Handbook of Enology
Volume 1
The Microbiology of Wine and Vinifications
2nd Edition

Handbook of Enology
Volume 1
The Microbiology of Wine and Vinifications
2nd Edition

Pascal Ribéreau-Gayon
Denis Dubourdieu
Bernard Donèche
Aline Lonvaud

Faculty of Enology
Victor Segalen University of Bordeaux II, Talence, France

Original translation by

Jeffrey M. Branco, Jr.
Winemaker
M.S., Faculty of Enology, University of Bordeaux II

Revision translated by

Christine Rychlewski
Aquitaine Traduction, Bordeaux, France

John Wiley & Sons, Ltd

Other Wiley Editorial Offices

John Wiley & Sons Inc., 111 River Street, Hoboken, NJ 07030, USA

Jossey-Bass, 989 Market Street, San Francisco, CA 94103-1741, USA

Wiley-VCH Verlag GmbH, Boschstr. 12, D-69469 Weinheim, Germany

John Wiley & Sons Australia Ltd, 42 McDougall Street, Milton, Queensland 4064, Australia

John Wiley & Sons (Asia) Pte Ltd, 2 Clementi Loop #02-01, Jin Xing Distripark, Singapore 129809

John Wiley & Sons Canada Ltd, 22 Worcester Road, Etobicoke, Ontario, Canada M9W 1L1

Wiley also publishes its books in a variety of electronic formats. Some content that appears in print may not be available in electronic books.

Library of Congress Cataloging-in-Publication Data:

Ribéreau-Gayon, Pascal.
  [Traité d'oenologie. English]
  Handbook of enology / Pascal Ribéreau-Gayon, Denis Dubourdieu, Bernard
Donèche ; original translation by Jeffrey M. Branco, Jr.—2nd ed. /
translation of updates for 2nd ed. [by] Christine Rychlewski.
        v. cm.
  Rev. ed. of: Handbook of enology / Pascal Ribéreau Gayon . . . [et al.].
c2000.
  Includes bibliographical references and index.
  Contents: v. 1. The microbiology of wine and vinifications
  ISBN-13: 978-0-470-01034-1 (v. 1 : acid-free paper)
  ISBN-10: 0-470-01034-7 (v. 1 : acid-free paper)
  1. Wine and wine making—Handbooks, manuals, etc. 2. Wine and wine
making—Microbiology—Handbooks, manuals, etc. 3. Wine and wine
making—Chemistry—Handbooks, manuals, etc. I. Dubourdieu, Denis. II.
Donèche, Bernard. III. Traité d'oenologie. English. IV. Title.
  TP548.T7613 2005
  663'.2—dc22
                                                    2005013973

British Library Cataloguing in Publication Data

A catalogue record for this book is available from the British Library

ISBN-13: 978-0-470-01034-1 (H/B)

Typeset in 10/12pt Times by Laserwords Private Limited, Chennai, India
Printed and bound in Great Britain by CPI Antony Rowe, Chippenham, Wiltshire

Contents

Remarks Concerning the Expression of Certain Parameters of Must and Wine Composition

UNITS

Metric system units of length (m), volume (l) and weight (g) are exclusively used. The conversion of metric units into Imperial units (inches, feet, gallons, pounds, etc.) can be found in the following enological work: *Principles and practices of winemaking*, R.B. Boulton, V.L. Singleton, L.F. Bisson and R.E. Kunkee, 1995, The Chapman & Hall Enology Library, New York.

EXPRESSION OF TOTAL ACIDITY AND VOLATILE ACIDITY

Although EC regulations recommend the expression of total acidity in the equivalent weight of tartaric acid, the French custom is to give this expression in the equivalent weight of sulfuric acid. The more correct expression in milliequivalents per liter has not been embraced in France. The expression of total and volatile acidity in the equivalent weight of sulfuric acid has been used predominantly throughout these works. In certain cases, the corresponding weight in tartaric acid, often used in other countries, has been given.

Using the weight of the milliequivalent of the various acids, the below table permits the conversion from one expression to another.

More particularly, to convert from total acidity expressed in H_2SO_4 to its expression in tartaric acid, add half of the value to the original value (4 g/l $H_2SO_4 \rightarrow 6$ g/l tartaric acid). In the other direction a third of the value must be subtracted.

The French also continue to express volatile acidity in equivalent weight of sulfuric acid. More generally, in other countries, volatile acidity is

Known Expression	Desired Expression			
	meq/l	g/l H_2SO_4	g/l tartaric acid	g/l acetic acid
meq/l	1.00	0.049	0.075	0.060
g/l H_2SO_4	20.40	1.00	1.53	1.22
g/l tartaric acid	13.33	0.65	1.00	
g/l acetic acid	16.67	0.82		1.00

Multiplier to pass from one expression of total or volatile acidity to another

expressed in acetic acid. It is rarely expressed in milliequivalents per liter. The below table also allows simple conversion from one expression to another.

The expression in acetic acid is approximately 20% higher than in sulfuric acid.

EVALUATING THE SUGAR CONCENTRATION OF MUSTS

This measurement is important for tracking grape maturation, fermentation kinetic and if necessary determining the eventual need for chaptalization.

This measurement is always determined by physical, densimetric or refractometric analysis. The expression of the results can be given according to several scales: some are rarely used, i.e. degree Baumé and degree Oechsle. Presently, two systems exist (Section 10.4.3):

1. The potential alcohol content (*titre alcoométraque potential* or TAP, in French) of musts can be read directly on equipment, which is graduated using a scale corresponding to 17.5 or 17 g/l of sugar for 1% volume of alcohol. Today, the EC recommends using 16.83 g/l as the conversion factor. The 'mustimeter' is a hydrometer containing two graduated scales: one expresses density and the other gives a direct reading of the TAP. Different methods varying in precision exist to calculate the TAP from a density reading. These methods take various elements of must composition into account (Boulton *et al.*, 1995).

2. Degree Brix expresses the percentage of sugar in weight. By multiplying degree Brix by 10, the weight of sugar in 1 kg, or slightly less than 1 liter, of must is obtained. A conversion table between degree Brix and TAP exists in Section 10.4.3 of this book. 17 degrees Brix correspond to an approximate TAP of 10% and 20 degrees Brix correspond to a TAP of about 12%. Within the alcohol range most relevant to enology, degree Brix can be multiplied by 10

and then divided by 17 to obtain a fairly good approximation of the TAP.

In any case, the determination of the Brix or TAP of a must is approximate. First of all, it is not always possible to obtain a representative grape or must sample for analysis. Secondly, although physical, densimetric or refractometric measurements are extremely precise and rigorously express the sugar concentration of a sugar and water mixture, these measurements are affected by other substances released into the sample from the grape and other sources. Furthermore, the concentrations of these substances are different for every grape or grape must sample. Finally, the conversion rate of sugar into alcohol (approximately 17 to 18 g/l) varies and depends on fermentation conditions and yeast properties. The widespread use of selected yeast strains has lowered the sugar conversion rate.

Measurements Using Visible and Ultraviolet Spectrometry

The measurement of optic density, absorbance, is widely used to determine wine color (Volume 2, Section 6.4.5) and total phenolic compounds concentration (Volume 2, Section 6.4.1). In these works, the optic density is noted as OD, OD 420 (yellow), OD 520 (red), OD 620 (blue) or OD 280 (absorption in ultraviolet spectrum) to indicate the optic density at the indicated wavelengths.

Wine color intensity is expressed as:

$$CI = OD\ 420 + OD\ 520 + OD\ 620,$$

Or is sometimes expressed in a more simplified form: CI = OD 420 + OD 520.

Tint is expressed as:

$$T = \frac{OD\ 420}{OD\ 520}$$

The total phenolic compound concentration is expressed by OD 280.

The analysis methods are described in Chapter 6 of *Handbook of Enology Volume 2, The Chemistry of Wine*.

Preface to the First Edition

Wine has probably inspired more research and publications than any other beverage or food. In fact, through their passion for wine, great scientists have not only contributed to the development of practical enology but have also made discoveries in the general field of science.

A forerunner of modern enology, Louis Pasteur developed simplified contagious infection models for humans and animals based on his observations of wine spoilage. The following quote clearly expresses his theory in his own words: 'when profound alterations of beer and wine are observed because these liquids have given refuge to microscopic organisms, introduced invisibly and accidentally into the medium where they then proliferate, how can one not be obsessed by the thought that a similar phenomenon can and must sometimes occur in humans and animals.'

Since the 19th century, our understanding of wine, wine composition and wine transformations has greatly evolved in function of advances in relevant scientific fields i.e. chemistry, biochemistry, microbiology. Each applied development has lead to better control of winemaking and aging conditions and of course wine quality. In order to continue this approach, researchers and winemakers must strive to remain up to date with the latest scientific and technical developments in enology.

For a long time, the Bordeaux school of enology was largely responsible for the communication of progress in enology through the publication of numerous works (Béranger Publications and later Dunod Publications):

Wine Analysis U. Gayon and J. Laborde (1912); *Treatise on Enology* J. Ribéreau-Gayon (1949);

Wine Analysis J. Ribéreau-Gayon and E. Peynaud (1947 and 1958); *Treatise on Enology* (2 Volumes) J. Ribéreau-Gayon and E. Peynaud (1960 and 1961); *Wine and Winemaking* E. Peynaud (1971 and 1981); *Wine Science and Technology* (4 volumes) J. Ribéreau-Gayon, E. Peynaud, P. Ribéreau-Gayon and P. Sudraud (1975–1982).

For an understanding of current advances in enology, the authors propose this book *Handbook of Enology Volume 1: The Microbiology of Wine and Vinifications* and the second volume of the *Handbook of Enology Volume 2: The Chemistry of Wine: Stabilization and Treatments*.

Although written by researchers, the two volumes are not specifically addressed to this group. Young researchers may, however, find these books useful to help situate their research within a particular field of enology. Today, the complexity of modern enology does not permit a sole researcher to explore the entire field.

These volumes are also of use to students and professionals. Theoretical interpretations as well as solutions are presented to resolve the problems encountered most often at wineries. The authors have adapted these solutions to many different situations and winemaking methods. In order to make the best use of the information contained in these works, enologists should have a broad understanding of general scientific knowledge. For example, the understanding and application of molecular biology and genetic engineering have become indispensable in the field of wine microbiology. Similarly, structural and quantitative physiochemical analysis methods such as chromatography,

NMR and mass spectrometry must now be mastered in order to explore wine chemistry.

The goal of these two works was not to create an exhaustive bibliography of each subject. The authors strove to choose only the most relevant and significant publications to their particular field of research. A large number of references to French enological research has been included in these works in order to make this information available to a larger non-French-speaking audience.

In addition, the authors have tried to convey a French and more particularly a Bordeaux perspective of enology and the art of winemaking. The objective of this perspective is to maximize the potential quality of grape crops based on the specific natural conditions that constitute their 'terroir'. The role of enology is to express the characteristics of the grape specific not only to variety and vineyard practices but also maturation conditions, which are dictated by soil and climate.

It would, however, be an error to think that the world's greatest wines are exclusively a result of tradition, established by exceptional natural conditions, and that only the most ordinary wines, produced in giant processing facilities, can benefit from scientific and technological progress. Certainly, these facilities do benefit the most from high performance installations and automation of operations. Yet, history has unequivocally shown that the most important enological developments in wine quality (for example, malolactic fermentation) have been discovered in ultra premium wines. The corresponding techniques were then applied to less prestigious products.

High performance technology is indispensable for the production of great wines, since a lack of control of winemaking parameters can easily compromise their quality, which would be less of a problem with lower quality wines.

The word 'vinification' has been used in this work and is part of the technical language of the French tradition of winemaking. Vinification describes the first phase of winemaking. It comprises all technical aspects from grape maturity and harvest to the end of alcoholic and sometimes malolactic fermentation. The second phase of winemaking 'winematuration, stabilization and treatments' is completed when the wine is bottled. Aging specifically refers to the transformation of bottled wine.

This distinction of two phases is certainly the result of commercial practices. Traditionally in France, a vine grower farmed the vineyard and transformed grapes into an unfinished wine. The wine merchant transferred the bulk wine to his cellars, finished the wine and marketed the product, preferentially before bottling. Even though most wines are now bottled at the winery, these long-standing practices have maintained a distinction between 'wine grower enology' and 'wine merchant enology'. In countries with a more recent viticultural history, generally English speaking, the vine grower is responsible for winemaking and wine sales. For this reason, the Anglo-Saxon tradition speaks of winemaking, which covers all operations from harvest reception to bottling.

In these works, the distinction between 'vinification' and 'stabilization and treatments' has been maintained, since the first phase primarily concerns microbiology and the second chemistry. In this manner, the individual operations could be linked to their particular sciences. There are of course limits to this approach. Chemical phenomena occur during vinification; the stabilization of wines during storage includes the prevention of microbial contamination.

Consequently, the description of the different steps of enology does not always obey logic as precise as the titles of these works may lead to believe. For example, microbial contamination during aging and storage are covered in Volume 1. The antiseptic properties of SO_2 incited the description of its use in the same volume. This line of reasoning lead to the description of the antioxidant related chemical properties of this compound in the same chapter as well as an explanation of adjuvants to sulfur dioxide: sorbic acid (antiseptic) and ascorbic acid (antioxidant). In addition, the on lees aging of white wines and the resulting chemical transformations cannot be separated from vinification and are therefore also covered in Volume 1. Finally, our understanding of phenolic compounds in red wine is based on complex chemistry. All aspects related to the nature of the corresponding substances, their properties and their evolution during grape maturation, vinification and aging are therefore covered in Volume 2.

These works only discuss the principles of equipment used for various enological operations and their effect on product quality. For example, temperature control systems, destemmers, crushers and presses as well as filters, inverse osmosis machines and ion exchangers are not described in detail. Bottling is not addressed at all. An in-depth description of enological equipment would merit a detailed work dedicated to the subject.

Wine tasting, another essential role of the winemaker, is not addressed in these works. Many related publications are, however, readily available. Finally, wine analysis is an essential tool that a winemaker should master. It is, however, not covered in these works except in a few particular cases i.e. phenolic compounds, whose different families are often defined by analytical criteria.

The authors thank the following people who have contributed to the creation of this work: J.F. Casas Lucas, Chapter 14, Sherry; A. Brugirard, Chapter 14, Sweet wines; J.N. de Almeida, Chapter 14, Port wines; A. Maujean, Chapter 14, Champagne; C. Poupot for the preparation of material in Chapters 1, 2 and 13; Miss F. Luye-Tanet for her help with typing.

They also thank Madame B. Masclef in particular for her important part in the typing, preparation and revision of the final manuscript.

Pascal Ribéreau-Gayon
Bordeaux

Preface to the Second Edition

The two-volume Enology Handbook was published simultaneously in Spanish, French, and Italian in 1999 and has been reprinted several times. The Handbook has apparently been popular with students as an educational reference book, as well as with winemakers, as a source of practical solutions to their specific technical problems and scientific explanations of the phenomena involved.

It was felt appropriate at this stage to prepare an updated, reviewed, corrected version, including the latest enological knowledge, to reflect the many new research findings in this very active field. The outline and design of both volumes remain the same. Some chapters have changed relatively little as the authors decided there had not been any significant new developments, while others have been modified much more extensively, either to clarify and improve the text, or, more usually, to include new research findings and their practical applications. Entirely new sections have been inserted in some chapters.

We have made every effort to maintain the same approach as we did in the first edition, reflecting the ethos of enology research in Bordeaux. We use indisputable scientific evidence in microbiology, biochemistry, and chemistry to explain the details of mechanisms involved in grape ripening, fermentations and other winemaking operations, aging, and stabilization. The aim is to help winemakers achieve greater control over the various stages in winemaking and choose the solution best suited to each situation. Quite remarkably, this scientific approach, most intensively applied in making the finest wines, has resulted in an enhanced capacity to bring out the full quality and character of individual *terroirs*. Scientific winemaking has not resulted in standardization or leveling of quality. On the contrary, by making it possible to correct defects and eliminate technical imperfections, it has revealed the specific qualities of the grapes harvested in different vineyards, directly related to the variety and *terroir*, more than ever before.

Interest in wine in recent decades has gone beyond considerations of mere quality and taken on a truly cultural dimension. This has led some people to promote the use of a variety of techniques that do not necessarily represent significant progress in winemaking. Some of these are simply modified forms of processes that have been known for many years. Others do not have a sufficiently reliable scientific interpretation, nor are their applications clearly defined. In this Handbook, we have only included rigorously tested techniques, clearly specifying the optimum conditions for their utilization.

As in the previous edition, we deliberately omitted three significant aspects of enology: wine analysis, tasting, and winery engineering. In view of their importance, these topics will each be covered in separate publications.

The authors would like to take the opportunity of the publication of this new edition of Volume 1 to thank all those who have contributed to updating this work:

— Marina Bely for her work on fermentation kinetics (Section 3.4) and the production of volatile acidity (Sections 2.3.4 and 14.2.5)

— Isabelle Masneuf for her investigation of the yeasts' nitrogen supply (Section 3.4.2)

— Gilles de Revel for elucidating the chemistry of SO_2, particularly, details of combination reactions (Section 8.4)

— Gilles Masson for the section on rosé wines (Section 14.1)

— Cornelis Van Leeuwen for data on the impact of vineyard water supply on grape ripening (Section 10.4.6)

— André Brugirard for the section on French fortified wines—*vins doux naturels* (Section 14.4.2)

— Paulo Barros and Joa Nicolau de Almeida for their work on Port (Section 14.4.3)

— Justo. F. Casas Lucas for the paragraph on Sherry (Section 14.5.2)

— Alain Maujean for his in-depth revision of the section on Champagne (Section 14.3).

March 17, 2005

Professor Pascal RIBEREAU-GAYON
Corresponding Member of the Institute
Member of the French Academy of Agriculture

1

Cytology, Taxonomy and Ecology of Grape and Wine Yeasts

1.1 INTRODUCTION

Man has been making bread and fermented beverages since the beginning of recorded history. Yet the role of yeasts in alcoholic fermentation, particularly in the transformation of grapes into wine, was only clearly established in the middle of the nineteenth century. The ancients explained the boiling during fermentation (from the Latin *fervere*, to boil) as a reaction between substances that come into contact with each other during crushing. In 1680, a Dutch cloth merchant, Antonie van Leeuwenhoek, first observed yeasts in beer wort using a microscope that he designed and produced. He did not, however, establish a relationship between these corpuscles and alcoholic fermentation. It was not until the end of the eighteenth century that Lavoisier began the chemical study of alcoholic fermentation. Gay-Lussac continued Lavoisier's research into the next century.

Handbook of Enology Volume 1 The Microbiology of Wine and Vinifications P. Ribéreau-Gayon, D. Dubourdieu, B. Donèche and A. Lonvaud
© 2006 John Wiley & Sons, Ltd

As early as 1785, Fabroni, an Italian scientist, was the first to provide an interpretation of the chemical composition of the ferment responsible for alcoholic fermentation, which he described as a plant–animal substance. According to Fabroni, this material, comparable to the gluten in flour, was located in special utricles, particularly on grapes and wheat, and alcoholic fermentation occurred when it came into contact with sugar in the must. In 1837, a French physicist named Charles Cagnard de La Tour proved for the first time that the yeast was a living organism. According to his findings, it was capable of multiplying and belonged to the plant kingdom; its vital activities were at the base of the fermentation of sugar-containing liquids. The German naturalist Schwann confirmed his theory and demonstrated that heat and certain chemical products were capable of stopping alcoholic fermentation. He named the beer yeast *zuckerpilz*, which means sugar fungus—*Saccharomyces* in Latin. In 1838, Meyen used this nomenclature for the first time.

This vitalist or biological viewpoint of the role of yeasts in alcoholic fermentation, obvious to us today, was not readily supported. Liebig and certain other organic chemists were convinced that chemical reactions, not living cellular activity, were responsible for the fermentation of sugar. In his famous studies on wine (1866) and beer (1876), Louis Pasteur gave definitive credibility to the vitalist viewpoint of alcoholic fermentation. He demonstrated that the yeasts responsible for spontaneous fermentation of grape must or crushed grapes came from the surface of the grape; he isolated several races and species. He even conceived the notion that the nature of the yeast carrying out the alcoholic fermentation could influence the gustatory characteristics of wine. He also demonstrated the effect of oxygen on the assimilation of sugar by yeasts. Louis Pasteur proved that the yeast produced secondary products such as glycerol in addition to alcohol and carbon dioxide.

Since Pasteur, yeasts and alcoholic fermentation have incited a considerable amount of research, making use of progress in microbiology,

biochemistry and now genetics and molecular biology.

In taxonomy, scientists define yeasts as unicellular fungi that reproduce by budding and binary fission. Certain pluricellular fungi have a unicellular stage and are also grouped with yeasts. Yeasts form a complex and heterogeneous group found in three classes of fungi, characterized by their reproduction mode: the sac fungi (Ascomycetes), the club fungi (Basidiomycetes), and the imperfect fungi (Deuteromycetes). The yeasts found on the surface of the grape and in wine belong to Ascomycetes and Deuteromycetes. The haploid spores or ascospores of the Ascomycetes class are contained in the ascus, a type of sac made from vegetative cells. Asporiferous yeasts, incapable of sexual reproduction, are classified with the imperfect fungi.

In this first chapter, the morphology, reproduction, taxonomy and ecology of grape and wine yeasts will be discussed. Cytology is the morphological and functional study of the structural components of the cell (Rose and Harrison, 1991).

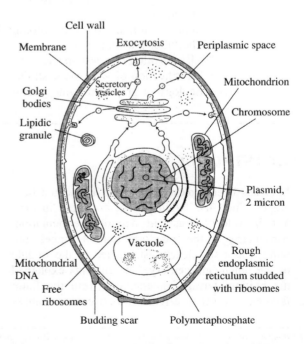

Fig. 1.1. A yeast cell (Gaillardin and Heslot, 1987)

Yeasts are the most simple of the eucaryotes. The yeast cell contains cellular envelopes, a cytoplasm with various organelles, and a nucleus surrounded by a membrane and enclosing the chromosomes. (Figure 1.1). Like all plant cells, the yeast cell has two cellular envelopes: the cell wall and the membrane. The periplasmic space is the space between the cell wall and the membrane. The cytoplasm and the membrane make up the protoplasm. The term protoplast or sphaeroplast designates a cell whose cell wall has been artificially removed. Yeast cellular envelopes play an essential role: they contribute to a successful alcoholic fermentation and release certain constituents which add to the resulting wine's composition. In order to take advantage of these properties, the winemaker or enologist must have a profound knowledge of these organelles.

1.2 THE CELL WALL

1.2.1 The General Role of the Cell Wall

During the last 20 years, researchers (Fleet, 1991; Klis, 1994; Stratford, 1999; Klis *et al.*, 2002) have greatly expanded our knowledge of the yeast cell wall, which represents 15–25% of the dry weight of the cell. It essentially consists of polysaccharides. It is a rigid envelope, yet endowed with a certain elasticity.

Its first function is to protect the cell. Without its wall, the cell would burst under the internal osmotic pressure, determined by the composition of the cell's environment. Protoplasts placed in pure water are immediately lysed in this manner. Cell wall elasticity can be demonstrated by placing yeasts, taken during their log phase, in a hypertonic (NaCl) solution. Their cellular volume decreases by approximately 50%. The cell wall appears thicker and is almost in contact with the membrane. The cells regain their initial form after being placed back into an isotonic medium.

Yet the cell wall cannot be considered an inert, semi-rigid 'armor'. On the contrary, it is a dynamic and multifunctional organelle. Its composition and functions evolve during the life of the cell, in response to environmental factors. In addition to its protective role, the cell wall gives the cell its particular shape through its macromolecular organization. It is also the site of molecules which determine certain cellular interactions such as sexual union, flocculation, and the killer factor, which will be examined in detail later in this chapter (Section 1.7). Finally, a number of enzymes, generally hydrolases, are connected to the cell wall or situated in the periplasmic space. Their substrates are nutritive substances of the environment and the macromolecules of the cell wall itself, which is constantly reshaped during cellular morphogenesis.

1.2.2 The Chemical Structure and Function of the Parietal Constituents

The yeast cell wall is made up of two principal constituents: β-glucans and mannoproteins. Chitin represents a minute part of its composition. The most detailed work on the yeast cell wall has been carried out on *Saccharomyces cerevisiae*—the principal yeast responsible for the alcoholic fermentation of grape must.

Glucan represents about 60% of the dry weight of the cell wall of *S. cerevisiae*. It can be chemically fractionated into three categories:

1. Fibrous β-1,3 glucan is insoluble in water, acetic acid and alkali. It has very few branches. The branch points involved are β-1,6 linkages. Its degree of polymerization is 1500. Under the electron microscope, this glucan appears fibrous. It ensures the shape and the rigidity of the cell wall. It is always connected to chitin.

2. Amorphous β-1,3 glucan, with about 1500 glucose units, is insoluble in water but soluble in alkalis. It has very few branches, like the preceding glucan. In addition to these few branches, it is made up of a small number of β-1,6 glycosidic linkages. It has an amorphous aspect under the electron microscope. It gives the cell wall its elasticity and acts as an anchor for the mannoproteins. It can also constitute an extraprotoplasmic reserve substance.

3. The β-1,6 glucan is obtained from alkali-insoluble glucans by extraction in acetic acid. The resulting product is amorphous, water soluble, and extensively ramified by β-1,3 glycosidic linkages. Its degree of polymerization is 140. It links the different constituents of the cell wall together. It is also a receptor site for the killer factor.

The fibrous β-1,3 glucan (alkali-insoluble) probably results from the incorporation of chitin on the amorphous β-1,3 glucan.

Mannoproteins constitute 25–50% of the cell wall of *S. cerevisiae*. They can be extracted from the whole cell or from the isolated cell wall by chemical and enzymatic methods. Chemical methods make use of autoclaving in the presence of alkali or a citrate buffer solution at pH 7. The enzymatic method frees the mannoproteins by digesting the glucan. This method does not denature the structure of the mannoproteins as much as chemical methods. Zymolyase, obtained from the bacterium *Arthrobacter luteus*, is the enzymatic preparation most often used to extract the parietal mannoproteins of *S. cerevisiae*. This enzymatic complex is effective primarily because of its β-1,3 glucanase activity. The action of protease contaminants in the zymolyase combine, with the aforementioned activity to liberate the mannoproteins. Glucanex, another industrial preparation of the β-glucanase, produced by a fungus (*Trichoderma harzianum*), has been recently demonstrated to possess endo- and exo-β-1,3 and endo-β-1,6-glucanase activities (Dubourdieu and Moine, 1995). These activities also facilitate the extraction of the cell wall mannoproteins of the *S. cerevisiae* cell.

The mannoproteins of *S. cerevisiae* have a molecular weight between 20 and 450 kDa. Their degree of glycosylation varies. Certain ones containing about 90% mannose and 10% peptides are hypermannosylated.

Four forms of glycosylation are described (Figure 1.2) but do not necessarily exist at the same time in all of the mannoproteins.

The mannose of the mannoproteins can constitute short, linear chains with one to five residues.

They are linked to the peptide chain by O-glycosyl linkages on serine and threonine residues. These glycosidic side-chain linkages are α-1,2 and α-1,3.

The glucidic part of the mannoprotein can also be a polysaccharide. It is linked to an asparagine residue of the peptide chain by an N-glycosyl linkage. This linkage consists of a double unit of N-acetylglucosamine (chitin) linked in β-1,4. The mannan linked in this manner to the asparagine includes an attachment region made up of a dozen mannose residues and a highly ramified outer chain consisting of 150 to 250 mannose units. The attachment region beyond the chitin residue consists of a mannose skeleton linked in α-1,6 with side branches possessing one, two or three mannose residues with α-1,2 and/or α-1,3 bonds. The outer chain is also made up of a skeleton of mannose units linked in α-1,6. This chain bears short side-chains constituted of mannose residues linked in α-1,2 and a terminal mannose in α-1,3. Some of these side-chains possess a branch attached by a phosphodiester bond.

A third type of glycosylation was described more recently. It can occur in mannoproteins, which make up the cell wall of the yeast. It consists of a glucomannan chain containing essentially mannose residues linked in α-1,6 and glucose residues linked in α-1,6. The nature of the glycan–peptide point of attachment is not yet clear, but it may be an asparaginyl–glucose bond. This type of glycosylation characterizes the proteins freed from the cell wall by the action of a β-1,3 glucanase. Therefore, *in vivo*, the glucomannan chain may also comprise glucose residues linked in β-1,3.

The fourth type of glycosylation of yeast mannoproteins is the glycosyl–phosphatidyl–inositol anchor (GPI). This attachment between the terminal carboxylic group of the peptide chain and a membrane phospholipid permits certain mannoproteins, which cross the cell wall, to anchor themselves in the plasmic membrane. The region of attachment is characterized by the following sequence (Figure 1.2): ethanolamine-phosphate-6-mannose-α-1,2-mannose-α-1,6-mannose-α-1,4-glucosamine-α-1,6-inositol-phospholipid. A C-phospholipase specific to phosphatidyl inositol and therefore capable of realizing this cleavage

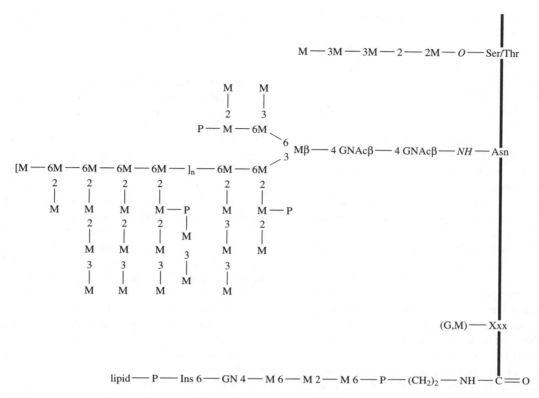

Fig. 1.2. The four types of glucosylation of parietal yeast mannoproteins (Klis, 1994). M = mannose; G = glucose; GN = glucosamine; GNAc = N-acetylglucosamine; Ins = inositol; Ser = Serine; Thr = threonine; Asn = asparagine; Xxx = the nature of the bond is not known

was demonstrated in the *S. cerevisiae* (Flick and Thorner, 1993). Several GPI-type anchor mannoproteins have been identified in the cell wall of *S. cerevisiae*.

Chitin is a linear polymer of *N*-acetylglucosamine linked in β-1,4 and is not generally found in large quantities in yeast cell walls. In *S. cerevisiae*, chitin constitutes 1–2% of the cell wall and is found for the most part (but not exclusively) in bud scar zones. These zones are a type of raised crater easily seen on the mother cell under the electron microscope (Figure 1.3). This chitinic scar is formed essentially to assure cell wall integrity and cell survival. Yeasts treated with D polyoxine, an antibiotic inhibiting the synthesis of chitin, are not viable; they burst after budding.

The presence of lipids in the cell wall has not been clearly demonstrated. It is true that cell walls

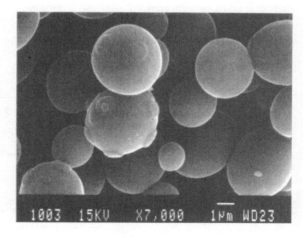

Fig. 1.3. Scanning electron microscope photograph of proliferating *S. cerevisiae* cells. The budding scars on the mother cells can be observed

prepared in the laboratory contain some lipids (2–15% for *S. cerevisiae*) but it is most likely contamination by the lipids of the cytoplasmic membrane, adsorbed by the cell wall during their isolation. The cell wall can also adsorb lipids from its external environment, especially the different fatty acids that activate and inhibit the fermentation (Chapter 3).

Chitin are connected to the cell wall or situated in the periplasmic space. One of the most characteristic enzymes is the invertase (*β*-fructofuranosidase). This enzyme catalyzes the hydrolysis of saccharose into glucose and fructose. It is a thermostable mannoprotein anchored to a *β*-1,6 glucan of the cell wall. Its molecular weight is 270 000 Da. It contains approximately 50% mannose and 50% protein. The periplasmic acid phosphatase is equally a mannoprotein.

Other periplasmic enzymes that have been noted are *β*-glucosidase, *α*-galactosidase, melibiase, trehalase, aminopeptidase and esterase. Yeast cell walls also contain endo- and exo-*β*-glucanases (*β*-1,3 and *β*-1,6). These enzymes are involved in the reshaping of the cell wall during the growth and budding of cells. Their activity is at a maximum during the exponential log phase of the population and diminishes notably afterwards. Yet cells in the stationary phase and even dead yeasts contained in the lees still retain *β*-glucanases activity in their cell walls several months after the completion of fermentation. These endogenous enzymes are involved in the autolysis of the cell wall during the

ageing of wines on lees. This ageing method will be covered in the chapter on white winemaking (Chapter 13).

1.2.3 General Organization of the Cell Wall and Factors Affecting its Composition

The cell wall of *S. cerevisiae* is made up of an outer layer of mannoproteins. These mannoproteins are connected to a matrix of amorphous *β*-1,3 glucan which covers an inner layer of fibrous *β*-1,3 glucan. The inner layer is connected to a small quantity of chitin (Figure 1.4). The *β*-1,6 glucan probably acts as a cement between the two layers. The rigidity and the shape of the cell wall are due to the internal framework of the *β*-1,3 fibrous glucan. Its elasticity is due to the outer amorphous layer. The intermolecular structure of the mannoproteins of the outer layer (hydrophobic linkages and disulfur bonds) equally determines cell wall porosity and impermeability to macromolecules (molecular weights less than 4500). This impermeability can be affected by treating the cell wall with certain chemical agents, such as *β*-mercaptoethanol. This substance provokes the rupture of the disulfur bonds, thus destroying the intermolecular network between the mannoprotein chains.

The composition of the cell wall is strongly influenced by nutritive conditions and cell age. The proportion of glucan in the cell wall increases

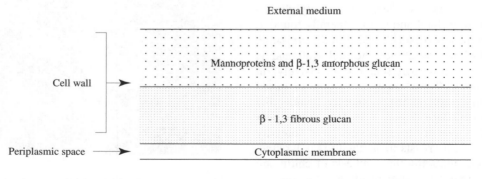

Fig. 1.4. Cellular organization of the cell wall of *S. cerevisiae*

with respect to the amount of sugar in the culture medium. Certain deficiencies (for example, in mesoinositol) also result in an increase in the proportion of glucan compared with mannoproteins. The cell walls of older cells are richer in glucans and in chitin and less furnished in mannoproteins. For this reason, they are more resistant to physical and enzymatic agents used to degrade them. Finally, the composition of the cell wall is profoundly modified by morphogenetic alterations (conjugation and sporulation).

1.3 THE PLASMIC MEMBRANE

1.3.1 Chemical Composition and Organization

The plasmic membrane is a highly selective barrier controlling exchanges between the living cell and its external environment. This organelle is essential to the life of the yeast.

Like all biological membranes, the yeast plasmic membrane is principally made up of lipids and proteins. The plasmic membrane of *S. cerevisiae* contains about 40% lipids and 50% proteins. Glucans and mannans are only present in small quantities (several per cent).

The lipids of the membrane are essentially phospholipids and sterols. They are amphiphilic molecules, i.e. possessing a hydrophilic and a hydrophobic part.

The three principal phospholipids (Figure 1.5) of the plasmic membrane of yeast are phosphatidylethanolamine (PE), phosphatidylcholine (PC) and phosphatidylinositol (PI) which represent 70–85% of the total. Phosphatidylserine (PS) and diphosphatidylglycerol or cardiolipin (PG) are less prevalent. Free fatty acids and phosphatidic acid are frequently reported in plasmic membrane analysis. They are probably extraction artifacts caused by the activity of certain lipid degradation enzymes.

The fatty acids of the membrane phospholipids contain an even number (14 to 24) of carbon atoms. The most abundant are C_{16} and C_{18} acids. They can be saturated, such as palmitic acid (C_{16}) and stearic acid (C_{18}), or unsaturated, as with oleic

acid (C_{18}, double bond in position 9), linoleic acid (C_{18}, two double bonds in positions 9 and 12) and linolenic acid (C_{18}, three double bonds in positions 9, 12 and 15). All membrane phospholipids share a common characteristic: they possess a polar or hydrophilic part made up of a phosphorylated alcohol and a non-polar or hydrophobic part comprising two more or less parallel fatty acid chains (Figure 1.6). In an aqueous medium, the phospholipids spontaneously form bimolecular films or a lipid bilayer because of their amphiphilic characteristic (Figure 1.6). The lipid bilayers are cooperative but non-covalent structures. They are maintained in place by mutually reinforced interactions: hydrophobic interactions, van der Waals attractive forces between the hydrocarbon tails, hydrostatic interactions and hydrogen bonds between the polar heads and water molecules. The examination of cross-sections of yeast plasmic membrane under the electron microscope reveals a classic lipid bilayer structure with a thickness of about 7.5 nm. The membrane surface appears sculped with creases, especially during the stationary phase. However, the physiological meaning of this anatomic character remains unknown. The plasmic membrane also has an underlying depression on the bud scar.

Ergosterol is the primary sterol of the yeast plasmic membrane. In lesser quantities, 24 (28) dehydroergosterol and zymosterol also exist (Figure 1.7). Sterols are exclusively produced in the mitochondria during the yeast log phase. As with phospholipids, membrane sterols are amphipathic. The hydrophilic part is made up of hydroxyl groups in C-3. The rest of the molecule is hydrophobic, especially the flexible hydrocarbon tail.

The plasmic membrane also contains numerous proteins or glycoproteins presenting a wide range of molecular weights (from 10 000 to 120 000). The available information indicates that the organization of the plasmic membrane of a yeast cell resembles the fluid mosaic model. This model, proposed for biological membranes by Singer and Nicolson (1972), consists of two-dimensional solutions of proteins and oriented lipids. Certain proteins are embedded in the membrane; they are called integral proteins (Figure 1.6). They interact

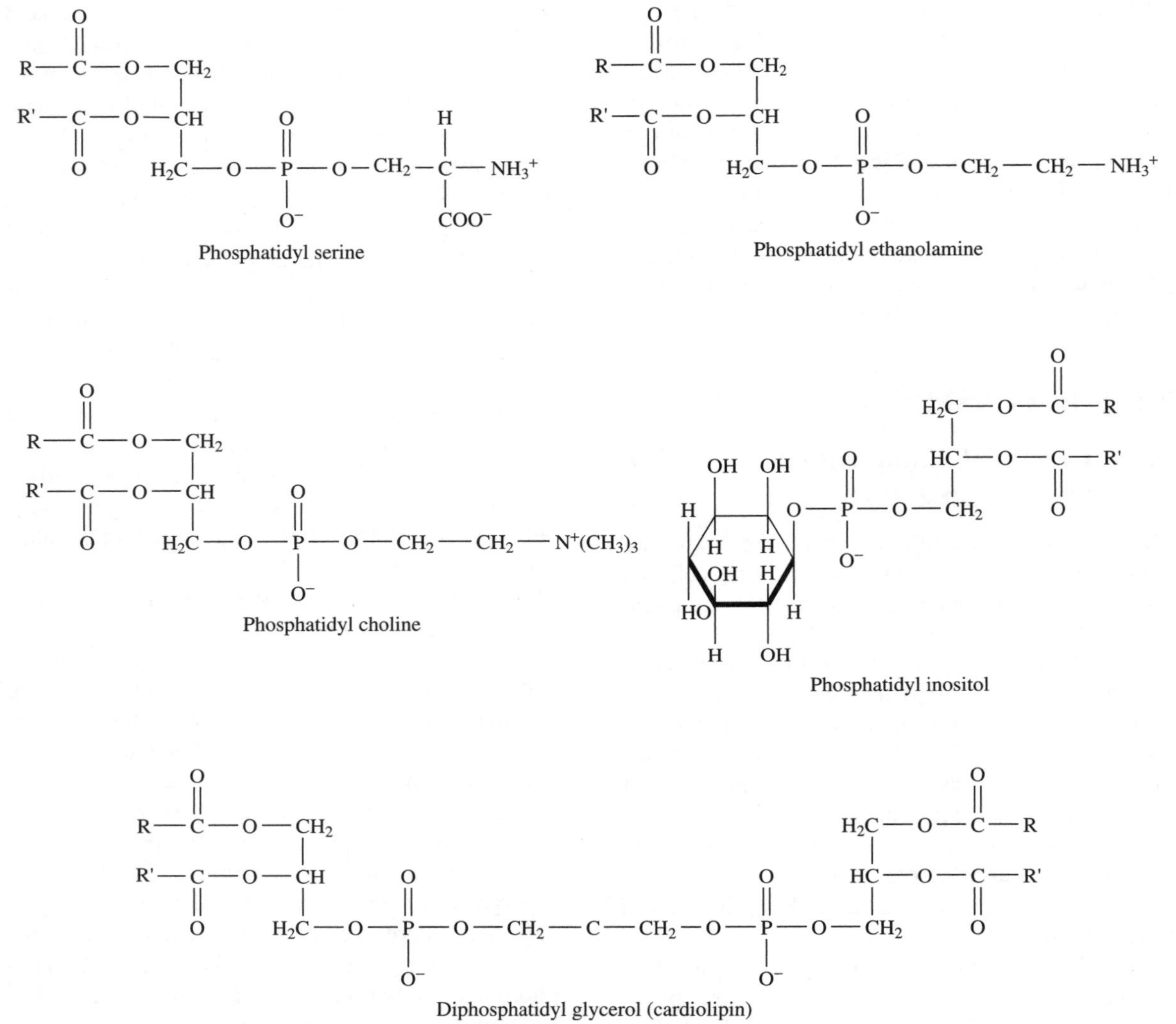

Fig. 1.5. Yeast membrane phospholipids

strongly with the non-polar part of the lipid bilayer. The peripheral proteins are linked to the precedent by hydrogen bonds. Their location is asymmetrical, at either the inner or the outer side of the plasmic membrane. The molecules of proteins and membrane lipids, constantly in lateral movement, are capable of rapidly diffusing in the membrane.

Some of the yeast membrane proteins have been studied in greater depth. These include adenosine triphosphatase (ATPase), solute (sugars and amino

acids) transport proteins, and enzymes involved in the production of glucans and chitin of the cell wall.

The yeast possesses three ATPases: in the mitochondria, the vacuole, and the plasmic membrane. The plasmic membrane ATPase is an integral protein with a molecular weight of around 100 Da. It catalyzes the hydrolysis of ATP which furnishes the necessary energy for the active transport of solutes across the membrane. (Note: an active

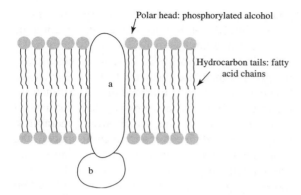

Polar head: phosphorylated alcohol

Hydrocarbon tails: fatty acid chains

a

b

Fig. 1.6. A membrane lipid bilayer. The integral proteins (a) are strongly associated to the non-polar region of the bilayer. The peripheral proteins (b) are linked to the integral proteins

transport moves a compound against the concentration gradient.) Simultaneously, the hydrolysis of ATP creates an efflux of protons towards the exterior of the cell.

The penetration of amino acids and sugars into the yeast activates membrane transport systems called permeases. The general amino acid permease (GAP) contains three membrane proteins and ensures the transport of a number of neutral amino acids. The cultivation of yeasts in the presence of an easily assimilated nitrogen-based nutrient such as ammonium represses this permease.

The membrane composition in fatty acids and its proportion in sterols control its fluidity. The hydrocarbon chains of fatty acids of the membrane phospholipid bilayer can be in a rigid and orderly state or in a relatively disorderly and fluid state. In the rigid state, some or all of the carbon bonds of the fatty acids are *trans*. In the fluid state, some of the bonds become *cis*. The transition from the rigid state to the fluid state takes place when the temperature rises beyond the fusion temperature. This transition temperature depends on the length of the fatty acid chains and their degree of unsaturation. The rectilinear hydrocarbon chains of the saturated fatty acids interact strongly. These interactions intensify with their length. The transition temperature therefore increases as the fatty acid chains become longer. The double bonds of the unsaturated fatty acids are generally *cis*, giving a curvature to the hydrocarbon chain (Figure 1.8). This curvature breaks the orderly

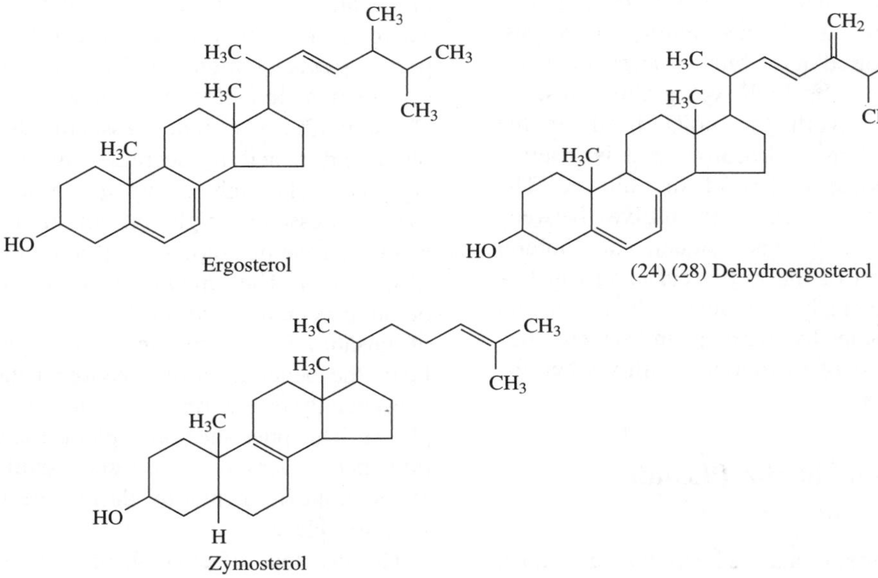

CH₃
H₃C
CH₃
H₃C
CH₃
H₃C
HO
Ergosterol

CH₂
H₃C
CH₃
H₃C
CH₃
H₃C
HO
(24) (28) Dehydroergosterol

H₃C
CH₃
H₃C
CH₃
H₃C
HO
H
Zymosterol

Fig. 1.7. Principal yeast membrane sterols

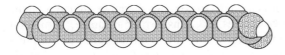

Stearic acid (C$_{18}$, saturated)

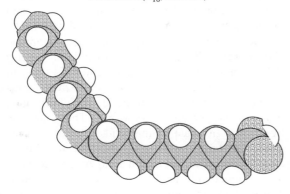

Oleic acid (C$_{18}$, unsaturated)

Fig. 1.8. Molecular models representing the three-dimensional structure of stearic and oleic acid. The *cis* configuration of the double bond of oleic acid produces a curvature of the carbon chain

stacking of the fatty acid chains and lowers the transition temperature. Like cholesterol in the cells of mammals, ergosterol is also a fundamental regulator of the membrane fluidity in yeasts. Ergosterol is inserted in the bilayer perpendicularly to the membrane. Its hydroxyl group joins, by hydrogen bonds, with the polar head of the phospholipid and its hydrocarbon tail is inserted in the hydrophobic region of the bilayer. The membrane sterols intercalate themselves between the phospholipids. In this manner, they inhibit the crystallization of the fatty acid chains at low temperatures. Inversely, in reducing the movement of these same chains by steric encumberment, they regulate an excess of membrane fluidity when the temperature rises.

1.3.2 Functions of the Plasmic Membrane

The plasmic membrane constitutes a stable, hydrophobic barrier between the cytoplasm and the environment outside the cell, owing to its phospholipids and sterols. This barrier presents a certain impermeability to solutes in function of osmotic properties.

Furthermore, through its system of permeases, the plasmic membrane also controls the exchanges between the cell and the medium. The functioning of these transport proteins is greatly influenced by its lipid composition, which affects membrane fluidity. In a defined environmental model, the supplementing of membrane phospholipids with unsaturated fatty acids (oleic and linoleic) promoted the penetration and accumulation of certain amino acids as well as the expression of the general amino acid permease (GAP), (Henschke and Rose, 1991). On the other hand, membrane sterols seem to have less influence on the transport of amino acids than the degree of unsaturation of the phospholipids. The production of unsaturated fatty acids is an oxidative process and requires the aeration of the culture medium at the beginning of alcoholic fermentation. In semi-anaerobic wine-making conditions, the amount of unsaturated fatty acids in the grape, or in the grape must, probably favor the membrane transport mechanisms of fatty acids.

The transport systems of sugars across the membrane are far from being completely elucidated. There exists, however, at least two kinds of transport systems: a high affinity and a low affinity system (ten times less important) (Bisson, 1991). The low affinity system is essential during the log phase and its activity decreases during the stationary phase. The high affinity system is, on the contrary, repressed by high concentrations of glucose, as in the case of grape must (Salmon *et al.*, 1993) (Figure 1.9). The amount of sterols in the membrane, especially ergosterol, as well as the degree of unsaturation of the membrane phospholipids favor the penetration of glucose in the cell. This is especially true during the stationary and decline phases. This phenomenon explains the determining influence of aeration on the successful completion of alcoholic fermentation during the yeast multiplication phase.

The presence of ethanol, in a culture medium, slows the penetration speed of arginine and glucose into the cell and limits the efflux of protons

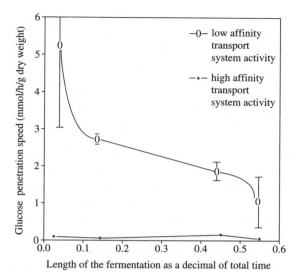

Fig. 1.9. Evolution of glucose transport system activity of *S. cerevisiae* fermenting a medium model (Salmon *et al.*, 1993). LF = Length of the fermentation as a decimal of total time GP = Glucose penetration speed (mmol/h/g of dry weight) 0 = Low affinity transport system activity * = High affinity transport system activity

resulting from membrane ATPase activity (Alexandre *et al.*, 1994; Charpentier, 1995). Simultaneously, the presence of ethanol increases the synthesis of membrane phospholipids and their percentage in unsaturated fatty acids (especially oleic). Temperature and ethanol act in synergy to affect membrane ATPase activity. The amount of ethanol required to slow the proton efflux decreases as the temperature rises. However, this modification of membrane ATPase activity by ethanol may not be the origin of the decrease in plasmic membrane permeability in an alcoholic medium. The role of membrane ATPase in yeast resistance to ethanol has not been clearly demonstrated.

The plasmic membrane also produces cell wall glucan and chitin. Two membrane enzymes are involved: β-1,3 glucanase and chitin synthetase. These two enzymes catalyze the polymerization of glucose and N-acetyl-glucosamine, derived from their activated forms (uridine diphosphates—UDP). The mannoproteins are essentially produced in the endoplasmic reticulum

(Section 1.4.2). They are then transported by vesicles which fuse with the plasmic membrane and deposit their contents at the exterior of the membrane.

Finally, certain membrane proteins act as cellular specific receptors. They permit the yeast to react to various external stimuli such as sexual hormones or changes in the concentration of external nutrients. The activation of these membrane proteins triggers the liberation of compounds such as cyclic adenosine monophosphate (cAMP) in the cytoplasm. These compounds serve as secondary messengers which set off other intercellular reactions. The consequences of these cellular mechanisms in the alcoholic fermentation process merit further study.

1.4 THE CYTOPLASM AND ITS ORGANELLES

Between the plasmic membrane and the nuclear membrane, the cytoplasm contains a basic cytoplasmic substance, or cytosol. The organelles (endoplasmic reticulum, Golgi apparatus, vacuole and mitochondria) are isolated from the cytosol by membranes.

1.4.1 Cytosol

The cytosol is a buffered solution, with a pH between 5 and 6, containing soluble enzymes, glycogen and ribosomes.

Glycolysis and alcoholic fermentation enzymes (Chapter 2) as well as trehalase (an enzyme catalyzing the hydrolysis of trehalose) are present. Trehalose, a reserve disaccharide, also cytoplasmic, ensures yeast viability during the dehydration and rehydration phases by maintaining membrane integrity.

The lag phase precedes the log phase in a sugar-containing medium. It is marked by a rapid degradation of trehalose linked to an increase in trehalase activity. This activity is itself closely related to an increase in the amount of cAMP in the cytoplasm. This compound is produced by a membrane enzyme, adenylate cyclase, in response to the stimulation of a membrane receptor by an environmental factor.

Glycogen is the principal yeast glucidic reserve substance. Animal glycogen is similar in structure. It accumulates during the stationary phase in the form of spherical granules of about 40 µm in diameter.

When observed under the electron microscope, the yeast cytoplasm appears rich in ribosomes. These tiny granulations, made up of ribonucleic acids and proteins, are the center of protein synthesis. Joined to polysomes, several ribosomes migrate the length of the messenger RNA. They translate it simultaneously so that each one produces a complete polypeptide chain.

1.4.2 The Endoplasmic Reticulum, the Golgi Apparatus and the Vacuoles

The **endoplasmic reticulum** (ER) is a double membrane system partitioning the cytoplasm. It is linked to the cytoplasmic membrane and nuclear membrane. It is, in a way, an extension of the latter. Although less developed in yeasts than in exocrine cells of higher eucaryotes, the ER has the same function. It ensures the addressing of the proteins synthesized by the attached ribosomes. As a matter of fact, ribosomes can be either free in the cytosol or bound to the ER. The proteins synthesized by free ribosomes remain in the cytosol, as do the enzymes involved in glycolysis. Those produced in the ribosomes bound to the ER have three possible destinations: the vacuole, the plasmic membrane, and the external environment (secretion). The presence of a signal sequence (a particular chain of amino acids) at the N-terminal extremity of the newly formed protein determines the association of the initially free ribosomes in the cytosol with the ER. The synthesized protein crosses the ER membrane by an active transport process called translocation. This process requires the hydrolysis of an ATP molecule. Having reached the inner space of the ER, the proteins undergo certain modifications including the necessary excising of the signal peptide by the signal peptidase. In many cases, they also undergo a glycosylation.

The yeast glycoproteins, in particular the structural, parietal or enzymatic mannoproteins, contain glucidic side chains (Section 1.2.2). Some of these are linked to asparagine by N-glycosidic bonds. This oligosaccharidic link is constructed in the interior of the ER by the sequential addition of activated sugars (in the form of UDP derivatives) to a hydrophobic, lipidic transporter called dolicholphosphate. The entire unit is transferred in one piece to an asparagine residue of the polypeptide chain. The dolicholphosphate is regenerated.

The **Golgi apparatus** consists of a stack of membrane sacs and associated vesicles. It is an extension of the ER. Transfer vesicles transport the proteins issued from the ER to the sacs of the Golgi apparatus. The Golgi apparatus has a dual function. It is responsible for the glycosylation of protein, then sorts so as to direct them via specialized vesicles either into the vacuole or into the plasmic membrane. An N-terminal peptidic sequence determines the directing of proteins towards the vacuole. This sequence is present in the precursors of two vacuolar-orientated enzymes in the yeast: Y carboxypeptidase and A proteinase. The vesicles that transport the proteins of the plasmic membrane or the secretion granules, such as those that transport the periplasmic invertase, are still the default destinations.

The **vacuole** is a spherical organelle, 0.3 to 3 µm in diameter, surrounded by a single membrane. Depending on the stage of the cellular cycle, yeasts have one or several vacuoles. Before budding, a large vacuole splits into small vesicles. Some penetrate into the bud. Others gather at the opposite extremity of the cell and fuse to form one or two large vacuoles. The vacuolar membrane or tonoplast has the same general structure (fluid mosaic) as the plasmic membrane but it is more elastic and its chemical composition is somewhat different. It is less rich in sterols and contains less protein and glycoprotein but more phospholipids with a higher degree of unsaturation. The vacuole stocks some of the cell hydrolases, in particular Y carboxypeptidase, A and B proteases, I aminopeptidase, X-propyl-dipeptidylaminopeptidase and alkaline phosphatase. In this respect, the yeast vacuole can be compared to an animal cell lysosome. Vacuolar proteases play an essential role in the turn-over of cellular proteins. In addition, the A protease is indispensable in the maturation of other vacuolar hydrolases. It excises a small peptide sequence and thus converts precursor forms (proenzymes) into active enzymes. The vacuolar proteases also autolyze the cell after its death. Autolysis, while ageing white wine on its lees, can affect wine quality and should concern the winemaker.

Vacuoles also have a second principal function: they stock metabolites before their use. In fact, they contain a quarter of the pool of the amino acids of the cell, including a lot of arginine as well as S-adenosyl methionine. In this organelle, there is also potassium, adenine, isoguanine, uric acid and polyphosphate crystals. These are involved in the fixation of basic amino acids. Specific permeases ensure the transport of these metabolites across the vacuolar membrane. An ATPase linked to the tonoplast furnishes the necessary energy for the movement of stocked compounds against the concentration gradient. It is different from the plasmic membrane ATPase, but also produces a proton efflux.

The ER, Golgi apparatus and vacuoles can be considered as different components of an internal system of membranes, called the vacuome, participating in the flux of glycoproteins to be excreted or stocked.

1.4.3 The Mitochondria

Distributed in the periphery of the cytoplasm, the mitochondria (mt) are spherically or rod-shaped organelles surrounded by two membranes. The inner membrane is highly folded to form cristae. The general organization of mitochondria is the same as in higher plants and animal cells. The membranes delimit two compartments: the inner membrane space and the matrix. The mitochondria are true respiratory organelles for yeasts. In aerobiosis, the *S. cerevisiae* cell contains about 50 mitochondria. In anaerobiosis, these organelles degenerate, their inner surface decreases, and the cristae disappear. Ergosterol and unsaturated fatty acids supplemented in culture media limit the degeneration of mitochondria in anaerobiosis. In

any case, when cells formed in anaerobiosis are placed in aerobiosis, the mitochondria regain their normal appearance. Even in aerated grape must, the high sugar concentration represses the synthesis of respiratory enzymes. As a result, the mitochondria no longer function. This phenomenon, catabolic glucose repression, will be described in Chapter 2.

The mitochondrial membranes are rich in phospholipids—principally phosphatidylcholine, phosphatidylinositol and phosphatidylethanolamine (Figure 1.5). Cardiolipin (diphosphatidylglycerol), in minority in the plasmic membrane (Figure 1.4), is predominant in the inner mitochondrial membrane. The fatty acids of the mitochondrial phospholipids are in C16:0, C16:1, C18:0, C18:1. In aerobiosis, the unsaturated residues predominate. When the cells are grown in anaerobiosis, without lipid supplements, the short-chain saturated residues become predominant; cardiolipin and phosphatidylethanolamine diminish whereas the proportion of phosphatidylinositol increases. In aerobiosis, the temperature during the log phase of the cell influences the degree of unsaturation of the phospholipids- more saturated as the temperature decreases.

The mitochondrial membranes also contain sterols, as well as numerous proteins and enzymes (Guerin, 1991). The two membranes, inner and outer, contain enzymes involved in the synthesis of phospholipids and sterols. The ability to synthesize significant amounts of lipids, characteristic of yeast mitochondria, is not limited by respiratory deficient mutations or catabolic glucose repression.

The outer membrane is permeable to most small metabolites coming from the cytosol since it contains porine, a 29 kDa transmembrane protein possessing a large pore. Porine is present in the mitochondria of all the eucaryotes as well as in the outer membrane of bacteria. The intermembrane space contains adenylate kinase, which ensures interconversion of ATP, ADP and AMP. Oxidative phosphorylation takes place in the inner mitochondrial membrane. The matrix, on the other hand, is the center of the reactions of the tricarboxylic acids cycle and of the oxidation of fatty acids.

The majority of mitochondria proteins are coded by the genes of the nucleus and are synthesized by the free polysomes of the cytoplasm. The mitochondria, however, also have their own machinery for protein synthesis. In fact, each mitochondrion possesses a circular 75 kb (kilobase pairs) molecule of double-stranded AND and ribosomes. The mtDNA is extremely rich in A (adenine) and T (thymine) bases. It contains a few dozen genes, which code in particular for the synthesis of certain pigments and respiratory enzymes, such as cytochrome *b*, and several sub-units of cytochrome oxidase and of the ATP synthetase complex. Some mutations affecting these genes can result in the yeast becoming resistant to certain mitochondrial specific inhibitors such as oligomycin. This property has been applied in the genetic marking of wine yeast strains. Some mitochondrial mutants are respiratory deficient and form small colonies on solid agar media. These '*petit*' mutants are not used in winemaking because it is impossible to produce them industrially by respiration.

1.5 THE NUCLEUS

The yeast nucleus is spherical. It has a diameter of 1–2 mm and is barely visible using a phase contrast optical microscope. It is located near the principal vacuole in non-proliferating cells. The nuclear envelope is made up of a double membrane attached to the ER. It contains many ephemeral pores, their locations continually changing. These pores permit the exchange of small proteins between the nucleus and the cytoplasm. Contrary to what happens in higher eucaryotes, the yeast nuclear envelope is not dispersed during mitosis. In the basophilic part of the nucleus, the crescent-shaped nucleolus can be seen by using a nuclear-specific staining method. As in other eucaryotes, it is responsible for the synthesis of ribosomal RNA. During cellular division, the yeast nucleus also contains rudimentary spindle threads composed of microtubules of tubulin, some discontinuous and others continuous (Figure 1.10). The continuous microtubules are stretched between the two spindle pole bodies (SPB). These corpuscles are permanently included in the nuclear membrane and

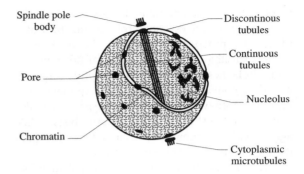

Fig. 1.10. The yeast nucleus (Williamson, 1991). SPB = Spindle pole body; NUC = Nucleolus; P = Pore; CHR = Chromatin; CT = Continuous tubules; DCT = Discontinuous tubules; CTM = Cytoplasmic microtubules

correspond with the centrioles of higher organisms. The cytoplasmic microtubules depart from the spindle pole bodies towards the cytoplasm.

There is little nuclear DNA in yeasts compared with higher eucaryotes—about 14 000 kb in a haploid strain. It has a genome almost three times larger than in *Escherichia coli*, but its genetic material is organized into true chromosomes. Each one contains a single molecule of linear double-stranded DNA associated with basic proteins known as histones. The histones form chromatin which contains repetitive units called nucleosomes. Yeast chromosomes are too small to be observed under the microscope.

Pulse-field electrophoresis (Carle and Olson, 1984; Schwartz and Cantor, 1984) permits the separation of the 16 chromosomes in *S. cerevisiae*, whose size range from 200 to 2000 kb. This species has a very large chromosomic polymorphism. This characteristic has made karyotype analysis one of the principal criteria for the identification of *S. cerevisiae* strains (Section 1.9.3). The scientific community has nearly established the complete sequence of the chromosomic DNA of *S. cerevisiae*. In the future, this detailed knowledge of the yeast genome will constitute a powerful tool, as much for understanding its molecular physiology as for selecting and improving winemaking strains.

The yeast chromosomes contain relatively few repeated sequences. Most genes are only present in a single example in the haploid genome, but the ribosomal RNA genes are highly repeated (about 100 copies).

The genome of *S. cerevisiae* contains transposable elements, or transposons—specifically, Ty (transposon yeast) elements. These comprise a central ε region (5.6 kb) framed by a direct repeated sequence called the δ sequence (0.25 kb). The δ sequences have a tendency to recombine, resulting in the loss of the central region and a δ sequence. As a result, there are about 100 copies of the δ sequence in the yeast genome. The Ty elements code for non-infectious retrovirus particles. This retrovirus contains Ty messenger RNA as well as a reverse transcriptase capable of copying the RNA into complementary DNA. The latter can reinsert itself into any site of the chromosome. The random excision and insertion of Ty elements in the yeast genome can modify the genes and play an important role in strain evolution.

Only one plasmid, called the 2 μm plasmid, has been identified in the yeast nucleus. It is a circular molecule of DNA, containing 6 kb and there are 50–100 copies per cell. Its biological function is not known, but it is a very useful tool, used by molecular biologists to construct artificial plasmids and genetically transform yeast strains.

1.6 REPRODUCTION AND THE YEAST BIOLOGICAL CYCLE

Like other sporiferous yeasts belonging to the class Ascomycetes, *S. cerevisiae* can multiply either asexually by vegetative multiplication or sexually by forming ascospores. By definition, yeasts belonging to the imperfect fungi can only reproduce by vegetative multiplication.

1.6.1 Vegetative Multiplication

Most yeasts undergo vegetative multiplication by a process called budding. Some yeasts, such as species belonging to the genus *Schizosaccharomyces*, multiply by binary fission.

Figure 1.11 represents the life cycle of *S. cerevisiae* divided into four phases: M, G1, S, and G2. M corresponds with mitosis, G1 is the period

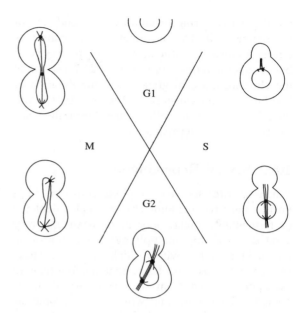

Fig. 1.11. *S. cerevisiae* cell cycle (vegetative multiplication) (Tuite and Oliver, 1991). M = mitosis; G1 = period preceding DNA synthesis; S = DNA synthesis; G2 = period preceding mitosis

preceding S, which is the synthesis of DNA and G2 is the period before cell division. As soon as the bud emerges, in the beginning of S, the splitting of the spindle pole bodies (SPB) can be observed in the nuclear membrane by electron microscopy. At the same time, the cytoplasmic microtubules orient themselves toward the emerging bud. These microtubules seem to guide numerous vesicles which appear in the budding zone and are involved in the reshaping of the cell wall. As the bud grows larger, discontinued nuclear microtubules begin to appear. The longest microtubules form the mitotic spindle between the two SPB. At the end of G2, the nucleus begins to push and pull apart in order to penetrate the bud. Some of the mitochondria also pass with some small vacuoles into the bud, whereas a large vacuole is formed at the other pole of the cell. The expansion of the latter seems to push the nucleus into the bud. During mitosis, the nucleus stretches to its maximum and the mother cell separates from the daughter cell. This separation takes place only after the construction of the separation cell wall and the deposit of a ring of chitin on the bud scar of the mother cell. The movement of chromosomes during mitosis is difficult to observe in yeasts, but a microtubule–centromere link must guide the chromosomes. In grape must, the duration of budding is approximately 1–2 hours. As a result, the population of the cells double during the yeast log phase during fermentation.

1.6.2 Sexual Reproduction

When sporiferous yeast diploid cells are in a hostile nutritive medium (for example, depleted of fermentable sugar, poor in nitrogen and very aerated) they stop multiplying. Some transform into a kind of sac with a thick cell wall. These sacs are called asci. Each one contains four haploid ascospores issued from meiotic division of the nucleus. Grape must and wine are not propitious to yeast sporulation and, in principal, it never occurs in this medium. Yet Mortimer *et al.* (1994) observed the sporulation of certain wine yeast strains, even in sugar-rich media. Our researchers have often observed asci in old agar culture media stored for several weeks in the refrigerator or at ambient temperatures (Figure 1.12). The natural conditions in which wild wine yeasts sporulate and the frequency of this phenomenon are not known. In the laboratory, the agar or liquid medium

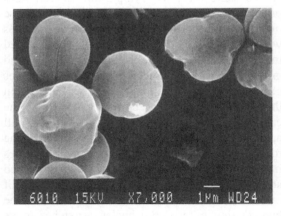

Fig. 1.12. Scanning electron microscope photograph of *S. cerevisiae* cells placed on a sugar-agar medium for several weeks. Asci containing ascospores can be observed

conventionally used to provoke sporulation has a sodium acetate base (1%). In *S. cerevisiae*, sporulation aptitude varies greatly from strain to strain. Wine yeasts, both wild and selected, do not sporulate easily, and when they do they often produce non-viable spores.

Meiosis in yeasts and in higher eucaryotes (Figure 1.13) has some similarities. Several hours after the transfer of diploid vegetative cells to a sporulation medium, the SPB splits during the DNA replication of the S phase. A dense body (DB) appears simultaneously in the nucleus near the nucleolus. The DB evolves into synaptonemal complexes—structures permitting the coupling and recombination of homologous chromosomes. After 8–9 hours in the sporulation medium, the two SPB separate and the spindle begins to form. This stage is called metaphase I of meiosis. At this stage, the chromosomes are not yet visible. Then, while the nuclear membrane remains intact, the SPB divides. At metaphase II, a second mitotic spindle stretches itself while the ascospore cell wall begins to form. Nuclear buds, cytoplasm and organelles migrate into the ascospores. At this point, edification of the cell wall is completed. The spindle then disappears when the division is achieved.

Placed in favorable conditions, i.e. nutritive sugar-enriched media, the ascospores germinate, breaking the cell wall of the ascus, and begin to multiply. In *S. cerevisiae*, the haploid cells have two mating types: **a** and α. The ascus contains two **a** ascospores and two α ascospores (Figure 1.14). Sign **a** (*MAT*a) cells produce a sexual pheromone **a**. This peptide made up of 12 amino acids is called sexual factor **a**. In the same manner, sign α cells produce the sexual factor α, a peptide made up of 13 amino acids. The **a** factor, emitted by the *MAT*a cells, stops the multiplication of *MAT*α cells in G1. Reciprocally, the α factor produced by α cells stops the biological cycle of **a** cells. Sexual coupling occurs between two cells of the opposite sexual sign. Their agglutination permits cellular and nuclear fusion and makes use of parietal glycoproteins and **a** and α agglutinins. The vegetative diploid cell that is formed (**a**/α) can no longer produce sexual pheromones and is insensitive to their action; it multiplies by budding.

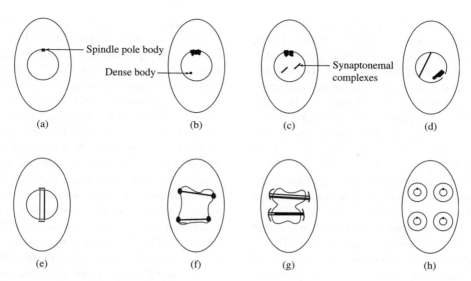

Fig. 1.13. Meiosis in *S. cerevisiae* (Tuite and Oliver, 1991). SPB = spindle pole body; DB = dense body; SC = synaptonemal complexes. (a) Cell before meiosis; (b) dividing of SPB; (c) synaptonemal complexes appear; (d) separation of the SPB; (e) constitution of spindle (metaphase I of meiosis); (f) dividing of the SPB; (g) metaphase. II of meiosis; (h) end of meiosis; formation of ascospores

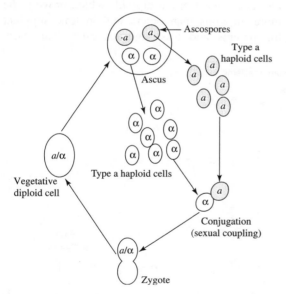

Fig. 1.14. Reproduction cycle of a heterothallic yeast strain (*a*, *α*: spore sexual signs)

Some strains, from a monosporic culture, can be maintained in a stable haploid state. Their sexual sign remains constant during many generations. They are **heterothallic**. Others change sexual sign

during a cellular division. Diploid cells appear in the descendants of an ascospore. They are **homothallic** and have an *HO* gene which inverses sexual sign at an elevated frequency during vegetative division. This changeover (Figure 1.15) occurs in the mother cell at the G1 stage of the

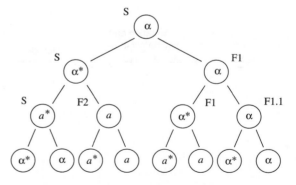

Fig. 1.15. Sexual sign commutation model of haploid yeast cells in a homothallic strain (Herskowitz *et al.*, 1992) (∗ designates cells capable of changing sexual sign at the next cell division, or cells already having undergone budding). S = initial cell carrying the *HO* gene; F1, F2 = daughter cells of S; F1.1. = daughter cell of F1

biological cycle, after the first budding but before the DNA replication phase. In this manner, a sign α ascospore S divides to produce two α cells (S and the first daughter cell, F1). During the following cellular division, S produces two cells (S and F2) that have become **a** cells. In the same manner, the F1 cell produces two α cells after the first division and two **a** cells during its second budding. Laboratory strains that are deficient or missing the *HO* gene have a stable sexual sign. Heterothallism can therefore be considered the result of a mutation of the *HO* gene or of genes that control its functioning (Herskowitz *et al.*, 1992).

Most wild and selected winemaking strains that belong to the *S. cerevisiae* species are diploid and homothallic. It is also true of almost all of the strains that have been isolated in vineyards of the Bordeaux region. Moreover, recent studies carried out by Mortimer *et al.* (1994) in Californian and Italian vineyards have shown that the majority of strains (80%) are homozygous for the *HO* character (*HO/HO*); heterozygosis (*HO/ho*) is in minority. Heterothallic strains (*ho/ho*) are rare (less than 10%). We have made the same observations for yeast strains isolated in the Bordeaux region. For example, the F10 strain fairly prevalent in spontaneous fermentations in certain Bordeaux growths is *HO/HO*. In other words, the four spores issued from an ascus give monoparent diploids, capable of forming asci when placed in a pure culture. This generalized homozygosis for the *HO* character of wild winemaking strains is probably an important factor in their evolution, according to the genome renewal phenomenon proposed by Mortimer *et al.* (1994) (Figure 1.16), in which the continuous multiplication of a yeast strain in its natural environment accumulates heterozygotic damage to the DNA. Certain slow-growth or functional loss mutations of certain genes decrease strain vigor in the heterozygous state. Sporulation, however, produces haploid cells containing different combinations of these heterozygotic characters. All of these spores become homozygous diploid cells with a series of genotypes because of the homozygosity of the *HO* character. Certain diploids which prove to be more vigorous than others will in time supplant the parents and less vigorous ones. This very

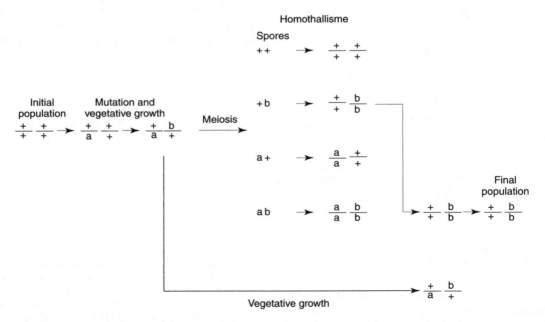

Fig. 1.16. Genome renewal of a homozygote yeast strain for the *HO* gene of homothallism, having accumulated recessive mutations during vegetative multiplication (Mortimer *et al.*, 1994) (a and b = mutation of certain genes)

tempting model is reaffirmed by the characteristics of the wild winemaking strains analyzed. In these, the spore viability rate is the inverse function of the heterozygosis rate for a certain number of mutations. The completely homozygous strains present the highest spore viability and vigor.

In conclusion, sporulation of strains in natural conditions seems indispensable. It assures their growth and fermentation performance. With this in mind, the conservation of selected strains of active dry yeasts as yeast starters should be questioned. It may be necessary to regenerate them periodically to eliminate possible mutations from their genome which could diminish their vigor.

1.7 THE KILLER PHENOMENON

1.7.1 Introduction

Certain yeast strains, known as killer strains (K), secrete proteinic toxins into their environment that are capable of killing other, sensitive strains (S). The killer strains are not sensitive to their toxin but can be killed by a toxin that they do not produce. Neutral strains (N) do not produce a toxin but are resistant. The action of a killer strain on a sensitive strain is easy to demonstrate in the laboratory on an agar culture medium at pH 4.2–4.7 at 20°C. The sensitive strain is inoculated into the mass of agar before it solidifies; then the strain to be tested is inoculated in streaks on the solidified medium. If it is a killer strain, a clear zone in which the sensitive strain cannot grow encircles the inoculum streaks (Figure 1.17).

This phenomenon, the killer factor, was discovered in *S. cerevisiae* but killer strains also exist in other yeast genera such as *Hansenula, Candida, Kloeckera, Hanseniaspora, Pichia, Torulopsis, Kluyveromyces* and *Debaryomyces*. Killer yeasts have been classified into 11 groups according to the sensitivity reaction between strains as well as the nature and properties of the toxins involved. The killer factor is a cellular interaction model mediated by the proteinic toxin excreted. It has given rise to much fundamental research (Tipper and Bostian, 1984; Young, 1987). Barre (1984, 1992), Radler (1988) and Van Vuuren and

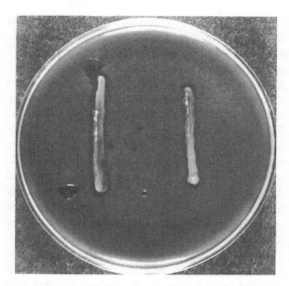

Fig. 1.17. Identification of the K2 killer phenotype in *S. cerevisiae.* The presence of a halo around the two streaks of the killer strain is due to the death of the sensitive strain cultivated on the medium

Jacobs (1992) have described the technological implications of this phenomenon for wine yeasts and the fermentation process.

1.7.2 Physiology and Genetics of the Killer Phenomenon

The determinants of the killer factor are both cytoplasmic and nuclear. In *S. cerevisiae*, the killer phenomenon is associated with the presence of double-stranded RNA particles, virus-like particles (VLP), in the cytoplasm. They are in the same category as non-infectious mycovirus. There are two kinds of VLP: M and L. The M genome (1.3–1.9 kb) codes for the K toxin and for the immunity factor (R). The L genome (4.5 kb) codes for an RNA polymerase and the proteinic capsid that encapsulates the two genomes. Killer strains (K^+R^+) secrete the toxin and are immune to it. The sensitive cells (K^-R^-) do not possess M VLP but most of them have L VLP. The two types of viral particles are necessary for the yeast cell to express the killer phenotype (K^+R^+), since the L mycovirus is necessary for the maintenance of the M type.

There are three kinds of killer activities in *S. cerevisiae* strains. They correspond with the K1, K2 and K3 toxins coded, respectively, by M1, M2 and M3 VLPs (1.9, 1.5 and 1.3 kb, respectively). According to Wingfield *et al.* (1990), the K2 and K3 types are very similar; M3 VLP results from a mutation of M2 VLP. The K2 strains are by far the most widespread in the *S. cerevisiae* strains encountered in wine. Neutral strains (K^-R^+) are insensitive to a given toxin without being capable of producing it. They possess M VLPs of normal dimensions that code only for the immunity factor. They either do not produce toxins or are inactive because of mutations affecting the M-type RNA.

Many chromosomal genes are involved in the maintenance and replication of L and M RNA particles as well as in the maturation and transport of the toxin produced.

The K1 toxin is a small protein made up of two sub-units (9 and 9.5 kDa). It is active and stable in a very narrow pH range (4.2–4.6) and is therefore inactive in grape must. The K2 toxin, a 16 kDa glycoprotein, produced by homothallic strains of *S. cerevisiae* encountered in wine, is active at between pH 2.8 and 4.8 with a maximum activity between 4.2 and 4.4. It is therefore active at the pH of grape must and wine.

K1 and K2 toxins attack sensitive cells by attaching themselves to a receptor located in the cell wall—a β-1,6 glucan. Two chromosomal genes, *KRE*1 and *KRE*2 (*Killer resistant*), determine the possibility of this linkage. The *kre*1 gene produces a parietal glycoprotein which has a β-1,6 glucan synthetase activity. The *kre*1 mutants are resistant to K1 and K2 toxins because they are deficient in this enzyme and devoid of a β-1,6 glucan receptor. The *KRE*2 gene is also involved in the fixation of toxins to the parietal receptor; the *kre*2 mutants are also resistant. The toxin linked to a glucan receptor is then transferred to a membrane receptor site by a mechanism needing energy. Cells in the log phase are, therefore, more sensitive to the killer effect than cells in the stationary phase. When the sensitive cell plasmic membrane is exposed to the toxin, it manifests serious functional alterations after a lag phase of about 40 minutes. These alterations include the interruption of the coupled transport of amino acids and protons, the acidification of the cellular contents, and potassium and ATP leakage. The cell dies in 2–3 hours after contact with the toxin because of the above damage, due to the formation of pores in the plasmic membrane.

The killer effect exerts itself exclusively on yeasts and has no effect on humans and animals.

1.7.3 The Role of the Killer Phenomenon in Winemaking

Depending on the authors and viticultural regions studied, the frequency of the killer character varies a lot among wild winemaking strains isolated on grapes or in fermenting grape must. In a work by Barre (1978) studying 908 wild strains, 504 manifested the K2 killer character, 299 were sensitive and 95 neutral. Cuinier and Gros (1983) reported a high frequency (65–90%) of K2 strains in Mediterranean and Beaujolais region vineyards, whereas none of the strains analyzed in Tourraine manifested the killer effect. In the Bordeaux region, the K2 killer character is extremely widespread. In a study carried out in 1989 and 1990 on the ecology of indigenous strains of *S. cerevisiae* in several tanks of red must in a Pessac-Léognan vineyard, all of the isolated strains manifested K2 killer activity, about 30 differentiated by their karyotype (Frezier, 1992). Rossini *et al.* (1982) reported an extremely varied frequency (12–80%) of K2 killer strains in spontaneous fermentations in Italian wineries. Some K2 killer strains were also isolated in the southern hemisphere (Australia, South Africa and Brazil). On the other hand, most of the killer strains isolated in Japan presented the K1 characteristic. Most research on the killer character of wine yeasts concerns the species *S. cerevisiae*. Little information exists on the killer effect of the alcohol-sensitive species which essentially make up grape microflora. Heard and Fleet (1987) confirmed Barre's (1980) observations and did not establish the existence of the killer effect in *Candida*, *Hanseniaspora*, *Hansenula* and *Torulaspora*. However, some killer strains of *Hanseniaspora uvarum* and *Pichia kluyveri* have been identified by Zorg *et al.* (1988).

Barre (1992) studied the activity and stability of the K2 killer toxin in enological conditions (Figure 1.18). The killer toxin only manifested a pronounced activity on cells in the log phase. Cells in the stationary phase were relatively insensitive. The amount of ethanol or SO_2 in the wine has practically no effect on the killer toxin activity. On the other hand, it is quickly destroyed by heat, since its half-life is around 30 minutes at 32°C. It is also quickly inactivated by the presence of phenolic compounds and is easily adsorbed by bentonite.

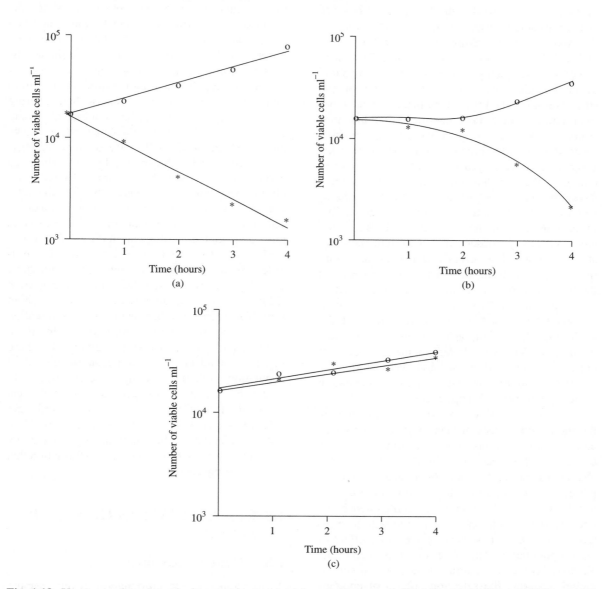

Fig. 1.18. Yeast growth and survival curves in a grape juice medium containing killer toxin (Barre, 1992): *, 10% K2 strain active culture supernatant; ○, 10% supernatant inactivated by heat treatment. (a) White juice, pH 3.4; cells in exponential phase introduced at time = 0. (b) Same juice, cells in stationary phase introduced at time = 0. (c) Red juice extracted by heated maceration, pH 3.4; cells in exponential phase introduced at time = 0

Scientific literature has reported a diversity of findings on the role of the killer factor in the competition between strains during grape must fermentation. In an example given by Barre (1992), killer cells inoculated at 2% can completely supplant the sensitive strain during the alcoholic fermentation of must. In other works, the killer yeast/sensitive yeast ratio able to affect the implantation of sensitive yeasts in winemaking varies between 1/1000 and 100/1, depending on the author. This considerable discrepancy can probably be attributed to implantation and fermentation speed of the strains present. The killer phenomenon seems more important to interstrain competition when the killer strain implants itself quickly and the sensitive strain slowly. In the opposite situation, an elevated percentage of killer yeasts would be necessary to eliminate the sensitive population. Some authors have observed spontaneous fermentations dominated by sensitive strains despite a non-negligible proportion of killer strains (2–25%). In Bordeaux, we have always observed that certain sensitive strains implant themselves in red wine fermentation, despite a strong presence of killer yeasts in the wild microflora (for example, 522M, an active dry yeast starter). In white winemaking, the neutral yeast VL1 and sensitive strains such as EG8, a slow-growth strain, also successfully implant themselves. The wild killer population does not appear to compete with a sensitive yeast starter and therefore is not an important cause of fermentation difficulties in real-life applications.

The high frequency of killer strains among the indigenous yeasts in many viticultural regions confers little advantage to the strain in terms of implantation capacity. In other words, this character is not sufficient to guarantee the implantation of a certain strain during fermentation over a wild strain equally equipped. On the other hand, under certain conditions, inoculating with a sensitive strain will fail because of the killer effect of a wild population. Therefore, the resistance to the K2 toxin (killer or neutral phenotype) should be included among the selection criteria of enological strains. The high frequency of the K2 killer character in indigenous wine yeasts facilitates this strategy.

A medium that contains the toxin exerts a selection pressure on a sensitive enological strain. Stable variants survive this selection pressure and can be obtained in this manner (Barre, 1984). This is the most simple strategy for obtaining a killer enological strain. However, the development of molecular genetics and biotechnology permits scientists to construct enological strains modified to contain one or several killer characters. Cytoduction can achieve these modifications. This method introduces cytoplasmic determinants (mitochondria, plasmids) issued from a killer strain into a sensitive enological strain without altering the karyotype of the initial enological strain. Seki *et al.* (1985) used this method to make the 522M strain K2 killer. By another strategy, new yeasts can be constructed by integrating the toxin gene into their chromosomes. Boone *et al.* (1990) were able to introduce the K1 character into K2 winemaking strains in this manner. The K1 killer character among wine yeasts is rare, and so the enological interest of this last application is limited. The acquiring of multi-killer character strains presents little enological advantage. Sensitive selected strains and current K2 killer strains can already be implanted without a problem. On the other hand, the dissemination of these newly obtained multi-killer strains in nature could present a non-negligible risk. These strains could adversely affect the natural microflora population, although we have barely begun to inventory its diversity and exploit its technological potentials. It would be detrimental to be no longer able to select wild yeasts because they have been supplanted in their natural environment by genetically modified strains—a transformation that has no enological interest.

1.8 CLASSIFICATION OF YEAST SPECIES

1.8.1 General Remarks

As mentioned in the introduction to this chapter, yeasts constitute a vast group of unicellular fungi—taxonomically heterogeneous and very complex. Hansen's first classification at the beginning of this century only distinguished between sporiferous and asporiferous yeasts. Since then, yeast taxonomy has incited considerable research. This research has been regrouped in successive works progressively creating the classification known today. The last enological treaty of the University of Bordeaux (Ribéreau-Gayon et al., 1975) was based on Lodder's (1970) classification. Between this monograph and the previous classification (Lodder and Kregger-Van Rij, 1952), the designation and classification of yeasts had already changed profoundly. In this book, the last two classifications by Kregger-Van Rij (1984) and Barnett et al. (2000) are of interest. These contain even more significant changes in the delimitation of species and genus with respect to earlier systematics.

According to the current classification, yeasts belonging to Ascomycetes, Basidiomycetes and imperfect fungi (Deuteromycetes) are divided into 81 genera, to which 590 species belong. Taking into account synonymy and physiological races (varieties of the same species), at least 4000 names for yeasts have been used since the nineteenth century. Fortunately, only 15 yeast species exist on grapes, are involved as an alcoholic fermentation agent in wine, and are responsible for wine diseases. Table 1.1 lists the two families to which enological yeasts belong: *Saccharomycetaceae* in the Ascomycetes (sporiferous) and *Cryptococcaceae* in the Deuteromycetes (asporiferous). Fourteen

genera to which one or several species of grape or wine yeasts belong are not listed in Table 1.1.

1.8.2 Evolution of the General Principles of Yeast Taxonomy and Species Delimitation

Yeast taxonomy (from the Greek *taxis*: putting in order), and the taxonomy of other microorganisms for that matter, includes classification and identification. Classification groups organisms into *taxa* according to their similarities and/or their ties to a common ancestor. The basic taxon is species. A species can be defined as a collection of strains having a certain number of morphological, physiological and genetic characters in common. This group of characters constitutes the standard description of the species. Identification compares an unknown organism to individuals already classed and named that have similar characteristics.

Taxonomists first delimited yeast species using morphological and physiological criteria. The first classifications were based on the phenotypic differences between yeasts: cell shape and size, spore formation, cultural characters, fermentation and assimilation of different sugars, assimilation of nitrates, growth-factor needs, resistance to cycloheximide. The treaty on enology by Ribéreau-Gayon et al. (1975) described the use of these methods on wine yeasts in detail. Since then, many rapid, ready-to-use diagnostic kits have been

Table 1.1. Classification of grape and wine sporogeneous and asporogeneous yeast genere (Kregger-Van Rij, 1984)

Saccharomycetaceae family (sporogeneous)			*Spermophtoracae* family (asporogeneous)	*Cryptococcaceae* family (asporogeneous)
Sub-family *Schizosaccharomycetoideae*	Sub-family *Nadsonioideae*	Sub-family *Saccharomycetoideae*		—
Genus *Schizosaccharomyces*	Genus *Saccharomycodes* *Hanseniaspora*	Genus *Saccharomyces* *Debaryomyces* *Dekker* *Hansenula* *Kluyveromyces* *Pichia* *Zygosaccharomyces* *Torulaspora*	Genus *Metschnikowia*	Genus *Brettanomyces* *Candida* *Kloeckera* *Rhodotorula*

proposed to determine yeast response to different physiological tests. Lafon-Lafourcade and Joyeux (1979) and Cuinier and Leveau (1979) designed the API 20 C system for the identification of enological yeasts. It contains eight fermentation tests, 10 assimilation tests and a cycloheximide resistance test. For a more complete identification, the API 50 CH system contains 50 substrates for fermentation (under paraffin) and assimilation tests. Heard and Fleet (1990) developed a system that uses the different tests listed in the work of Barnett *et al.* (1990).

Due to the relatively limited number of yeast species significantly present on grapes and in wine, these phenotypic tests identify enological yeast species in certain genera without difficulty. Certain species can be identified by observing growing cells under the microscope. Small apiculated cells, having small lemon-like shapes, designate the species *Hanseniaspora uvarum* and its imperfect form *Kloeckera apiculata* (Figure 1.19). *Saccharomycodes ludwigii* is characterized by much larger (10–20 μm) apiculated cells. Since most yeasts multiply by budding, the genus *Schizosaccharomyces* can be recognized because of its vegetative reproduction by binary fission (Figure 1.20). In modern taxonomy, this genus only contains the species *Schizosacch. pombe.* Finally, the budding of *Candida stellata* (formerly known as *Torulopsis stellata*) occurs in the shape of a star.

According to Barnett *et al.* (1990), the physiological characteristics listed in Table 1.2 can be used to distinguish between the principal grape and wine yeasts. Yet some of these characters (for example, fermentation profiles of sugars) vary within the species and are even unstable for a given strain during vegetative multiplication. Taxonomists realized that they could not differentiate species based solely on phenotypic discontinuity criteria. They progressively established a delimitation founded on the biological and genetic definition of a species.

In theory, a species can be defined as a collection of interfertile strains whose hybrids are themselves fertile—capable of producing viable spores. This biological definition runs into several problems

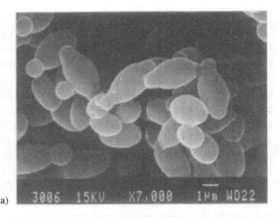
(a)

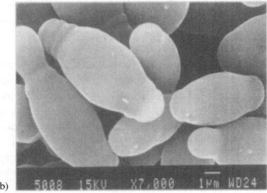
(b)

Fig. 1.19. Observation of two enological yeast species having an apiculated form. (a) *Hanseniaspora uvarum.* (b) *Saccharomycodes ludwigii*

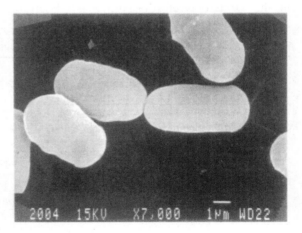

Fig. 1.20. Binary fission of *Schizosaccharomyces pombe*

Table 1.2. Physiological characteristics of the principal grape and wine yeasts (Barnett *et al.*, 1990)

	F1 D-Glucose fermentation	F2 D-Galactose fermentation	F3 Maltose fermentation	F4 Me α-D-glucoside fermentation	F5 Sucrose fermentation	F5 α,α-Trehalose fermentation	F7 Melibiose fermentation	F8 lactose Fermentation	F9 Cellobiose fermentation	F10 Melezitose fermentation	F11 Raffinose fermentation	F12 Inulin fermentation	F13 Starch fermentation	C1 D-Glucose growth	C2 D-Galactose growth	C3 L-Sorbose growth	C4 D-Glucosamine growth	C5 D-Ribose growth	C6 D-Xylose growth	C10 Sucrose growth	C11 Maltose growth	C12 α,α-Trehalose growth	C13 Me α-D-Glucoside growth	C14 Cellobiose growth	C15 Salicin growth	C16 Arbutin growth	C17 Melibiose growth	C18 Lactose growth	C19 Raffinose growth	C20 Melezitose growth	C21 Inulin growth	C22 Starch growth	C23 Glycerol growth	C24 Erythritol growth	C25 Ribitol growth	C26 Xylitol growth	C28 D-Glucitol growth	C29 D-Mannitol growth
Candida stellata	+	−	−	−	+	−	−	−	−	−	v	−	−	+	−	v	−	−	−	+	−	−	−	−	−	−	−	−	+	−	−	−	−	−	−	−	−	−
Candida vini	v	−	−	−	−	−	−	−	−	−	−	−	−	+	−	−	−	−	−	−	−	−	−	−	−	−	−	−	−	−	−	−	v	−	v	v	+	+
Candida famata	v	v	v	−	v	v	v	−	v	v	v	−	−	+	+	v	v	v	+	+	+	+	+	v	v	v	v	v	+	+	v	v	+	v	+	+	+	+
Dekkera anomala	+	v	v	v	+	v	−	v	v	v	v	−	−	v	v	−	v	−	−	v	v	v	v	v	v	v	−	v	v	v	−	−	v	−	−	−	−	−
Dekkera bruxellensis	v	v	v	+	+	+	−	−	v	v	−	−	−	v	v	−	v	v	−	v	v	v	v	v	v	v	−	−	v	v	v	−	v	−	−	−	−	−
Hanseniaspora uvarum	+	−	−	−	−	−	−	−	v	−	−	−	−	+	−	−	−	−	−	−	−	v	−	+	+	+	−	−	−	−	−	−	−	−	−	−	v	−
Metschnikowia pulcherrima	+	v	−	−	−	−	−	−	−	−	−	−	−	+	+	v	v	v	v	+	+	+	+	+	+	+	−	−	−	+	−	−	+	−	v	+	+	+
Pichia anomala	+	v	v	v	+	v	−	−	v	−	v	−	v	+	v	−	−	v	v	+	+	v	+	v	+	+	−	−	v	v	−	+	+	v	v	v	+	+
Pichia fermentans	+	−	−	−	−	−	−	−	−	−	−	−	−	+	−	−	v	−	+	−	−	−	−	−	−	−	−	−	−	−	−	−	+	−	−	−	−	−
Pichia membranefaciens	v	−	−	−	−	−	−	−	−	−	−	−	−	+	−	v	v	−	v	−	−	−	−	−	−	−	−	−	−	−	−	−	v	−	−	v	−	−
*Saccharomyces cerevisiae	+	v	v	v	v	v	v	−	−	v	v	−	v	+	v	−	−	−	−	v	v	v	v	−	−	−	v	−	v	v	−	v	v	−	−	−	v	v
Saccharomycodes ludwigii	+	−	−	−	+	−	−	−	−	−	v	−	−	+	−	−	−	−	−	+	−	−	−	+	+	+	−	−	v	−	−	−	v	−	−	−	−	−
Kluyveromyces thermolerens	+	v	v	v	+	v	−	−	−	v	+	v	−	+	v	+	−	−	−	+	+	+	+	−	−	−	−	−	+	+	v	−	+	−	v	+	v	v
Schizosaccharomyces pombe	+	v	v	v	+	−	v	−	−	−	v	v	v	+	v	−	−	−	−	+	+	v	v	−	−	−	v	−	+	−	v	v	−	−	−	−	−	−
Zygosaccharomyces bailii	+	−	−	−	v	v	−	−	−	−	v	−	−	+	v	v	−	−	−	v	−	v	−	−	−	−	−	−	v	−	−	−	v	−	v	v	v	v

	C30 Calactitol growth	C31 *myo*-Inositol growth	C32 D-Glucono-1,5-lactone growth	C33 2-Keto-D-gluconate growth	C34 5-Keto-D-gluconate growth	C35 D-Gluconate growth	C38 DL-Lactate growth	C39 Succinate growth	C40 Citrate growth	C42 Ethanol growth	N1 Nitrate growth	N2 Nitrite growth	N3 Ethylamine growth	N4 L-Lysine growth	N5 Cadaverine growth	N6 Creatine growth	N7 Creatinine growth	N8 Glucosamine growth	V1 Growth W/O vitamins	V2 Growth W/O myo-Inositol	V3 Growth W/O Pantothenate	V4 Growth W/O Biotin	V5 Growth W/O Thiamin	V6 Growth W/O Biotin & Thiamin	V7 Growth W/O Pyridoxine	V8 Growth W/O Pyridoxine & Thiamin	V9 Growth W/O Niacin	T2 Growth at 30°C	T3 Growth W/O at 35°C	T4 Growth W/O at 37°C	T5 Growth W/O at 40°C	O1 0.01% Cycloheximide growth	O3 1% Acetic acid growth	O4 50% D-Glucose growth	M2 Acetic acid production	M3 Urea hydrolysis
Candida stellata	−	−	−	−	−	−	−	−	−	−	−	−	−	+	−	−	−	−	−	−	v	−	−	−	+	−	v	+	v	−	−	−	−	v	−	−
Candida vini	−	−	−	v		−	v	+	−	+	−	−	+	+	+	−	−	−	−	+	+	v	v	v	+		+	v	−	−	−	−	−	−	−	−
Candida famata	v	−	v	+		v	v	+	+	+	−	v	+	+	+	v	−	−	v	+	+	v	+	v	+		+	+	v	v	−	v	−	+	−	−
Dekkera anomala	−	−	v	−	−	v	v	v	−	v	v	+	+	+	+	−	−	−	−	v	+	−	−	−	+	−	+	+	+	v	−	+	−	−	+	−
Dekkera bruxellensis	−	−	v	v	−	−	v	−	−	v	v	v	+	+	+	−	−	−	−	+	+	−	−	−	+	−	+	+	v	v	v	+	v	−	+	−
Hanseniaspora uvarum	−	−	+	+	−	v	−	−	−	−	−	−	+	+	+	−	−	−	−	−	−	v	v	−	−	−	−	+	−	−	−	+	−	v	−	−
Metschnikowia pulcherrima	−	−	+	+		+	v	+	+	+	−	−	+	+	+	−	−	−	−	+	+	−	+	−	+		+	v	v	−	−	−	−	v	−	−
Pichia anomala	−	−	+	−	−	v	+	+	+	−	+	+	+	+	+	−	−	−	+	+	+	+	+	+	+	+	+	+	v	v	−	−	−	v	−	−
Pichia fermentans	−	−	v	−	−	−	+	+	+	−	−	−	+	+	+	−	−	+	−	+	+	+	−	−	+	−	+	+	+	+	−	−	−	v	−	−
Pichia membranefaciens	−	−	v	−	−	−	v	v	−	+	−	−	+	+	+	−	−	v	v	+	+	v	v	v	+	v	+	+	v	v	−	−	−	v	−	−
*Saccharomyces cerevisiae	−	−	v	−	−	−	v	v	−	v	−	−	−	−	−	−	−	−	v	v	v	v	v	v	v	v	+	+	v	v	v	−	−	v	−	−
Saccharomycodes ludwigii	−	−	v	−	−	−	v	−	−	v	−	−	+	+	+	−	−	+	−	+	−	−	v	−	−	−	−	+	+	v	−	−	−	−	−	−
Kluyveromyces thermolerens	−	−	v	v	−	v	−	v	−	v	−	−	+	+	+	−	−	−	−	−	v	v	v	v	v		v	+	+	v	−	−	−	+	−	−
Schizosaccharomyces pombe	−	−	v	v	−	v	−	−	−	−	−	−	v	v	v	−	−	−	−	−	v	v	v	−	v	−	−	+	+	+	v	v	+	+	−	+
Zygosaccharomyces bailii	−	−	v	v	−	−	−	−	−	+	−	−	+	+	+	−	−	−	−	+	+	−	+	−	+		+	+	v	v	−	−	+	+	−	−

+: test positive; −: test negative; v: variable result.
*With these tests they cannot be differentiated from *S. bayanus, S. paradoxus* and *S. pastorianus*.

when applied to yeasts. First of all, a large number of yeasts (Deuteromycetes) are not capable of sexual reproduction. Secondly, a lot of Ascomycetes yeasts are homothallic; hybridization tests are especially fastidious and difficult for routine identification. Finally, certain wine yeast strains have little or no sporulation aptitude, which makes the use of strain infertility criteria even more difficult.

To overcome these difficulties, researchers have developed a molecular taxonomy over the last 15 years based on the following tests: DNA recombination; the similarity of DNA base composition; the similarity of enzymes; ultrastructure characteristics; and cell wall composition. The DNA recombination tests have proven to be effective for delimiting yeast species. They measure the recombination percentages of denatured nuclear DNA (mono-stranded) of different strains. An elevated recombination rate between two strains (80–100%) indicates that they belong to the same species. A low recombination percentage (less than 20% of the sequences in common) signifies that the strains belong to different and very distant species. Combination rates between these extremes are more difficult to interpret.

1.8.3 Successive Classifications of the Genus *Saccharomyces* and the Position of Wine Yeasts in the Current Classification

Due to many changes in yeast classification and nomenclature since the beginning of taxonomic studies, enological yeast names and their positions in the classification have often changed. This has inevitably resulted in some confusion for enologists and winemakers. Even the most recent enological works (Fleet, 1993; Delfini, 1995; Boulton *et al.*, 1995) use a number of different epithets (*cerevisiae, bayanus, uvarum*, etc.) attached to the genus name *Saccharomyces* to designate yeasts responsible for alcoholic fermentation. Although still in use, this enological terminology is no longer accurate to designate the species currently delimited by taxonomists.

The evolution of species classification for the genus *Saccharomyces* since the early 1950s

(Table 1.3) has created this difference between the designation of wine yeasts and current taxonomy. By taking a closer look at this evolution, the origin of the differences may be understood.

In Lodder and Kregger-Van Rij (1952), the names *cerevisiae, oviformis, bayanus, uvarum*, etc. referred to a number of the 30 species of the genus *Saccharomyces*. Ribéreau-Gayon and Peynaud (1960) in the *Treatise of Œnology* considered that two principal fermentation species were found in wine: *S. cerevisiae* (formerly called *ellipsoideus*) and *S. oviformis*. The latter was encountered especially towards the end of fermentation and was considered more ethanol resistant. The difference in their ability to ferment galactose distinguished the two species. *S. cerevisiae* (Gal^+) fermented galactose, whereas *S. oviformis* (Gal^-) did not. According to the same authors, the species *S. bayanus* was rarely found in wines. Although it possessed the same physiological fermentation and sugar assimilation characters as *S. oviformis*, its cells were more elongated, its fermentation was slower, and it had a particular behavior towards growth factors. The species *S. uvarum* was identified in wine by many authors. It differed from *cerevisiae, oviformis* and *bayanus* because it could ferment melibiose.

In Lodder's following edition (Lodder 1970), the number of species of the genus *Saccharomyces* increased from 30 to 41. Some species formerly grouped with other genera were integrated into the genus *Saccharomyces*. Moreover, several species names were considered to be synonyms and disappeared altogether. Such was the case of *S. oviformis*, which was moved to the species *bayanus*. The treatise of Ribéreau-Gayon *et al.* (1975) considered, however, that the distinction between *oviformis* and *bayanus* was of enological interest because of the different technological characteristics of these two yeasts. Nevertheless, by the beginning of the 1980s most enological work had abandoned the name *oviformis* and replaced it with *bayanus*. This name change began the confusion that currently exists.

The new classification by Kregger-Van Rij (1984), based on Yarrow's work on base percentages of guanine and cytosine in yeast DNA,

Table 1.3. Evolution of the nomenclature for the *Saccharomyces* genus, 1952–1990

1952: *The Yeasts, a Taxonomic Study*—I (Lodder and Kreger-Van Rij)	1970: *The Yeasts, a Taxonomic Study*—II (Lodder)	1984: *The Yeasts, a Taxonomic Study*—III (Yarrow)	2000: *Yeasts* (Barnett *et al.*)	
Saccharomyces cerevisiae ⇑	Saccharomyces cerevisiae		Saccharomyces cerevisiae	Group I Saccharomyces sensu stricto
Saccharomyces pastorianus	Saccharomyces aceti			
Saccharomyces bayanus ⇑	Saccharomyces bayanus		Saccharomyces bayanus	
Saccharomyces oviformis	Saccharomyces capensis			
Saccharomyces logos	Saccharomyces prostoserdovii	Saccharomyces cerevisiae		
Saccharomyces chevalieri ⇑	Saccharomyces chevalieri			
Saccharomyces fructuum	Saccharomyces coreanus			
Saccharomyces lactis	Saccharomyces diastaticus			
Saccharomyces elegans	Saccharomyces globosus		Saccharomyces paradoxus	
Saccharomyces heterogenicus	Saccharomyces heterogenicus			
Saccharomyces fermentati	Saccharomyces hienipiensis			
Saccharomyces mellis	Saccharomyces inusitatus			
Saccharomyces italicus ⇑	Saccharomyces italicus			
Saccharomyces steineri	Saccharomyces norbensis		Saccharomyces pastorianus	
Saccharomyces pastori	Saccharomyces oleaceus			
Saccharomyces carlsbergensis	Saccharomyces oleaginosus			
Saccharomyces uvarum ⇑	Saccharomyces uvarum			
	Saccharomyces incompspicuus			
Saccharomyces acidifaciens	Saccharomyces amurcae			
Saccharomyces bailii ⇑	Saccharomyces bailii			
Saccharomyces fragilis	Saccharomyces cidri			
	Saccharomyces dairensis ⇑	Saccharomyces dairensis	Saccharomyces dairensis	
Saccharomyces delbrueckii ⇑	Saccharomyces delbrueckii			
Saccharomyces marxianus	Saccharomyces eupagycus			
Saccharomyces exiguus ⇑	Saccharomyces exiguus ⇑	Saccharomyces exiguus	Saccharomyces exiguus	Group II Saccharomyces sensu lato
Saccharomyces veronae	Saccharomyces fermentati			
Saccharomyces florentinus ⇑	Saccharomyces florentinus ⇑			
Saccharomyces bisporus ⇑	Saccharomyces bisporus ⇑		Saccharomyces unisporus	
Saccharomyces williamus	Saccharomyces kloeckerianus			
	Saccharomyces kluyveri ⇑			
Saccharomyces microellipsodes ⇑	Saccharomyces microellipsodes			
	Saccharomyces montanus		Saccharomyces servazzi	
	Saccharomyces mrakii			
	Saccharomyces pretoriensis			
Saccharomyces rosei ⇑	Saccharomyces rosei			
Saccharomyces rouxii ⇑	Saccharomyces rouxii		Saccharomyces castelli	
	Saccharomyces saitoanus			
	Saccharomyces telluris ⇑	Saccharomyces telluris		
	Saccharomyces transvaalensis			
	Saccharomyces unisporus ⇑	Saccharomyces unisporus	Saccharomyces kluyveri	Group III
	Saccharomyces vafer	Saccharomyces servazzi		

brought forth another important change in the designation of the *Saccharomyces* species. Only seven species continued to exist, while 17 names became synonyms of *S. cerevisiae*. Certain authors considered them to be races or physiological varieties of the species *S. cerevisiae*. As with the preceding classification, these races of *S. cerevisiae* were differentiated by their sugar utilization profile (Table 1.4). However, this method of classification was nothing more than an artificial taxonomy without biological significance. Enologists took to the habit of adding the varietal name to *S. cerevisiae* to designate wine yeasts: *S. cerevisiae* var. *cerevisiae*, var. *bayanus*, var. *uvarum*, var. *chevalieri*, etc. In addition, two species, *bailii* and *rosei*, were removed from the genus *Saccharomyces* and integrated into another genus to become *Zygosaccharomyces bailii* and *Torulaspora delbrueckii*, respectively.

The latest yeast classification (Barnett *et al.*, 2000) is based on recent advances in genetics and molecular taxonomy—in particular, DNA recombination tests reported by Vaughan Martini and Martini (1987) and hybridization experiments

between strains (Naumov, 1987). It has again thrown the species delimitation of the genus *Saccharomyces* into confusion. The species now number 10 and are divided into three groups (Table 1.3). The species *S. cerevisiae*, *S. bayanus*, *S. paradoxus* and *S. pastorianus* cannot be differentiated from one another by physiological tests but can be delimited by measuring the degree of homology of their DNA (Table 1.5). They form the group *Saccharomyces sensu stricto*. *S. pastorianus* replaced the name *S. carlsbergensis*, which was given to brewer's yeast strains used for bottom fermentation (lager) and until now included in the species *cerevisiae*. The recently delimited *S. paradoxus* species includes strains initially isolated from tree exudates, insects, and soil (Naumov *et al.*, 1998). It might constitute the natural common ancestor of three other yeast species involved in the fermentation process. Recent genomic analysis (Redzepovic *et al.*, 2002) identified a high percentage of *S. paradoxus* in Croatian grape microflora. The occurrence of this species in other vineyards around the world and its winemaking properties certainly deserve further investigation.

Table 1.4. Physiological Races of *Saccharomyces cerevisiae* regrouped under a single species *Saccharomyces cerevisiae* by Yarrow and Nakase (1975)

	Fermentation					
	Ga	Su	Ma	Ra	Me	St
Saccharomyces						
aceti	−	−	−	−	−	−
bayanus	−	+	+	+	−	−
capensis	−	+	−	+	−	−
cerevisiae	+	+	+	+	−	−
chevalieri	+	+	−	+	−	−
coreanus	+	+	−	+	+	−
diastaticus	+	+	+	+	−	+
globosus	+	−	−	−	−	−
heterogenicus	−	+	+	−	−	−
hienipiensis	−	−	+	−	+	−
inusitatus	−	+	+	+	+	−
norbensis	−	−	−	−	+	−
oleaceus	+	−	−	+	+	−
oleaginosus	+	−	+	+	+	−
prostoserdovii	−	−	+	−	−	−
steineri	+	+	+	−	−	−
uvarum	+	+	+	+	+	−

Ga = D-galactose; Su = saccharose; Ma = maltose; Ra = raffinose; Me = melibiose; St = soluble starch.

Table 1.5. DNA/DNA reassociation percentages between the four species belonging to genus Saccharomyces *sensu stricto* (Vaughan Matini and Martini, 1987)

	S. cerevisiae	S. bayanus	S. pastorianus	S. paradoxus
S. cerevisiae	100			
S. bayanus	20	100		
S. pastorianus	58	70	100	
S. paradoxus	53	24	24	100

A second group, *Saccharomyces sensu largo*, is made up of the species *exiguus, castelli, servazzi* and *unisporus*. The third group consists only of the species *kluyveri*. Only the first group comprises species of enological interest: *S. cerevisiae, S. bayanus*, and, possibly, *S. paradoxus*, if its suitability for winemaking is demonstrated. This new classification has created a lot of confusion in the language pertaining to the epithet *bayanus*. For taxonomists, *S. bayanus* is a species distinct from *S. cerevisiae*. For enologists and winemakers, *bayanus* (ex *oviformis*) designates a physiological race of *S. cerevisiae* that does not ferment galactose and possesses a stronger resistance to ethanol than *Saccharomyces cerevisiae* var. *cerevisiae*.

By evaluating the infertility of strains (a basic species delimitation criterion), Naumov *et al.* (1993) demonstrated that most strains fermenting melibiose (Mel[+]) isolated in wine, and until now classed as *S. cerevisiae* var. *uvarum*, belong to the species *S. bayanus*. Some strains, however, can be crossed with a reference *S. cerevisiae* to produce fertile descendants. They are therefore attached to *S. cerevisiae*. These results confirm, but nevertheless put into perspective, the past works of Rossini *et al.* (1982) and Bicknell and Douglas (1982), which were based on DNA recombination tests. The DNA recombination percentages are low between the *uvarum* and *cerevisiae* strains tested, but they are elevated between these same *uvarum* strains and the *S. bayanus* strain (CBS 380). In other words, most enological strains formerly called *uvarum* belong to the species *S. bayanus*. This relationship, however, is not complete. Certain *Saccharomyces* Mel[+] found in the spontaneous fermentations of grapes belong to *cerevisiae*. The yeasts that enologists commonly called *S. cerevisiae* var. *bayanus*, formerly *S. oviformis*, were studied to determine if they belong to

the species *bayanus* or to the species *cerevisiae*, as the majority of *uvarum* strains. In this case, their designation only leads to confusion.

All of the results of molecular taxonomy presented above show that the former phenotypic classifications, based on physiological identification criteria, are not even suitable for delimiting the small number of fermentative species of the genus *Saccharomyces* found in winemaking. Moreover, specialists have long known about the instability of physiological properties of *Saccharomyces* strains. Rossini *et al.* (1982) reclassified a thousand strains from the yeast collection of the Microbiology Institute of Agriculture at the University of Perouse. During this research, they observed that 23 out of 591 *S. cerevisiae* strains conserved on malt agar lost the ability to ferment galactose. Twenty three strains 'became' *bayanus*, according to Lodder's (1970) classification. They found even more frequently that, over time, strains acquired the ability to ferment certain sugars. For example, 29 out of 113 strains of *Saccharomyces oviformis* became capable of fermenting galactose, thus 'becoming' *cerevisiae*. According to these authors, this physiological instability is a specific property of strains from the *Saccharomyces* group *sensu stricto*. In the same collection, no noticeable change in fermentation profiles was observed in 150 strains of *Saccharomyces rosei* (today *Torulaspora delbrueckii*) or in 300 strains of *Kloeckera apiculata*. Genetic methods are therefore indispensable for identifying wine yeasts. Yet DNA recombination percentage measures or infertility tests between homothallic strains, a long and fastidious technique, are not practical for routine microbiological controls. The amplification of genome segments by polymerization chain reaction (PCR) is a quicker and easier method which has recently proved to be an excellent tool for discrimination of wine yeast species.

1.8.4 Delimitation of Winemaking Species of *S. cerevisiae* and *S. bayanus* by PCR

Since its discovery by Saiki *et al.* (1985), PCR has often been used to identify different plant and bacteria species. This technique consists of enzymatically amplifying one or several gene fragments *in vitro*. The reaction is based on the hybridization of two oligonucleotides which frame a target region on a double-stranded DNA or template. These oligonucleotides have different sequences

and are complementary to the DNA sequences which frame the strand to amplify. Figure 1.21 illustrates the different stages of the amplification process. The DNA is first denatured at a high temperature (95°C). The reactional mixture is then cooled to a temperature between 37 and 55°C, permitting the hybridization of these oligonucleotides on the denatured strands. The strands serve as primers from which a DNA polymerase permits the stage-by-stage addition of desoxyribonucleotidic units in the 5′–3′ direction. The DNA polymerase (Figure 1.22) requires four deoxyribonucleoside-5′

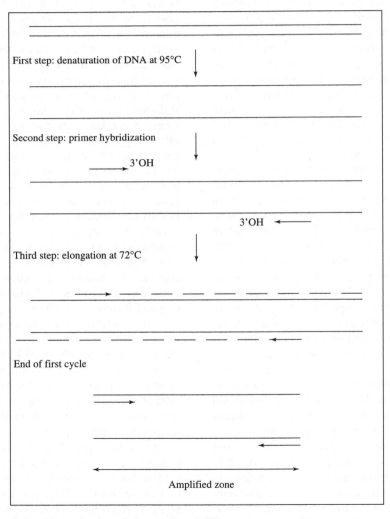

Fig. 1.21. Principle of the polymerization chain reaction (PCR)

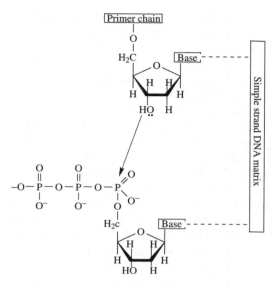

Fig. 1.22. Mode of action of a DNA polymerase

triphosphates (dATP, dGTP, dTTP, dCTP). A phosphodiester bond is formed between the 3'-OH end of the primer and the innermost phosphorus of the activated deoxyribonucleoside; pyrophosphate is thus liberated. The newly synthesized strand is formed on the template model. A thermoresistant enzyme, the TAQ DNA polymerase, is derived from the thermoresistant bacteria *Thermus aquaticus*. It permits a large number of amplification cycles (25–40) *in vitro* without having to add the DNA polymerase after each denaturation. In this manner, the DNA fragment amplified during the first cycle serves as the template for the following cycles. In consequence, each successive cycle doubles the target DNA fragment—amplified by a factor of 10^5 to 10^6 during 25–30 amplification cycles.

Hansen and Kielland-Brandt (1994) proposed *MET*2 gene PCR amplification to differentiate between *S. cerevisiae* and *S. bayanus*, while working on strain types of the species *cerevisiae* and *bayanus* and on a strain of the variety *S. uvarum*. This gene, which codes for the synthesis of the homoserine acetyltransferase, has different sequences in the two species. Part of the gene is initially amplified by using two complementary oligonucleotides of the sequences which border the

fragment to be amplified. The amplificat obtained, about 600 b.p, is the same size for the strains of the species *cerevisiae* and *bayanus* tested, as well as for the isolate designated *S. uvarum*. Different restriction endonucleases, which recognize certain specific DNA sequences, then digest the amplified fragment. Figure 1.23 gives an example of the mode of action of the EcoRI restriction endonuclease. This enzyme recognizes the base sequences GAATTC and cuts at the location indicated by the arrows. Electrophoresis is used to separate the obtained fragments. As a result, the restriction fragment length polymorphism (RFLP) can be appreciated. The restriction profiles obtained differ between *cerevisiae* and *bayanus*. They are identical for the strain types *bayanus* and *uvarum* tested.

This PCR–RFLP technique associated to the *MET*2 gene has been developed and adapted for rapid analysis. The whole cells are simply heated in water (95°C), 10 minutes before amplification. Only two restriction enzymes are used: EcoRI and PstI (Masneuf *et al.*, 1996a,b). The *MET*2 amplificat (580 bp) is cut into two fragments (369 and 211 bp) by EcoRI in *S. cerevisiae*. PstI restriction creates two fragments for strain type *S. bayanus*. EcoRI does not cut the *MET*2 amplificat of *S. bayanus*, nor does PstI cut the *S. cerevisiae* amplificat (Figure 1.24). Masneuf (1996) demonstrated that *S. paradoxus* can be identified by this method. Its *MET*2 gene amplificat produces one fragment of the same size as with the two other species. This one, however, is not cleaved by EcoRI or PstI, but rather by Mae III.

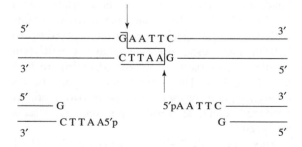

Fig. 1.23. Recognition site and cutting mode of an *ECO*R1 restriction endonuclease

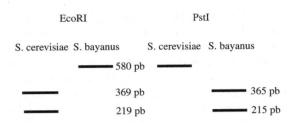

Fig. 1.24. Identification principles for the species *S. cerevisiae* and *S. bayanus* by the *MET2* gene PCR-RFLP technique, after cutting the amplificat by *Eco*RI and *Pst*I restriction enzymes

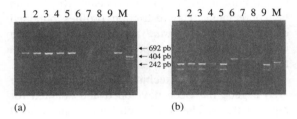

Fig. 1.25. Agar gel electrophoresis (1.8%) of (a) *Eco*R1 and P*st*I digestions of the *MET2* gene amplificats of the Mel+ strains studied by Naumov *et al.* (1993). Band 1: *S. bayanus* SCU 11; band 2: *S. bayanus* SCU 13; band 3: *S. bayanus* SCU 73; band 4: *S. bayanus* L19; band 5: *S. bayanus* L490: band 6: *S. cerevisiae* L 579; band 7: *S. cerevisiae* L 1425; band 8: *S. cerevisiae* VKM-Y 502; band 9: *S. bayanus* VKM-Y 1146. M = molecular weight markers

By applying this relatively simple and quick technique to different enological strains of *S. uvarum* studied by Naumov *et al.* (1993), strains attached to the species *bayanus* by hybridization tests have been clearly demonstrated to present the same profile characteristic as *bayanus* (two bands after restriction with PstI, no restriction with EcoRI). On the other hand, the *uvarum* strains included in the species *cerevisiae*, according to hybridization tests, effectively have a restriction profile characteristic of *S. cerevisiae* (Figure 1.25 and Table 1.6). The delimitation of the species *cerevisiae* and *bayanus* by these two methods produced identical results for the 12 strains analyzed.

This type of PCR–RFLP analysis of the *MET2* gene has been extended to different selected yeast strains available in the trade and currently used

in winemaking. Depending on their ability to ferment galactose, wine professionals in the entire world still call these strains *cerevisiae* or *bayanus* (Table 1.7). For all of these strains, restriction profile characteristics of the species *S. cerevisiae* have been obtained.

In the same manner, the species of 82 indigenous *Saccharomyces* strains isolated in wines in fermentation and on grapes has been determined (Table 1.8). For the eight Gal⁺Mel⁻ strains and the 47 Gal⁻Mel⁻ strains analyzed, called *cerevisiae* and *bayanus* respectively by enologists, the restriction profiles of the *MET2* gene amplificat are characteristic of the species *S. cerevisiae*. Similar results were obtained for 2 *chevalieri* strains fermenting galactose but not maltose (Ma⁻), as well as for the *capensis* strain (Gal⁻Ma⁻). Most of the Mel⁺ strains, called *uvarum* until now, (11 out of 12 for the isolates from Sauternes and 11 out of 11 for the isolates from Sancerre), belong to the species *S. bayanus*. Yet certain Mel⁺ are *S. cerevisiae* (one strain from Sauternes and two strains from the Lallemand collection). To summarize, the classification of the main winemaking yeasts (Section 1.8.3) has gone through three stages. Initially, several separate species were envisaged: *S. cerevisiae*, *S. bayanus* and/or *S. oviformis*, and *S. uvarum*. Subsequently, all of these were thought to belong to a single species: *S. cerevisiae*. The current classification identifies three distinct species on the basis of molecular biological data: *S. cerevisiae*, *S. bayanus*, and *S. paradoxus*. As the strains of *S. bayanus* used in winemaking belong exclusively to the *S. uvarum* variety (or sub-species), the remainder of this Handbook will consider just two species of winemaking yeasts, *S. cerevisiae* and *S. uvarum*. The involvement of *S. paradoxus* in grape fermentation microflora has yet to be confirmed.

Finally, PCR–RFLP associated with the *MET2* gene can be used to demonstrate the existence of hybrids between the species *S. cerevisiae* and *S. bayanus*. This method has been used to prove the existence (Masneuf *et al.*, 1998) of such a natural hybrid (strain S6U var. *uvarum*) among dry yeasts commercialized by Lallemand Inc. (Montreal, Canada). Ciolfi (1992, 1994) isolated

Table 1.6. Characterization by PCR–RFLP of the *MET2* gene of 12 *S. uvarum* (Mel⁺) reclarified, after hybridization testy by Naumov *et al.* (1993), as the species *S. cerevisiae* and *S. bayanus* (Masneuf, 1996)

Strain	CLIB number	Origin	Author	Hybridization test	PCR–RFLP of the *MET2* gene
VKM Y-502	219	Russia	Naumov	*S. cerevisiae* (control)	*S. cerevisiae*
VKM Y-1146	218	Russia grape	Naumov	*S. bayanus* (control)	*S. bayanus*
58 1	—	FŒB must	Sapis-Domercq	*S. bayanus*	*S. bayanus*
SCU 11	101	ITVN wine	Poulard	*S. bayanus*	*S. bayanus*
SCU 13	102	ITVN wine	Poulard	*S. bayanus*	*S. bayanus*
SCU 74	103	ITVN wine	Poulard	*S. bayanus*	*S. bayanus*
L 19	108	ITVT wine	Cuinier	*S. bayanus*	*S. bayanus*
L 99	109	ITVT wine	Cuinier	*S. bayanus*	*S. bayanus*
L 490	110	ITVT wine	Cuinier	*S. bayanus*	*S. bayanus*
DBVPG 1642	113	UPG grape	Vaughan Martini	*S. bayanus*	*S. bayanus*
DBVPG 1643	114	UPG grape	Vaughan Martini	*S. bayanus*	*S. bayanus*
DBVPG 1689	115	UPG grape	Vaughan Martini	*S. bayanus*	*S. bayanus*
L 579	94	ITVT wine	Cuinier	*S. cerevisiae*	*S. cerevisiae*
L 1425	95	ITVT wine	Cuinier	*S. cerevisiae*	*S. cerevisiae*

CLIB: Collection de Levures d'Intérêt Biotechnologique (collection of yeasts of Biotechnological Interest) INA-PG, Grignon, France.
FŒB: Faculté Œnologie de l'Université de Bordeaux II, Talence, France.
ITVN: Institut Technique du Vin (Institut of Wine Technology), Centre d'expérimentation de Nantes, France.
ITVT: Institut Technique du Vin (Institut of Wine Technology), Centre d'expérimentation de Tours, France.
UPG: Univeraita degli Studi de Perugia, Italy.

Table 1.7. Characterization by PCR–RFLP of the *MET2* gene of various selected commercial strains used in winemaking

Strains	Commercial brand	Origin	Enological designation	Species
VL1	Zymaflore VL1	FŒB	*S. cerevisiae*	*S. cerevisiae*
VL3c	Zymaflore VL3	FŒB	*S. cerevisiae*	*S. cerevisiae*
WET 136	Siha levactif 3	Dormstadt	*S. cerevisiae*	*S. cerevisiae*
71B	Actiflore primeur	INRA Narbonne	*S. cerevisiae*	*S. cerevisiae*
F10	Zymaflore F10	FŒB	*S. bayanus*	*S. cerevisiae*
R2	Vitlevure KD	n-a	*S. bayanus*	*S. cerevisiae*
BO213	Actiflore bayanus	Institut Pasteur	*S. bayanus*	*S. cerevisiae*
CH158	Siha levactif 4	n-a	*S. bayanus*	*S. cerevisiae*
QA23	Lalvin QA23	UTM	*S. bayanus*	*S. cerevisiae*
IOC182007	IOC 182007	IŒC	*S. bayanus*	*S. cerevisiae*
DV10	Vitlevure DV10	CIVC	*S. bayanus*	*S. cerevisiae*
O16	Lalvin O16	UB	*S. bayanus*	*S. cerevisiae*
Epemay2	Uvaferm CEG	n-a	*S. bayanus*	*S. cerevisiae*

CLIB: Collection de Levures d'Intérêt Biotechnologique (Collection of yeast of biotechnological interest), INA-PG, Grignon, France.
FŒB: Faculté d' Œnologie de l'Université de Bordeeux II, Talence, France.
UTM: Université de Trasos Montes, Portugal.
ŒC: Institut Œnologique de Champagne, France.
CIVC: Comité Interprofessionnel des vins de Champagne (Interprofessional Champagne Committee), Epernay, France.
UB: Université de Bourgogne, Dijon, France.
na: not available.

Table 1.8. Characterization by PCR–RFLP of the *MET2* gene of various species of wild *Saccharomyces* isolated on the grape and in wine (Masneuf, 1996)

Number of different strain analyzed	Origin	Collection	Enological designation	Species
8	Sauternes wines	FŒB	*cerevisiae*	*S. cerevisiae*
2	Dry white Bordeaux wines	FŒB	*bayanus*	*S. cerevisiae*
9	Sauternes wines	FŒB	*bayanus*	*S. cerevisiae*
2	Dry white Bordeaux wines	FŒB	*chevalieri*	*S. cerevisiae*
1	Sauternes wines	FŒB	*capensis*	*S. cerevisiae*
36	Unknown	Lallemand	*bayanus*	*S. cerevisiae*
11	Sauternes wines	FŒB	*uvarum*	*S. bayanus*
1	Sauternes wines	FŒB	*uvarum*	*S. cerevisiae*
10	Sancerre and Pouilly/Loire Valley wines	FŒB	*uvarum*	*S. bayanus*
1	Sancerre grapes	FŒB	*uvarum*	*S. bayanus*
2	Unknown	Lallemand	*uvarum*	*S. cerevisiae*

FŒB: Faculté dŒnologie de l'Université de Bordeaux II, Talence, France. Lallemand: Lallemand Inc. Montreal, Quebec, Canada.

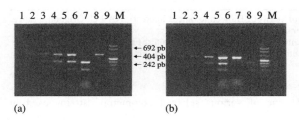

(a)   (b)

Fig. 1.26. Electrophoresis in agarose gel (1.8%) of (a) *Eco*RI and (b) P*st*I digestions of the amplificats of the *MET2* gene of the strain hybrid. Bands 1, 2, 3: sub-clones of the hybrid strain; band 4: hybrid strain; band 5: *S. cerevisiae* VKM-Y 502; band 6: *S. bayanus* VKM-Y 1146. M = molecular weight marker

this yeast in an Italian winery. It was selected for certain enological properties, in particular its aptitude to ferment at low temperatures, its low production of acetic acid, and its ability to preserve must acidity. The *MET2* gene restriction profiles of this strain by EcoRI and PstI, constituted by three bands, are identical (Figure 1.26). In addition to the amplified fragment, two bands characteristic of *S. cerevisiae* with EcoRI and two bands characteristic of the species *S. bayanus* with PstI are obtained. The bands are not artefacts due to an impurity in the strain, because the amplification of the *MET2* gene carried out on subclones (obtained from the multiplication of unique cells isolated by a micromanipulator) produces identical

results. Furthermore, after sporulation of the strain in the laboratory, 10 tetrads were equally isolated by a micromanipulator after the digestion of the ascus cell wall. None of the 40 ascospores analyzed could germinate. The non-viability of the ascospores concurs with the hypothesis that this strain is an interspecific hybrid. Hansen of the Carlsberg laboratory (Denmark) sequenced two of the MET2 gene alleles from this strain. The sequence of one of the alleles is identical to that of the *S. cerevisiae* MET2 gene, with the exception of one nucleotide. The sequence of the other allele is 98.5% similar to that of *S. bayanus*. The presence of this allele is thus probably due to an interspecific cross.

Recent research (Naumov *et al.*, 2000b) has shown that the S6U strain is, in fact, a tetraploid interspecific hybrid. Indeed, the percentage germination of spores from 24 tetrads, isolated using a micromanipulator, was very high (94%), whereas it would have been very low for a "normal" diploid interspecific hybrid. The monospore clones in this first generation (F1) were all capable of sporulating, while none of the ascospores of the second-generation tetrads were viable. The hybrid nature of the monospore clones produced by F1 was confirmed by the presence of the *S. cerevisiae* and *S. uvarum* MET2 gene, identified by PCR/RFLP. Finally, measuring the DNA content per cell using flux cytometry estimation confirmed that the descendants of S6U were interspecific diploids and that S6U itself was an allotetraploid. Natural *S. cerevisiae/S. uvarum* hybrids have been isolated on grapes and in spontaneously fermenting musts in Alsace (Lejeune and Masneuf, unpublished results).

Several other methods using PCR/RFLP have been applied to typing *Saccharomyces* itself. The fragments amplified were ribosomal DNA sequences (DNAr) (Guillamon *et al.*, 1998; Nguyen *et al.*, 2000).

1.9 IDENTIFICATION OF WINE YEAST STRAINS

1.9.1 General Principles

The principal yeast species involved in grape must fermentation, particularly *S. cerevisiae*, comprise a very large number of strains with varied technological properties. The yeast strains which are involved during winemaking influence fermentation speed, the nature and quantity of secondary products formed during alcoholic fermentation, and the aromatic characters of the wine. The ability to differentiate between the different strains of *S. cerevisiae* is required for the following fields: the ecological study of wild yeasts responsible for the spontaneous fermentation of grape must; the selection of strains presenting the best enological qualities; production and marketing controls; the verification of the implantation of selected yeasts used as yeast starter; and the constitution and maintenance of wild or selected yeast collections.

Bouix *et al.* (1981) (cited in Van Vuuren and Van Der Meer, 1987) conducted the initial research on infraspecific differentiation within *S. cerevisiae*. They attempted to distinguish strains by electrophoretic analysis of their exocellular proteins and later (1987) used the separation of intracellular proteins. Other teams proposed identifying the strains by the analysis of long-chain fatty acids using gas chromatography (Tredoux *et al.*, 1987; Augustyn *et al.*, 1991; Bendova *et al.*, 1991; Rozes *et al.*, 1992). Although these different techniques differentiate between certain strains, they are irrefutably less discriminating than genetic differentiation methods. They also present the major inconvenience of depending on the physiological state of the strains and the cultural conditions, which must always be identical.

In the late 1980s, owing to the development of genetics, certain techniques of molecular biology were successfully applied to characterize wine yeast strains. They are based on the clonal polymorphism of the mitochondrial and genomic DNA of *S. cerevisiae*. These genetic methods are independent of the physiological state of the yeast, unlike the previous techniques based on the analysis of metabolism byproducts.

1.9.2 Mitochondrial DNA Analysis

The mtDNA of *S. cerevisiae* has two remarkable properties: it is extremely polymorphous, depending on the strain; and its is stable (it mutates very little) during vegetative multiplication. Restriction endonucleases (such as *Eco*R5) cut this DNA at specific sites. This process generates fragments of variable size which are few in number and can be separated by electrophoresis on agarose gel.

Aigle *et al.* (1984) first applied this technique to brewer's yeasts. Since 1987, it has been used for the characterization of enological strains of *S. cerevisiae* (Dubourdieu *et al.*, 1987; Hallet *et al.*, 1988).

The extraction of mtDNA comprises several stages. The protoplasts obtained by enzymatic digestion of the cell walls are lysed in a hypotonic buffer. The mtDNA is then separated from the chromosomic DNA by ultracentrifugation in a cesium chloride gradient, in the presence of bisbenzimide which acts as a fluorescent intercalating agent. This agent amplifies the difference in density between chromosomic and mtDNA. The mtDNA has an elevated amount of adenine and thymine base pairs for which the bisbenzimide has a strong affinity. Finally, the mtDNA is purified by a phenolchloroform-based extraction and an ethanol-based precipitation.

Defontaine *et al.* (1991) and Querol *et al.* (1992) simplified this protocol by separating the mitochondria from the other cellular constituents before extracting the DNA. In this manner, they avoided the ultracentrifugation step. The coarse cellular debris is eliminated from the yeast lysate by centrifuging at 1000 g. The supernatant is recentrifuged at 1500 g to obtain the mitochondria. The mitochondria are then lysed in a suitable buffer to liberate the DNA.

Unlike industrial brewer strains analyzed by Aigle *et al.* (1984), which have the same mtDNA restriction profile, implying that they are of common origin, the enological yeast strains have a large mtDNA diversity. This method differentiates between most of the selected yeasts used in winemaking as well as wild strains of *S. cerevisiae* found in spontaneous fermentations (Figure 1.27).

This technique is very discriminating and not too expensive, but it is long and requires several complex manipulations. It is useful for the subtle characterization of a small number of strains. Inoculation effectiveness can also be verified by this method. To verify an inoculation, a sample is taken during or towards the end of alcoholic fermentation. In the laboratory, the lees are placed in a liquid medium culture. The mtDNA restriction profile of this total biomass and of the yeast starter strain are compared. If the restriction profile of the sample has no supernumerary bands with respect to the yeast starter strain profile, the yeast starter has been properly implanted, with an accuracy of 90%. In fact, in the case of a binary mixture, the minority strain must represent around 10% of the total population to be detected (Hallet *et al.*, 1989).

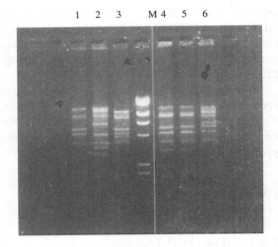

Fig. 1.27. Restriction profile by *Eco*R5 of mtDNA of different strains of *S. cerevisiae.* Band 1: F10; band 2: BO213; band 3: VLI; band 4: 522; band 5: Sita 3; band 6: VL3c. M = marker

1.9.3 Karyotype Analysis

S. cerevisiae has 16 chromosomes with a size range between 250 and 2500 Kb. Its genomic DNA is very polymorphic; thus it is possible to differentiate strains of the species according to the size distribution of their chromosomes. Pulse-field electrophoresis is used to separate *S. cerevisiae* chromosomes and permits the comparison of the karyotypes of the strains. This technique uses two electric fields oriented differently (90 to 120 degrees). The electrodes placed on the sides of the apparatus apply the fields alternately (Figure 1.28).

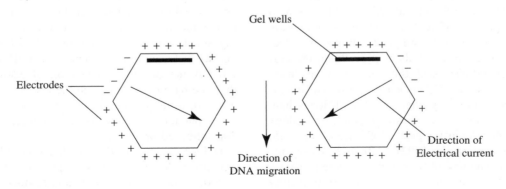

Fig. 1.28. CHEF pulsed field electrophoresis device (contour clamped electrophoresis field)

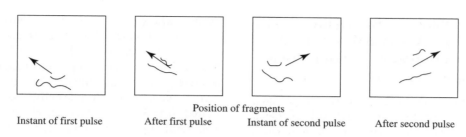

Instant of first pulse   After first pulse   Instant of second pulse   After second pulse

Position of fragments

Fig. 1.29. Principle of DNA molecule separation by pulsed field electrophoresis

The user can define the duration of the electric current that will be applied in each direction (pulse). With each change in direction of the electric field, the DNA molecules reorientate themselves. The smaller chromosomes reorientate themselves more quickly than the larger ones (Figure 1.29).

Blondin and Vezhinet (1988), Edersen *et al.* (1988) and Dubourdieu and Frezier (1990) applied this technique to identify enological yeast strains. Sample preparation is relatively easy. The yeasts are cultivated in a liquid medium, collected during the log phase, and then placed in suspension in a warm agarose solution that is poured into a partitioned mold to form small plugs.

Figure 1.30 gives an example of the identification of *S. cerevisiae* strains isolated from a grape must in spontaneous fermentation by

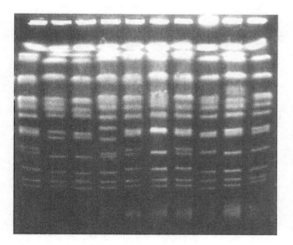

Fig. 1.30. Example of electrophoretic (pulsed field) profile of *S. cerevisiae* strain caryotypes

this method. Vezhinet *et al.* (1990) have shown that karyotype analysis can distinguish between strains of *S. cerevisiae* as well or better than the use of mtDNA restriction profiles. Furthermore, karyotype analysis is much quicker and easier to use than mtDNA analysis. In the case of ecological studies of spontaneous fermentation microflora, pulse-field electrophoresis of chromosomes is extensively used today to characterize strains of *S. cerevisiae* (Frezier and Dubourdieu, 1992; Versavaud *et al.*, 1993, 1995).

Very little research on the chromosomic polymorphism in other species of grape and wine yeasts is currently available. Naumov *et al.* (1993) suggested that *S. bayanus* and *S. cerevisiae* karyotypes can be easily distinguished. Other authors (Vaughan-Martini and Martini, 1993; Masneuf, 1996) have confirmed his results. In fact, a specific chromosomic band systematically appears in *S. bayanus*. Furthermore, there are only two chromosomes whose sizes are less than 400 kb in *S. bayanus* but generally more in *S. cerevisiae*, in all of the strains that we have analyzed.

Species other than *Saccharomyces*, in particular apiculated yeasts (*Hanseniaspora uvarum*, *Kloeckera apiculata*), are present on the grape and are sometimes found at the beginning of fermentations. These species have fewer polymorphous karyotypes and fewer bands than in *Saccharomyces*. Versavaud *et al.* (1993) differentiated between strains of apiculated yeast species and *Candida famata* by using restriction endonucleases at rare sites (Not I and Sfi I). The endonucleases cut the chromosomes into a limited number of fragments, which were then separated by pulse-field electrophoresis.

1.9.4 Genomic DNA Restriction Profile Analysis Associated with DNA Hybridization by Specific Probes (Fingerprinting)

The yeast genome contains DNA sequences which repeat from 10 up to 100 times, such as the δ or Y1 sequences of the chromosome telomeres. The distribution, or more specifically, the number and location of these elements, has a certain intraspecific variability. This genetic fingerprint is used to identify strains (Pedersen, 1986; Degre et al., 1989).

The yeasts are cultivated in a liquid medium. Samples are taken during the log phase, as in the preceding techniques. The entire DNA is isolated and digested by restriction endonucleases. The generated fragments are separated by electrophoresis on agarose gel and then transferred to a nylon membrane (Southern, 1975). Complementary radioactive probes (nucleotide sequences taken from δ and Y1 elements) are used to hybridize with fragments having homologous sequences. The result gives a hybridization profile containing several bands.

Genetic fingerprinting is a more complicated and involved method than mtDNA or karyotype analysis. It is, however, without doubt the most discriminating strain identification method and may even discriminate too well. It has correctly indicated minor differences between very closely related strains. In fact, in the Bordeaux region, *S. cerevisiae* clones isolated from spontaneous fermentations in different wineries have been encountered which have the same karyotype and the same mtDNA restriction profile. Yet their hybridization profiles differ according to sample origin (Frezier, 1992). These strains, probably descendants of the same mother strain, have therefore undergone minor random modifications, maintained during vegetative multiplication.

1.9.5 Polymerization Chain Reaction (PCR) Associated to δ Sequences

This method consists of using PCR to amplify certain sequences of the yeast genome (Section 1.8.4),

occurring between the repeated δ elements, whose separation distance does not exceed a certain value (1 kb). Ness et al. (1992) and Masneuf and Dubourdieu (1994) developed this method to characterize *S. cerevisiae* strains. The amplification is carried out directly on whole cells. They are simply heated to make the cellular envelopes permeable. The resulting amplification fragments are separated according to their size by electrophoresis in agarose gel and viewed using ultraviolet fluorescence (Figure 1.31).

PCR profile analysis associated with δ sequences can distinguish between most *S. cerevisiae* active dry yeast strains (ADY) used in winemaking (Figure 1.32): 25 out of the 26 selected commercial yeast strains analyzed. Lavallée et al. (1994) also observed excellent discriminating power with this method while analyzing industrially produced commercial strains from Lallemand Inc. (Montreal, Canada). In addition, this method permits the identification of 25 to 50 strains per day; it is the quickest of the different strain identification techniques currently available. When used for

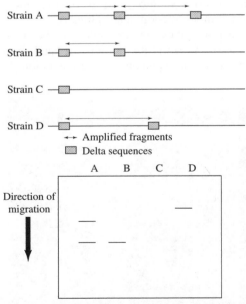

Fig. 1.31. Principle of identification of *S. cerevisiae* strains by PCR associated with *delta* elements

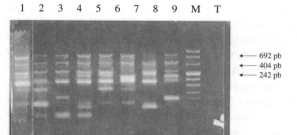

Fig. 1.32. Electrophoresis in agar gel (at 1.8%) of amplified fragments obtained from various commercial yeast strains. Band 1: F10; band 2: BO213; band 3: VL3c; band 4: UP30Y5; band 5: 522 D; band 6: EG8; band 7: L-1597; band 8: WET 136. M = molecular weight marker; T = negative control

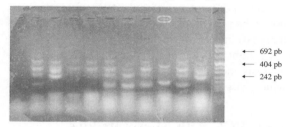

Fig. 1.33. Electrophoresis in agar gel (1.8%) of amplified fragments illustrating examples of verifying yeast implantation (successful: yeasts B and C; unsuccessful: yeasts A, D and E). Band 1: negative control; band 2: Lees A; band 3: ADY A; band 4: Lees B; band 5: ADY B; band 6: Lees C; band 7: ADY C; band 8: Lees D; band 9: ADY D; band 10: Lees E; band 11: ADY E. M = molecular weight marker

indigenous strain identification in a given viticultural region, however, it seems to be less discriminating than karyotype analysis. PCR profiles of wild yeasts isolated in a given location often appear similar. They have several constant bands and only a small number of variable discriminating bands. Certain strains have the same PCR amplification profile while having different karyotypes. In a given location, the polymorphism witnessed by PCR associated with δ sequences is less important than that of the karyotypes. This method is therefore complementary to other methods for characterizing winemaking strains. PCR permits a rapid primary sort of an indigenous population. Karyotype analysis refines this discrimination.

S. bayanus strains cannot be distinguished by this technique because their genome contains only a few Ty elements.

Finally, because of its convenience and rapidity PCR associated with δ sequences facilitates verification of the implantation of yeast starters used in winemaking. The analyses are effectuated on the entire biomass derived from lees, placed beforehand in a liquid medium in a laboratory culture. The amplification profiles obtained are compared with inoculated yeast strain profiles. They are identical with a successful implantation, and different if the inoculation fails. Figure 1.33 gives examples of successful (yeasts B and C) and unsuccessful

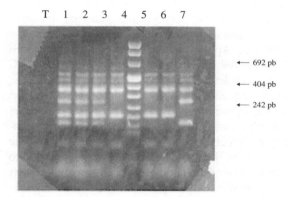

Fig. 1.34. Determination of the detection threshold of a contaminating strain. Band 1: strain A 70%, strain B 30%; band 2: strain A 80%, strain B 20%; band 3: strain A 90%, strain B 10%; band 4: strain A 99%, band B 1%; band 5: strain A 99.9%, strain B 0.1%; band 6: strain A; band 7: strain B. M = molecular weight marker; T = negative control

(yeasts A, D, and E) implantations. Contaminating strains have a different amplification profile than the yeast starter. The detection threshold of a contaminating strain was studied in the laboratory by analyzing a mixture of two strains in variable proportions. In the example given in Figure 1.34, the contaminating strain is easily detected at 1%. In winery fermentations, however, several minority indigenous strains can coexist with the inoculated strain. When must in fermentation or lees is analyzed by PCR, the yeast implantation rate is at least 90% when the amplification profiles of the lees and the yeast starter are identical.

In light of various research, different DNA analysis methods should be combined to identify wine yeast strains.

1.9.6 PCR with Microsatellites

Microsatellites are tandem repeat units of short DNA sequences (1–10 nucleotides), i.e. in the same direction and dispersed throughout the eukaryote genome (Field and Wills, 1998). The number of motif repetitions is extremely variable from one individual to another, making these sequences highly polymorphous in size. These regions are easily identified, thanks to the full sequence of the *S. cerevisiae* genome, available on the Internet in the *Saccharomyces* Genome Database. Approximately 275 sequences have been listed, mainly AT dinucleotides and AAT and AAC trinucleotides (Perez *et al.*, 2001). Furthermore, these sequences are allelic markers, transmitted to the offspring in a Mendelian fashion. Consequently, these are ideal genetic markers for identifying specific yeast strains, making it possible not only to distinguish between strains but also to arrange them in related groups. This technique has many applications in man: paternity tests, forensic medicine, etc. In viticulture, this molecular identification method has already been applied to *Vitis vinifera* grape varieties (Bowers *et al.*, 1999).

The technique consists of amplifying the region of the genome containing these microsatellites, then analyzing the size of the amplified portion to a level of detail of one nucleotide by electrophoresis on acrylamide gel. This varies by a certain number of base pairs (approximately 8–40) from one strain to another, depending on the number of times the motif is repeated. A yeast strain may be heterozygous for a given locus, giving two different-sized amplified DNA fragments. Using 6 microsatellites, Perez *et al.* (2001) were able to identify 44 different genotypes within a population of 51 strains of *S. cerevisiae* used in winemaking. Other authors (Gonzalez *et al.*, 2001; Hennequin *et al.*, 2001)

have shown that the strains of *S. cerevisiae* used in winemaking are weakly heterozygous for the loci studied. However, interstrain variability of the microsatellites is very high. The results are expressed in numerical values for the size of the microsatellite in base pairs or the number of repetitions of the motifs on each allele. These digital data are easy to interpret, unlike the karyotype images on agarose gel, which are not really comparable from one laboratory to another. Microsatellite analysis has also been used to identify the strains of *S. uvarum* used in winemaking (Masneuf and Lejeune, unpublished). As the *S. uvarum* and *S. cerevisiae* microsatellites have different amplification primers, this method provides an additional means of distinguishing between these species and their hybrids.

In future, this molecular typing method will certainly be a useful tool in identifying winemaking yeast strains, ecological surveys, and quality control of industrial production batches.

Finally, another technique has recently been proposed for identifying *Saccharomyces* strains with PCR by amplifying introns of the COX1 mitochondrial DNA gene, which varies in number and position in different strains. It is possible to amplify either purified DNA or fermenting must. This technique has been used to monitor yeast development during fermentation (Lopez *et al.*, 2003).

1.10 ECOLOGY OF GRAPE AND WINE YEASTS

1.10.1 Succession of Grape and Wine Yeast Species

Until recently, a large amount of research focused on the description and ecology of wine yeasts. It concerned the distribution and succession of species found on the grape and then in wine during fermentation and conservation (Ribéreau-Gayon *et al.* 1975; Lafon-Lafourcade 1983).

The ecological study of grape and wine yeast species represents a considerable amount of research. De Rossi began his research in the 1930s (De Rossi, 1935). Castelli (1955, 1967) pursued yeast ecology in Italian vineyards. Peynaud and Domercq (1953) and Domercq (1956) published the first results on the ecology of enological yeasts in France. They described not only the species found on the grape and during alcoholic fermentation, but also contaminating yeasts and diseases. Among the many publications on this theme since the 1960s in viticultural regions around the world, the following works are worth noting: Brechot *et al.* (1962), Minarik (1971), Barnett *et al.* (1972), Park (1975), Cuinier and Guerineau (1976), Soufleros (1978), Belin (1979, 1981), Poulard *et al.* (1980), Poulard and Lecocq (1981), Bureau *et al.* (1982), Rossini *et al.* (1982).

Yeasts are widespread in nature and are found in soils, on the surface of vegetables and in the digestive tract of animals. Wind and insects disseminate them. They are distributed irregularly on the surface of the grape vine; found in small quantities on leaves, the stem and unripe grapes, they colonize the grape skin during maturation. Observations under the scanning electron microscope have identified the location of yeasts on the grape. They are rarely found on the bloom, but multiply preferentially on exudates released from microlesions in zones situated around the stomatal apparatus. *Botrytis cinerea* and lactic acid and acetic acid bacteria spores also develop in the proximity of these peristomatic fractures (Figure 1.35).

The number of yeasts on the grape berry, just before the harvest, is between 10^3 and 10^5, depending on the geographical situation of the vineyard,

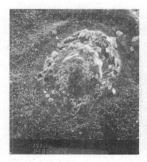

Fig. 1.35. Grape surface under scanning electron microscope, with detail of yeast peristomatic zones. Department of Electronic Microscopy, University of Bordeaux I

climatic conditions during maturation, the sanitary state of the harvest, and pesticide treatments applied to the vine. The most abundant yeast populations are obtained in warm climatic conditions (lower latitudes, elevated temperatures). Insecticide treatments and certain fungicidal treatments can contribute to the rarefaction of indigenous grape microflora. Quantitative results available on this subject, however, are few. After the harvest, transport and crushing of the crop, the number of cells capable of forming colonies on an agar medium generally attains 10^6 cells/ml of must.

The number of yeast species significantly present on the grape is limited. Strictly oxidative metabolism yeasts, which belong to the genus *Rhodotorula* and a few alcohol-sensitive species, are essentially found there. Among the latter, the apiculated species (*Kloeckera apiculata* and its sporiferous form *Hanseniaspora uvarum*) are the most common. They comprise up to 99% of the yeasts isolated in certain grape samples. The following are generally found but in lesser proportions: *Metschnikowia pulcherrima*, *Candida famata*, *Candida stellata*, *Pichia membranefaciens*, *Pichia fermentans*, *Hansenula anomala*.

All research confirms the extreme rarity of *S. cerevisiae* on grapes. Yet these yeasts are not totally absent. Their existence cannot be proven by spreading out diluted must on a solid medium prepared in aseptic conditions, but their presence on grapes can be proven by analyzing the spontaneous fermentative microflora of grape samples placed in sterile bags, then aseptically crushed and vinified in the laboratory in the absence of all contamination. Red and white grapes from the Bordeaux region were treated in this manner. At mid-fermentation in the majority of cases, *S. cerevisiae* represented almost all of the yeasts isolated. In some rare cases, no yeast of this species developed and apiculated yeasts began the fermentation.

Ecological surveys carried out at the Bordeaux Faculty of Enology from 1992 to 1999 (Naumov *et al.*, 2000a) demonstrated the presence of *S. uvarum* yeasts on grapes and in spontaneously fermenting white musts from the Loire Valley, Jurançon, and Sauternes. The frequency of the presence of this species alongside *S. cerevisiae* varies from 4–100%. On one estate in Alsace, strains of *S. uvarum* were identified on grapes, in the press, and in vats, where they represented up to 90% of the yeasts involved throughout fermentation in two consecutive years (Lejeune, unpublished work). More recently, Naumov *et al.* (2002) showed that *S. uvarum*, identified on grapes and in fermenting must, was involved in making Tokay wine.

The adaptation of *S. uvarum* to relatively low temperatures (6–10°C) certainly explains its presence in certain ecological niches: northerly vineyards, late harvests, and spontaneous "cool" fermentation of white wines. In contrast, this strain is sensitive to high temperatures and has not been found in spontaneous fermentations of red Bordeaux wines.

Recent observations also report the presence of natural *S. cerevisiae/S. uvarum* hybrids on grapes and in wineries where both species are present (Lejeune, unpublished work).

Between two harvests, the walls, the floors, the equipment and sometimes even the winery building are colonized by the alcohol-sensitive species previously cited. Winemakers believe, however, that spontaneous fermentations are more difficult to initiate in new tanks than in tanks which have already been used. This empirical observation leads to the supposition that *S. cerevisiae* can also survive in the winery between two harvests. Moreover, this species was found in non-negligible proportions in the wooden fermenters of some of the best vineyards in Bordeaux during the harvest, just before they were filled.

In the first hours of spontaneous fermentations, the first tanks filled have a very similar microflora to that of the grapes. There is a large proportion of apiculated yeasts and *M. pulcherrima*. After about 20 hours, *S. cerevisiae* develops and coexists with the grape yeasts. The latter quickly disappear at the start of spontaneous fermentation. In red winemaking in the Bordeaux region, as soon as must density drops below 1.070–1.060, the colony samples obtained by spreading out diluted must on a solid medium generally isolate exclusively *S. cerevisiae* (10^7 to 10^8 cells/ml). This species

plays an essential role in the alcoholic fermentation process. Environmental conditions influence its selection. This selection pressure is exhibited by four principal parameters: anaerobiosis; must or grape sulfiting; the sugar concentration; and the increasing presence of ethanol. In winemaking where no sulfur dioxide is used, such as white wines for the production of spirits, the dominant grape microflora can still be found. It is largely present at the beginning of alcoholic fermentation (Figure 1.36). Even in this type of winemaking, the presence of apiculated yeasts is almost nonexistent at mid alcoholic fermentation.

During dry white winemaking, the separation of the marc after pressing combined with clarification by racking strongly reduces yeast populations, at least in the first days of the harvest. The yeast population of a severely racked must rarely exceeds 10^4 to 10^5 cells/ml.

A few days into the harvest, the alcogeneous *S. cerevisiae* yeasts contaminate the harvest material, grape transport machinery and especially the harvest receiving equipment, the crusher-stemmer, and the wine press. For this reason, it is already largely present at the time of filling the tanks (around 50% of yeasts isolated during the first homogenization pumping-over of a red-grape tank). Fermentations are initiated more rapidly in the course of the winemaking campaign because

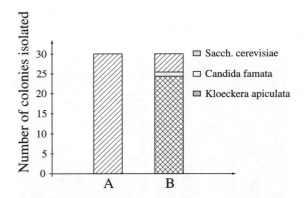

Fig. 1.36. Comparison of yeast species present at the start of alcoholic fermentation (d = 1.06). (A) in a tank of sulfited red grapes in Bordeaux (Frezier, 1992); (B) in a tank of unsulfited white must, for the elaboration of Cognac (Versavaud, 1994)

of this increased percentage of *S. cerevisiae*. In fact, the last tanks filled often complete their fermentations before the first ones. Similarly, static racking in dry white winemaking is becoming more and more difficult to achieve, even at low temperatures, from the second week of the harvest onward, especially in hot years. The entire installation inoculates the must with a sizeable alcogeneous yeast population. General weekly disinfection of the pumps, the piping, the wine presses, the racking tanks, etc. is therefore strongly recommended.

During the final part of alcoholic fermentation (the yeast decline phase), the population of *S. cerevisiae* progressively decreases while still remaining greater than 10^6 cells/ml. In favorable winemaking conditions, characterized by a rapid and complete exhaustion of sugars, no other yeast species significantly appears at the end of fermentation. In poor conditions, spoilage yeasts can contaminate the wine. One of the most frequent and most dangerous contaminations is due to the development of *Brettanomyces intermedius*, which is responsible for serious olfactive flaws (Volume 2, Section 8.4.5).

In the weeks that follow the completion of alcoholic fermentation, the viable populations of *S. cerevisiae* drop rapidly, falling below a few hundred cells/ml. In many cases, other yeast species (spoilage yeasts) can develop in wines during ageing or bottle storage. Some yeasts have an oxidative metabolism of ethanol and form a veil on the surface of the wine, such as *Pichia* or *Candida*, or even certain strains of *S. cerevisiae*—sought after in the production of specialty wines. By topping off regularly, the development of these respiratory metabolism yeasts can be prevented. Some other yeasts, such as *Brettanomyces* or *Dekkera*, can develop in anaerobiosis, consuming trace amounts of sugars that have been incompletely or not fermented by *S. cerevisiae*. Their population can attain 10^4 to 10^5 cells/ml in a contaminated red wine in which alcoholic fermentation is otherwise completed normally. These contaminations can also occur in the bottle. Refermentation yeasts can develop significantly in sweet or botrytized sweet wines during ageing or bottle storage; the

principal species found are *Saccharomycodes ludwigii*, *Zygosaccharomyces bailii*, and also some strains of *S. cerevisiae* that are particularly resistant to ethanol and sulfur dioxide.

1.10.2 Recent Advances in the Study of the Ecology of *S. cerevisiae* Strains

The ecological study of the clonal diversity of yeasts, and in particular of *S. cerevisiae* during winemaking, was inconceivable for a long time because of a lack of means to distinguish yeast strains from one another. Such research has become possible with the development of molecular yeast strain identification methods (Section 1.9). This Section focuses on recent advances in this domain.

The alcoholic fermentation of grape must or grapes is essentially carried out by a single yeast species, *S. cerevisiae*. Therefore, an understanding of the clonal diversity within this species is much more important for the winemaker than investigations on the partially or non-fermentative grape microflora.

The analysis in this Section of *S. cerevisiae* strains in practical winemaking conditions in particular intends to answer the following questions:

- Is spontaneous fermentation carried out by a dominant strain, a small number or a very large number of strains?

- Can the existence of a succession of strains during alcoholic fermentation be proven? If so, what is their origin: grape, harvest material, or winery equipment?

- During winemaking and from one year to another in the same winery or even the same vineyard, is spontaneous alcoholic fermentation carried out by the same strains?

- Can the practice of inoculating with selected strains modify the wild microflora of a vineyard?

During recent research (Dubourdieu and Frezier 1990; Frezier 1992; Masneuf 1996), many samples of yeast microflora were taken at the vineyard and the winery from batches of white and red wines spontaneously fermenting or inoculated with active dry yeasts. Several conclusions can be drawn from this research, carried out on several thousand wild strains of *S. cerevisiae*.

In the majority of cases, a small number of major strains (one to three) representing up to 70–80% of the colonies isolated, carry out the spontaneous fermentations of red and dry white wines. These dominant strains are found in comparable proportions in all of the fermentors from the same winery from start to end of alcoholic

fermentation. This phenomenon is illustrated by the example given in Figure 1.37, describing the indigenous microflora of several tanks of red must in a Pessac-Léognan vineyard in 1989. The strains of *S. cerevisiae*, possessing different karyotypes, are identified by an alphanumeric code comprising the initial of the vineyard, the tank number, the time of the sampling, the isolated colony number and the year of the sample. Two strains, FzIb1 (1989) and FzIb2 (1989), are encountered in all of the tanks during the entire alcoholic fermentation process.

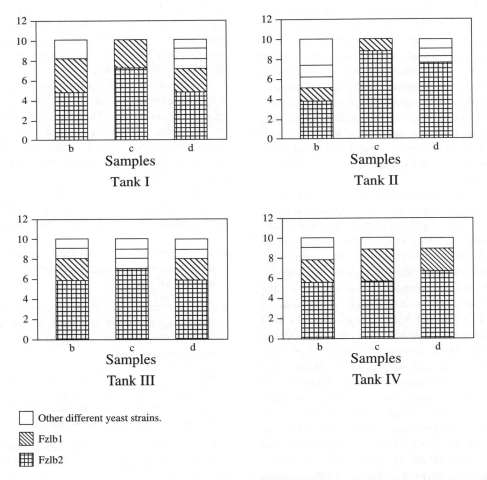

Fig. 1.37. Breakdown of *S. cerevisiae* caryotypes during alcoholic fermentation in red grape tanks in Fz vineyard (Pessac-Léognan, France) in 1989 (Frezier, 1992) (b, c, and d designate the start, middle, and end of alcoholic fermentation, respectively). Tanks I and II (Merlot) and III and IV (Cabernet-Sauvignon) are filled on the 1st, 3rd, 7th and 23rd day of the harvest, respectively

The spontaneous fermentation of dry white wines in the same vineyard is also carried out by the same dominant yeast strains in all of the barrels.

The tank filling order and the grape variety have little effect on the clonal composition of the populations of S. cerevisiae spontaneously found in the winery. The daily practice of pumping-over the red grape must with pumping equipment used for all of the tanks probably ensures the dissemination of the same strains in the winery. In white winemaking, the wine press installation plays the same role as an inoculator.

In addition, in Figure 1.37, all of the strains analyzed are K2 killer. The two dominant strains do not ferment galactose (phenotype Gal⁻). Their former denomination was therefore S. oviformis or S. cerevisiae (race bayanus) in previous classifications. Domercq (1956) observed a lesser proportion of S. oviformis in the spontaneous microflora of Bordeaux region fermentations in the 1950s (one-fifth at the beginning of fermentation to one-third at the end). In the indigenous fermentative microflora of Bordeaux musts, certain strains of S. cerevisiae Gal⁻ which dominate from the start of alcoholic fermentation were selected over time. The causes of this change in the microflora, remain unknown. On the other hand, a systematic increase in the proportion of Gal⁻ strains during red or dry white wine fermentation has not been observed (Table 1.9). In botrytized sweet wines from Sauternes, the succession of strains is more distinct.

The same major strain is frequently encountered for several consecutive vintages in the same vineyard in spontaneous-fermentation red-grape must tanks. In 1990, one of the major strains was the same as the previous year in the red grape must tank of the Fz vineyard. Other strains appeared, however, which had not been isolated in 1989.

When sterile grape samples are taken, pressed sterilely, sulfited at winemaking levels and fermented in the laboratory in sterile containers, one or several dominant strains responsible for spontaneous fermentations in the winery exist in some samples. These strains are therefore present at the vineyard. In practice, they probably begin

Table 1.9. Example of physiological race breakdown (%) of *Saccharomyces cerevisiae* during spontaneous alcoholic fermentation (Frezier, 1992)

Stage of fermentation	Physiological race			
	cerevisiae	*oviformis*	*capensis*	*chevalieri*
Red wine[a]				
start	23	77	—	—
middle	10	90	—	—
end	14	84	2	—
White wine[b]				
start	23	62	—	14
middle	35	60	—	4
end	32	62	—	6
Sweet wine[c]				
start	37	51	4	8
middle	40	56	4	3
end	23	73	4	4

[a]Isolation of 100 colonies from six tanks of a Pessac-Léognan vineyard (four tanks in 1989 and two tanks in 1990).
[b]Isolation of 100 colonies in three barrels of a Pessac-Léognan vineyard.
[c]Isolation of 100 colonies in two barrels of a Sauternes vineyard in 1990.

Table 1.10. Rate of occurrence of the dominant FZIB2-89 caryotype in microvinifications carried out on sterile grape samples (I, II, III) in the FZ vineyard

Year	Number of clones analyzed	Sample I	Sample II	Sample III
1990	30	—	70%	—
1991	60	—	—	—
1992	85	25%	31%	3%
1993	74	—	—	—
1994	79	87%	—	40%

to multiply as soon as the grapes arrive at the winery. A few days into the harvest, they infest the winery equipment which in turn ensures a systematic inoculation of the fresh grape crop.

The presence each year of the same dominant strain in the vineyard is not systematic (Table 1.10). In the Fz vineyard, the FzIb2-89 strain could not be isolated in 1991 although it was present in certain vineyard samples in 1990, 1992 and 1994. In 1993, another strain proved to be dominant in spontaneous fermentations of sterile grape samples.

The spontaneous microflora of *S. cerevisiae* seems to fluctuate. At present, the factors involved in this fluctuation have not been identified. In a given vineyard, spontaneous fermentation is not systematically carried out by the same strains each year; strain specificity does not exist and therefore does not participate in vineyard characteristics. Ecological observations do not confirm the notion of a vineyard-specific yeast. Furthermore, some indigenous strains, dominant in a given vineyard, have been found in other nearby or distant vineyards. For example, the FzIb2-89 strain, isolated for the first time in a vineyard in Pessac-Léognan, was later identified not only in the spontaneous fermentation of dry white and red wines of other vineyards in the same appellation, but also in relatively distant wineries as far away as the Médoc. This strain has since been selected and commercialized under the name Zymaflore F10.

In some cases (Figure 1.38), *S. cerevisiae* populations with a large clonal diversity carry out spontaneous must fermentation. Many strains coexist. Their proportions differ from the start to the end of fermentation and from one winery to another. In the Bordeaux region, this diversity causes slow fermentations and sometimes even stuck fermentations. No strain is capable of asserting itself. On the other hand, the presence of a small number of dominant strains generally characterizes complete and rapid spontaneous fermentations. These dominant strains are found from the start to the end of the fermentation.

In normal red winemaking conditions, the inoculation of the first tanks in a winery influences the wild microflora of non-inoculated tanks. The strain(s) used for inoculating the first tanks are frequently found in majority in the latter. Figure 1.39 provides an example comparing the microflora of a tank of Merlot from Pomerol, inoculated with an active dry yeast strain (522M) on the first day of the harvest, with a non-inoculated tank filled later. From the start of alcoholic fermentation, the selected strain is successfully implanted in the inoculated tank. Even in the non-inoculated tank, the same strain is equally implanted throughout the fermentation. It is therefore difficult to select the dominant wild strains in red winemaking tanks, when some of the tanks have been inoculated. An early and massive inoculation of the must, however, permits the successful implantation of

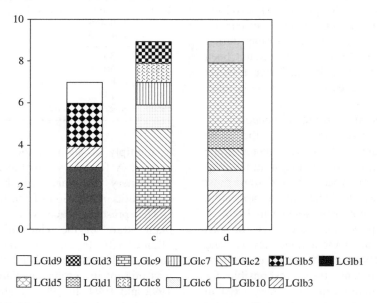

Fig. 1.38. Breakdown of *S. cerevisiae* caryotypes in tank I of red grapes at LG vineyard (Pomerol, France) in 1989 (Frezier, 1992) (b, c, and d designate the start, middle, and end of alcoholic fermentation, respectively)

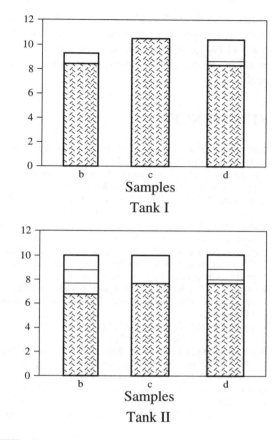

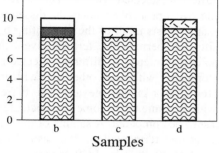

Tank I, inoculated with F5

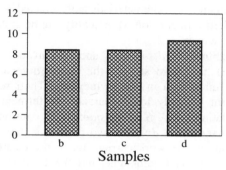

Tank II, inoculated with F10

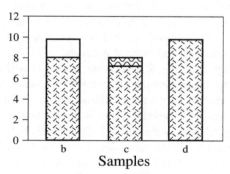

Tank III, inoculated with 522M

Fig. 1.39. Breakdown of *S. cerevisiae* caryotypes in tank I and tank II P vineyard in 1990 (b, c, and d designate the start, middle, and end of alcoholic fermentation, respectively). Tank I is inoculated with 522M dry yeasts and tank II underwent spontaneous fermentation (b, c and d designate the start, middle and end of alcoholic fermentation, respectively)

Fig. 1.40. Breakdown of *S. cerevisiae* caryotypes in tanks I, II and III in F vineyard in 1990, with massive early inoculation with F5, F10 and 522M, respectively (Frezier, 1992) (b, c, and d designate the start, middle, and end of alcoholic fermentation, respectively)

different selected yeasts in several tanks at the same winery (Figure 1.40).

In white winemaking, inoculating rarely influences the microflora of spontaneous fermentations in wineries. For the most part, dominant indigenous strains in non-inoculated barrels of fermenting dry white wine are observed, whereas in the same wine cellar, other batches were inoculated with different selected yeasts. The absence of pumping-overs probably hinders the dissemination of the same yeasts in all of the fermenting barrels. This situation permits the fermentative behavior and enological interest of different selected strains to be compared with each other and with indigenous strains in a given vineyard. The barrels are filled with the same must; some are inoculated with the yeast to be compared. A sample of the biomass is taken at mid fermentation. The desired implantation is then verified by PCR associated with δ sequences. Due to the ease of use of this method, information on characteristics of selected strains and their influence on wine quality can be gathered at the winery.

Vezhinet *et al.* (1992) and Versavaud *et al.* (1995) have also studied the clonal diversity of yeast microflora in other vineyards. Their results confirm the polyclonal character of fermentative populations of *S. cerevisiae*. The notion of dominant strains (one to two per fermentation) is obvious in the work carried out in the Charentes region. As in Champagne and the Loire Valley, some Charentes region strains are found for several years in a row in the same winery. The presence of these dominant strains on the grape has been confirmed before any contact with winery equipment during several harvests.

Why do some *S. cerevisiae* strains issued from a very heterogeneous population become dominant during spontaneous fermentation? Why can they be found several years in a row at the same vineyard and wine cellar? Despite their practical interest, these questions have not often been studied and there are no definitive responses. It seems that these strains rapidly start and complete alcoholic fermentation and have a good resistance to sulfur dioxide (up to 10 g/hl). Furthermore, during mixed inoculations in the laboratory of either 8% ethanol or non-fermented musts, these strains rapidly become dominant when placed in the presence of other wild non-dominant strains of *S. cerevisiae* isolated at the start and end of fermentation. This subject merits further research. Without a doubt, it would be interesting to compare the genetic characteristics of dominant and non-dominant

yeasts and their degree of heterozygosity. Considering the genome renewal theory of Mortimer *et al.* (1994) (Section 1.6.2), dominant strains are possibly more homozygous than non-dominant strains.

REFERENCES

Aigle M., Erbs D. and Moll M. (1984) *Am. Soc. of Brewing Chemists*, 42 (1), 1–7.

Alexandre E.H., Rousseaux I. and Charpentier C. (1994) *Biotechnol. Appl. Biochem.*, 19.

Augustyn O.P.H., Kock J.L.F. and Ferreira D. (1991) *System. Appl. Microbiol.*, 15, 105–115.

Barnett J.A., Delaney M.A., Jones E., Magson A.B. and Winch B. (1972) *Arch. Microbiol.*, 83, 52–55.

Barnett J.A., Payne R.W. and Yarrow D. (2000) Yeasts: *Characteristics and Identification*, 3rd edn. Cambridge University Press Cambridge, UK.

Barre P. (1978) Killer factor activity under vinification conditions. *Vth International Symposium on Yeasts*, Montpellier.

Barre P. (1980) *Bull. OIV*, 53, 560–567.

Barre P. (1984) *Bull. OIV*, 57, 635–643.

Barre P. (1992) Le facteur Killer. In *Les acquisitions récentes de la microbiologie du vin* (ed. B. Doneche), pp. 63–69. Tec & Doc Lavoisier, Paris.

Belin J.M. (1979) *Mycopathologia*, 67 (2), 67–81.

Belin J.M. (1981) Biologie des levures liées à la vigne et au vin. Thèse Docteur ès Sciences Naturelles, Université de Dijon.

Bendova O., Richter V., Janderova B. and Haüsler J. (1991) *Appl. Microbiol. Biotechnol.*, 35, 810–812.

Bicknell J.N. and Douglas H.C. (1982) *J. Bacteriol.*, 101, 505–512.

Bisson L.F. (1991) Yeasts–metabolism of sugars. In *Wine Microbiology and Biotechnology*, pp. 55–75. Harwood Academic Publishers.

Blondin B. and Vezhinet F. (1988) *Rev. Fr. Oenol.*, 115, 7–11.

Boone C., Sdicu A.M., Wagner J., Degre R., Sanchez C. and Bussey H. (1990) *Am. J. Enol. Vitic.*, 41, 37–42.

Bouix M., Leveau J.Y. and Cuinier C. (1981) *Conn. Vigne Vin*, 15, 41–52.

Boulton R.B., Singleton V.L., Bisson L.F. and Kunkee R. (1995) *Principles and Practices of Winemaking*. Chapman & Hall Enology Library, New York.

Bowers J., Boursiquot J.M., This P., Chu K., Johansson H. and Meredith C. (1999) *Science*, 285, 1562–1565.

Brechot P., Chauvet J. and Girard H. (1962) *Ann. Techn. Agric.*, 11 (3), 235–244.

Bureau G., Brun O., Vigues A., Maujean A., Vesselle G. and Feuillat A. (1982) *Conn. Vigne Vin*, 16 (1), 15–32.

Carle G.F. and Olson M.V. (1984) *Nucl. Acids Res.*, 12, 5647–5664.

Castelli T. (1955) *Am. J. Enol. Vitic.*, 6, 18–20.

Castelli T. (1967) Ecologie et systématique des levures du vin. *IIéme Symposium International d'Enologie*, Bordeaux-Cognac, INRA Ed. Paris.

Charpentier Cl. (1995) *Rev. Enolo.* 73 S, 25–28.

Ciolfi G. (1992) *L'enotechnico*, XXVIII, (11) 87.

Ciolfi G. (1994) *L'enotechnico*, Nov., 71–76.

Cuinier C. and Gros C. (1983) *Vignes Vins*, 318, 25–27.

Cuinier C. and Guerineau L. (1976) Evolution de la microflore au cours de la vinification des vins de Chinon. *Vignes Vins*, 269, 29–41.

Cuinier C. and Leveau J.Y. (1979) *Vignes Vins*, 283, 44–49.

Defontaine A., Lecocq F.M. and Hallet J.N. (1991) *Nucl. Acids Res.*, 19 (1), 185.

Degre R., Thomas D.Y., Frenette J. and Mailhot K. (1989). *Rev. Fr. Oenol.*, 119, 23–26.

Delfini C. (1995) *Scienza e technica di microbioloia enologica*. Il Lievito, Asti.

De Rossi G. (1935) Il lieviti della fermentazione nella regione umbra. *IVème Congrès International de la Vigne et du Vin*, Lausanne.

Domercq S. (1956) Etude et classification des levures de vin de la Gironde. Thèse de Docteur-Ingénieur de l'Université de Bordeaux.

Dubourdieu D. and Frezier V. (1990) *Rev. Fr. Oenol.*, 30, 37–40.

Dubourdieu D. and Moine V. (1995) Rôle des conditions d'élevage sur la stabilisation protéique des vins blancs. In *Actualités enologiques 95*, Compte-rendus du 5ème Symposium International d'Enologie de Bordeaux.

Dubourdieu D., Sokol A., Zucca J., Thalouarn P., Datee A. and Aigle M. (1987) *Conn. Vigne Vin*, 4, 267–278.

Field D. and Wills C. (1998) *Proc. Natl. Acad. Sci.*, 95, 1647–1652.

Fleet G.H. (1991) Cell wall. In *The Yeasts*, Vol. 4: *Yeasts Organelles*, pp. 199–277. Academic Press, London.

Fleet G.H. (1993) *Wine Microbiology and Biotechnology*. Harwood Academic Publishers, Chur, Switzerland.

Flick J.S. and Thorner J. (1993) *Mol. Cell. Biol.* 13, 5861–5876.

Frezier V. (1992) Recherche sur l'écologie des souches de *Saccharomyces cerevisiae* au cours des vinifications bordelaises. Thèse de Doctorat de l'Université de Bordeaux II.

Frezier V. and Dubourdieu D. (1992) *Am. J. Enol. Vitic.* 43, 375–380.

Gaillardin C. and Heslot H. (1987) *La levure. La Recherche*, 188 (18) 586–596.

Gonzalez Techera A., Jubany S., Carrau F.M. and Gaggero C. (2001) *Lett. Appl. Microbiol.*, 33, 71–75.

Guerin B. (1991) Mitochondria. In *The Yeasts*, Vol. 4: *Yeasts Organelles*, pp. 541–589. Academic Press, London.

Guillamon J.M., Sabate J., Barrio E., Cano J. and Querol A. (1998) *Arch. Microbiol.*, 169 (5), 387–392.

Hallet J.N., Craneguy B., Zucca J. and Poulard A. (1988) *Prog. Agric. Vitic.*, 105, 328–333.

Hallet J.N., Craneguy B., Daniel P. and Poulard A. (1989) Caractérisation des souches levuriennes des mouts et des lies par le polymorphism de restriction de leur ADN mitochondrial. In *Actualités Enologiques 89*, Comptes rendus du 4ème Symposium d'Enologie de Bordeaux. Dunod, Paris.

Hansen J. and Kielland-Brandt M.C. (1994) *Gene*, 140, 33–40.

Heard G.M. and Fleet G.H. (1987) *Appl. Environ. Microbiol.* 51, 539–45.

Heard G.M. and Fleet G. (1990) *J. Appl. Bacteriol.*, 68, 445–447.

Hennequin C., Thierry A., Richard G.F., Lecointre G., Nguyen H.V., Gaillardin Cl., Dujon B. (2001), *J. Clin. Microbiol.*, 39 (2), 551–559.

Henschke P.A. and Rose A.H. (1991) Plasma membrane. In *The Yeasts*, Vol. 4: *Yeasts Organelles*, pp. 297–345. Academic Press, London.

Herskowitz I., Rine J. and Strathern J.N. (1992) Mating-type determination and mating-type interconversion in *Saccharomyces cerevisiae*. In Jones, (eds E.W. J.N. Strathern and J.R. Broach). Broach J.R. (eds) *The Molecular Biology of the Yeast* Saccharomyces: *Gene Expression* (eds E.W. Jones, J.N. Strathern and J.R. Broach), pp. 583–656. Cold Spring Harbor, Laboratory Press, New York.

Klis F.M. (1994) *Yeast*, 10, 851–869.

Kregger-Van Rij N.J.W. (1984) *The Yeasts, a Taxonomic Study*. Elsevier Science Publishers, B.V. Amsterdam.

Lafon-Lafourcade S. (1983) Wine and brandy. In *Biotechnology* Vol. V: *Food and Feed Production with Microorganisms* (eds H.J. Rehm and G. Reed). Verlag Chemie, Weinheim.

Lafon-Lafourcade S. and Joyeux A. (1979) *Conn. Vigne Vin*, 4, 295–310.

Lavallée F., Salves S., Lamy S., Thomas D.Y., Degre R. and Dulau L. (1994) *Am. J. Enol. Vitic.* 45, 86–91.

Lodder J. (1970) *The Yeasts, a Taxonomic Study*, 2nd edn. Elsevier Science Publishers, B.V. Amsterdam.

Lodder J. and Kregger-Van Rij N.J.W. (1952) *The Yeasts, a Taxonomic Study*. Elsevier Science Publishers, B.V. Amsterdam.

Lopez V., Fernandez-Espinar M.T., Barrio E., Ramon D. and Querol A. (2003) *Int. J. Food Microbiol.*, 81 (1), 63–71.

Masneuf I. (1996) Recherches sur l'identification génétique des levures de vinification. Applications enologiques. Thèse de Doctorat de l'Université de Bordeaux II.

Masneuf I., Aigle M. and Dubourdieu D. (1996a) *FEMS Microbiol. Letters*,.

Masneuf I., Aigle M. and Dubourdieu D. (1996b) *J. Int. Sci. Vigne Vin*, 30 (1), 15–21.

Masneuf I. and Dubourdieu D. (1994) *J. Int. Sci. Vigne Vin*, 28 (2), 153–160.

Masneuf I., Hansen J., Groth C., Piskpur J. and Dubourdieu D. (1998) *Appl. Environ. Microbiol.*, 64 (10), 3887–3892.

Minarik E. (1971) *Conn. Vigne Vin*, 2, 185–198.

Mortimer R.K., Romano P., Suzzi G. and Polsinelli M. (1994) *Yeast*, 10, 1543–1552.

Naumov G.I. (1987) Genetic basis for classification and identification of the ascomycetous yeasts. *Studies in Mycology*, 30, 469–475.

Naumov G.I., Masneuf I., Naumova E.S., Aigle M. and Dubourdieu D. (2000a) *Res. Microbiol.*, 151, 683–691.

Naumov G.I., Naumova E.S., Masneuf I., Aigle M., Kondratieva V.I. and Dubourdieu D. (2000b) *Syst. Appl. Microbiol.*, 23, 442–449.

Naumov G.I., Naumova E.S., Antunovics Z. and Sipiczki M. (2002) *Appl. Microbiol. Biotechnol.*, 59 (6), 727–30.

Naumov G., Naumova E. and Gaillardin C. (1993) Genetic and karyotypic identification of wine *Saccharomyces bayanus* yeasts isolated in France and Italy. *System. Appl. Microbiol.*, 16, 274–279.

Naumov G.I., Naumova E.S. and Sniegowski P.D. (1998) *Can. J. Microbiol.*, 44, 1045–1050.

Nguyen H.V., Lepingle A. and Gaillardin C.A. (2000) *Syst. Appl. Microbiol.*, 23 (1), 71–85.

Ness F., Lavallée F., Dubourdieu D., Aigle M. and Dulau L. (1992) *J. Sci. Food Agric.*, 62, 89–94.

Park Y.H. (1975) *Conn. Vigne Vin*, 3, 253–278.

Pasteur L. (1866) *Etudes sur le vin*. Imprimerie Impériale, Paris.

Pasteur L. (1876) *Etudes sur la bière*. Gauthier-Villars, Paris.

Pedersen M.B. (1986) *Carlsberg Res. Commun.* 51, 163–183.

Perez M.A., Gallego F.J., Martinez I. and Hidalgo P. (2001) *Lett. Appl. Microbiol.*, 33, 461–466.

Peynaud E. and Domercq S. (1953) *Ann. Technol. Agric.*, 4, 265–300.

Poulard A. and Lecocq M. (1981) *Rev. Fr. Oenol.*, 82, 31–35.

Poulard A., Simon L. and Cuinier Cl. (1980) *Conn. Vigne Vin*, 14, 219–238.

Querol A., Barrio E. and Ramon D. (1992) *System. Appl. Microbiol.*, 15, 439–446.

Radler F. (1988) in *Application à l'oenologie des progrès récents en microbiologie et en fermentation*. Office international de la vigne et du vin Éd., Paris. pp. 273–282.

Redzepovic S., Orlic S., Sikora S., Madjak A. and Pretorius I.S. (2002) *Lett. Appl. Microbiol.*, 35, 305–310.

Ribéreau-Gayon J., Peynaud E., Ribéreau-Gayon P. and Sudraud P. (1975) *Traité d'Enologie. Sciences et techniques du vin*, Vol 2. Dunod Paris.

Rose A.H. and Harrison J.S. (1969) *The Yeasts*, Vol. 1: *Biology of Yeasts*. Academic Press, London.

Rose A.H. and Harrison J.S. (1991) *The Yeasts*, Vol. 4: *Yeasts Organelles*. Academic Press, London.

Rossini G., Federici F. and Martini A. (1982) *Microbiol. Ecol.*, 8, 83–89.

Rozes N., Garcia-Jares C., Larue F. and Lonvaud-Funel A. (1992) *J. Sci. Food Agric.*, 59, 351–357.

Saiki R.K., Sharf S., Falcona F. and Lonvaud-Funel A. (1992) *J. Sci. Food Agric.*, 59, 315–357.

Salmon J.M., Vincent O., Mauricio J.C. and Bely M. (1993) *Am. J. Enol. Vitic.*, 44 (1), 56–64.

Schwartz D.C. and Cantor C.R. (1984) *Cell*, 37, 67–75.

Seki T., Choi E.H and Ryu D. (1985) *Appl. Environ. Microbiol.* 49, 1211.

Singer S.J. and Nicolson G.L. (1972) *Science*, 175, 720–731.

Soufleros E. (1978) Les levures de la région viticole de Naoussa (Grèce). Identification et classification. Etude des produits volatils formés au cours de la fermentation. Thèse Docteur-Ingénieur Université de Bordeaux II.

Southern E. (1975) *J. Biol. Chem.*, 98, 503–517.

Stratford M. (1994) *Yeast*, 10, 1741–1752.

Tipper D.J. and Bostian K.A. (1984) Double stranded ribonucleic acid killer system in Yeasts. *Microbiol. Rev.* 48 (2), 125–136.

Tredoux H.G., Kock J.L.F., Lategan P.L. and Muller H.B. (1987) *Am. J. Enol. Vitic.*, 38, 161–164.

Tuite M.F. and Oliver S.G. (1991) *Saccharomyces Biotechnology Handbooks*. Plenum Press, New York.

Van Vuuren H.J.J. and Jacobs C.J. (1992) *Am. J. Enol. Vitic.*, 43 (2), 119–128.

Van Vuuren H.J.J. and Van Der Meer L. (1987) *Am. J. Enol. Vitic.*, 38, 49–53.

Vaughan Martini A. and Martini A. (1987) Three newly delimited species of *Saccharomyces sensu stricto*. *Antonie van Leeuwenhoek*, 53, 77–84.

Vaughan Martini A. and Martini A. (1993) *System. Appl. Microbiol.*, 16, 113–119.

Versavaud A., Dulau L. and Hallet J.N. (1993) *Rev. Fr. Oenol.*, 142, 20–28.

Versavaud A., Courcoux Ph., Roulland Cl., Dulau L. and Hallet J.N. (1995) *Appl. Environ. Microbiol.*, 61 (10), 3521–3529.

Vezhinet F., Blondin B. and Hallet J.N. (1990) *Appl. Micro-Biotecn.*, 32, 568–571.

Vezhinet F., Hallet J.N., Valade M. and Poulard A. (1992) *Am. J. Enol. Vitic.*, 43 (1), 83–86.

Williamson D.H. (1991) Nucleus, chromosomes and plasmids. In *The Yeasts*, Vol. 4: *Yeasts organelles*, (eds

A.H. Rose and J.S. Harison), pp. 433–482. Academic Press, London.

Wingfield B.D., Van Der Meer L.J., Pretorius S. and Van Vuuren H.J.J. (1990) *Mycol. Res.*, 94, 901–906.

Yarrow D. and Nakase T. (1975) *Antonie van leeuwenhoek*, 41 81–88.

Young T.W. (1987) Killer yeasts. In *The Yeasts*, Vol. 2, *Yeasts and the Environment* (eds A.H. Rose and J.S. Harrison), pp. 131–164. Academic Press, New York.

Zorg J. and Kilian J. and Radler F. (1988) *Archiv. Microbiol.*, 149, 261–267.

2

Biochemistry of Alcoholic Fermentation and Metabolic Pathways of Wine Yeasts

2.1 INTRODUCTION

The synthesis of living material is endergonic, requiring the consumption of energy. Chlorophyllous plants, called phototrophs, collect solar energy. Some bacteria obtain energy from the oxidation of minerals: they are chemolithotrophs. Like most animals and bacteria, fungi, including yeast, are chemoorganotrophs: they draw their necessary energy from the degradation of organic nutrients.

In a growing organism, energy produced by degradation reactions (catabolism) is transferred to the chain of synthesis reactions (anabolism). Conforming to the laws of thermodynamics, energy furnished by the degradation of a substrate is only partially converted into work; this is called free energy (the rest is dissipated in the form of heat). Part of this free energy can be used for transport, movement, or synthesis. In most cases, the free energy transporter particular to biological systems is adenosine triphosphate (ATP). This molecule is rich in energy because its triphosphate unit contains two phosphoanhydride bonds (Figure 2.1). The hydrolysis of ATP into adenosine diphosphate (ADP) results in the liberation of a large quantity of free energy (7.3 kcal/mol). Biosynthesis and the active transport of metabolites make use of free energy.

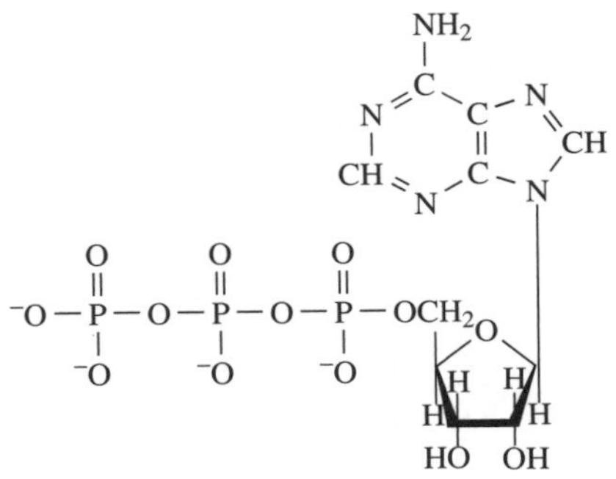

Fig. 2.1. Structure of adenosine triphosphate (ATP)

$$ATP + H_2O \rightleftharpoons ADP + Pi + H^+ \Delta G^{\circ\prime}$$

$$= -7.3 \text{ kcal/mol} \qquad (2.1)$$

In this reaction, $\Delta G^{\circ\prime}$ is the change in free energy. ATP is considered to be 'the universal money of free energy in biological systems' (Stryer, 1992). In reality, microorganism growth or, in this case, yeast growth is directly related to the quantity of ATP furnished by metabolic pathways used for degrading a substrate. It is indirectly related to the quantity of substrate degraded.

In the living cell, there are two processes which produce ATP: substrate-level phosphorylation and oxidative phosphorylation. Both of these pathways exist in wine yeasts.

Substrate-level phosphorylation can be either aerobic or anaerobic. During oxidation by electron loss, an ester–phosphoric bond is formed. It is an energy-rich bond between the oxidized carbon of the substrate and a molecule of inorganic phosphate. This bond is then transferred to the ADP by transphosphorylation, thus forming ATP. This process takes place during glycolysis.

Oxidative phosphorylation is an aerobic process. The production of ATP is linked to the transport of electrons to an oxygen molecule by the cytochromic respiratory chain. This oxygen molecule is the final acceptor of the electrons. These reactions occur in the mitochondria.

This chapter describes the principal biochemical reactions occurring during grape must fermentation by wine yeasts. It covers sugar metabolisms, i.e. the biochemistry of alcoholic fermentation, and

nitrogen metabolisms. Volatile sulfur-containing compounds and volatile phenol formation mechanisms will be discussed in Volume 2, Chapter 8 in the section concerning olfactory flaws. The influence of yeasts on varietal wine aromas will be covered in Volume 2, Chapter 7.

2.2 SUGAR DEGRADATION PATHWAYS

Depending on aerobic conditions, yeast can degrade sugars using two metabolic pathways: alcoholic fermentation and respiration. These two processes begin in the same way, sharing the common trunk of glycolysis.

2.2.1 Glycolysis

This series of reactions, transforming glucose into pyruvate with the formation of ATP, constitutes a quasi-universal pathway in biological systems. The elucidation of the different steps of glycolysis is intimately associated with the birth of modern biochemistry. The starting point was the fortuitous discovery by Hans and Eduard Buchner, in 1897, of the fermentation of saccharose by an acellular yeast extract. Studying possible therapeutic applications for their yeast extracts, the Buchners discovered that the sugar used to preserve their yeast extract was rapidly fermented into alcohol. Several years later, Harden and Young demonstrated that inorganic phosphate must be added to acellular yeast extract to assure a constant glucose fermentation speed. The depletion of inorganic phosphate during *in vitro* fermentation led them to believe that it was incorporated into a sugar phosphate. They also observed that the yeast extract activity was due to a non-dialyzable component, denaturable by heat, and a thermostable dialyzable component. They named these two components 'zymase' and 'cozymase'. Today, zymase is known to be a series of enzymes and cozymase is composed of their cofactors as well as metal ions and ATP. The complete description of glycolysis dates back to the 1940s, due in particular to the contributions of Embden, Meyerhoff and Neuberg.

For that reason, glycolysis is often called the Embden–Meyerhoff pathway.

The transport of must hexose (glucose and fructose) across the plasmic membrane activates a complex system of proteinic transporters not fully explained (Section 1.3.2). This mechanism facilitates the diffusion of must hexoses in the cytoplasm, where they are rapidly metabolized. Since solute moves in the direction of the concentration gradient, from the concentrated outer medium to the diluted inner medium, it is not an active transport system requiring energy.

Next, glycolysis (Figure 2.2) is carried out entirely in the cytosol of the cell. It includes a first stage which converts glucose into fructose 1,6-biphosphate, requiring two ATP molecules. This transformation itself comprises three steps: an initial phosphorylation of glucose into glucose 6-phosphate, the isomerization of the latter into fructose 6-phosphate and a second phosphorylation forming fructose 1,6-biphosphate. These three reactions are catalyzed by hexokinase, phosphoglucose isomerase and phosphoglucokinase, respectively.

In fact, *Saccharomyces cerevisiae* has two hexokinases (PI and PII) capable of phosphorylating glucose and fructose. Hexokinase PII is essential and is active predominantly during the yeast log phase in a medium with a high sugar concentration. Hexokinase PI, partially repressed by glucose, is not active until the stationary phase (Bisson, 1991).

Mutant strains devoid of phosphoglucoisomerase have been isolated. Their inability to develop on glucose suggests that glycolysis is the only catabolic pathway of glucose in *Saccharomyces cerevisiae* (Caubet *et al.*, 1988). The oxidative pentose phosphate pathway, by which some organisms utilize sugars, serves only as a means of synthesizing ribose 5-phosphate, incorporated in nucleic acids and in reduced nicotinamide-adenine dinucleotide phosphate (NADPH) in *Saccharomyces*.

The second stage of glycolysis forms glyceraldehyde 3-phosphate. Under the catalytic action of aldolase, fructose 1,6-biphosphate is cleaved thus forming two triose phosphate isomers: dihydroxyacetone phosphate and glyceraldehyde 3-phosphate. The triose phosphate isomerase catalyzes the isomerization of these two compounds. Although at equilibrium the ketonic form is more abundant, the transformation of dihydroxyacetone phosphate into glyceraldehyde 3-phosphate is rapid, since this compound is continually eliminated by the ensuing glycolysis reactions. In other words, a molecule of glucose leads to the formation of two molecules of glyceraldehyde 3-phosphate.

The third phase of glycolysis comprises two steps which recover part of the energy from glyceraldehyde 3-phosphate (G3P). Initially, GA3P is converted into 1,3-biphosphoglycerate (1,3-BPG). This reaction is catalyzed by glyceraldehyde 3-phosphate dehydrogenase. It is an oxidation coupled with a substrate-level phosphorylation. Nicotinamide–adenine dinucleotide (NAD$^+$) is the cofactor of the dehydrogenation. At this stage, it is in its oxidized form; nicotinamide is the reactive part of the molecule (Figure 2.3). Simultaneously, an energy-rich bond is established between the oxidized carbon of the substrate and the inorganic phosphate. The NAD$^+$ accepts two electrons and a hydrogen atom lost by the oxidized substrate. Next, phosphoglyceratekinase catalyzes the transfer of the phosphoryl group of the acylphosphate from 1,3-BPG to ADP; and 3-phosphoglycerate and ATP are formed.

The last phase of glycolysis transforms 3-phosphoglycerate into pyruvate. Phosphoglyceromutase catalyzes the conversion of 3-phosphoglycerate into 2-phosphoglycerate. Enolase catalyzes the dehydration of the latter, forming phosphoenolpyruvate. This compound has a high phosphoryl group transfer potential. By phosphorylation of ADP, pyruvic acid and ATP are formed; the pyruvate kinase catalyzes this reaction. In this manner, glycolysis creates four ATP molecules. Two are immediately used to activate a new hexose molecule, and the net gain of glycolysis is therefore two ATP molecules per molecule of hexose metabolized. This stage marks the end of the common trunk of glycolysis; alcoholic fermentation, glyceropyruvic fermentation or respiration follow, depending on various conditions.

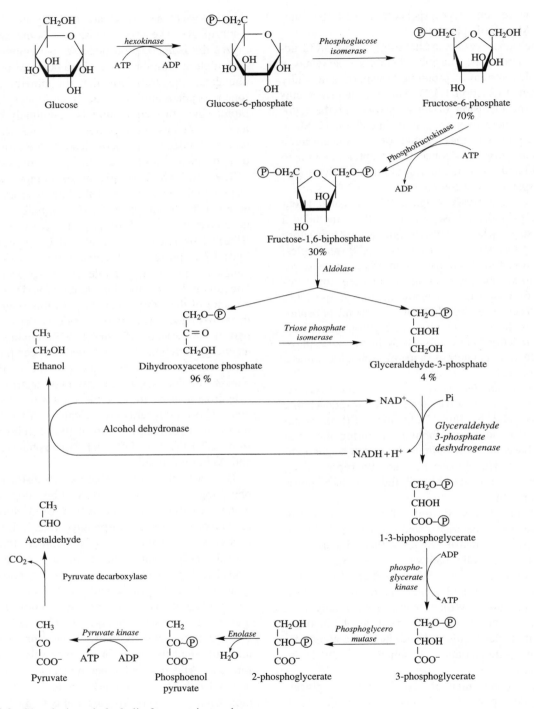

Fig. 2.2. Glycolysis and alcoholic fermentation pathway

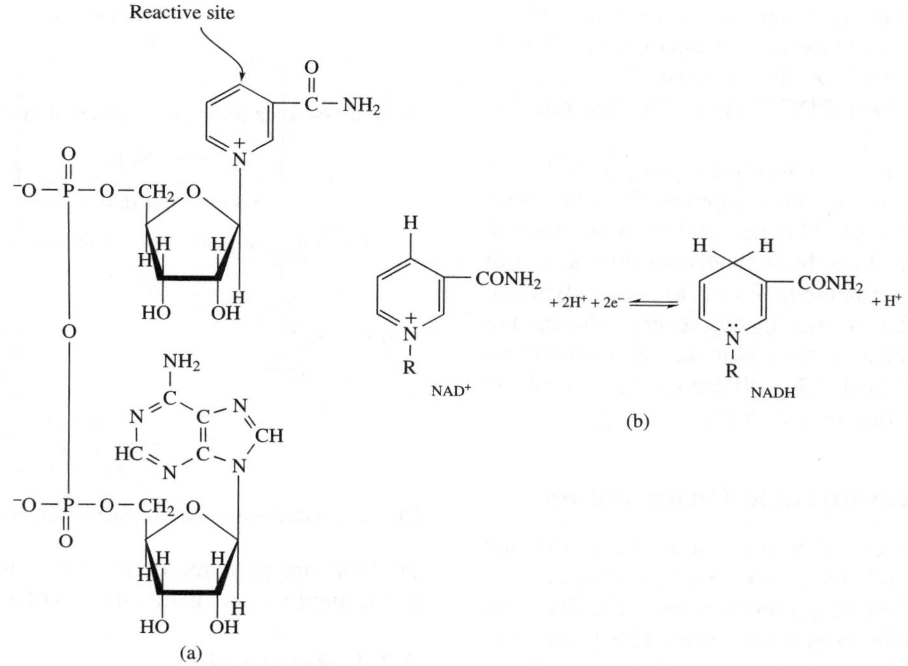

Fig. 2.3. (a) Structure of nicotinamide adenine dinucleotide in the oxidized form (NAD^+). (b) Equilibrium reaction between the oxidized (NAD^+) and reduced (NADH) forms

2.2.2 Alcoholic Fermentation

The reducing power of NADH, produced by glycolysis, must be transferred to an electron acceptor to regenerate NAD^+. In alcoholic fermentation, it is not pyruvate but rather acetaldehyde, its decarboxylation product, that serves as the terminal electron acceptor. With respect to glycolysis, alcoholic fermentation contains two additional enzymatic reactions, the first of which (catalyzed by pyruvate decarboxylase), decarboxylates pyruvic acid.

The cofactor is thiamine pyrophosphate (TPP) (Figure 2.4). TPP and pyruvate form an intermediary compound. More precisely, the carbon atom located between the nitrogen and the sulfur of the TPP thiazole cycle is ionized. It forms a carbanion, which readily combines with the pyruvate carbonyl group. The second step reduces acetaldehyde into ethanol by NADH. This reaction is catalyzed by the alcohol dehydrogenase whose active site contains a Zn^{2+} ion.

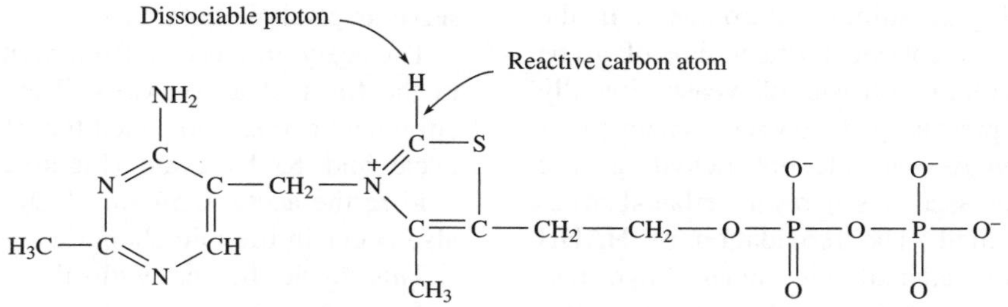

Fig. 2.4. Structure of thiamine pyrophosphate (TPP)

S. cerevisiae pyruvate decarboxylase (PDC) comprises two isoenzymes: a major form, PDC1, representing 80% of the decarboxylase activity, and a minor form, PDC5, whose function remains uncertain.

From an energy viewpoint, glycolysis followed by alcoholic fermentation supplies the yeast with two molecules of ATP per molecule of glucose degraded, or 14.6 biologically usable kcal/mol of glucose fermented. From a thermodynamic viewpoint, the change in free energy during the degradation of a mole of glucose into ethanol and CO_2 is -40 kcal. The difference (25.4 kcal) is dissipated in the form of heat.

2.2.3 Glyceropyruvic Fermentation

In the presence of sulfite (Neuberg, 1946), the fermentation of glucose by yeasts produces equivalent quantities of glycerol, carbon dioxide, and acetaldehyde in its bisulfitic form. This glyceropyruvic fermentation takes place in the following manner. Since the acetaldehyde combined with sulfite cannot be reduced into ethanol, dihydroxyacetone-1-phosphate becomes the terminal electron acceptor. It is derived from the oxidation of glyceraldehyde 3-phosphate and reduced to glycerol 3-phosphate, which is itself dephosphorylated into glycerol. This mechanism was used for the industrial production of glycerol. In this fermentation, only two molecules of ATP are produced for every molecule of hexose degraded. ATP is required to activate the glucose in the first step of glycolysis (Figure 2.5). Glyceropyruvic fermentation, whose net gain in ATP is nil, does not furnish biologically assimilable energy for yeasts.

Glyceropyruvic fermentation does not occur uniquely in a highly sulfitic environment. In the beginning of the alcoholic fermentation of grape must, the inoculum consists of yeasts initially grown in the presence of oxygen. Their pyruvate decarboxylase and alcohol dehydrogenase are weakly expressed. As a result, ethanal accumulation is limited. The reoxidation of NADH does not involve ethanal, but rather dihydroxyacetone. Glycerol, pyruvate and some secondary fermentation products are formed. These secondary

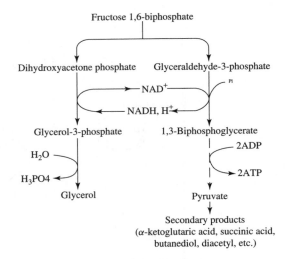

Fig. 2.5. Glyceropyruvic fermentation pathway

products are pyruvate derivatives—including, but not limited to, succinate and diacetyl.

2.2.4 Respiration

When sugar is used by the respiratory pathway, pyruvic acid (originating in glycolysis) undergoes an oxidative decarboxylation in the presence of coenzyme A (CoA) (Figure 2.6) and NAD^+. This process generates carbon dioxide, NADH and acetyl-CoA:

$$\text{pyruvate} + \text{CoA} + \text{NAD}^+ \longrightarrow$$
$$\text{acetyl CoA} + CO_2 + \text{NADH} + H^+ \quad (2.2)$$

The enzymatic system of the pyruvate dehydrogenase catalyzes this reaction. It takes place in the interior of the mitochondria. Thiamine pyrophosphate (TPP), lipoamide and flavin–adenine dinucleotide (FAD) participate in this reaction and serve as catalytic cofactors.

The acetyl unit issued from pyruvate is activated in the form of acetyl CoA. The reactions of the citric acid cycle, also called the tricarboxylic acids cycle and Krebs cycle (Figure 2.7), completely oxidize the acetyl CoA into CO_2. These reactions also occur in the mitochondria.

This cycle begins with the condensation of a 2-carbon acetyl unit with a 4-carbon compound, oxaloacetate, to produce a tricarboxylic

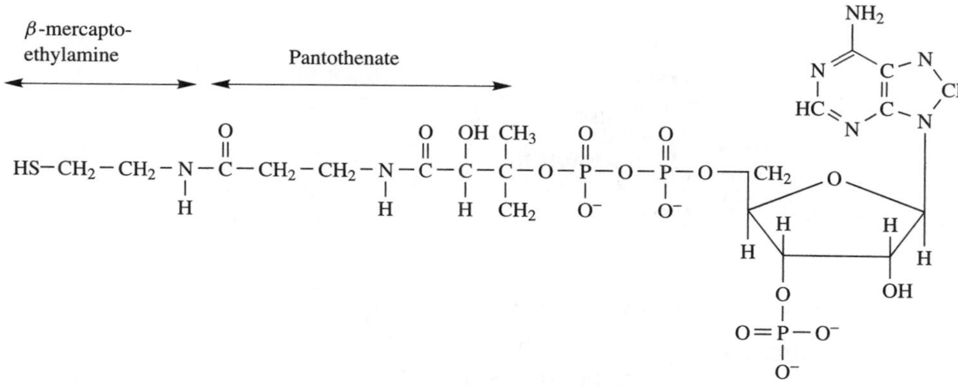

Fig. 2.6. Structure of coenzyme A. The reactive site is the terminal thiol group

acid with 6 carbon atoms: citric acid. Four oxidation–reduction reactions regenerate the oxaloacetate. The oxidative pathway decarboxylates isocitrate, an isomer of citrate, into α-ketoglutarate. The isocitrate dehydrogenase catalyzes this reaction. The α-ketoglutarate, a 5-carbon atom compound, undergoes an oxidative decarboxylation to become succinate by the α-ketoglutarate dehydrogenase. In these two reactions, NAD$^+$ is the hydrogen acceptor. The fumarate dehydrogenase is responsible for the reduction of succinate into fumarate; FAD is the hydrogen acceptor (Figure 2.8). Finally, fumarate is hydrated into L-malate. The latter is reduced into oxaloacetate by the malate dehydrogenase. In this case, the NAD$^+$ is an electron acceptor once again.

From acetate, each complete cycle produces two CO_2 molecules, three hydrogen ions transferred to three NAD$^+$ molecules (six electrons) and a pair of hydrogen atoms (two electrons) transferred to an FAD molecule. The cytochrome chain transports these electrons towards the oxygen. ATP is formed during this process. This oxidative phosphorylation (Figure 2.9) takes place in the mitochondria. This process makes use of three enzymatic systems (the NADH-Q reductase, the cytochrome reductase and the cytochrome oxidase). Two electron transport systems (ubiquinone, or coenzyme Q, and cytochrome c) link these enzymatic systems.

Oxidative phosphorylation yields three ATP molecules per pair of electrons transported between the NADH and the oxygen—two ATP molecules with FADH$_2$. In the Krebs cycle, substrate-level phosphorylation forms an ATP molecule during the transformation of succinyl CoA into succinate.

The respiration of a glucose molecule (Table 2.1) produces 36 to 38 ATP molecules. Two originate from glycolysis, 28 from the oxidative

Table 2.1. Energy balance of oxidation of glucose in respiration

Stage	Reduction coenzyme	Number of molecules of ATP formed
Glycolysis	2NADH	4 or 6
Net gain in ATP from glycolysis		2
Pyruvate $\longrightarrow$ acetyl CoA	NADH	6
Isocitrate $\longrightarrow$ α-Ketoglutarate	NADH	6
α-Ketoglutarate $\longrightarrow$ succinyl CoA	NADH	6
Succinyl CoA $\longrightarrow$ succinate		2
Succinate $\longrightarrow$ fumarate	FADH$_2$	4
Malate $\longrightarrow$ oxaloacetate	NADH	6
Net yield from glucose		36–38

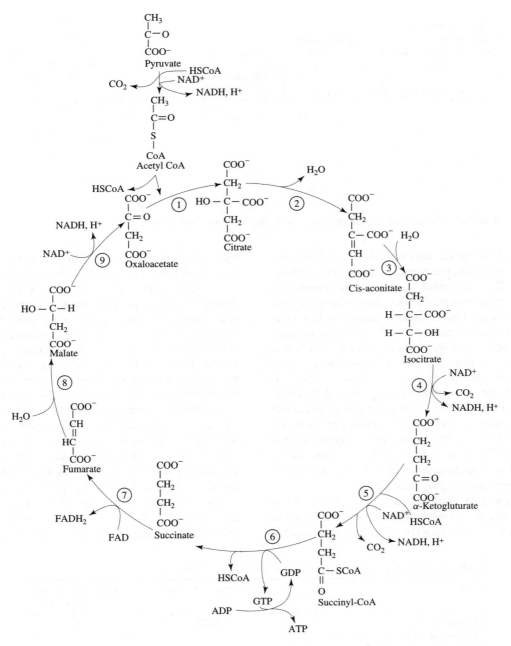

Fig. 2.7. Tricarboxylic acid or Krebs cycle. 1 = citrate synthase; 2–3 = aconitase; 4 = isocitrate dehydrogenase; 5 = complex α-ketoglutarate dehydrogenase; 6 = succinyl-CoA synthetase, 7 = succinate dehydrogenase; 8 = fumarase; 9 = malate dehydrogenase; GTP = guanosine triphosphate; GDP = guanosine diphosphate

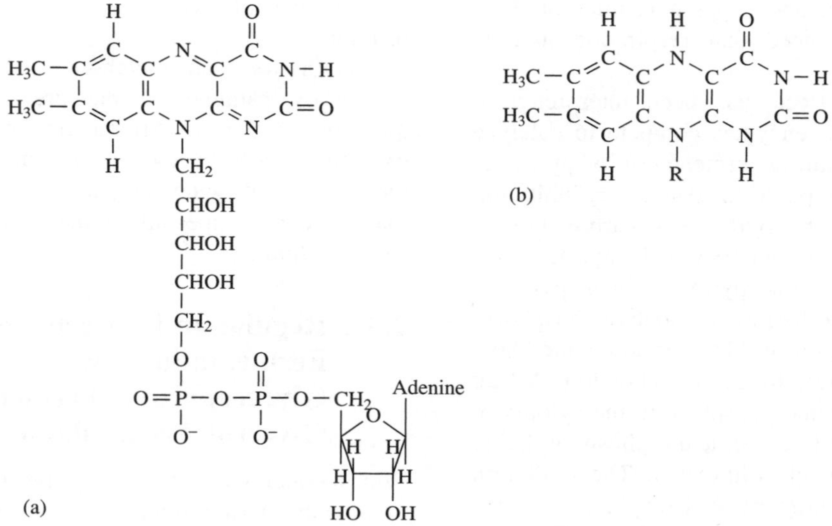

Fig. 2.8. Structure of flavin adenine dinucleotide (FAD): (a) oxidized form (FAD); (b) reduced form (FADH₂)

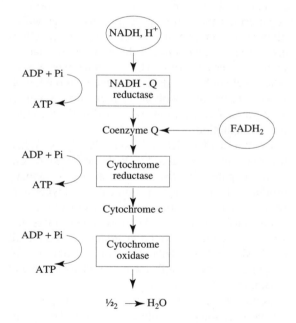

Fig. 2.9. Oxidative phosphorylation during electron transport in the respiratory chain

phosphorylation of NADH and FADH₂ generated by the Krebs cycle, and two from substrate-level phosphorylation during the formation of succinate. Four to six ATP molecules result from

the oxidative phosphorylation of two NADH molecules produced in glycolysis. The precise number depends on the transport system used to move the electrons of the cytosolic NADH to the respiratory chain in the mitochondria. The respiration of the same amount of sugar produces 18 to 19 times more biologically usable energy available to yeasts than fermentation. Respiration is used for industrial yeast production.

2.3 REGULATION OF SUGAR-UTILIZING METABOLIC PATHWAYS

2.3.1 Regulation Between Fermentation and Respiration: Pasteur Effect and Crabtree Effect

Pasteur was the first to compare yeast growth in aerobiosis and anaerobiosis. For low concentrations of glucose on culture media, yeasts utilize sugars through either respiration or fermentation. Aeration induces an increase in biomass formed (total and per unit of sugar degraded) and a decrease in alcohol production and sugar consumption. Pasteur therefore deduced that respiration inhibits fermentation.

The 'Pasteur effect' has been interpreted in several ways. Two enzymes compete to catalyze either the respiration or fermentation of pyruvate. This competition explains the respiratory inhibition of fermentation. The pyruvate decarboxylase is involved in the fermentative pathway. It has a lower affinity towards pyruvate than pyruvate dehydrogenase. Furthermore, oxidative phosphorylation consumes a lot of ADP and inorganic phosphate, which migrate to the mitochondria. A lack of ADP and inorganic phosphate in the cytoplasm ensues. This deficit can limit the phosphorylation and thus slow the glycolitic flux. The inhibition of glycolysis enzymes by ATP explains the Pasteur effect for the most part. The ATP issued from oxidative phosphorylation inhibits phosphofructokinase in particular. Phosphorylated hexoses accumulate as a result. The transmembrane transport of sugars and thus glycolysis is slowed.

For high glucose concentrations—for example, in grape must—S. cerevisiae only metabolizes sugars by the fermentative pathway. Even in the presence of oxygen, respiration is impossible. Discovered by Crabtree (1929) on tumoral cells, this phenomenon is known by several names: catabolic repression by glucose, the Pasteur contrary effect and the Crabtree effect. Yeasts manifest the following signs during this effect: a degeneration of the mitochondria, a decrease in the proportion of cellular sterols and fatty acids, and a repression of both the synthesis of Krebs cycle mitochondrial enzymes and constituents of the respiratory chain. With S. cerevisiae, there must be at least 9 g of sugar per liter for the Crabtree effect to occur. The catabolic repression exerted by glucose on wine yeasts is very strong. In grape must, at any level of aeration, yeasts are only capable of fermenting because of the high glucose and fructose concentrations. From a technological viewpoint, yeasts consume sugars by the respiratory pathway for the industrial production of dry yeast, but not in winemaking. If must aeration helps the alcoholic fermentation process (Section 3.7.2), the fatty acids and sterols synthesized by yeasts, proliferating in the presence of oxygen, are responsible but not respiration.

S. cerevisiae can metabolize ethanol by the respiratory pathway in the presence of small quantities of glucose. After alcoholic fermentation, oxidative yeasts develop in a similar manner on the surface of wine as part of the process of making certain specialty wines (Sherry, Yellow Wine of Jura).

2.3.2 Regulation Between Alcoholic Fermentation and Glyceropyruvic Fermentation; Glycerol Accumulation

Wines contain about 8 g of glycerol per 100 g of ethanol. During grape must fermentation, about 8% of the sugar molecules undergo glyceropyruvic fermentation and 92% undergo alcoholic fermentation. The fermentation of the first 100 g of sugar forms the majority of glycerol, after which glycerol production slows but is never nil. Glyceropyruvic fermentation is therefore more than an inductive fermentation which regenerates NAD^+ when acetaldehyde, normally reduced into ethanol, is not yet present. Alcoholic fermentation and glyceropyruvic fermentation overlap slightly throughout fermentation.

Pyruvic acid is derived from glycolysis. When this molecule is not used by alcoholic fermentation, it participates in the formation of secondary products. In this case, a molecule of glycerol is formed by the reduction of dihydroxyacetone.

Glycerol production therefore equilibrates the yeast endocellular oxidation–reduction potential, or $NAD^+/NADH$ balance. This 'relief valve' eliminates surplus NADH which appears at the end of amino acid and protein synthesis.

Some winemakers place too much importance on the organoleptic role of glycerol. This compound has a sugary flavor similar to glucose. In the presence of other constituents of wine, however, the sweetness of glycerol is practically imperceptible. For the majority of tasters, even well trained, the addition of 3–6 g of glycerol per liter to a red wine is not discernible and so the pursuit of winemaking conditions that are more conducive to glyceropyruvic fermentation has no enological interest. On the contrary, the winemaker should favor a pure alcoholic fermentation and should limit glyceropyruvic fermentation. The production of glycerol is accompanied by the formation of other secondary products, derived from pyruvic acid, the increased presence of which (such as carbonyl function compounds and acetic acid) decreases wine quality.

2.3.3 Secondary Products Formed from Pyruvate by Glyceropyruvic Fermentation

When a molecule of glycerol is formed, a molecule of pyruvate cannot be transformed into ethanol following its decarboxylation into ethanal. In anaerobic conditions, oxaloacetate is the means of entry of pyruvate into the cytosolic citric acid cycle. Although the mitochondria are no longer functional, the enzymes of the tricarboxylic acids cycle are present in the cytoplasm. Pyruvate carboxylase (PC) catalyzes the carboxylation of pyruvate into oxaloacetate. The prosthetic group of this enzyme is biotin; it serves as a CO_2 transporter. The reaction makes use of an ATP molecule:

$$biotin-PC + ATP + CO_2 \longrightarrow$$

$$CO_2-biotin-PC + ADP + [iP] \quad \textbf{(2.3)}$$

$$CO_2-biotin-PC + pyruvate \longrightarrow$$

$$biotin-PC + oxaloacetate \quad \textbf{(2.4)}$$

In these anaerobic conditions, the citric acid cycle cannot be completed since the succinodehydrogenase activity requires the presence of FAD, a strictly respiratory coenzyme. The chain of reactions is therefore interrupted at succinate, which accumulates (0.5−1.5 g/l). The NADH generated by this portion of the Krebs cycle (from oxaloacetate to succinate) is reoxidized by the formation of glycerol from dihydroxyacetone.

The α-ketoglutarate dehydrogenase has a very low activity in anaerobiosis; some authors therefore believe that the oxidative reactions of the Krebs cycle are interrupted at α-ketoglutarate. In their opinion, a reductive pathway of the

citric acid cycle forms succinic acid in anaerobiosis: oxaloacetate → malate → fumarate → succinate. Bacteria have a similar mechanism. In yeast, this is probably a minor pathway since only the oxidative pathway of the Krebs cycle can maintain the $NAD^+/NADH$ redox balance during fermentation (Oura, 1977). Furthermore, additional succinate is formed during alcoholic fermentation on a glutamate-enriched medium. The glutamate is deaminated to form α-ketoglutarate, which is oxidized into succinate.

Among secondary products, ketonic function compounds (pyruvic acid, α-ketoglutaric acid) and acetaldehyde predominantly combine with sulfur dioxide in wines made from healthy grapes. Their excretion is significant during the yeast proliferation phase and decreases towards the end of fermentation. Additional acetaldehyde is liberated in the presence of excessive quantities of sulfur dioxide in must. An elevated pH and fermentation temperature, anaerobic conditions, and a deficiency in thiamine and pantothenic acid increase production of ketonic acids. Thiamine supplementing of must limits the accumulation of ketonic compounds in wine (Figure 2.10).

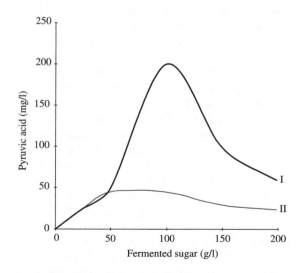

Fig. 2.10. Effect of a thiamine addition on pyruvic acid production during alcoholic fermentation (Lafon-Lafourcade, 1983). I = control must; II = thiamine supplemented must

Other secondary products of fermentation are also derived from pyruvic acid: acetic acid, lactic acid, butanediol, diacetyl and acetoin. Their formation processes are described in the following paragraphs.

2.3.4 Formation and Accumulation of Acetic Acid by Yeasts

Acetic acid is the principal volatile acid of wine. It is produced in particular during bacterial spoilage (acetic spoilage and lactic disease) but is always formed by yeasts during fermentation. Beyond a certain limit, which varies depending on the wine, acetic acid has a detrimental organoleptical effect on wine quality. In healthy grape must with a moderate sugar concentration (less than 220 g/l), *S. cerevisiae* produces relatively small quantities (100–300 mg/l), varying according to the strain. In certain winemaking conditions, even without bacterial contamination, yeast acetic acid production can be abnormally high and become a problem for the winemaker.

The biochemical pathway for the formation of acetic acid in wine yeasts has not yet been clearly identified. The hydrolysis of acetyl CoA can produce acetic acid. The pyruvate dehydrogenase produces acetyl CoA beforehand by the oxidative decarboxylation of pyruvic acid. This reaction takes place in the matrices of the mitochondria but is limited in anaerobiosis. Aldehyde dehydrogenase can also form acetic acid by the oxidation of acetaldehyde (Figure 2.11). This enzyme, whose cofactor is NADP$^+$, is active during alcoholic fermentation. The NADPH thus formed can be used to synthesize lipids. When pyruvate dehydrogenase is repressed, this pathway forms acetyl CoA through the use of acetyl CoA synthetase. In anaerobiosis on a model medium, yeast strains producing the least amount of acetic acid have the highest acetyl CoA synthetase activity (Verdhuyn *et al.*, 1990).

The acetaldehyde dehydrogenase in *S. cerevisiae* has five isoforms, three located in the cytosol (Section 1.4.1) (Ald6p, Ald2p, and Ald3p) and the remaining two (Ald4p and Ald5p) in the mitochondria (Section 1.4.3). These enzymes differ by their specific use of the NAD$^+$ or NADP$^+$ cofactor (Table 2.2).

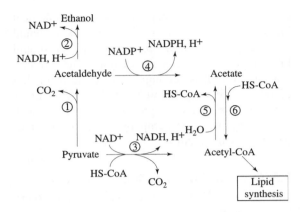

Fig. 2.11. Acetic acid formation pathways in yeasts. 1 = pyruvate decarboxylase; 2 = alcohol dehydrogenase; 3 = pyruvate dehydrogenase; 4 = aldehyde dehydrogenase; 5 = acetyl-CoA hydrolase; 6 = acetyl-CoA synthetase

Table 2.2. Isoforms of acetaldehyde dehydrogenase in *S. cerevisiae* (Navarro-Avino *et al.*, 1999)

Chromosome	Gene	Location	Cofactor
XIII	ALD2	Cytosol	NAD$^+$
XIII	ALD3	Cytosol	NAD$^+$
XV	ALD4	Mitochondria	NAD$^+$ and NADP$^+$
V	ALD5	Mitochondria	NADP$^+$
XVI	ALD6	Cytosol	NADP$^+$

Remize *et al.* (2000) and Blondin *et al.* (2002) studied the impact of the deletion of each gene and demonstrated that the NADP-dependent cytoplasmic isoform *ALD6* played a major role in the formation of acetic acid during the fermentation of dry wines, while the *ALD5* isoform was also involved, but to a lesser extent (Figure 2.12).

Practical winemaking conditions likely to lead to abnormally high acetic acid production by *S. cerevisiae* are well known. As is the case with glycerol formation, acetic acid production is closely dependent on the initial sugar level of the must, independent of the quantity of sugars fermented (Table 2.3). The higher the sugar content of the must, the more acetic acid (and glycerol) the yeast produces during fermentation. This is due to the yeast's mechanism for adapting to a medium with a high sugar concentration:

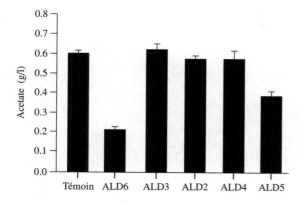

Fig. 2.12. Acetate production by strains of *S. cerevisiae* (V5) following deletion of different genes coding for isoforms of acetaldehyde dehydrogenase (Blondin *et al.*, 2002)

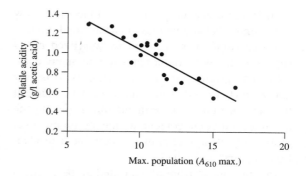

Fig. 2.13. Correlation between volatile acidity production and the maximum cell population in high-sugar, botrytized musts

Table 2.3. Effect of initial sugar concentration of the must on the formation of secondary products of the fermentation (Lafon-Lafourcade, 1983)

Initial sugar (g/l)	Fermented sugar (g/l)	Secondary products		
		Acetic acid (g/l)	Glycerol (g/l)	Succinic acid (g/l)
224	211	0.26	4.77	0.26
268	226	0.45	5.33	0.25
318	211	0.62	5.70	0.26
324	179	0.84	5.95	0.26
348	152	1.12	7.09	0.28

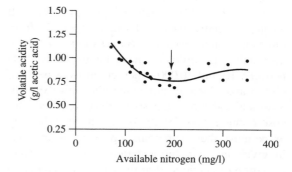

Fig. 2.14. Effect of the available nitrogen content in must (with or without ammonium supplements) on the production of volatile acidity (initial sugar content: 350 g/l)

S. cerevisiae increases its intracellular accumulation of glycerol to counterbalance the osmotic pressure of the medium (Blomberg and Alder, 1992). This regulation mechanism is controlled by a cascade of signal transmissions leading to an increase in the transcription level of genes involved in the production of glycerol (*GPD1*), but also of acetate (*ALD2* and *ALD3*) (Attfield *et al.*, 2000). Acetate formation plays an important physiological role in the intracellular redox balance by regenerating reduced equivalents of NADH. Thus, it is clear that an increase in acetate production is inherent to the fermentation of high-sugar musts. However, Bely *et al.* (2003) demonstrated that it was possible to reduce acetate production by supplying more NADH to the redox balance process. This may be done indirectly by stimulating biomass formation, which generates an excess of NADH during amino acid synthesis. Available nitrogen in the must plays a key role in this process. Thus, in high-sugar musts, acetate production is inversely correlated with the maximum cell population (Figure 2.13), which is, in turn, related to the available nitrogen content of the must. It is strongly recommended to monitor the available nitrogen content of botrytized musts and supplement them with ammonium sulfate, if necessary. The optimum available nitrogen concentration in this type of must to minimize acetic acid production is approximately 190 mg/l (Figure 2.14). The best time for adding nitrogen supplements is at the very beginning of fermentation, as later additions are less effective and may even increase acetate production. Indeed, in view of the unpredictable increase in acetic acid production that sometimes occurred in botrytized musts supplemented with ammonium sulfate, many enologists had given up the practice entirely. It is now known that, provided the supplement is added at the very beginning of fermentation, adjusting the available nitrogen content to the optimum level (190 mg/l) always minimizes acetic acid production in botrytized wines.

In wines made from noble rotted grapes, certain substances in the must inhibit yeast growth and increase the production of acetic acid and glycerol during fermentation. *Botrytis cinerea* secretes these 'botryticine' substances (Ribéreau-Gayon *et al.*, 1952, 1979). Fractional precipitation with ethanol partially purifies these compounds from must and culture media of *Botrytis cinerea*. These heat stable glycoproteins have molecular weights between 10 and 50 000. They comprise a peptidic (10%) and glucidic part containing mostly mannose and galactose and some rhamnose and glucose (Dubourdieu, 1982). When added to healthy grape must, these compounds provoke an increase in glyceropyruvic fermentation and a significant excretion of acetic acid at the end of fermentation (Figure 2.15). The mode of action of these glycoproteins on yeasts has not yet been identified. The physiological state of yeast populations at the time of inoculation seems to play an important role in the fermentative development of botrytized grape must. Industrial dry yeast preparations are much more sensitive to alcoholic fermentation inhibitors than yeast starters obtained by preculture in healthy grape musts.

Other winemaking factors favor the production of acetic acid by *S. cerevisiae*: anaerobiosis, very low pH (<3.1) or very high pH (>4), certain amino acid or vitamin deficiencies in the must, and too high of a temperature (25–30°C) during the yeast multiplication phase. In red winemaking, temperature is the most important factor, especially when the must has a high sugar concentration. In hot climates, the grapes should be cooled when filling the vats. The temperature should not exceed 20°C at the beginning of fermentation. The

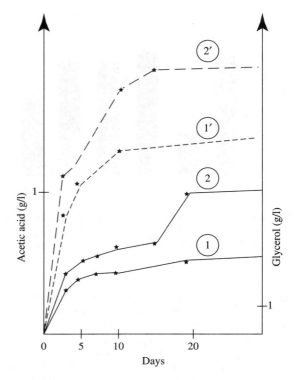

Fig. 2.15. Effect of an alcohol-induced precipitate of a botrytized grape must on glycerol and acetic acid formation during the alcoholic fermentation of healthy grape must (Dubourdieu, 1982). (1) Evolution of acid acetic concentration in the control must; (2) evolution of acid acetic concentration in the must supplemented with the freeze-dried precipitate; (1′) evolution of glycerol concentration in the control must; (2′) evolution of glycerol concentration in the must supplemented with the freeze-dried precipitate

same procedure should be followed in thermovinification immediately following the heating of the grapes.

In dry white and rosé winemaking, excessive must clarification can also lead to the excessive production of volatile acidity by yeast. This phenomenon can be particularly pronounced with certain yeast strains. Therefore, must turbidity should be adjusted to the lowest possible level which permits a complete and rapid fermentation (Chapter 13). Solids sedimentation (must lees) furnishes long-chain unsaturated fatty acids (C18:1, C18:2). Yeast lipidic alimentation greatly

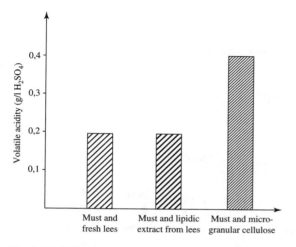

Fig. 2.16. Effect of the lipidic fraction of must lees on acetic acid production by yeasts during alcoholic fermentation (Lavigne, 1996)

influences acetic acid production during white and rosé winemaking.

The experiment in Figure 2.16 illustrates the important role of lipids in acetic acid metabolism (Delfini and Cervetti, 1992; Alexandre *et al.*, 1994). The volatile acidity of three wines obtained from the same Sauvignon Blanc must was compared. After filtration, must turbidity was adjusted to 250 Nephelometric turbidity units (NTU) before inoculation by three different methods: reincorporating fresh lees (control); adding cellulose powder; and supplementing the same quantity of lees adsorbed on the cellulose powder with a lipidic extract (methanol-chloroform). The volatile acidities of the control wine and the wine that was supplemented with a lipidic extraction of lees before fermentation are identical and perfectly normal. Although the fermentation was normal, the volatile acidity of the wine made from the must supplemented with cellulose (therefore devoid of lipids) was practically twice as high (Lavigne, 1996). Supplementing the medium with lipids appears to favor the penetration of amino acids into the cell, which limits the formation of acetic acid.

During the alcoholic fermentation of red or slightly clarified white wines, yeasts do not continuously produce acetic acid. The yeast metabolizes a large portion of the acetic acid secreted in must during the fermentation of the first 50–100 g of sugar. It can also assimilate acetic acid added to must at the beginning of alcoholic fermentation. The assimilation mechanisms are not yet clear. Acetic acid appears to be reduced to acetaldehyde, which favors alcoholic fermentation to the detriment of glyceropyruvic fermentation. In fact, the addition of acetic acid to a must lowers glycerol production but increases the formation of acetoin and butanediol-2,3. Yeasts seem to use the acetic acid formed at the beginning of alcoholic fermentation (or added to must) via acetyl CoA in the lipid-producing pathways.

Certain winemaking conditions produce abnormally high amounts of acetic acid. Since this acid is not used during the second half of the fermentation, it accumulates until the end of fermentation. By refermenting a tainted wine, yeasts can lower volatile acidity by metabolizing acetic acid. The wine is incorporated into freshly crushed grapes at a proportion of no more than 20–30%. The wine should be sulfited or filtered before incorporation to eliminate bacteria. The volatile acidity of this mixture should not exceed 0.6 g/l in H_2SO_4. The volatile acidity of this newly made wine rarely exceeds 0.3 g/l in H_2SO_4. The concentration of ethyl acetate decreases simultaneously.

2.3.5 Other Secondary Products of the Fermentation of Sugars

Lactic acid is another secondary product of fermentation. It is also derived from pyruvic acid, directly reduced by yeast L(+) and D(−) lacticodehydrogenases. In anaerobiosis (the case in alcoholic fermentation), the yeast synthesizes predominantly D(−) lacticodehydrogenase. Yeasts form 200–300 mg of D(−) lactic acid per liter and only about a dozen milligrams of L(+) lactic acid. The latter is formed essentially at the start of fermentation. By determining the D(−) lactic acid concentration in a wine, it can be ascertained whether the origin of acetic acid is yeast or lactic bacteria (Section 14.2.3). Wines that have undergone malolactic fermentation can contain several grams per liter of exclusively L(+) lactic acid. On

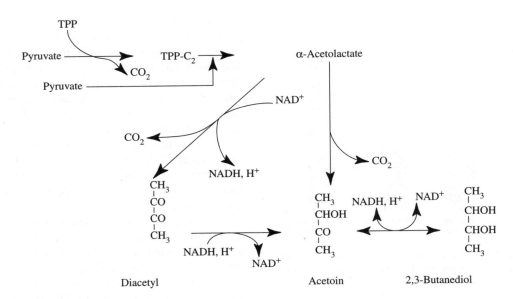

Fig. 2.17. Acetoin, diacetyl and 2,3-butanediol formation by yeasts in anaerobiosis. TPP = thiamine pyrophosphate; TPP-C2 = active acetaldehyde

the other hand, the lactic fermentation of sugars (lactic disease) forms D(−) lactic acid. Lactic bacteria have transformed substrates other than malic acid, when D(−) lactic acid concentrations exceed 200 to 300 mg/l.

Yeasts also make use of pyruvic acid to form acetoin, diacetyl and 2,3-butanediol (Figure 2.17). This process begins with the condensation of a pyruvate molecule and active acetaldehyde bound to thiamine pyrophosphate, leading to the formation of α-acetolactic acid. The oxidative decarboxylation of α-acetolactic acid produces diacetyl. Acetoin is produced by either the non-oxidative decarboxylation of α-acetolactic acid or the reduction of diacetyl. The reduction of acetoin leads to the formation of 2,3-butanediol; this last reaction is reversible.

From the start of alcoholic fermentation, yeasts produce diacetyl, which is rapidly reduced to acetoin and 2,3-butanediol. This reduction takes place in the days that follow the end of alcoholic fermentation, when wines are conserved on the yeast biomass (de Revel *et al.*, 1996). Acetoin and especially diacetyl are strong-smelling compounds which evoke a buttery aroma. Above a certain

concentration, they have a negative effect on wine aroma. The concentration of wines that have not undergone malolactic fermentation is too low (a few milligrams per liter for diacetyl) to have an olfactory influence. On the other hand, lactic bacteria can degrade citric acid to produce much higher quantities of these carbonyl compounds than yeasts (Section 5.3.2).

Finally, yeasts condense acetic acid (in the form of acetyl CoA) and pyruvic acid to produce citramalic acid (0–300 mg/l) and dimethylglyceric acid (0–600 mg/l) (Figure 2.18). These compounds have little organoleptic incidence.

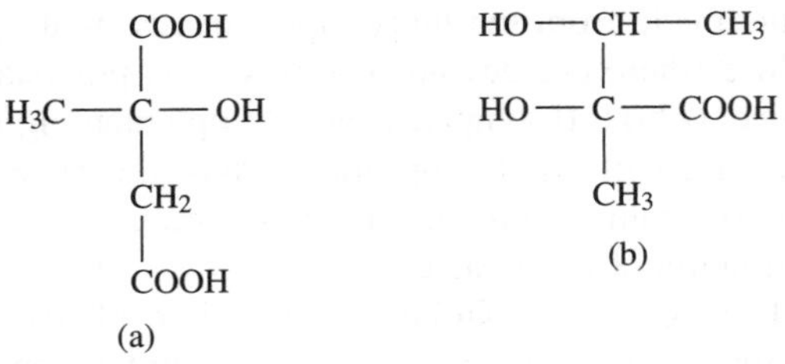

Fig. 2.18. (a) Citramalic acid and (b) dimethylglyceric acid

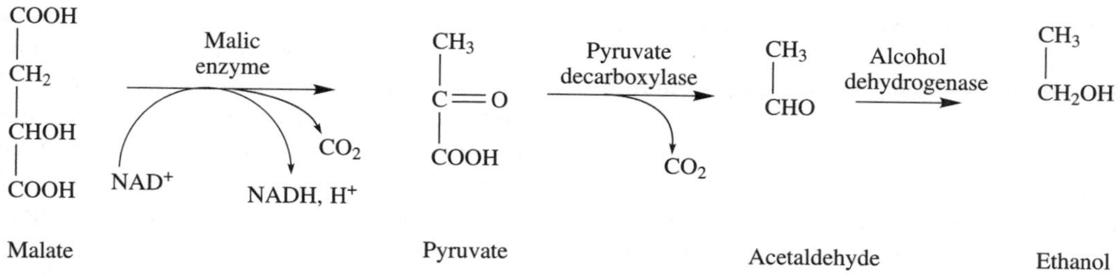

Fig. 2.19. Decomposition of malic acid by yeasts during alcoholic fermentation

2.3.6 Degradation of Malic Acid by Yeast

Saccharomyces cerevisiae partially degrades must malic acid (10–25%) during alcoholic fermentation. Different strains degrade varying amounts of this acid, and degradation is more significant when the pH is low. Alcoholic fermentation is the principal pathway degrading malic acid. The pyruvic acid resulting from this transformation is decarboxylated into ethanal, which is then reduced to ethanol. The malic enzyme is responsible for the transformation of malic acid into pyruvic acid (Figure 2.19). This oxidative decarboxylation requires NAD^+ (Fuck and Radler, 1972). This maloalcoholic fermentation lowers wine acidity significantly more than malolactic fermentation.

Schizosaccharomyces differs from wine yeasts. The alcoholic fermentation of malic acid is complete in yeasts of this genus, which possess an active malate transport system. (In *S. cerevisiae*, malic acid penetrates the cell by simple diffusion.) Yet at present no attempts to use *Schizosaccharomyces* in winemaking have been successful (Peynaud *et al.*, 1964; Carre *et al.*, 1983). First of all, the implantation of these yeasts in the presence of *S. cerevisiae* is difficult in a non-sterilized must. Secondly, their optimum growth temperature (30°C) higher than for *S. cerevisiae*, imposes warmer fermentation conditions. Sometimes, the higher temperature adversely affects the organoleptical quality of wine. Finally, some grape varieties fermented by *Schizosaccharomyces* do not express their varietal aromas. The acidic Gros Manseng variety produces a very fruity wine when correctly vinified with *S. cerevisiae*, but has no varietal aroma when fermented by *Schizosaccharomyces*. To resolve these problems, some researchers have used non-proliferating populations of *Schizosaccharomyces* enclosed in alginate balls. These populations degrade the malic acid in wines having already completed their alcoholic fermentation (Magyar and Panyid, 1989; Taillandier and Strehaiano, 1990). Although no organoleptical defect is found in these wines, the techniques have not yet been developed for practical use.

Today, molecular biology permits another strategy for making use of the ability of *Schizosaccharomyces* to ferment malic acid. It consists of integrating *Schizosaccharomyces* malate permease genes and the malic enzyme (mae 1 and mae 2) in the *S. cerevisiae* genome (Van Vuuren *et al.*, 1996). The technological interest of a wine yeast genetically modified in this manner is not yet clear, nor are the risks of its proliferation in wineries and nature.

2.4 METABOLISM OF NITROGEN COMPOUNDS

The nitrogen requirements of wine yeasts and the nitrogen supply in grape musts are discussed in Chapter 3 (Section 3.4.2). The following section covers the general mechanisms of assimilation, biosynthesis, and degradation of amino acids in yeasts. The consequences of these metabolisms, which occur during alcoholic fermentation and affect the production of higher alcohols and their associated esters in wine, are also discussed.

2.4.1 Amino Acid Synthesis Pathways

The ammonium ion and amino acids found in grape must supply the yeast with nitrogen. The yeast can also synthesize most of the amino acids necessary for constructing its proteins. It fixes an ammonium ion on a carbon skeleton derived from the metabolism of sugars. The yeast uses the same reactional pathways as all organisms. Glutamate and glutamine play an important role in this process (Cooper, 1982; Magasanik, 1992).

The NADP$^+$ glutamate dehydrogenase (NADP$^+$-GDH), product of the GDH1 gene, produces glutamate (Figure 2.20) from an ammonium ion and an α-ketoglutarate molecule. The latter is an intermediary product of the citric acid cycle. The yeast also possesses an NAD$^+$ glutamate dehydrogenase (NAD$^+$-GDH), product of the GDH2 gene. This dehydrogenase is involved in the oxidative catabolism of glutamate. It produces the inverse reaction of the precedent, liberating the ammonium ion used in the synthesis of glutamine. NADP$^+$-GDH activity is at its maximum when the yeast is cultivated on a medium containing exclusively ammonium as its source of nitrogen. The NAD$^+$-GDH activity, however, is at its highest level when the principal source of nitrogen is glutamate. Glutamine synthetase (GS) produces glutamine from glutamate and ammonium. This amination requires the hydrolysis of an ATP molecule (Figure 2.21).

Through transamination reactions, glutamate then serves as an amino group donor in the biosynthesis of different amino acids. Pyridoxal phosphate is the transaminase cofactor (Figure 2.22); it is derived from pyridoxine (vitamin B$_6$).

The carbon skeleton of amino acids originates from glycolysis intermediary products (pyruvate, 3-phosphoglycerate, phosphoenolpyruvate), the

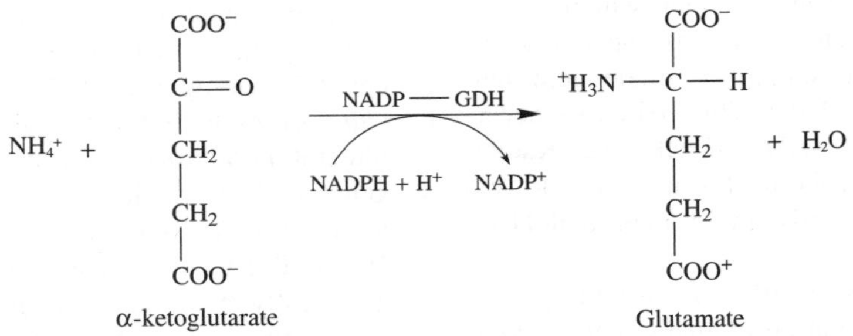

Fig. 2.20. Incorporation of the ammonium ion in α-ketoglutarate catalyzed by NADP glutamate dehydrogenase (NADP-GDP)

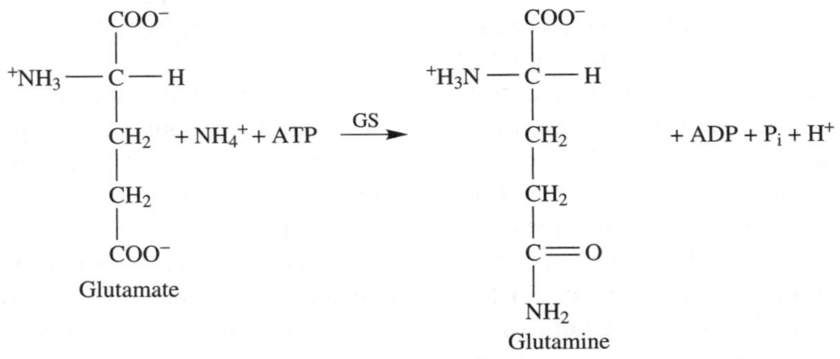

Fig. 2.21. Amidation of glutamate into glutamine by glutamine synthetase (GS)

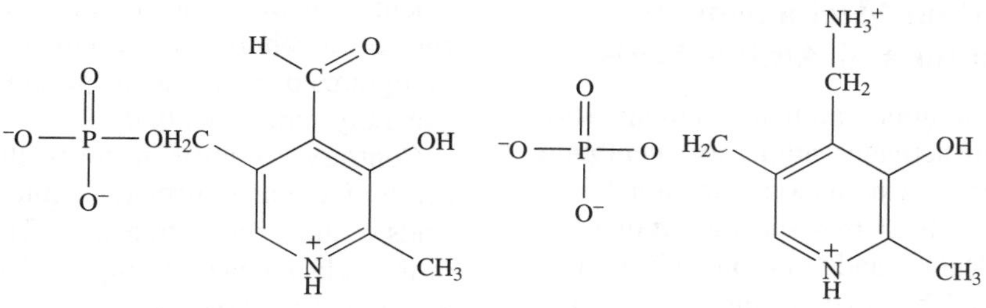

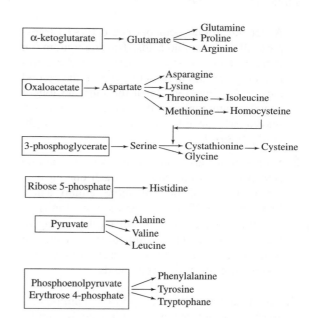

Fig. 2.22. Pyridoxal phosphate (PLP) and pyridoxamine phosphate (PMP)

citric acid cycle (α-ketoglutarate, oxaloacetate) or the pentose phosphate cycle (ribose 5-phosphate, erythrose 4-phosphate). Some of these reactions are very simple, such as the formation of aspartate or alanine by transamination of glutamate into oxaloacetate or pyruvate:

oxaloacetate + glutamate $\longrightarrow$

$$\text{aspartate} + \alpha\text{-ketoglutarate} \qquad (2.5)$$

pyruvate + glutamate $\longrightarrow$

$$\text{alanine} + \alpha\text{-ketoglutarate} \qquad (2.6)$$

Other biosynthetic pathways are more complex, but still occur in yeasts as in the rest of the living world. The amino acids can be classified into six biosynthetic families depending on their nature and their carbon precursor (Figure 2.23):

1. In addition to glutamate and glutamine, proline and arginine are formed from α-ketoglutarate.

2. Asparagine, methionine, lysine, threonine and isoleucine are derived from aspartate, which is issued from oxaloacetate. ATP can activate methionine to form *S*-adenosylmethionine, which can be demethylated to form *S*-adenosylhomocysteine, the hydrolysis of which liberates adenine to produce homocysteine.

3. Pyruvate is the starting point for the synthesis of alanine, valine and leucine.

4. 3-Phosphoglycerate leads to the formation of serine and glycine. The condensation of homocysteine and serine produces cystathionine, a precursor of cysteine.

Fig. 2.23. General biosynthesis pathways of amino acids

5. The imidazole cycle of histidine is formed from ribose 5-phosphate and adenine of ATP.

6. The amino acids possessing an aromatic cycle (tyrosine, phenylalanine, trytophan) are derived from erythrose 4-phosphate and phosphoenolpyruvate. These two compounds are intermediaries of the pentose cycle and glycolysis, respectively. Their condensation forms shikimate. The condensation of this compound with another molecule of phosphoenolpyruvate produces chorismate, a precursor of aromatic amino acids.

2.4.2 Assimilation Mechanisms of Ammonium and Amino Acids

The penetration of ammonium and amino acids into the yeast cell activates numerous membrane proteinic transporters or permeases (Section 1.3.2). *S. cerevisiae* has at least two specific ammonium ion transporters (Dubois and Grenson, 1979). Their activity is inhibited by several amino acids, in a non-competitive manner.

Two distinct categories of transporters ensure amino acid transport:

1. A general amino acid permease (GAP) transports all of the amino acids. The ammonium ion inhibits and represses the GAP. The GAP therefore only appears to be active during the second half of fermentation, when the must no longer contains ammonium. It acts as a 'nitrogen scavenger' towards amino acids (Cartwright *et al.*, 1989).

2. *S. cerevisiae* also has many specific amino acid permeases (at least 11). Each one ensures the transport of one or more amino acids. In Contrast to GAP, the ammonium ion does not limit their activity. From the beginning of the yeast log phase during the first stages of fermentation, these transporters ensure the rapid assimilation of must amino acids.

Glutamate and glutamine, crossroads of amino acid synthesis, are not the only amino acids rapidly assimilated. Most of the amino acids are practically depleted from the must by the time the first 30 g of sugar have been fermented. Alanine and arginine are the principal amino acids found in must. Yeasts make use of these two compounds and ammonium slightly after the depletion of other amino acids. Furthermore, yeasts massively assimilate arginine only after the disappearance of ammonium from the medium. Sometimes, yeasts do not completely consume γ-aminobutyric acid. Yeasts do not utilize proline during fermentation, although it is one of the principal amino acids found in must.

During fermentation, yeasts assimilate between 1 and 2 g/l of amino acids. Towards the end of fermentation, yeasts excrete significant but variable amounts of different amino acids. Finally, at the end of alcoholic fermentation, a few hundred milligrams of amino acids per liter remain; proline generally represents half.

Contrary to must hexoses that penetrate the cell by facilitated diffusion, ammonium and amino acids require active transport. Their concentration in the cell is generally higher than in the external medium. The permease involved couples the transport of an amino acid molecule (or ammonium ion) with the transport of a hydrogen ion. The hydrogen ion moves in the direction of the concentration gradient: the concentration of protons in the must is higher than in the cytoplasm. The amino acid and the proton are linked to the same transport protein and penetrate the cell simultaneously. This concerted transport of two substances in the same direction is called symport (Figure 2.24). Obviously, the proton that penetrates the cell must then be exported to avoid acidification of the cytoplasm. This movement is made against the concentration gradient and requires energy. The membrane ATPase ensures the excretion of the hydrogen ion across the plasmic membrane, acting as a proton pump.

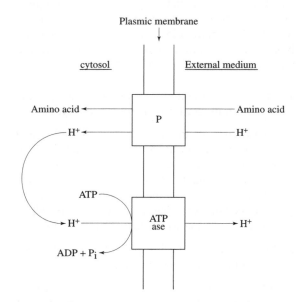

Fig. 2.24. Active amino acid transfer mechanisms in the yeast plasma membrane. P = protein playing the role of an amino acid 'symporter'

Ethanol strongly limits amino acid transport. It modifies the composition and the properties of the phospholipids of the plasmic membrane. The membrane becomes more permeable. The H^+ ions of the medium massively penetrate the interior of the cell by simple diffusion. The membrane ATPase must increase its operation to control the intracellular pH. As soon as this task monopolizes the ATPase, the symport of the amino acids no longer functions. In other words, at the beginning of fermentation, and for as long as the ethanol concentration in the must is low, yeasts can rapidly assimilate amino acids and concentrate them in the vacuoles for later use, according to their biosynthesis needs.

2.4.3 Catabolism of Amino Acids

The ammonium ion is essential for the synthesis of amino acids necessary for building proteins, but yeasts cannot always find sufficient quantities in their environment. Fortunately, they can obtain ammonium from available amino acids through various reactions.

The most common pathway is the transfer of an α-amino group, originating from one of many different amino acids, onto α-ketoglutaric acid to form glutamate. Aminotransferases or transaminases catalyze this reaction, whose prosthetic group is pyridoxal phosphate (PLP). Glutamate is then deaminated by oxidative pathways to form NH_4^+ (Figure 2.25). These two reactions can be summarized as follows:

$$\alpha\text{-amino acid} + NAD^+ + H_2O \longrightarrow$$

$$\alpha\text{-ketonic acid} + NH_4^+ + NADH + H^+ \quad (2.7)$$

During transamination, pyridoxal phosphate is temporarily transformed into pyridoxamine phosphate (PMP). The PLP aldehydic group is linked to a lysine residue ε-amino group on the active site of the aminotransferase to form an intermediary product (E-PLP) (Figure 2.26). The α-amino group of

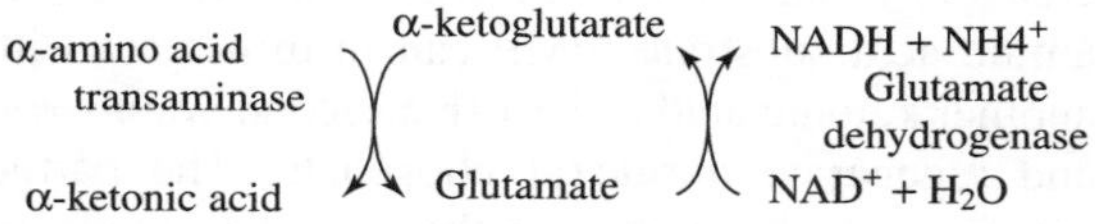

Fig. 2.25. Oxidative deamination of an amino acid, catalyzed by a transaminase and glutamate dehydrogenase

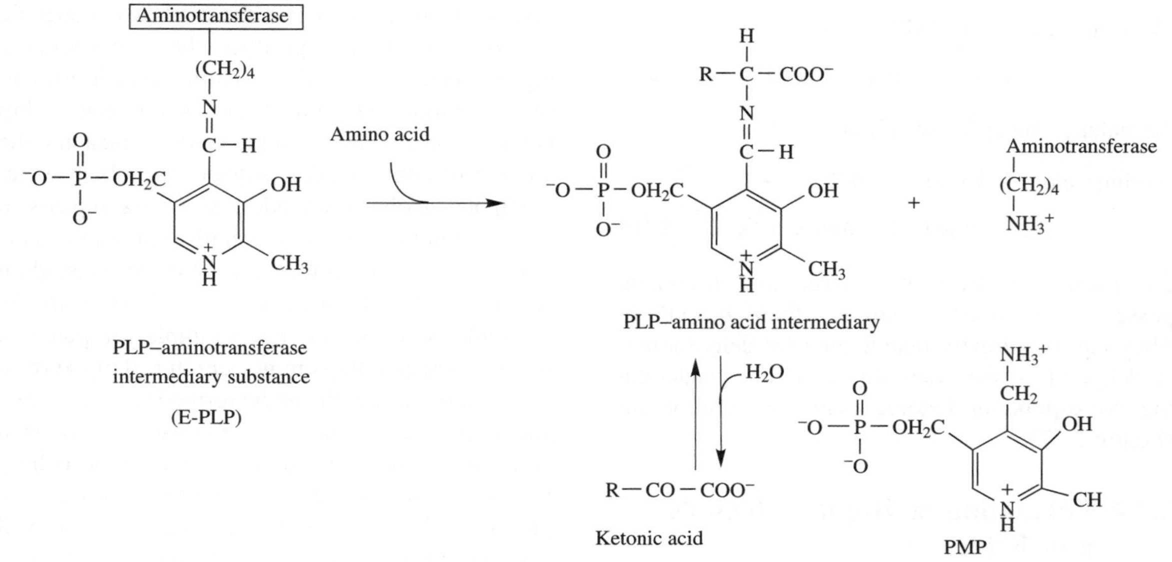

Fig. 2.26. Mode of action of pyridoxal phosphate (PLP) in transamination reactions. Formation of intermediary products between PLP and aminotransferase or the amino acid substrate

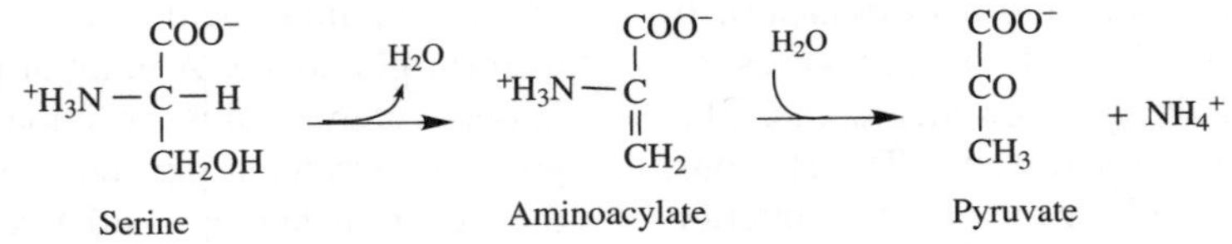

Fig. 2.27. Deamination of serine by a dehydratase

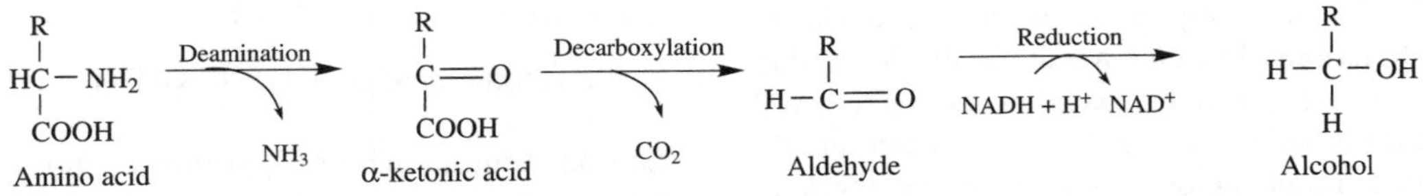

Fig. 2.28. Formation of higher alcohols from amino acids (Ehrlich reactions)

the amino acid substrate of the transamination displaces the lysine residue ε-amino group linked to PLP. The cleavage of this intermediary product liberates PMP and ketonic acid, corresponding to the amino acid substrate. PMP can in turn react with another ketonic acid to furnish a second amino acid and regenerate pyridoxal phosphate. The partial reactions can be written in the following manner:

$$\text{amino acid } 1 + \text{E-PLP} \longrightarrow$$
$$\text{ketonic acid } 1 + \text{E-PMP} \quad (2.8)$$

$$\text{ketonic acid } 2 + \text{E-PMP} \longrightarrow$$
$$\text{amino acid } 2 + \text{E-PLP} \quad (2.9)$$

the balance sheet for which is:

$$\text{amino acid } 1 + \text{ketonic acid } 2 \longrightarrow$$
$$\text{ketonic acid } 1 + \text{amino acid } 2 \quad (2.10)$$

Some amino acids, such as serine and threonine, possess a hydroxyl group on their β carbon. They can be directly deaminated by dehydration. A dehydratase catalyses this reaction, producing the corresponding ketonic acid and ammonium (Figure 2.27).

2.4.4 Formation of Higher Alcohols and Esters

Yeasts can excrete ketonic acids originating from the deamination of amino acids only after their

decarboxylation into aldehyde and reduction into alcohol (Figure 2.28). This mechanism, known as the Ehrlich reaction, explains in part the formation of higher alcohols in wine. Table 2.4 lists the principal higher alcohols and their corresponding amino acids, possible precursors of these alcohols.

Several experiments clearly indicate, however, that the degradation of amino acids is not the only pathway for forming higher alcohols in wine. In fact, certain ones, such as propan-1-ol and butan-1-ol, do not have amino acid precursors. Moreover, certain mutants deficient in the synthesis of specific amino acids do not produce the corresponding higher alcohol, even if the amino acid is present in the culture medium. There is no relationship between the amount of amino acids in must and the amount of corresponding higher alcohols in wine.

Higher alcohol production by yeasts appears to be linked not only to the catabolism of amino acids but also to their synthesis via the corresponding ketonic acids. These acids are derived from the metabolism of sugars. For example, propan-1-ol has no corresponding amino acid. It is derived from α-ketobutyrate which can be formed from pyruvate and acetyl coenzyme A. α-Ketoisocaproate is a precursor of isoamylic alcohol and an intermediary product in the synthesis of leucine. It too can be produced from α-acetolactate, which is derived from pyruvate. Most higher alcohols in wine can also be formed by the metabolism of glucose without the involvement of amino acids.

Table 2.4. The principal alcohols found in wine and their amino acid precursors

Higher alcohol	Concentration in wine (mg/l)	Amino acid precursor
$CH_3-\overset{\underset{\mid}{CH_3}}{CH}-CH_2-CH_2OH$ *3-methyl-butan-1-ol* *or isoamyl alcohol*	80–300	$CH_3-\overset{\underset{\mid}{CH_3}}{CH}-CH_2-\overset{\underset{\mid}{NH_2}}{CH}-COOH$ *Leucine*
$CH_3-CH_2-\overset{\underset{\mid}{CH_3}}{CH}-CH_2OH$ *2-methyl-butan-2-ol* *or active amyl alcohol*	30–100	$CH_3-CH_2-\overset{\underset{\mid}{CH_3}}{CH}-\overset{\underset{\mid}{NH_2}}{CH}-COOH$ *Isoleucine*
$CH_3-\overset{\underset{\mid}{CH_3}}{CH}-CH_2OH$ *2-methyl-propan-1-ol* *or isobutyl alcohol*	50–150	$CH_3-\overset{\underset{\mid}{CH_3}}{CH}-\overset{\underset{\mid}{NH_2}}{CH}-COOH$ *Valine*
(phenyl ring)$-CH_2-CH_2OH$ *Phenylethanol*	10–100	(phenyl ring)$-CH_2-\overset{\underset{\mid}{NH_2}}{CH}-COOH$ *Phenylalanine*
$HO-$(phenyl ring)$-CH_2-CH_2OH$ *Tyrosol*	20–50	$HO-$(phenyl ring)$-CH_2-\overset{\underset{\mid}{NH_2}}{CH}-COOH$ *Tyrosine*
$CH_3-CH_2-CH_2OH$ *Propan-1-ol*	10–50	?
$CH_3-CH_2-CH_2-CH_2OH$ *Butan-1-ol*	1–10	?
(indole ring)$-CH_2-CH_2OH$ *Thyptophol*	0–1	(indole ring)$-CH_2-\overset{\underset{\mid}{NH_2}}{CH}-COOH$ *Tryptophane*
$CO-CH_2-CH_2-CH_2$ with O bridge *γ-Butyrolatone*	0–5	$COOH-CH_2-CH_2-\overset{\underset{\mid}{NH_2}}{CH}-COOH$ *Glutamic acid*
$CH_3-S-CH_2-CH_2-CH_2OH$ *Methionol*	0–5	$CH_3-S-CH_2-CH_2-\overset{\underset{\mid}{NH_2}}{CH}-COOH$ *Methionine*

The physiological function of higher alcohol production by yeasts is not clear. It may be a simple waste of sugars, a detoxification process of the intracellular medium, or a means of regulating the metabolism of amino acids.

With the exception of phenylethanol, which has a rose-like fragrance, higher alcohols smell bad. Most, such as isoamylic alcohol, have heavy solvent-like odors. Methionol is a peculiar alcohol because it contains a sulfur atom. Its cooked-cabbage odor has the lowest perception threshold (1.2 mg/l). It can be responsible for the most persistent and disagreeable olfactory flaws of reduction, especially in white wines. In general, the winemaker should avoid excessive higher alcohol odors. Fortunately, their organoleptic impact is limited at their usual concentrations in wine, but it depends on the overall aromatic intensity of the wine. Excessive yields and rain at the end of maturation can dilute the must, in which case the wine will have a low aromatic intensity and the heavy, common character of higher alcohols can be pronounced.

The winemaking parameters that increase higher alcohol production by yeasts are well known: high pH, elevated fermentation temperature, and aeration. In red winemaking, the extraction of pomace constituents and the concern for rapid and complete fermentations impose aeration and elevated temperatures, and in this case higher alcohol production by yeast cannot be limited. In white winemaking, a fermentation temperature between 20 and 22°C limits the formation of higher alcohols.

Ammonium and amino acid deficiencies in must lead to an increased formation of higher alcohols. In these conditions, the yeast appears to recuperate all of the animated nitrogen available by transamination. It abandons the unused carbon skeleton in the form of higher alcohols. Racking white must also limits the production of higher alcohols (Chapter 13).

The nature of the yeast (species, strain) responsible for fermentation also affects the production of higher alcohols. Certain species, such as *Hansenula anomala*, have long been known to produce a lot, especially in aerobiosis (Guymon *et al.*, 1961). Yet production by wine yeasts is limited, even in spontaneous fermentation. More recently, various researchers have shown that most *S. bayanus (ex uvarum)* produce considerably more phenylethanol than *S. cerevisiae*. Finally, higher alcohol production in *S. cerevisiae* depends on the strain. A limited higher alcohol production (with the exception of phenylethanol) should be among selection criteria for wine yeasts.

Due to their esterase activities, yeasts form various esters (a few milligrams per liter). The most important acetates of higher alcohols are isoamyl acetate (banana aroma) and phenylethyl acetate (rose aroma). Although they are not linked to nitrogen metabolism, ethyl esters of medium-chain fatty acids are also involved. They are formed by the condensation of acetyl coenzyme A. These esters have more interesting aromas than the others. Hexanoate has a flowery and fruity aroma reminiscent of green apples. Ethyl decanoate has a soap-like odor. In white winemaking, the production of these esters can be increased by lowering the fermentation temperature and increasing must clarification. Certain yeast strains (71B) produce large quantities of these compounds, which contribute to the fermentation aroma of young wines. They are rapidly hydrolyzed during their first year in bottle and have no long-term influence on the aromatic character of white wines.

REFERENCES

Alexandre H., Nguyen Van Long T., Feuillat M. and Charpentier C., 1994, *Rev. Fr. Œnol.,* cahiers scientifiques, 146, 11–20.

Bely M., Rinaldi A. and Dubourdieu D., 2003, *J. Biosci. Bioeng.* 96 (6), 507.

Bisson L.F., 1993, Yeasts-Metabolism of sugars, in *Wine microbiology and biotechnology*, p. 55–75, G.H. Fleet., Harwood Academic Publishers, Chur, Switzerland.

Blomberg A. and Alder L., 1992, *Adv. Microbiol. Physiol.*, 33, 145–212.

Blondin B., Dequin S., Saint Prix F. and Sablayrolles J.M., 2002, *La formation d'acides volatils par les levures, dans 13*ième *Symposium international d'œnologie 09–10 Juin 2002*, INSA, INRA, Montpellier, France.

Carre E., Lafon-Lafourcade S. and Bertrand A., 1983, *Conn. Vigne Vin*, 17, 43–53.

Cartwright C.P., Rose A.H., Calderbank J. and Keenan M.H.J., 1989, Solute transport, in *The Yeasts*, vol. 3, p. 5–55. A.H. Rose, J.S. Harrison, Academic Press, London.

Caubet R., Guérin B. and Guérin M., 1988, *Arch. Microbiol.*, 149, 324–329.

Cooper T.G., 1982, Nitrogen metabolism in *Saccharomyces cerevisiae*, in *The molecular biology of the yeast Saccharomyces: metabolism and gene expression* p. 39–99, J.N. Strathern, E.W. Jones, J.R. Broach., Cold Spring Harbor Laboratory, New York.

Crabtree H.G., 1929, *Biochem. J.*, 23, 536–545.

De Revel G., Lonvaud-Funel A. and Bertrand A., 1996, Étude des composés dicarbonylés au cours des fermentations alcoolique et malolactique, in *Œnologie 95, 5ᵉ Symposium international d'œnologie*, p. 321–325, A. Lonvaud, Tec et Doc, Pairs.

Delfini C. and Cervetti F., 1992, *Vitic. Enol. Sci.*, 46, 142–150.

Dubois E. and Grenson M., 1979, *Molecular and general genetics*, 175, 67–76.

Dubourdieu D., 1982, Recherches sur les polysaccharides sécrétés par *Botrytis cinerea* dans la baie de raisin, *Thése Doctorat és Sciences*, Université de Bordeaux II.

Fuck E. and Radler F., 1972, *Archiv, für Mikrobiologie*, 87, 149–164.

Guymon J.F., Ingraham J.L. and Crowell E.A., 1961, *Arch. Biochem. Biophys.*, 95, 163–168.

Kishimoto M., 1994, *J. of Fermentation and Bio. engineering*, 77, 4, 432–435.

Lafon-Lafourcade S., 1983, Wine and Brandy, in *Biotechnology*, 83–163 H.J. Rehm et G. Reed, Verlag Chemie, Weinheim.

Lavigne V., 1996, *Recherche sur les composés volatils formés par la levure au cours de la vinification et de l'élevage des vins blancs secs*, Thése de Doctorat, Université de Bordeaux II.

Magasanik B., 1992, Regulation of nitrogen utilization, in *The molecular biology of the yeast Saccharomyces.*

Gene expression, p. 283–317, E.W. Jones, J.R. Pringle, J.R. Broach., Cold Spring Harbor Laboratory, New York.

Magyar I. and Panyik I., 1989, *Am. J. Oenol. Vitic.* 40, 233–240.

Masneuf I., 1996, *Recherches sur l'identification génétique des levures de vinification. Applications œnologiques*, Thése de Doctorat de l'Université de Bordeaux II.

Navarro-Avino J.P., Prasad R., Miralles V.J., Benito R.M. and Serrano R., 1999, *Yeast*, 15, 929–842.

Neuberg C., 1946, *Ann. Rev. Biochem.*, 15, 435.

Oura E., 1977, *Process Biochem.*, 12, 19–21.

Peynaud E., Domercq S. and Boidron A., Lafon-Lafourcade S., Guimberteau G., 1964, *Archiv. für Mikrobiologie*, 48, 150–165.

Remize F., Andrieu E. and Dequin S., 2000, *Appl. Environ. Microbiol.*, 66, 3151–3159.

Ribéreau-Gayon J., Peynaud E. and Lafourcade S., 1952, *C. R. Acad. Sc.* 234, 478.

Ribéreau-Gayon P., Lafon-Lafourcade S., Dubourdieu D., Lucmaret V. and Larue F., 1979, C. R. Acad. Sc., 289 D, 441.

Stryer L., 1992, Métabolisme: concepts et vue d'ensemble, in *La biochimie de Lubert Stryer*, p. 315–328, Flammarion, Paris.

Taillandier P., Strehaiano P., 1990, Dégradation de l'acide malique par *Schizosaccharomyces*, in *Actualités œnologiques 89, C. R. du 4ᵉ Symposium international d'œnologie, Bordeaux 1989*, P. Ribéreau-Gayon, A. Lonvaud, Dunod-Bordas, Paris.

Van Vuuren H.J.J., Viljoen M., Grobler J., Volschenk H., Bauer F. and Subden R.E., 1996, Genetic analysis of the *Schizosaccharomyces pombe* malate permeases, *Mael* and malic enzyme, *Maell*, genes and their expression in *Saccharomyces cerevisiae*, in *Œnologie 95, 5ᵉ Symposium international d'œnologie*, p. 195–197, A. Lonvaud, Tech et Doc, Paris.

Verdhuyn C., Postma E., Scheffers W.A. and Van Dijken J.P., 1990, *J. Gen. Micro.* 136, 359–403.

3

Conditions of Yeast Development

3.1 INTRODUCTION

Grape must is a highly fermentable medium in which yeasts find the necessary substances to ensure their vital functions. Carbohydrates (glucose and fructose) are used as carbon and energy sources—the yeasts deriving ethanol. Ethanol gives wines their principal character. Organic acids (tartaric and malic acid) and mineral salts (phosphate, sulfate, chloride, potassium, calcium and magnesium) ensure a suitable pH. Nitrogen compounds exist in several forms: ammonia, amino acids, polypeptides and proteins. Grape must also contains substances serving as growth (vitamins) and survival factors. Other grape constituents (phenolic compounds, aromas) contribute to wine character, but do not play an essential role in fermentation kinetics.

In general, with an adequate inoculation (10^6 cells/ml), fermentation is easily initiated in grape must. It is complete, if the initial carbohydrate concentration is not excessive, but different factors can disrupt yeast growth and fermentation kinetics. Some factors have a chemical nature and correspond with either nutritional deficiencies or the presence of inhibitors formed during fermentation (ethanal, fatty acids, etc.). Others have a physicochemical nature—for example, oxygenation, temperature and must clarification. Finally, fermentation difficulties can lead to the development of undesired microorganisms. They can antagonize the desired winemaking yeast strain.

The successful completion of fermentation depends on all of these factors. A perfect mastery of fermentation is one of the primary responsibilities of an enologist, who must use the necessary means to avoid microbial deviations and lead the fermentation to its completion—the complete depletion of sugars in dry wines. Stuck fermentations are a serious problem. They are not only often difficult to restart but can also lead to bacterial spoilage such as lactic disease, in sugar-containing media. These issues have been discussed in connection with practical winemaking applications (Ribéreau-Gayon et al., 1975a, 1976). Theoretical information concerning yeast physiology can be found in more fundamental works (Fleet, 1992).

3.2 MONITORING AND CONTROLLING FERMENTATIONS

3.2.1 Counting Yeasts

Different methods (described elsewhere) permit the industrial characterization of yeast strains involved in fermentation. Tracking the yeast population according to fermentation kinetics can be useful. Lafon-Lafourcade and Joyeux (1979) described enumeration and identification techniques for different microorganisms in must and wine (yeast, acetic and lactic bacteria).

After the appropriate dilution of fermenting must, the total number of yeast cells can be estimated under the microscope, using a Malassez cell. After calibration, this determination can also be made by measuring the optic density of the fermentation medium at 620 nm. This measurement permits a yeast cell count by interpretation of the medium's cloudiness, provoked by yeasts.

The total cells counted in this manner include both 'dead' yeast and 'live' yeast. The two must be differentiated in enology. In fact, counting 'viable microorganisms' is preferable. When placed in a suitable, solid nutritive medium, viable cells are capable of developing and forming a microscopic cluster, visible to the naked eye, called a colony. The number of viable yeast cells can be determined by counting the colonies formed on this medium after 3–4 days. The culture medium is prepared by adding 1 ml of correctly diluted yeast solution to 5 ml of nutritive medium. This mixture is transferred in its liquid form at 40°C into a Petri dish; it solidifies as it cools. The nutritive medium consists of equal volumes of a 20 g/l agar solution and grape must (180 g sugar/l and 3.2 pH) diluted to half its initial concentration. The average of two population counts having between 100 and 300 colonies is calculated.

Viable yeast populations can also be estimated directly by counting under the microscope using specific coloration or epifluorescence techniques. Viable populations can also be determined with ATP measurements using bioluminescence. Bouix et al. (1997) proposed the use of immunofluorescence to detect bacterial contaminations during winemaking.

Microbiological control is useful for research work, but these control methods are relatively long and difficult. For this reason, fermentations are generally followed by more simple methods at the winery.

3.2.2 Monitoring Fermentation Kinetics

Winemakers must closely monitor wine fermentation in each tank of the winery. This close supervision allows them to observe transformations, anticipate their evolution, and act quickly if necessary. They should effectuate both fermentation and temperature controls daily (Figure 3.1).

Fermentation kinetics can be tracked by measuring the amount of sugar consumed, alcohol formed, or carbon dioxide released, but the measurement of the mass per unit volume (density) is a simpler-method for tracking its evolution. Mass per unit volume constitutes an approximate measure of the amount of sugar contained in the grape must. Since a relationship exists between the amount of alcohol produced during fermentation and the

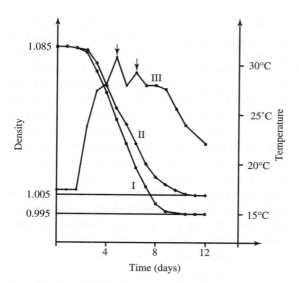

Fig. 3.1. Example of daily fermentation monitoring in two tanks (initial must density = 1.085). (I) Normal fermentation curve: after a latency period, fermentation initiates, accelerates and then slows before stopping on the 10th day at a density below 0.995, when all sugar is fermented. (II) Fermentation curve leading to a stuck fermentation: fermentation stops on the 11th day at a density of 1.005; unfermented sugar remains. Fermentation slows early enough to take preventive action. (III) Evolution of the temperature (red wine fermentation): arrows indicate cooling

initial concentration of sugar in the must, must density can directly give an approximate potential alcohol. The density and potential alcohol are generally indicated on the stem of the hydrometer: approximately 17 g of sugar produces 1% alcohol in volume (Section 10.3.2, Table 10.6). Expressing potential alcohol is without doubt the best solution.

During fermentation, sugar depletion and ethanol formation result in decreased density. A hydrometer is used to monitor must density. Samples are taken from the middle of the tank by using the tasting faucet. Before taking a sample, the faucet should be cleared by letting a few centiliters flow. The density is then corrected according to must temperature. No other conversions or interpretations are necessary. Plotting the points in the form of a graph permits the winemaker to evaluate fermentation kinetics. The regularity of the fermentation can also be evaluated. More

importantly, fermentation difficulties, which lead to stuck fermentations, can be anticipated. Due to the heterogeneity of fermentation kinetics in a red winemaking tank (the fermentation is most active under the pomace cap), homogenizing the tank is recommended before taking samples.

3.2.3 Taking the Temperature

The daily monitoring of tank temperature during fermentation is indispensable, but this measurement must be taken properly. In red winemaking, in particular, the tank temperature is never homogeneous. The temperature is highest in the pomace cap and lowest at the bottom of the tank. In the first hours of fermentation, abrupt temperature increases occur in the pomace and are sometimes very localized. As a result, the must temperature against the tank lining is always less than in the center of the tank. The temperature taken in these conditions, even after properly clearing the tasting faucet, is not representative of the entire tank. The temperature should be taken after a pumping-over, which homogenizes tank temperature. In this manner, the average temperature can be obtained, but the maximum temperature, also important, remains unknown.

The must temperature can be taken with a dial thermometer having a 1.5 m probe. This effective method can measure must temperature directly in different areas, especially just below the pomace cap in the hottest part of the tank. This zone has the most significant fermentation activity. The temperature can also be taken by thermoelectric probes judiciously placed in each tank. The probes are linked to a measurement system in the winery laboratory. With this system, the winemaker can verify the temperature of the tanks at any moment. Certain temperature control systems automatically regulate tank temperature when the temperature reaches certain value.

3.2.4 Fermentation Control Systems

Various automated systems simplify the monitoring of fermentation and make it more rigorous. These systems can also automatically heat and cool the must according to temperature.

Some of these systems can be very sophisticated. For example, when the temperature exceeds the set limit, the apparatus initially homogenizes the must in the tank by pumping. If the temperature is still too high, the refrigeration unit of the system cools the must. Owing to the seasonal character of winemaking, some winemakers must make do with manual control systems. Automation can, however, be fully justified, when it permits greater precision in winemaking. For example, a temperature gradient can be produced in this manner during red winemaking. At the beginning, a moderate temperature (18–20°C) favors cell growth and vitality; at the end, a higher temperature (>30°C) facilitates the extraction of pomace constituents.

New approaches to automated systems could be further developed (Flanzy, 1998). For the moment at least, they influence temperature, modulating it according to fermentation kinetics. In addition to 'on-line' temperature control, a system should be developed to monitor fermentation kinetics. Various methods have been tested: measurement of carbon dioxide given off, the decrease in weight, and the decrease in density (measured as the difference in pressure between the top and the bottom of the tank). Measurement of gas released seems to be the most reliable method (by weighing in the laboratory or by using a domestic gas counter for large capacity tanks; El Halaoui et al., 1987). This process assumes that the tanks are completely airtight. The method, however, is not particularly recommended, especially in red winemaking where pumping-over is indispensable.

Sablayrolles et al. (1990) stated that automated systems should modulate temperature in order to control fermentation kinetics better and limit the use of cooling systems.

Automated control systems should be linked to fermentation speed and therefore yeast activity, a parameter certainly as important as the temperature. When the speed of CO_2 production exceeds a previously established limit, the obtained temperature should be maintained. As soon as the speed decreases, the apparatus should let the temperature rise in order to revive the fermentation. This operation would permit the fermentation to be completed more quickly. Temperature modulation should be related to must fermentability. Simultaneously letting the temperature increase would result in a decrease in the total energy demand (Table 3.1). Of course, the apparatus should maintain the temperature within compatible enological limits.

Another approach consists of creating a model of the alcoholic fermentation process. Different calculations concerning time, temperature, and alcohol and sugar concentrations are used to predict fermentation behavior—especially the risk of a stuck fermentation (Bovée et al., 1990). In this way, the need and moment of a certain operation (especially temperature control) could be anticipated.

Finally, these control system rules must be adapted to the particular needs of each tank in terms of quantifiable data and the enologist's experience. A highly advance automated system optimizing alcoholic fermentations in winemaking would have to make use of artificial intelligence to take into account the enologist's own experience (Grenier et al., 1990).

3.2.5 Avoiding Foam Formation

During fermentation, foam can be formed as carbon dioxide is released. This can result in the tank overflowing. To avoid the problem, tanks are

Table 3.1. Comparison of a temperature-controlled (17–22°C) fermentation and isothermal fermentations at 17°C and 22°C (Sablayrolles et al., 1990)

Temperature	Isothermal 17°C	Isothermal 22°C	Temperature-controlled 17°C–22°C
Duration of fermentation (h)	263	174	183
Maximum rate (g CO_2/l/h)	0.69	1.12	0.72
Total frigorific units needed (kcal/l)	18.2	16.1	10
Maximum frigorific unit demand (kcal/l/h)	0.174	0.257	0.179

sometimes filled to only half their capacity. This constraint is not acceptable. Factors that influence foam formation include must nitrogen composition (especially proteic concentration), fermentation temperature, and the nature of the yeast strain. Attempts to create fermentation conditions (for example, eliminating proteins by using bentonite) capable of limiting this phenomenon have not led to satisfactory results.

For this reason, some American wineries have adopted the use of products that increase surface tension. This process reduces foam formation and stability. Two anti-foaming agents are gaining popularity: dimethyl polysiloxane and a mixture of oleic acid mono- and diglyceride. They are used at a concentration of less than 10 mg/l and do not leave a residue in wine, especially after filtration. Due to their efficiency, red wine tanks can be filled to 75–80% capacity and white wine tanks to 85–90%. These products are not toxic. The Office International de la Vigne et du Vin recommends the exclusive use of the mixture of oleic acid mono- and diglyceride.

3.3 YEAST GROWTH CYCLE AND FERMENTATION KINETICS

In an unsulfited and non-inoculated must, contamination yeasts can begin to develop within a few hours of filling the tank. Apiculated yeasts (*Kloechera, Hanseniaspora*) are the most frequently encountered. Aerobic yeasts also develop (*Candida, Pichia, Hansenula*), producing acetic acid and ethyl acetate. *Brettanomyces* and its characteristic animal-like odors are rare in must. Although such yeasts can be relatively resistant to sulfur dioxide (Fleet, 1992), sulfiting followed by inoculating with a selected strain of *Saccharomyces cerevisiae* constitute, in practice, an effective means of avoiding contamination (Section 3.5.4).

In general, *S. cerevisiae* inoculated at 10^6 cells/ml, either naturally or by a selected strain inoculation, induces grape must fermentation.

The yeast growth cycle and grape must fermentation kinetics are depicted in Figure 3.2 (Lafon-Lafourcade, 1983). In order to accentuate certain phenomena, the figure concerns a must containing

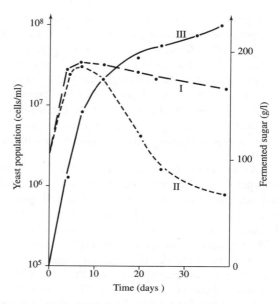

Fig. 3.2. Yeast growth cycle and fermentation kinetics of grape must containing high sugar concentrations (320 g/l) (Lafon-Lafourcade, 1983). (I) Total yeast population. (II) Viable yeast population. (III) Fermented sugar

particularly high sugar concentrations, which cannot be completely fermented. Analysis of this figure prompts the following remarks.

1. The growth cycle has three principal phases: a limited growth phase (2–5 days) increases the population to between 10^7 and 10^8 cells/ml; a quasi-stationary phase follows and lasts about 8 days; finally, the death phase progressively reduces the viable population to 10^5 cells/ml. The final phase can last for several weeks.

2. During this particularly long cycle, growth is limited to four or five generations.

3. The stopping of growth is not the result of a disappearance of energy nutrients.

4. The duration of these different phases is not equal. The death phase, in particular, is three to four times longer than the growth phase.

5. Fermentation kinetics are directly linked to the growth cycle. The fermentation speed is at its maximum and practically constant for

a little over 10 days. This time period corresponds with the first two phases of the growth cycle. The fermentation speed then progressively slows but fermentation nevertheless lasts several weeks. At this stage, the yeast population is in the survival phase. Finally, the stopping of fermentation is not simply the result of insufficient yeast growth. The metabolic activity of non-proliferating cells can also be inhibited.

The cessation of metabolic activity has been interpreted as the depletion of cellular ATP and the accumulation of ethanol in the cell—most likely due to transport difficulties across the membranes because of cellular sterol depletion. The cell enzymatic systems still function during this survival phase but the intracellular sugar concentration decreases gradually (Larue *et al.*, 1982).

These phenomena have several technological consequences. For a limited concentration in sugar (less than 200 g/l), fermentation occurs during the first two phases of the cycle. It takes place rapidly and most often without a problem. On the other hand, if there is an elevated amount of sugar, a population in its death phase carries out the last part of fermentation. In this case, its metabolic activity continues to decrease throughout the process. The total transformation of sugar into alcohol depends on the survival capacity of this population; it is as important as in the initial growth phase.

Excessive temperatures and sugar concentrations can provoke sluggish or stuck fermentations. Nutritional deficiencies and inhibition phenomena can also be involved. All of them have either chemical or physicochemical origins. Fermentation kinetics can be ameliorated by different methods which influence these phenomena. Early action appears to increase their effectiveness; yeasts in their growth phase and in a medium containing little ethanol are more receptive to external stimuli. The winemaker should anticipate fermentation difficulties: the possible operations are much less effective after they occur.

Moreover, certain operations, intended to activate fermentation, affect yeast growth and improve fermentation kinetics at its beginning but do not always affect yeast survival or the final stages of fermentation, at least in musts with high sugar

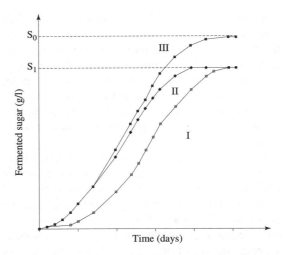

Fig. 3.3. Example of a theoretical activation of grape must fermentation (S_0: initial sugar content; S_1: sugar content at the end of fermentation). (I) Control: fermentation stops, leaving unfermented sugar ($S_0 - S_1$). (II) Activation at the initiation of fermentation, but the completion of fermentation in musts having high sugar concentrations is not improved. (III) Activation acting on yeast population growth and survival; fermentation is complete

concentrations (the most difficult to ferment). Increasing must temperature is a classic example of an operation that increases the fermentation speed at the beginning but leads to stuck fermentations (see Section 3.7.1, Figure 3.8). In other cases, the activation of the fermentation influences both growth and survival; the duration of fermentation is prolonged. Figure 3.3 gives an example: activation (curve II) apparently improved the fermentation kinetics of musts with relatively low sugar levels as compared to the control (curve I). However, in a high-sugar must, an activator that did not enhance survival was unable to prevent stuck fermentation. In the case of curve III, where both survival and growth factors were added, fermentation was completed smoothly, even in a high-sugar must.

3.4 NUTRITION REQUIREMENTS

3.4.1 Carbon Supply

In grape must, yeasts find glucose and fructose—sources of carbon and energy. The total sugar concentration in must is between 170 and 220 g/l, corresponding to wines between 10 and 13% vol.ethanol after fermentation. The amount of dissolved sugar can be even higher in grapes for the production of sweet wines—up to 350 g/l in Sauternes musts. The must sugar concentration can influence the selection of the yeast strains, ensuring fermentation (Fleet, 1992).

Fermentation is slow in a medium containing a few grams of sugar per liter. Its speed increases in musts which have 15–20 g/l and remains stable until about 200 g/l. Above this concentration, fermentation slows. In fact, alcohol production can be lower in a must containing 300 g/l than in another containing only 200 g/l. From 600 to 650 g of sugar per liter, the concentrated grape must becomes practically unfermentable. The presence of sugar, as well as alcohol, contributes to the stability of fortified wine.

Thus, an elevated amount of sugar hinders yeast growth and decreases the maximum population. Consequently, fermentation slows even before the production of a significant quantity of ethanol—which normally has an antiseptic effect (Section 3.6.1).

Fermentation slows down in the same way when the high sugar concentration is due to the addition of sugar (chaptalization) or concentrated must, or the elimination of water by reverse osmosis or vacuum evaporation (Section 15.5.1). The effect is exacerbated if sugar is added when fermentation is already well advanced and alcohol has started to inhibit yeast development, although intervening at this stage has the advantage of avoiding overheating the must. When sugar is added, it is advisable to wait until the second day after fermentation starts, i.e. the end of the yeast growth phase. In these conditions, the population reaches a higher value, because it grows in a medium with a relatively low sugar concentration. Next, the addition of sugar in a medium containing yeast in full activity increases the fermentation capacity of the cells and therefore the transformation of sugar. Of course, a refrigeration system is necessary to compensate for the corresponding temperature increase of the must.

3.4.2 Nitrogen Supply

Henschke and Jiranek (1992) analyzed many theoretical works on this subject in detail. The research for these different works was carried out under a large range of conditions and so the results were not always applicable to the winemaking conditions analyzed by Ribéreau-Gayon et al. (1975a).

Grape must contains a relatively high concentration of nitrogen compounds (0.1 to 1 g of soluble nitrogen per liter), although only representing about a quarter of total berry nitrogen. These constituents include the ammonium cation (3–10% of total nitrogen), amino acids (25–30%), polypeptides (25–40%) and proteins (5–10%). The grape nitrogen concentration depends on variety, rootstock, environment and growing conditions—especially nitrogen fertilization. It decreases when rot develops on the grapes or the vines suffer from drought conditions. Water stress, however, is generally a positive factor in quality red wine production. Planting cover crop in the vineyard to control yields also reduces the grapes nitrogen levels. The effects are variable, e.g. 118 mg/l nitrogen in grapes from the control plot and 46 mg/l nitrogen for the plot with cover crop in one case and 354 and 210 mg/l nitrogen, respectively, in another. The nitrogen content of overripe grapes may increase due to concentration of the juice.

In dry white winemaking, juice extraction methods influence the amino group compound and protein concentration in must. Slow pressing and skin maceration, which favor the extraction of skin constituents, increase their concentration (Dubourdieu et al., 1986).

Yeasts find the nitrogen supply necessary for their growth in grape must. The ammonium cation is easily assimilated and can satisfy yeast nitrogen needs, in particular, for the synthesis of amino acids. Polypeptides and proteins do not participate in S. cerevisiae growth, since this yeast cannot hydrolyze these substances. S. cerevisiae does not need amino acids as part of its nitrogen supply, since it is capable of synthesizing them individually, but their addition stimulates yeasts more than ammoniacal nitrogen. A mixture of amino acids and ammoniacal nitrogen is an even more effective stimulant. Yeasts use amino acids according to three mechanisms (Henschke and Jiranek, 1992):

1. Direct integration without transformation into proteins.

2. Decomposition of the amino group, which is used for the biosynthesis of different amino constituents. The corresponding carbon molecule is excreted. Such a reaction is one of the pathways of higher alcohol formation present in wine:

$$R-CHNH_2-COOH + H_2O \longrightarrow$$

$$R-CH_2OH + CO_2 + NH_3$$

Yeasts are probably capable of obtaining ammoniacal nitrogen from amino acids through other pathways.

3. The amino acid molecule can be used as a source of carbon in metabolic reactions. The yeast simultaneously recuperates the corresponding ammoniacal nitrogen.

The assimilation of different amino acids depends on the functioning of transport systems and the regulation of metabolic systems. Several studies have been published on this subject (Castor 1953; Ribéreau-Gayon and Peynaud, 1966). Due to the diversity of must composition, the results are not identical. The assimilation of amino acids by yeasts does not always improve growth. The most easily assimilated amino acids are not necessarily the most significant in cell composition, but are instead the most easily transformed by yeasts. Yeasts have difficulty assimilating arginine when it is the only amino acid in the environment, but arginine is easily assimilated when furnished in a mixture. Yeasts do not use it when ammonium is present.

To avoid the difficulty of precisely defining grape must composition, Henschke and Jiranek (1992) carried out their experiments using a well-known medium model. Their results have given researchers a new understanding of this subject.

Although complex mixtures of ammonium salts and amino acids are more effective for promoting yeast growth and fermentation speed, ammonium salts are used almost exclusively to increase

nitrogen concentrations in must, for reasons of simplicity. Positive results have been obtained in laboratory tests but their effectiveness is less spectacular in practical conditions. Moreover, the addition of assimilable nitrogen is not always sufficient for resolving difficult final stages of fermentation, although it accelerates fermentation in the early stages.

For a long time, diammonium phosphate was the exclusive form of ammonium salts used. The phosphate ion involved in glucidic metabolic reactions was also thought to favor fermentation. In reality, the must is sufficiently rich in the phosphate ion (incidentally participating in the iron casse of white wines) and for this reason it is preferable to use diammonium sulfate. EU regulations authorize the addition of 30 g of one of the these two salts per hectoliter, corresponding to 63 mg/l nitrogen. In the USA, the limit is set at 95 g of diammonium phosphate per hectoliter. In Australia, its addition must not lead to a concentration of inorganic phosphate greater than 40 g/hl. The standard dose is between 10 and 20 g/hl. (Note that 100 g of diammonium phosphate or sulfate contains approximately 27 mg of ammonium and 73 mg of phosphate or sulfate ions). The addition of this form nitrogen to must increases acidity, due to the contribution of the anion. For 10 g of diammonium salt per hectoliter, the must acidity can increase by 0.35 g/l (in H_2SO_4) or 0.52 g/l (in tartaric acid).

The initial concentration of ammonium cations and amino acids in the must is one of the most important elements in determining the need for supplements. When the NH_4^+ concentration is less than 25 mg/l, nitrogen addition is necessary. It is useful for concentrations to be between 25 and 50 mg/l. Above this concentration, supplementing has no adverse effects. It is, however, unlikely that an addition will activate the fermentation. If the values are expressed in free amino nitrogen (FAN), detectable by ninhydrin, between 70 and 140 mg/l are necessary to have a complete fermentation of musts containing between 160 and 250 g of sugar per liter (Henschke and Jiranek, 1992).

Bely *et al.* (1990) determined that adding nitrogen was effective if the available nitrogen content (NH_4^+ + amino acids, except proline) in the must was below 130 mg/l, but was unnecessary and could even be harmful at initial concentrations of 200–350 mg/l. Aerny's (1996) formol index provides a simple estimate of the free amino acid and ammonium cation content. A formol index of 1 corresponds to 14 mg of amino nitrogen per liter. According to Lorenzini (1996), the addition of nitrogen in Swiss varietal musts is indispensable if the index is less than 10, and is recommended if the index is between 10 and 14. The formol method provides a quick, simple assessment of available nitrogen deficiency by assaying NH_4^+ and free amino acids, except proline. It should be more widely used to monitor ripeness and fermentation. However, the nitrogen requirements of *S. cerevisiae* vary from one strain to another. Julien *et al.* (2000) recently proposed a test for comparing yeast nitrogen requirements, estimated during the stationary phase of alcoholic fermentation. This test measures the quantity of nitrogen required to maintain a constant fermentation rate during this stationary phase. Nitrogen requirements varied by a factor of 2 among the 26 winemaking yeast strains tested. An assessment of the nitrogen requirement is certainly an important criterion in selecting yeast strains to use in nitrogen-deficient musts.

For example, in a study of musts from Bordeaux vineyards in the 1996 to 2000 vintages (Table 3.2), Masneuf *et al.* (2000) found nitrogen levels of 36–270 mg/l in white musts, with deficiencies in 22% of the samples (nitrogen concentrations under 140 mg/l). In reds, levels ranged from 46 to 354 mg/l, with deficiencies in 49%; in rosés, levels ranged from 42 to 294 mg/l, with deficiencies in 60%, while 89% of botrytized musts were nitrogen-deficient.

Choné (2000) analyzed Cabernet Sauvignon musts in 1997 and found significant variations in nitrogen levels, from 95 to 218 mg/l.

Finally, different nitrogen concentrations have also been found in individual plots within the same vineyard, e.g. 25–45 mg/l in vines with less vegetative growth and 152–294 mg/l in grapes from more vigorous vines.

All these analytical findings show the extent and frequency of nitrogen deficiencies, which are much more common than was generally thought in the past, perhaps due to changes in vineyard management techniques. It was generally accepted previously that nitrogen concentrations in musts from northerly vineyards in the northern hemisphere (temperate, oceanic climate) were sufficiently high.

Adding nitrogen to musts containing insufficient levels is extremely useful in achieving good fermentation kinetics. In some cases of severe nitrogen deficiency, it may even be opportune to add as much as 40 g/l of ammonium sulfate, which would require a change to EU legislation, as the current limit is 30 g/l, corresponding to 63 mg/l of nitrogen. Some observations suggest that an excessive increase in nitrogen content may have a negative effect on fermentation kinetics, so it is advisable to modulate nitrogen additions according to the natural level in the must and ensure that the total never exceeds 200 mg/l.

Adding excessive amounts of nitrogen may also result in the presence of non-assimilated residual nitrogen at the end of fermentation. Although there are no specific data on this issue, residual nitrogen may have a negative impact on a wine's microbiological stability. An excess of ammoniacal

Table 3.2. Available nitrogen content (NH_4^+ and free amino acids expressed in mg/l) in musts from Bordeaux vineyards (1996–1999 vintages) determined by the formol method (Masneuf *et al.*, 2000)

	White	Red	Rosé	Botrytized
Number of samples	32	55	48	9
Minimum value	36	46	42	22
Maximum value	270	354	294	157
Mean	181.9	157	119	82.8
Standard deviation	32	55	48	9
Deficient musts (%) <140 mg/l	22	49	60	89

nitrogen can also lead to a modification of the aromatic characters of wine. Since the yeast no longer needs to deaminate amino acids, it forms less secondary products (higher alcohols and their esters). This modifies wine aroma, especially white wines. Finally, the nitrogen supply affects ethyl carbamate production. This undesirable constituent has carcinogenic properties and is controlled by legislation.

The must sugar concentration also affects the impact of the nitrogen supply on fermentation kinetics, especially the successful completion of fermentation. For moderate concentrations of sugar (less than 200 g/l), the addition of nitrogen increases the biomass of yeast formed and in consequence the fermentation speed; the fermentation is completed a few days in advance. For high concentrations of sugar, the fermentation is accelerated at the beginning with respect to the control sample but, as the fermentation continues, the gap between the control sample and the supplemented sample decreases. Finally, their fermentations spontaneously stop with similar quantities of residual sugar remaining. Curve II in Figure 3.3 depicts the effect of supplemental nitrogen (or other activator effect) on a must with a high sugar concentration, having a normal nitrogen concentration. On the other hand, if fermentation sluggishness is due to a nitrogen deficiency, the addition of ammonium salts manifestly stimulates it (Curve III, Figure 3.3). Stuck fermentations can sometimes be avoided in this manner.

Other factors affect the assimilation of nitrogen during fermentation. Yeasts have strain-specific capabilities. Henschke and Jiranek (1992) reported that different *S. cerevisiae* strains fermenting grape must assimilated quantities of nitrogen varying from 329 to 451 mg/l at 15.5°C and from 392 to 473 mg/l at 20°C. These last figures also show, among other things, that temperature increases nitrogen assimilation. Julien *et al.* (2000) compared the nitrogen and oxygen requirements of several yeast strains used in winemaking.

Oxygen, however, has the most effect on the assimilation of nitrogen. Yeasts have long been known to use considerably more nitrogen in the presence of oxygen (Ribéreau-Gayon *et al.*,

1975a). It has been observed that yeasts fermenting in the complete absence of oxygen assimilate 200 mg of nitrogen per liter. When they develop in the presence of oxygen, their assimilation increases to 300 mg/l. In aerobiosis, they can assimilate up to 735 mg/l without a proportional increase in cellular multiplication.

The impact of oxygen on fermentation kinetics, irrespective of any addition of NH_4^+, is apparently complex and dependent on several factors (Sablayrolles *et al.*, 1996a and 1996b), as well as the type of must (sugar content and possible nitrogen deficiency). It is accepted that adding nitrogen accelerates fermentation, resulting in faster completion. It is, however, more difficult to identify the conditions under which adding nitrogen can prolong sugar conversion by the yeasts and prevent fermentation from becoming stuck, at least in musts rich in sugar.

In an experiment carried out by Rozès *et al.* (1988), using a must containing 222 g/l of sugar with a normal nitrogen content (35 mg/l NH_4^+ corresponding to approximately 200 mg/l nitrogen), fermentation stopped prematurely in the absence of air. Adding NH_4^+ (0.15 or 0.50 g/l $(NH_4)_2SO_4$) initially accelerated fermentation but did not increase the amount of sugar fermented. With aeration on the 3rd day, fermentation was faster and all the sugar was fermented. Adding NH_4^+ did not improve fermentation kinetics, but, on the contrary, after an initial acceleration, yeast activity stopped when 9 g/l of sugar was still unfermented. Of course, these experimental results must be interpreted in the light of the specific conditions (sugar and nitrogen content of the must). The results would not necessarily have been the same under different conditions, particularly if there had been a significant nitrogen deficiency in the must. In any case, this experiment shows quite clearly that adding nitrogen does not necessarily eliminate problems with the end of fermentation. Further experiments using nitrogen-deficient must are required to identify a possible improvement.

The timing of the addition of ammonium salts appears to be important. Ribéreau-Gayon *et al.* (1975a) had suggested their addition in must before the initiation of fermentation. Yeasts react best to stimuli during the growth phase in a medium containing little ethanol. They witnessed an assimilation of ammoniacal nitrogen supplement (100 mg/l) varying between 100 and 50% when the addition was made before the initiation of fermentation or on the fourth day. Enhanced nitrogen assimilation did not necessarily increase the yeasts' fermentation potential. This explains why adding nitrogen has no significant impact on accelerating a sluggish final stage in fermentation and is even less effective in restarting a stuck fermentation.

Sablayrolles *et al.* (1996a and 1996b) reported slightly different findings. According to these authors, nitrogen supplements were most effective in mid-fermentation, together with aeration. This combined operation had more impact on fermentation kinetics than aeration alone and provided an optimum solution for avoiding prematurely stuck fermentation (Sablayrolles and Blateyron, 2001).

In conclusion, supplementing musts with naturally low nitrogen levels ($N_{total} \leqslant 140$ mg/l) with nitrogen salts is likely to improve fermentation kinetics, with varying effects on yeast growth and sugar conversion. For maximum effectiveness, total nitrogen after supplementation should not exceed 200 mg/l. Some experimental findings indicate that fermentation may slow down following the addition of excessive amounts of nitrogen. If the must already had a sufficiently high nitrogen content, further supplementation was likely to cause an initial acceleration in fermentation, but the effect wore off gradually. Adding nitrogen cannot be expected to remedy problems in the final stages of fermentation (high-sugar musts or strictly anaerobic conditions). It is, however, true that nitrogen deficiencies (old vines or vineyards with cover crop) have not been given sufficient consideration in the past, and that completion of fermentation is facilitated in these cases by adding ammonium salts. Total nitrogen in must should be analyzed in vat before the start of fermentation as a matter of course, together with sugar and acidity levels.

Adding oxygen at the start of fermentation (Section 3.7.2) when the yeast population is in the growth phase is still the most effective way of accelerating fermentation and preventing premature stoppages. Opinions diverge on the correct time to add ammonium salts, varying from the beginning of fermentation to halfway through. In any case, nitrogen supplements are more effective at accelerating fermentation than preventing it from becoming stuck with unwanted residual sugar.

3.4.3 Mineral Requirements

The yeasts that Pasteur cultivated in the following medium proliferated well: water, 1000 ml; sugar, 100 g; ammonium tartrate, 1 g; ashes of 10 g of yeast. Yeast ashes supply the yeast with all of its required minerals. Dry yeast contains 5–10% mineral matter, whose average composition (in percentage weight of ashes) is as follows:

K_2O	23–48
Na_2O	0.06–2.2
CaO	1.0–4.5
MgO	3.7–8.5
Fe_2O_3	0.06–7.3
P_2O_5	45–59
SO_3	0.4–6.3
SiO_2	0–1.8
Cl	0.03–1.0

Other minerals not listed above are present in trace amounts: Al, Br, Cr, Cu, Pb, Mn, Ag, Sr, Ti, Sn, Zn, etc. These are called trace elements. Not all of them are indispensable but some are essential enzyme constituents.

The precise function of only a few minerals is known. Grape must contains, both qualitatively and quantitatively, a sufficient mineral supply to ensure yeast development.

3.5 FERMENTATION ACTIVATORS

3.5.1 Growth Factors

Growth factors affect cellular multiplication and activity, even in small concentrations. They are indispensable to microorganisms and a deficiency in these substances disturbs the metabolism. Microorganisms behave differently in relation to growth factors. Some can totally or partially synthesize them; others cannot and must find them in their environment.

The substances that are growth factors for microorganisms are also necessary vitamins for higher organisms (Figure 3.4). They are essential components of coenzymes and are involved in metabolic reactions. Grape must has an ample supply of growth factors (Table 3.3) but alcoholic fermentation alters its vitamin composition. For example, thiamine disappears almost entirely: yeasts are capable of consuming greater amounts of thiamine (600–800 µg/l) than the must contains;

but yeasts form riboflavin. The concentration of nicotinamide remains constant in red wines and musts, but only 60% remains in white wines. Pantothenic acid, pyridoxine and biotin are used by yeasts and then released; their concentrations are nearly identical in musts and wines. Mesoinositol is practically untouched.

Although musts contain sufficient amounts of growth factors to ensure yeast development and alcoholic fermentation, natural concentrations do not necessarily correspond with optimal concentrations. For this reason, supplementing must with certain growth factors is recommended.

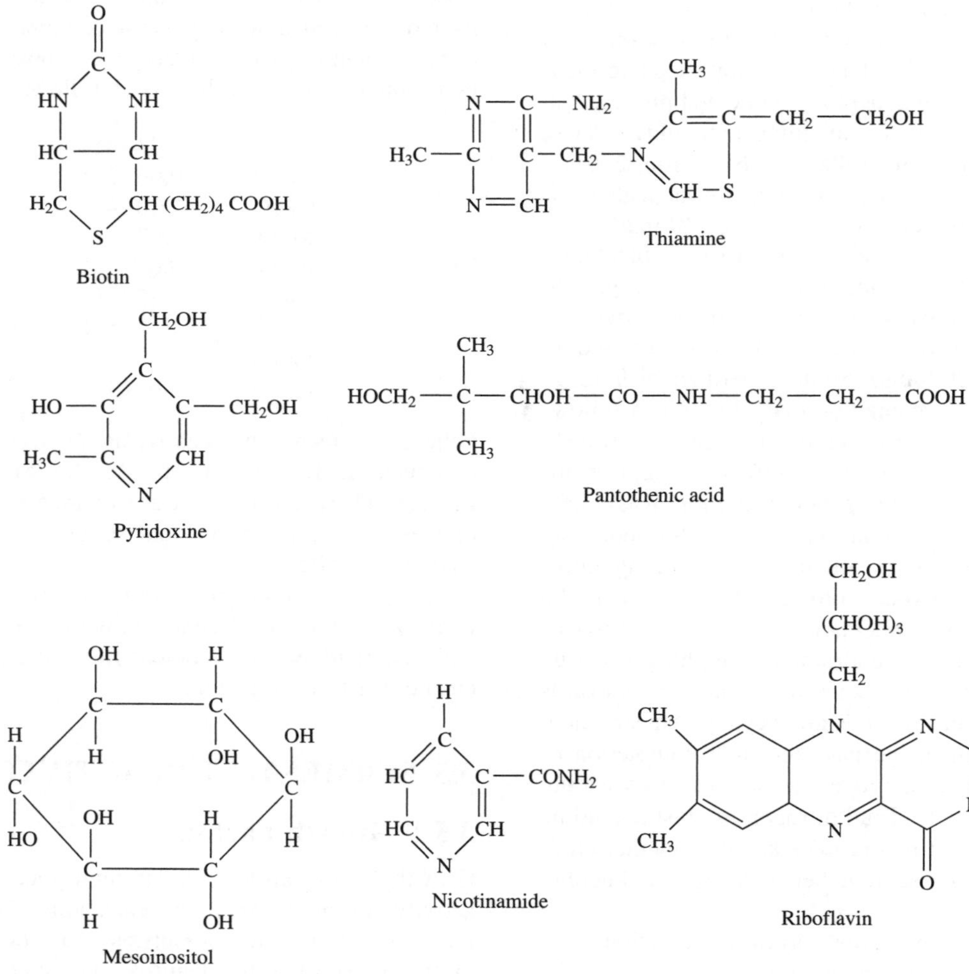

Fig. 3.4. Yeast growth factors

Table 3.3. Maximum and minimum growth factor concentrations (μg/l) in musts and wines (Ribéreau-Gayon *et al.*, 1975a)

Vitamins	Grape musts	Wines Whites	Reds
Thiamine	160–450	2–58	103–245
Riboflavin	3–60	8–133	0.47–1.9
Pantothenic acid	0.5–1.4	0.55–1.2	0.13–0.68
Pyridoxine	0.16–0.5	0.12–0.67	0.13–0.68
Nicotinamide	0.68–2.6	0.44–1.3	0.79–1.7
Biotin	1.5–4.2	1–3.6	0.6–4.6
Mesoinositol (mg/l)	380–710	220–730	290–334
Cobalamine	0	0–0.16	0.04–0.10
Choline	19–39	19–27	20–43

A deficiency in pantothenic acid causes the yeast to accumulate acetic acid but it has not been proven that the (unauthorized) addition of pantothenic acid to a fermenting must lowers the wine's volatile acidity originating from yeast. The production by yeasts of abnormally high levels of volatile acidity is probably due to the must's deficiencies in certain lipids. These deficiencies are most likely linked to deficiencies in pantothenic acid, which is involved in the formation of acetyl coenzyme A, responsible for fatty acid and lipid synthesis.

The supplementing of biotin and especially thiamine improved the must fermentation kinetics in numerous experiments. An addition of 0.5 mg of thiamine per liter can increase the viable population by 30%; the fermentation of sugar is also quicker. These results, although regularly observed in laboratory experiments, are not always obtained under practical conditions. The natural concentration of thiamine may or may not be a limiting factor of fermentation kinetics, depending on the nature of the grape and on maturation conditions.

The addition of thiamine is legal in several countries (EU, at a dose of 50 mg/hl) but it is rarely used to accelerate fermentation in winemaking. It effectively decreases significant ketonic acid concentrations by decarboxylation (pyruvic and α-ketoglutaric acid). Large quantities of these acids bind to sulfur dioxide in botrytized sweet wines (Section 8.4.2).

3.5.2 Survival Factors

The idea of survival factors is derived from the interpretation of the mode of action of sterols and certain long-chain fatty acids on yeast activity and fermentation kinetics. The first works on this subject (Andreasen and Stier, 1953; Bréchot *et al.*, 1971) were analyzed by Ribéreau-Gayon *et al.* (1975a). The growth factor activity of ergosterol in complete anaerobiosis is optimal at a concentration of 7 mg/l; it is solubilized with Tween 80. For example, in a must with a high sugar concentration (260 g/l), *S. cerevisiae* ferments in complete anaerobiosis 175 g of sugar per liter in 10 days in the control sample and 258 g/l in the presence of 5 mg/l of ergosterol. In aerobiosis, on the other hand, a slight inhibition of the fermentation is observed when ergosterol is added. The authors concluded that these sterols are indispensable to yeasts in complete anaerobiosis, because they cannot be synthesized in these conditions. Sterols are necessary for ensuring cell membrane permeability. In the presence of oxygen, yeasts are capable of producing sterols. In anaerobiosis, ergosterol is in some ways an oxygen substitute for yeasts.

Other sterols and long-chain fatty acids share most of the properties of ergosterol. Some are constituents of grape bloom and cuticular wax, such as oleanolic acid—especially when associated with oleic acid (Figure 3.5). These constituents explain the results of past experiments, indicating an acceleration of the fermentation speed of grape must in complete anaerobiosis when grape skins and seeds were added in suspension.

Later works (Larue *et al.*, 1980; Lafon-Lafourcade, 1983) showed that the action mechanism of sterols is in fact more complex. These authors confirmed the growth factor action in a strictly anaerobic fermentation: the maximum population increases. They also witnessed the inhibitory effect of sterols on a fermentation with permanent aeration. Neither of these two conditions correspond exactly to winemaking conditions.

In the winery, large-volume fermentations are certainly anaerobic, but the must is aerated during extraction and inoculated with a yeast starter which was precultivated in aerobiosis; the yeasts

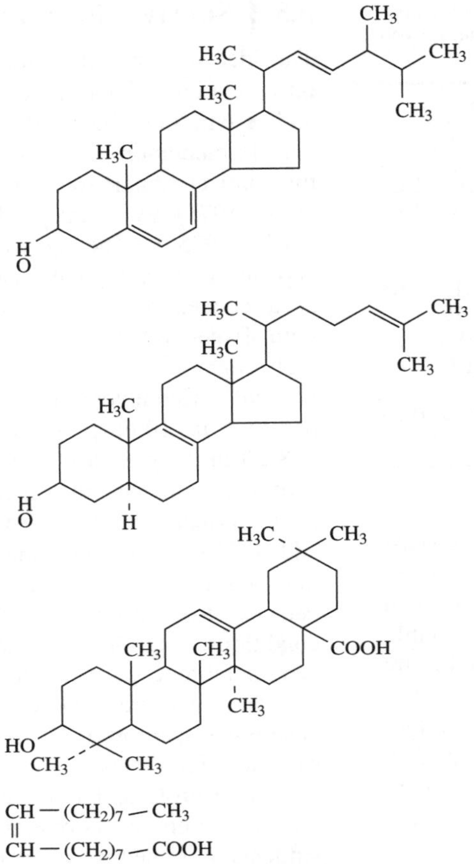

Ergosterol $C_{26}H_{44}O$

Zymosterol $C_{27}H_{42}O$

Oleanolic acid $C_{30}H_{46}O_3$

Oleic acid $C_{18}H_{34}O_2$

Fig. 3.5. Structure of some steroids and fatty acids playing a role in yeast growth

are therefore well equipped in sterols. Both commercial active dry yeasts and indigenous yeasts, which develop on the surface of material in contact with the harvest, initially develop in aerobiosis. In these conditions, the addition of ergosterol or oleanolic acid does not increase the maximum population. The fermentation speed is also not affected during the first 10 days. Yet the yeast cells well equipped in sterols maintain their fermentation activity for a longer time. At the end of fermentation, they will have degraded a larger amount of sugar than non-supplemented cells (Table 3.4). The term 'survival factor' has been proposed for this action that does not correspond with an increase in growth. The evolution of yeast populations during fermentation in the presence of sterols is represented in Figure 3.6; the incidence of the temperature is also indicated.

The notion of survival factors complements the notion of growth factors. They are especially interesting in the case of difficult fermentations—for example, musts containing high sugar concentrations. Of course, the direct addition of sterols to tanks in fermentation should not be considered. Winemaking can, however, be orientated towards methods which promote sterol synthesis. Moreover, their existence in the solid parts of the grape should be taken into account: crushed red grapes ferment better than white grape musts because of solids contact during fermentation. In addition, the elimination of sterols during the excessive clarification of white grape

Table 3.4. Sterol concentrations in yeasts during alcoholic fermentation of grape must (Larue *et al.*, 1980)

Conditions	Constant aeration			Anaerobiosis		
	C	+E	+OA	C	+E	+OA
Day 2						
Fermented sugar (g/l)	30	27	24	37	36	23
Sterols (% of dry weight)	2.70	2.80	2.30	1.60	1.40	1.70
Viable cells (10^6/ml)				22	20	17
Day 5						
Fermented sugar (g/l)	116	101	95	113	111	105
Sterols (% of dry weight)	1.90	1.90		0.60	1.10	0.40
Viable cells (10^6/ml)				13	10	12
Day 9						
Fermented sugar (g/l)	187	175		164	169	154
Sterols (% of dry weight)	1.20	1.10	0.7	0.40	1.00	0.30
Viable cells (10^6/ml)				5	7	5
End of fermentation						
Fermented sugar (g/l)	256	234	211	170	199	185
Sterols (% of dry weight)	1.00	0.80	0.40	0.30	0.60	0.20
Viable cells (10^6/ml)				0.05	0.5	0.1

C = control must; +E = ergosterol (25 mg/l); +OA = oleanolic acid (50 mg/l).
Must sugar concentration after addition: 250 g/l; active dry yeast: *Saccharomyces cerevisiae*; initial sterol concentration: 1.5%; initial population: 2.2×10^6 cells/ml.

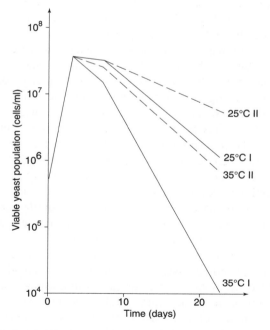

Fig. 3.6. Influence of sterols on yeast survival during their death phase at different temperatures (Lafon-Lafourcade, 1983). (I) Control. (II) Plus ergosterol

must can result in extremely difficult fermentations (Section 3.7.3). This concept can also explain past experiments which show increased fermentation speeds with the addition of ground grape skins and seeds.

3.5.3 Other Fermentation Activators

Ribéreau-Gayon *et al.* (1975a) examined other fermentation activators. These activators generally help the yeast to make better use of must nitrogen. Incidentally, the same phenomenon is observed each time that the fermentation is accelerated by the presence of air or by the addition of yeast extract nutriments or survival factors (Table 3.5).

Hydrolyzed yeast extracts are rich in assimilable nitrogen, survival factors and mineral salts. They have often been used (at high concentrations, up to 4 g/l) to accelerate fermentation in the food industry. Some can impart foreign odors and tastes.

Among the activator formulas proposed (Ribéreau-Gayon *et al.*, 1975a), 200 mg/l of the following mixture may facilitate fermentation in musts with vitamin and nitrogen deficiencies:

Table 3.5. Effect of air and survival factors on grape must nitrogen assimilation (adapted from Ingledew and Kunkee, 1985, by Cantarelli, 1989, cited by Henschke and Jiranek, 1992)

Atmosphere	Additions	Nitrogen utilization (mg of FAN/l)	Number of cells ($\times 10^7$/l)	Fermentation speed (g of sugar per 100 ml/24 h)
Air	0	87	35.0	1.80
Nitrogen	0	65	1.3	0.50
Air	YE	314	36.0	4.00
Nitrogen	YE	107	3.0	1.22
Nitrogen	YE + SF	293	32.0	3.60

FAN = free amino nitrogen; YE = yeast extract; SF = survival factors (Tween 80, ergosterol). Musts without and with addition of YE contain 99 and 386 mg FAN/l, respectively.

100 g of diammonium sulfate, 250 mg of thiamine, 250 mg of calcium pantothenate, and 2 mg of biotin.

Other products, extracted from fungi, are also alcoholic fermentation activators. Some have been commercialized in the past. One of the most effective is prepared from a culture medium of *Aspergillus niger*. It modifies the concentration of secondary products by promoting the glyceropyruvic fermentation of sugar. A yeast activator is also obtained from the mycelium of *Botrytis cinerea*. These activators are not authorized by viticultural legislation, at least in the EU.

In white winemaking, suspended solids activate fermentation (Section 3.7.3). Certain constituents, probably sterols and fatty acids, are involved in this phenomenon (Ribéreau-Gayon *et al.*, 1975b). Although these substances are not very soluble, yeasts are capable of using them to improve fermentation kinetics. They probably act in conjunction with other factors, such as oxygenation and possibly nitrogen additions. Yeast hulls have a similar effect independent of their ability to eliminate inhibition (Section 3.6.2).

As with nitrogen or growth factor supplementation, fermentation is not activated to the same degree in the winery as in the laboratory. Oxygenation, yeast starter preparation and the fermentation medium play an essential role, in addition to the must's possible nutritional deficiencies.

3.5.4 Adding Yeast Starter

Winemakers have always been interested in improving fermentation kinetics and wine quality by inoculating with activated yeast starter. This practice has certainly become more widespread since relatively economical, easy-to-use Dried Activated Yeast (DAY) became available on the market. DAY are simply reactivated in water or a mixture of equal volumes water and must at a temperature of 35–40°C. There are around a hundred commercial yeast preparations on the market, and each one should be prepared for use according to the manufacturer's instructions. DAY also makes it possible to eliminate apiculated yeasts and select strains with a high sugar–alcohol conversion rate. Together with other winegrowing practices, the use of DAY has contributed to the general increase in average alcohol content in wines and to the corresponding decrease in the need for chaptalization (adding sugar) in certain regions. On the contrary, in some situations, there is an interest in using yeast strains with a lower conversion rate. This mainly concerns hot areas where there may be a high sugar level in the grape flesh although the other ripeness indicators (skin) have not reached optimum levels. In this case, it is necessary to delay the harvest, with the risk of obtaining very high-sugar musts that are difficult to ferment in order to produce high-alcohol wines.

An initial inoculum of 10^6 cells/ml is generally considered necessary to obtain good fermentation kinetics. In view of the current constraints on white winemaking, this initial level is rarely achieved; so the use of yeast starter has become practically compulsory. In red winemaking, there may be insufficient inoculum in the first few vats filled, but grape and must handling operations in the winery rapidly result in proliferation of the yeasts. The use of DAY is mainly justified in the first vats. Later vats either require no starter at all, or DAY can be replaced with 2–5% must from a vat where fermentation is going well.

There has, however, been interest for some time in the possibility of improving a wine's quality by selecting the appropriate strain for fermentation. It is certainly true that the composition of grapes and other natural factors (e.g. *terroir*) are the main elements of the specific characteristics recognized as the basis of quality, especially in the concept of *appellation d'origine contrôlée*. It is also true that positive results have been obtained, especially with white wines (Section 13.7). Several strains capable of fermenting musts with low turbidity without producing excessive volatile acidity have been isolated. Strains have also been identified that do not produce vinyl phenols, with their unpleasant chemical odor and other undesirable characteristics, due to high levels of fermentation esters. These neutralize varietal aromas and can only be recommended for wines made from non-aromatic varieties. It is clear that using yeast starter is a good way of avoiding these types of defects and, consequently, making the most of the grapes' intrinsic quality. Another example is the development of anaerobic yeasts likely to produce ethyl acetate as soon as vats are filled with non-sulfited grapes or must. Reports in the literature indicate that it is possible to avoid these defects by using appropriate yeast strains after the grapes/must have been sulfited.

Today, there is increasing interest in selecting yeast strains capable of enhancing the varietal aromas of various grape varieties by releasing variable quantities of odoriferous molecules from their odorless precursors. Research has focused on Muscat varieties and, above all, Sauvignon Blanc (Section 13.7.2). Although different yeast strains have varying impacts on wine aromas, it is impossible to say that the use of yeast starter leads to the development of a uniform character that depends mainly on the grapes' composition in terms of aroma precursors.

In the case of red wines, it has been reported that the use of specific yeast starters has an impact on color intensity and the aromatic character of some grape varieties. These effects may occur during fermentation itself or result from the autolysis of dead yeast cells, which justifies the practice of aging wine on the lees. These observations, however, require a more detailed theoretical investigation.

A dose of 10^6 cells/ml yeast starter is generally recommended, which corresponds to 10–20 g/hl of DAY. As the yeast population in strongly fermenting must is of the order of 10^8 cells/ml, a 1% inoculum is theoretically sufficient, but 2% is more commonly used, or even 5%, to offset any potential difficulties. Experiments carried out with higher doses of yeast starter (20–25 g/hl of DAY) indicated that there was a lower risk of fermentation becoming sluggish towards the end, but some off-aromas could be produced. In any case, the most important selection criteria for winemaking yeast starters are temperature resistance and the ability to complete fermentation in high-sugar musts. These properties are characteristic of yeasts formerly known as *Saccharomyces bayanus* (Section 1.8.4).

When the yeast starter is added, it is important to avoid antagonism with other strains naturally present in the must. Antagonistic reactions may reduce the fermentation rate and contribute to causing stuck fermentation (Section 3.8.1). For inoculation with DAY to be successful, the yeast starter must be more abundant and more active than the indigenous yeasts, which must be inhibited by proper hygiene, sufficiently low temperatures, and appropriate use of sulfite.

3.6 INHIBITION OF THE FERMENTATION

This section covers the phenomenon of inhibition in grape must fermentation. A large number of substances exist that may hinder yeast multiplication: chemical antiseptics and antibiotics and fungicides (Ribéreau-Gayon *et al.*, 1975a). Inhibitors used for the conservation of wines (in particular sulfur dioxide) are described in Chapters 8 and 9.

3.6.1 Inhibition by Ethanol

Ethanol produced by fermentation slows the assimilation of nitrogen and paralyzes the yeast. Ethanol

Table 3.6. Effect of ethanol addition to must on fermentation (in limited aerobiosis at 25°C) (Ribéreau-Gayon *et al.*, 1975a)

Alcohol addition (% vol.)	Delay for initiation of fermentation	Yeast population (10^6/ml)	Alcohol content attained (% vol.)	Alcohol formed (% vol.)	Residual sugar (g/l)	Nitrogen assimilated (mg/l)	Glycerol (mmol/l)
+0	1 day	80	14.0	14.0	2	252	57
+2	2 days	67	15.6	13.6	6	233	65
+6	4 days	62	18.2	12.2	15	194	72
+10	12 days	30	16.0	6.0	125	81	80

acts by modifying cell active transport systems across the membrane (Henschke and Jiranek, 1992). The quantity of alcohol necessary to block fermentation depends on many factors, including yeast strain, temperature and aeration.

The presence of ethanol at the time of inoculation prolongs the latent phase and reduces cellular multiplication. An elevated temperature increases this inhibitory action. This effect of ethanol on yeast growth and fermentation speed occurs even at low concentrations from the start of fermentation. The difficulty of restarting a stuck fermentation is, therefore, understandable.

The experiment in Table 3.6 shows the effect of the addition of alcohol to grape must. It slows the initiation of fermentation and limits the assimilation of nitrogen and the formation of alcohol. Yet the yeasts can continue their activity up to a higher alcohol content, as long as the inhibitory action of ethanol is not excessive. In this experiment, the variation of the glycerol concentration represents significant metabolic modification. As seen in Section 3.4.1, ethanol intensifies the inhibitory effect of an elevated sugar concentration in must.

3.6.2 Inhibition by Fermentation By-Products: the Use of Yeast Hulls

Past observations have indicated the possibility of the formation of substances other than ethanol, during fermentation, having an inhibitory action on yeast. Geneix *et al.* (1983) confirmed this hypothesis (Figure 3.7). Synthetic fermentation media containing variable concentrations of ethanol were

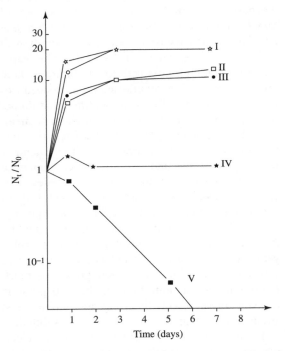

Fig. 3.7. Evolution of *Saccharomyces cerevisiae* population in fermenting media containing different alcohol concentrations (A = 1.7% vol.; B = 7.0% vol.; C = 9.5% vol., obtained by fermentation or alcohol additions) (Geneix *et al.*, 1983). N_t = cell count at time t; N_0 = cell count at start (approximately 10^7 cells/ml). (I) non-fermented media A and B. (II) non-fermented medium C. (III) pre-fermented medium A. (IV) pre-fermented medium B. (V) pre-fermented medium C

inoculated with *S. cerevisiae*. A first series consisted of non-fermented and entirely synthetic media. A second series consisted of pre-fermented media. The ethanol in the second series came from a fermentation stopped by double centrifugation in the growth, stationary and death phases of the population growth cycle.

The composition of the non-fermented and pre-fermented media was as follows:

A: alcohol content = 1.7% vol.; sugar = 160 g/l

B: alcohol content = 7.0% vol.; sugar = 65 g/l

C: alcohol content = 9.5% vol.; sugar = 23 g/l

Figure 3.7 shows that yeasts grow in all the non pre-fermented media. In the pre-fermented media, on the other hand, growth is only possible in medium A, which has a low alcohol concentration. Population decline is significant in pre-fermented medium C, which has an elevated alcohol content. According to this experiment, fermentation creates other substances besides ethanol which inhibit yeast growth and alcoholic fermentation. A complementary experiment indicated that these substances are eliminated by charcoal, confirming past observations. For a stuck fermentation, charcoal helps to restart yeast activity by removing yeast metabolism products from wine.

Research into the impact of various fermentation by-products on yeast demonstrated the inhibiting effect of C_6, C_8, and C_{10} short-chain fatty acids found in wine at concentrations of a few milligrams per liter. They affect cell membrane permeability and hinder exchanges between the inside of the cell and the fermenting medium. When fermentation stops, the yeast enzymatic systems still function, but the sugars can no longer penetrate the cell to be metabolized (Larue et $al.$, 1982). Salmon et $al.$ (1993) confirmed that loss of activity

of $S.$ $cerevisiae$ in enological conditions was linked to inhibition of the transport of sugar. Fermentation is inhibited by these C_6, C_8 and C_{10} saturated fatty acids—hexanoic (caproic), octanoic (caprylic) and decanoic (capric) acids. Other unsaturated, long-chain fatty acids (C_{18}), however, are activators in certain conditions: oleic acid, one double bond; and linoleic acid, two double bonds (Section 3.5.2). The term 'fatty acid' used in both cases can lead to confusion.

The preceding facts lead to the use of yeast hulls in winemaking. They are currently the most effective fermentation activators known for winemaking (Lafon-Lafourcade et $al.$, 1984). Yeast hulls eliminate the inhibition of the fermentation by fixing the toxic fatty acids (Lafon-Lafourcade et $al.$, 1984). The permeability of the cellular membranes is re-established in this manner. Munoz and Ingledew (1989) confirmed the toxic effect of C_6, C_8 and C_{10} fatty acids and the activation of fermentation by different varieties of yeast hulls. According to these authors, in addition to their properties of adsorption of fatty acids, yeast hulls contribute sterols and unsaturated, long-chain fatty acids to the medium. These constituents are considered to be 'oxygen substitutes' or survival factors (Section 3.5.2). Whatever their mode of action, yeast hulls are universally recognized as fermentation activators. Table 3.7 gives an example of the activation of the fermentation of a grape must containing high sugar concentrations. The numbers show the superiority of yeast hulls with respect to ammonium salts for the activation of the fermentation. During the final stage of fermentation, the

Table 3.7. Stimulation of wine fermentation by the addition of $(NH_4)_2SO_4$ or yeast hulls to grape must before fermentation (results at the end of fermentation) (Lafon-Lafourcade et $al.$, 1984)

	Control (no addition)	Addition of yeast hulls (g/l)		Addition of $(NH_4)_2SO_4$ (0.2 g/l)
		0.2	1	
Sugar fermented (g/l)	206	247	257	212
Total population (10^7 cells/ml)	9	11	14	10
Viable population (10^7 cells/ml)	3.5	10	26	1.7

Initial sugar concentration: 260 g/l; initial viable yeast population: 10^6 cells/ml; dry yeast: $Saccharomyces$ $cerevisiae$; fermentation temperature: 19°C. Yeast populations are counted at the end of fermentation.

total cell population has not greatly increased but the viable yeast population is clearly more significant in the presence of yeast hulls. This characteristic survival factor effect does not exist when only ammonium salts are added. The addition of yeast hulls on the fifth day following the initiation of fermentation, after the growth phase, has a more pronounced effect on population survival than an addition before fermentation.

Yeast hulls have proven to be effective in musts that are difficult to ferment; for example, those containing high sugar concentrations or containing pesticide residues. Yeasts are also more temperature resistant in their presence (Table 3.8). They may be used, although less effectively, in cases of stuck fermentation (Section 3.8.3).

Yeast hulls must be perfectly purified to avoid an organoleptical impact on wines. The industrial preparation of yeast extract results in pronounced odors or tastes in the product and these must be removed from the envelopes before use in wine. Moreover, if the hulls are not sufficiently purified, a souring (due to the presence of residual lipids) may occur during storage in certain conditions. This souring leads to an unfavorable development in the wine's organoleptic characters. Such circumstances have incited excessive criticism concerning the use of yeast hulls.

The involvement of yeast hulls in fermentation processes is also accompanied by variation in the concentration of secondary products (higher alcohols, fatty acids and their esters). As a result, wine aromas and tastes can be modified. All operations that affect fermentation kinetics affect the wine—temperature, oxygenation, addition of ammonium salts, etc.—and yeast hulls have no more of an impact on the fermentation than these other factors, and certainly less than temperature,

for example. Whatever the case, yeast hulls should be used with prudence for the fermentation of wines having a simple structure, such as certain dry white wines. In this case, their effect can be more significant.

3.6.3 Inhibition from Different Origins

Some vine treatment spray residues (e.g. Folpel) are well known to inhibit fermentation. Sulfur- and chloride-based compounds are the most harmful to yeasts. Inoculation with fresh yeast once the inhibiting residue has broken down is generally sufficient to reactivate fermentation in the must (see Hatzidimitriou et al., 1997). However, certain difficult final stages of fermentation can be attributed to the presence of these residues. The minimum time between the last application of a product and the harvest date indicated by the manufacturer is not always sufficient.

Elevated concentrations of tannins and colored matter found in certain varieties of red wines can hinder yeast activity. They bind to the cell wall by a kind of tanning process. The effect of these substances is not clear: some activate fermentation, while others inhibit it.

Carbon dioxide produced by fermentation is known to have an inhibitory effect. This occurs during fermentations under pressure (sparkling wines). A slight internal pressure in the tank is sufficient to slow the fermentation, and above 7 bars fermentation becomes impossible. In normal winemaking conditions, carbon dioxide is released freely and exercises no inhibition on the fermentation.

The difference in fermentability of various grapes and musts is linked with many poorly controlled factors. In the same way that specific

Table 3.8. Stimulation of red wine fermentation by the addition of yeast hulls (maximum fermentation temperature of 34°C attained on day 7) (Lafon-Lafourcade et al., 1984)

	Residual sugars during fermentation (g/l)			
	Day 4	Day 8	Day 9	Day 16
Control	194	22	15	15
Addition of yeast hulls (0.3 g/l)	206	15	2	

activators must exist, the presence of natural inhibitors in grapes has been considered. Their interaction would affect must fermentability.

More specifically, the preliminary development of yeasts, lactic bacteria or *Botrytis cinerea* can hinder the alcoholic fermentation process. In the case of alcoholic fermentation difficulties, bacterial growth and malolactic fermentation exacerbate these difficulties. Stuck fermentations often result (Sections 3.8.1; 6.4.1).

Among these causes of fermentation difficulties, the involvement of *Botrytis cinerea* has been the most studied. Must derived from parasitized grapes (noble or gray rot) is more difficult to ferment than must originating from healthy grapes. Past works, summarized by Ribéreau-Gayon *et al.* (1975a), identified a substance with antibiotic properties; the authors named it botryticin. Sulfiting and prolonged heating at 120°C destroy this substance. Ethanol at 80% can precipitate it. Subsequent work showed that this fungistatic substance is an either partially or completely mannose-based neutral polysaccharide (Ribéreau-Gayon *et al.*, 1979). The phytotoxic properties of such substances are known and this polysaccharide affects fermentation kinetics. It is also the cause of certain metabolic deviations induced by *Botrytis cinerea*—in particular an increase in the production of glycerol and acetic acid (Section 2.3.4; Volume 2, Section 3.7.2).

3.7 PHYSICOCHEMICAL FACTORS AFFECTING YEAST GROWTH AND FERMENTATION KINETICS

3.7.1 Effect of Temperature

Alcoholic fermentation, depicted by the following chemical equation:

$$C_6H_{12}O_6 \longrightarrow 2C_2H_5OH + 2CO_2,$$

liberates 40 kcal of free energy per molecule. Yeasts use part of this energy to ensure their vital functions, in particular their growth and multiplication, and to form two ATP molecules from a sugar molecule. These ATP molecules have a high energy potential:

$$2ADP + 2H_3PO_4 \longrightarrow 2ATP + 2H_2O$$

It takes 7.3 kcal of energy to form one ATP molecule. The difference $(40 - 14.6 = 25.4 \text{ kcal})$ is non-utilized energy that is dissipated in the form of heat, causing the fermentation tanks to heat.

This estimate of non-utilized energy is open to discussion. In situations where part of the ATP formed is not needed by the yeast, it is hydrolyzed by the corresponding enzymes. Yet 25 kcal corresponds fairly well with past thermodynamic measures of dissipated heat by the fermentation of one molecule of sugar.

The fermentation of must containing 180 g (one molecule) of sugar per liter therefore liberates 25 kcal in the form of heat. This liberation of heat theoretically could raise the temperature from 20 to 45°C. Such an increase in temperature would kill the yeast. Fortunately, this increase is the hypothetical case of an explosive, instantaneous fermentation or a fermentation in a fully insulated tank. In reality, fermentation takes place over several days. During this time, the calories produced are dissipated by several phenomena: by being entrained with the large quantity of carbon dioxide released during fermentation; by cooling resulting from the evaporation of water and alcohol; and by exchanges across the tank wall.

Temperature increases in fermenters depend on several factors.

1. The must sugar concentration, which determines the amount of calories liberated.

2. The initial must temperature.

3. The fermentation speed, which depends on must composition (nitrogen-based substances) and yeast inoculation conditions. Operations such as aeration, chaptalization and inoculation will increase the fermentation speed, limit the dissipation of calories and increase the maximum temperature. Reciprocally, not crushing the red harvest in carbonic maceration (Section 12.9.1) will slow fermentation kinetics and lower the maximum temperature.

4. Tank dimensions. When the volume increases, the surface of the walls and their thermic exchange capacity decrease when the same proportions are maintained.

5. Tank material. The global thermic exchange coefficient (K, expressed in cal/h/m^2 for each degree difference in temperature) is from 0.7 to 0.78 for a concrete wall, 10 cm thick, from 1.46 to 1.49 for a 5 cm wooden wall, and from 5.34 to 32.0 for a 3 mm stainless steel wall. The coefficient varies the most in the case of stainless steel. A stainless steel tank is sensitive to the external conditions (temperature, aeration) in which it is placed.

6. Aeration and cellar temperature. The ventilation of the winery limits fermenting tank temperatures by dissipating heat.

The maximum tank temperature is related to all of these factors by complex laws and is difficult to predict. Depending on the circumstances, the maximum temperature can be compatible with red winemaking. In this case, a maximum temperature between 25 and 30°C ensures sufficient extraction of phenolic compounds from the solid parts during maceration. In other cases, refrigeration is necessary to avoid exceeding the maximum temperature limit. Refrigeration is always necessary for white wines: their fermentation must be carried out at around 20°C to retain their aromas.

Current refrigeration methods include circulating cold water or other cold fluid through the double lining of metallic tanks or through a temperature exchanger submerged in the tank. In certain cases, spraying the exterior of a metallic tank can be sufficient. The must can also be sent through tubular exchangers cooled by circulating water, itself refrigerated by an air exchanger.

Temperature has an impact on yeast development and fermentation kinetics. According to Fleet and Heard (1992), temperature can affect indigenous yeast ecology. The authors suggest that different strains are more or less adapted to different temperatures, ranging from 10 to 30°C. More precisely, the growth rate varies for each strain according to temperature—for instance,

with *Kloechera apiculata* and *S. cerevisiae*. The possible enological consequences of this phenomenon merit further research.

Numerous overlapping factors make it difficult to anticipate the impact of temperature on fermentation kinetics. Ough (1964, 1966) developed equations for the estimation of the impact of temperature on fermentation kinetics as a function of numerous parameters. Bordeaux enologists dedicated much research to this subject, summarized by Ribéreau-Gayon *et al.* (1975a). Temperature profoundly affects yeast respiratory and fermentation intensity (Table 3.9): the fermentation intensity doubles for every 10°C temperature increase. It is at its maximum at 35°C and begins to decrease at 40°C. These numbers show the importance of the fermentation temperature. The fermentation of sugar is twice as fast at 30°C as at 20°C, and for each temperature increase of 1°C the yeast transforms 10% more sugar in the same elapsed time. The optimal fermentation temperature varies according to the yeast species.

Temperature influences fermentation kinetics. The alcohol yield is generally lower at elevated temperatures, in which case some of the alcohol may be entrained with the intense release of carbon dioxide. Additionally, most of the secondary products of glyceropyruvic fermentation are found in greater concentrations. Fatty acids, higher alcohols and their esters are the most affected: their formation is at its maximum at about 20°C and then progressively diminishes. Low fermentation

Table 3.9. Average fermentation and respiratory intensities (mm^3 of O_2 consumed or of CO_2 released/g of dry yeasts/hour) of various *Saccharomyces cerevisiae* species according to temperature (Ribéreau-Gayon *et al.*, 1975a)

Temperature	Respiratory intensity (Q_{O_2})	Fermentation intensity $\left(\dfrac{CO_2}{QCO_2}\right)$
15°C	4.2	118
20°C	6.7	168
25°C	9.6	229
30°C	11.4	321
35°C	6.2	440
40°C	3.0	376

temperatures are justified when these products are desired in white winemaking.

In addition to its influence on yeast activity, temperature affects fermentation speed and limits. Between 15 and 35°C, the duration of the latent phase and the delay before the initiation of fermentation become shorter as the temperature increases. Simultaneously, yeast consumption of nitrogen increases (Section 3.4.2).

For example, a grape must with a limited sugar concentration (less than 200 g/l) takes several weeks to ferment at 10°C, 15 days at 20°C and 3 to 4 days at 30°C. For musts with higher sugar concentrations, the fermentation becomes more limited as the temperature increases; in fact, fermentation can stop, leaving non-fermented sugar.

Table 3.10 concerns a must from Sauternes containing more than 300 g of sugar per liter. The same phenomenon occurred in Müller-Thurgau's experiment in 1884 (cited in Ribéreau-Gayon *et al.*, 1975a). He fermented the same must with increasing concentrations of sugar (Table 3.11).

The initial sugar concentration of must and excessive temperatures limit ethanol production. Other factors, such as the amount of oxygen present, can limit or stop the fermentation at relatively low alcohol concentrations (11–12% vol.) and temperatures (less than 30°C). An excessively high temperature (25–30°C), during the yeast multiplication phase affects their viability and favors stuck fermentations. The impact of temperature on fermentation kinetics also varies from one yeast strain to another.

These facts are important for winery practices and show the difficulty of determining a maximum acceptable temperature limit.

The impact of temperature on fermentation is depicted in Figure 3.8. The latency time decreases and the initial fermentation speed increases as the temperature rises. The risks and severity of a stuck fermentation also increase with temperature. Of course, if the initial sugar concentration had been lower in this example, the fermentation would have been complete at 35°C. On the other hand, a higher sugar concentration would have resulted in a stuck fermentation even at 25°C. This shows that fermentation speed increases as the temperature rises but that the fermentation is also increasingly limited.

Table 3.10. Fermentation initiation speed and limits according to temperature (Ribéreau-Gayon *et al.*, 1975a)

Temperature	Initiation of fermentation	Alcohol content attained (% vol.)
10°C	8 days	16.2
15°C	6 days	15.8
20°C	4 days	15.2
25°C	3 days	14.5
30°C	36 hours	10.2
35°C	24 hours	6.0

(Initial sugar concentration: approximately 300 g/l).

Table 3.11. Alcohol formation (% vol.) according to fermentation temperature (Müller-Thurgau, 1884)

Sugar concentration (g/l)	Potential alcohol (% vol.)	Alcohol produced at			
		9°C	18°C	27°C	36°C
127	7.2	7.0	6.9	6.9	4.2
217	12.4	11.8	11.0	9.4	4.8
303	17.3	9.9	9.1	7.7	5.1

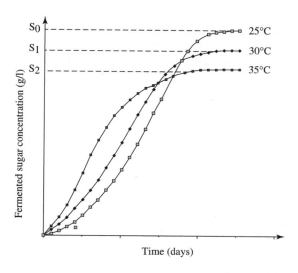

Fig. 3.8. Influence of temperature on fermentation speed and limit (S_0 = initial sugar concentration). At 25°C, fermentation is slower, but complete. At 30°C, and especially 35°C, it is more rapid, but stops at fermented sugar concentrations S_1 and S_2, respectively, below S_0

An abrupt temperature change during fermentation can lead to thermic shock. This phenomenon is different from the notion of shock or stress that is used in microbiology (Section 6.2.5). In certain wineries, white wine fermentation tanks are situated outdoors due to space limitations. These wines, fermenting at moderate temperatures (20°C), have difficulty withstanding the abrupt temperature variations between day and night when autumn cold first arrives. The fermentation progressively slows and finally stops. A second inoculation is not effective. If these wines are transferred to a tank at a constant temperature before the fermentation has completely stopped and after a consequent inoculation, fermentation will be completed. Laboratory tests confirm that an abrupt temperature change, in one direction or the other, affects yeast activity. This phenomenon merits more in-depth study.

The data in Table 3.12 are taken from a laboratory experiment. At 12°C, the fermentation is slow but complete. At 19°C, the fermentation is quicker and sugar transformation is also complete. If the fermentation begins at 19°C and is abruptly lowered to 12°C, it stops, leaving non-degraded sugar. The same is true for a fermentation whose temperature abruptly increases from 12 to 19°C. If thermic shock occurs during the final stage of fermentation (after the fermentation of 120 g of

Table 3.12. Effect of temperature variations (thermal shocks) on grape must fermentation (Larue *et al.*, 1987)

Fermentation temperature	Duration of fermentation (days)	Residual sugar concentration at end of fermentation (g/l)
12°C	93	<2
12°C, transferred at 19°C after transformation of 40 g sugar/l.	56	27
19°C	50	<2
19°C, transferred at 12°C after transformation of 40 g sugar/l	21	108

Initial sugar concentration: 220 g/l.

sugar per liter, for example), its effect is less significant.

To recapitulate, an increase in temperature accelerates the fermentation but also adversely affects its limit. A stuck fermentation can also occur if other limiting factors add their effects (richness in sugar, anaerobiosis). For this reason, there is no fixed temperature limit above which fermentation is no longer possible. In red winemaking, the fermentation temperature should never exceed 30°C: above this temperature, the risks are certain. Of course, this does not mean that a fermentation cannot be complete at temperatures at or above 35°C.

The yeast is most sensitive to temperature at the beginning of its development; it is more resistant during the final stage of fermentation. For this reason, in red winemaking, the fermentation should begin at 18–20°C and be allowed to increase progressively to 32°C or even a little higher. The higher final temperature favors maceration phenomena. An excessively high temperature at the initiation of fermentation (around 30°C) can result in a difficult final stage of fermentation.

3.7.2 Influence of Oxygen—Effect of Must Aeration

Yeasts use energy derived from the degradation of sugars. This degradation is carried out by either the respiratory or the fermentation pathway. In grape must, due to the catabolic repression of respiration exerted by must glucose in *S. cerevisiae*, sugar degradation is carried out exclusively by alcoholic fermentation.

Yeast respiratory capacity is put to good use in enology for the production of flor wines. In this case, yeasts oxidize ethanol into aldehyde in dry wines. Oxidative yeasts can also develop during winemaking: they oxidize ethanol into carbon dioxide and are considered to be spoilage yeasts.

Pasteur spoke of must fermentation as a type of 'life without air'. Yeast development and fermentation have long been known to be impossible in the complete absence of oxygen (Ribéreau-Gayon *et al.*, 1951). The complete absence of oxygen supposes the fermentation of a must devoid of oxygen in complete anaerobiosis. The must would also have to be inoculated by a yeast-starter cultivated in the absence of oxygen. Experimental conditions in complete anaerobiosis are difficult to maintain. For this reason, some authors might have thought that oxygen was not absolutely necessary for fermentation kinetics.

On the contrary, oxygen has a considerable impact on the fermentation kinetics of wine. The addition of oxygen is probably the most effective method available to the winemaker for controlling must fermentation. For this reason, the terms 'complete anaerobiosis', 'semi-anaerobiosis' or 'limited aerobiosis' are sometimes used to explain the amount of oxygen added during fermentation. In laboratory experiments, samples are sealed with a sterile wad which permits the controlled introduction of oxygen and ensures an exchange in both directions between the interior and exterior environment. Complete anaerobiosis is obtained by obturating the opening of the samples with a fermentation lock.

In the winery, open tanks leaving the wine in contact with air permit a permanent aeration. They are not recommended, because of the risk of bacterial development—closed tanks are preferable but a controlled amount of oxygen should be added to the fermenting wine in these tanks during pumping-over, for example. This technique has been widely used in red winemaking for many years. The fermenting must is easily saturated in oxygen (6–8 mg/l). Pumping-over, however, is less used in white winemaking because of fears of oxidating the must and modifying the aromas. In fact, this fear has not been confirmed in practice, since the yeast can absorb a large amount of

oxygen during fermentation. The aromas of must before fermentation and in white wine separated from its lees after fermentation are susceptible to oxidation, but the aromas of must during fermentation are probably less affected by aeration.

Aeration accelerates fermentation and as a result increases the demand for nitrogen-containing nutrients. Oxygen favors the synthesis of sterols and unsaturated fatty acids, improving the cell membrane permeability and consequently glucide penetration. The addition of oxygen has an effect similar to the addition of sterols, which are considered to be oxygen substitutes.

The data in Table 3.13, taken from an experiment carried out many years ago in the laboratory of Ribéreau-Gayon (Ribéreau-Gayon et al., 1951), show the effects of controlled aeration on the fermentation kinetics of a relatively high-sugar must. In complete anaerobiosis, fermentation stops on the 14th day, leaving 75 g of sugar per liter. On the 21st day, the yeast population was 5×10^7 cells/ml. In limited aerobiosis, the fermentation is complete with a yeast population twice as great. Moreover, the initiation of fermentation and its initial speed are greater in the presence of oxygen. Of course, if the initial sugar concentration had been lower in the experiment in Table 3.13, the fermentation would have been complete in both cases, although slower in anaerobiosis. A higher sugar concentration would have led to a stuck fermentation in both cases before the complete depletion of sugar.

In winemaking, permanent aeration is rarely possible, but momentary aerations are a suitable replacement. The data in Table 3.14 are also

Table 3.13. Evolution of the fermentation of a grape must containing a high sugar concentration (270 g/l) according to aeration conditions (Ribéreau-Gayon et al., 1951)

Time	Limited aerobiosis (cotton stoppered flasks)		Anaerobiosis (bubbler stoppered flasks)	
	Residual sugars (g/l)	Total cells (10^7/ml)	Residual sugars (g/l)	Total cells (10^7/ml)
7 days	86		140	
14 days	2		75	
21 days	2	10.7	75	5

Initial inoculation: 10^6 cells/ml.

Table 3.14. Effect of oxygen addition at different stages of grape must fermentation containing 228 sugar/gl (measurement made on day 14 of fermentation) (Ribéreau-Gayon *et al.*, 1951)

Fermentation aeration	Oxygen added[a] (ml/l)	Total yeast cells (10^7/ml)	Fermented sugars (g/l)
In contact with air (limited permanent aeration)		9.3	225
Without air contact			
Without aeration	0	5.1	164
Brief aeration before fermentation	6.0	6.2	164
Brief aeration on day 2	0.15	5.8	190
	0.75	6.1	196
	1.5	6.3	205
	6.0	7.5	223
Brief aeration on day 4	0.75	5.3	184
	1.5	6.0	202
	6.0	6.0	202
Brief aeration on day 8	6.0	5.2	173

[a]Oxygen can be expressed in mg/l by multiplying the values by 1.43 (density of oxygen with respect to air).

taken from a past experiment (Ribéreau-Gayon *et al.*, 1951). They confirm the difference in fermentability between musts in the presence and in the absence of oxygen. The effect of oxygen on fermentation kinetics increases with the quantity of oxygen introduced. The timing of the addition of oxygen appears to be especially important. The acceleration of the fermentation is most significant when oxygen is added on the second day following the initiation of fermentation, during the growth phase of the yeast population. In this case, the fermentation is nearly as (if not as) rapid as a fermentation with limited permanent aeration; the same amount of sugar is also transformed. In fact, it is not the must that needs to be aerated but rather the yeasts fermenting the must—especially the population in the growth phase. Other experiments confirm yeast use of oxygen primarily during the first stages of fermentation. They do not benefit from an aeration when the fermentation is in its advanced stages and the alcohol concentration is excessive.

These observations are significant for practical purposes and they should be taken into account during winemaking—especially red winemaking (Section 12.4.2). The risks of winemaking

in open tanks, or permanent aerobiosis, are known (Section 12.5.1). Anaerobiosis in tanks is recommended if oxygen is added at the right moment. The acceleration of the fermentation induced by a momentary aeration must nevertheless be anticipated. This acceleration results in a more significant heating of the wine.

The experiments cited in this section date back to Ribéreau-Gayon *et al.* (1951). Even if no longer cited in recent works, the results are still valid (Fleet, 1992). In particular, Sablayrolles and Barre (1986) retained the same values for yeast oxygen needs, i.e around 10 mg/l; they also confirmed the significant influence of oxygen and the moment of its addition on fermentation kinetics. Other research has shown that aeration, combined with the addition of nitrogen in mid-fermentation, is more effective than aeration alone (Sablayrolles *et al.*, 1996a and 1996b) (Section 3.4.2). This effect is apparently more marked in certain media under certain fermentation conditions.

3.7.3 Effect of Must Clarification on White Grapes

Must clarification before the initiation of fermentation has long been known to affect the quality of white wine (Section 13.5.1). Yeasts fermenting clear must form more higher alcohols, fatty acids and corresponding esters. In addition, must suspended solids can impart heavy and disagreeable vegetal odors. The racking of must is therefore essential.

Must clarification also affects fermentation phenomena. It eliminates some of the wild yeasts, along with the natural vegetal sediment, but inoculation compensates for this loss, and is also often recommended to compensate for the small population present at the time of filling the tank. In this manner, fermentation is carried out by selected strains that best express must quality, without the development of olfactory flaws.

During juice settling, certain conditions useful for sedimentation (such as low temperature and sulfiting) can promote the development of certain strains resistant to these conditions. Of course, this growth must not become a fermentation; otherwise, it would put the sediments back in suspension. Nevertheless, these strains can develop preferentially during fermentation even after an active yeast inoculation (Fleet, 1992).

Clarification essentially modifies must fermentability. Clear must is known to ferment with more difficulty than cloudy must (though the elimination of yeasts is not the only reason for this fermentation difficulty, as was once thought). Yet clarification simultaneously favors the aromatic finesse of the wine. White wine quality is thought to be enhanced by a somewhat difficult and slow fermentation. In general, all operations that

accelerate fermentation will lower wine quality, and vice versa. A compromise permitting complete fermentation and satisfactory wine quality must be sought.

Several researchers have studied the effect of grape must lees and other solid materials on fermentation kinetics (Ough and Groat, 1978; Delfini and Costa, 1993). Ribéreau-Gayon et al. (1975b) evaluated the effect of several related factors on must fermentability: the elimination of yeast, the elimination of nutritive elements released by grape must sediment and a possible support effect which would permit a greater yeast activity, possibly by the fixation of toxic compounds (short-chain fatty acids). A number of fermentations were carried out in the laboratory under different conditions. For this experiment, the medium was heated at 100°C for 5 minutes to destroy the yeasts and ensure the solubilization of the nutritive elements likely to be involved in fermentation. In view of the destruction of the yeasts under certain conditions, the medium was systematically inoculated, using a strain of S. cerevisiae with good fermentation potential at a relatively low concentration (10^5/ml) to assess any possible effect of natural elimination of the yeast. In this manner, the subsequent effect of the natural elimination of yeasts can be appreciated. By comparing the different samples, the contribution of each of the three parameters to the loss of fermentability could be evaluated. The results of two tests with different concentrations of sugar are given in Table 3.15. Fermentability loss due to

Table 3.15. Analysis of the effect of different must clarification operations on fermentability loss[a] (Ribéreau-Gayon et al., 1975b)

Cause of fermentability loss	Trial A (initial sugar concentration 220 g/l) Measured on day 2	Trial B (initial sugar concentration 285 g/l) Measured on day 5
Total fermentability loss due to clarification	**62%**	**37%**
Elimination of yeasts in lees	27%	10%
Elimination of support effect of lees	9%	13%
Elimination of nutritive elements released by lees	23%	21%
Total fermentability loss	**59%**	**44%**

[a]Fermentability loss is defined as the percentage of sugar remaining unfermented, in comparison to a non-clarified control, all other conditions being the same.

juice clarification appeared to be significant. The measurements were taken on the second and fifth day of fermentation and it should be taken into account that the slowing of the fermentation due to juice settling and clarification is most obvious at this time. The differences tend to become less apparent over time. In test A, the fermentation is complete even after juice settling. In test B, juice settling results in a stuck fermentation. In spite of the necessary approximations made in carrying out this experiment, it shows the multitude of effects of juice settling. Furthermore, the sum of the individual fermentability losses corresponds fairly well with the total loss.

The nature of the sediment and its effect on must fermentability are related to grape origin, grape sanitary conditions and must extraction conditions. Lafon-Lafourcade et al. (1980) demonstrated the essential role of the crushing and pressing conditions of white grapes. In identical clarification conditions, musts extracted after the energetic crushing of grapes ferment less well than musts originating from pressing without crushing. This difference in fermentation can result in a stuck fermentation when associated with other unfavorable conditions (elevated sugar concentrations, complete anaerobiosis, etc.). Pressing without crushing maintains the juice in contact with the skins for a certain period, and this contact seems to permit the diffusion of grape skin steroids. In the same manner, pre-fermentation skin contact generally results in a must with a good fermentability, even after careful juice settling. Delfini et al. (1992) noticed that must clarification eliminates long-chain fatty acids. Their elimination has been linked to the increased production of acetic acid frequently reported in white musts with fermentation problems, particularly those that have been highly clarified.

Must lees particles and even glucidic macromolecules, making up part of the colloidal turbidity of musts such as yeast hulls, can adsorb short-chain fatty acids (C8 and C10) (Section 3.6.2) (Ollivier et al., 1987). In consequence, the level of must clarification should be controlled for each type of white winemaking by measuring must cloudiness or turbidity, expressed in NTU

(Sections 13.5.2; 13.5.3). For vineyards in the Bordeaux region, a turbidity of less than 60 NTU can lead to serious fermentation difficulties. Above 200 NTU, the risk of olfactive deviations due to the presence of must lees is certain.

3.8 STUCK FERMENTATIONS

3.8.1 Causes of Stuck Fermentations

Stuck fermentations have always been a major problem in winemaking. French enological literature has mentioned them since the beginning of the 20th century. The production of fortified wines was definitely a response to difficult final stages of fermentation and the ensuing microbial accidents, especially in countries with warm climates. These wines were rapidly stabilized by the addition of pure alcohol. Around the world, many of these wines disappeared as progress in microbiology permitted the elaboration of dry wines.

Stuck fermentation continues to be a much discussed subject. In some cases, a modification of vine varieties has produced grapes that have high sugar concentrations. These grapes are more difficult to ferment than past varieties. In other cases, winemakers have recently realized that sluggish fermentations spread over several months are not ideal for making wine.

As red winemaking techniques used in Bordeaux (grapes with a relatively high sugar content and long vatting requiring closed vats) are particularly conducive to this problem, it has been studied here for many years. In the 1950s in-depth research in Bordeaux resulted in important discoveries concerning temperature regulation and aeration. The ubiquity of stuck fermentations in other viticultural regions led to new research confirming past work.

The slowing of fermentation can be monitored by tracking the mass per unit volume. If a density below 1.005 decreases by only 0.001 or 0.002 per day, a stuck fermentation can be anticipated before the complete depletion of sugar. From past experience, the consequences of a stuck fermentation are not too serious if it occurs with at least 15 g of sugar per liter and a moderate alcohol content (<12% vol.). In this case, the restarting of fermentation does not pose any major problems. On the other hand, with a sugar concentration of less than 10 g/l, it is often very difficult to restart stuck fermentations, particularly when malolactic fermentation is initiated.

A stuck fermentation can result from a particular cause. For example, an excessive must sugar concentration makes a complete fermentation impossible; in this case, only a sweet wine can be made. Stuck fermentations can be expected in the case of excessive temperatures, and this type of stuck fermentation is generally the result of several causes. The effects are cumulative, although sometimes individually without consequence. Winemakers do not always understand possible cumulative risks and the precautions that must be taken.

To summarize, the following factors can be involved in stuck fermentations:

1. The must sugar concentration has an inhibitory effect which compounds the toxicity of the alcohol formed. The addition of sugar to must (chaptalization), when it is too late, requires yeasts to pursue their metabolic activity although already hindered by the alcohol formed.

2. An excessive temperature results from the initial temperature, the quantity of sugar fermented, and the type of tank used (dimensions and material). All operations that accelerate the transformation speed of sugar increase the maximum temperature. The temperature becomes a limiting factor at about 30°C. The effect is more pronounced when the temperature is elevated in the early stages of fermentation. Normally, the fermentation should begin at a moderate temperature (20°C).

3. Conversely, too low of an initial temperature can limit yeast growth and lead to an insufficient yeast population. At moderate temperatures, yeasts have difficulty supporting extreme temperature changes (thermic shocks).

4. Complete anaerobiosis does not permit satisfactory yeast activity (growth and survival). Aeration increases fermentation speed. It must take place in the early stages of fermentation, during the population growth phase. Oxygen substitutes such as steroids and long-chain fatty acids can also improve fermentation kinetics.

5. Yeast activity can be affected by nutritional deficiencies: nitrogen compounds, growth factors, and possibly minerals. Combined additions of oxygen and ammoniacal nitrogen appear to be particularly effective. These nitrogen deficiencies probably occur in specific situations that we are now capable of predicting. The effectiveness of the addition of nutritive elements, observed in laboratory work, should be interpreted with respect to other fermentation conditions (sugar concentration and aeration). Certain grape growing conditions, such as hydric stress, old vines, and cover cropping vineyards to decrease vine vigor, can lead to less fermentable musts. Under these conditions, stuck fermentation is probably due to nutrient deficiencies (nitrogen), which probably require further investigation.

6. Metabolic by-products (C6, C8 and C10 saturated fatty acids) inhibit yeast growth, intensifying alcohol toxicity.

7. Anti-fungal substances can be present in must—pesticide residues used to protect the vine or compounds produced by *Botrytis cinerea* in rotten grapes.

8. In white winemaking, must extraction conditions have a significant influence: grape crushing, conditions of juice draining, pressing of the crushed grapes, and especially the level of must clarification (juice settling). These operations may result in the excessive elimination of steroids, which act as survival factors for the yeast.

In the acidity range of the must, a high acidity does not seem to favor fermentation, but an elevated pH can make the consequences of a stuck fermentation much more serious. A low pH combines with the effect of sulfiting to inhibit bacterial growth. In this case, antagonistic phenomena between bacteria and yeasts diminish and the fermentation is more steady. Another unfortunate consequence of low pH is that it promotes the formation of volatile acidity by the yeast.

In addition to chemical and physicochemical causes of stuck fermentations, microbial phenomena also are involved. First of all, the quantity of the initial yeast inoculum can be insufficient (Section 3.5.4) if, for example, must temperature is exaggeratedly low. Antagonistic phenomena between different yeast strains can also occur, and the killer factor (Section 1.7) explains this fairly widespread antagonism. Fermentation can be rapid in some tanks while being slower in others, and strain identification techniques have shown that fermentation is carried out almost exclusively by one strain in the first case, whereas several strains ferment the must in the second case (Section 1.10.2). These antagonistic phenomena can affect an inoculation. In certain conditions (for example, a significant natural inoculum in full activity), inoculating with dry commercial yeast leads to a slower fermentation than not inoculating. Yeast strains must therefore be selected according to the type of wine being made, ensuring that they are more active and numerous than the indigenous yeasts. The necessary conditions for controlling fermentation include: cleanliness; inhibiting natural yeasts sufficiently early by maintaining low temperatures; sulfiting appropriately; and inoculating with an active yeast starter as soon as the tank is filled to ensure its rapid implantation.

Antagonistic phenomena between yeasts and lactic acid bacteria can also cause fermentation difficulties (Section 6.4.1), especially in red winemaking (Section 12.4.3). The initial sulfiting of the grapes must temporarily inhibit the bacteria while at the same time permitting yeast development and sugar fermentation. Bacteria do not develop as long as yeast activity is sufficient, but if alcoholic fermentation slows for some reason, bacteria can begin to grow—especially if the initial sulfiting was insufficient. This bacterial development aggravates yeast difficulties and increases the risks of a premature, stuck alcoholic fermentation. The bacterial risk is an additional justification for red grape sulfiting (5 g/hl) before fermentation (Section 8.7.4). The addition of lysozyme

(200–300 mg/l), extracted from egg whites, has been suggested to reinforce the inhibitory effect of sulfuring on bacteria in difficult fermentations (Section 9.5.2). In addition, the inoculation of lactic acid bacteria (*Oenococcus oeni*) before alcoholic fermentation to activate a subsequent malolactic fermentation is not recommended; in the case of difficult alcoholic fermentations, this operation increases the risk of bacterial spoilage (Section 3.8.2). Yet this is standard practice in some vineyards. The relationship between this practice and an increase in volatile acidity should be considered.

For a long time, difficult final stages of fermentation and stuck fermentations were a real problem during red winemaking. Temperature control systems and the general practice of pumping-over with aeration limited these incidents. White musts, however, have become increasingly difficult to ferment because of excessive clarification and mechanized must extraction conditions. This excessive clarification removes must constituents essential for a complete fermentation. In white winemaking, the must is often not aerated to avoid oxidation, yet a lack of aeration during fermentation also contributes to fermentation difficulties. Today, controlled aeration of white musts is recommended during fermentation as the CO_2 being released protects them from oxidation.

Human error is another factor that certainly has an impact on stuck fermentations, although it is difficult to prove. It is not unusual to find wineries where stuck fermentations occur with some regularity, as though there was a specific, technical cause that could be identified and corrected, then disappear completely following a change in winemaker.

3.8.2 Consequences of Stuck Fermentations

Residual sugar is not acceptable in dry white and most red wines. A stuck fermentation therefore requires the restarting of yeast activity in a hostile medium. Evidently, if the alcohol content is already elevated (13% vol.), the chances of restarting the fermentation are slim.

The risk of bacterial spoilage is the principal danger of a stuck fermentation.

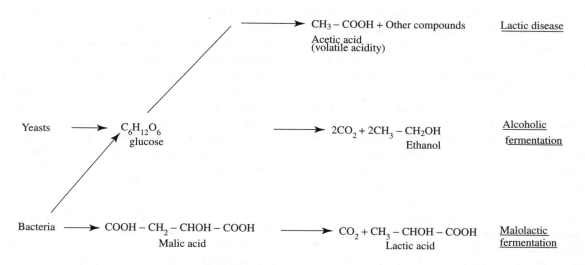

Fig. 3.9. Effect of different microbial phenomena during primary and secondary fermentations (Ribéreau-Gayon 1999)

Figure 3.9 schematizes the involvement of different microbial phenomena, (red winemaking is specifically represented since malolactic fermentation is taken into account) (Ribéreau-Gayon, 1999). The various stages of fermentation are understood. Yeasts ferment sugars, and yeast activity should stop only when all of the sugar molecules have been consumed. The lactic acid bacteria then assert themselves, exclusively decomposing malic acid molecules in a process called malolactic fermentation. If yeast activity stops before the complete depletion of must sugar, bacteria can develop. Bacterial development depends on several factors, including the initial sulfiting of the grapes and the possible addition of lysozyme (Section 9.5.2). The inoculation of malolactic fermentation bacteria in the must also promotes their development. Acetic acid is formed when lactic bacteria, mainly heterofermentative *Oenococcus*, are present in a medium containing sugar. In these situations, the volatile acidity can rapidly increase to unacceptable levels even if there is a relatively small residual sugar concentration. In fact, bacteria form acetic acid from sugar after their growth phase, during which malic acid is assimilated. In consequence, in the case of a stuck alcoholic fermentation, the winemaker can let the malolactic fermentation continue until its completion before inhibiting the bacteria.

The understanding of the processes represented in Figure 3.9 constituted an unquestionable progress in wine microbiology. As a result, certain operations were initiated to control the alcoholic fermentation and avoid stuck fermentations. The volatile acidity of top-ranked red wines decreased substantially, with a corresponding improvement in quality. In the 1930s in the Bordeaux region, the volatile acidity of these wines was often 1.0 g/l (expressed in H_2SO_4) or 1.2 g/l (expressed in acetic acid). The content has been decreased half this value and today's higher figures are due to various problems during fermentation and storage, which can and must be avoided.

3.8.3 Action in Case of a Stuck Fermentation

Many stuck fermentations result from winemaking errors. Moreover, systematic stuck fermentations have been observed in certain wineries year after year. They disappear without any apparent reason at the same time that the winemaker changes. More often than not, the necessary operations are known but not carried out properly. In red winemaking, stuck fermentations often result from excessive temperatures at the initiation of fermentation and a poor control of tank temperature during fermentation. Insufficient dissolution of oxygen during pumping-over can also contribute. In white winemaking, excessive must clarification, temperature variations and the absence of air contribute to fermentation problems. Of course, a higher sugar concentration in the must increases the risks, but in certain situations there is no satisfactory explanation for stuck fermentations.

A density that remains stable during 24 or 48 hours confirms a stuck fermentation. In this case, different procedures exist to restart the fermentation while avoiding bacterial spoilage.

Restarting a fermentation is often a difficult operation. First of all, this medium is rich in alcohol and poor in sugar—it is not conducive to the development of a second fermentation. In addition, the yeasts are exhausted from the first fermentation and they react poorly to the different stimuli employed. They benefit more from different operations such as nitrogen supplementation and oxygenation at the beginning of their development, when the medium contains high sugar concentrations and does not contain ethanol. Therefore, an operation after a stuck fermentation cannot compensate for a winemaking error. All operations beneficial to the fermentation should be employed from the start of the winemaking process to avoid stuck fermentations. The prevention of stuck fermentations is essential to winemaking and it should take into account all of the recommendations previously stated.

In spite of all precautions, a stuck fermentation may still occur. In this case, white wines must be treated differently from reds which undergo malolactic fermentation. At the time of the stuck fermentation, the red wine tank contains must and pomace rich in bacteria. The wine should be drained rapidly, even if the skin and seed maceration is not complete. Draining eliminates part of the bacterial contamination and introduces oxygen, which favors the restarting of fermentation and decreases the temperature. The wine can be sulfited at the same time, to inhibit bacterial development. In some cases, the fermentation restarts spontaneously.

Even if the stuck fermentation results from the combination of several elementary causes, each has

its own effect on the ease of restarting the fermentation. Excessive temperatures destroy yeasts but do not make the medium unfermentable, as does fermentation in complete anaerobiosis.

If the fermentation does not restart on its own, an inoculation with active yeast is required. At present, commercial dry yeasts are inactive in media containing more that 8–9% vol. of alcohol, due to manufacturing conditions. In the future, industrially prepared yeast capable of developing in a medium containing alcohol would be desirable. Bacteria with this property have now been developed for malolactic fermentation.

An active yeast starter must be prepared using the stuck fermentation medium adjusted to 9% vol. alcohol and 15 g of sugar per liter; 3 g of SO_2 per hectoliter is also added. The active dry yeasts are added at a concentration of 20 g/hl. Their growth at 20°C requires several days and it is monitored by measuring the density or measuring the sugars. When all of the sugar has been consumed, the yeasts are at the peak of their growth phase. This yeast starter, rich in activated yeasts and no longer containing sugar, is inoculated into the stuck fermentation medium at a concentration of 5–10%. Several days are required for the complete exhaustion the last few remaining grams of sugar. It is a long and painstaking operation. The volume of the yeast starter can also be progressively increased by adding larger and larger quantities of the stuck fermentation wine to it.

In choosing a yeast strain, the yeasts should certainly be resistant to ethanol. Yeasts commercialized under the name *S. bayanus* could be recommended but they seem to have a propensity to form volatile acidity in these conditions. Commercially available *S. cerevisiae* yeasts (formerly *S. bayanus*), known for their resistance to ethanol and low probability of producing volatile acidity, are recommended for this purpose.

Effectively, the volatile acidity of the wine tends to increase during the restarting of a stuck fermentation. This generally occurs when the yeasts encounter unfavorable conditions. Certain yeast strains are more predisposed to forming it than others. The addition of 50 mg of pantothenic acid per hectoliter (not authorized by EU legislation) not only can limit its formation but also can contribute to the disappearance of an excess of acetic acid.

The temperature for restarting the fermentation must be considered. A slightly elevated temperature favors cellular multiplication but the antiseptic properties of ethanol increase with temperature. The risk of an increase in volatile acidity also seems to be in function of temperature. For these reasons, the restarting of the fermentation should be carried out at a temperature between 20 and 25°C.

Existing activated yeasts in the winery can also be used to restart a stuck fermentation. If there is a large volume of fresh harvest available at the right moment, the tank with the stuck fermentation can be drowned with it. This operation, however, is in conflict with the legitimate desire to select cuvées. The practice of adding 5–20% of a medium in full fermentation to a stopped tank should be carried out with prudence. Active yeasts are certainly added but sugar is too. In this situation, the fermentation has sometimes been observed to restart and then stop again, leaving about the same amount of sugar that existed before the operation. The lees of a tank that has normally completed its fermentation can also be used as a yeast starter to restart a stuck fermentation. Supplemental sugar is not introduced into the medium, but yeast in their death phase are no longer very active. The correct restarting of a fermentation requires the introduction of active yeasts in this alcoholic medium without introducing supplemental sugar. The preparation of a yeast starter using dry yeast gives the most satisfactory results.

In white winemaking, at least when malolactic fermentation is not sought, the wine with a stuck fermentation should be lightly sulfited to protect against bacterial development. The fermentation can then be restarted using a yeast starter prepared according to the preceding instructions.

Many possible adjuvants helping to restart a stuck fermentation have been proposed. The addition of ammonium salts does not raise any counterindications, but no appreciable improvement of the second fermentation has been observed. The addition of ammonium sulfate should be limited to 5 g/hl due to the limited use of nitrogen by yeasts.

Flash-pasteurization (heating between 72 and 76°C for 20 seconds) seems to be effective. It improves the fermentability of wines with stuck fermentation (Dubernet, 1994). This operation is valid for red, rosé and dry white wines and should be carried out before inoculating. Its heating effect can be likened to the effect observed during thermo-vinification (Section 12.8.3). In spite of the destruction of yeasts, the heated musts ferment especially well. The effects of this process merit further study but several explanations can be proposed: fermentation by a sole strain avoiding microbial antagonisms; addition of nutritive elements due to yeast lysis; elimination of toxic substances; and modification of the colloidal structure.

Active charcoal has also been used for a long time to reactivate fermentations (10–20 g/hl). Such an addition is hardly conceivable in red wines, but its effectiveness for stuck fermentations in white wines is recognized. It works by eliminating yeast inhibitors (fatty acids) (Section 3.6.2).

The addition of yeast hulls is certainly the most effective way of restarting a stuck fermentation, although less so than in preventing fermentation from stopping in the first place. (Section 3.6.2). They can be added to the yeast starter preparation or directly to the medium with the stuck fermentation.

In results of an experiment given in Table 3.16, the first fermentation of a must initially containing 250 g of sugar per liter stops at 67 g of nonfermented sugar per liter. The second fermentation is conducted after an inoculation at 10^6 cells of *S. cerevisiae* per milliliter, without the addition of yeast hulls in the control sample and with an addition of 0.5 g/l in the test sample 24 hours later. This addition permits a complete fermentation

Table 3.16. Restarting fermentation (after a spontaneous stuck fermentation) by addition of yeast hulls (Lafon-Lafourcade *et al.*, 1984)

	Residual sugars (g/l)		
	Day 9	Day 16	Day 36
Control	57	36	13
+50 g/hl of yeast hulls	53	23	1.4

in 36 days, which is not possible in the control sample.

A massive addition of yeast hulls combined with an inoculation of active yeast can result in olfactive modifications of light wines such as whites and rosés. Doses between 20 and 30 g/hl (maximum) are therefore recommended.

In white as well as red winemaking, the restarting of a stuck fermentation should be closely monitored, especially by measuring volatile acidity to ensure that the alcoholic fermentation is pure. The smallest increase in volatile acidity represents a bacterial contamination, which should absolutely be avoided. A judicious sulfiting should prevent contamination; without a doubt, it slows the fermentation. Yet if the doses are adapted to the situation (3–5 g/hl), the fermentation will not be definitively compromised; it will restart after inoculating. It must also prevent all bacterial development before the complete depletion of sugars, even though its addition can make malolactic fermentation more difficult.

Tanks with stuck fermentations must be restarted as soon as possible. In the middle of winter, this operation can become impossible and in these situations it is preferable to wait until the following spring, when fermentation may restart spontaneously.

REFERENCES

Aerny J. (1996) *Revue Suisse Arboric. Hortic.*, 28 (3), 161.

Andreasen A.A. and Stier T.J.B. (1953) *J. Cell. Comp. Physiol.*, 41, 23.

Bely M., Sablayrolles J.M. and Barre P. (1990) *J. Ferment. Bioeng.*, 70, 246.

Bouix M., Busson C., Charpentier M., Leveau J.Y., Dutertre B. (1997) *J. Int. Sci. Vigne Vin*, 31 (1), 11.

Bovée J.P., Blouin J., Marion J.M. and Strehaiano P. (1990) In *Actualités Enologiques 90* (Eds P. Ribéreau-Gayon and A. Lonvaud). Dunod, Paris. pp. 281–286.

Bréchot P., Chauvet J., Dupuy P., Crosson M. and Rabatu U. (1971) *CR Acad. Sci.*, 272, 890.

Cantarelli C. (1989) In *Biotechnology Application in Beverage Production* (Eds C. Cantarelli and G. Lanzarem). Elsevier Applied Sciences, London.

Castor J.G.B. (1953) *J. Food Res.*, 18, 146.

Choné X. (2000) *Contribution à l'étude des terroirs bordelais: étude de l'influence des déficits hydriques modérés et de l'alimentation en azote sur le potentiel aromatique des raisins de vitis vinifera, cv. sauvignon blanc*. Thèse de doctorat, Université Victor Segalen Bordeaux 2.

Delfini C., Costa A. (1993) *Am. J. Enol. Vitic.*, 44 (1), 86.

Delfini C., Conterno L., Giacosa D., Cocito C., Ravaglia S. and Bardi L. (1992) *Vitic. Enol. Sci.*, 47, 69.

Dubernet M. (1994) *La Vigne*, Dec., 52.

Dubourdieu D., Ollivier C. and Boidron J.N. (1986) *Conn. Vigne Vin*, 20 (1), 53.

El Halaoui N., Ricque D. and Corrieu, G. (1987) *Sci. Alim.*, 7, 241.

Flanzy C. (1998) *OEnologie. Fondements Scientifiques et Technologiques*. Lavoisier, Technique & documentation, Paris.

Fleet G.H. (1992) *Wine Microbiology and Biotechnology*. Harwood Academic Publishers, Chur, Switzerland.

Fleet G.H. and Heard G.M. (1992) In *Wine Microbiology and Biotechnology* (Ed. G.M. Fleet). Harwood Academic Publishers, Chur, Switzerland, pp. 27–54.

Geneix C., Lafon-Lafourcade S. and Ribéreau-Gayon P. (1983) *C.R. Acad. Sci., Série III*, 296, 943.

Grenier P., Feuilloley P. and Sablayrolles J.M. (1990) In *Actualités Enologiques 89*, (Eds P. Ribéreau-Gayon and A. Lonvaud). Dunod, Paris. pp. 521–526.

Guinand G. (1989) *Groupe d'experts technologie du vin*. Office International de la Vigne et du Vin, Paris.

Hatzidimitriou E., Darriet P., Bertrand A. and Dubourdieu D. (1997) *J. Int. Sci. Vigne Vin*, 31 (1), 51.

Henschke P.A. and Jiranek V. (1992) In *Wine Microbiology and Biotechnology* (Ed. G.H. Fleet). Harwood Academic Publishers, Chur, Switzerland. pp. 77–169.

Ingledew W.M. and Kunkee R. (1985) *Am. J. Enol. Vit.*, 36, 65.

Julien A., Roustan J.L., Dulau L. and Sablayrolles J.M. (2000) *Am. J. Viti. Oenol.*, 51 (3), 215.

Lafon-Lafourcade S. (1983) Wine and Brandy, in *Biotechnology*, Vol. 5 (Eds H.J. Rehm and G. Read): Verlag Chemie, Weinheim. pp. 81–163.

Lafon-Lafourcade S. and Joyeux A. (1979) *Conn. Vigne Vin*, 13 (4), 295.

Lafon-Lafourcade S., Larue F., Bréchot P. and Ribéreau-Gayon P. (1977) *CR Acad. Sci.* 284, 1939.

Lafon-Lafourcade S., Dubourdieu D., Hadjinicolaou D. and Ribéreau-Gayon P. (1980) *Conn. Vigne Vin*, 14 (2), 127.

Lafon-Lafourcade S., Geneix C. and Ribéreau-Gayon P. (1984) *Appl. Envir. Microbiol.*, 47 (6), 1246.

Larue F., Lafon-Lafourcade S. and Ribéreau-Gayon P. (1980) *Appl. Envir. Microbiol.*, 39 (4), 808.

Larue F., Lafon-Lafourcade S. and Ribéreau-Gayon P. (1982) *CR Acad. Sci., Série III*, 294, 587.

Larue F., Couralet M. and Ribéreau-Gayon P. (1987) In *Compte-rendu des activités de recherches (1984–1986) de l'Institut d'Enologie, Université de Bordeaux II.*

Lorenzini F. (1996) *Revue Suisse Arboric. Hortic.*, 28 (3), 169.

Masneuf I., Murat M.L., Choné X. and Dubourdieu D. (2000) *Viti*, 249, 41.

Müller-Thurgau M. (1884) Reported by Ribéreau-Gayon et al. (1975a).

Munoz E. and Ingledew W.M. (1989) *Am. J. Enol. Vitic.*, 40 (1), 61.

Ollivier C., Stonestreet T., Larue F. and Dubourdieu D. (1987) *Conn. Vigne Vin*, 21, 59.

Ough C.S. (1964) *Am. J. Enol. Vitic.*, 15 (4), 167.

Ough C.S. (1966) *Am. J. Enol. Vitic.*, 17 (1), 20 and 74.

Ough C.S. and Groat M.L. (1978) *Appl. Envir. Microbiol.*, 35, 881.

Pavlenko N.M. (1988) *Groupe d'experts technologie du vin.* Office International de la Vigne et du Vin. Paris.

Ribéreau-Gayon J. and Peynaud E. (1966) *Am. Nutrit. Alim.*, 20, 1.

Ribéreau-Gayon J., Peynaud E. and Lafourcade S. (1951) *Ind. Agri. Alim.* 68, 141.

Ribéreau-Gayon J., Peynaud E., Ribéreau-Gayon P. and Sudraud P. (1975a) *Sciences et Techniques du Vin.* Vol 2: *Caractères des Vins, Maturation du raisin, Levures et bactéries.* Dunod, Paris.

Ribéreau-Gayon P. (1999) *J. Int. Sci. Vigne Vin*, 33 (1), 39.

Ribéreau-Gayon P., Lafon-Lafourcade S. and Bertrand A. (1975b) *Conn. Vigne Vin*, 2, 117.

Ribéreau-Gayon J., Peynaud E., Ribéreau-Gayon P. and Sudraud P. (1976) *Sciences et Techniques du Vin*, Vol. 3: *Vinifications. Transformations du vin.* Dunod, Paris.

Ribéreau-Gayon P., Lafon-Lafourcade S., Dubourdieu D., Lucmaret V. and Larue F. (1979) *CR Acad. Sci., Série D*, 289, 491.

Rozès N., Cuzange B., Larue F. and Ribéreau-Gayon P. (1988) *Conn. Vigne Vin*, 22 (2), 163.

Sablayrolles J.M. and Barre P. (1986) *Sci. Alim.*, 6, 177 and 373.

Sablayrolles J.M. and Blateyron L. (2001) *Bull. OIV*, 845–846, 464.

Sablayrolles J.M., Grenier P. and Corrieu G. (1990) In *Actualités Enologiques 89* (Eds P. Ribéreau-Gayon and A. Lonvaud). Dunod, Paris. pp. 275–280.

Sablayrolles J.M., Dubois C., Manginot C., Roustan J.L. and Barre P. (1996a) *J. Ferm. Bioeng.*, 82 (4), 377.

Sablayrolles J.M., Salmon J.M. and Barre P. (1996b) *Progrès agric. vitic.*, 113, 339.

Salmon J.M., Vincent O., Mauricio J.C., Bely M. and Barre P. (1993) *Am. J. Enol. Vitic.*, 44 (1), 58.

4

Lactic Acid Bacteria

Lactic acid bacteria are present in all grape musts and wines. Depending on the stage of the winemaking process, environmental conditions determine their ability to multiply. When they develop, they metabolize numerous substrates. Lactic acid bacteria therefore play an important role in the transformation of grape must into wine. Their impact on wine quality depends not only on environmental factors acting at the cellular level but also on the selection of the best adapted species and strains of bacteria.

All the strains have a similar cellular organization, but their physiological differences account for their specific characteristics and varying impact on wine quality. They are classified according to their morphological, genetic, and biochemical traits.

4.1 THE DIFFERENT COMPONENTS OF THE BACTERIA CELL

Bacteria are procaryotic cells with an extremely simple organization. They can be distinguished from eucaryotes (to which yeast belong) by their small size and a lack of a nuclear membrane delimiting a nucleus.

It is impossible to distinguish between such different bacteria as *Escherichia coli* and *Oenococcus oeni* (*O. oeni*, formerly known as *Leuconostoc oenos* or *L. oenos*) by simple microscopic examination. In fact, the structure of all bacteria is very similar. It can be divided into three principal elements (Figure 4.1):

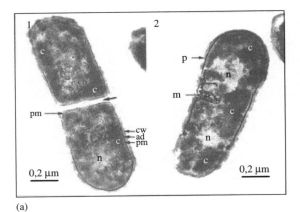

(a)

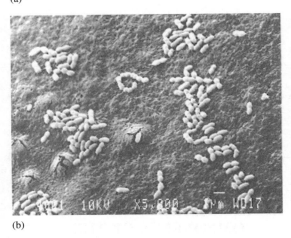

(b)

Fig. 4.1. Lactic acid bacteria isolated in wine under a scanning electron microscope (Departement de Microscopie Electronique, University of Bordeaux I). (a) Photograph of *Lactobacillus plantarum* cells (transmission, Lonvaud, 1975): c = cytoplasm; pm = plasma membrane; cw = cell wall; s = septum; m = mesosome; n = nucleus. (b) Photograph of *Leuconostoc oenos* (*Oenococcus oeni*) (scanning electron microscope).

- Cellular envelopes, including the cell wall and the membrane. The cell is delimited by the cytoplasmic membrane doubled towards the exterior by the cell wall. Between the cell wall and the membrane, the periplasmic space is a more or less fluid gel wherein proteins move about.

- The cytoplasm.

- The nucleus.

4.1.1 The Cell Wall

The cell wall of Gram-positive bacteria, such as lactic acid bacteria, is essentially composed of a peptidoglycan that is only found in procaryotes (Figure 4.2). This polymer wraps the bacterial cell with a kind of meshwork made up of polysaccharidic chains linked by peptides. The oses are glucose derivatives: N-acetylmuramic acid and N-acetylglucosamine (Figure 4.2). They alternate along the entire length of the chain, linked by β-type (1–4) glycosidic bonds that can be hydrolyzed by lysozyme or mutanolysine.

A chain of four amino acids is linked to muramic acid; L-alanine, D-alanine and D-glutamic acid are in majority. A peptide bond links the tetrapeptide of another polysaccharidic chain to the third amino acid (Figure 4.3). The peptidic chains vary depending on the species of the bacteria. The sequence of their amino acids can be used in taxonomy.

The cell walls of lactic acid bacteria, like those of nearly all Gram-positive bacteria, also contain

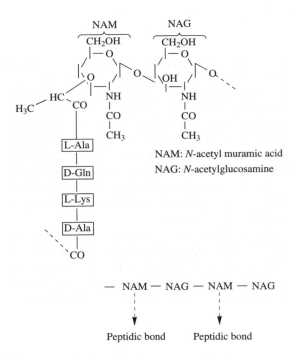

Fig. 4.2. Polysaccharidic chain of bacterium peptidoglycan

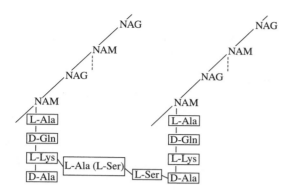

Fig. 4.3. Structure diagram of the peptidoglycan of *Leuconostoc oenos* (*Oenococcus oeni*) bacteria

ribitol phosphate or glycerol phosphate polymers called teichoic acids. Phosphodiester linkages can fix amino acids and oses to these chains. Glycerol based teichoic acids contain a glycolipid by which they attach themselves to the external layer of the plasmic membrane. They pass through the peptidoglycan and are at the surface of the cell wall acting as the antigenic sites of bacteria. The proportion of peptidoglycans and teichoic acids varies depending on the species and also the phase of the cell development cycle. Teichoic acids can represent up to 50% of the weight of the cell wall.

The cell wall is rigid and gives the cell its form: round for cocci, elongated for bacilli. It permits the cell to resist very high internal osmotic pressures (up to 20 bars). The culture of cells in the presence of penicillin, inhibiting the synthesis of the cell wall, leads to the formation of protoplasts: they are only viable in isotonic media. Similarly, lysozyme hydrolyzes the glycosidic linkages of peptidoglycan, provoking the bursting of the cell in a hypotonic medium.

Water, mineral ions, substrates and metabolic products diffuse freely across the cell wall. At this level, proteases also release amino acids from proteins and peptides which are used for cellular metabolism.

Observations under the electron microscope have also proven the existence of a protein layer on the cell wall surface (S-layer) in several lactic acid bacteria species. The study of this S-layer in wine bacteria has not yet been attempted. Finally, the accumulation of polysaccharides piled upon these proteins can form a more or less distinct capsule. Its thickness varies according to environmental conditions. In enology, *Pediococcus damnosus* gives the best example of this phenomenon. In certain conditions, strains of this species synthesize significant quantities of polysaccharides which make the wine viscous (ropiness). These cells are easily recognized under an optical microscope by the refringent halo that surrounds them.

4.1.2 The Plasmic Membrane

The membrane is situated against the cell wall, delimiting a periplasmic space. Folds are sometimes visible in the interior of the cell: these are mesosomes.

The membrane of lactic acid bacteria has the classic structure of all biological membranes: a lipid bilayer creating a central hydrophobic zone (Chapter 1, Figure 1.6). The proteins are more or less tightly joined to it. Among them, the hydrosoluble proteins are only fixed to the surface by ionic or hydrogen bonds (peripheral proteins, 30% of the proteins). The others are lodged in the membrane by hydrophobic bonds (integral proteins). The peripheral proteins have a certain mobility in the periplasmic space between the peptidoglycan and the membrane, whereas the integral proteins are almost immobile. Some protrude from the membrane while others only appear on the surface. Hydrophobic bonds between aliphatic lipidic and protein chains create the framework of the membrane. The high number of these bonds ensures the solidity of this structure, but there are no covalent bonds and so the framework created remains fluid. The biochemical functions ensured by the membrane depend on this fluidity, i.e. lipid–protein interactions. The structure can be destroyed by organic solvents and detergents. It is also disturbed by wine components. Finally, on the surface, the hydrophilic parts of the lipids and the ionized groups of the proteins establish ionic bonds between themselves.

Membranous lipids represent nearly all (95–99%) bacteria cell lipids. They essentially include

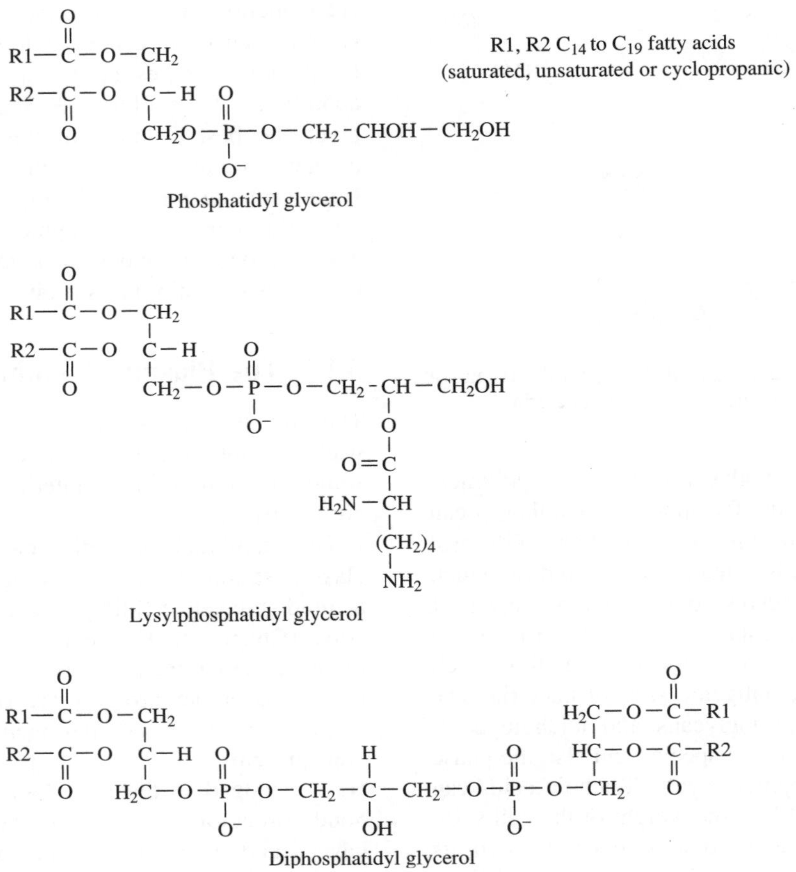

$$
\begin{array}{l}
\quad\quad\; O \\
\quad\quad\; \| \\
R1\!-\!C\!-\!O\!-\!CH_2 \\
\quad\quad\quad\quad | \\
R2\!-\!C\!-\!O\quad C\!-\!H\quad O \\
\quad\quad\; \|\quad\quad\; |\quad\quad\; \| \\
\quad\quad\; O\quad\quad CH_2O\!-\!P\!-\!O\!-\!CH_2\text{-}CHOH\!-\!CH_2OH \\
\quad\quad\quad\quad\quad\quad\quad | \\
\quad\quad\quad\quad\quad\quad\quad O^-
\end{array}
$$

R1, R2 C_{14} to C_{19} fatty acids
(saturated, unsaturated or cyclopropanic)

Phosphatidyl glycerol

$$
\begin{array}{l}
\quad\quad\; O \\
\quad\quad\; \| \\
R1\!-\!C\!-\!O\!-\!CH_2 \\
\quad\quad\quad\quad | \\
R2\!-\!C\!-\!O\quad C\!-\!H\quad\quad O \\
\quad\quad\; \|\quad\quad\; |\quad\quad\quad \| \\
\quad\quad\; O\quad\quad\; CH_2\!-\!O\!-\!P\!-\!O\!-\!CH_2\text{-}CH\!-\!CH_2OH \\
\quad\quad\quad\quad\quad\quad\quad |\quad\quad\quad\quad\; | \\
\quad\quad\quad\quad\quad\quad\quad O^-\quad\quad\quad\; O \\
\quad\quad\quad\quad\quad\quad\quad\quad\quad\quad\quad\quad | \\
\quad\quad\quad\quad\quad\quad\quad\quad\quad\quad O\!=\!C \\
\quad\quad\quad\quad\quad\quad\quad\quad\quad\quad\quad\quad | \\
\quad\quad\quad\quad\quad\quad\quad\quad H_2N\!-\!CH \\
\quad\quad\quad\quad\quad\quad\quad\quad\quad\quad\quad\quad | \\
\quad\quad\quad\quad\quad\quad\quad\quad\quad\quad\; (CH_2)_4 \\
\quad\quad\quad\quad\quad\quad\quad\quad\quad\quad\quad\quad | \\
\quad\quad\quad\quad\quad\quad\quad\quad\quad\quad\quad NH_2
\end{array}
$$

Lysylphosphatidyl glycerol

$$
\begin{array}{l}
O \quad\quad\quad\quad\quad\quad\quad\quad\quad\quad\quad\quad\quad\quad\quad\quad\quad\quad\quad\quad\quad\quad\quad\; O \\
\| \quad\quad\quad\quad\quad\quad\quad\quad\quad\quad\quad\quad\quad\quad\quad\quad\quad\quad\quad\quad\quad\quad\quad\; \| \\
R1\!-\!C\!-\!O\!-\!CH_2 \quad\quad\quad\quad\quad\quad\quad\quad\quad\quad\quad\quad\quad H_2C\!-\!O\!-\!C\!-\!R1 \\
\quad\quad\quad\quad | \quad\quad\quad\quad\quad\quad\quad\quad\quad\quad\quad\quad\quad\quad\quad\quad | \\
R2\!-\!C\!-\!O\quad C\!-\!H\quad O\quad\quad\quad\quad\quad H\quad\quad\quad O\quad HC\!-\!O\!-\!C\!-\!R2 \\
\quad\quad \|\quad\quad | \quad\quad\| \quad\quad\quad\quad\quad | \quad\quad\quad\| \quad\quad\quad\quad\quad\| \\
\quad\quad O\quad H_2C\!-\!O\!-\!P\!-\!O\!-\!CH_2\!-\!C\!-\!CH_2\!-\!O\!-\!P\!-\!O\!-\!CH_2\quad O \\
\quad\quad\quad\quad\quad\quad\quad | \quad\quad\quad\quad\quad | \quad\quad\quad\quad | \\
\quad\quad\quad\quad\quad\quad\quad O^-\quad\quad\quad\; OH\quad\quad\; O^-
\end{array}
$$

Diphosphatidyl glycerol

Fig. 4.4. Chemical formulae of some membrane phospholipids

phospholipids and glycolipids. Phospholipids are most abundant; they consist of a glycerol molecule which has a primary alcohol function and a secondary alcohol function esterified by fatty acids. The other primary function is esterified by phosphoric acid, which is esterified by glycerol, forming phosphatidyl glycerol. Lactic acid bacteria also contain diphosphatidyl glycerol (cardiolipid), amino esters of phosphatidyl glycerol with alanine (*Oenococcus oeni*) and lysine (*Lactobacillus plantarum*) (Figure 4.4). Bacteria phospholipid concentrations vary according to growth stage and cultural conditions.

Glycolipids—generally glycosides of diglycerides—are formed by glycosidic bonds between a mono or disaccharide (glucose, fructose, galactose, rhamnose) and the primary alcohol function of a diglyceride (Figure 4.5).

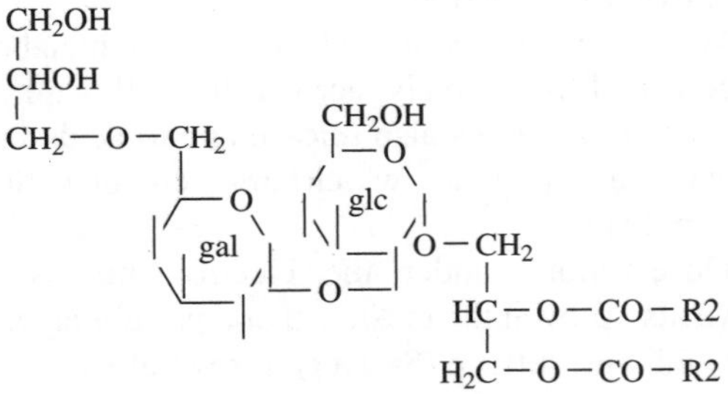

$$
\begin{array}{l}
CH_2OH \\
| \\
CHOH \\
| \\
CH_2\!-\!O\!-\!CH_2
\end{array}
$$

Fig. 4.5. Formula of a glycolipid

Fatty acids possess a long hydrocarbon chain and a terminal carboxylic acid function. These molecules are characterized by the length of their chain, their level of unsaturation, the *cis* or *trans* conformation of the double bonds, and (for Gram-positive) the *iso* or *anti-iso* ramification:

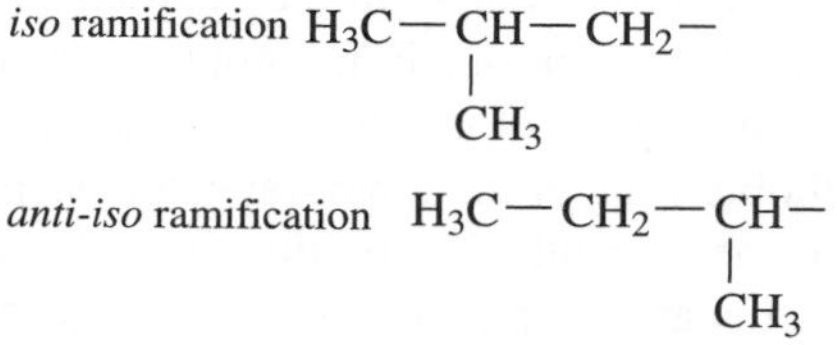

iso ramification $H_3C - CH - CH_2 -$
$\qquad\qquad\qquad\qquad |$
$\qquad\qquad\qquad\quad CH_3$

anti-iso ramification $H_3C - CH_2 - CH -$
$\qquad\qquad\qquad\qquad\qquad\qquad |$
$\qquad\qquad\qquad\qquad\qquad\quad CH_3$

In bacteria, most fatty acids have 14 to 20 carbon atoms and are saturated or mono-unsaturated. Lactic acid bacteria also contain a characteristic cyclopropanic acid: lactobacillic acid (*cis*-11,12-methylene-octodecanoic). Table 4.1 lists the principal fatty acids of lactic acid bacteria found in wine—notably *Oenococcus oeni* (Lonvaud-Funel and Desens, 1990). Malonyl CoA and acetate condense to form fatty acids with an even number of carbon atoms. For an odd number of carbon atoms, fatty acids are synthesized by the condensation of malonyl CoA and propionate. Anaerobic bacteria synthesize unsaturated acids by the action of a dehydratase on hydroxydecanoate which is formed by the addition of a malonyl unit on an octanoic acid molecule.

The following reactions are given very schematically:

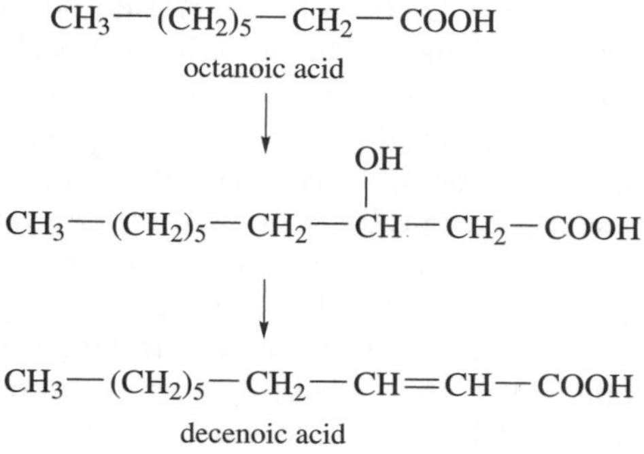

$$CH_3 - (CH_2)_5 - CH_2 - COOH$$
octanoic acid

$$\downarrow$$

$$CH_3 - (CH_2)_5 - CH_2 - \overset{\displaystyle OH}{\overset{\displaystyle |}{CH}} - CH_2 - COOH$$

$$\downarrow$$

$$CH_3 - (CH_2)_5 - CH_2 - CH = CH - COOH$$
decenoic acid

The progressive elongation of this acid leads to the formation of *cis*-vaccenic acid (C_{18}), a precursor of lactobacillic acid (C_{19}). In this last step, the double bond of the unsaturated acid (the precursor) is methylated to form the corresponding cyclopropanic acid. The fatty acid composition of the bacteria lipids varies during the physiological cycle and is also strongly influenced by several environmental factors.

Finally, besides polar lipids, the bacteria membranes contain neutral lipids, analogous to sterols in eucaryotes. These triterpenic and pentacyclic molecules are called hapanoids. They are formed by the cyclization of squalene in an anaerobic process. They have not been clearly identified in lactic acid bacteria.

The membrane is even more vital to bacteria than the cell wall. Numerous proteins in the membrane ensure essential enzymatic functions such as substrate and metabolic product transfers and the

Table 4.1. Principal fatty acids of lactic acid bacteria

		Chain
Myristic acid	tetradecanoic	C14:0
Palmitic acid	hexadecanoic	C16:0
Palmitoleic acid	*cis*-9-hexadecanoic	C16:1 Δ 9
Stearic acid	octadecanoic	C18:0
Oleic acid	*cis*-9-octadecanoic	C18:1 Δ 9
cis-Vaccenic acid	*cis*-11-octadecanoic	C18:1 Δ 11
Hydrosterculic acid*	*cis*-9-10-methylene octadecanoic	C19 cyc-9
Lactobacillic acid*	*cis*-11-12-methylene octadecanoic	C19 cyc-11

*These two acids contain a cyclopropanic cycle $\quad - CH - CH -$
$$\qquad\qquad\qquad\qquad\qquad\qquad\qquad\qquad\qquad\quad \underset{\displaystyle CH_2}{\backslash \;\; /}$$

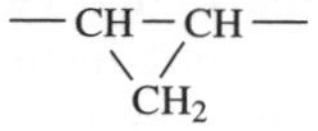

ATPase system. Lactic acid bacteria do not have a respiratory system. The selective permeability ensured by the membrane creates a transmembrane electrochemical proton gradient between the inside and outside of the cell. This difference generates electrochemical energy used in the synthesis of ATP. Moreover, the membrane maintains an optimum cellular pH for the functioning of numerous reactions of the cellular metabolism. It constitutes a barrier whose optimal functioning is guaranteed by the fluidity. The fluidity determines the specific activity of the proteins according to the lipidic environment, but this fluidity must be controlled for the membrane to remain an effective barrier between the cytoplasm and the environment. During the cell growth cycle and in response to multiple external parameters such as temperature, pH and the presence of toxic substances (ethanol), the cell manages to modify membrane composition to adapt to and resist environmental effects. The physical properties of the membrane are maintained at least as long as the stress factor remains within certain limits. The mechanisms put into play act together on the same properties. They affect the average length and the unsaturation, ramification and cyclization level of fatty acid chains, the proportion of neutral and polar lipids and the quantity of proteins. In this way, from the growth phase until the stationary phase, cis-vaccenic acid diminishes greatly to the point where it represents less than 10% of the total fatty acids, whereas lactobacillic acid attains a proportion of 55% in Oenococcus oeni, Lactobacillus plantarum and Pediococcus damnosus (Lonvaud-Funel and Desens, 1990).

The effect of temperature on membrane composition is one of the most understood effects. At low temperatures, the fatty acid unsaturation rate increases as does the proportion of acids with ramified chains. At the same time, the length of the chains decreases. In this manner, palmitic acid (C_{16}) increases and cis-vaccenic and lactobacillic acid decrease in Oenococcus oeni and Lactobacillus plantarum when the temperature of the culture increases from 25 to 30°C. The introduction of a methyl group, the formation of a propanic cycle, has the same effect on the physical properties of bacteria as a double bond. The inverse phenomena occur when the culture temperature is higher. The unsaturated fatty acids are less abundant. Neutral lipids also participate in cell adaptation to the medium by increasing membrane viscosity.

The presence of ethanol in the medium provokes significant modifications in membrane structure. It exerts a detergent effect by intercalating in the hydrophobic zone of the membrane, whose polarity increases as a result. The fluidity is increased and the proteins are denatured. In general, an increase in the unsaturated/saturated fatty acid ratio is observed. In Oenococcus oeni, this ratio increases from 0.4 to 2.1, when bacteria are cultivated in the presence of 9% ethanol. The results are the same for the species Lactobacillus hilgardii whose strains, like Oenococcus oeni, are capable of growing better than other species in an alcoholic medium (Desens, 1989).

The membrane proteins also participate in cell response to an environmental change. The stress proteins in microorganisms are becoming better known. Their synthesis is increased, for example, by temperature, acidity or the concentration in ethanol. Certain proteins also change when the cell enters the stationary phase. Several families of these proteins have been constituted and the specific functions of some of them have been identified. Their overexpression in the cell is related to a better resistance to stress factors. Their induction by heat shock protects the cell not only against the toxic effect of heat but also against the effect of other factors, such as ethanol and acidity. In certain wine lactic acid bacteria, especially Oenococcus oeni, the proteins exist but their role is not known. Their synthesis is increased when wine is added to their culture medium or when the cells are directly inoculated into wine. The concentrations found in Oenococcus oeni have been found to be up to five times higher than in other species. (Garbay, 1994). Among these, two proteins have been identified and coded by the Omr A and Fts H genes (Bourdineaud et al., 2003a, 2003b).

4.1.3 The Cytoplasm

The cytoplasm contains the main elements for cell operation: enzymes, nuclear material and sometimes reserve substances. The entire metabolism—both degradation reactions (catabolism) and synthesis reactions (anabolism)—is carried out in a programmed manner according to exchanges with the external environment, to produce the energy necessary for cell growth.

Coded by the genome, the cytoplasmic proteins are always the same for any given bacterial strain, but for some of them their level of expression varies with cultural conditions. Stress proteins, produced by drastic changes in conditions, have also been identified. One of those produced in *O. oeni, Lo18*, has been particularly studied (Delmas *et al.*, 2001). The electrophoretic profile of the soluble proteins of the cell can therefore be used as an identification method by comparison with established strains.

Cytoplasmic granulations can be revealed by specific coloration techniques. They are insoluble reserve substances of an organic nature: polymers of glucose or of the polyester of β-hydroxybutyric acid. These reserve substances accumulate in the event of a nitrogen deficiency, when a source of carbon is still present. Inclusions of volutin (a polymer of insoluble, inorganic phosphate) are characteristic in lactic acid bacteria, especially certain species of the genus of strictly homofermentative *Lactobacillus*. Volutin comprises a phosphate reserve available for the synthesis of phosphorylated molecules such as nucleic acids.

Under the transmission electron microscope, the interior of the bacterial cell appears granular. This is due to the ribosomes, which are essential players in protein synthesis. They ensure, along with the t-RNA, the translation of the genetic code. The ribosomes consist of two parts characterized by their sedimentation speed, expressed in Svedberg values (S). These two sub-units are different in size: 30 S and 50 S in procaryotes. The assembled ribosome has a sedimentation constant of 70 S. The 30 S sub-unit contains a 16 S ribosomal RNA molecule (1542 nucleotides) and 21 different protein molecules; its molecular mass is 900 KDa. The larger 50 S sub-unit contains two ribosomal RNA molecules, a 23 S and a 5 S molecule (2904 and 120 nucleotides, respectively). It also contains 35 proteins and has a molecular mass of 1600 KDa. During protein synthesis at the translation step, the sub-units (at first separated) reassemble. The genes encoding proteins and ribosomal RNA are known for the bacteria *Escherichia coli*. They are organized in operons, ensuring the control of the synthesis of ribosomal components. The operon of genes encoding the rRNA have the following structure:

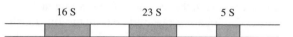

16 S   23 S   5 S

The nucleotide sequences of these genes, especially those of rRNA16S, are known for many species and identified in gene banks. Sequence comparison forms the basis of molecular identification methods.

4.1.4 The Nucleus and Genetic Material

The bacteria nucleus consists of a single circular chromosome of double stranded DNA suspended in the cytoplasm without any separation. Its size varies depending on the species. In *Lactobacillus plantarum*, its length is about 2400 kb. It is much smaller in *Oenococcus oeni* (about 1400 kb) and *Pediococcus pentosaceus* (1200 kb) (Daniel, 1993). The chromosome carries the essential genetic information of a cell.

Other more or less vital functions are determined by plasmids. These small, circular DNA molecules are completely independent of the chromosome. They vary in size and number depending on the species and strain of bacteria. In *Oenococcus oeni*, plasmids of 2 to 40 kb are often identified and one of them has been sequenced (Fremaux *et al.*, 1994). So far, no function of enological or physiological interest has been attributed to them. In general, the plasmids determine functions such as the fermentation of certain sugars, the hydrolysis of proteins, resistance to phages, antibiotics, heavy metals, etc.

In wine lactic acid bacteria of the species *Pediococcus damnosus*, a plasmid has been identified as a determinant of polysaccharidic synthesis. Strains that contain it are responsible for ropiness in wines (Lonvaud-Funel *et al.*, 1993). This plasmid has been entirely sequenced. It has three coding regions, one of which is probably responsible for synthesizing the expolysaccharide (Section 5.4.4) as it probably codes for a glucosyl transferase (Walling, 2003). Characteristically, plasmids are relatively unstable from one generation to the next, but some, in *Oenococcus oeni* manifest an immense stability. Others are easily lost in the absence of environmental pressures. Conjugative plasmids can naturally transfer from one strain to another, though this property has never been demonstrated for the lactic acid bacteria of wine. A strain's plasmidic profile can therefore vary.

4.1.5 Multiplication of Bacteria

All bacteria multiply by binary division (Figure 4.1). A cell gives two completely identical daughter cells. Multiplication supposes, on the one hand, division of nuclear material, and on the other hand, synthesis for the construction of new cellular envelopes and cytoplasmic elements, in particular ribosomes and enzymes.

The genetic material is transmitted after the duplication of the chromosomal DNA and the potentially existing plasmids. DNA replication, according to the semi-conservative mechanism, leads to the formation of two molecules that are identical to the parental chromosome or plasmid. The replication occurs almost during the entire cellular cycle at the mesosomes. When it is finished, the scission of the cytoplasm begins.

A septum is formed in the middle of the cell as a result of the synthesis of portions of the membrane and the cell wall. It separates the mother cell little by little into two daughter cells. The genetic material and the other cellular components are simultaneously distributed between them. Finally, when the septum is completely formed, the two daughter cells separate. Cell and nucleus division are not synchronous; replication is quicker. Moreover, a replication cycle can start before cell division is completed. For this reason, bacteria cells in their active growth phase contain more than one chromosome per cell. During division, plasmids (much smaller than the chromosome) are not always correctly distributed between the cells after their replication, hence their instability over generations.

4.2 TAXONOMY

The objective of taxonomy is to identify, describe and class microorganisms. Classification is made according to several hierarchical levels. For bacteria, the highest level corresponds with their classification among procaryotes. The lowest level is species. In a species of bacterium, strains grouped together share a number of identical characters. These characters radically differentiate them from other strains.

Lactic acid bacteria belong to the Gram-positive group, based on color tests (Section 4.3.2). The primary product of their metabolism of glucose is lactic acid.

4.2.1 Phenotypic Taxonomy, Molecular Taxonomy and Phylogeny

Phenotypes include morphological, physiological, biochemical and immunological characters as a whole and the composition of certain cellular components. Certain phenotypic characters appear to vary in a given strain—for example, the assimilation of certain sugars. Certain strains having different phenotypes but belonging to the same species are atypical strains.

Progress in molecular biology provides new classification criteria based on genome analysis. Molecular taxonomy consist of classifying bacteria according to similarities in their genome. Diverse methods exist, permitting several levels of classification.

A first level takes into account the percentage of guanine and cytosine bases in the DNA—the $(G + C)\%$ with respect to the total number. Two strains are not necessarily related because they have the same $(G + C)\%$. In fact, the base composition does not give any indication of the DNA sequence. Among Gram-positives, lactic acid bacteria belong to the phylum *Clostridium*. The

Table 4.2. Sub-division of Gram-positive bacteria according to (G + C)%

(G + C)%	<50%	>50%
Phylum	*Clostridium*	*Actinomycetes*
Genera	*Lactobacillus* *Leuconostoc* *Pediococcus* *Oenococcus* *Weissella*	*Bifidobacterium* *Brevibacterium*
	Carnobacterium *Lactococcus* *Streptococcus* *Vagococcus* *Enterococcus* *Carnobacterium*	*Corynebacterium* *Microbacterium* *Propionibacterium*

(G + C)% of this phylum is less than 50%. The *Actinomycetes* whose (G + C)% is greater than 50% include other bacteria that are also important to the food and beverage industries (Table 4.2). The *Clostridium* branch consists of three groups: the first includes the *Lactobacillus, Pediococcus, Leuconostoc, Oenococcus,* and *Weissella* genera; the second, *Streptococcus* and *Lactococcus*; and the third, *Carnobacterium, Vagococcus and Enterococcus* (Gasser *et al.*, 1994).

Dicks *et al.* (1995) proposed a new species, *Oenococcus oeni*, for bacteria previously known as *Leuconostoc oenos*, currently the only species in the *Oenococcus* genus. This proposition was based on the phylogenetic distance of *O. oeni* with respect to other lactic acid bacteria.

The homology of genomic DNA permits the definition of bacterial species by their nucleotide sequence. The homology is measured by the reassociation percentage between strands of DNA from the strain to be classed and a type strain of the species. The strands are isolated by DNA denaturation. Two bacterial strains belong to the same species if the hybridization percentage is greater than 70%. For a lesser value, the strains are part of the same genus, on the condition that the hybridization remains measurable.

Nucleotide sequence comparison can be carried out on portions of the genome rather than the entire genome. The chosen portions correspond with essential functions, common to the strains to be compared. The genes encoding the synthesis of ribosomes, particularly for the ribosomal RNA (a conserved molecule), are a noteworthy example. Several types of analysis are possible. In one of them, the 16 S RNA is affected by the action of a restriction endonuclease. The tiny oligonucleotide fragments smaller than 20 bp are separated by electrophoresis and then sequenced, permitting the creation of a data bank. Sequences specific to groups of bacteria can be identified in this manner. The similarity coefficient between strains can also be defined.

Another type of analysis consist of sequencing the 16 S, 23 S, and 5 S RNA. The 16 S RNA sequence can provide the most interesting indications. It contains conserved zones and variable zones: the comparison of conserved zones is valid for distantly related bacteria; variable zone comparison can be used on closely related bacteria.

Sequencing of the 16 S rRNA divides lactic acid bacteria into three phylogenetic groups:

1. The group *Lactobacillus delbrueckii* contains this species and other strictly homofermentative lactobacilli.

2. The group *Leuconostoc* is divided into two subgroups: one containing *L. paramesenteroides* and heterofermentative lactobacilli; the other containing *Leuconostoc sensu stricto*, to which *Oenococcus oeni* belongs (although individualized).

3. The group *Lactobacillus casei—Pediococcus* is a more heterogeneous group since it comprises strictly and facultatively heterofermentative species and strictly homofermentative species.

Grouping by means of 16 S rRNA sequences is based on phylogenetic relationships between bacteria. It does not support the grouping realized by using phenotypes, such as morphology and physiology. At present, therefore, it is difficult to class lactic acid bacteria if referring to both the phenotype and the genome. Physiological and biochemical criteria remain useful, but the contribution of molecular taxonomy is considerable and seems more absolute since it is directly related to the genetic heritage of a strain.

4.2.2 Classification of Wine Lactic Acid Bacteria. Description of Genera

The lactic acid bacteria of grape must and wine belong to the genera *Lactobacillus, Leuconostoc, Oenococcus* and *Pediococcus*. Besides their morphology in coccal or rod-like forms, the homofermentative or heterofermentative character is a deciding factor in their classification. Homofermentative bacteria produce more than 85% lactic acid from glucose. Heterofermentative bacteria produce carbon dioxide, ethanol and acetic acid in addition to lactic acid.

Among the cocci, the bacteria from the genus *Pediococcus* are homofermenters and those from the genera *Leuconostoc* and *Oenococcus* are heterofermentative. The lactobacilli can present the two behaviors. They are divided into three groups:

- Group I: strict homofermenters (this group has never been identified in wine).

- Group II: facultative heterofermenters.

- Group III: strict heterofermenters.

The strictly homofermentative lactobacilli do not ferment pentose, and form two molecules of lactic acid from one molecule of glucose by the Embden–Meyerhoff pathway.

In facultative heterofermenters (Group II), one glucose molecule, as in the case of Group I, leads to two molecules of lactic acid, but the pentoses are fermented into lactic and acetic acid by the heterofermentative pentose phosphate pathway. The strictly heterofermentative bacteria in Group III do not possess the fructose 1,6-diphosphate aldolase that is characteristic of the Embden–Meyerhoff pathway. They ferment glucose into CO_2, lactic and acetic acid, and ethanol by the pentose phosphate pathway, and pentose into lactic and acetic acid in the same manner as bacteria from Group II.

Table 4.3 lists the lactic acid bacteria most often encountered in grape must and wine. *Oenococcus oeni* is known for ensuring malolactic fermentation in the great majority of cases. So far, the strictly homofermentative lactobacilli of Group I have not been isolated in must or wine. The species are therefore divided into facultative and strict heterofermenters for lactobacilli and into homofermenters (*Pediococcus*) and heterofermenters (*Leuconostoc*) for cocci. It is likely that this classification will be modified—on one hand due to anticipated progress in the identification of new species in wine, and on the other hand due to eventual reclassifications of lactobacilli in the groups described above.

No lactic acid bacteria possess cytochrome. The catalase activity is generally assumed not to exist, but several species of bacteria (*Lactobacillus, Pediococcus* and *Leuconostoc*) can synthesize a manganese-dependent, non-hemic pseudocatalase. A hemic catalase activity has been identified in many strains.

The following is a general description of three genera of lactic acid bacteria in wine.

Table 4.3. List of the most widespread lactic acid bacteria species in grape must and wine

Lactobacilli	Facultative heterofermenters (Group II)	*Lactobacillus casei* *Lactobacillus plantarum*
	Strict heterofermenters (Group III)	*Lactobacillus brevis* *Lactobacillus hilgardii*
Cocci	Homofermenters	*Pediococcus damnosus* *Pediococcus pentosaceus*
	Heterofermenters	*Leuconostoc oenos* (*Oenococcus oeni*) *Leuconostoc mesenteroides* subsp. *mesenteroides*

Genus *Leuconostoc Oenococcus*

- Non-mobile, non-sporulating, spherical or slightly elongated cells, assembled in pairs or small chains; diameter 0.5–0.7 μm, a length 0.7–1.2 μm.

- Facultative anaerobiosis.

- Chemo-organotroph: requires a rich medium and fermentable sugars.

- Optimum growth temperature 20–30°C.

- Metabolic products of glucose: CO_2, lactic acid and ethanol.

- Arginine is metabolized by certain strains of *Oenococcus oeni*, whereas other *Leuconostoc* species respond negatively to this test.

- (G + C)% from 38 to 44%

- No teichoic acid.

Genus *Pediococcus*

- Non-mobile, non-sporulating, sometimes isolated, spherical (never elongated) cells; diameter 1–2 μm; division in two right-angled planes which leads to the formation of tetrads—no chains.

- Facultative anaerobiosis.

- Chemo-organotroph: requires a rich medium and fermentable sugars.

- Metabolic products of glucose: DL or L lactic acid, no CO_2.

- (G + C)% from 34 to 42%.

- No teichoic acid.

Genus Lactobacillus

- Non-mobile, non-sporulating, regular elongated cells. 0.5–1.2 μm by 1.0–10 μm, often long rod-like forms. Some are very small (nearly the same dimensions as *Leuconostoc*). Assembled in pairs or in variably sized chains.

- Facultative anaerobiosis.

- Chemo-organotroph: requires a rich medium and fermentable sugars.

- Fermentative metabolism: at least half of the products of the metabolism of glucose is lactic acid. The homofermentative metabolism leads to this sole molecule. The heterofermentative metabolism also produces acetic acid, ethanol and CO_2.

- (G + C)% from 36 to 47%.

- Many species contain teichoic acid in the cell wall.

4.3 IDENTIFICATION OF LACTIC ACID BACTERIA

4.3.1 General Principles

Since the beginning of microbiology, the identification of bacteria has been based on their phenotypic characters (Section 4.3.2). Besides by its morphology, which gives little information, a strain is identified essentially by the substrates and products of its metabolism. When more discriminating analytical methods appeared, the chemical composition of microorganisms (fatty acids and proteins, Section 4.3.8) also participated in their identification.

More recently, and in a spectacular manner, the tools of molecular biology have made the identification even more precise at the genus and species level and even within the same species. For a long time, lactic acid bacteria of wine were identified by their phenotypes. Now, with DNA analysis, more reliable results are obtained (Sections 4.3.3–4.3.6).

Identification by phenotypic analysis of clones isolated in wine often poses two kinds of problems. First, these clones are difficult to multiply in laboratory conditions. Numerous sub-cultures are needed to obtain a sufficient biomass to carry out all of the tests. For the same reasons, the response to biochemical tests in the API tests (Section 4.3.2) can be ambiguous. The change of color of the indicator is not distinct if the strain does not multiply sufficiently in the microtube. Second, the phenotypic character, such as the assimilation of a substrate or the formation of a particular product, represents the result of a metabolic chain that depends on the entirety of cell enzymatic activity. For a phenotype to be positive, all of the genes encoding the enzymes of the particular chain must be expressed; the enzymes must also be functional. The induction or repression of enzyme synthesis as well as inhibitions due to certain medium conditions can therefore modify a phenotype.

The DNA composition of strains is strictly specific. It is not influenced by culture conditions. It can, however, undergo punctual mutations over generations. At the laboratory culture scale, these mutations do not significantly affect the genomic DNA characteristics. Strain identification by genomic analysis therefore appears to be the most reliable approach. Several types of analysis exist which permit different levels of identification: strain, species, genus.

The general principle consists of looking for similarities between the DNA of the unidentified strain and the DNA of the reference strain. There are several methods based on various tools and properties of the DNA molecule. The study of restriction polymorphism is based on the specific action of restriction enzymes. Hybridization is based on the ability of single-strand DNA chains to reassemble in double-strand chains. Combining these two methods and varied uses of each method considerably broaden the possibilities of analysis.

Finally, polymerization chain reaction (PCR) permits the amplification of portions of the genome delimited by markers. These markers are primers (oligonucleotides) which must hybridize with the DNA matrix for amplification to start. Depending on the primers chosen, the electrophoretic profile of the amplicon obtained can, permit different levels of classification within the genus or the species.

4.3.2 Phenotypic Analysis

Phenotypic analysis encompasses morphology, the assimilation of diverse substrates and the nature of metabolic products.

Morphological observations can be made with fresh cells but they are more precise with fixed preparations. Microscopic observation can be coupled with the Gram coloration test, which is used to verify that bacteria are Gram-positive. After the bacteria are placed on the slide and dried by the flame of a Bunsen burner, the preparation is dipped first in a violet colorant, then in an alcohol–acetone solution, and finally in a rose colorant. The cell wall of Gram-positive bacteria is not altered by the organic solvent: these bacteria retain the violet coloration. Conversely, Gram-negative bacteria are rose colored. Cell form, whether coccal or rod-like, is easy to identify, as is cell arrangement (pairs, tetrads, small chains).

Secondly, the homofermentative or heterofermentative character is determined. The unidentified strain is cultivated in a medium with glucose as the energy source. After cell growth, the metabolic products are characterized and measured. A release of CO_2 manifests the heterofermentative character of the strain. This result is regularly confirmed by measuring acetic acid and ethanol concentrations. Their presence is also proof of a heterofermentative metabolism. Conversely, the exclusive formation of lactic acid attests to a homofermentative character. In culture conditions, facultative heterofermentative bacilli (for example, *Lactobacillus plantarum*) have a homofermentative metabolism with respect to glucose.

During the same test, it is interesting to determine the optical nature of lactic acid formed from glucose. This analysis makes use of an enzymatic process. The two stereoisomers of lactic acid: (L and D) are analyzed separately. This form of analysis is particularly adapted to the identification of heterofermentative cocci (*Oenococcus oeni*, *Leuconostoc mesenteroides*), which only form the D isomer, and of *Lactobacillus casei*, which only forms L-lactic acid.

These initial investigations permit bacterial identification at the genus level: *Lactobacillus* by morphology, and *Pediococcus* and *Leuconostoc* by morphology and determination of their homofermentative or heterofermentative character. Classification at the species level makes use of the analysis of the fermentation profiles of a large number of sugars.

For this type of analysis, the API 50 CHL identification system (Bio-Mérieux) is commonly used. In this system, the classic tests that were once performed in test tubes are miniaturized. The unidentified strain is inoculated in to a nutritive medium that contains all of the nitrogen-based nutrients, vitamins and salts necessary for its growth. Different carbohydrate energy sources are represented in each microtube of the system. In this manner, 49 substances are tested, including hexoses, pentoses, disaccharides, etc. An indicator in the culture medium, which changes color, facilitates the reading of results. Fermentation in a microtube acidifies the medium, provoking the indicator to change color.

To carry out the API test, 0.1 ml of bacterial suspension is deposited in each of the 50 microtubes. The tubes are sealed with a drop of paraffin to ensure anaerobiosis. Generally, the system is incubated at 25°C for 24 hours. Tubes in which the color changes from blue to yellow indicate positive characters. In this manner, the fermentation profile of the examined strain can be established (Figure 4.6).

This method is well adapted for the identification of numerous lactic acid bacteria, but it must be carried out very carefully with bacteria isolated in wine. Experience has shown that the strain should undergo several successive sub-cultures in the standard laboratory medium beforehand.

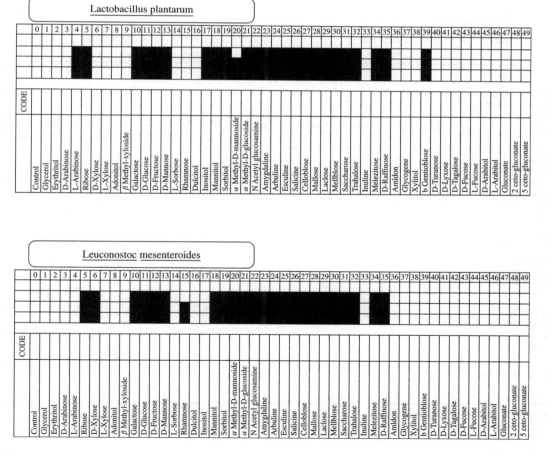

Fig. 4.6. Biochemical substrate assimilation profiles (API 50 CHL gallery) of two lactic acid bacteria species

This preparation is essential for obtaining profile stability, which is indispensable before referring to the identification key. In any case, a strain cannot be identified solely on these results. The method's discriminating powers are not sufficient and the similarity in profiles of *Lactobacillus plantarum* and *Leuconostoc mesenteroides* demonstrates this point. All of the other phenotypes previously described should also be taken into consideration at the same time. These characters as a whole make the determination of a species possible without too much ambiguity, by referring to *Bergey's Manual of Determinative Bacteriology* (1986).

Tables 4.4 and 4.5 summarize the phenotypic characters used to determine genus and species. Most of the strains for each species have

Table 4.4. Determination of the genus of lactic acid bacteria isolated in wine

Morphology	Round cells (cocci)		Elongated cells (bacilli)
Cell arrangement	Pairs Small chains	Tetrads	Pairs or small chains
Glucose fermentation	Heterofermentative	Homofermentative	Heterofermentative or homofermentative
Lactic acid stereo isomer	D	L or D, L	L, D or D, L
Genus	*Leuconostoc* *(Oenococcus)*	*Pediococcus*	*Lactobacillus*

Table 4.5. Determination of wine lactic acid bacteria species and similar species (*Bergey's Manual of Systematic Bacteriology,* 1986)

Species	Fructose	Glucose	Galactose	Glycerol	Arabinose	Ribose	Xylose	Esculine hydrolysis	Lactic acid isomer	Arginine dihydrolase
Lb. casei	+	+	+	−	−	+	−	+	L	−
Lb. homohiochii	+	+	−		−	+	−	−	L	−
Lb. pentosus	+	+	+	+	+	+	v	+	DL	−
Lb. plantarum	+	+	+	−	v	+	v	+	L	−
Lb. rhamnosus	+	+	+	(+)	v	+	−	+	DL	−
Lb. sake	+	+	+	−	+	+	−	+	DL	−
Lb. brevis	+	+	v	−	+	+	+	v	DL	+
Lb. fructivorans	+	+	−	−	−	+	−	−	DL	+
Lb. fructosus	+	+	+	−	−	−	−		D or L	+
Lb. hilgardii	+	+	v	−	−	−	+	−	DL	+
Lb. kefir	+	+	−		v	+	−	−	DL	+
Lb. sanfrancisco	−	+	+		−	−	−		DL	−
Pc. acidilactici	+	+	+	−	v	+	+		L ou DL	+
Pc. damnosus	+	+	+	−	−	−	−		DL	−
Pc. parvulus	+	+	+	−	v	−	−		DL	v
Pc. pentosaceus	+	+	+	−	+	+	v	+	DL	+
Ln. mesenteroides subsp. *mesenteroides*	+	+	+		+	+	v	+	D	−
O. oeni	+	v	v		v		v	+	D	−

⁺, positive; ⁻, negative; ᵛ, variable.
Lb. = *Lactobacillus.*
Ln. = *Leuconostoc.*
O. = *Oenococcus.*
Pc. = *pediococcus.*

fermentative profiles which correspond with those listed in Table 4.5. Nevertheless, besides differences that can be introduced by the pre-culture of the strain before the test, the more pronounced intraspecific variability of certain characters must be taken into account. For example, *Leuconostoc* (in particular *Oenococcus oeni*) was long thought not to possess the arginine hydrolysis character, but recent studies have proven that numerous strains of *Oenococcus oeni* possess all of the necessary enzymatic equipment to hydrolyze arginine, leading to the production of citrulline, ornithine and carbamyl phosphate (Liu *et al.*, 1994; Makaga, 1994). The hydrolysis activity depends on environmental conditions which determine not only the enzymatic activity but also its synthesis. Furthermore, the "arc" operon, containing genes coding for the various enzymes in the metabolic pathway, has been identified and sequenced in strains of *O. oeni* (Tonon *et al.*, 2001).

The use of fermentative profiles as an identification method should therefore be standardized. Bacterial characters can then be expressed in the most reproducible manner possible. Finally, these tests place bacteria in optimal growth and metabolic conditions—they give no indication of their true metabolism in wine. The fermentation of a carbohydrate found in the API 50 CHL system can be totally impossible in wine: its nutritional conditions are far from those in the synthetic medium. Conversely, a substrate that cannot be metabolized in optimal conditions can be metabolized in wine because of the totally different metabolic regulations in play.

4.3.3 Extraction and Visualization of DNA for Genomic Study

Before analysis, the entire genomic DNA of bacteria must be separated from the lipids, glucides and proteins constituting the cell. The extraction protocols for lactic acid bacteria include all of the stages of cell lysis, deproteinization and DNA precipitation. There are slight differences between the protocols (Lonvaud-Funel *et al.*, 1989). Lysis of Gram-positive cells is obtained by the action of lysozyme. The peptidoglycans are hydrolyzed

to form protoplasts, which are submitted to the action of SDS—a powerful detergent that destroys the membranes and liberates the cellular contents. The addition of phenol, most often mixed with chloroform and isoamylic alcohol, precipitates the proteins. The organic and aqueous phases are separated by centrifugation. The denatured proteins assemble together at the interphase. The lower phenolic phase contains the lipids and proteins; the upper phase contains the dissolved DNA.

The phenol is eliminated from this phase by successive extractions with a mixture of chloroform, alcohol, and isoamylic alcohol. The DNA is precipitated by ethanol in the presence of salts. It can be stored at −20°C after being dissolved in a buffer.

Electrophoresis is the most popular, simple and reliable technique for analyzing DNA extract. At an alkaline pH, the DNA phosphate groups are ionized. The molecules placed in an electric field therefore migrate towards the anode. In a viscous gel (most often agarose), the electrophoretic mobility depends on the size and conformation of the molecules. The smaller the linear molecules are, the faster they migrate. Circular molecules of equal size are less mobile than linear molecules. Plasmids, for example, exist in circular and coiled circular form. The size of linear DNA is easily calculated from the migration distance of DNA weight markers. There is an inverse relationship between the mobility and the logarithm of the molecule size. Molecules separated by electrophoresis are revealed by an ethidium bromide based coloration. This compound is an analogue of an aromatic base; it intercalates in the DNA and fluoresces orange under ultraviolet light.

4.3.4 Identification Based on Restriction Fragment Length Polymorphism

This method consists of hydrolyzing the DNA with the help of restriction enzymes. These enzymes produce different sized fragments which are separated by electrophoresis. The electrophoretic profile differs depending on the strain. The enzymes, in fact, act on specific sites, recognizing palindromic sequences of generally four to seven nucleotides on the two DNA strands.

The *Eco* RI enzyme, for example, hydrolyzes DNA according to the following schema:

$$5' - G \mid AA \ TT \quad C - 3'$$
$$3' - C \quad TT \ AA \mid G - 5'$$

Depending on the sequence recognized, the number of sites cut on the polynucleotide varies, but it is always identical for a given enzyme and nucleotide. On the one hand, different fragments varying in size and number can be obtained from the same DNA by using a variety of restriction enzymes available. On the other hand, the restriction of different DNA by the same enzyme leads to different sized fragments.

The restriction products are analyzed by electrophoresis. The characteristic profiles are obtained after revelation by coloration. Considering the number and sequence of the nucleotides, enzymes that statistically cut the DNA too often will produce complex profiles that are difficult to study. If the number of cut sites is very low, the profiles are simpler but the length of the fragments produced requires the use of pulse-field electrophoresis in order to separate them. This technique is very reliable and well adapted for the identification of yeasts but remains very difficult to use for bacteria (Daniel, 1993).

Restriction polymorphism is not relevant for the identification of bacterial species. No profile type exists for *Oenococcus oeni*, for example, nor for each of the other species that are of interest in winemaking. This method seems better adapted to the differentiation of strains of the same species. It is thus easy to identify strains of *O. oeni*, following hydrolysis of their DNA by the *NotI* enzyme with rare restriction sites. This method is used to monitor development of inoculated selected strains, used during winemaking to promote malolactic fermentation. The restriction profile of the biomass collected in wine during malolactic fermentation is compared with that of the selected strain that was added (Gindreau *et al.*, 1997; 2003).

4.3.5 Identification by Specific Probe DNA/DNA Hybridization

Hybridization is a technique often used in molecular genetics and it is very well adapted for the identification of species and even strains. The technique is based on the ability of double-strand DNA to separate, reversibly, into two single strands, in certain conditions that destroy their hydrogen bonds. Along with the ionic force of the medium, the temperature is the determining parameter of DNA denaturation. A temperature increase provokes the separation of the two strands, which reassociate when the temperature diminishes once again.

In favorable environmental conditions, the chains can reassociate if the nucleotide sequences present are complementary. For example, the oligonucleotides of the following sequences recombine: 5'-ATGCAATTGGCC-and 3'-TACGTTAACCGG-.

A hybridization consists of two single strands, each coming from different cells, reassociating due to their complementary sequences. This property is used for identification and strains are considered to belong to the same species if they have a 70% homology of their genomic DNA sequence.

The DNA of a reference strain is needed to identify a strain by DNA hybridization. One of the two fragments (most often the reference DNA) constitutes the isotopic probe or the chemically derived base analogue labeled probe. A probe is a single-strand DNA fragment that combines with the complementary sequence of the target DNA.

These operations are schematized in Figure 4.7. The target DNA (the DNA of the unidentified strain) is fixed and denatured on a nylon membrane. The membrane is then placed in a hybridization buffer without probes. During this pre- hybridization step, all of the non-specific sites on the nylon are saturated by a mixture of macromolecules. These molecules have no affinity for the probe. The hybridization takes place in the same physicochemical conditions after the addition of the marked DNA probe. At the end of this step, the single strands of the DNA probe will have strongly combined with the complementary target DNA, but also more weakly with DNA having less similar sequences. Revelation is used to localize the target DNA that truly corresponds with

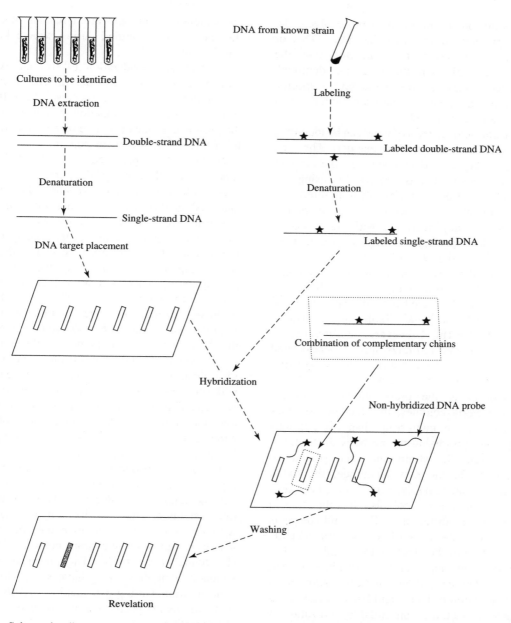

Fig. 4.7. Schematic diagram of the general principle for the identification of lactic acid bacteria using specific DNA/DNA probe

hybrids of strains of the same species. Therefore, this step is performed after the elimination of the probe and of the DNA strands that present little homology. Successive washes with ionic buffers of decreasing force are very important and participate in the specificity and the sensitivity of this method. The revelation process makes use of autoradiography for the isotopic probes and immunoenzymatic reactions for the non-isotopic probes most often used.

Depending on the problem to be resolved, the probe is prepared from either the entire DNA or a specific DNA fragment. In the first case, the species of an unknown strain can be identified. In the second case, strains possessing a specific gene and in consequence a characteristic functional property can be identified.

Most lactic acid bacteria of wine can be identified at the species level in this manner. The first working probe was developed for the species *Oenococcus oeni*. It was created by using the total DNA of different strains taken as references (Lonvaud-Funel *et al.*, 1989). The DNA probes of *O. oeni* do not hybridize with the genomic DNA of other species; the inverse is also true. The presence of the *O. oeni* species of bacterium can be identified even in a mixture containing other bacteria. Subsequently, this method also proved to be well adapted for other species found in grape must and wine (Table 4.6) (Lonvaud-Funel *et al.*, 1991b). However, the similarity of the *L. hilgardii* and *L. brevis* species necessitated the development of a more specific probe targeting *L. hilgardii* (Sohier *et al.*, 1999).

By hybridizing the probe directly with the DNA of bacteria colonies, this method becomes considerably more interesting than the previous technique of hybridizing the probe with the DNA extracted from the strain and then placed on a membrane. In the new technique, the nylon membrane is placed on the surface of a Petri dish after the development of colonies. It is then treated successively in different buffers and reagents which provoke the lysis of the bacteria and the immobilization of the DNA on the membrane. The prehybridization steps follow; hybridization and washes are then carried out. In these conditions, a mixed population of lactic acid bacteria can easily be studied. In fact, after an initial hybridization of the membrane with a given probe, dehybridization and then rehybridization with a second probe permit the localization of clones belonging to another species. At least five different species can successively be detected in a mixture with this system (Figure 4.8) (Lonvaud-Funel *et al.*, 1991a). Thanks to this method, by preparing probes representing the eight species most often

encountered in enology, the dynamics of each of the species were studied during winemaking for the first time.

DNA/DNA hybridization is also an excellent tool for identifying strains that differ in phenotype but belong to the same species. The difference rests on a metabolic function which depends on the presence of one or more enzymes and therefore the presence of the corresponding genes. In this case, the probe is prepared from a DNA fragment representing all or part of the gene.

At present in enology, two particular cases are analyzed in this manner: strains of *Pediococcus damnosus*, responsible for ropiness disease, and strains which produce histamine, notably *O. oeni*. Preliminary studies have shown that *P. damnosus* strains capable of synthesizing the 'ropy wine' polysaccharide possess a supplementary plasmid, contrary to normal strains. The ropy character is linked to the presence of this plasmid. A fragment was cloned in *E. coli* and now constitutes the base material for preparing the probe. In this manner, colony hybridization permits the identification of 'ropy' clones even when mixed with other *Pediococcus* clones or other species of bacteria. This method is routinely used to identify this undesirable population in the microflora of wines at the end of winemaking and during aging (Lonvaud-Funel *et al.*, 1993).

The other cloned specific probe is prepared from a gene fragment of histidine decarboxylase. This enzyme catalyzes the decarboxylation of histidine into histamine. The hybridization of a bacterium with this probe, to more than half of the gene length, signifies that the strain possesses the gene (Le Jeune *et al.*, 1995). The presence of these strains in wine most likely increases the histamine concentration. During winemaking, and also aging, these strains in specific can be counted by colony hybridization.

4.3.6 Identification by Polymerization Chain Reaction (PCR)

PCR consists of using polymerization to amplify one or more DNA fragments, located by specific sequences. The obtained product exists in sufficient

Table 4.6. Identification of wine lactic acid bacteria by probe specific DNA hybridization (Lonvaud-Funel *et al.*, 1991b).

n°	DNA Probes of reference strains	strain	1	2	3	4	5	6	7	8	9	10	11	12	13	14	15	16	17	18	19	20	21	22	23	24
1	Lactobacillus plantarum	CHL	+	−	−	−	−						−	−	−	−										
	Lactobacillus casei																										
2	subsp. rhamnosus	393		+	+	+	−							−	−	−										
5	subsp. casei	7469			+	+	+						−	−	−	−										
6	Lactobacillus homohiochii	15434						−	+	+	+	+												−		
7	Lactobacillus hilgardii	8290						+	−	−	−	−														
15	Leuconostoc mesenteroides	8293															+	+	−	−	−					
17	Oenococcus oeni	23279															−	−	+	+	+					
20	Pediococcus dextrinicus	33087																				+	−	−		
21	Pediococcus pentosaceus	33316																				−	+	+	−	−
23	Pediococcus damnosus	25248																					−	−	+	+

Target DNA (Strains to be identified)

+: hybridization; −: no hybridization

n° 1: *Lactobacillus plantarum*; n° 2, 3, 4, 5: *Lactobacillus casei*; n° 6: *Lactobacillus hilgardii*; n° 7, 8, 9, 10: *Lactobacillus homohiochii*; n° 11, 12, 13, 14: *Lactobacillus brevis*;
n° 15, 16: *Leuconostoc mesenteroides*; n° 17, 18, 19: *Oenococcus oeni*; n° 20: *Pediococcus dextrinicus*; n° 21, 22: *Pediococcus pentosaceus*; n° 23, 24: *Pediococcus damnosus*.

Sources: *L. plantarum* CHL (Chr. Hansen A/S) the ATCC (American Type Culture Collection).
Source of target strains: the IŒB collection (Faculté d'Œnologie de Bordeaux).

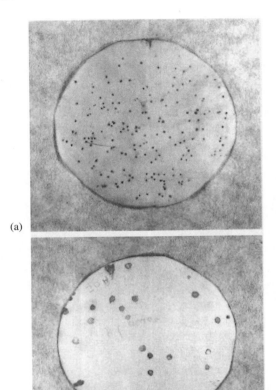

(a)

(b)

Fig. 4.8. Specific lactic acid bacteria population counts by hybridization on colonies with reference DNA probes. (a) hybridization on colonies of cultured *Oeno-coccus oeni*. (b) detection of *L. hilgardii* colonies in a mix of 5 species (*O. oeni*, *L. mesenteroides*, *P. damnosus*, *L. plantarum*, *L. hilgardii*). Result obtained after 4 successive hybridizations and dehybridizations with probes from four other species

of one of the primers serves as a template for the other after denaturation. The repetition of cycles comprising primer annealing, extension reactions and denaturation leads to an accumulation of identical neosynthesized molecules, flanked by the chosen oligonucleotides (Chapter 1, Figure 1.21).

For the three steps of the amplification to be successful and to avoid complicated manipulations of the sample during multiplication, a temperature-resistant enzyme is necessary. The use of the Taq polymerase resolved this problem; it is thermoresistant and functions at elevated temperatures up to 72°C. The cycles are therefore repeated, generally 30 to 40 times. A large quantity of specific DNA fragments determined by the primers are produced in this manner. Theoretically, 2^n target fragments are obtained after n cycles. The automatic equipment currently in use permits the different parameters of the three steps of each cycle to be programmed. From a single copy of a DNA fragment, a sufficient number of copies can be obtained in order to view easily after ethidium bromide coloration. The products of PCR are analyzed by electrophoresis. If a fluorescent band is revealed with the expected size, this means that the DNA matrix contained the sequence identified by the primers. Thus, within a genome with approximately two million nucleotides, like that of *Oenococcus oeni*, it is possible to determine whether a gene or gene fragment several hundred nucleotides long is present or not.

The specificity of PCR is based on the level of hybridization between the oligonucleotides and the template. It therefore depends on the primer sequence and length, the ionic force of the medium, the temperature and the concentration of Mg^{2+} ions. The choice of primers can be very precise when the sequence of the regions bordering the zone to be amplified are known. The use of random primers also gives valuable results, when the bacteria genome is totally unknown.

The primary value of PCR is its sensitivity. The presence of a gene in a small number of cells can generate a quantity of DNA which is easily analyzable by gel electrophoresis. In enology, PCR is also used to identify the two problems previously described, through the use

quantities to be easily revealed by electrophoresis. This method includes an enzymatic reaction for the synthesis of DNA which requires primers and a template. The polymerase copies the DNA target starting from the primer at 3′ towards 5′. The PCR technique uses two oligonucleotide primers, chosen for their complementary sequences: each one is complementary to a single strand of the DNA target. Synthesis is carried out between the two primers by a polymerase. The extensional product

of 'ropy' strain and histamine-producing strain-specific probes. The amplification reaction is very specific; the size of the amplified fragment is verified after electrophoresis. These undesirable strains of bacteria are therefore detected in a mixture of other bacteria, even if they are few in number PCR amplification using a region of the histidine decarboxylase gene makes it possible to identify bacteria likely to produce histamine, irrespective of the species (Coton *et al.*, 1998). Sequencing the plasmid of bacteria that produced glucane led to the identification of the *dps* gene responsible for this effect. Primers were selected for PCR detection of these bacteria directly in wine (Gindreau *et al.*, 2001). It is now also possible to detect lactic bacteria that break down glycerol to produce acrolein (Claisse and Lonvaud-Funel, 2001) as well as those that decarboxylate tyrosine to form tyramine.

PCR will soon have another application in our domain for the differentiation of strains of the same species. Random primers are used for the moment. In this case, the reactions amplify several zones of the bacterium genome. After electrophoresis, the amplification products furnish a profile that can be characteristic of the strain. The difficulty lies in finding the primers. The best adapted ones for recognizing strains must give a profile for each strain in a reproducible manner. Among the lactic bacteria in wine, this process has only been applied to strains of *O. oeni*. The main application is in monitoring selected bacteria for malolactic fermentation.

PCR is a useful tool, especially due to its great sensitivity and speed. This method complements the colony hybridization method by specific probes, and the two methods permit the early detection (PCR) and quantification (specific probes) of bacterial strains that alter wine. However, in the near future, the more recently developed quantitative PCR in real time is likely to provide quantitative data with all the accuracy and speed of PCR. In the future, other methods of genome analysis will probably permit the identification of species less common to wine with greater certainty. These species are interesting because of their metabolism or their resistance to wine

stabilization processes. Ribotyping, for example, consists of hybridizing the genomic DNA with a probe prepared from the DNA encoding ribosomal genes. Before this process, the genomic DNA must first be submitted to the action of restriction enzymes and undergo separation by electrophoresis. This method, which permits the analysis of simplified hybridization patterns, has been used to study a few strains of *L. hilgardii* and *L. brevis*. Profile types permit the classification of strains into two species that truly correspond with the habitual phenomena described (Le Jeune and Lonvaud-Funel, 1994).

4.3.7 Identification by Fatty Acid and Protein Composition

Besides their phenotypic characteristics, the protein and fatty acid composition of bacteria is also determined by the mass of information in the genome and can, therefore, be used for identification purposes. In both cases, these components result from a succession of genetically determined syntheses. Differences in the fatty acid and protein composition therefore reflect differences between strains. They can possibly even lead to identification of genus and species.

The total fatty acids are dosed in the form of esters after saponification. The analysis makes use of gas phase chromatography (Rozès, 1993). Even if this analysis is reliable, it must be used with caution for identification; in fact, several studies have clearly proven that, for a given lactic bacterium, the same fatty acids are always represented, but their proportion varies significantly according to the cellular cycle phase and even more so the physicochemical growth conditions. Modifications essentially concern the level of saturation and the length of the carbon chains. Moreover, for a given species, strains are capable of synthesizing very long-chain fatty acids (more than 20 carbon atoms) to adapt to growth in an alcoholic environment (Desens, 1989; Kalmar, 1995).

Bacteria can therefore only be identified by their composition in total fatty acids when the culture of the cells to be analyzed is standardized. Even if this method does not seem easy to use, it merits being mentioned. In the genus *Pediococcus*, it was used to characterize three groups in which six species are classified. *P. damnosus* and *P. pentosaceus*, encountered in enology, belong to two of these groups. The authors of this work (Uchida and Mogi, 1972) observed that culture age and environmental conditions modify the proportions of fatty acids without affecting the separation of the groups.

Bacteria cell proteins constitute another level of genomic expression. The amino acid sequence of proteins is the result of a direct translation of genes. It is therefore normal to distinguish bacteria from one another by the proteins that they contain. The primary structure determines molecule mobility in an electrophoretic gel in conditions where the secondary, tertiary and quaternary structures are denatured. This identification method therefore involves subjecting the total cell contents of bacteria to electrophoresis. After staining, the protein profiles are compared either visually or by computer-assisted analysis. The electrophoretic profiles are reproducible. They are standardized by markers which are required to compare several gels.

According to Kersters (1985), protein profiles of strains are identical when their DNA presents a homology greater than 90%; they are very similar up to 70%. Strains can therefore be identified in this manner at the species level. Nevertheless, as with the method using fatty acids, all of the following conditions must be rigorously standardized: bacteria culture conditions; the moment of sampling; extraction; and electrophoretic protocol. Recently, lactic acid bacteria spoiling fortified wines have been discovered in this manner: *O. oeni* (Dicks *et al.*, 1995), and diverse species of lactobacilli (*L. hilgardii, L. fructivorans, L. collinoides* and *L. mali*—the last three being rare) (Couto and Hogg, 1994).

REFERENCES

Bourdineaud J.P., Nehmé B., Tessé S. and Lonvaud-Funel A. (2003a) *Appl. Environ. Microbiol.*, 69, 2512.

Bourdineaud J.P., Nehmé B., Tessé S. and Lonvaud-Funel A. (2003b) *Int. J. Food Microbiol.*, 92, 1–14.

Claisse O. and Lonvaud-Funel A. (2001) *J. Food Prot.*, 64, 833.

Coton E., Rolan G.C., Bertrand A. and Lonvaud-Funel A. (1998) *Am. J. Enol. Vitic.*, 49, 199.

Couto J.A. and Hogg T. (1994) *J. Appl. Bacteriol.*, 76, 487.

Daniel P. (1993) Thèse de Doctorat, Université de Nantes.

Delmas F., Pierre F., Coucheney F., Divies C. and Guzzo J. (2001) *J. Mol. Microbiol. Biotechnol.*, 3, 601.

Desens C. (1989) Thèse de Doctorat, Université de Bordeaux II.

Dicks L.M.T., Dellaglio F. and Collins M.D. (1995) *Int. J. System. Bacteriol.*, 45, 395.

Dicks L.M.T., Loubser P.A. and Augustyn O.P.H. (1995) *J. Appl. Bacteriol.*, 79, 43.

Fremaux C., Aigle M. and Lonvaud-Funel A. (1993) *Plasmids*, 30, 212.

Garbay S. (1994) Thèse de Doctorat, Université de Bordeaux II.

Gasser F., Martel M.C., Talon R. and Champomier M. (1994) in *Les Bactéries Lactiques* (eds H. de Roissart and F.M. Luquet). Lorica, Uriage, France.

Gindreau E., Joyeux A., de Revel G., Claisse O. and Lonvaud-Funel A. (1997) *J. Int. Sci. Vigne Vin*, 31, 197.

Gindreau E., Kheim H., de Revel G., Bertrand A. and Lonvaud-Funel A. (2003) *J. Int. Sci. Vigne Vin*, 37, 51.

Gindreau E., Walling E. and Lonvaud-Funel A. (2001) *J. Appl. Microbiol.*, 90, 535.

Kalmar Z. (1995) Thèse de Doctorat, Université de Bordeaux II.

Kersters K. (1985) In *Computer-assisted Bacterial Systematics* (eds M. Goodfellow and D.E. Minnkin). Academic Press, London.

Le Jeune C. (1994) Thèse de Doctorat, Université de Bordeaux II.

Le Jeune C. and Lonvaud-Funel A. (1994) *Food Microbiol.*, 11, 195.

Le Jeune C., Lonvaud-Funel A., Ten Brink B., Hofstra H. and Van der Vossen J.M.B.M. (1995) *J. Appl. Microbiol.*, 78, 316–326.

Liu S.Q., Pritchard G.G., Hardman M.J. and Pilone G.J. (1994) *Amer. J. Enol. Vitic.*, 45, 235.

Lonvaud-Funel A. (1986) Thèse de Doctorat ès Sciences, Université de Bordeaux II.

Lonvaud-Funel A., Biteau N. and Frémaux C. (1989) *Sci. Alim.*, 9, 533.

Lonvaud-Funel A. and Desens C. (1990) *Sci. Alim.*, 10, 817.

Lonvaud-Funel A., Joyeux A. and Ledoux O. (1991a) *J. Appl. Bacteriol.*, 71, 501.

Lonvaud-Funel A., Fremaux C., Biteau N. and Joyeux A. (1991b) *Food Microbiol.*, 8, 215–222.

Lonvaud-Funel A., Guilloux Y. and Joyeux A. (1993) *J. Appl. Bacteriol.*, 74, 41.

Makaga E. (1994) *Thèse de Doctorat*, Université de Reims.

Rozès N., Garbay S., Denayrolles M. and Lonvaud-Funel A. (1993) *Lett. Appl. Microbiol.*, 17, 126.

Sohier D., Coulon J. and Lonvaud-Funel A. (1999) *Int. J. Syst. Bacteriol.*, 49, 1075.

Tonon T., Bourdineaud J.P. and Lonvaud-Funel A. (2001) *Res. Microbiol.*, 152, 653.

Uchida K. and Mogi K. (1972) *J. Gen. Appl. Microbiol.*, 18, 109.

Walling E. (2003) Thèse Doctorat, Université de Bordeaux.

5

Metabolism of Lactic Acid Bacteria

5.1 GENERALITIES—A REVIEW

Metabolism represents the biochemical reactions of degradation and synthesis carried out by the bacteria cell during multiplication. Catabolic reactions provide energy, transforming substrates from the environment or reserve substances of the cell; anabolic reactions guarantee cellular synthesis from environmental substrates and intermediary catabolism products.

Lactic acid bacteria are chemotrophic: they find the energy required for their entire metabolism from the oxidation of chemical compounds. The oxidation of substrates represents the loss of electrons that must be accepted by another molecule, which is reduced. Most oxidations, simultaneously liberate protons and electrons. The transport of these two particles to the final acceptor can activate a chain of successive oxidation–reductions.

Thus the biological oxidation of a substrate is always coupled with the reduction of another molecule. In the following oxidation–reduction reaction the oxidized substance is noted as DH_2 and the final electron and proton acceptor as A:

$$DH_2 \longrightarrow D + 2H^+ + 2e^- + \text{energy} \quad (5.1)$$

$$A + 2H^+ + 2e^- \longrightarrow AH_2 \quad (5.2)$$

The overall reaction is:

$$DH_2 + A \longrightarrow AH_2 + D + \text{energy}. \quad (5.3)$$

The nature of the final electron acceptor A determines the type of metabolism: fermentative or respiratory. The presence of oxygen also distinguishes aerobic and anaerobic microorganisms.

In aerobiosis, the electrons and protons are transported to oxygen, which is most often reduced to water. This process is called aerobic respiration. The transport system consists of a group of cytochromes. The proton flux creates a proton motive force, which permits the synthesis of ATP molecules. The conservation of the oxidation energy is ensured by the synthesis of the pyrophosphate bond of ATP. This bond generates energy when it is hydrolyzed. This system does not exist in lactic acid bacteria, although some species can synthesize cytochromes from precursors.

Some lactic acid bacteria reduce oxygen from the environment by forming hydrogen peroxide according to the following reaction:

$$O_2 + 2e^- + 2H^+ \longrightarrow H_2O_2 \quad \textbf{(5.4)}$$

Hydrogen peroxide must be eliminated since it is toxic. Cells that are not capable of eliminating it cannot develop in the presence of oxygen: they are strict anaerobes. Depending on their behavior with respect to oxygen, lactic acid bacteria are classed as strict anaerobes, facultative anaerobes, microaerophiles or aerotolerants. The distinction between these different categories is often difficult to establish for a given strain.

Most lactic acid bacteria tolerate the presence of oxygen but do not use it in energy-producing mechanisms. Depending on the species, they use different pathways to eliminate the toxic peroxide, activating peroxidases which use NADH as a reducer: a superoxide dismutase, a pseudo catalase and sometimes Mn^{2+} ions (Desmazeaud and Roissart, 1994). To date, this subject has not been specifically studied for species isolated in wine.

If the final electron and proton acceptor is a mineral ion (sulfate, nitrate), the microorganism functions in anaerobiosis, but a respiratory mechanism is still involved. This process is called anaerobic respiration.

In anaerobiosis, the reduced molecule can also be an endogenic substance—one of the products of metabolism. This is the case with fermentation.

In lactic acid bacteria, this molecule is pyruvate. It is reduced into lactate in the reaction which characterizes lactic fermentation:

$$\text{pyruvate} + NADH + H^+ \longrightarrow \text{lactate} + NAD^+ \quad \textbf{(5.5)}$$

Contrary to the reoxidation of the coenzyme by the respiratory chain, this reaction is not energy producing.

Other kinds of reactions can lead to ATP synthesis. They occur in aerobiosis or anaerobiosis. During these reactions, the oxidation of the substrate accompanies the creation of an energy-rich bond between the oxidized carbon and a phosphate molecule:

$$X H_2 + [Pi] \longrightarrow X \sim P + 2H^+ + 2e^- \quad \textbf{(5.6)}$$

The energy of the esterphosphoric bond is then stored in a pyrophosphate bond:

$$X \sim P + ADP \longrightarrow X + ATP \quad \textbf{(5.7)}$$

In this manner the phosphoenolpyruvate and the acetylphosphate can transfer their phosphate group to the ADP in the same type of reaction. The two intermediary molecules in the catabolism of sugar are therefore very important from an energetic viewpoint.

5.2 METABOLISM OF SUGARS BY LACTIC ACID BACTERIA

The oxidation of sugars constitutes the principal energy-producing pathway. This energy is essential for bacterial growth. In lactic acid bacteria, fermentation is the pathway for the assimilation of sugars. For a given species, the type of sugar fermented and environmental conditions (the presence of electron acceptors, pH, etc.) modify the energy yield and the nature of the final products.

The cytoplasmic membrane is an effective barrier separating the external environment from the cellular cytoplasm. Although permeable to water, salts and low molecular weight molecules, it is impermeable to many organic substances. Various works describe the different active sugar transport systems in lactic acid bacteria. They are for the most part ATP dependent and activate enzymatic systems—sometimes complex. These systems are specific to the sugars being transported. Heterofermentative bacteria, particularly the species that interest enologists, have not been studied in depth, but the existence of active transport systems using ATP-dependent permeases is highly probable.

Lactic acid bacteria of the genera *Lactobacillus, Leuconostoc* and *Pediococcus* assimilate sugars by either a homofermentative or heterofermentative pathway. Among the cocci, *Pediococcus* bacteria are homofermentative, while *Leuconostoc* and *Oenococcus* are heterofermentative. In lactobacilli, heterofermenters and homofermenters are distinguished according to the pathway used for hexose degradation. Pentoses, when degraded, are metabolized by heterofermentation.

5.2.1 Homofermentative Metabolism of Hexoses

Homofermentative bacteria transform nearly all of the hexoses that they use, especially glucose, into lactic acid. Depending on the species, either the L or D lactic isomer is formed (see Chapter 4). The homofermentative pathway or the Embden-Meyerhof pathway includes a first phase containing all of the reactions of glycolysis that lead from hexose to pyruvate. During this stage, the oxidation reaction takes place generating the reduced coenzyme NADH + H$^+$. This pathway is used by numerous cells. For aerobic organisms, this pathway is followed by the citric acid or Krebs cycle.

In lactic acid bacteria, the reaction of the second phase characterizes lactic fermentation. The reduced coenzyme is oxidized into NAD$^+$ during the reduction of pyruvate into lactate.

The reactions of glycolysis are listed in Figure 5.1. In the first stage, the glucokinase phosphorylates glucose into glucose 6-P (glucose 6-phosphate). This molecule then undergoes an isomerization to become fructose 6-P. Another phosphorylation leads to the formation of fructose 1,6-diphosphate. At this stage, the two most

important reactions have already occurred. They activate the kinases which require bivalent ions (Mg^{2+}, Mn^{2+}) and use an ATP molecule each time. One of them, the phosphofructokinase, an allosteric enzyme controlled by ATP, determines the speed of glycolysis.

The fructose 1,6-diphosphate is then split into two molecules of triosephosphate. This reaction is catalyzed by aldolase, a key enzyme of the glycolytic pathway. Homofermentative bacteria present a high fructose 1,6-diphosphate aldolase activity. The products of this reaction are glyceraldehyde 3-P and dihydroxyacetone-P.

Only glyceraldehyde 3-P pursues the transformation pathway. The dihydroxyacetone-P is rapidly isomerized into glyceraldehyde 3-P. In reality, the equilibrium between these two molecules favors dihydroxyacetone-P, but it is continually reversed, since the glyceraldehyde 3-P is eliminated by the reaction which follows. In the next stage, energy production processes begin. The glyceraldehyde 3-P is oxidized into 1,3-diphosphoglycerate. A phosphorylation from inorganic phosphate accompanies the oxidation. The NAD$^+$ coenzyme is reduced to NADH + H$^+$. These reactions permit the synthesis of an acyl–phosphate bond—a high energy potential bond. During the hydrolysis of this bond, the reaction immediately following recuperates the energy by the synthesis of an ATP molecule.

The 1,3-diphosphoglycerate is transformed into 3-P glycerate. This molecule undergoes a rearrangement: its phosphate group passes from position 3 to position 2, esterifying in this manner the secondary alcohol function of the glycerate. An internal dehydration of the molecule then occurs. The important reaction which follows generates an enolphosphate, a high energy potential molecule, called phosphoenolpyruvate. Finally, this energy is used for the synthesis of ATP from ADP in a reaction which forms pyruvate.

From the moment when the triose molecules are utilized, the second part of glycolysis comprises the most important energy-producing phases. Two reactions ensure the synthesis of ATP for each of the glyceraldehyde-P molecules coming from hexose. The total reaction energy from the

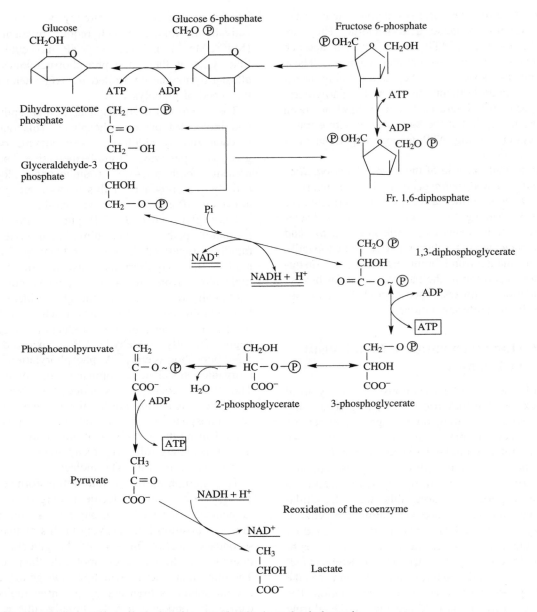

Fig. 5.1. Metabolic pathway of glucose fermentation by homolactic bacteria

transformation of a glucose molecule is therefore the synthesis of two ATP molecules and incidentally the reduction of NAD$^+$.

For each hexose molecule assimilated, the cell requires an NAD$^+$ molecule. The cell must therefore make use of a system that maintains an acceptable NAD$^+$ level. Lactic acid bacteria use the pyruvate formed by glycolysis as an electron acceptor to oxidize NADH. This character defines lactic fermentation. In general, bacteria therefore transform a hexose molecule into two lactate molecules by the homolactic pathway.

5.2.2 Heterofermentative Metabolism of Hexoses

Bacteria using the heterofermentative pathway transform hexoses principally but not exclusively into lactate. The other molecules produced by this metabolism are essentially CO_2, acetate and ethanol; this is the pentose phosphate pathway. After being transported into the cell, a glucokinase phosphorylates the glucose into glucose 6-P (glucose 6-phosphate). Its destination is completely different from the glucose 6-P of the homofermentative pathway. Two oxidation reactions occur successively: the first leads to gluconate 6-P; the second, accompanied by a decarboxylation, forms ribulose 5-P (Figure 5.2). In each of these reactions, a molecule of the coenzyme NAD^+ or $NADP^+$ is reduced. The ribulose 5-P is then epimerized into xylulose 5-P.

The xylulose 5-P phosphoketolase is the key enzyme of this pathway: it catalyzes the cleavage of the pentulose 5-P molecule into acetyl-P

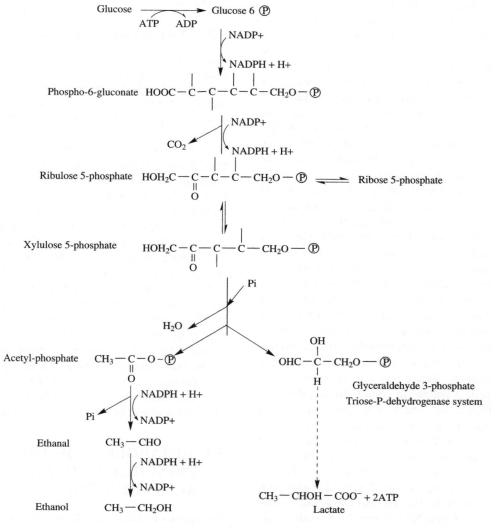

Fig. 5.2. Metabolic pathway of glucose fermentation by heterolactic bacteria (pentose phosphate pathway)

and glyceraldehyde 3-P. This reaction requires phosphate. The glyceraldehyde 3-P is metabolized into lactic acid by following the same pathway as in the homofermentative pathway. The acetyl-P has two possible destinations, depending on environmental conditions. This molecule can be successively reduced into ethanal and then ethanol, in which case the molecules of the coenzyme NADH + H$^+$ or NADPH + H$^+$, formed during the two oxidation reaction of hexose at the beginning of the heterofermentative pathway, are reoxidized. This reoxidation is essential for regenerating the coenzymes necessary for the assimilation of sugar.

In certain conditions, when the cell makes use of other coenzyme reoxidation systems, the acetate kinase catalyzes a reaction that leads to the formation of acetate from acetyl-P. This reaction simultaneously recuperates the bond energy of the P group of acetyl-P by the synthesis of an ATP molecule. In this case, the coenzyme reoxidation systems activate NADH or NADPH oxidases, when the cells are in aerobiosis or reduction reactions such as the transformation of fructose into mannitol. When acetyl-P leads to the formation of acetate, there is a definite energetic advantage. A supplementary ATP molecule is formed for each hexose molecule transformed.

The final quantity of glucose metabolism products (presence of ethanol and acetate) from heterofermentative bacteria demonstrates that this pathway is nearly always used. Yet the use of this pathway varies more or less depending on the degree of aeration and the presence of other proton and electron acceptors. In this way, bacteria of the genus *Leuconostoc* preferentially produce lactate and ethanol in a slightly aerated environment and, on the contrary, lactate and acetate in an aerated environment. Changes in conditions therefore not only influence the nature of the products formed but also the energy yield and thus growth.

Heterofermentative bacteria produce acetic acid from hexoses, but regulation mechanisms modify production. In anaerobic conditions, the NADH oxidase cannot regenerate NAD. Glucose preferentially leads to the formation of lactic acid and ethanol. When NADH can be reoxidized by another process, the amount of ethanol formed

decreases, resulting in an increase in acetic acid. This occurs in aerobic conditions or in the presence of another substance that can be reduced. Homolactic bacteria ferment glucose almost exclusively into lactic acid. In an anaerobic environment with a limited glucose concentration, homofermentative bacteria such as *Lactobacillus casei* form less lactic acid; the primary products can become acetic acid, formic acid and ethanol. The change is linked to the regulation of the L-LDH by fructose 1,6-diphosphate. The change is less obvious when the homofermentative species possess the two LDH types, L and D. FDP does not regulate the D-LDH.

5.2.3 Metabolism of Pentoses

Certain strains of *Lactobacillus, Pediococcus* or *Leuconostoc* ferment pentoses such as ribose, arabinose and xylose, whether they are homofermenters or heterofermenters, according to the same schema (Figure 5.3). The pentoses are phosphorylated by reactions activating kinases and using ATP. Specific isomerases then lead to the formation of the xylulose 5-P molecule. The following reactions are described in the heterofermentative pathway for glucose assimilation. In spite of glyceraldehyde 3-P having the same fate in this case, acetyl-P exclusively leads to the formation of the acetate molecule, generating an ATP molecule in this manner. In fact, a reduced coenzyme molecule is not available to reduce acetyl-P into ethanol. The pathway furnishes two ATP molecules for each pentose molecule fermented. This pathway therefore has a greater yield than the fermentation of a hexose by the pentose phosphate pathway.

The study of the homofermentative and heterofermentative metabolic pathways of sugars therefore permits the prediction of the nature of the products formed. Pentoses are always at the origin of acetic acid and of course lactic acid production.

5.3 METABOLISM OF THE PRINCIPAL ORGANIC ACIDS OF WINE

Bacteria essentially degrade two organic acids of wine: malic and citric acid. Other acids can of

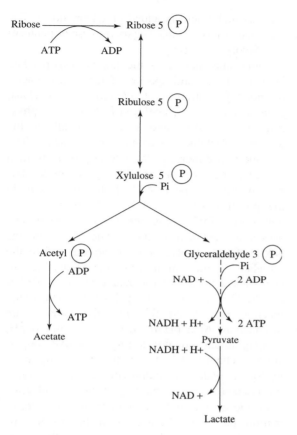

Fig. 5.3. Pentose fermentation pathway by lactic acid bacteria

course be degraded but are of less interest in enology—with the exception of tartaric acid, which has rarely been studied. Since the initial research of lactic acid bacteria and their role in winemaking, malic acid has been the focus of a large number of studies. Yet the degradation of citric acid also plays an important role in winemaking. The majority of bacterial species preponderant in wine after alcoholic fermentation degrade these two acids. This degradation is evidently the source of many organoleptical changes noted after their development. The enologist may consider the transformation of malic acid to be the most important phenomenon of the malolactic fermentation phase, but other transformations, of citric acid in particular, should also be taken into account.

5.3.1 Transformation of Malic Acid

In the case of non-proliferating cells in a laboratory medium and during winemaking, lactic acid bacteria of wine transform L-malic acid exclusively into L-lactic acid. Seifert (1901) established the reaction of the malolactic transformation according to the following equation:

$$\text{malic acid} \longrightarrow \text{lactic acid} + CO_2. \quad (5.8)$$

This equation was confirmed when the stereoisomers could be separately determined for each of the two acids.

This reaction therefore involves a decarboxylation without an intermediary product capable of following another metabolic pathway. Several authors have reported that certain bacterial strains form other molecules from malic acid, suggesting in this manner the existence of other reactions. Even if their existence cannot be ruled out, malolactic transformation is the only reaction that exists in the lactic acid bacteria involved in winemaking.

Alizade and Simon (1973) studied the stereochemistry of this transformation. Enzymatic methods were used to determine the specific quantities of the stereoisomers. In addition, the fermentation of radioactively labeled glucose and malic acid permitted the study of their products. The heterofermentative cocci (*Oenococcus*), abundant or exclusive during winemaking, were found to present several properties. They form exclusively D-lactic acid from glucose (Chapter 4) and exclusively L-lactic acid from L-malic acid (Figure 5.4).

This observation suggests that the transformation of malic acid does not pass by the intermediary of pyruvic acid. Peynaud (1968) concluded that the substrate was decarboxylated directly. A lot of

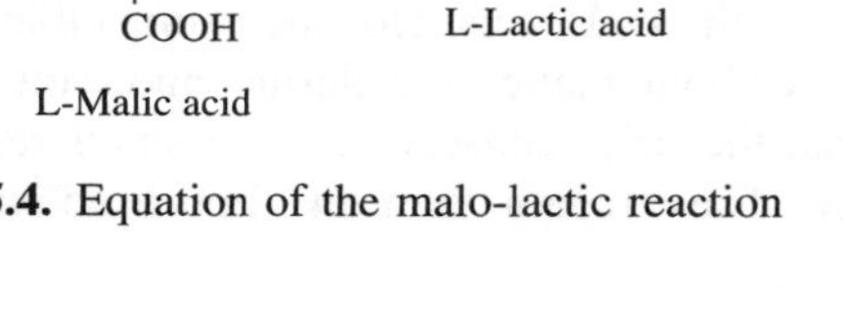

$$\begin{array}{ccc}
\text{COOH} & & \text{COOH} \\
| & & | \\
\text{HO}-\text{C}-\text{H} & \longrightarrow & \text{H}-\text{C}-\text{OH} \quad + \text{ CO}_2 \\
| & & | \\
\text{H}-\text{C}-\text{H} & & \text{CH}_3 \\
| & & \text{L-Lactic acid} \\
\text{COOH} & & \\
\text{L-Malic acid} & &
\end{array}$$

Fig. 5.4. Equation of the malo-lactic reaction

research was carried out to elucidate this mechanism. It naturally leads to the examination of the enzymatic aspect of this transformation.

At that time only the malate dehydrogenase (MDH) and the malic enzymes were, known to be capable of fixing and catalyzing a reaction whose substrate is L-malic acid. These two enzymes were described in numerous vegetal and animal cells and in diverse microorganisms. They catalyze the following reactions:

$$\text{MDH: L-malate} \xrightleftharpoons{\text{Mn}^{2+}}$$

$$\text{oxaloacetate} + \text{NADH} + \text{H}^+ + \text{NAD}^+ \tag{5.9}$$

$$\text{malic enzyme: L-malate} + \text{NAD}^+ \xrightleftharpoons{}$$

$$\text{pyruvate} + \text{CO}_2 + \text{NADH} + \text{H}^+ \tag{5.10}$$

Since oxaloacetate is easily decarboxylated into pyruvate and CO_2, these two reactions lead to the formation of pyruvate from L-malate. Since the final product of the malolactic transformation in wine is L-lactic acid, MDH or the malic enzyme would be associated to an LDH catalyzing the reduction of pyruvate into L-lactate in this metabolic pathway. At least for wine bacteria, this concept is not acceptable since these bacteria only possess a D-LDH. Malic acid would only lead to the formation of D-lactic acid.

Therefore, the hypothesis of the existence of an enzyme catalyzing the direct decarboxylation of L-malic acid into L-lactic acid was made. The enzyme, called the malolactic enzyme, was isolated for the first time in *Lactobacillus plantarum* (Lonvaud, 1975; Schütz and Radler, 1974). From acellular bacterial extracts and thanks to successive purification stages, the authors obtained purified fractions responding to the functional criteria of the malolactic enzyme. L-Malic acid is transformed stoichiometrically into L-lactic acid. These fractions do not have an LDH activity.

At least in *L. mesenteroides* and *L. oenos* (*O. oeni*), the malolactic enzyme is inducible. Cultivated without malic acid during numerous generations, the cells conserve a very small residual activity. They regain their maximum activity as

soon as malic acid is added (1 g/l or more). The presence of fermentable sugars (hexose or pentose) also favors its activity.

Some time later, the same enzyme was purified in other strains and species of lactic acid bacteria, notably in strains of *L. plantarum, L. murinus, L. mesenteroides, O. oeni* and *L. lactis*. The physical characteristics and kinetics of all of the described malolactic enzymes are the same. The enzyme is a dimeric or tetrameric protein formed by the association of a 60 kDa polypeptide. The pH_i of the enzyme is 4.35. It functions only in the presence of the NAD^+ cofactor and bivalent ions, Mn^{2+} being the most effective, and uses a sequential mechanism. The Mn^{2+} and the NAD^+ fix themselves on the protein before the L-malate. At the optimum pH, the Michaelis constants are 2×10^{-3} M for malate and 4×10^{-5} M for NAD. The optimum pH of the enzymatic reaction is 5.9. At this pH, the kinetics are Michaelian. At a pH far from the optimum pH, it is sigmoidal—demonstrating a positive cooperative mechanism which signifies a growing affinity for the malate. Homopolymeric enzymes share this characteristic: the binding of the first substrate molecule on the first promoter transmits a deformation, increasing the affinity of the others. This cooperativeness permits an increase in the effectiveness of the system in unfavorable conditions. Evidently, in winemaking, bacteria are in far from optimal conditions (Lonvaud-Funel and Strasser de Saad, 1982).

The carboxylic acids of wine—succinic, citric and L-tartaric acid—are competitive inhibitors with the following respective inhibition constants: 8×10^{-2} M, 1×10^{-2} M and 0.1 M. L-Lactic acid, a product of the reaction, is an inefficient, noncompetitive inhibitor whose inhibition constant of 0.3 M indicates a weak affinity.

Although this enzyme is becoming better known, a question still remains unanswered: what is the real role of NAD^+ in the oxidation–reduction exchange? The indispensable coenzyme of the reaction is not involved, at least in a conventional manner.

The malolactic enzyme purified from *Lactococcus lactis*, a lactic bacterium of milk origin, has exactly the same characteristics as the enological strain enzyme. It was used to study the structure of the gene. The protomer N-terminal end was sequenced on 20 amino acids (IBMC Laboratory, University of Bordeaux II). The corresponding nucleotide sequences were deduced from the five first and five last amino acids of this portion of the protein (Denayrolles *et al.*, 1994). These oligonucleotide sequences were used as primers in a PCR amplification reaction with bacterial DNA as templates. In this manner, a 60-nucleotide fragment was isolated and used to produce a probe, permitting the identification of the malolactic gene in the bacterial chromosome. This fragment, and progressively the entire gene, was sequenced. Ansanay *et al.* (1993) obtained the same result by another method, starting with the same purified enzyme preparation.

The nucleotide sequence encoding the malolactic enzyme is therefore known and shows a strong resemblance to the malic enzyme. The binding sites of the coenzyme on the protein have also been located (Figure 5.5). Finally, after having been inserted into a vector, this gene was transferred into *E. coli* and also into laboratory strains of *S. cerevisiae* yeast; the gene was expressed in these conditions (Ansanay *et al.*, 1993; Denayrolles *et al.*, 1995).

In the late 1980s, the program aimed at developing a "malolactic yeast" capable of carrying out the malolactic transformation during alcoholic fermentation was supported by winemakers in France and abroad. The first stage consisted of cloning the gene of the malolactic enzyme and expressing it in an yeast. Unfortunately, it very rapidly became obvious that the system was limited by the fact that malic acid entered the yeast. To overcome this problem, the Stellenbosch team (South Africa) decided to clone the malate permease gene from *Schizosaccharomyces pombe*, another yeast found in wine (Grobler *et al.*, 1995). Having demonstrated that a yeast could be transformed by a vector bearing the gene coding for permease and another bearing the malolactic enzyme, the same team inserted these two genes into a yeast chromosome to stabilize the desired genetic data. A yeast strain with malolactic activity now exists,

can be produced on an industrial scale, and has shown a certain level of performance (Van Vuuren and Husnik, 2003). There is no question of carrying out full malolactic fermentation as this yeast does not exhibit all the other bacterial activities involved in enhancing the gustatory qualities of wine. It may be useful as an organic agent for deacidifying wines when the winemaker wishes to preserve the characteristic aromas revealed by yeast. However, this yeast is a genetically modified microorganism and, as such, is far from gaining universal acceptance.

The production of the enzyme for direct use in wines is of no use, since this protein is rapidly inhibited by diverse substances in wine—acids, alcohol and polyphenols. The malolactic reaction takes place at the interior of the bacterium in a medium protected from inhibitors by the bacterial membrane. The degradation rate of malic acid is limited by its transport speed in the interior of the cell. Although the optimal pH for enzyme activity is around 6.0, it is around 3.0 to 3.5 for whole cells of *O. oeni*. At this pH, malic acid penetrates more easily into the bacterium than at higher pHs.

The inhibitory action of tartaric and succinic acid is even stronger on whole cells than on proteins. Citric acid at a concentration of 0.5 g/l, normally not reached in wine, only slows cellular activity by around 5% (Lonvaud-Funel and Strasser de Saad, 1982).

Finally, among the questions raised as early as the period of initial research on malolactic fermentation, the physiological role of malic acid remains to be interpreted. The addition of malic acid in a culture medium of lactic acid bacteria simultaneously increases the yield and the growth rate. A partial explanation of this observation was discovered only recently. The malolactic reaction itself is not very exergonic, yet it indirectly constitutes a real energy source for the cell. Poolman (1993) demonstrated that, following the decarboxylation reaction, the increase of the internal pH (which imposes an influx of protons), the uptake of malic acid and the efflux of lactic acid combine to create a proton motor force, permitting the conservation of energy via the membrane ATPase.

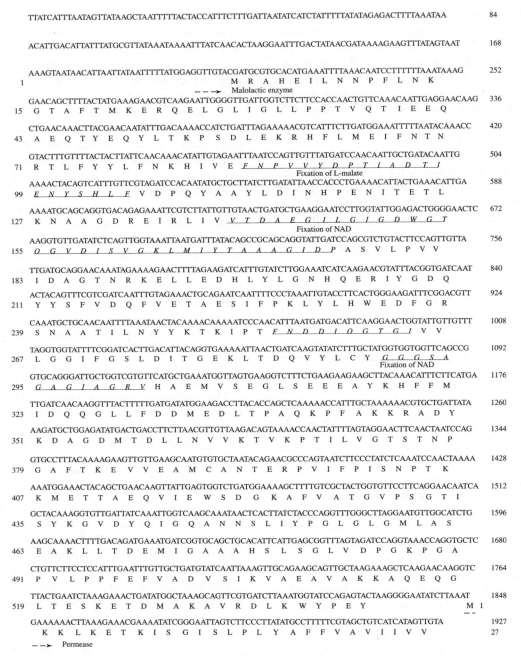

TTATCATTTAATAGTTATAAGCTAATTTTTACTACCATTTCTTTGATTAAATATCATCTATTTTTATATAGAGACTTTTAAATAA   84

ACATTGACATTATTTATGCGTTATAAATAAAATTTATCAACACTAAGGAATTTGACTATAACGATAAAAGAAGTTTATAGTAAT   168

AAAGTAATAACATTAATTATAATTTTTATGGAGGTTGTACGATGCGTGCACATGAAATTTTAAACAATCCTTTTTTAAATAAAG   252
1   M R A H E I L N N P F L N K
                                           – – –→   Malolactic enzyme

GAACAGCTTTTACTATGAAAGAACGTCAAGAATTGGGGGTTGATTGGTCTTCTTCCACCAACTGTTCAAACAATTGAGGAACAAG   336
15   G T A F T M K E R Q E L G L I G L L P P T V Q T I E E Q

CTGAACAAACTTACGAACAATATTTGACAAAACCATCTGATTTAGAAAAACGTCATTTCTTGATGGAAATTTTTAATACAAACC   420
43   A E Q T Y E Q Y L T K P S D L E K R H F L M E I F N T N

GTACTTTGTTTTACTACTTATTCAACAAACATATTGTAGAATTTAATCCAGTTGTTTATGATCCAACAATTGCTGATACAATTG   504
71   R T L F Y Y L F N K H I V E *F N P V V Y D P T I A D T*
                                                          Fixation of L-malate

AAAACTACAGTCATTTGTTCGTAGATCCACAATATGCTGCTTATCTTGATATTAACCACCCTGAAAACATTACTGAAACATTGA   588
99   *E N Y S H L F* V D P Q Y A A Y L D I N H P E N I T E T L

AAAATGCAGCAGGTGACAGAGAAATTCGTCTTATTGTTGTAACTGATGCTGAAGGAATCCTTGGTATTGGAGACTGGGGAACTC   672
127 K N A A G D R E I R L I V *V T D A E G I L G I G D W G* T
                                                          Fixation of NAD

AAGGTGTTGATATCTCAGTTGGTAAATTAATGATTTATACAGCCGCAGCAGGTATTGATCCAGCGTCTGTACTTCCAGTTGTTA   756
155 *O G V D I S V G K L M I Y T A A A G I D* P A S V L P V V

TTGATGCAGGAACAAATAGAAAAGAACTTTTAGAAGATCATTTGTATCTTGGAAATCATCAAGAACGTATTTACGGTGATCAAT   840
183 I D A G T N R K E L L E D H L Y L G N H Q E R I Y G D Q

ACTACAGTTTCGTCGATCAATTTGTAGAAACTGCAGAATCAATTTTTCCCTAAATTGTACCTTCACTGGGAAGATTTCGGACGTT   924
211 Y Y S F V D Q F V E T A E S I F P K L Y L H W E D F G R

CAAATGCTGCAACAATTTTAAATAACTACAAAACAAAAATCCCAACATTTAATGATGACATTCAAGGAACTGGTATTGTTGTTT   1008
239 S N A A T I L N Y Y K T K I P T *F N D D I O G T G* I V V

TAGGTGGTATTTTCGGATCACTTGACATTACAGGTGAAAAATTAACTGATCAAGTATATCTTTGCTATGGTGGTGGTTCAGCCG   1092
267 L G G I F G S L D I T G E K L T D Q V Y L C Y *G G G S A*
                                                              Fixation of NAD

GTGCAGGGATTGCTGGTCGTGTTCATGCTGAAATGGTTAGTGAAGGTCTTTCTGAAGAAGAAGCTTACAAACATTTCTTCATGA   1176
295 *G A G I A G R V* H A E M V S E G L S E E E A Y K H F F M

TTGATCAACAAGGTTTACTTTTTGATGATATGGAAGACCTTACACCAGCTCAAAAACCATTTGCTAAAAAACGTGCTGATTATA   1260
323 I D Q Q G L L F D D M E D L T P A Q K P F A K K R A D Y

AAGATGCTGGAGATATGACTGACCTTCTTAACGTTGTTAAGACAGTAAAACCAACTATTTTAGTAGGAACTTCAACTAATCCAG   1344
351 K D A G D M T D L L N V V K T V K P T I L V G T S T N P

GTGCCTTTACAAAAGAAGTTGTTGAAGCAATGTGTGCTAATACAGAACGCCCAGTAATCTTCCCTATCTCAAATCCAACTAAAA   1428
379 G A F T K E V V E A M C A N T E R P V I F P I S N P T K

AAATGGAAACTACAGCTGAACAAGTTATTGAGTGGTCTGATGGAAAAAGCTTTTGTCGCTACTGGTGTTCCTTCAGGAACAATCA   1512
407 K M E T T A E Q V I E W S D G K A F V A T G V P S G T I

GCTACAAAGGTGTTGATTATCAAATTGGTCAAGCAAATAACTCACTTATCTACCCAGGTTTGGGCTTAGGAATGTTGGCATCTG   1596
435 S Y K G V D Y Q I G Q A N N S L I Y P G L G L G M L A S

AAGCAAAACTTTTGACAGATGAAATGATCGGTGCAGCTGCACATTCATTGAGCGGTTTAGTAGATCCAGGTAAACCAGGTGCTC   1680
463 E A K L L T D E M I G A A A H S L S G L V D P G K P G A

CTGTTCTTCCTCCATTTGAATTTGTTGCTGATGTATCAATTAAAGTTGCAGAAGCAGTTGCTAAGAAAGCTCAAGAACAAGGTC   1764
491 P V L P P F E F V A D V S I K V A E A V A K K A Q E Q G

TTACTGAATCTCAAAGAAACTGATATGGCTAAAGCAGTTCGTGATCTTAAATGGTATCCAGAGTACTAAGGGGAAATATCTTAAAT   1848
519 L T E S K E T D M A K A V R D L K W Y P E Y   M 1
                                                                                – –

GAAAAAAACTTAAAGAAACGAAAATATCGGGAATTAGTCTTCCCTTATATGCCTTTTTCGTAGCTGTCATCATAGTTGTA   1927
     K K L K E T K I S G I S L P L Y A F F V A V I I V V   27
     – – –→   Permease

Fig. 5.5. Nucleotidic sequence of the DNA fragment carrying the malolactic enzyme (*mLeS*) gene and proteic sequences coded by this fragment. Certain proteic zones particularly well conserved between the malolactic enzyme and malic enzymes have been underlined. A potential function has been specified for some

5.3.2 Metabolism of Citric Acid

Certain lactic acid bacteria (heterofermentative cocci and homofermentative bacilli) degrade citric acid. Among the species found in wine, *L. plantarum*, *L. casei*, *O. oeni* and *L. mesenteroides* rapidly use citric acid. Strains of the genus *Pediococcus* and of the species *L. hilgardii* and *L. brevis* cannot.

In certain dairy industry bacteria, the lack of utilization of citric acid is linked to the loss of the plasmid encoding the citrate permease, essential for the uptake of the acid. In bacteria isolated in wine, the citrate permease may exist, but its role is inconsequential, since at the pH of wine, the non-dissociated substrate diffuses across the membrane

without needing the permease. The species and the strains that do not degrade citric acid are therefore at least deficient in the first enzyme of the metabolic pathway: the citrate lyase. This enzyme was studied in wine lactic acid bacteria and more particularly in a strain of *L. mesenteroides* (Weinzorn, 1985).

Within bacteria, citric acid is split into an oxaloacetate molecule and an acetate molecule by the lyase (Figure 5.6). The largest quantities of this enzyme are synthesized in low sugar concentration media containing citric acid. Glucose acts as a repressor. The protein is active in an acetylated form. The inactive deacetylated form can be reacetylated *in vivo* by the citrate lyase ligase with acetyl CoA or acetate and ATP. This

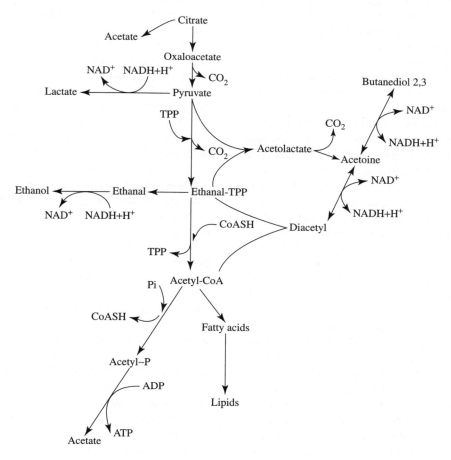

Fig. 5.6. Metabolic pathway for citric acid degradation by lactic acid bacteria

first degradation stage leads to the formation of an acetate molecule for each molecule of the substrate.

Oxaloacetate is then decarboxylated into pyruvate in *Oenococcus*, the most important bacteria in enology. In certain *lactobacillus*, it can also lead to a partial formation of succinate and formate. Pyruvate is the source of acetoin compounds: diacetyl, acetoin and 2,3-butanediol. The first is particularly important organoleptically. It is the very aromatic molecule that gives butter its smell. The olfactive intensity of acetoin and butanediol, which are derived from diacetyl by reduction of the ketonic functions, is much lower (Figure 5.6).

Two diacetyl synthesis pathways have been proposed. In one, diacetyl results from the reaction of acetyl CoA with ethanal-TPP (active acetaldehyde), catalyzed by a diacetyl synthetase, which has never been isolated. The other pathway supposes that from two pyruvate molecules, α-acetolactate synthetase produces α-acetolactate which is then decarboxylated into acetoin. The diacetyl is derived from it by oxidation. This is an aerobic pathway.

In addition to acetoinic substances, the pyruvate molecules coming from citrate have other destinations. First of all, if the coenzyme NADH, produced by other pathways, is available, it leads to the formation of lactate. Next comes the decarboxylation of pyruvate and then a reduction produces ethanol. Finally, the pyruvate derived from citrate participates in the synthesis of fatty acids and lipids via acetyl CoA. The radioactivity of labeled citrate supplied to the bacteria is incorporated into the cellular material. In this pathway, part of the acetyl CoA can also generate acetate molecules (Figure 5.6).

The citric acid metabolism products are therefore very diverse. Whatever the conditions, more than a molecule of acetic acid is surely formed from a substrate molecule. The production of others is largely determined by factors influenced by growth conditions. In limited glucose concentration conditions, a low pH and the presence of growth inhibitors, citric acid preferentially leads to the formation of acetoinic substances. In fact, due to conditions, pyruvate is orientated neither

towards the synthesis of cellular material, since growth is difficult, nor towards lactate and ethanol, because of a lack of reduced coenzymes. The acetoinic substance synthesis pathway is considered to be a detoxification process of the cell. In order to maintain its intracellular pH, it must eliminate pyruvate. Conversely, when growth is easy, pyruvate is utilized by fatty acid synthesis pathways; acetic acid is produced in larger quantities. In a laboratory medium, *Oenococcus* forms more than two acetic acid molecules from one citric acid molecule at a pH of 4.8 and only 1.2 molecules at a pH of 4.1. Conversely, the production of acetoinic molecules is four times higher (Lonvaud-Funel *et al.*, 1984).

It is therefore not surprising that some wines contain more than 10 mg of diacetyl per liter. Yet it has been determined that several milligrams of diacetyl per liter (2–3 mg/l for white wines and about 5 mg/l for red wines) contribute favorably to the bouquet. Above these concentrations, the buttery aroma, distinctly perceived, diminishes wine quality. Malolactic fermentation conditions and the quantity of citric acid degraded (from 0.2 to 0.3 g/l) and also without doubt the species of *O. oeni* involved determines the quantity of diacetyl produced. Several other reactions contribute to its final concentration. First of all, yeasts also synthesize diacetyl during alcoholic fermentation by a completely different pathway linked to the metabolism of amino acids. Diacetyl is then reduced into acetoin by the diacetyl reductase, an enzyme present in yeasts and lactic acid bacteria. The diacetyl concentration attains two maxima in this manner: one during alcoholic fermentation, the other during the degradation of citric acid by bacteria. It diminishes between the two fermentations and at the final stage of bacterial activity. Maintaining wine on yeast and bacteria lees at the end of fermentation ensures this reduction and also determines the final diacetyl level (De Revel *et al.*, 1989). Finally, sulfur dioxide further diminishes its concentration by combining with the ketonic functions.

Citric acid is always metabolized during fermentation because in nearly every case the species *O. oeni* is involved. Its degradation begins at the same time as malic acid degradation, but it is much slower—so much so that at the end of malolactic fermentation, citric acid often remains, up to 0.15 g/l or sometimes even more. It represents an additional energy source. In fact, ATP is formed from acetyl-P, derived from pyruvate, which is directed towards the production of cell components. In the presence of residual sugars (glucose or fructose) degraded by the heterofermentative pathway, part of the pyruvate derived from citric acid acts as an electron and proton acceptor. Part of the acetyl-P originating from sugars leads to the formation of acetate in producing ATP. In this manner the presence of citric acid in a wine favors bacterial growth and survival, notably in the presence of residual sugars. This metabolism therefore participates, along with the malic acid metabolism, in the microbiological stabilization of wine by eliminating energy sources (Lonvaud-Funel, 1986).

5.3.3 Metabolism of Tartaric Acid

Wine lactic acid bacteria can degrade tartaric acid, but this metabolism differs from malic and citric acid metabolisms. It is a veritable bacterial spoilage. Pasteur described it in the last century and named it *tourne* disease. It is dangerous since the disappearance of tartaric acid, an essential acid in wine, lowers the fixed acidity and is accompanied by an increase in the volatile acidity. The degradation can be total or partial, depending on the level of bacterial development, but it always lowers wine quality.

This spoilage is rare since the strains capable of degrading tartaric acid seem to be relatively few in number. Studies carried out on this subject in the 1960s and 1970s showed that this property is not linked to a particular bacterial species. Strains belonging to different species have been isolated by various authors, but they are most often lactobacilli. Among them, Radler and Yannissis (1972) found four strains of *L. plantarum* and one strain of *L. brevis* having this trait out of the 78 strains examined. Peynaud (1967) discovered 30 or so strains capable of partially or totally using tartaric acid in a study carried out on more than 700 strains. The scarcity of this property constitutes in sum the first protection against this disease.

Since a high pH is always propitious to the multiplication of a larger number of bacteria, higher acidity wines are less affected. Moreover, these bacteria are sensitive to SO_2. Therefore, respecting the current rules of hygiene in the winecellar and in wine should be sufficient to avoid this problem.

Few studies exist examining the metabolic pathways of the transformation of tartaric acid. The only results existing describe different pathways for *L. plantarum* and *L. brevis* (Radler and Yannissis, 1972). From a molecule of tartaric acid, *L. plantarum* produces 0.5 molecules of acetic acid, 0.3 of succinic acid and 1.3 of CO_2. *L. brevis* forms 0.7 molecules of acetic acid, 0.3 of succinic acid and 1.3 of CO_2 (Figure 5.7).

5.4 OTHER TRANSFORMATIONS LIKELY TO OCCUR IN WINEMAKING

5.4.1 Degradation of Glycerol

Glycerol is one of the principal components of wine, both in its concentration (5–8 g/l) and in its contribution to taste. Yeasts form glycerol by glyceropyruvic fermentation at the beginning of fermentation. The degradation of glycerol harms wine quality, partly because of the decrease in its concentration and partly because of the resulting products of the metabolism.

Certain bacterial strains produce bitterness in wine—a fact known since the time of Pasteur. Lactic acid bacteria make use of a glycerol dehydratase to transform glycerol into β-hydroxypropionaldehyde (Figure 5.8). This molecule is the precursor of acrolein, which is formed in wine by heating, or slowly during aging. The combination of wine tannins and acrolein, or its precursor, gives a bitter taste.

Like *tourne*, this spoilage is not widespread, due to the rarity of strains capable of degrading glycerol by this pathway. No single species of bacteria is responsible for degrading glycerol in wine. Little research has been devoted to this problem, but strains of two species of bacteria, *Lactobacillus hilgardii* and *Lactobacillus diolivorans*, have been isolated from wine following degradation of the

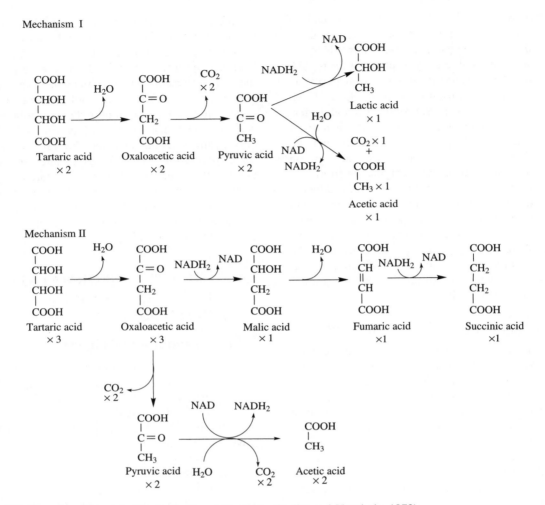

Fig. 5.7. Tartaric acid metabolism by lactic acid bacteria (Radler and Yannissis, 1972)

glycerol (Claisse, 2002). A key enzyme, glycerol dehydrogenase, has been studied in several strains. In strains of *L. brevis* and *L. buchneri*, Schutz and Radler (1984) demonstrated the degradation of glycerol by the glycerol dehydratase to 1,3-propanediol via β-hydroxypropionaldehyde when the medium also contained glucose or fructose. NADH (or NADPH), produced by the fermentation of sugar, reduces aldehyde to 1,3-propanediol. As described earlier, this co-metabolism leads to a deviation of acetyl-P derived from sugar towards the production of additional ATP and acetate. It therefore facilitates bacterial growth.

Some strains degrade glycerol in wine by the glycerol dehydratase pathway, while others also use the 3-P-glycerol dehydrogenase pathway. The genes coding the enzymes for the first pathway have been studied. They are organized in a set including a total of 13 genes, probably all necessary for the functioning of glycerol dehydratase, which consists of three protein subunits and propane-1,3-diol-dehydrogenase, leading, finally, to propane-1,3-diol. *L. hilgardii* and *L. diolivorans* are organized in the same way, as are strains of *L. collinoides* isolated from cider affected by acrolein spoilage (Gorga

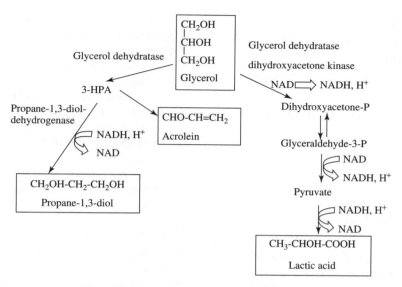

Fig. 5.8. Glycerol degradation pathways by lactic acid bacteria (Ribéreau-Gayon *et al.*, 1975)

et al., 2002). Oligonucleotide primers for detecting these bacteria have been selected from the gene sequence coding for one of the glycerol dehydratase subunits (Claisse and Lonvaud-Funel, 2001). Degradation of glycerol results not only in the production of hydroxy-3-propionaldehyde, the precursor of acrolein, but also, by metabolic coupling, to an increase in volatile acidity produced from the L-lactic acid in the wine. The other pathway consists of glycerokinase phosphorylating the glycerol and 3-P-glycerol dehydrogenase resulting in 3-P-dehydroxyacetone. This molecule enters into glycolysis reactions by oxidation into dihydroxyacetone-P which result in the formation of pyruvate. The final products of this pathway are those previously described from pyruvate degradation, notably acetic acid and acetoinic substances. The quantity of products varies depending on environmental conditions, in particular the amount of fermentable sugar and aeration.

In particular, large amounts of lactic acid are formed, increasing the wine's total and fixed acidity and causing its pH to drop. Acidification was as high as 0.8 g/l (expressed in tartaric acid) in wines where the pH dropped by 0.25 to 0.30.

Acetoinic molecules may also be formed from pyruvate. The fate of the pyruvate molecules is probably, as usual, determined by the availability of NADH/NAD coenzymes, i.e. the intracellular redox condition. If NADH is available, lactate is produced. If not, the pyruvate is eliminated in the form of acetoinic compounds. Interactions between the metabolic pathways and the bacteria's environment are the decisive factor. In any case, the presence of bacteria capable of degrading glycerol is a risk to the extent that, even if it is not totally metabolized, its metabolic by-products may spoil the wine to varying degrees.

5.4.2 Decarboxylation of Histidine

Histidine, an amino acid in wine, is decarboxylated into histamine, whose toxicity, although low, is additive to the toxicity of other biogenic amines (tyramine, phenyl ethylamine, putrescine and cadaverine) (Figure 5.9). Tyrosine is decarboxylated to form tyramine by a similar reaction.

In general, biogenic amines—histamine in particular—are more abundant in wines after malolactic fermentation. This explains the results presented in various works: red wines appeared to be richer in amines than white wines. Some researchers (Aerny, 1985) also proved that histamine is formed mainly at the end of malolactic fermentation, and even later.

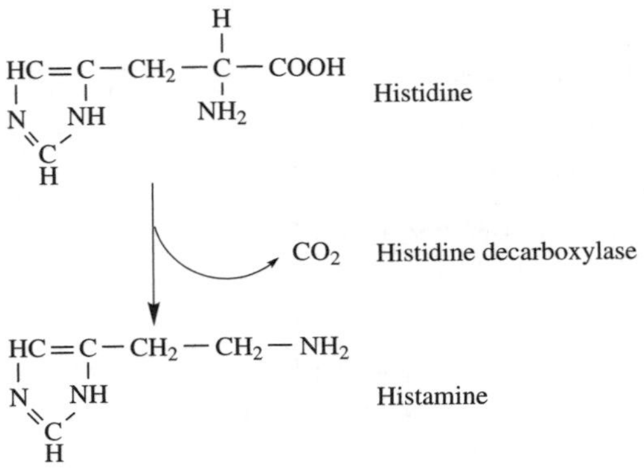

H
|
HC＝C－CH₂－C－COOH
| | | Histidine
N   NH   NH₂
  ＼C／
   ‖
   H

CO_2 Histidine decarboxylase

HC＝C－CH₂－CH₂－NH₂
| |
N   NH   Histamine
  ＼C／
   ‖
   H

Fig. 5.9. Histidine decarboxylation reaction

For a long time, many attempts were made to isolate bacteria capable of decarboxylating histidine, and they led to the conclusion that only contamination strains belonging to the genus *Pediococcus* had this property. According to some authors, the presence of histamine indicated a lack of winemaking skill and hygiene in the winery.

Yet this phenomenon occurs even during the malolactic fermentations of wines whose microflora is almost exclusively *O. oeni*. This fact contradicted the above results. An in-depth study of the microflora of wines rich in histamine finally led to the isolation of *O. oeni* strains that produce histamine from histidine (Lonvaud-Funel and Joyeux, 1994). In laboratory media, the formation of histamine by these strains increases as the growth conditions become less favorable: the absence of other substrates (in particular, sugar) at a low pH and in the presence of ethanol. The histamine concentration also, of course, depends on the histidine concentration. The addition of yeast lees, which progressively liberate amino acids and peptides, increases the concentration in the medium. Bacteria that also possess peptidases find a significant histidine source there. It is, therefore, not surprising that the histamine concentration increases toward the end and even after fermentation, as long as lactic acid bacteria are present. A precise study of the histidine carboxylase activity of a strain of *O. oeni* shows that even non-viable cells conserve their activity for a very long time in wine.

This enzyme has been fully purified. Its characteristics and properties are similar to those of the enzyme purified from a *Lactobacillus* of non-enological origin. The protein comprises two different subunits grouped in a type $[\alpha\beta]_6$ hexamer. The activity is optimal at a pH of 4.5; it is the same for whole cells. The enzyme is synthesized in the form of an inactive precursor Π. It is then activated by the scission of Π into α and β. Like the malolactic reaction, the decarboxylation of histidine does not directly generate cellular energy, but the strains of *O. oeni* that are capable of using it profit by its presence in the medium and grow more quickly than strains that do not have this capability. The energetic advantage is explained (as in *L. buchneri*) according to the process described for malolactic fermentation by Poolman (1993). The exchange between histidine and histamine at the membrane level creates a proton gradient and a proton motive force, generating ATP.

The strains of *O. oeni* that use histidine, therefore, have an additional advantage which can be a deciding factor after winemaking when the medium is poor in nutrients. Their survival can be facilitated in this manner.

Not all *O. oeni* strains possess the histidine decarboxylase; in fact, there are only a few. The detection of these particular strains was made possible by the use of molecular tools. DNA amplification by PCR, by using the appropriate oligonucleotide primers, permits the identification of these particular strains in a mixture. The labeling of a gene fragment, or a whole gene, that encodes the enzyme supplies the specific probe. Thus, the detection and enumeration of strains capable of forming histamine no longer present a problem. For example, the analysis of 250 samples of Bordeaux wines showed that nearly 50% contained such strains. This method can be extended to other bacteria, since it is specific only to the presence of the gene and not the bacterial species (Le Jeune *et al.*, 1995; Coton, 1996).

Until now, the only strains isolated from wine capable of producing tyramine were identified as *L. brevis* and *L. hilgardii* (Moreno-Arribas *et al.*, 2000). The tyrosine decarboxylase enzyme was studied in a particularly active strain of

L. brevis, then extracted from it (Moreno-Arribas and Lonvaud-Funel, 2001). As the aim was to produce detection tools specific to these strains, the protein was sequenced to find the coding sequence of the gene and, finally, the oligonucleotide primers required for PCR amplification (Lucas and Lonvaud-Funel, 2002; Landete *et al.*, 2003).

In future, tools for the rapid, accurate identification of strains producing biogenic amines will enable winemakers to assess the risks and also to carry out studies of the ecology of these strains, particularly their distribution and conditions encouraging their presence in some wineries.

5.4.3 Metabolism of Arginine

Among grape must amino acids, arginine is the most rapidly and completely consumed at the beginning of alcoholic fermentation. It is then secreted by the yeast and liberated during autolysis. The use of this amino acid by bacteria during fermentation was not closely studied until recently, notably as part of research on the origins of ethyl carbamate (Makaga, 1994; Liu *et al.*, 1994).

The microorganisms employ several metabolic pathways for using arginine. In lactic acid bacteria, the most widely used is the arginine deiminase pathway. The enzyme of this first step catalyzes the deamination of arginine into citrulline (Figure 5.10). Next, ornithine transcarbamylase and carbamate kinase lead to the formation of ornithine, CO_2, ammonium and the synthesis of ATP. Strictly heterofermentative *lactobacilli* were long considered to be the only ones capable of these transformations. Heterofermentative lactobacilli were long considered to be the only ones capable of effecting these transformations. This is the case of *L. hilgardii*, which has a highly active arginine metabolism that provides a high-level energy source (Tonon and Lonvaud-Funel, 2002).

However, some studies also identified this pathway in optionally heterofermentative *lactobacilli*, such as *Lactobacillus plantarum* (Arena *et al.*, 1999). Interestingly, *Oenococcus oeni* (*O. oeni*) was classified in Bergey's reference manual for the identification of bacteria as being among those that do not hydrolyze arginine. Recent works of authors

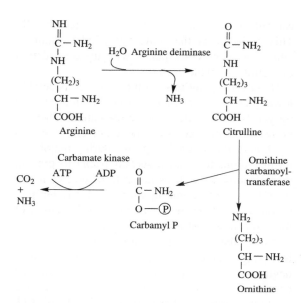

Fig. 5.10. Arginine degradation mechanism by *Oenococcus oeni*

cited earlier have, however, definitively confirmed that certain strains of the species *O. oeni* are capable of degrading arginine by the arginine deiminase pathway. The frequency of this character is not currently known.

The synthesis of three enzymes—arginine deiminase, ornithine transcarbamylase and carbamate kinase—is induced by the presence of arginine in the medium. A strain of *O. oeni* studied by Liu *et al.* (1994) transforms arginine stoichiometrically into a mole of ornithine and two moles of ammonium. In *O. oeni*, the genes that code for the enzymes of this metabolic pathway are organized in an operon on the chromosome including *arc* A, B, and C coding for arginine deiminase, ornithine transcarbamylase, and carbamate kinase, preceded by a regulation gene, *arc* R, and followed by two genes coding for transport proteins, *arc* D1 and *arc* D2 (Tonon *et al.*, 2001; Divol *et al.*, 2003).

In a collection of *O. oeni*, the strains do not systematically have this complete region of the chromosome. Of course, only those that have the complete region are capable of assimilating arginine via this pathway.

A small quantity of citrulline is still liberated in the medium (up to 16%); it is not taken up in the catabolism of *O. oeni* as opposed to lactobacilli. This molecule is the source of ethyl carbamate, whose presence corresponds with the degradation of arginine by wine bacteria.

However, quantities formed are considerably lower than those originating from the urea released by yeasts during alcoholic fermentation. There is no need to worry about the development of strains of *O. oeni* that degrade arginine during malolactic fermentation. It is, however, advisable to prevent these strains, as well as heterofermentative *lactobacilli*, from proliferating after malolactic fermentation. It is advisable not to risk the formation of these precursors of ethyl carbamate.

As is the case in decarboxylation reactions involving malic acid, histidine, and tyrosine, degrading arginine provides yeasts with additional energy resources. The net energy gain via the arginine pathway consists of a molecule of ATP produced from carbamyl-P.

Nevertheless, as with strains that decarboxylate histidine, strains that degrade arginine have an advantage over other strains. The net energy gain is an ATP molecule produced from carbamyl-P. In lactobacilli, it has been demonstrated that the uptake of arginine is coupled with the excretion of ornithine by an antiport system; therefore it does not require energy. At least in lactobacilli, arginine stimulates growth. The same effect has been demonstrated in *O. oeni* (Tonon and Lonvaud-Funel, 2000).

5.4.4 Synthesis of Exocellular Polysaccharides

The synthesis of exocellular polysaccharides by lactic acid bacteria is a very widespread character. *L. mesenteroides* and *Streptococcus mutans* produce glucose homopolymers such as dextran and glucan; fructose homopolymers (levans) and heteropolymers are also synthesized. Dextran of *L. mesenteroides* is the best known, as much for its different structures and its biosynthesis as for its various applications.

The production of exocellular polysaccharides increases viscosity and it can be measured or evaluated visually by the ropy character of the medium.

Exocellular polysaccharides are of interest to the milk industry and for industrial and medical applications, but much less so in enology. They give rise to ropiness and the *graisse* disease, studied by Pasteur. A lot less rare than *tourne* and *amertume*, this spoilage has incited new research.

In the literature, increased viscosity in ciders and beers is attributed to different lactic acid bacteria species, notably *P. damnosus* and *L. brevis* — which are also found in wine (Williamson, 1959; Beech and Carr, 1977). Luthi (1957) established that the symbiosis between lactic acid bacteria producing polysaccharides and acetic bacteria accelerates the increase in viscosity of the medium.

In the early 1980s, this spoilage reappeared with increasing frequency. Isolations carried out since then have demonstrated the involvement of the species *P. damnosus*. This fact does not exclude the possible participation of other species, but they are generally in much smaller proportions. Polysaccharide production was studied both by measuring medium viscosity and by determining polysaccharide concentrations. For a given wine, the viscosity increase corresponds directly to the production of the polysaccharide.

It is not simply a matter of measuring viscosity: it is the visual aspect of the wine that is the determining criterion for characterizing this problem. For example, a wine with a viscosity of 1.637 centistoke (cst) and a polysaccharide concentration of 95 mg/l is not ropy, as opposed to a wine with a viscosity of 1.615 and a polysaccharide concentration of 300 mg/l. Many other medium factors contribute to wine viscosity, notably the alcoholic content.

Compared with non-ropy strains of *P. damnosus*, the ropy strains are distinguished by the existence of a sort of refringent capsule around the cell, clearly visible under the microscope. The colonies formed on a solid medium are also easily identified by the formation of a thick fiber when picked with a platinum fiber. At the physiological level, the ropy strains demonstrate the occasional ability to adapt to growth in wine. They develop with the same ease whatever the alcoholic content, even greater than 12%. Their growth rate is hardly reduced at a pH of 3.5 compared with a pH of 4.5, and is not much affected by sulfur dioxide.

Growth remains normal at pH 3.7 with a free SO_2 concentration of 30 mg/l (Lonvaud-Funel and Joyeux, 1988).

Ropy *P. damnosus* strains increase wine viscosity when they multiply in a medium containing glucose. The disease is clearly visible when the population exceeds 10^7 units forming colony (UFC)/ml. Wine sugars such as fructose or pentoses do not permit the synthesis of the polysaccharide. It is a glucan, a glucose homopolymer with a structure comprising a β 1-3 chain on which a glucose unit is attached in β 1-2 every two units (Llaubères *et al.*, 1990):

$$-[glc - glc]-_n$$
$$|$$
$$glc$$

The particular structure of this glucan does not permit an enzymatic treatment of affected wines with currently known enzymes.

In wine, the polysaccharide is therefore formed from residual glucose. Several dozen milligrams per liter suffice to increase the viscosity. The spoilage can occur in the tank at the end of fermentation, but most problems are posed by spoilage in the bottle—mainly a few months after bottling. In-depth studies have shown that these strains adopt physiological forms that ensure not only their survival but also their growth. Furthermore, glucan production is greater in a nutrient-poor medium, such as wine. For the same amount of bacterial biomass, two to three 3 times more glucan is formed in a medium containing 0.1 g of glucose per liter than in a medium with 2 g/l. Similarly, for the same glucose concentration, the production is greater in a nitrogen-deficient medium. Independent of the survival and growth rate, the strain physiology in extreme media directs their metabolism towards the synthesis of glucan.

Laboratory studies of several strains have shown the great instability of the ropy phenotype. Strains transferred to a medium without ethanol rapidly lose this property. This result led to the comparison of ropy strains and their mutants. The presence of a plasmid distinguishes the strains. On this is a small plasmid with 5.5 kbp; three coding genes have been identified by homology, one probably for replicase, the second for a mobilization protein, and the third for a glucosyl transferase. This third gene is probably the key to its property of synthesizing exopolysaccharide (Walling *et al.*, 2001). Knowledge of this plasmid has made it possible to develop tools for detecting these strains, either by hybridization with a probe or by PCR (Lonvaud-Funel *et al.*, 1993; Gindreau *et al.*, 2001). They make it possible to identify and count "ropy" strains in a heterogeneous population of wine bacteria, including "non-ropy" *P. damnosus* or other species.

If spoilage occurs at the winery or a warehouse, the first precaution is evidently the disinfection of all of the tanks and winery material to avoid future contaminations. In general, a ropy wine does not present any other organoleptic defects and it can therefore be commercialized after the appropriate treatments. The viscosity can generally be lowered by the mechanical action of shaking the wine. Sulfiting at a minimum of 30 mg of free SO_2 per liter and progressive filtrations up to a sterile filtration ensure the preparation of the wine for a risk-free re-bottling. Heat treatment of the wine, just before bottling, is another a reliable solution for these fragile wines.

5.5 EFFECT OF THE METABOLISM OF LACTIC ACID BACTERIA ON WINE COMPOSITION AND QUALITY

Unless the appropriate inhibitory treatments are applied, lactic acid bacteria are part of the normal microflora of all white and red wines. From the start to the end of fermentation, and even during aging and storage, they alternate between successive growth and regression periods depending on the species and the strains. All multiplication or survival involves a metabolism that is perhaps very active or, on the contrary, hardly perceptible and even impossible to detect with current analytical methods. Substrates are transformed and consequently organoleptic characters are modified. Some metabolic activities are favorable and others are without consequence, while some are totally detrimental to wine quality (Volume 2, Section 8.3).

The main substrates for wine bacteria known to date are simple molecules: sugars and organic acids. Although their transformation is not currently verified, other more complex wine components, such as phenolic compounds, aromatic compounds or aroma precursors, present in small quantities, are without doubt partially metabolized. The repercussion of these minor transformations on organoleptic characters can be (depending on the molecules concerned) at least as important as the principal reactions.

The only substrate always metabolized by the same pathway by all species of wine bacteria is L-malic acid. Cellular activity is modulated by the presence of other compounds acting on the transport level or on the enzyme activity. The growth of lactic acid bacteria in wine is sought after because of this activity; indeed, it is the only activity truly desired. It permits the softening of wine provoked by deacidification and by the replacement of malic acid with lactic acid, a compound with a less aggressive flavor.

Bacteria degrade must and wine sugars with a different affinity depending on the species and perhaps even the strain. Hexoses are fermented into L- or D-lactic acid, or a mixture of the two forms, depending on the species. In general, bacterial development occurs after yeast development. Therefore, the lactic acid formed from sugars is in negligible quantities compared with the amount coming from malic acid. Several bacterial species produce D-lactic acid but it is the exclusive form for heterofermentative cocci, and thus *O. oeni*, the most important bacterium to enology. Among the sugar fermentation products of *O. oeni*, acetic acid is significant because of its contribution to the volatile acidity of wines. Like D-lactic acid, it is produced in small quantities as long as the bacteria do not ferment too much residual sugar. An increase in volatile acidity can therefore be attributed to lactic acid bacteria, if an abnormal amount (>0.3 g/l) of D-lactic acid is simultaneously formed. In this case, *O. oeni* fermented a significant quantity of sugars (a few grams per liter). This situation is called lactic disease.

Acetic acid is also one of the unavoidable metabolic products of citric acid, produced by homofermentative lactobacilli and especially by heterofermentative cocci. The fermentation of a few hundred milligrams of sugars per liter increases the volatile acidity during malolactic fermentation. Although carried out on a small quantity of the substrate, the degradation of citric acid is certainly important on account of the production of diacetyl.

Diacetyl, like the other α-dicarbonylated compounds in wine, glyoxal, methylglyoxal, and pentanedione, produced partly by the metabolism of lactic bacteria, are highly reactive. Reactions, in particular those with cysteine in wine, produce heterocycles such as thiazole, described as smelling of popcorn, toast, and hazelnuts, and thiophene, and furan, with aromas of coffee and burnt rubber (Marchand *et al.*, 2000).

Methionine and cysteine are metabolized into volatile sulfur compounds. The *O. oeni* species is particularly active in converting cysteine into hydrogen sulfide and 2-sulfanyl ethanol, and methionine into dimethyl disulfide, 3-(methasulfanyl) propanol, 3-(methasulfanyl) propan-1-ol and 3-(methasulfanyl) propionic acid. The most interesting of these compounds from a sensory point of view is 3-(methasulfanyl) propionic acid, with its earthy, red-berry fruit nuances (Pripis-Nicolau, 2002).

As for the other known metabolisms of lactic acid bacteria in wine, they all participate in one way or another in the spoilage of the wine. The degradation of essential wine components such as tartaric acid and glycerol into volatile acidity and bitter-tasting substances, respectively, completely destroys the organoleptic quality of the wine. The metabolism of amino acids (arginine, histidine, etc.) does not affect taste, but at a toxicological level it creates a problem by increasing the concentrations of biogenic amine and ethyl carbamate precursors in the wine. All things considered, ropiness seems to be the most widespread and spectacular disease, but even if it causes economic loss, the damages can be limited since the spoiled wine can be treated and commercialized.

In contrast to the metabolisms of malic acid, sugars and citric acid, these last transformations are carried out by certain strains belonging to normally inoffensive species. Bacterial spoilage can no longer be attributed to a specific bacterial species, as in the past. Certain strains of *O. oeni* form biogenic amines, and other strains form citrulline—precursor of ethyl carbonate.

REFERENCES

Aerny J. (1985) *Bull. OIV*, 656–657, 1016–1019.

Alizade M.A. and Simon H. (1973) *Hoppe-Seyler's Z. Physiol. Chem.*, 354, 163–168.

Ansanay V., Dequin S., Blondin B. and Barre P. (1993) *FEBS Lett.*, 332, 74–80.

Arena M.E., Saguir F.M. and Manca de Nadra M.C. (1999) *Int. J. Food Microbiol.*, 47, 203.

Beech F.W. and Carr J.G. (1977) in *Alcoholic Beverages*, (ed. A.H. Rosa) pp. 139–313. Economic Microbiology. Academic Press, London.

Claisse O. (2002) *Diplôme Expérimentation et Recherche en Oenologie*. Université Victor Segalen, Bordeaux 2.

Claisse O. and Lonvaud-Funel A. (2001) *J. Food Prot.*, 64, 833.

Coton E. (1996) Thèse de Doctorat, Université de Bordeaux II.

Denayrolles M., Aigle M. and Lonvaud-Funel A. (1994) *FEMS Microbiol. Lett.*, 116, 79–86.

Denayrolles M., Aigle M. and Lonvaud-Funel A. (1995) *FEMS Microbiol. Lett.*, 125, 37–44.

De Revel G., Bertrand A. and Lonvaud-Funel A. (1989) *Conn. Vigne Vin*, 23, 39–45.

Desmazeaud M. and de Roissart H. (1994) *in Les Bactéries Lactiques*, Vol. 1, (eds H. de Roissart and F.M. Luquet) pp. 194–198. Lorica, Uriage, France.

Divol B., Tonon T., Morichon S., Gindreau E. and Lonvaud-Funel A. (2003) *J. Appl. Microbiol.*, 94, 738.

Gindreau E., Walling E. and Lonvaud-Funel A. (2001) *J. Appl. Microbiol.*, 90, 535.

Gorga A., Claisse O. and Lonvaud-Funel A. (2002) *Sci. Aliment.*, 22, 113.

Grobler J., Bauer F., Subden R.E. and Van Vuuren H.J.J. (1995) *Yeast*, 11, 613.

Landete J.M., Ferrer S., Lucas P., Lonvaud-Funel A. and Pardo I. (2003) ≪ Oenologie 2003 ≫. 7ème *Symposium International D'œnologie, Bordeaux-Arcachon Tec et Doc*, Lavoisier Paris (à paraître).

Le Jeune C., Lonvaud-Funel A., Ten Brink B., Hofstra H. and Van der Voosen J.M.B.M. (1995) *J. Appl. Bacteriol.*, 78, 316–326.

Liu S.Q., Pritchard G.G., Hardman M.J. and Pilone G.J. (1994) *Am. J. Enol. Vitic.*, 45, 235–242.

Llaubères R.M., Richard B., Lonvaud-Funel A. and Dubourdieu D. (1990) *Carbohyd. Res.*, 203, 103–107.

Lonvaud M. (1975) Thèse Doctorat 3ème cycle, Université de Bordeaux II.

Lonvaud-Funel A. (1986) Thèse de Doctorat ès Sciences, Universitè de Bordeaux II.

Lonvaud-Funel A. and Joyeux A. (1988) *Sci. Alim.*, 8, 33–49.

Lonvaud-Funel A. and Joyeux A. (1994) *J. Appl. Bacteriol.*, 77, 401–407.

Lonvaud-Funel A. and Strasser de Saad A.M. (1982) *Appl. Environ. Microbiol.*, 43, 357–361.

Lonvaud-Funel A., Zmirou-Bonnamour C. and Weinzorn F. (1984) *Sci. Alim.*, 4 (HS III), 81–85.

Lonvaud-Funel A., Guilloux Y. and Joyeux A. (1993a) *J. Appl. Bacteriol.*, 74, 41–47.

Lonvaud-Funel A., Guilloux Y. and Joyeux A. (1993b) *J. Appl. Microbiol.*, 74, 41.

Lucas P. and Lonvaud-Funel A. (2002) *FEMS Microbiol. Lett.*, 21, 85.

Luthi H. (1957) *Am. J. Enol. Vitic.*, 8, 176–181.

Makaga E. (1994) Thèse de Doctorat, Université de Reims.

Marchand S., de Revel G. and Bertrand A. (2000) *J. Agric. Food Chem.*, 48, 4890.

Moreno-Arribas V., Torlois S., Joyeux A., Bertrand A. and Lonvaud-Funel A. (2000) *J. Appl. Microbiol.*, 88, 584.

Peynaud E. (1967) Etudes rècentes sur les bactèries lactiques du vin. *IIème Symposium International d'Œnologie*, Bordeaux.

Peynaud E. (1968) *CR Acad. Sci.*, 267D, 121–122.

Poolman B. (1993) *FEMS Microbiol. Rev.*, 12, 125–148.

Pripis-Nicolau L. (2002) Thèse de Doctorat, Université Victor Segalen Bordeaux 2.

Radler F. and Yannissis C. (1972) *Arch. Microbiol.*, 82, 219–239.

Ribéreau-Gayon J., Peynaud E., Ribéreau-Gayon P. and Sudraud P. (1995) *Traité d'Œnologie, Sciences et Techniques du Vin*, Vol. 2 Dunod, Paris.

Schütz M. and Radler F. (1974) *Arch. Microbiol.*, 96, 329–339.

Schütz H. and Radler F. (1984) *Arch. Microbiol.*, 139, 366–370.

Seifert W. (1901) *Z. Landwirstch. Versuchsu. Dent. Oest.*, 4, 980–992.

Tonon T. and Lonvaud-Funel A. (2000) *J. Appl. Microbiol.*, 89, 526.

Tonon T. and Lonvaud-Funel A. (2002) *Food Microbiol.*, 19, 451.

Tonon T., Bourdineaud J.P. and Lonvaud-Funel A. (2001) *Res. Microbiol.*, 152, 653.

Van Vuuren H.J.J. and Husnik J. (2003) Oenologie 2003. 7ème *Symposium International D'Oenologie, Bordeaux-Arcachon, Tec et Doc*, Paris (à paraître).

Walling E., Gindreau E. and Lonvaud-Funel A. (2001) *Dairy Sci. Technol.*, 81, 289.

Weinzorn F. (1985) Thèse de Doctorat 3ème cycle, Universitè de Bordeaux II.

Williamson D.H. (1959) *J. Appl. Bacteriol.*, 22, 392–402.

6

Lactic Acid Bacteria Development in Wine

6.1 LACTIC ACID BACTERIA NUTRITION IN WINE

Like all microorganisms, lactic bacteria cells multiply when conditions are favorable: presence of nutritional factors, absence of toxic factors, and adequate temperature. All of the principal reactions of its metabolism are directed towards the biosynthesis of cellular components: nucleic acids for the transmission of genetic heritage, carbohydrates, lipids, structure proteins and of course biologically active proteins. To ensure these syntheses, the cell must first find the necessary chemical elements in the medium: carbon, nitrogen and minerals—in usable forms. Since all of these synthesis reactions are endergonic, the medium must

also supply molecules capable of liberating the necessary energy. Most of the energy is supplied by the assimilation of various substrates. In addition, the cell receives energy from sophisticated systems which activate electron and proton transport phenomena. Although these systems cannot ensure the totality of cell growth, they contribute to it, very actively in certain cases, particularly when the cells are in nutritionally limited conditions.

6.1.1 Energy Sources

Most of the energy comes from the assimilation of numerous organic substrates, sugars, amino acids and organic acids. Lactic acid bacteria are chemo-organotrophic organisms. The oxidation of these substrates is principally represented by the fermentation of sugars. Heterofermentative and homofermentative lactic acid bacteria degrade hexoses and pentoses. At different stages of their metabolism, exergonic reactions permit the stocking of energy in ATP molecules (Section 5.2). The oxidation of sugars is always coupled with the reduction of coenzymes. In anaerobiosis, the lactic fermentation process is responsible for their reoxidation. In the metabolism of other substrates, the liberation of energy by reactions can be accompanied by the synthesis of ATP: this is the case with the degradation of citric acid and arginine.

The energetic importance of the proton motive force created at the membrane level has been demonstrated in several lactic acid bacteria species. The bacterial membrane in fact has a dual role: on the one hand, it is a barrier opposing the free diffusion of components of the medium and the cytoplasm; on the other hand, it is the site of proton and electron exchange. The proton motive force has two components: a difference in electric potential (negative inside) and a proton gradient (of pH). Maintaining a proton motive force requires an H^+-ATPase of the membrane, which functions reversibly. An influx of protons leads to the synthesis of ATP; conversely, the efflux of protons consumes energy. In lactic acid bacteria, the efflux of lactate from the metabolism is associated with the efflux of two protons (symport). In this manner, the efflux of protons does not require energy.

During malolactic fermentation, the use of malate produces a sufficient proton motive force for the synthesis of ATP. The influx of negatively charged malate into the cell is coupled with the efflux of neutral lactate; a difference in potential is created. Furthermore, the decarboxylation provokes the alkalinization of the cytoplasm and thus increases the pH gradient. All of this leads to the creation of the proton motive force. The energy indirectly furnished by the malolactic transformation is therefore conserved. This same process explains the energy gain by the decarboxylation of histidine and tyrosine. The histidine/histamine exchange, accompanied by the transfer of a negative charge, and the decarboxylation reaction provoke the alkalinization of the internal environment, ensuring the conservation of energy as in the previous case (Poolman, 1993).

6.1.2 Nutrients, Vitamins and Trace Elements

Apart from water (the most important component), cells draw carbon, nitrogen and mineral elements such as phosphorus and sulfur from their environment. These substances enter into the composition of cellular components.

Carbon essentially comes from sugars and sometimes organic acids. Glucose and fructose are the most represented sugars in wine after alcoholic fermentation (a few hundred milligrams per liter). Mannose, galactose, pentoses (arabinose, xylose, ribose), rhamnose and a few disaccharides are also present in small concentrations (a dozen milligrams of each per liter). The sugar degradation capacity depends on the bacterial species and (for example, for glucose) on environmental factors.

Oenococcus oeni degrades fructose more easily than glucose. Its presence in a mixture with glucose is beneficial to growth. Its reduction into mannitol regenerates coenzyme molecules necessary for the oxidation of glucose. Through a lack of reduced coenzymes, acetylphosphate does not lead to the formation of ethanol, but rather to acetic acid and ATP.

The energy obtained by the fermentation of residual sugars largely suffices to ensure the necessary growth for successfully starting and completing malolactic fermentation. According to Radler (1967), less than 1 g of glucose per liter covers the needs of the bacteria to form the biomass necessary for malolactic fermentation. In fact, much less than 1 g of glucose suffices, since other sugars in the medium are also used. The available sugars not only come directly from grape must but probably also from the hydrolysis of some of its components, notably polysaccharides.

Amino acids and sometimes peptides supply lactic acid bacteria with their assimilable nitrogen. Amino acid requirements vary with respect to the species and even the strain. These acids can be strictly indispensable or simply growth activators.

According to Ribéreau-Gayon *et al.* (1975), the following amino acids are necessary as a whole or in part, depending on the strain: Ala, Arg, Cys, Glu, His, Leu, Phe, Ser, Trp, Tyr and Val. Cocci have stricter demands than bacilli. The results of auxotrophic studies are, however, difficult to obtain and interpret. In a more recent study on *O. oeni*, Frémaux (1990) demonstrated their auxotrophy for Ile, Leu and Val. The synthesis pathways for these acids have enzymes in common for the production of aromatic acids (Phe, Trp, Tyr), derived from the same precursor, chorismic acid, and for Arg, His, Ser and Met. New observations suggest that His is a stimulant and not an essential.

Although these data remain very imprecise, an amino acid deficiency does not appear to be responsible for growth difficulties of lactic acid bacteria in wine. Temporary deficiencies can be noted at the beginning of alcoholic fermentation during the rapid yeast multiplication phase, but at the end this is no longer the case. The metabolism and then the autolysis of yeasts release a large variety and quantity of amino acids into the environment. The culture of *Oenococcus* and *Lactobacillus* in a synthetic laboratory medium shows that all of the amino acids of the medium can be consumed during growth. In wine, certain amino acids diminish while others increase in concentration, probably because of the simultaneous hydrolysis of peptides or proteins. In addition, the ammonium concentration increases following the deamination reaction (Ribéreau-Gayon *et al.*, 1975). Amino acids are essentially used for protein synthesis. Depending on the strain, some can be catabolized and serve as energy sources (arginine, histidine, and tyrosine).

Among nitrogen compounds, puric and pyrimidic bases play an important role in activating growth. In this case, the needs for adenine, guanine, uracile, thymine and thymidine are also dependent on the strain. They are not always essential.

Minerals such as Mg^{2+}, Mn^{2+}, K^+, and Na^+ are necessary. The first two are often used as key enzyme cofactors of the metabolism (kinases, malolactic enzyme). The following trace elements are involved in the nutrition of lactococci: Cu^{2+}, Fe^{3+}, Mo^{4+} and Se^{4+}. Yet the role of these metal ions is not yet established for wine lactic acid bacteria.

Vitamins are coenzymes or coenzyme precursors. Lactic acid bacteria are incapable of synthesizing B-group vitamins, in particular nicotinic acid, thiamin, biotin and pantothenic acid. A glycosyled derivative of pantothenic acid was identified in grape juice; it had been initially purified from tomato juice (Tomato Juice Factor: Amachi, 1975).

Finally, among the important chemical elements, phosphorus plays a primordial role in lactic acid bacteria, as in all cells, in the composition of nucleic acids, phospholipids and in the stocking of energy in the form of ATP.

All of the minerals and vitamins cited, as well as carbon substrates and nitrogen nutriments, are found in sufficient quantities in wine. Only in exceptional cases, are developmental difficulties of bacteria after alcoholic fermentation likely to be due to nutritional deficiencies. A simple experiment suffices to prove this statement: a favorable modification of one of the physicochemical factors that will be studied later (temperature, pH) usually permits the multiplication of the population. Independent of these physicochemical factors, the absence of growth must be considered to be caused by inhibitors.

6.2 PHYSICOCHEMICAL FACTORS OF BACTERIAL GROWTH

Four parameters very distinctly determine the growth rate of lactic acid bacteria in wine: pH, temperature, alcohol content and SO_2 concentration. Other factors are also in play but to a lesser degree and can only be determinant in some conditions.

These four essential factors have been known for a long time. They permitted the establishment of "enological rules". Progress in winery equipment has made these rules progressively easier to follow (Section 12.7.4). None of these factors can be considered independently of the others: the four act together as a unit. A favorable level of one compensates an unfavorable value of one or several others. It is also rather difficult to give an exact limit for each of them. In this way, bacteria tolerate higher alcohol contents and SO_2 concentrations in wines with favorable pHs than in wines with low pHs.

6.2.1 Influence of pH

The variation in growth rate related to the pH presents an optimum value and extreme limits. Most bacteria develop better at a pH near neutrality. This is not the case with acidogenic bacteria such as lactic acid bacteria: their acidophily permits their active development in wine at low pHs, around 3.5. At pHs as low as 2.9–3.0, growth remains possible but slow. At the upper pH limits of wine (3.7–3.8), it is much quicker. Stopped growth due to environmental acidity occurs when the intracellular pH attains a certain limit (pH_i). It not only depends on the environmental pH but also on the nature of the acids (McDonald *et al.*, 1990). In fact, the fraction of acids that freely penetrate in non-dissociated form is dissociated inside the cell, resulting in a decrease in pH. Consequently, the intracellular enzyme activity is more or less inhibited with respect to the optimum pH of their activity. The proton motive force and the dependent transports are also slowed, interfering with the global metabolism of the cell and thus multiplication. The lower limit tolerated for pH_i varies depending on the species. It is approximately 4.7 and 5.5, respectively, for *L. plantarum* and *L. mesenteroides*, according to McDonald *et al.* (1990). At pH 3.5, *O. oeni* maintains a higher pH_i than *L. plantarum* (Henick-Kling, 1986). The strains of this species adapt better to acidity than other species. Moreover, when cultivated in an acidic environment, they have a higher pH_i and thus a greater proton motive force—linked to the higher proton gradient.

Acidity adaptation mechanisms are not known but actively participate in the natural selection of this species in wine. It has been established for a long time that wines with relatively high pHs present not only a more abundant lactic microflora but also a much more varied one with respect to acidic wines. These wines are more microbiologically fragile as some of the bacteria

are spoilage factors, and as a broader range of substrates is metabolized. High pH facilitates the growth of bacteria in wine, as well as promotes their survival, not only directly but also by reducing the effectiveness of free sulfur dioxide. Spoilage may develop several months, or even years, after fermentation. The pH also has an impact on the malolactic activity of the entire cell.

Besides growth, the pH affects the malolactic activity of the entire cell. Although the optimum pH of the purified enzyme is 5.9, it is not the same for cells. The malolactic activity of *O. oeni* strains is optimum at a pH between 3.0 and 3.2 and around 60% of its maximum activity at pH 3.8. The usual pH range of wines, therefore, corresponds well with the maximum malolactic activity of the bacterial cell. Yet the malolactic fermentation rate depends on not only the activity but also the quantity of cells. Finally, at usual wine values, the pH affects both in the same way. Consequently, when all other conditions are equal, malolactic fermentation is quicker at higher pHs. For example, malolactic fermentation lasts 164 days for a wine adjusted to pH 3.15 and 14 days for a wine adjusted to 3.83 (Bousbouras and Kunkee, 1971).

According to Ribéreau-Gayon *et al.* (1975), the pH also conditions the nature of the substrates transformed. The authors defined the threshold pH for malic acid and sugar assimilation. It corresponds to the lowest pH at which the substrate is transformed and it varies according to the strain. The threshold pH for malic acid is lower than the threshold pH for sugars. In the zone between these two pHs, bacteria degrade malic acid without fermenting a large quantity of sugars and thus without producing volatile acidity. The larger the zone, the better adapted for winemaking is the strain. The average threshold pH of 400 heterofermentative coccus strains tested is 3.23 for malic acid and 3.51 for sugars. These values are respectively 3.38 and 3.32 for the 250 heterofermentative lactobacilli strains tested. The presence of the latter therefore does not guarantee a malolactic fermentation without the risk of volatile acidity production.

The pH is therefore very important and comes into play at several levels: in the selection of the best adapted strains; in the growth rate and yield; in the malolactic activity; and even in the nature of the substrates transformed.

The role of pH has diverse practical consequences in the control of the malolactic fermentation. First of all, the malolactic fermentation is initiated more easily and rapidly in press wines than in the corresponding free run wine. A partial chemical deacidification of wine may be advisable in the most difficult cases. It is especially recommended in the preparation of a malolactic fermentation starter—used for the inoculation of recalcitrant wine tanks. Finally, particular attention must be paid to musts and wines with elevated pHs. They sustain a more or less anarchic bacterial growth of a large variety of bacteria and are thus subject to spoilage. A sensible sulfiting is the only tool for controlling these microorganisms.

6.2.2 Effect of Sulfur Dioxide

In wine, sulfur dioxide (SO_2) is in equilibrium between its free and bound forms. Its effectiveness as a germicide and as an antioxidant is directly linked to wine composition and pH (Section 8.3.1). The active form, in fact, is molecular SO_2 which depends on the concentration of free SO_2 and the pH. To calculate it, the Sudraud and Chauvet (1985) formula can be used which gives the percentage of molecular SO_2 in function of the pH.

$$\%\text{molecular } SO_2 = 100/10^{pH-1.81} + 1 \quad (6.1)$$

For example, at pH 3.2 this percentage is 3.91%. It is, respectively, 2.00% and 1.01% at pH 3.5 and pH 3.8. These numbers demonstrate the influence of pH. Four times more free SO_2 is necessary at pH 3.8 than at pH 3.2 to obtain the same effectiveness.

The mechanism of the action of SO_2 was studied in yeasts in particular, but it is most likely very similar in bacteria. According to Romano and Suzzi (1992), SO_2 penetrates into the cell in molecular form by diffusion. In the cytoplasm where the pH is highest, it dissociates and reacts with essential biological molecules: enzymes with their disulfur bonds, coenzymes and vitamins. The

result is cessation of growth and, finally, cell death. The inhibitory action of SO_2 on the malolactic enzyme of *Oenococcus* is in addition to its effect on cellular growth.

For the same concentration of total SO_2, bacteria inhibition depends on the binding power of the wine (Section 13.3.2), which in turn determines the free SO_2 remaining and the pH. This establishes the amount of active molecular SO_2. It is only possible to give an approximation of the quantity of SO_2 necessary to inhibit bacterial development. As a general rule, lactic acid bacteria have difficulty in developing at concentrations ≥ 100 mg of total SO_2 per liter and 10 mg of free SO_2 per liter. Evidently, the result is not the same at pH 3.2 and at pH 3.8. Their sensitivity also varies according to the strain. Finally, for a given strain, the sensitivity varies according to environmental growth conditions and physiological adaptation possibilities.

Lafon-Lafourcade and Peynaud (1974), found that cocci seem less resistant than lactobacilli. Thus, *O. oeni* growth is hindered more than *L. hilgardii* growth, for example. The effect is also connected to the strain. *Pediococcus damnosus* is a useful example: the ropy strains are insensitive to SO_2 doses that inhibit or kill other strains. After 2 months of bottle storage, ropy type bacteria can maintain populations between 10^4 and 10^6 UFC/ml in wines containing 50 mg of free SO_2 per liter (Lonvaud-Funel and Joyeux, 1988).

Bound SO_2 also exerts a growth inhibitor effect, demonstrated by Fornachon (1963) (Section 8.6.3). Lactic acid bacteria may be capable of metabolizing the aldehyde fraction of the combination and liberating SO_2. The SO_2 then exerts its activity on the cell, but it is less effective. From their tests, Lafon-Lafourcade and Peynaud (1974) concluded that bound SO_2 is 5 to 10 times less active than free SO_2. Other authors have observed that its concentration in wine can easily be 5 to 10 times more elevated.

Technological consequences can be drawn from these results. When the elaborated wine must undergo malolactic fermentation, it is important to sulfite the grapes judiciously. The sulfiting must exert a transitory inhibitory effect on the lactic acid bacteria. At the end of alcoholic fermentation, the bound SO_2 persists and can delay bacterial growth. Obviously, sulfiting the wine during running off is not recommended, except in very unusual cases (Section 12.6.2).

6.2.3 Influence of Ethanol

Like most microorganisms, lactic acid bacteria are sensitive to ethanol. Generally, in laboratory conditions, bacteria isolated from wine are inhibited at an alcoholic strength of around 8–10% volume. Results vary according to the genus, species, and strain. Ribéreau-Gayon *et al.* (1975), found that cocci are altogether more sensitive to ethanol than are lactobacilli. At an alcohol content of 13% volume, more than 50% of the lactobacilli resist as opposed to only 14% of the cocci.

The growth of *O. oeni* strains isolated from wine and cultivated in the laboratory is activated at around 5–6% volume of ethanol; it is inhibited in environments richer in ethanol and difficult at or above 13–14% volume. The ethanol tolerance of laboratory strains is much less than for the same strains cultivated in wine. Bacteria that multiply in wine adapt to the presence of ethanol but also probably to the wine environment as a whole. In addition to the intrinsic strain tolerance of ethanol, their adaptation capacity varies. It is therefore difficult to set a limit above which lactic acid bacteria no longer multiply.

Strains of *Lactobacillus fructivorans, L. brevis* and *L. hilgardii* (heterofermentative bacilli) are frequently isolated from fortified wines with alcoholic strengths from 16 to 20% volume. They seem to be naturally adapted to ethanol but lose this adaptation after isolation. Strains of *L. fructivorans* nevertheless remain very tolerant of ethanol, which has an activator role in their case (Kalmar, 1995). *P. damnosus* bacteria are not particularly resistant to alcohol, but the ropy strains multiply at the same rate and with the same yield in the presence or absence of 10–12% volume of alcohol. The adaptation phenomena are definitely dissimilar in nature. In most cases, they are the result of a structural (fatty acid, phospholipid and protein composition) and functional modification of the membrane. In the case of ropy *P. damnosus*

strains, the polysaccharidic capsule possibly acts as a supplementary protector.

6.2.4 Effect of Temperature

Temperature influences the growth rate of all microorganisms. As with chemical reactions, it accelerates biochemical reactions. Cellular activity (resulting from all of the involved enzyme activities) and consequently growth vary with temperature according to a bell curve. At the optimum temperature, generation time is the quickest. This curb not only varies with the species and strains but also with the environment in which the bacteria multiply.

In a laboratory culture medium, lactic acid bacteria strains isolated from wine multiply between 15 and 45°C but their optimum growth range is between 20 to 37°C. The optimum growth temperature for *O. oeni* is from 27 to 30°C. but it is not the same in an alcoholic medium, especially in wine. The optimum temperature range is more limited: from 20 to 23°C. When the alcohol content increases to 13–14% volume of alcohol, the optimum temperature decreases. Growth slows as the temperature decreases, becoming nearly impossible around 14–15°C.

The ideal temperature for lactic acid bacteria growth (notably *O. oeni*) and for malic acid degradation in wine is around 20°C. An excessive temperature of 25°C or above always slows malolactic fermentation—principally by inhibiting the bacterial biomass. Additionally, an excessive temperature increases the risk of bacterial spoilage and increased volatile acidity. In practice, therefore, maintaining a wine at 20°C is recommended. It should not be allowed to cool too much after alcoholic fermentation. If the temperature of the winery decreases, the wine should be warmed.

When the temperature is less than 18°C, the initiation of malolactic fermentation is delayed and its duration is longer. A malolactic fermentation under way can continue even in a wine with a temperature between 10 and 15°C. In these cases, the bacterial biomass was normally constituted under favorable conditions. The cooling blocks the multiplication of bacteria but does not eliminate them. The cellular activity, however, is slower. Malolactic fermentation of a wine therefore continues after its initiation even in the case of being cooled, but the duration is much longer. The time frame for degrading all of the malic acid can range from 5–6 days to several weeks or months.

Along with pH, temperature is certainly the factor that most strongly influences the malolactic fermentation speed of a properly vinified wine not excessively sulfited. This factor is also the most easily monitored and controlled.

6.2.5 Other Factors Involved in Lactic Acid Bacteria Activity and Growth; Adaptation of Bacteria to Growth in Wine

The action of phenolic compounds on lactic acid bacteria growth remains relatively unknown. Past results have shown that polyphenols tested alone or in a mixture had an inhibitory effect. Saraiva (1983) noticed, on the contrary, that gallic acid stimulates yeasts and lactic acid bacteria. Conversely, different phenolic acids (coumaric, protocatechic acid, etc.) and condensed anthocyanins inhibited them. Enological tannins were found to have an antibacterial effect (Ribéreau-Gayon *et al.*, 1975). The effect of phenolic compounds on lactic acid bacteria growth remains unclear.

Nevertheless, a systematic study of several types of molecules clearly demonstrated the inhibitory effect of vanillic acid, seed procyanidins and oak ellagitannins and, at the same time, the stimulating effect of gallic acid and free anthocyanins (Vivas *et al.*, 1995). These results pertain to *O. oeni* growth, but may also be valid for other bacterial species. By favoring growth, gallic acid and anthocyanins activate malolactic fermentation. Bacteria degrade these two compounds. The transformation of anthocyanins seems to activate a β-glucosidase—freeing the aglycon fraction and the glucose, which is metabolized by bacteria.

Polyphenols, along with wine components as a whole, affect bacteria. Some are favorable and others unfavorable to bacterial growth and activity, but they play a secondary role compared with the other four parameters examined earlier. These

elements, among many others (most of them unknown), determine the malolactic fermentability of wines.

Similarly, oxygen can influence the multiplication of lactic acid bacteria in wine but its effect is not clear. In fact, the behavior of bacterial species present in wine can be diverse with respect to oxygen. They can be indifferent to its presence, adapt better in its absence (facultative anaerobiosis), tolerate oxygen at its partial pressure in air but be incapable of using it (aerotolerant), or finally can require a small oxygen concentration for optimal growth (microaerophiles). Furthermore, the behavior of a given strain can vary with its environment. In a laboratory culture medium, growth is activated in an inert gas atmosphere: CO_2 and N_2.

It is therefore difficult to specify the possible oxygen needs of lactic acid bacteria in wine. Current observations indicate that a limited aeration, after running off or racking wine, can strongly favor the initiation of malolactic fermentation.

Wine is an extremely complex environment and it is not possible to elucidate the effects of all of its components on lactic acid bacteria. In any case, this would not help the enologist, since these individual effects are cumulative—acting in synergy or, on the contrary, compensating each other. In this medium, lactic acid bacteria, particularly *O. oeni*, develop in extreme conditions. Acidity and ethanol combine with other molecules to inhibit the growth of the isolated strains.

It has long been known that a strain of *O. oeni* isolated from a wine undergoing malolactic fermentation, therefore capable of multiplying, then cultivated in a laboratory medium, loses its viability when re-inoculated in wine. Many observations, both in the laboratory and in the winery, suggest the existence of adaptation phenomena that ensure the survival and growth of bacteria in these extreme conditions. Isolated cells cultivated in a laboratory medium with wine added have a generally higher tolerance to low pH, SO_2, ethanol and wine than isolated cells cultivated in the absence of wine (Table 6.1).

The plasmic membrane probably participates actively in these adaptation phenomena, which have been shown to exist in *O. oeni* and other

Table 6.1. Influence of culture medium on the population (UFC/ml) of four *O. oeni* strains (A, B, C and D) after inoculation in red wine (Garbay, 1994)

Time (days)	Strains							
	A		B		C		D	
	M	MW	M	MW	M	MW	M	MW
0	3×10^6	6×10^6	4×10^5	3×10^6	1×10^7	6×10^6	1×10^7	3×10^6
1	$<10^4$	6×10^6	$<10^4$	4×10^6	4×10^5	3×10^6	3×10^5	2×10^6
3	$<10^4$	5×10^6	$<10^4$	6×10^6	$<10^4$	2×10^6	5×10^4	4×10^5
8	$<10^3$	9×10^7	$<10^4$	1×10^7	$<10^3$	6×10^7	1×10^5	1×10^7

M: Cells cultivated in laboratory medium.
MW: Cells cultivated in laboratory medium with wine added (V:V).

bacterial species. The first adaptation mechanism to be discovered was modification of the fatty acid composition of the membrane. All stress by the medium (addition of ethanol or wine, a temperature change, etc.) capable of provoking a modification of membrane fluidity, and thus membrane function, is compensated by an adjustment of the length and unsaturation level of the fatty acid chains. For all species studied (*O. oeni, Pediococcus damnosus, L. plantarum, L. hilgardii and L. fructivorans*), the presence of ethanol in the medium, for example, greatly increases the proportion of unsaturated fatty acids (Desens, 1989; Garbay *et al.*, 1995; Kalmar, 1995).

A second phenomenon, quantitatively and qualitatively significant, concerns membrane proteins. Their concentration increases following a shock—whether physical (cold or heat) or chemical (acidity, addition of ethanol, fatty acids or wine) (Table 6.2). In this manner, as with all living cells, *O. oeni* and the other lactic acid bacteria react to a shock by inducing the synthesis of "shock proteins". These proteins participate in the reaction of the cell against environmental stress (Garbay and Lonvaud-Funel, 1996). The Lo18 protein is induced in *O. oeni* by heat, ethanol, and acidity. It is associated with the membrane and maintains its integrity, following induction by a change in fluidity (Guzzo *et al.*, 1997; Delmas *et al.*, 2001).

Lactic acid bacteria are extremely exacting in their development in laboratory media. Contrary to all expectations, these microorganisms develop spontaneously in wine. Their development is due

Table 6.2. Influence of different types of stress on the protein concentration of the *O. oeni* plasmic membrane (Garbay, 1994)

Stress[a]	Proteins (mg per 10^{12} cells)
Control	2.2
Heating to 37°C	2.6
Heating to 42°C	2.8
Heating to 50°C	6.0
Incubation 10% ethanol	2.7
Incubation 10% ethanol + fatty acids	4.8

[a]Shock exposure time = 30 min.

to their complex group of adaptation phenomena—notably the induction of proteins whose functions remain unknown.

6.3 EVOLUTION OF LACTIC ACID BACTERIA MICROFLORA DURING FERMENTATION AND AGING, AND INFLUENCE ON WINE COMPOSITION

6.3.1 Evolution of the Total Lactic Acid Bacteria Population

In the production of wines requiring malolactic fermentation, the bacterial microflora passes through several phases (Figure 6.1). During the first days of fermentation, as soon as the tanks are filled, they are present in very variable quantities—most often from 10^2 to 10^4 UFC/ml. The extent of the population depends on climatic conditions during the

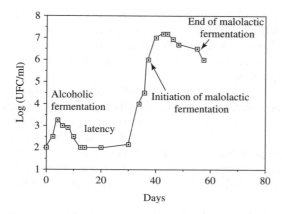

Fig. 6.1. Evolution of lactic acid bacteria population during alcoholic and malolactic fermentation

last days of maturation. It is generally lower when the conditions are propitious for healthy grapes and in these situations a single bacterial colony can be impossible to isolate on the berry. However, during the diverse operations from harvest to filling the tank, the must is inoculated very rapidly—probably by the equipment. During the harvest period the bacteria, like the yeasts, progressively colonize the winery. In general, the last tanks filled present the highest populations.

During the first days of alcoholic fermentation, the bacteria and yeasts multiply. The latter, better adapted to grape must, rapidly invade the medium with elevated populations. During this time, the bacteria multiply but their growth remains limited, with a maximum population of 10^4 to 10^5 UFC/ml. To a large extent their behavior at this time depends on the pH of the medium and the grape sulfiting level.

Normally, the doses of SO_2 added (about 5 g/hl) at pHs between 3.2 and 3.4 do not prevent their growth, but simply limit it. Then, from the most active phase of alcoholic fermentation to the depletion of sugars, the bacteria rapidly regress to 10^2 to 10^3 UFC/ml. This level also depends on environmental conditions (pH and SO_2).

Following alcoholic fermentation, the bacterial population remains in a latent phase for a varying period, which can last several months when the pH, ethanol and temperature parameters are at their lower limits. Usually, this phase lasts only for a

few days, and in certain cases, it does not occur at all. In the most frequently encountered situation, the multiplication phase takes place after the wine has been run off.

One microorganism follows the other: the yeasts first and then the lactic acid bacteria. These are ideal winemaking conditions, in which all of the fermentable sugars are depleted before the bacteria invade the medium. In the opposite case, the bacteria multiply actively towards the end of alcoholic fermentation: they ferment sugars using the heterofermentative pathway and increase the volatile acidity of the wine.

The growth phase lasts for several days and raises the population to around 10^7 UFC/ml or more. Evidently, its duration also depends on the composition of the medium. The subsequent stationary phase also varies. The bacteria then begin the decline phase. As soon as the malic acid is completely transformed, sulfiting is used to eliminate the bacteria as quickly as possible.

The malolactic fermentation phase begins during the growth phase, as soon as the total population exceeds 10^7 UFC/ml. It continues and is completed during the stationary phase, or sometimes at the beginning of the death phase. In very favorable conditions with a limited concentration of malic acid, malolactic fermentations are often completed even before the end of the growth phase. The optimum population in these cases exceeds 10^8 UFC/ml. As soon as a sufficient biomass is formed, malic acid is degraded. The malolactic acid bacterial activity is always present but depends on various conditions, especially the temperature. The transformation of 2 g of malic acid per liter can take more time than 4 g/l if the population level attained is lower.

If wine is not sulfited after malolactic fermentation, bacteria continue to survive for months. Carre (1982) observed a small decrease from 10^7 UFC/ml to 10^5 UFC/ml after 6 months of conservation in a wine stored at 19°C with a pH of 3.9 and an ethanol volume of 11.25%. Sulfiting immediately after the end of malolactic fermentation is intended to accelerate this death phase. No significant viable population should be left in the wine. Even if they can no longer multiply very actively, cells can metabolize diverse substrates to ensure their survival. These transformations have not all been explained but they increase the wine's concentrations of undesirable substances from a sensory or health standpoint (biogenic amines, ethyl carbamate, etc.).

Sulfiting at the end of secondary fermentation is practiced to adjust the free SO_2 concentration to 30–40 mg/l. At this concentration, nearly all of the lactic acid bacteria disappear within a few days (≤1–10 UFC/ml). The results also depend on the composition of the medium (Figure 6.2). Additionally, numerous observations have shown that the lactic population is maintained more easily in the barrel than in the tank. During 18 months of barrel aging, a decrease from only 10^6 UFC/ml to 10^3 UFC/ml was noted in spite of a free SO_2 concentration of between 20 and 30 mg/l. The last fining realized with egg whites effectively helps to eliminate bacteria.

In fact, the drop in the bacterial population assessed by counting the colonies developed on a nutrient medium does not apparently provide an accurate representation of the situation after sulfiting. Counting the bacteria by epifluorescence shows that part of the population retains some

metabolic activity, although the cells are incapable of multiplying in a nutrient medium. This physiological condition is described as "viable but non-cultivable" (Millet and Lonvaud-Funel, 2000) (Section 6.3.2).

According to Coton (1996), histamine may be accumulated in wines by this type of cell, which still has a high histidine decarboxylase activity. It is, therefore, possible that bacteria may be responsible for other transformations in wine constituents, even after sulfiting and during aging. These effects are not necessarily entirely negative.

6.3.2 Viable but Non-cultivable Bacteria

Viable bacteria are usually counted as bacterial colonies, on the principle that bacteria placed on a nutrient gel will multiply. After an incubation period, the resulting population becomes so large it is visible to the naked eye and is, thus, easily counted. Counting by epifluorescence is based on a completely different principle. Bacteria cells on a microscope slide or filter membrane are placed in contact with a substrate that is transformed by passing through each cell. The most common substrate is a fluorescence ester hydrolyzed by an esterase in the cell, which makes the bacterial content fluoresce under UV light. All cells that fluoresce under these conditions are considered viable as hydrolysis of the ester indicates the existence of enzyme activity.

Both methods give the same results for bacterial suspensions in the growth phase. When they move into the stationary, and then the decline phase, the difference between the results increases. While counting by epifluorescence shows a slight decrease in the number of cells, there is a sharp drop in the number of colonies visible. This difference may be explained by the fact that part of the population of fluorescent cells is still biologically active but is incapable of the metabolic and physiological functions necessary for multiplication. They are described as "viable, non-cultivable" (VNC) cells. Table 6.3 shows the lactic bacteria count after sulfiting a wine.

As expected, sulfiting eliminated the viable (and revivable) population. However, there were still

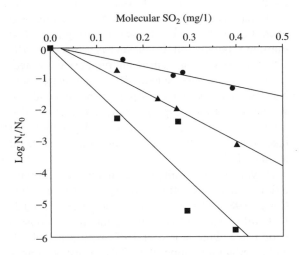

Fig. 6.2. Influence of molecular SO_2 concentration on lactic acid bacteria survival. (Lonvaud-Funel, unpublished results.) N_t = viable bacteria population 20 days after sulfiting; N_o = initial bacteria population, (■), wine A; (▲) wine B; (●) wine C

Table 6.3. Lactic bacteria populations counted by epifluorescence (cells.m/l) and by visible colonies (UFC. m/l) after sulfiting (Millet and Lonvaud-Funel, 2000)

Free SO$_2$ (mg/l)	Epifluorescence	Colonies
30	$(44 \pm 6) \times 10^5$	<1
50	$(44 \pm 5) \times 10^4$	<1

large numbers of VNC cells (Millet and Lonvaud-Funel, 2000).

This phenomenon is not exclusive to lactic bacteria, but certainly applies to many other microorganisms. It is easily demonstrated for acetic bacteria in winemaking. As soon as they are deprived of oxygen, the difference between the viable and VNC populations increases rapidly, then disappears completely as soon as the wine is aerated (Millet and Lonvaud-Funel, 2000). The same experiments showed that yeast and bacteria in VNC state decrease in size and some of them may pass through filters intended to eliminate them.

The extent of this phenomenon and its importance in winemaking have yet to be assessed. Further research is required to determine what happens to VNC cells and their capacity to recover viability, i.e. to multiply and produce colonies.

6.3.3 Evolution of Various Bacterial Species

During fermentation, the lactic microflora evolves not only in number but also in variety of species. Carre (1982) isolated bacteria on grapes before the harvest belonging to the following species: *Lactobacillus plantarum*, *L. hilgardii* and *L. casei*. The species *O. oeni*, which becomes the most significant later, is barely present at the beginning of fermentation. Just after its arrival in the tank, grape must contains a very diverse microflora, generally belonging to the eight usual lactobacilli and cocci species: *L. plantarum, L. casei, L. hilgardii, L. brevis, P. damnosus, P. pentosaceus, L. mesenteroides* and *O. oeni*) (Table 6.4).

All species are not always represented, or at least cannot be identified by current analytical methods, but a natural selection has been confirmed which takes place progressively during alcoholic fermentation. The lactic population regresses after reaching its optimum. At the same time, the homofermentative then heterofermentative lactobacilli disappear, to the benefit of *O. oeni*. Afterwards, the homofermentative cocci and *L. mesenteroides* also give way to *O. oeni* (see Table 6.4).

Certain species may also subsist at very low residual populations—less than 10 or 10^2 UFC/ml. Molecular methods, such as PCR and PCR-DGGE should enhance our knowledge of total residual microflora. As these methods amplify the signal specific to a particular microorganism, they make it possible to identify minority species. Furthermore, PCR-DGGE reveals the presence of unexpected microorganisms, as a region common to all bacteria is amplified, then each species is identified individually (Claisse and Lonvaud-Funel, 2003). Some species may proliferate at a later stage, once fermentation is completed, if the wine is not properly protected. After fermentation,

Table 6.4. Population (UFC/ml) of the different lactic acid bacteria species during the alcoholic fermentation of Cabernet Sauvrignon must (Lonvaud-Funel *et al.*, 1991)

Day	Alcohol Content (% vol.)	*Oenococcus oeni*	*Leuconostoc mesenteroides*	*Pediococcus damnosus*	*Lactobacillus hilgardii*	*Lactobacillus brevis*	*Lactobacillus plantarum*
0	0	nd	2.9×10^2	6.0×10^2	1.1×10^3	nd	7.5×10^1
3	7	nd	1.7×10^4	3.8×10^4	8.0×10^4	2.0×10^4	2.0×10^4
6	9	nd	9.6×10^4	3.7×10^4	4.0×10^4	4.5×10^3	nd
10	13	4.2×10^3	3.2×10^3	4.9×10^3	4.4×10^3	nd	nd
18	13	3.4×10^6	nd	nd	nd	nd	nd

nd: not detected.

some can multiply if the wine is poorly protected. In fact, species other than *O. oeni* are most often responsible for wine spoilage.

The spontaneous evolution of a mixture of species corresponds with the selection of those best adapted to wine—which is a hostile acidic and alcoholic environment. The composition of the plasmic membrane, and the various mechanisms, that permit it to react to the aggressiveness of the medium, seem to influence this adaptation. Certain species or strains may also differ in their ability to carry out these transformations. Strains of *L. fructivorans* adapt better to ethanol than *L. plantarum* and *L. hilgardii*, due to a more effective modification of their fatty acids (unsaturation and chain length) (Kalmar, 1995). Unsurprisingly, strains of this species are often identified in fortified wines tainted by lactic disease with an alcohol content between 15 and 20% volume.

6.3.4 Evolution of Wine Composition in the Different Phases of Bacterial Development

As soon as lactic acid bacteria multiply, they inevitably modify wine composition. In fact, their growth requires the assimilation of substrates to supply the cell with energy and carbon and nitrogen. The division of a bacterium into two daughter cells evidently supposes the neosynthesis of all of the structural components and molecules having a biological activity. In wine, bacteria transform sugars, organic acids and a multitude of other components. The type of reactions and the nature and concentration of these substances more or less profoundly modify the wine—improving or, on the contrary, spoiling it.

The only bacterial intervention truly sought after in winemaking is the transformation of malic acid into lactic acid (Section 12.7.2). It is the source of the most manifest organoleptic change, resulting from malolactic fermentation: the deacidification and the softening of wine. Malic acid, a dicarboxylic acid, is transformed molecule for molecule into lactic acid, monocarboxylic. The loss of an acid function per molecule is intensified by the replacement of an acid with

a particularly aggressive taste by a much softer acid. This transformation is carried out on 1.5 g/l to 8 g/l maximum, depending on the variety and grape maturation conditions.

Bacteria do not transform all of the malic acid contained in the grape. From the start, during alcoholic fermentation, yeasts metabolize a maximum of 30% of the malic acid. The product, pyruvate, then enters one of many yeast metabolic pathways—notably leading to the formation of ethanol. This "malo-alcoholic" fermentation is catalyzed at the first stage by the malic enzyme. The bacteria must develop a sufficient population before malolactic fermentation can truly start. The production of L-lactic acid is coupled with the decrease in malic acid (Figure 6.3).

The degradation of citric acid is also very important in enology. First of all, its disappearance from the medium contributes to the natural microbiological stabilization of wine by eliminating a potentially energetic substrate. Additionally, the organoleptical impact of the products of its metabolism, fairly well known at present, has been proven (Section 5.3.2), (De Revel *et al.*, 1996). Diacetyl is certainly involved and at low concentrations it gives wine an aromatic complexity that is much appreciated. In certain kinds of wine, tasters even prefer wine that has a very pronounced odor of this component. The degradation of citric acid also increases volatile acidity—a maximum of approximately 70 mg/l (H_2SO_4). Organoleptical deviations due to an excess of volatile acidity or diacetyl coming from the degradation of citric acid, however, are rare; they can have other origins. Recently, enologists and researchers have shown increased interest in this subject, but opinions differ to such a point that some currently advocate avoiding the degradation of citric acid as much as possible while others look for ways of ensuring it. Entirely different approaches have been considered, including the use of transformed bacterial strains to ensure or prevent this transformation or even for accumulating diacetyl.

In any case, citric acid is always degraded during malolactic fermentation, since *O. oeni* species have all of the necessary enzyme equipment. Its transformation is nevertheless slower than malic

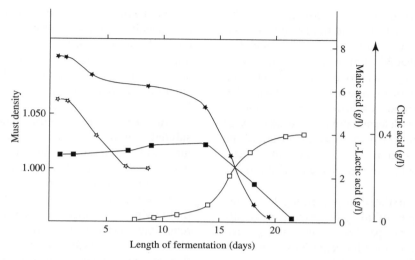

Fig. 6.3. Evolution of different compound concentrations during fermentation. (Lonvaud-Funel, 1986.) Filled star, malic acid; open star, density; filled square, citric acid; open square, L-lactic acid

acid, and several dozen mg of citric acid per liter often remain at the time of sulfiting. Yet since lactic acid bacteria are not eliminated immediately and remain active for several days (sometimes several weeks), only traces of citric acid remain in wine.

During winemaking, lactic disease is a dreaded bacterial spoilage. By definition, it corresponds with the increase of volatile acidity caused by the heterofermentative fermentation of sugars. Normally, lactic acid bacteria multiply only after the completion of alcoholic fermentation. The residual sugars—glucose, fructose and pentose—are in small but sufficient concentrations to ensure the essential energy needs of the bacteria. If bacterial growth occurs before the end of alcoholic fermentation, when more than 4–5 g of reducing sugar per liter remains in the wine, lactic disease can result. In fact, heterofermentative bacteria (*O. oeni*) ferment sugars not only into lactic acid but also into acetic acid. Moreover, the study of metabolisms has shown that the fermentation of glucose in the presence of fructose, which is the case in wine, preferentially directs the acetyl phosphate molecules towards acetic acid—fructose being reduced into mannitol (Section 5.2.2).

Lactic disease, therefore, occurs when environmental conditions are favorable to bacterial growth, even though yeasts have not yet completely fermented the sugars. This category essentially comprises wines derived from particularly ripe harvests. In fact, the sugar concentration is elevated, the medium is often poor in nitrogen and the pH is high. A stuck alcoholic fermentation (Section 3.8.2) or simply a sluggish fermentation should be expected. These phenomena by may lead to a rapid multiplication of lactic acid bacteria.

Lactic disease is also a widespread form of bacterial spoilage in fortified wines. These wines are elaborated by the addition of alcohol to grape must that has been slightly (or not at all) fermented. They are generally stabilized due to their high alcohol content. Yet lactic acid bacteria, and most often heterofermentative lactobacilli, are particularly resistant to ethanol. They develop easily in this very sugar-rich medium. Not only is there malolactic fermentation (which is not a real problem); there is also, lactic disease. The volatile acidity of these wines frequently attains 1–1.5 g/l (H_2SO_4). This phenomenon often occurs in the bottle, producing carbon dioxide and a cloudy wine.

A posteriori, the diagnosis of lactic disease is based on the nature of the products of the bacterial metabolism (Section 14.2.3). Wines presenting an elevated volatile acidity can also have been the site of acetic bacterial multiplication or a metabolic deviation of yeasts. Yet when they produce acetic acid, lactic acid bacteria (heterofermenters by definition) also form lactic acid from sugars—more precisely, exclusively D-lactic for cocci and DL-lactic for lactobacilli. Lafon-Lafourcade (1983) demonstrated that yeasts produce little D-lactic and concluded that lactic disease has occurred when the D-lactic concentration exceeds 0.2 g/l in wine. A simple enzymatic determination of the quantity of D-lactic acid therefore determines the origin of the wine spoilage. A limit of 0.3–0.4 g/l of D-lactic acid per liter would seem to ensure a more reliable diagnosis.

In wines, the first step in preventing lactic disease is the proper sulfiting of grapes, especially when they are very ripe. The corresponding musts are more subject to stuck fermentations than others (Section 3.8.1). The winemaker must react accordingly and, if need be, use additives such as nitrogen, vitamins and yeast hulls whose effectiveness is clearly established. Of course, elemental operations, notably aeration and temperature control, must also be scrupulously respected.

In the particular case of fortified wines, studies are under way to propose the best solutions for microbiological stabilization. The hygiene of the winery and barrels used in their production is essential. Sulfiting can resolve some of the problems but is not authorized for certain fortified wines. Heat treatment just before bottling is probably also a suitable solution for these wines.

The catabolism of sugars, malic acid and citric acid are normal occurrences during fermentation. Lactic disease only exists if *O. oeni* multiplies prematurely. Many other transformations also occur and some depend on the nature of the strain. Malolactic fermentation has been confirmed to cause chromatic changes in wines and a decrease in their color, while stabilizing it.

The *Met* and *Cys* sulfur-based amino acids as well as, probably, other precursor compounds are converted into volatile odoriferous compounds. They contribute to the increasing complexity in a wine's aromas and bouquet after malolactic fermentation De Revel *et al.*, 1999. *O. oeni* produces methanethiol, dimethyl sulfide, 3 (methylsulfanyl) propanol-1-ol, and 3(methylsulfanyl) propionic acid. Synthetic solutions of

3(methylsulfanyl) propionic acid, described as having chocolate and toasty aromas, have a perception threshold of 50 μg/l. Concentrations increase significantly after malolactic fermentation, and interaction with other components of wine produces an aroma reminiscent of red-berry fruit (Pripis-Nicolau, 2002).

After the secondary fermentation, the wine is sulfited. Sulfiting stabilizes the wine by eliminating viable bacteria and definitively blocking all microbial growth. Even then, profiting from a weak sulfitic protection or more often due to a natural resistance, spoilage strains sometimes succeed in multiplying. Wine diseases such as ropiness, *amertume* and *tourne* can be triggered (Section 5.4). Nonviable lactic acid bacteria, or at least those that are no longer capable of multiplying, can also still modify wine composition. Some strains produce histamine in this manner—these *O. oeni* bacteria decarboxylate histidine from the must, the metabolism and later yeast autolysis. The determination of histamine concentrations has shown an increase in concentrations during aging. The histidine decarboxylase enzyme maintains an elevated level of activity for several months in nonviable or at least non-cultivable cells (Coton, 1996). Consequently, these residual populations can be responsible for other minor, unidentified transformations of the wine during aging.

6.4 MICROBIAL INTERACTIONS DURING WINEMAKING

When the must arrives in the tank, it contains an extensive variety of microorganisms, fungi, yeasts, and lactic and acetic bacteria. Initially they come from the grape and from harvest equipment and then later from equipment that transports whole and crushed grapes to the tank. From this mixture, the microorganisms involved in winemaking are selected naturally—very quickly at first and afterwards more progressively. This selection takes place due to changes in environmental conditions (composition, oxidation–reduction potential) and specific antagonistic and synergistic interactions between the different microorganisms.

Successively or simultaneously, yeasts and bacteria interact not only with the different types of microorganisms (yeasts and bacteria) but also at the species and strain level. Due to the great diversity of microorganisms and their varying adaptation ability in the medium, a multitude of interactions between them can be imagined, depending on the winemaking stage. Only a few are well known. Some, on the contrary, are very difficult to identify and study. The yeasts/bacteria interactions during fermentation seem to be the most important.

6.4.1 Interactions Between Yeasts and Lactic Acid Bacteria

Yeasts are well adapted to growth in grape must. From the first days of fermentation, their multiplication is very rapid. Lactic acid bacteria also multiply very easily when inoculated alone in this same environment. Yet in practical conditions, yeasts and bacteria are mixed; yeasts are always observed to dominate bacteria. The experimental inoculation of grape must with *S. cerevisiae* and diverse lactobacilli or *S. cerevisiae* and a mixture of *O. oeni* clearly shows a behavioral difference between the bacteria.

When yeasts and lactic acid bacteria are inoculated in approximately equal concentrations (7×10^5 UFC/ml), lactobacilli are completely eliminated after 8 days. *O. oeni* disappears more slowly and subsists at a very low concentration. If the same must is inoculated with 10–100 times more bacteria, they remain viable for a longer period but eventually disappear—with the exception of *O. oeni*. This species is better adapted than the others to winemaking.

The interactions between *S. cerevisiae* and *O. oeni* have therefore been studied in greater detail. Grape must (220 g of sugar per liter) has been simultaneously inoculated with both microorganisms. Figure 6.4 illustrates their evolution in must at pH 3.4. In an initial phase, corresponding with the explosive growth of yeasts, the bacterial population regresses. After a transitory phase, the inverse phenomenon occurs. The yeast death phase coincides with the rapid growth phase of the bacteria. This evolution can be interpreted as

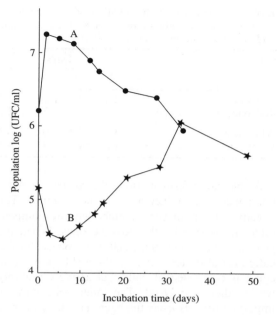

Fig. 6.4. Evolution of yeast and lactic acid bacteria (*Oenococcus oeni*) populations mixed inoculated in grape must (Lonvaud-Funel *et al.*, 1988b). Grape must pH = 3.4; concentration in sugars = 220 g/l. (A) Yeasts; (B) lactic acid bacteria

an antagonism exerted by yeasts on the *O. oeni* population. The bacteria not only do not multiply but also are partially eliminated. At this stage, nutritional deficiencies may also be responsible in part. Moreover, during the rapid growth period at the beginning of fermentation, yeasts have been proven to deplete the medium of amino acids. Arginine can be totally consumed. These deficiencies, hindering bacterial multiplication, are combined with the toxic effects of metabolites liberated by yeasts. In the first 3–4 days, the alcohol formed cannot explain this effect. Moreover, at low concentrations (5–6% volume,) it activates bacterial growth. Other substances are involved among the following: fatty acids liberated by yeasts, such as hexanoic, octanoic, decanoic and especially dodecanoic acid (Table 6.5) (Lonvaud-Funel *et al.*, 1988a). These acids target and alter the bacterial membrane. The incubation of whole cells in the presence of these fatty acids results in an ATP leak and a loss of malolactic activity.

Table 6.5. Influence of the addition of fatty acids on the malolactic fermentation rate (concentration in malic acid g/l) (Lonvaud-Funel *et al.*, 1988a)

Lot	Days			
	2	4	6	8
Red wine	2.5	2.1	1.7	1.4
Red Wine $+C_{10}$(23 μM)	2.8	2.7	2.5	2.5
White wine	3.5	2.4	0.6	0.2
White wine $+C_{12}$(2.5 μM)	4.5	4.5	4.5	4.5

As alcoholic fermentation takes place, the alcohol concentration increases in the medium. The negative effects of yeast metabolism are compensated in the end by the positive ones. When the yeast population enters the stationary phase, the situation is not static: in reality, the viable population count is composed of cells that actively multiply while others are lyzed. The latter cells play an important role *vis-à-vis* the bacteria—they liberate vitamins, nitrogen bases, peptides and amino acids. All of these components act as growth factors for the bacteria.

Therefore, in the final stage of alcoholic fermentation, yeasts stimulate bacterial growth. This effect is also combined with a lesser known phenomenon corresponding to an inhibition of yeasts by bacteria (Section 3.8.1) (Figure 6.5). More precisely, the bacteria accelerate the yeast death phase

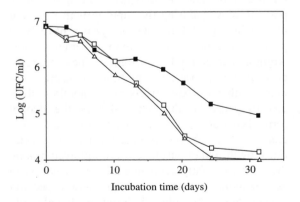

Fig. 6.5. Effect of lactic acid bacteria on the evolution of the yeast population after alcoholic fermentation (Paraskevopoulos, 1988). (■), pure yeast culture; (□), mixed culture (bacteria 10^6 UFC/ml); (▲), mixed culture (bacteria 10^7 UFC/ml)

(Lonvaud-Funel *et al.*, 1988b). Glucosidase and bacterial protease activities are certainly responsible for the hydrolysis of the yeast cell wall and lead to the lysis of the entire cell.

At the end of alcoholic fermentation, bacteria therefore accelerate yeast autolysis. Their growth is equally stimulated by the released products. These phenomena amplify each other and finally lead to a rapid decrease in yeast activity and viability. They contribute to slow or even stuck alcoholic fermentations. Yet bacteria probably also produce yeast inhibitors. In fact, grape must precultivated by bacteria (cocci or lactobacilli), is less fermentable by yeast than the control must. The wines obtained conserve several dozen grams of non-fermented sugar per liter. Among the species tested, *O. oeni* has the highest incidence (unpublished results). The role of *L. plantarum*, a species very common in musts, nevertheless needs to be emphasized. A strain of this species inhibits not only bacteria but also a large proportion of yeasts from the genera *Saccharomyces, Zygosaccharomyces* and *Schizosaccharomyces*. The inhibitory substance is an extracellular protein that is stable but inactivated by heat (Rammelsberg and Radler, 1990).

Environmental conditions, in particular pH and grape sulfiting, play an important role in the evolution of these mixed cultures (Figure 6.6). An elevated pH is favorable to bacterial growth. Evidently, the inverse is true of low pHs. But sulfiting considerably limits bacterial survival and growth at the beginning of the primary fermentation. Its role is essential. Yeasts should be allowed to multiply without leaving room for the bacteria. They must regress but remain in the medium, all the same, to take advantage of the yeast death phase and then multiply. These observations illustrate the importance of sulfiting grapes correctly. By taking the pH into account, winemaking incidents caused by the competition between yeasts and bacteria—such as lactic disease or, on the contrary, malolactic fermentation difficulties—can be avoided.

The nature and quantity of peptides, polysaccharides and other macromolecules in wine released by yeasts are different depending on winemaking techniques and the yeast strain. As a result, the

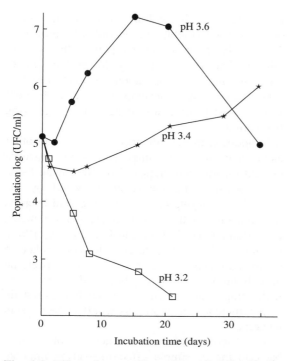

Fig. 6.6. Effect of must pH on the evolution of lactic acid bacteria populations in the presence of yeasts: grape must 220 g of sugar per liter (Lonvaud-Funel *et al.*, 1988b)

Table 6.6. Influence of the yeast strain on wine malolactic fermentability (Lonvaud-Funel, unpublished results)

Yeast number	Wine concentrations		Average duration of malolactic fermentation[a] (days)
	SO_2 (mg/l)	Dodecanoic acid (mg/l)	
1	8	0.27	12
2	10	0.42	9.25
3	3	0.22	6.25
4	21	0.32	9.25
5	35	0.28	8.5
6	51	0.44	21
7	16	0.31	10.25
8	2	0.42	7

[a]Calculated for the eight wines inoculated with four strains of *O. oeni*.

malolactic fermentability of wines obtained from the same must but fermented by different yeast strains varies greatly. The inhibitory contribution of yeast fatty acids, however, remains certain. Wines that are richest in these toxic substances are often the least propitious for bacterial development. Yet macromolecules, particularly polysaccharides, are capable of adsorbing these fatty acids and carry out a veritable detoxification of the medium. Yeast autolysis, autolysis rate (Lonvaud-Funel *et al.*, 1985; Guilloux-Benatier and Feuillat, 1993) and the nature of the molecules, which vary with the strain, influence their liberation in the medium.

For example, different wines were obtained from fermentation of the same must by eight commercial yeast strains. The difference in the duration of malolactic fermentation was then compared (Table 6.6). The wines had similar ethanol contents and pHs. After sterile filtration

and inoculation by four different *O. oeni* strains, they underwent malolactic fermentation. Their fermentation speeds varied from 6 to 21 days on average. Determinations of the quantity of sulfur dioxide and dodecanoic acid showed extremely varied concentrations, depending on the wine. A relationship between the duration of the malolactic fermentation and the concentration of these known bacterial inhibitory substances appears. The role of lees is very important: when added to a synthetic medium containing only sugars, malic acid and salts, they permit bacterial activity and growth.

6.4.2 Interactions Between Lactic Acid Bacteria

The succession of bacterial species during alcoholic fermentation can be explained by a difference in the sensitivity of bacteria to interactions with yeasts. Interactions between lactic acid bacteria must also exist, simultaneously. Like other microorganisms, they can synthesize and liberate substances with antimicrobial activities. This problem has been examined closely in the milk industry, where the consequences are more serious. These substances are simple (hydrogen peroxide, organic acids, etc.) or more complex. Bacteriocins are a class of proteins whose bactericidal activity generally has a narrow range of action. It is sometimes even limited to the same species as the producing strain. Fundamental and applied research on bacteriocins is on the increase and a large range of these substances produced by a large variety of lactobacilli and cocci is now known. So many have been discovered, in fact, that it could be imagined that each strain produces a specific bacteriocin. The key to proving their existence rests in finding sensitive strains.

Rammelsberg and Radler (1990), Lonvaud-Funel and Joyeux (1993) and Strasser de Saad et al. (1996) tackled this problem for wine lactic acid bacteria. The first of these works reported the discovery of two bacteriocins: brevicin from an L. brevis strain and caseicin from an L. casei strain. The first has a large range of action and inhibits O. oeni and P. damnosus strains in addition to L. brevis. Caseicin is only active on L. casei. Brevicin is a small thermostable protein (3 kDa) and is stable in a large pH range. Caseicin is less stable, with a much higher molecular weight (40–42 kDa). The same authors observed that a strain of L. plantarum has an antibacterial activity towards many bacterial species, including lactobacilli and cocci, notably O. oeni. The active protein synthesized by this strain has not yet been isolated. In a P. pentosaceus strain, Strasser de Saad et al. (1996) demonstrated the production of a bactericidal protein vis à vis several strains of L. hilgardii, P. pentosaceus and O. oeni. This bacteriocin, produced in large quantities in grape juice, is stable in the acidic conditions and ethanol concentrations of wine.

In the same way, various strains belonging to all of the species of the FOEB (Faculté d'Oenologie de Bordeaux) collection and isolated in wine were tested to look for possible reactions. Several associations were clearly demonstrated to create reactions in the liquid medium. The most obvious effects were recorded for P. pentosaceus and L. plantarum, both strongly inhibiting the growth of O. oeni and L. mesenteroides. This inhibition not only exists in mixed cultures but also when a culture medium pre-fermented by these two strains is added to O. oeni culture medium (Lonvaud-Funel and Joyeux, 1993). Different experiments have permitted the characterization of the possible roles of hydrogen peroxide, pH and lactic acid. For two strains, the inhibitory molecules which

accumulate in the culture medium are small peptides, thermostable and degraded by proteases. Their toxic effect is only temporary; they do not kill the bacteria but merely lower the growth rate and the final population. A more resistant sub-population may develop in the end or, more simply, these peptides are degraded by the growing population.

In addition to the influence of yeasts and other lactic acid bacteria, fungi and acetic bacteria present on infected grapes also affect wine lactic acid bacteria. The media precultivated by the above have varying effects on lactic acid bacteria multiplication with respect to the control media (San Romao, 1985; Lonvaud-Funel et al., 1987). Organic acids and polysaccharides accumulate in the medium and either impede or activate bacterial growth, but in practice they have little effect. Even if the grapes are tainted, these metabolites remain in insufficient concentrations to affect lactic acid bacteria.

The discovery of these few active molecules—bacteriocins or simple effectors—gives only an indication of the true situation in wine. They are specific not only to genera but also to species and especially strains. It is therefore impossible to try to identify them all. Nevertheless, they exist and carry out the selection of the strains observed in all winemaking. In the majority of cases, conditions ensure that the undesirable strains are swept aside during winemaking.

6.5 BACTERIOPHAGES

Bacteriophages are viruses capable of massively destroying cultures of sensitive bacterial strains. For lactic acid bacteria, bacteriophages were first discovered in the milk and cheese industry: they provided explanations of incidents during cheese production. Phage accidents increased in this industry with the use of unique strain ferments. Considerable research led to the use of mixed fermentation starters, which minimized these problems. In the future, phage-resistant strains will be developed genetically.

The bacteriophage must infect a bacterium in order to multiply. Inside the cell, it uses its own genome as code and the enzyme equipment of the cell to ensure the necessary syntheses. Depending on whether the phage is moderate or virulent, the multiplication cycle does not have the same effect on the development of the culture. With a moderate phage, the genome remains integrated in the bacterial chromosome in the form of a prophage and is replicated and transmitted altogether normally to the daughter cells. With a virulent phage, the virus multiplies into many copies—liberated in the medium after cell lysis. Each one of these copies then infects another cell, and so the destruction of the culture is massive. In certain conditions, the prophage carried by the lysogen can excise itself from the chromosome and start another lytic cycle.

In enology, the Suisse de Sozzi team carried out the first research on bacteriophages of lactic acid bacteria of the species *O. oeni* (Sozzi *et al.*, 1976; 1982). The phages were first discovered under electron microscope after centrifugation of the wine (Figure 6.7). Subsequently, identification was simplified by isolating sensitive indicator strains. Plaques could be observed on the indicator strain. The phages were then isolated and purified. According to the Sozzi team, abrupt stoppages of malolactic fermentation are caused by a phage attack, which destroys the total *O. oeni* population. Other authors, such as Davis *et al.* (1985), Henick-Kling *et al.* (1986) and Arendt *et al.* (1990), also demonstrated the existence of bacteriophages, without linking them to winemaking incidents.

The DNA extracts of all of the *O. oeni* phages hybridize together, and the marking of any of them furnishes a probe. By DNA/DNA hybridization, this probe permits the detection of lysogenic strains in a mixture. In this manner, we have established that nearly 90% of the *O. oeni* strains from our collection, isolated during malolactic fermentation, are lysogenic (Poblet and Lonvaud-Funel, 1996). The restriction profiles of isolated phages are not all identical, which confirms that several *O. oeni* strains coexist in wine during malolactic fermentation. Due to diverse interactions and variable phage sensitivity, these strains succeed each other.

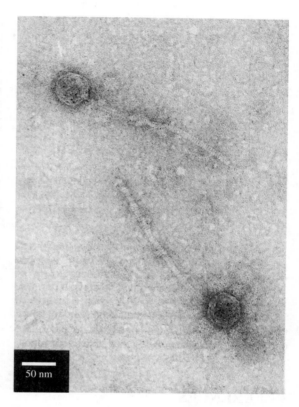

Fig. 6.7. Electron microscope photograph of *Oenococcus oeni* phages. (Photograph from Centre de Microscopie Électronique, Université de Bordeaux I.)

Bacteria and phage counts in two tanks during malolactic fermentation showed that both populations developed in a similar way. Phage populations appeared after a short time lag, decreased as the bacteria populations did, and reached a maximum two logarithmic units lower than the viable bacteria count (Figure 6.8). This result is normal, since the phage appears when bacteria multiply as a result of the excision of the prophage.

The diversity of *O. oeni* strains present in wine ensures against stuck malolactic fermentations caused by the phage destruction of bacteria. None of the strains is likely to have the same sensitivity to the phages. The elimination of one strain by phage attack is probably followed by the multiplication of other strains. In fact, a natural bacterial strain rotation can occur during winemaking.

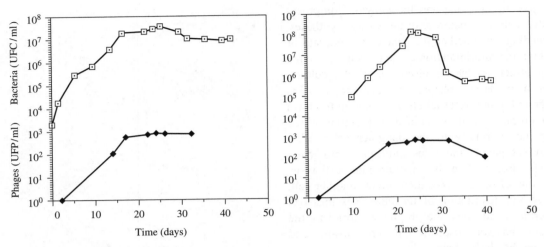

Fig. 6.8. Evolution of phage and lactic acid bacteria populations during malolactic fermentation (Poblet and Lonvaud-Funel, 1996)

A stuck malolactic fermentation can be feared only in exceptional circumstances when the phage and bacterial population reach the same number.

REFERENCES

Amachi T. (1975) *In Lactic Acid Bacteria in Beverages and Food* (eds J.G. Carr, C.V. Cutting and G.C. Whitings). Academic Press, London.

Arendt E., Neve H. and Hammes W.P. (1990) *Appl. Microbiol. Biotechnol.*, 34, 220–224.

Bousbouras G.E. and Kunkee R.E. (1971) *Am. J. Enol. Vitic.*, 22, 121.

Carre E. (1982) Thèse de Doctorat, Institut d'Œnologie, Université de Bordeaux II.

Coton E. (1996) Thèse de Doctorat, Faculté d'Œnologie, Université de Bordeaux II.

Claisse O. and Lonvaud-Funel A. (2003) *In œnologie* Tec et Doc. Lavoisier, (à paraître).

Davis C.R., Silveira N.F.A. and Fleet G.H. (1985) *Appl. Environ. Microbiol.*, 50, 872–876.

Delmas F., Pierre F., Coucheney F., Divies C. and Guzzo J. (2001) *J. Mol. Microbiol. Biotechnol.*, 3, 601.

De Revel G., Lonvaud-Funel A. and Bertrand A. (1996) *In Œnologie 95*. Tec. & Doc Lavoisier, Paris.

De Revel G., Martini N., Pripis-Nicolau L., Lonvaud-Funel A. and Bertrand A. (1999) *J. Agri. Food Chem.*, 47, 4003.

Desens C. (1989) Thèse de Doctorat, Institut d'Œnologie, Université de Bordeaux II.

Fornachon J.-C.M. (1963) *J. Sci. Food Agric.*, 14, 857.

Fremaux C. (1990) Thèse de Doctorat, Institut d'Œnologie, Université de Bordeaux II.

Garbay S. and Lonvaud-Funel A. (1996) *J. Appl. Bacteriol.*, 81, 613–625.

Garbay S., Rozès N. and Lonvaud-Funel A. (1995) *Food Microbiol.*, 12, 387–395.

Guilloux-Benatier M. and Feuillat M. (1993) *J. Int. Sci. Vigne Vin*, 27, 299–311.

Guzzo J., Delmas F., Pierre F., Jobin M., Samyn B., Van Beeumen J., Carvin J.F. and Divies C. (1997) *Lett. Appl. Microbiol.*, 214, 393.

Henick-Kling T. (1986) PhD Thesis, University of Adelaide, Australia.

Henick-Kling T., Lee T.H. and Nicholas D.J.D. (1986) *J. Appl. Bacteriol.*, 61, 525–534.

Joyeux A. (1988) *Sci. Nutriments*, 8, 33–49.

Kalmar Z.P. (1995) Thèse de Doctorat, Institut d'Œnologie, Université de Bordeaux II.

Lafon-Lafourcade S. (1983) *In Biotechnology*, Vol. 5 (eds H.J. Rehm and G. Reed) Verlag-Chemie, Weinheim.

Lafon-Lafourcade S. and Peynaud E. (1974) *Conn. Vigne Vin*, 8, 187.

Lonvaud–Funel A. (1986) Thèse de Doctorat ès Sciences, Institut d'Œnologie, Université de Bordeaux II.

Lonvaud-Funel A. and Joyeux A. (1988) *Sci. Aliments*, 8, 33–49.

Lonvaud-Funel A. and Joyeux A. (1993) *Food Microbiol.*, 10, 411–419.

Lonvaud-Funel A., Zmirou-Bonnamour C. and Weinzorn F. (1984) *Sci. Alim.*, 4, (H.S. III), 81–85.

Lonvaud-Funel A., Desens C. and Joyeux A. (1985) *Conn. Vigne Vin*, 19, 229–240.

Lonvaud-Funel A., San Romao M.V., Joyeux A. and Chauvet S. (1987) *Sci. Nutriments*, 7, 267–274.

Lonvaud-Funel A., Joyeux A. and Desens C. (1988a) *J. Sci. Food Agric.*, 44, 183–192.

Lonvaud-Funel A., Masclef J.-Ph., Joyeux A. and Paraskevopoulos Y. (1988b) *Conn. Vigne Vin*, 22, 11–24.

Lonvaud-Funel A., Joyeux A. and Ledoux O. (1991) *J. Appl. Bacteriol.*, 71, 501–508.

McDonald L.C., Fleming H.P. and Hassan H.M. (1990) *Appl. Environ. Microbiol.*, 56, 2120–2124.

Millet V. and Lonvaud-Funel A. (2000) *Lett. Appl. Microbiol.*, 30, 136.

Paraskevopoulos Y. (1988) Thèse de Docteur-Ingénieur, Institut d'Œnologie, Université de Bordeaux II.

Poblet M. and Lonvaud-Funel A. (1996) *in Œnologie 95*, pp. 313–316. Tec & Doc Lavoisier, Paris.

Poolman B. (1993) *FEMS Microbiol. Rev.*, 12, 125–148.

Pripis-Nicolau L. (2002) Thèse Doctorat de l'Université Victor Segalen, Bordeaux 2.

Radler F. (1967) *Conn. Vigne Vin*, 1, 73.

Rammelsberg M. and Radler F. (1990) *J. Appl. Microbiol.*, 69, 177–184.

Ribéreau-Gayon J., Peynaud E. and Ribéreau-Gayon P. (1975) *In Traité d'Œnologie. Sciences et Techniques du Vin*, vol. 2. Dunod, Paris.

Romano P. et and Suzzi G. (1992) *In Wine Microbiology and Biotechnology* (ed. G.H. Fleet). Harwood Academic Publishers, Chur, Switzerland.

San Romao M.V. (1985) *Conn. Vigne Vin*, 19, 109–116.

Saraiva R. (1983) Thèse Doctorat 3ème cycle, Institut d'Œnologie, Université de Bordeaux II.

Sozzi T., Maret R. and Poulin J.M. (1976) *Experientia*, 32, 568–569.

Sozzi T., Gnaegi F., d'Amico N. and Hose H. (1982) *Rev. Suisse Vitic. Arboric. Hortic.*, 14, 2–8.

Strasser de Saad A.M., Pasteris S. and Manca de Nadra (1996) *In Œnologie 95*, pp. 329–331. Tec. & Doc Lavoisier, Paris.

Sudraud P. et Chauvet S. (1985) *Conn. Vigne Vin*, 19, 31–40.

Vivas N., Bellemère L., Lonvaud-Funel A. and Glories Y. (1995) *Rev. Fr. Œnol.*, 151, 39–45.

7

Acetic Acid Bacteria

7.1 PRINCIPAL CHARACTERISTICS AND CYTOLOGY

Acetic acid bacteria are very prevalent in nature and are well adapted to growth in sugar-rich and alcohol-rich environments. Wine, beer and cider are natural habitats of these bacteria when production and storage conditions are not correctly controlled. Their quality is clearly lowered, except in the case of certain very particular beers.

Acetic acid bacteria cells generally have an ellipsoidal or rod-like form, with dimensions of 0.6–0.8 μm by 1–4 μm. They can be either single or organized in pairs or small chains. Some

are equipped with cilia, surrounding the cell or at its ends. These locomotive organs give the cells a mobility that is visible under the microscope. These bacteria, like lactic bacteria, do not sporulate. Their metabolism is strictly aerobic. Cellular oxidations of sugars, ethanol or other substrates are coupled with respiratory chain electron transport mechanisms. Oxygen is the ultimate acceptor of electrons and protons (coming from oxidation reactions).

The cellular structure of an acetic bacterium is similar to that of other bacteria: a cytoplasm containing genetic material (chromosome, plasmids), ribosomes and all of the enzymatic equipment, a plasmic membrane and cell wall. At the structural level, only the cell wall clearly distinguishes it from lactic bacteria. Acetic acid bacteria are Gram-negative, whereas lactic bacteria are Gram-positive (Section 4.1.1).

The Gram coloration reflects a significant structural difference of the cell wall of the two types of bacteria. Peptidoglycan is the principal constituent of Gram-positive cell walls, but it is much less present in Gram-negatives. In the latter, an essentially lipidic external membrane is present. It is destroyed by ethanol which acts as a solvent in the Gram test, resulting in the washing away of the violet dye. The external membrane is composed of phospholipids, lipoproteins and lipopolysaccharides. Like the plasmic membrane, it is organized into a lipid bilayer; a hydrophobic zone is contained between the layers. The lipopolysaccharides comprise a lipidic zone integrated into the external layer of the membrane, an oligosaccharide and a polysaccharidic chain at the exterior of the membrane. This chain carries the antigenic specificity of the bacterium. The lipoproteins join the thin peptidoglycan layer to the external membrane. Buried in the lipidic layers, crossing the entire membrane, proteins called porines form canals that permit exchanges across the cell wall.

7.2 CLASSIFICATION AND IDENTIFICATION

7.2.1 Classification

Acetic acid bacteria belong to the *Acetobacteraceae* family. Besides the previously mentioned characters, their principal property is to oxidize ethanol into acetic acid. Their (G + C) DNA base composition is from 51 to 65%. They are chemo-organotropic.

The bacteria of this family are separated into two genera: *Acetobacter* and *Gluconobacter*. The key distinguishing features according to *Bergey's Manual.* (De Ley *et al.*, 1984) are as follows:

- Genus *Acetobacter:* oxidize lactic and acetic acid into CO_2; non-mobile or peritrichous.

- Genus *Gluconobacter:* do not oxidize lactic or acetic acid; non-mobile or polar flagella.

De Ley *et al.* (1984) referenced a total of five species: *A. aceti, A. liquefasciens, A. pasteurianus* and *A. hansenii* for the genus *Acetobacter* and only *G. oxydans* for the genus *Gluconobacter*. Later studies on acetic acid bacteria led to the identification of new species: *A. diazotrophicus, A. methanolicus, A. xylinum, G. asaii, G. cerinus* (Swings, 1992) and more recently *A. europaeus*. This last species is clearly separated from other *Acetobacter* and *Gluconobacter* by its very low DNA/DNA hybridization percentage of between 0 and 22% (Sievers *et al.*, 1992). This species is pre-eminently used in vinegar production, due to its high ethanol and acetic acid tolerance. In *Gluconobacter*, a fourth species *G. frateurii* cannot be differentiated from *G. cerinus* by phenotypic comparison, but from its low DNA/DNA hybridization percentage it is very distant from it at the genetic level (Sievers *et al.*, 1995).

The classification has been updated still more recently on the basis of molecular phylogenic criteria to include the *Acetobacter, Gluconobacter, Gluconoacetobacter, Acidomonas, Asai,* and *Kozakia* genera in the *Acetobacteraceae* family of bacteria (Yamada *et al.*, 2002). Bacteria in the *A. liquefasciens, A. hansenii, A. methanolicus, A. xylinum, A. diatrophicus,* and *A. europaeus* species according to the previous classification (Table 7.1) are now included in the *Gluconoacetobacter* genus.

Three species unaffected by recent changes in the classification, *G. oxydans, A. aceti,* and *A. pasteurianus,* are the ones that are most frequently found in the course of winemaking, as well as, to a lesser extent, *Gluconoacetobacter liquefasciens* and *Gluconoacetobacter hansenii*. The three species succeed each other during winemaking. The *G. oxydans* present on the grape disappears and gives way to *Acetobacter*, which subsists in wine (Lafon-Lafourcade and Joyeux, 1981).

7.2.2 Isolation and Identification

The isolation of acetic acid bacteria from grape must or wine is carried out by culture on a solid nutritive medium. The composition of the medium varies, depending on the researcher. Nevertheless, taking into account their nutritional demands,

Table 7.1. Principal distinctive characteristics of Acetobacter species (Swings, 1992; Sievers *et al.*, 1992)

Characteristic	A. aceti	A. lique-fasciens	A. pasteu-rianus	A. hansenii	A. xylinum	A. methano-licus	A. diazo-trophicus	A. europaeus
Growth on acetic acid	−	−	−	−	−	−	nd	+
Produced from glucose:								
5-ketogluconic acid	+	d	−	d	+	−	−	d
2,5-diketogluconic acid	−	+	−	−	−	−	+	nd
Ketonic acid from glycerol	+	+	−	+	+	(+)	d	nd
Growth on ethanol	+	+	d	−	−	(+)	+	+
Growth on methanol	−	−	−	−	−	+	−	−

+: Positive. −: Negative. d: 11 to 89% of the strains are positive. (+): Low positive reaction. nd: Not determined.

these bacteria only develop well on rich media containing yeast extract, amino acids and glucose as the principal energy source. Swings (1992) described diverse media for isolating bacteria from different ecological niches. To isolate acetic bacteria from wine, the same medium may be used as for lactic bacteria (Ribéreau-Gayon *et al.*, 1975): 5 g/l of yeast extract, 5 g/l of amino acids from casein, 10 g/l of glucose, and 10 ml/l of tomato juice, with the pH adjusted to 4.5. It is also possible to use grape juice diluted with an equal amount of water plus 5 g/l yeast extract. The medium is solidified by adding 20 g/l of agar. To ensure that the medium supports only the growth of acetic bacteria, 0.2 ml 0.5% pimaricin and 0.1 ml 0.125% penicillin are added per 10 mL culture medium to eliminate yeasts and lactic bacteria. The culture must be incubated under aerobic conditions.

After isolation, the colonies put into pure cultures are identified by a group of tests and identification keys in *Bergey's Manual* (De Ley *et al.*, 1984). The first test is Gram coloration. Researchers also depend on the aptitude of the strain for developing on diverse constituents and on its metabolism in relation to different substrates. According to Swings (1992) and Sievers *et al.* (1992), Table 7.1 presents the identification keys for *Acetobacter* and Table 7.2 for *Gluconobacter* (Swings, 1992).

Gluconobacter and *Acetobacter* are differentiated by their ability to oxidize lactate: 20 g of lactate per liter is added to the medium, already constituted of yeast extract at 5 g/l. *Acetobacter*

Table 7.2. *Gluconobacter* species differentiation (Swings, 1992)

	Growth on ribitol	Growth on arabitol	Growth on nicotinic acid
G. oxydans	−	−	−
G. cerinus	+	+	+
G. asaii	−	−	+

oxidizes lactate; a cloudy zone is formed by the precipitation of calcium around the colony.

Ethanol oxidation by the two genera of bacteria is verified by culture in a medium containing 5 g of yeast extract per liter and 2–3% ethanol. The acidification of the medium is demonstrated either by titration or by the addition of a color-changing indicator (bromocresol green).

7.3 PRINCIPAL PHYSIOLOGICAL CHARACTERISTICS

The bacteria of the two genera *Acetobacter* and *Gluconobacter* are obligatory aerobic microorganisms with an exclusively respiratory metabolism. Their growth, at the expense of substrates that they oxidize, is therefore determined by the presence of dissolved oxygen in the environment. All of these species develop on the surface of liquid media and form a halo or haze, less often a cloudiness and a deposit.

Although present in the two genera, the characteristic metabolism of *Acetobacter* is the oxidation of ethanol into acetic acid with a high transformation yield. This is not the case for

Gluconobacter, which are characterized by a high oxidation activity of sugars into ketonic compounds (this activity is low in *Acetobacter*). In sum, bacteria of the genus *Acetobacter* prefer ethanol to glucose for their growth; the inverse is true for bacteria of the genus *Gluconobacter*. In addition, ethanol tolerance is parallel. In consequence, *Acetobacter* bacteria are more common in alcoholized environments (partially fermented musts and wines) than *Gluconobacter*, which are more present on the grape and in the must.

Some *Acetobacter* strains form cellulose in non-agitated culture media. Certain *Gluconobacter* produce other polysaccharides (glucans, levan, etc.), which make the medium viscous.

The vitamin demands are approximately identical for all acetic acid bacteria. Growth is only possible in environments enriched in yeast extract and peptone, which furnish the necessary carbon substrates. In order of preference, the best substrates for *Acetobacter* are ethanol, glycerol and lactate; for *Gluconobacter* they are mannitol, sorbitol, glycerol, fructose and glucose. Acetic acid bacteria are not known to require a specific amino acid. Certain *Acetobacter* and *Gluconobacter* are capable of using ammonium from its environment as a nitrogen source.

The optimum pH range for growth is from 5 to 6, but the majority of strains can easily multiply in acidic environments as low as pH 3.5.

Although they oxidize ethanol, acetic acid bacteria are not especially resistant to it. On average, *Gluconobacter* do not tolerate more than 5% ethanol, and few *Acetobacter* develop at above 10%. Evidently, adaptation phenomena (probably similar to those described for lactic bacteria) occur, ensuring their ethanol tolerance in wine. Acidity and ethanol concentration simultaneously influence the physiology and the resistance of acetic acid bacteria.

7.4 METABOLISMS

7.4.1 Metabolism of Sugars

The direct incomplete oxidation of sugars without phosphorylation leads to the formation of the corresponding ketones. The aldoses are oxidized into aldonic acids. The aldehydic function of this sugar is transformed into a carboxylic acid function. Glucose is oxidized into gluconic acid in this manner. The glucose oxidase catalyzes the reaction, which is coupled with the reduction of FAD. In acetic acid bacteria, electrons and protons are transported by the cytochrome chain to oxygen, which is the final acceptor.

Bacteria of the genus *Gluconobacter* in particular also have the property of oxidizing gluconic acid, leading to the formation of keto-5 gluconic, keto-2 gluconic and diketo-2,5 gluconic acids (Figure 7.1). These different molecules energetically bind with sulfur dioxide (Section 8.4.3). Wines made from grapes tainted by *Gluconobacter* are therefore very difficult to conserve.

Certain *Acetobacter* strains also form diketonic acid. Similarly, other aldoses, mannose and galactose, lead to the formation of mannonic and galactonic acid.

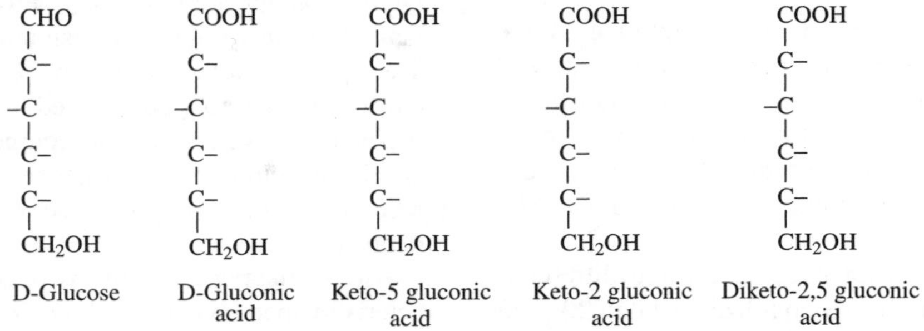

Fig. 7.1. Oxidation of gluconic acid by *Gluconobacter* bacteria

Ketoses are less easily oxidized by acetic acid bacteria. The oxidation of fructose can lead to the formation of gluconic acid and keto-5 fructose. The carbon chain of the sugar can also be divided, resulting in the accumulation of glyceric, glycolic and succinic acid. Especially for the *Acetobacter*, the final oxidation products of hexose are gluconate and ketogluconate.

The complete oxidation of sugars, however, furnishes the necessary energy for bacterial growth. The hexose monophosphate pathway is the metabolic pathway for the utilization of sugars. In *Acetobacter*, the tricarboxylic acid cycle is also used, but is absent in *Gluconobacter*. The enzymes of glycolysis either do not exist or only partially exist in acetic acid bacteria.

Oxidation by the hexose monophosphate pathway begins with the phosphorylation of sugar, followed by two successive oxidation reactions. The second is accompanied by a decarboxylation. The xylulose 5-P enters a group of transketolization and transaldolization reactions (Figure 7.2). The overall reaction is the degradation of a glucose molecule into six molecules of CO_2. In parallel, 12

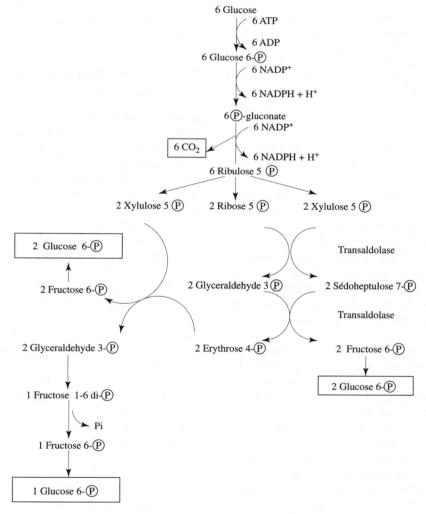

Fig. 7.2. Degradation of glucose by acetic acid bacteria (hexose monophosphate pathway)

coenzyme molecules are reduced. The transfer of electrons and protons by the cytochromic chains in turn reoxidizes the coenzymes. The transfer generates three molecules of ATP per pair of H^+ and e^-, 36 ATPs for the oxidation of a molecule of glucose into CO_2. This metabolic pathway is regulated by the pH of the environment and the glucose concentration. It is significantly inhibited by a low pH (<3.5) and a glucose concentration above 2 g/l. In these conditions, gluconic acid accumulates in the medium.

7.4.2 Metabolism of Ethanol

Among the transformations carried out by acetic acid bacteria, enologists are most interested in the transformation of ethanol. It is the source of an increase in volatile acidity in many cases. In fact, the oxidation of ethanol leads to the formation of acetic acid. The transformation takes place in two steps; the intermediary product is ethanal (acetaldehyde):

$$CH_3-CH_2OH \longrightarrow CH_3CHO \longrightarrow CH_3COOH$$
$$\text{ethanol} \qquad\qquad \text{ethanal} \qquad\qquad \text{acetic acid}$$
$$(7.1)$$

Acetobacter are also capable of oxidizing acetic acid, but this reaction is inhibited by ethanol. It therefore does not exist in enological conditions. Acetic acid slows the second step, when it accumulates in the medium, in which case the ethanal concentration of the wine may increase. According to Asai (1968), this second step is a dismutation of ethanal into ethanol and acetic acid. In aerobiosis, up to 75% of the ethanal leads to the formation of acetic acid. In intense aeration conditions, the oxidation and the dismutation convert all of the ethanol into acetic acid. When the medium grows poorer in oxygen, ethanal accumulates in the medium. Furthermore, a pH-dependent metabolic regulation preferentially directs the pathway towards oxidation rather than towards dismutation in an acidic environment.

The enzymes involved are, successively, alcohol dehydrogenase (ADH) and acetaldehyde dehydrogenase (ALDH). These two enzymes were proven to exist in *Acetobacter* and *Gluconobacter*. Two kinds have been distinguished: an NADP

coenzyme-dependent ADH and ALDH and a soluble coenzyme-independent ADH and ALDH. The first are soluble and cytoplasmic; the second are linked to the plasmic membrane. For the latter, the electrons generated in the oxidation reaction are conveyed to oxygen by an electron transport system integrated in the membrane. These membrane enzymes are incapable of reducing the NADP coenzyme (or NAD) but *in vitro* they reduce electron acceptors such as ferrocyanure and methylene blue. They are probably the most involved in the oxidation of ethanol in wine since they function at low pHs. Conversely, cytoplasmic enzymes, which function at the interior of the cell, have an elevated optimum pH of approximately 8.0.

7.4.3 Metabolism of Lactic Acid and Glycerol

In vitro, all of the species of the genus *Acetobacter* oxidize D- and L-lactic acid. Certain strains completely oxidize it into CO_2 and H_2O, but most stop at the acetic acid stage. The two isomers are transformed, but the activity is exerted more effectively on the D isomer. Pyruvate is the first intermediary. It is first decarboxylated into ethanal, which is oxidized into acetic acid by the ALDH.

Two types of enzymes have been identified, one in the membrane and the other in the cytoplasm (Asai, 1968). The D- and L-lactate oxydases are membranal enzymes which do not require a cofactor but function with the cytochrome chain. The membranes also contain the pyruvate decarboxylase, catalyzing the transformation into ethanal. In the cytoplasm, D- and L-lactate dehydrogenases ensure the oxidation of lactate into pyruvate. The pyruvate decarboxylase ensures ethanal production. Finally, the NADP-dependent ALDH leads to the formation of acetic acid. This metabolism does not seem to be particularly active in wine; it has never been proven that it is the source of wine spoilage.

The oxidation of glycerol leads to the formation of dihydroxyacetone (DHA: CH_2OH-CO-CH_2OH). Acetic acid bacteria, except *A. pasteurianus*, produce this compound. This reaction requires an intense oxygenation of the environment and is inhibited by ethanol. It is unlikely that it occurs in wine, but the conditions are more favorable on spoiled grapes. In fact, acetic acid bacteria are present on the grape alongside *Botrytis cinerea*, glycerol being one of its principal metabolites.

7.4.4 Formation of Acetoin

Acetoin is the compound at the intermediary oxidation level in the group of three acetoinic molecules present in wine: diacetyl, acetoin and butanediol. It is formed from pyruvate, itself a metabolic intermediary product having different origins in microorganisms.

Acetic acid bacteria produce acetoin from lactic acid that is oxidized beforehand. For most strains, this pathway is not very significant, but some form up to 74% of the theoretical maximum quantity.

Two synthesis pathways are believed to exist (Asai, 1968):

1. Pyruvate is decarboxylated in the presence of thiamine pyrophosphate (TPP) and leads to the formation of ethanal-TPP. The reaction of ethanal and ethanal-TPP forms acetoin:

$$2CH_3\text{--}CO\text{--}COOH + 2TPP \longrightarrow$$
$$2CH_3CHO\text{--}TPP + 2CO_2 \quad \textbf{(7.2)}$$

$$CH_3CHO\text{--}TPP \longrightarrow CH_3\text{--}CHO + TPP \quad \textbf{(7.3)}$$

$$CH_3CHO\text{--}TPP + CH_3\text{--}CHO \longrightarrow$$
$$CH_3CHOH\text{--}CO\text{--}CH_3 + TPP \quad \textbf{(7.4)}$$
$$\text{acetoin}$$

2. The other synthesis pathway bypasses the intermediary step of forming α-acetolactate (identical to the lactic acid bacteria pathway), by the reaction of ethanal-TPP and pyruvate:

$$CH_3CHO\text{--}TPP + CH_3\text{--}CO\text{--}COOH$$
$$\longrightarrow CH_3\text{--}CO\text{--}(CH_3)COH\text{--}COOH \quad \textbf{(7.5)}$$
$$\alpha\text{-acetolactate}$$

$$CH_3\text{--}CO\text{--}(CH_3)COH\text{--}COOH$$
$$\longrightarrow CO_2 + CH_3CHOH\text{--}CO\text{--}CH_3 \quad \textbf{(7.6)}$$
$$\text{acetoin}$$

G. oxydans can also oxidize butanediol into acetoin (Swings, 1992). The presence of acetoin in wine is less problematic than that of diacetyl, which has a much more pronounced aroma. Moreover, it has never been proven that acetic acid bacteria are capable of oxidizing acetoin into diacetyl.

7.5 ACETIC ACID BACTERIA DEVELOPMENT IN GRAPE MUST

Acetic acid bacteria are present on a ripe grape. The populations vary greatly according to grape health. On healthy grapes, the population level is low—around 10^2 UFC/ml, and it is almost entirely made up of *Gluconobacter oxydans*. Rotten grapes, however, are very contaminated: populations can reach upwards of 10^5 to 10^6 UFC/ml and are mixed, comprising varying proportions of *Gluconobacter* and *Acetobacter* (Lafon-Lafourcade and Joyeux, 1981).

This bacterial microflora modifies must composition by metabolizing sugars and sometimes organic acids. *Acetobacter* partially degrades citric and malic acids (Joyeux *et al.*, 1984). However, the most significant activity of these bacteria, especially those in the *Gluconoacetobacter oxydans* species, is producing substances that combine strongly with SO_2 (Sections 8.4.3; 8.4.6). They transform glucose into gluconic acid and its lactone derivatives, γ and δ gluconolactone, may combine up to 135 mg/l SO_2, in a must containing 24 g/l gluconic acid with a free SO_2 content of 50 mg/l (Barbe *et al.*, 2000). These bacteria also oxidize glycerol to form dihydroxyacetone, which combines with SO_2 in an unstable manner. This compound is metabolized by yeast during fermentation and is no longer a factor in forming combinations in wine.

The most abundant SO_2 combination due to *Gluconobacter* results in 5-oxofructose, which is not metabolized by yeast, so it remains unchanged in the wine. It is formed by oxidation of any fructose in the medium, as is the case in grape must. In a botrytized must where the fungus has developed to its most advanced stage, this compound alone accounts for 60% of all combinations. In wines made from this type of must, 5-oxofructose and γ- and δ-gluconolactones are involved in 50% of the SO_2 combinations, while ethanal and ketoacids formed by yeast account for most of the remainder (Barbe *et al.*, 2000).

In addition to acetic acid bacteria, yeasts contaminate grapes. Although alcohol production is limited, these strains do produce small quantities of ethanol directly on extremely rotten grapes or immediately following crushing and pressing. This alcohol is immediately oxidized by acetic acid bacteria. Some musts can therefore have a relatively high volatile acidity before fermentation.

Furthermore, acetic acid bacteria produce yeast-inhibiting substances. Lafon-Lafourcade and Joyeux (1981) demonstrated this fact for *Gluconobacter*. Cultivated during 3, 7 and 14 days in a must then inoculated by *S. cerevisiae*, this bacterium stopped alcoholic fermentation. There remained 1.5 g, 9 g and 18 g of non-fermented sugar per liter on average, respectively, as opposed to 0.5 g/l in the control. Gilliland and Lacey (1964) identified the same inhibitive effect of *Acetobacter* toward strains from seven yeast genera possibly present in must, including *Saccharomyces*. As a general rule, however, this effect is very limited. In fact, acetic acid bacteria activity in grape must is obligatorily short-lived: it stops at almost the same time that alcoholic fermentation begins.

In summary, the principal inconvenience of grape contamination by acetic acid bacteria is the production of volatile acidity and ketonic substances. This contamination results from sugars liberated through fissures on the grape caused by fungi during their proliferation on the berry. Part of the sugar is fermented into ethanol, which is oxidized into acetic acid; the rest is oxidized directly. In all cases, the winemaker's task is complicated: the volatile acidity can be high before fermentation, and musts altered in this manner strongly bind with sulfur dioxide.

7.6 EVOLUTION OF ACETIC ACID BACTERIA DURING WINEMAKING AND WINE AGING, AND THE IMPACT ON WINE QUALITY

The principal physiological characteristic of acetic acid bacteria is their need for oxygen to multiply. In wine, *Acetobacter aceti* and *A. pasteurianus* draw their energy from the oxidation of ethanol. Acetic acid concentrations indicate their activity. Finished wines contain around 0.3–0.5 g of volatile acidity (H_2SO_4) per liter, resulting from yeast and lactic acid bacteria metabolisms. Above this concentration, acetic acid accumulation most often comes from acetic acid bacteria; this problem is called acetic spoilage. This contamination must be avoided not only because of its negative effect on wine quality but also because of the legal limits on the concentration of volatile acidity permitted in wine. Acetic spoilage is accompanied by an increase in ethyl acetate. The perception threshold of this ester is around 160–180 mg/l. Yeasts also form it in concentrations of up to 50 mg/l. An excessive temperature during wine storage accelerates this spoilage.

Acetic acid bacteria multiply easily in aerobiosis, i.e. in grape must or wine at the surface in contact with air, but this is not the case during fermentation. As soon as alcoholic fermentation begins, the environment grows poor in oxygen and the oxidation–reduction potential falls.

Lafon-Lafourcade and Joyeux (1981) observed the evolution of bacteria during the production of two kinds of wine. In a white grape must parasitized by *B. cinerea*, the initial population of 2×10^6 UFC/ml fell to 8×10^4 UFC/ml, five days after harvest. At this stage, 90 g of sugar per liter was fermented. The population was less than 10^3 UFC/ml on the 12th day, after the fermentation of 170 g of sugar per liter. Similarly in the red grape must, the initial population of 2×10^4 UFC/ml progressively diminished to

20 UFC/ml by the time the wine was run off. The acetic acid bacteria are therefore not involved in alcoholic fermentation. The same is true during malolactic fermentation. Yet in all cases, they never totally disappear.

During barrel or tank aging, the wine should be protected from air to avoid both chemical and biological oxidation. In addition to oxidative yeasts, acetic acid bacteria still viable after both fermentations are capable of multiplying in the presence of air. To avoid this problem, the containers (tanks or barrels) should be filled as completely as possible. Topping off should be practiced with a wine of excellent microbiological quality to avoid contamination. An inert gas may also be used to replace the atmosphere present at the top of the tanks.

Aging also entails racking for clarifying and aerating the wine—causing limited oxidations that are indispensable to wine evolution. In the absence of air between rackings, acetic acid bacteria remain present in the entire wine mass at concentrations of 10^2 to 10^4 UFC/ml. During traditional barrel maturation, the dissolving of oxygen is more significant than in tanks, due to diffusion across the wood and the bung (when bung is on top). This slight oxygen dissolution suffices to ensure an oxidation–reduction level compatible with bacteria survival.

At the time of racking, the conditions are radically modified. The transfer from one tank or barrel to another is accompanied by the dissolution of 5–6 mg of oxygen per liter in the wine, unless very careful precautions are taken. The gas dissolves more quickly when air contact is favored and the temperature is lowered. This oxygen is at first rapidly and then more progressively consumed by the oxidizable substances in wine. The oxidation–reduction potential follows the same evolution. Table 7.3 illustrates the evolution of the dissolved oxygen concentration and the acetic acid bacteria population—the growth of which is very active just after racking. Afterwards, the bacteria slowly lose their viability until the next racking, several months later. The same

Table 7.3. Evolution of dissolved oxygen and acetic acid bacteria concentration when racking wine from one barrel to another

Stage	Dissolved oxygen (mg/l)	Acetic acid bacteria (UFC/ml)
Before racking	0.2	1.0×10^3
During racking	6.0	—
After 3 days	0.8	1.2×10^4
After 20 days	0.6	2.0×10^3
After 60 days	0.3	10

phenomenon occurs at each racking during the 18 months of aging.

Acetic acid is always synthesized during each growth phase of the bacteria. In a series of observations, its concentration increased from 0.03 to 0.04 g/l following rackings. These values vary greatly. They are linked to the bacteria population level and multiplication rate. For example, 0.02 g of acetic acid per liter was formed when a population was doubled from 3.5×10^4 UFC/ml to 7.0×10^4 UFC/ml. The acetic acid concentration increased by 0.08 g/l when the initial population of 50 UFC/ml multiplied to 1.5×10^3 UFC/ml. These observations prove that acetic bacteria play a key role in the increase in volatile acidity during aging (Millet *et al.*, 1995).

The principal factors affecting acetic acid bacteria development (as with lactic bacteria) are the alcoholic content, the pH, the SO_2 concentration, the temperature, and the oxidation–reduction potential. The more the pH and temperature are increased, the more easily the bacteria survive. Their multiplication is quicker in the case of aeration.

There is no effective method for eliminating acetic acid bacteria. Current observations show that even when protected by 25–30 mg of free SO_2 per liter, wines always conserve a viable bacteria population—up to 10^3 to 10^4 UFC/ml during barrel aging. Only a relatively low temperature of around 15°C can eventually limit this problem.

To avoid spoilage related to acetic acid bacteria, the winemaker should first of all concentrate on winery hygiene in order to eliminate potential contamination sources. Furthermore, all the other parameters (alcohol, pH, etc.) being equal, the influence of storage conditions on the oxidation–reduction potential is a deciding factor. In large-capacity tanks, the increase in volatile acidity is lower than in barrels. Similarly, even when the population is around 10^3 UFC/mL at the time of bottling, it decreases slowly but inexorably during bottle aging as the redox potential becomes very restrictive.

REFERENCES

Asai T. (1968) In *Acetic Acid Bacteria. Classification and biochemical activities*. University of Tokyo Press, Tokyo.

Barbe J.C., de revel G., Joyeux A., Lonvaud-Funel A. and Bertrand A. (2000) *J. Agric. Food Chem.*, 48, 3413.

De Ley J. Gillis M. and Swings J. (1984) In *Bergey's Manual of Systematic Bacteriology*. eds N.R. Krieg and J.G. Holt, pp. 267–268. Williams and Wilkins, Baltimore.

Gilliland R.P. and Lacey J.P. (1964) *Nature*, 202, 727–728.

Joyeux A., Lafon-Lafourcade S. and Ribéreau-Gayon P. (1984) *Sci. Aliments*, 4, 247–255.

Lafon-Lafourcade S. and Joyeux A. (1981) *Bull. OIV* 608, 803–829.

Millet V., Vivas N. and Lonvaud-Funel A. (1995) *J. Sci. Tech. Tonnellerie*, 1, 123.

Ribéreau-Gayon J., Peynaud E., Ribéreau-Gayon P. and Sudraud P. (1975) In *Traité d'Œnologie. Sciences et Techniques du Vin*, Vol. 2. Dunod, Paris.

Sievers M., Ludwig W. and Teuber M. (1992) *Syst. Appl. Microbiol.*, 15, 386–392.

Sievers M., Gaberthüel C., Boesch C., Ludwig W. and Teuber M. (1995) *FEMS Microbiol. Lett.* 126, 123–126.

Swings J. (1992) In *The Procaryotes* eds A. Balows, H.G. Trüpper, M. Dworkin, W. Harder and K.H. Schleifer pp. 2268–2286. Springer Verlag, New York.

Yamada Y., Katsura K., Kawasaki H., Widyastuti Y., Saono S., Seki T., Yamada Y., Uchimura T. and Komagata K. (2002) *Int. J. Syst. Evol. Microbiol.*, 52, 813.

8

The Use of Sulfur Dioxide in Must and Wine Treatment

8.1 INTRODUCTION

The general use of sulfur dioxide (SO_2) appears to date back to the end of the 18th century. Its many properties make it an indispensable aid in winemaking. Perhaps some wines could be made in total or near-total absence of SO_2 but it would certainly be presumptuous to claim that all of the wines produced in the various wineries throughout the world could be made in this manner. It must also be taken into account that yeasts produce small quantities of SO_2 during fermentation. In general, the amount formed is rarely more than 10 mg/l, but in certain cases it can exceed 30 mg/l. Consequently, the total absence of sulfur dioxide in wine is rare, even in the absence of sulfiting.

Its principal properties are as follows:

1. Antiseptic: it inhibits the development of microorganisms. It has a greater activity on bacteria than on yeasts. At low concentrations, the inhibition is transitory. High concentrations destroy a percentage of the microbial population. The effectiveness of a given

concentration is increased by lowering the initial population, by filtration for example. During storage, SO_2 hinders the development of all types of microorganisms (yeasts, lactic bacteria, and, to a lesser extent, acetic bacteria), preventing yeast haze formation, secondary fermentation of sweet white wines (Section 8.6.2), *Brettanomyces* contamination and the subsequent formation of ethyl-phenols (Volume 2, Section 8.4.4), the development of mycodermic yeast (flor) (Volume 2, Section 8.3.4), and various types of bacteria spoilage (Volume 2, Sections 8.3.1 and 8.3.3).

2. Antioxidant: in the presence of catalyzers, it binds with dissolved oxygen according to the following reaction:

$$SO_2 + \tfrac{1}{2}O_2 \longrightarrow SO_3 \qquad (8.1)$$

This reaction is slow. It protects wines from chemical oxidations, but it has no effect on enzymatic oxidations, which are very quick. SO_2 protects wine from an excessively intense oxidation of its phenolic compounds and certain elements of its aroma. It prevents madeirization. It also contributes to the establishment of a sufficiently low oxidation–reduction potential, favoring wine aroma and taste development during storage and aging.

3. Antioxidasic: it instantaneously inhibits the functioning of oxidation enzymes (tyrosinase, laccase) and can ensure their destruction over time. Before fermentation, SO_2 protects musts from oxidation by this mechanism. It also helps to avoid oxidasic casse in white and red wines made from rotten grapes.

4. Binding ethanal and other similar products, it protects wine aromas and makes the flat character disappear.

Adding SO_2 to wine raises a number of issues. Excessive doses must be avoided, above all for health reasons, but also because of their impact on aroma. High doses neutralize aroma, while even larger amounts produce characteristic aroma defects, i.e. a smell of wet wool that rapidly becomes suffocating and irritating, together with

a burning sensation on the aftertaste. However, an insufficient concentration does not ensure the total stability of the wine. Excessive oxidation or microbial development can compromise its presentation and quality.

It is not easy to calculate the precise quantities required, because of the complex chemical equilibrium of this molecule in wine. It exists in different forms that possess different properties in media of different composition.

The concentration of sulfur dioxide in wine is habitually expressed in mg SO_2 per liter (or ppm) although this substance exists in multiple forms in wine (Section 8.3).

The words sulfur dioxide, sulfur anhydride or sulfurous gas can all be used equally, or even sulfurous acid, though the corresponding molecule cannot be isolated. The expression 'sulfur', however, is fundamentally incorrect. Additions made to wine are always expressed in the anhydrous form, in mg/l or in g/hl, regardless of the form effectively employed—sulfur dioxide gas or liquid solution, potassium bisulfite ($KHSO_3$) or potassium metabisulfite ($K_2S_2O_5$). The effect of the addition to wine is the same, regardless of the form used. The equilibrium established between the various forms is identical. It depends on the pH and the presence of molecules that bind with the sulfur dioxide.

Substantial progress in the understanding of the chemistry of sulfur dioxide and its properties have permitted the winemaker to reason its use in wine. As a result, the concentrations of SO_2 employed in wine have considerably decreased. Simultaneously, this technological progress has led to a decrease in authorized concentrations. In 1907, French legislation set the legal limit in all wines at 350 mg/l increased, in 1926, to 450 mg/l. Today, French wines are subject to EU legislation (Table 8.1), which has gradually reduced the permitted level to 160 mg/l for most red wines and 210 mg/l for the majority of white wines. Higher doses may only be used in wines with very high sugar content. They are generally premium wines produced in small volumes and consumed in moderate quantities.

In practice, the concentration used is even lower. For white French wines (excluding special wines)

Table 8.1. Maximum sulfur dioxide concentrations depending on wine type. EU regulations and OIV recommendations (values expressed in mg/l)

A. EU regulations no: 1493/1999 and 1622/2000, modified in 1655/2001

Types of wine	Sugar content <5 g/l	Sugar content = or >5 g/l
Red wines	160 (+40)*	210 (+40)*
White and rosé wines	210 (+40)*	260 (+40)*
Red *vins de pays*	125	150
White and rosé *vins de pays*	150	175
Dessert wines	150	200
Vins de pays (TAV >15% vol.; sugar >45 g/l)		
White AOC wines		
Bordeaux superieur, Graves de Vayres, Côtes de Bordeaux Saint-Macaire, Premieres Côtes de Bordeaux, Sainte-Foy Bordeaux, Côtes de Bergerac suivie ou non de la denomination Côtes de Saussignac, Haut Montravel, Côtes de Montravel et Rosette, Gaillac		
White DO wines		
Allela, La Mancha, Navarra, Penedes, Rioja, Reuda, Tarragona et Valencia		
Alto Adlge, Trentino "passito" "vendemmia tardiva"		
Vqprd Moscato di Pantelleria naturale and Moscato di Pantelleria		
United Kingdom Vqprd described as follows:		
botrytis, noble harvest, noble late harvested		
German wines		
Spätlese		300
Auslese and some Rumanian white wines		350
Beerenauslese, Ausbruch, Ausbruch-wein, Trockenbeerenauslese, Elswein		400
White AOC wines		
Sauternes, Barsac, Cadillac, Cérons, Louplac, Sainte-Crolx-du-Mont, Graves supérieurs, Monbazillac, Jurançon, Pacherenc du Vic Bilh. Anjou-Coteaux de la Loire, Bonnezeaux, Quarts de Chaume, Coteaux de l'Aubance, Coteaux du Layon sulvi du nom de la commune d'origine, Coteaux du Layon suivi du nom de Chauma, Coteaux de Saumur		
Alsace et Alsace grand cru suivi de la mention "vendanges tardives" ou "selection de grains nobles"		
Sweet wines from Greece (sugar = or >45 g/l)		
Samos, Rhodes, Patras, Rio Patron, Cephalonie, Limnos, Sitia, Santorin, Néméa, Daphines		
Certain Canadian white wines (Icewine)		400

*When required due to weather conditions in certain vineyard areas.

B. OIV—maximum acceptable limits: International Code of Winemaking Practices and Collection of International Wine Analysis Methods, 2001.

Types of wine	Sugar content = or <4 g/l	Sugar content >4 g/l
Red wines	150	300
White and rosé wines	200	300
Certain sweet white wines		400

the average concentration is 105 mg/l; for red wines it is 75 mg/l. The Office International de la Vigne et du Vin (OIV) recommends slightly higher values than those advocated by the EU in its member countries. In certain countries, the regulation of sulfur dioxide dictates a common limit for all wines. For example, this value is 350 mg/l in the USA, in Canada, in Japan and in Australia.

Due to the fluctuating equilibrium between free and bound forms of SO_2, in general, the legislation of different countries exclusively refers to the total sulfur dioxide concentration. Certain countries, however, have regulations for the free fraction.

Today, especially for health reasons, the possibility of further reducing the authorized concentrations in different kinds of wines is sought after. Such an approach consists of optimizing the conditions and perfecting the methods of using this product. This supposes more in-depth knowledge of the chemical properties of the sulfur dioxide molecule and its enological role. Substitute products can also be considered. Due to the various effects of sulfur dioxide in wine, the existence of another substance performing the same roles without the disadvantages seems very unlikely, but, the existence of adjuvants, complementing the effect of SO_2 in some of its properties, is perfectly conceivable. Enological research has always been preoccupied by the quest for such a product or substitution process (Chapter 9).

In conclusion, sulfur dioxide permits the storage of many types of wine known, today that would not exist without its protection. In particular, it permits extended barrel maturation and bottle aging. In view of its involvement in a wide variety of chemical reactions, it is not easy to determine the optimum dose to obtain all the benefits of SO_2 without any of its unfortunate side-effects. The adjustment should be made within plus or minus 10 mg/l.

8.2 PHYSIOLOGICAL EFFECTS

The addition of sulfur dioxide to wine raises health-related objections. These should be taken into account, although this product boasts a long history of use. Its use has always been regulated and enological techniques have always sought methods of lowering its concentrations. Since the beginning of the century, the possible toxicity of sulfur dioxide has been the subject of much research (Vaquer, 1988).

Acute toxicity has been studied in animals. The absorption of a single dose of sulfites is slightly toxic. Depending on the animal species, the LD_{50} (lethal dose for 50% of individuals) is between 0.7 and 2.5 g of SO_2 per kilogram of body weight. Sodium sulfite would therefore have an acute toxicity similar to inoffensive products such as sodium bicarbonate or potassium chloride.

Chronic toxicity has also been studied in animals (Til et al., 1972). During several generations, a diet containing 1.5 g of SO_2/kg was regularly absorbed. Three kinds of complications resulted: a thiamine deficiency linked to its destruction by sulfur dioxide; a histopathological modification of the stomach; and slowed growth. This study permitted the establishment of a maximum nontoxic concentration for rats at 72 mg/kg of body weight. This value led the World Health Organization to set the RDA (recommended daily allowance) at 0.7 mg of SO_2/kg of body weight.

Concerning its toxicity in humans, studies carried out indicate the appearance of intoxication symptoms such as nausea, vomiting and gastric irritation at significantly high absorbed concentrations (4 g of sodium sulfite in a single concentration). No secondary effects were observed with a concentration of 400 mg of sulfur dioxide during 25 days. In humans, its possible toxicity has often been attributed to the well-known destruction of thiamine or vitamin B1 by sulfites, but the corresponding reaction has been observed to be very limited at a pH of around 2, which corresponds to stomach pH.

In 1973, allergic reactions to sulfites were proven to exist. They occur at very low ingested concentrations (around 1 mg) and primarily concern asthmatics (4–10% of the human population). Asthmatics are therefore urged to abstain from drinking wine. Although SO_2 sensitivity has not been clearly demonstrated for non-asthmatics, these allergic reactions led the US FDA (Food and Drug Administration) to require the mention of the presence of sulfites on wine labels in the United States when the concentration exceeds 10 mg/l.

Considering an RDA of 0.7 mg/kg/day, the acceptable concentration for an individual is between 42 and 56 mg per day, depending on body weight (60 and 80 kg, respectively). The consumption of half a bottle of wine per day (375 ml) can supply a quantity of SO_2 higher than the RDA. If the total SO_2 concentration is at the maximum limit authorized by the EU (160 mg/l for red wines and 210 mg/l for white wines), the quantity of SO_2 furnished by half a bottle is 60 mg for reds and 79 mg for whites. The average SO_2 concentrations observed in France are much lower: 75 mg/l for red wines and 105 mg/l for white wines. Therefore, the daily consumption of half a bottle furnishes 28 and 39 mg of SO_2, respectively.

In any case, the figures clearly indicate that, with respect to World Health Organization norms, wines can supply a non-negligible quantity of SO_2. It is therefore understandable that national and international health authorities recommend additional decreases in the accepted legal limits.

Experts from the OIV estimate that the concentrations recommended by the EC can be decreased by 10 mg/l, at least for the most conventional wines. In this perfectly justified quest for lowering SO_2 concentrations, specialty wines such as botrytized wines must be taken into account. Due to their particular chemical composition, they possess a significant combining power with sulfur dioxide. Consequently, their stabilization supposes extensive sulfiting. The EU legislation authorizing 400 mg/l is perfectly reasonable, but this concentration is not always sufficient. In particular, it does not guarantee the stability of some batches of botrytized wines and will not prevent them from secondary fermentation.

8.3 CHEMISTRY OF SULFUR DIOXIDE

8.3.1 Free Sulfur Dioxide

During the solubilization of SO_2, equilibria are established:

$$SO_2 + H_2O \overset{K_1}{\rightleftharpoons} HSO_3^- + H^+ \quad (8.2)$$

$$HSO_3^- \overset{K_2}{\rightleftharpoons} SO_3^{2-} + H^+ \quad (8.3)$$

The H_2SO_3 acid molecule would not exist in a solution. It nevertheless possesses two acid functions whose pKs are 1.81 and 6.91, respectively at 20°C. The neutralization of an acid begins at approximately pH = pK − 2. The absence of neutral sulfites (SO_3^{2-}) at the pH of wine can therefore be deduced. But the first function is partially neutralized according to the pH. Knowing the proportion of free acid (active SO_2) and bisulfite (HSO_3^-) is important, since the essential enological properties are attributed to the first. The calculation is made by applying the mass action law:

$$\frac{[H^+][HSO_3^-]}{[SO_2][H_2O]} = K_1 \quad (8.4)$$

The water concentration can be treated as a constant or very near to 1:

$$\frac{[H^+][HSO_3^-]}{[SO_2]} = K_1 \quad (8.5)$$

which results in:

$$\mathrm{Log}\, \frac{[HSO_3^-]}{[SO_2]} = pH - pK_1 \quad (8.6)$$

Table 8.2 indicates the results for the pH range corresponding to various kinds of wine. The proportion of molecular SO_2, approximately corresponding to active SO_2, varies from 1 to 10. This explains the need for more substantial sulfiting when the must or wine pH is high.

Table 8.2. Molecular SO_2 and bisulfite percentages according to pH (at 20°C) in aqueous solution

pH	Molecular SO_2	Bisulfite (HSO_3^-)
3.00	6.06	94.94
3.10	4.88	95.12
3.20	3.91	96.09
3.30	3.13	96.87
3.40	2.51	97.49
3.50	2.00	98.00
3.60	1.60	98.40
3.70	1.27	98.73
3.80	1.01	98.99
3.90	0.81	99.19
4.00	0.64	99.36

Table 8.3. Sulfur dioxide pK1 value according to alcoholic strength and temperature (Usseglio-Tomasset, 1995)

alcohol (% vol.)	Temperature (°C)							
	19	22	25	28	31	34	37	40
0	1.78	1.85	2.00	2.14	2.25	2.31	2.37	2.48
5	1.88	1.96	2.11	2.24	2.34	2.40	2.47	2.56
10	1.98	2.06	2.21	2.34	2.44	2.50	2.57	2.66
15	2.08	2.16	2.31	2.45	2.54	2.61	2.67	2.76
20	2.18	2.26	2.41	2.55	2.64	2.72	2.78	2.86

The pK value is also influenced by temperature and alcoholic strength (Table 8.3), and equally by ionic force—the concentration in salts. Usseglio-Tomasset (1995) calculated the effect of these factors on the proportion of sulfur dioxide in the form of active SO_2 (Table 8.4).

The bisulfite ion (HSO_3^-) represents the corresponding fraction of the acid neutralized by bases, thus almost entirely in the form of ionized salts. Active SO_2 (or sulfurous acid in the free acid state) represents free sulfur dioxide as defined in enology. The difference between the chemical notion of a free acid and a salified acid should be taken into account.

As a result, the antiseptic properties of a given concentration of free SO_2 towards yeasts or bacteria vary in function of pH, even if the HSO_3^- form is attributed with a certain activity. In the same manner, the disagreeable taste and odor of sulfur dioxide, for the same value of free SO_2, increase the more acidic the wine. The disagreeable odor of SO_2 is sometimes less the result of an exaggerated SO_2 addition than the nature of the wine—inferior quality, an absence of character and aroma, and very high acidity.

Table 8.4. Percentage of active molecular SO_2 at pH 3.0 according to alcoholic strength and temperature (Usseglio-Tomasset, 1995)

alcohol (% vol.)	Temperature (°C)		
	19	28	38
0	4.88		
10	7.36	15.40	27.55
20	10.95		

8.3.2 Bound Sulfur Dioxide

Bisulfites possess the property of binding molecules which contain carbonyl groups according to the following reversible reaction:

$$R\text{—}CHO + HSO_3^- \rightleftharpoons R\text{—}\underset{\underset{H}{|}}{\overset{\overset{OH}{|}}{C}}\text{—}SOH_3^- \quad (8.7)$$

$$R\text{—}\underset{\underset{O}{\|}}{C}\text{—}R' + HSO_3^- \rightleftharpoons R\text{—}\underset{\underset{OH}{|}}{\overset{\overset{R'}{|}}{C}}\text{—}SO_3^- \quad (8.8)$$

These additional forms represent bound sulfur dioxide, or bound SO_2 as it is defined in enology. The sum of free SO_2 plus bound SO_2 is equal to total SO_2. With respect to free SO_2, bound SO_2 has much less significant (even insignificant), antiseptic and antioxidant properties (Section 8.6).

In the reactions forming these combinations, the equilibrium point is given by the formula in Eqn (8.9), for the reaction in Eqn (8.7). This formula presents the molar concentration relationship between the different molecules:

$$\frac{[R\text{–}CHO][HSO_3^-]}{[R\text{–}CHOH\text{–}SO_3^-]} = K \quad (8.9)$$

K is a constant characteristic of each substance, with aldehydic or ketonic functions, able to bind SO_2.

This relationship can be written as follows:

$$\frac{[R\text{–}CHOH\text{–}SO_3^-]}{[R\text{–}CHO]} = \frac{[HSO_3^-]}{K} \quad (8.10)$$

For example, a concentration of 20 mg of free SO_2 per liter represents 25 mg of HSO_3^- per liter (molecular weights 64 and 81, respectively). The molar concentration is therefore:

$$[HSO_3^-] = \frac{25}{81} * 10^{-3} = \frac{10^{-3}}{3.24} \qquad (8.11)$$

The relationship in Eqn (8.10) becomes:

$$\frac{[C]}{[A]} = \frac{[R\text{--}CHO\text{--}SO_3^-]}{[R\text{--}CHO]} = \frac{10^{-3}}{3.24 \times K} \qquad (8.12)$$

It expresses the proportion of carbonyl group molecules bound to SO_2 (C) and in their free form (A).

First case: K has a low value equal to or less than 0.003×10^{-3} M, at equilibrium:

$$\frac{[C]}{[A]} = \frac{10^{-3}}{3.24 \times 0.003 \times 10^{-3}} = \frac{1}{0.01} = 100 \qquad (8.13)$$

In this case, there exists 100 times more of the bound form than the free form. The binding molecule is considered to be almost entirely in the combined form. Free SO_2 can only exist when all of the molecules in question are completely bound. Furthermore, this combination is stable and definitive; the depletion of free SO_2 by oxidation does not cause an appreciable displacement of the equilibrium.

Second case: K has an elevated value equal to or greater than 30×10^{-3} M:

$$\frac{[C]}{[A]} = \frac{10^{-3}}{3.24 \times 30 \times 10^{-3}} = \frac{1}{100} \qquad (8.14)$$

In this case, there exists 100 times more of the free form than the combined form. The binding molecule is considered to be slightly combined and the corresponding combination is not very stable. When free SO_2 is depleted by oxidation, the dissociation of this combination, necessary for reestablishing the equilibria, regenerates free SO_2.

Of course, [C] plus [A] represents the total molar concentration of the combining molecule as given by analysis, expressed in millimole per liter. It is therefore possible to establish overall reaction values of bound SO_2 for different free SO_2 values. In fact, by determining the quantity of each combining molecule, the amount of bound SO_2 can be calculated using the value of K and the concentration of free SO_2 (see Figure 8.3). The sum of the individual combinations must correspond with the total bound SO_2 determined by analysis (Section 8.4.3).

Figure 8.1 gives SO_2 combination curves for different values of K and for a combining molecular concentration of 10^{-3} M. The maximum bound SO_2 concentration is also 10^{-3} M, 64 mg/l.

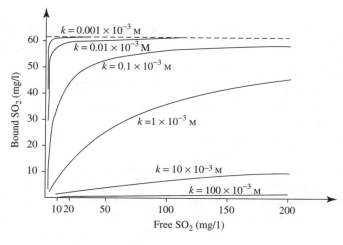

Fig. 8.1. Sulfur dioxide combination curves in accordance with the chemical dissociation constant K (concentration of carbonyled substance $= 10^{-3}$ M (Blouin, 1965)

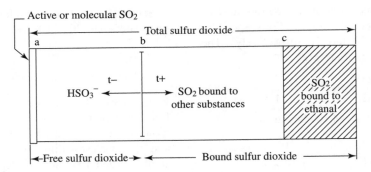

Fig. 8.2. The different states of sulfur dioxide in wine (Ribéreau-Gayon *et al.*, 1977)

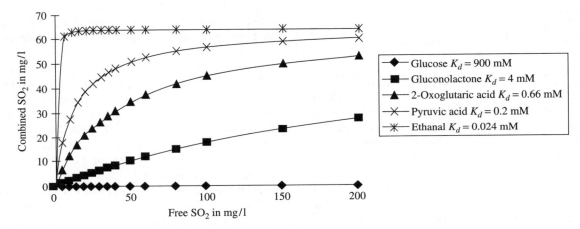

Fig. 8.3. Sulfur dioxide combination curves for various compounds at a concentration of 1 mM, in function of their K_d (Barbe, 2000)

In conclusion, the different forms of sulfur dioxide existing in wine are summarized in the Figure 8.2. Active SO_2 is located to the left; its separation (a) with HSO_3^- varies according to the pH. To the right, sulfurous aldehydic acid represents the SO_2 fraction combined with ethanal. Since K is low, this combination is very stable and depends on the ethanal concentration. The (c) separation line is definitive. On the other hand, the (b) separation between sulfur dioxide and sulfur dioxide combined with other substances varies, moving in one direction or the other according to temperature and the free SO_2 concentration.

8.4 MOLECULES BINDING SULFUR DIOXIDE

8.4.1 Ethanal

The reaction:

$$CH_3-CHO + HSO_3^- \rightleftarrows CH_3-CHOH-SO_3^- \tag{8.15}$$

generally represents the most significant portion of bound SO_2 in wine. The value of K is extremely low (0.0024×10^{-3}) and corresponds to a combination rate of greater than 99%. The ethanal concentrations between 30 and 130 mg/l correspond to possible bound SO_2 values between 44 and 190 mg/l.

In wine no longer containing free SO_2, a weak dissociation of sulfurous aldehydic acid liberates a trace of ethanal. This ethanal is said to be responsible for the flat character in wine, but the presence of free ethanal is considered to be impossible in wine containing free SO_2.

The combination is rapid. At pH 3.3, 98% of it is combined in 90 minutes and the combination is total in 5 hours. Within normal limits, the combination is independent of temperature. The amount of free SO_2 liberated by raising the temperature is very small. Concentrations in botrytized musts are of the order of 10 mg/l, up to a maximum of 20 mg/l. These concentrations may explain a mean combination of under 10 mg/l SO_2.

Alcoholic fermentation is the principal source of ethanal in wine. It is an intermediary product in the formation of ethanol from sugars. Its accumulation is linked to the intensity of the glyceropyruvic fermentation. It principally depends on the level of aeration, but the highest values are obtained when yeast activity occurs in the presence of free SO_2. The formation of sulfurous aldehydic acid is a means of protection for the yeasts against this antiseptic. Consequently, the level of grape sulfiting controls the ethanal and ethanal bound to SO_2 concentration.

Considering these phenomena, the addition of SO_2 to a fermenting must should be avoided. It would immediately be combined without being effective. When the grapes are botrytized, the variation in the ethanal content of different wines when 50 mg/l of SO_2 is added to the must accounts for a combining power approximately 40 mg/l higher than that of non-sulfited control wines. When stopping the fermentation of a sweet wine, a sufficient concentration should be added which stops all yeast activity. This concentration can be decreased by initially reducing the yeast population, using centrifugation or cold stabilization ($-4°C$), for example. The highest ethanal concentrations are encountered when successive fermentations occur. The necessary multiple sulfitings progressively increase bound SO_2 concentrations.

The chemical oxidation of ethanol, by oxidation–reduction in the presence of a catalyzer, may also increase the ethanal concentration during storage—for example, during rackings. The combining power of the wine therefore also increases.

8.4.2 Ketonic Acids

Pyruvic acid and 2-oxoglutaric acid (formerly α-ketoglutaric acid) are generally present in wine (Table 8.5). They are secondary products of alcoholic fermentation. Considering their low K value, they can play an important role in the SO_2 combination rate. For example, a wine containing 200 mg of pyruvic acid and 100 mg of 2-oxoglutaric acid per liter has 93 mg of SO_2 per liter bound to these acids for 20 mg of free SO_2.

Those two substances may combine with very different amounts of SO_2. In wines made from botrytized grapes, for a free SO_2 content of 50 mg/l, 2-oxoglutaric acid is likely to combine with an average of 43 mg/l and pyruvic acid with 58 mg/l (Barbe, 2000). The average percentages of pyruvic and 2-oxoglutaric acids in the SO_2 combination balance are 20.7% and 16.7%, respectively.

It is therefore interesting to understand the formation and accumulation conditions of these acids during alcoholic fermentation. They are formed at the beginning of the fermentative process. After initially increasing, their concentration decreases towards the end of fermentation. This explains the higher concentration of these molecules in sweet

Table 8.5. The ketonic acids of wine (Usseglio-Tomasset, 1995)

Name	Formula	K	Average concentrations in wine
Pyruvic acid	$CH_3–CO–COOH$	0.3×10^{-3} M	10–500 mg/l
2-Oxoglutaric acid	$COOH–CO–CH_2–CH_2–COOH$	0.5×10^{-3} M	2–350 mg/l

Table 8.6. Action of thiamine on ketonic acids and free sulfur dioxide concentrations (mg/l), calculated for 250 mg of total SO_2 per liter (Ribereau-Gayon *et al.*, 1977)

Origin of wines	Control			+Thiamine		
	Pyruvic acid	2-Oxoglutaric acid	Free SO_2 for total 250 mg/l	Pyruvic acid	2-Oxoglutaric acid	Free SO_2 for total 250 mg/l
Monbazillac	10	traces	136	12	traces	134
Barsac	traces	128	104	traces	107	108
Cerons	traces	108	113	traces	82	111
Sauternes	264	121	44	40	73	108
Monbazillac	330	273	20	51	74	109
Sauternes	61	205	52	10	100	88
Cerons	108	72	48	41	70	81

wines with respect to dry wines. Elevated temperatures and pHs, along with aerations, favor the synthesis and accumulation of ketonic acids. In numerous fermentations, thiamine (at a concentration of 0.5 mg/l) has been shown to diminish the concentration of these acids and consequently sulfur dioxide combinations. The effect of thiamine is not surprising. It is an essential element of the carboxylase which assures decarboxylation of pyruvic acid into ethanal. This is an essential step of alcoholic fermentation. The accumulation of ketonic acids appears to result from a thiamine deficiency.

The figures in Table 8.6 show the effect of thiamine on the accumulation of ketonic acids and the corresponding combining power. In the first three wines made from slightly rotten grapes, the sulfur dioxide equilibrium is not modified after the addition of thiamine. In the other cases, the presence of thiamine decreases the ketonic acid concentration and often improves the sulfur dioxide equilibrium.

To be effective, thiamine needs to be added to clarified and sulfited must sufficiently early. It has no action on the accumulation of ethanal. In certain cases, useful secondary effects are observed: activation of the fermentation and diminution of volatile acidity. On average, in eight cases out of 10, thiamine increases the free SO_2 concentration in sweet wines by 20 mg, for the same bound SO_2 concentration.

8.4.3 Sugars and Sugar Derivatives

Considering the existence of aldehydic and ketonic functions in different sugar molecules, they can

be expected to have a combining power with sulfur dioxide. Fructose and saccharose, however, practically do not react.

Arabinose binds SO_2 at a rate of approximately 8 mg of SO_2 per gram of arabinose for 50 mg of free SO_2 per liter. Since the concentration of arabinose in wine is low (less than 1 g/l), this combination is not generally taken into account. Glucose has a much lower combining power. One gram combines 0.3 mg of SO_2 for 50 mg of free SO_2 per liter. Due to the high concentration of glucose in musts and sweet wines, this combination should be taken into account and it is included in the interpretation of the decrease in free SO_2 after sulfiting the grapes or the must.

Burroughs and Sparks (1964 and 1973) identified the following substances: keto-5-fructose (5-oxofructose), xylosone, keto-2-gluconic (2-oxogluconic) and diketo-2,5-gluconic (2,5-dioxogluconic) acids (Table 8.7). Due to their concentrations in some wines (capable of attaining several dozen milligrams per liter), and their K values, some of them can play a significant role in binding with sulfur dioxide. These substances exist naturally in healthy, ripe grapes and they are also formed in large quantities by *Botrytis cinerea* and acetic acid bacteria (*Acetobacter* and *Pseudomonas*). Their development frequently accompanies various forms of rot.

According to more recent findings (Barbe *et al.*, 2002), among all the previously-mentioned compounds, 2-oxo and 2,5-dixogluconic acids always present in a ratio of 2.5/1, do not have a significant affinity for SO_2. In contrast, at the pH of botrytized

Table 8.7. Sulfur-dioxide binding sugar derivatives (based on Burroughs and Sparks, 1964 and 1973)

	Uronic acid		Sugar oxidation products			
	Galacturonic acid	Glucuronic acid	Keto-2-gluconic acid	Diketo-2,5-gluconic acid	Keto-5-fructose	Xylosone
Formulae	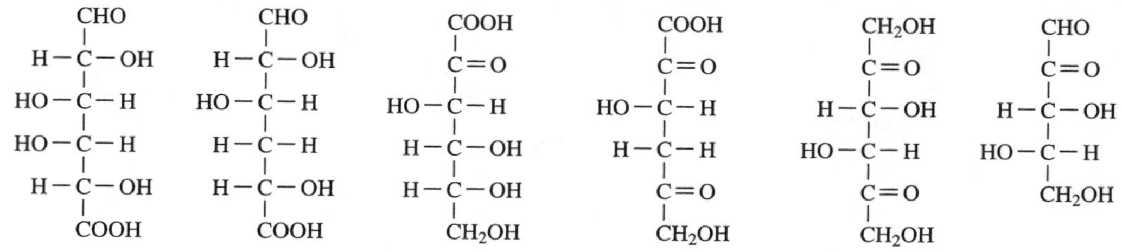 CHO H−C−OH HO−C−H HO−C−H H−C−OH COOH	CHO H−C−OH HO−C−H H−C−H H−C−OH COOH	COOH C=O HO−C−H H−C−OH H−C−OH CH_2OH	COOH C=O HO−C−H H−C−H C=O CH_2OH	CH_2OH C=O H−C−OH HO−C−H C=O CH_2OH	CHO C=O H−C−OH HO−C−H CH_2OH
K	20×10^{-3} M	20×10^{-3} M	0.4×10^{-3} M	0.4×10^{-3} M	0.3×10^{-3} M	0.15×10^{-3} M
Combination rate[a]	4.4	1.5	66	66	72	84

[a]Percentage of combined substances per 50 mg of free SO_2.

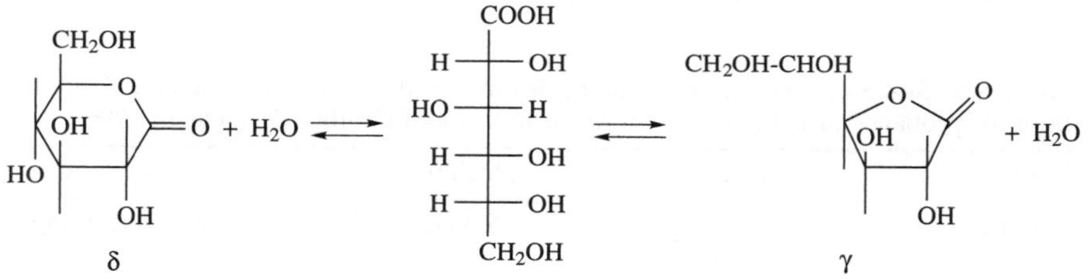

Fig. 8.4. Formation of γ- and δ-gluconolactone from D-gluconic acid

musts and wines, gluconic acid (20 g/l) is in equilibrium with two lactones, γ- and δ-gluconolactone (Figure 8.4), representing about 10% of the concentration of the acid. The affinity corresponds to that of a monocarbonyl compound with a bisulfite combination dissociation constant $K = 4.22$ mM. Thus, the lactones of gluconic acid are likely to combine with up to 135 mg/l SO_2 for a free SO_2 content of 50 mg/l.

The 5-oxofructose content is also frequently of the order of 100 mg/l in wines made from botrytized grapes (Barbe, 2000). Concentrations increase with the combining power (Figure 8.5). According to Barbe (2000), 5-oxofructose may account for the combination of 4–78% of the sulfur dioxide. Concentrations of this compound are not altered by alcoholic fermentation or any other aspect of yeast metabolism. Excessive concentrations can, therefore, only be avoided by monitoring grape quality. In the special case of must made from grapes affected by rot in the mature stage, it contributes, on average, over 60% to the combination balance. This compound is produced from fructose by acetic bacteria in the genus *Gluconobacter* (Section 7.5).

8.4.4 Dicarbonyl Group Molecules

In grapes affected by rot, Guillou-Largeteau (1996) identified molecules with two carbonyl groups (Table 8.8). They are probably formed during the development of *Botrytis cinerea* and other microorganisms involved in various types of rot. In view of the fact that concentrations do not

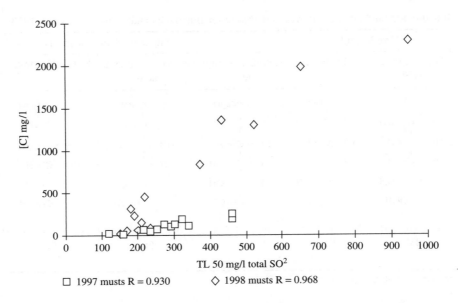

Fig. 8.5. Changes in the 5-oxofructose content according to the combining power of the must (Barbe *et al.*, 2000)

Table 8.8. Some dicarbonyl group molecules involved in sulfur dioxide combinations (hydroxypropanedial is a tautomer form of reductone) (Guillou-Largeteau, 1996)

Name	Chemical formula	Healthy grapes	Botrytized grapes
Glyoxal		Several mg/l	Several dozen mg/l
Methylglyoxal		Several mg/l	Several dozen mg/l

exceed 3 mg/l, the contribution of glyoxal to the SO_2 combination balance is practically negligible. Methylglyoxal makes a more significant contribution and may be responsible for combining over 50 mg/l SO_2 for a free SO_2 content of 50 mg/l (Barbe, 2000). Glyoxal and, especially, methylglyoxal, concentrations decrease during alcoholic fermentation, so these two α-dicarbonyl compounds

are only responsible for insignificant amounts of combined SO_2 in wine.

8.4.5 Other Combinations

Other substances likely to fix small amounts of sulfur dioxide have been identified: glucuronic, galacturonic acid and xylosone (Table 8.7), glyoxylic acid, oxaloacetic acid, glycolic aldehyde, acetoine, diacetyl, 5-(hydroxymethyl)furfural, etc. Their individual contribution is insignificant. In the case of dihydroxyacetone, 100 mg/l accounts for the combination of approximately 16 mg/l for 50 mg/l free SO_2, although this value may be as high as 72 mg/l in certain types of must (Barbe et al., 2001c). While glyceraldehyde has a greater affinity for SO_2 ($K = 0.4$ mM, Blouin, 1995), it is only present in tiny amounts, so it makes a negligible contribution to the SO_2 combination balance.

SO_2 can also bind with phenolic compounds. In the case of proanthocyanic tannins, a solution of 1 g/l binds with 20 mg/l of SO_2 per liter. The combinations are significant with anthocyanins. These reactions are directly visible by the decoloration produced. The combination is reversible; the color reappears when the free sulfur dioxide disappears. This reaction is related to temperature (Section 8.5.2) and acidity (Section 8.5.1), which affect the quantity of free SO_2. The SO_2 involved in these combinations is probably titrated by iodine along with the free SO_2. In fact, due to their low stability, they are progressively dissociated to reestablish the equilibrium as the free SO_2 is oxidized by iodine.

8.4.6 The Sulfur Dioxide Combination Balance in Wines Made from Botrytized Grapes

Burroughs and Sparks (1973) calculated the SO_2 combination balance for two wines on the basis of the concentrations of the various constituents involved, determined by chemical assay and expressed in millimoles per liter (Section 8.3.2). The combined SO_2 calculated by this method was in good agreement with the combined SO_2 assay results, so it would appear that the SO_2 combinations were fully known in that case.

Blouin (1965) had previously demonstrated the particular importance of ketonic acids in this type of combination. In spite of all these findings, the sulfur dioxide combination balance cannot be considered complete and satisfactory. Progress has been made in establishing the combination balance for wines made from botrytized grapes by finding out about other compounds, such as dihydroxyacetone, which is in balance with glyceraldehydes (Blouin, 1995; Guillou-Largeteau, 1996), and work on neutral carbonyl compounds in wines (Guillou-Largeteau, 1996). Finally, more recent research by Barbe and colleagues (2000; 2001a; b; and c; 2002) has improved control of sulfur dioxide concentrations by adding to knowledge of the origins of these compounds.

In wines made from botrytized grapes with high or low combination capacities, almost all of these combinations are accounted for by the concentrations of 5-oxofructose, dihydroxyacetone, γ- and δ-gluconolactone, ethanal, pyruvic and 2-oxoglutaric acid, glyoxal, methylglyoxal, and glucose (Table 8.9). In contrast, in must made from the same type of grapes, the high combining power is precisely accounted for by the quantities of SO_2 combined by 5-oxofructose, dihydroxyacetone, and gluconic acid lactones (Table 8.10).

Carefully-controlled fermentation of botrytized musts minimizes the accumulation of yeast metabolic products combining SO_2, although much higher concentrations of these compounds are implicated in stopping fermentation than those present in dry wines. Various technological parameters during fermentation make it possible to reduce the quantities of sulfur dioxide, by affecting only those combining compounds produced by fermentation yeasts. Wines with a lower sulfur dioxide combining power may be obtained by not sulfiting must, adding 0.5 mg/l of thiamine to must, choosing a yeast strain known to produce little ethanal or 2-oxoacids, and delaying mutage until the yeast metabolism has been completely shut down (e.g. by filtering or chilling the wine) (Barbe et al., 2001c).

These compounds are produced due to the presence of microorganisms in botrytized grapes. Although yeasts represent a preponderant part of the microorganisms present, acetic bacteria, especially those in the Gluconobacter genus, are responsible for producing large amounts of these compounds, which act as intermediaries in their metabolism of the two main sugars in botrytized grapes (Barbe et al., 2001a).

Table 8.9. Combining powers of compounds in a wine. SO_2 combinable by all the compounds assayed or only 6 of them (ethanal, pyruvic acid, 2-oxoglutaric acid, γ- and δ-gluconolactone, and 5-oxofructose) in 9 wines (Barbe, 2000)

Wines	CL50 mg/l Total SO_2	Total combinable SO_2		SO_2 combinable by the 6 compounds accounting for the largest contributions	
		in mg/l	in % CL50	in mg/l	in % CL50
1	100	92	92	76	76
2	180	175	98	164	92
3	215	217	102	196	92
4	245	228	94	201	83
5	260	261	101	240	93
6	290	258	90	232	81
7	340	306	90	288	85
8	350	339	97	328	94
9	450	449	100	437	98

Table 8.10. Average quantities (mg/l) of sulfur dioxide combined by the compounds under study in different musts (Barbe *et al.*, 2001)

Compound	Musts ($n = 24$) with low combining power CL50 = 171 mg/l total SO_2	Musts ($n = 7$) with high combining power CL50 = 498 mg/l total SO_2
5-oxofructose	24	258
γ- and δ-gluconolactone	17	55
dihydroxyacetone	7	60
glucose	48	45
methylglyoxal	12	9
glyoxal	2	2
ethanal	10	14
2-oxoglutaric acid	14	15
pyruvic acid	5	9
other	32	31

The SO_2 combination balance varies considerably between different musts (metabolism of the acetic bacteria) and wines (fermentation parameters). Furthermore, the total content of these combinant compounds in wine may result from both sources, as shown in Table 8.11.

Finally, *Botrytis cinerea* indirectly plays two major roles in the accumulation of substances that combine with SO_2. Firstly, it causes in-depth modifications in the grape skins, which become permeable, thus facilitating access to the various substrates for acetic bacteria. Secondly, noble rot causes glycerol to accumulate in the grapes and is thus indirectly responsible for dihydroxyacetone production.

Table 8.12 recapitulates all the substances that combine SO_2 identified in musts and wines made from botrytized grapes.

8.5 PRACTICAL CONSEQUENCES: THE STATE OF SULFUR DIOXIDE IN WINES

8.5.1 Equilibrium Reactions

In a sulfited wine, an equilibrium exists between the free sulfur dioxide and the bound sulfur dioxide—more precisely, the bound sulfur dioxide with a high dissociation constant K. Sulfur dioxide

Table 8.11. Sulfur dioxide combination balance (in %) in two wines with similar TL50 (Barbe *et al.*, 2001b)

Wine A: TL50 = 310 mg/l total SO_2 (i.e. CL50 = 260 mg/l total SO_2)
Wine A: TL50 = 340 mg/l total SO_2 (i.e. CL50 = 290 mg/l total SO_2)

	Wine A	Wine B
5-oxofructose	2	32
γ- and δ-gluconolactone	11	16
trioses (glyceraldehyde + DHA)	1	2
ethanal	45	16
pyruvic acid	18	9
2-oxoglutaric acid	16	12
α-dicarbonyls (methylglyoxal + glyoxal)	2	1
glucose	5	6
other	—	6

Table 8.12. Concentrations found and K_d calculated for the main molecules identified in botrytized musts and wines (Burroughs and Sparks, 1964, 1973; Blouin, 1965, 1995; Guillou-Largeteau 1996; Barbe, 2000)

Molecules	Concentrations in wine min−max (mg/l)	K_d (Mm)	Combination ratio (mg/l)*
ethanal	20−100	0.0024	99.7
pyruvic acid	20−330	0.3	28
2-oxoglutaric acid	50−330	0.5	25
glyoxal	0.2−2.5	—	81
methylglyoxal	0.7−6	0.017	87
galacturonic acid	100−700	17	10
glucuronic acid	traces−60	50	1
5-oxofructose	traces−2500	0.48	22
dihydroxyacetone	traces−20	2.65	16
glyceraldehyde	traces−10	0.4	26
gluconic acid	1 000−25 000	20	—
2-oxogluconic acid	traces−1200	1.8	—
5-oxogluconic acid	traces−500	—	—
γ- and δ-gluconolactone	6% and 4% of the gluconic acid	4.22	5.6
glucose	±100 g/l	800	0.03

*For a 100 mg/l concentration of the compound, the combination value in SO_2 for a free SO_2 content of 50 mg/l.

bound to ethanal does not participate in this equilibrium, since its combination has a very low K value and thus is very stable.

Any addition of sulfur dioxide to a wine results in the combination of a part of this sulfur dioxide. Conversely, the depletion of free sulfur dioxide by oxidation results in a decrease of the bound fraction to such a degree that the loss of free sulfur dioxide is less than the amount oxidized. This liberation mechanism is advantageous, since it automatically prolongs the effectiveness of a given concentration of sulfur dioxide.

When the free sulfur dioxide concentration of a wine decreases to a very low level, it rarely falls completely to zero, unless yeasts are involved or other factors modify wine composition. The decombination of bound SO_2 progressively replaces the missing free sulfur dioxide.

As a result of these equilibria, the total sulfur dioxide concentrations of different wines cannot be

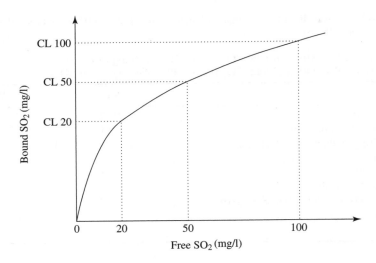

Fig. 8.6. Graphical representation of the binding of sulfur dioxide in wine at CL20, CL50 and CL100

compared if they do not have the same free sulfur dioxide concentration. For example, if a sweet wine has a total SO_2 concentration close to the legal limit, the consequence is not at all the same if it only contains 10 mg of free SO_2 per liter (insufficient for ensuring its stability) or 50 mg (largely sufficient).

To remedy this difficulty, Blouin (1965) recommended the use of the expressions 'CL20' and 'CL50' (Figure 8.6) which represent, respectively, the quantities of bound SO_2 necessary to have 20 or 50 mg of free SO_2 per liter. Known sulfur dioxide additions are used to obtain these numbers experimentally (Kielhöfer and Würdig, 1960). These considerations are most important in the case of sweet wines (Sauternes, Monbazillac, Coteaux de Layon, and Tokay), which require relatively high free SO_2 concentrations to ensure their stability. In practice, the combining power (TL50), or the amount of total SO_2 necessary in a must or wine to obtain 50 mg/l of free SO_2, is calculated by drawing a graph of total SO_2 against free SO_2 (Barbe, 2000).

8.5.2 Influence of Temperature

The determination of the free sulfur dioxide concentration in samples of a botrytized sweet white wine with a strong binding power varies according to temperature, although the total SO_2 concentration remains constant (Table 8.13). The results for determining free SO_2 concentrations are therefore variable. Depending on the conditions, the results obtained can differ by as much as 20 mg/l.

The storage temperature of the wine must also be taken into account in the evaluation of the effectiveness of sulfiting, at least in the case of sweet wines. Finally, the influence of temperature becomes particularly important when heating wine. The SO_2 concentration can double, or even more. This liberation of sulfur dioxide singularly reinforces the effectiveness of heating.

Table 8.13. Influence of temperature on the state of sulfur dioxide (mg/l) in a botrytized sweet wine (sugar 74 g/l; ethanal 70 mg/l)

Sulphur dioxide	Temperature		
	0°C	15°C	30°C
Total sulfur dioxide	412	412	412
Free sulfur dioxide	68	85	100
Bound sulfur dioxide (SO_2C)	344	327	312
SO_2C (to ethanal)	104	104	104
SO_2C (to other substances)	240	223	208

At the time of bottling, wines can be sterilized at relatively low temperatures (between 45 and 50°C, for example), due in part to this phenomenon.

8.5.3 Empirical Laws of Combination

For a long time, enology has tried to determine applicable combination rules, both for sulfiting a new wine immediately following fermentation and for adjusting the free SO_2 during storage.

The most satisfactory solution consists of adding increasing concentrations of SO_2 to various samples of the same wine to produce a curve as in Figure 8.6. This operation is long and difficult; consequently, it is not always feasible. Laboratory tests are, however, recommended before the first sulfiting of unknown wines immediately following fermentation. Due to the diversity of harvests, a standard SO_2 concentration can lead to an insufficient free SO_2 concentration for ensuring stability, or, on the contrary, an excessively high concentration that would be difficult to lower.

The combination curve (Figure 8.6) clearly reveals that the bound SO_2 increases with the free SO_2. Yet the increase is slower and slower as the free sulfur dioxide concentration increases.

To increase the free SO_2 concentration of a wine already containing some, the combination of the added concentration must be taken into account. The lower the free SO_2 concentration, the more the added concentration combines. As a general rule in standard wines already containing free SO_2, two-thirds of the supplementary concentration remains in a free state and one-third combines. As a result, 3 g/hl are necessary to increase the free SO_2 concentration by 20 mg/l. Eventual abnormal cases must also be anticipated, corresponding with a much higher combination rate.

In practice, a few days after the addition of SO_2 to wine, the free SO_2 concentration should be verified to ensure that it corresponds with the desired concentration and that the stabilization conditions are obtained.

8.6 ANTIMICROBIAL PROPERTIES OF SULFUR DIOXIDE

8.6.1 Properties of the Different Forms

The enological properties of sulfur dioxide were summarized at the beginning of this chapter (Section 8.1). It is essentially a multifaceted antiseptic and a powerful reducing agent that protects against oxidation. Its antifungal and antibacterial activities will be covered in Sections 8.6.2 and 8.6.3; the antioxidizing and antioxidasic properties will be covered in Section 8.7.2. The various forms of sulfur dioxide do not share these properties to the same extent (Table 8.14).

Its various properties can make sulfur dioxide seem indispensable in winemaking. The goal of enology is not to eliminate this substance completely but rather to establish responsible

Table 8.14. Wine conservation properties of the different forms of sulfur dioxide (Ribéreau-Gayon *et al.*, 1977)

Property	SO_2	HSO_3^-	$R-SO_3^-$
Fungicidal	+	low	0
Bactericidal	+	low	low
Antioxidant	+	+	0
Antioxidasic	+	+	0
Gustatory amelioration:			
Reduction-oxidation potential	+	+	0
Neutralization of ethanal	+	+	+
Gustatory role of SO_2	biting odor, SO_2 taste	odorless, salty, bitter taste	odorless, tasteless at normal concentrations

concentration limits. This supposes a sufficient knowledge of its properties and conditions of use.

8.6.2 Antifungal Activities

The antiseptic action of SO_2 with respect to yeasts can appear in different ways. On one hand, it can be used to stop the fermentation of sweet wines (mutage) (Section 14.2.5b). It effectively destroys the existing population (fungicidal action). On the other hand, it protects these same sweet wines from possible refermentations—evaluated by the growth of a small residual population. It effectively inhibits cellular multiplication (fungistatic activity). Moderate sulfiting is also known to inhibit yeast growth temporarily without their total destruction. The subsequent disappearance of free SO_2 permits the revival of yeast activity. In practice, in the winery, new yeast activity may also come from new contaminations resulting from contact with non-sterile equipment and containers.

For these different reasons, the results concerning the action of SO_2 on wine yeasts cited in various research work and obtained in different conditions are not always easily compared. Moreover, the data on this subject seems incomplete.

Bound sulfur dioxide does not have an antiseptic action on yeasts. Yeasts make use of the formation of this combination to inactivate SO_2. HSO_3^- also possesses a low but undetermined antiseptic activity. Table 8.15 indicates the concentrations of free SO_2, titratable by iodine, that must be added to wines (according to their pHs) to have an antiseptic activity equal to 2 mg of active molecular SO_2 per liter. The antiseptic activity of the bisulfite form HSO_3^- is more or less significant, depending on the various hypotheses being considered. According to experience obtained on wine stability, HSO_3^- seems to be 20 times less active than SO_2^-, notably in wines containing reducing sugars.

Sulfur dioxide is fungistatic at high pHs and at low concentrations, and it is a fungicide at low pHs and high concentrations. The HSO_3^- form is exclusively fungistatic. Each yeast strain

Table 8.15. Free sulfur dioxide concentrations necessary in wines to maintain an antiseptic activity equal to 2 mg of active molecular SO_2 per liter (Ribéreau-Gayon et al., 1977)

Wine pH	Hypothesis: $H-SO_3^-$ activity			
	None	100 times less than SO_2	20 times less than SO_2	10 times less than SO_2
2.8	22	20	14	11
3.0	34	29	19	14
3.2	54	43	24	16
3.4	87	61	28	18
3.6	134	81	31	19
3.8	200	100	33	20

probably has a specific sensitivity to the different forms of sulfur dioxide. Romano and Suzzi (1992) considered possible mechanisms that could explain these differences. According to these same authors (Suzzi and Romano, 1982), sulfiting must before fermentation increases yeast resistance to SO_2. Yeasts from a non-sulfited must, isolated after fermentation, are more sensitive to SO_2 than those coming from the same must which is sulfited before fermentation.

Concerning the *mutage* of sweet wines (fungicidal activity), the fermentation seems to stop abruptly after the addition of 100 mg of SO_2 per liter. The concentration of sugar remains constant, although carbon dioxide continues to be released for about an hour. During this time, the yeasts do not seem to be affected by the sulfiting—they are still capable of multiplying (Table 8.16), whatever the concentration used. It is necessary to wait at least 5 hours, and more often 24 hours, to observe a decrease in cell viability.

To ensure a complete cessation of fermentation, Sudraud and Chauvet (1985) estimated that 1.50 mg of molecular SO_2 per liter must be added to wine. According to the same authors, after the elimination of yeasts by different treatments, 1.20 mg of molecular SO_2 per liter seems sufficient for ensuring the proper storage of wines containing residual sugars (fungistatic activity). Lower concentrations could be recommended for wines

Table 8.16. Sulfiting to inhibit yeasts in a sweet wine at the end of fermentation (values are number of viable cells, capable of producing colonies in Petri dishes, per ml; initial population 58×10^6/ml) (Ribéreau-Gayon et al., 1977)

SO$_2$ concentration added (mg/l)	Time		
	1 hour	5 hours	24 hours
100	58×10^6	8×10^6	10^5
150	58×10^6	3×10^6	0
300	58×10^6	10^6	0

stored at low temperatures having a low yeast population.

Romano and Suzzi (1992) summarized the current understanding of the action of the sulfur dioxide molecule on yeasts. Molecular SO$_2$ penetrates the cell by either active transport or simple diffusion. Considering the intracellular pH, it must exist in the cell in the form of HSO$_3^-$. Once inside the cell, it reacts with numerous constituents such as coenzymes (NAD, FAD, FMN), cofactors and vitamins (thiamine). It would also have an effect on numerous enzymatic systems and on nucleic acids. Finally, a significant decrease in ATP is also attributed to it.

8.6.3 Antibacterial Activities

The activity of free SO$_2$ on lactic acid bacteria is well known. It is even more influenced by pH than the activity with respect to yeasts. Yet the fraction combined with ethanal or pyruvic acid is also now known to possess an antibacterial activity. The combined SO$_2$ molecule has a direct action on bacteria. The mechanism is not explained by the decomposition of the combination by bacteria, resulting in the liberation of free SO$_2$.

The sulfur dioxide combined with ethanal (or pyruvic acid) seems to possess an antibacterial activity 5–10 times weaker than free SO$_2$, yet it can be 5–10 times more abundant.

A large number of bacteria are eliminated by 5 mg of free SO$_2$ per liter. The same concentration in the combined form lowers the population by 50%. *Oenococcus oeni* is less resistant to sulfur dioxide than *Lactobacillus* and *Pediococcus*.

Significant technical applications for controlling malolactic fermentation and storing wines have resulted from these observations. Sulfiting the grapes does not only act rapidly on bacteria in the pre-fermentation period; it acts by leaving a certain concentration of combined sulfur dioxide which effectively protects and retards bacterial growth until completion of alcoholic fermentation. In this manner, the medium that still contains sugar is protected from an untimely bacterial development which could lead to the production of volatile acidity (Section 3.8.1).

When malolactic fermentation is not sought (in dry white wines, for example), it should be noted that wine stability is not due solely to the bactericidal action of free SO$_2$ but rather to the concentration of combined SO$_2$ that the wine conserves after fermentation; its action is long-lasting during storage. In certain types of wine with too low a pH, combined SO$_2$ concentrations of 80–120 mg/l can make malolactic fermentation impossible.

Sulfur dioxide is also active on acetic acid bacteria but additional studies on this subject are needed. These bacteria resist relatively high concentrations. In the winery, acetic acid bacteria are most effectively prevented by avoiding contact with oxygen in the air and controlling temperature in the winery.

8.7 THE ROLE OF SULFUR DIOXIDE IN WINEMAKING

8.7.1 Advantages and Disadvantages

Although the use of sulfur dioxide in the storage of wine seems to be fairly ancient, its use in winemaking is more recent. It was recommended at the beginning of the 20th century—essentially for avoiding oxidasic casse. The very appreciable improvement in wine quality by sulfiting rotten grapes was an essential factor in the gain in popularity of this process. Its antiseptic properties and its role in the prevention of bacterial spoilage were discovered later.

Nevertheless, the generalization of sulfiting in winemaking, or at least the establishment of a precise and homogeneous doctrine from one vinicultural region to another, took a long time to come about. Besides its many advantages, sulfiting also presented some disadvantages; therefore, a sufficiently precise understanding of the properties of sulfur dioxide had to be obtained before defining the proper conditions of its use. These conditions permit the winemaker to profit fully from its advantages while avoiding its disadvantages.

When used in excessively high concentrations, this product has a disagreeable odor and a bad taste which it imparts to the wine; the taste of hydrogen sulfide and mercaptans in young wines can also appear when they are stored too long on their lees. The most serious danger of improper sulfiting is the slowing or definitive inhibition of the malolactic fermentation of red wines. Incidentally, for a long time sulfited grapes were observed to produce red wines with higher acidities. Before the understanding of malolactic fermentation, this observation was attributed to an acidifying effect of sulfur dioxide or an acidity fixation.

8.7.2 Protection Against Oxidation

The chemical consumption of oxygen by SO_2 is slow. It corresponds to the following reaction:

$$SO_2 + \tfrac{1}{2}O_2 \longrightarrow SO_3 \qquad (8.16)$$

In a synthetic medium, SO_2 has been shown to take several days to consume 8.0–8.6 mg of oxygen per liter (this amount corresponds with the saturation of this medium). Such oxidation requires the presence of catalyzers, notably iron and copper ions. Yet musts are very oxidizable and should therefore be rapidly and effectively protected against oxidation. Sulfiting accomplishes this. Sulfur dioxide, however, cannot act by its anti-oxygen effect, that is to say by combining with oxygen which is no longer available for the oxidation of other must constituents.

Dubernet and Ribéreau-Gayon (1974) confirmed this hypothesis. The experiment consisted of saturating a white grape must with oxygen and measuring the oxygen depletion rate electrometrically (Figure 8.7). In the absence of sulfiting, the depletion of this oxygen is very rapid and is complete within a few minutes (4 to 20 on average). This phenomenon demonstrates the extremely high oxidability of grape must. If at a given moment

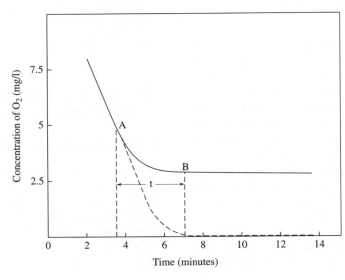

Fig. 8.7. Oxygen consumption in musts following sulfiting (Dubernet and Ribéreau-Gayon, 1974). (A) Addition SO_2 of (B) Stopping point of oxygen consumption (t = time necessary for oxygen consumption to stop)

Table 8.17. Protection of color of red wine made from botrytized grapes by sulfiting (Sudraud, 1963)

Level of harvest sulfiting	Composition of the wines obtained		
	Total phenolic compounds (index)	Color intensity[a]	Oxidasic casse potential
Without SO$_2$	32	0.53	++++
+10g SO$_2$/hl	41	0.63	++
+20g SO$_2$/hl	45	0.83	0

[a]Color intensity = OD 420 + OD 520 (under 1 mm thickness).

the must is sulfited, oxygen is no longer consumed and its concentration remains constant after a given time t, which varies depending on the conditions but is always fairly short. As an initial approximation, the value t varies between 1 and 6 minutes when sulfiting varies between 100 and 10 mg/l. The value t is much greater for must obtained from rotten grapes.

In summary, although the anti-oxygen effect of sulfur dioxide is involved in wine storage, its role is insignificant during winemaking. In this case, SO$_2$ protects against oxidations by destroying oxidases (laccase) or, at least, blocking their activity, if destruction is not total. The enzymatic oxidation phenomena are inhibited in this manner until the start of fermentation. From this point, the reductive character of the fermentation continues to ensure the protection. Yet oxidative phenomena can resume at the end of fermentation insofar as active oxidases remain after the depletion of free SO$_2$. The oxidasic casse test, or, even better, the determination of laccase activity, permits the evaluation of the risk and the necessary precautions to be taken.

In must, enzymatic oxidations are more significant than chemical oxidations because they are more rapid. In wine, however, chemical oxidations play an unquestionable role, since oxidative enzymes no longer exist. In this case, SO$_2$ reacts with oxygen to protect the wine.

Rot is responsible for the most serious oxidative phenomena. In fact, *Botrytis cinerea* secretes a laccase more active and stable than the tyrosinase of grapes. It is responsible for the oxidasic casse in red wines derived from rotten grapes. An appropriate sulfiting can protect against this phenomenon

to some extent. The figures in Table 8.17 show that intense sulfiting of rotten grapes (since they could be used in the past) increases the total phenolic compound concentration and the color intensity while decreasing the risk of oxidasic casse. Progress in phytosanitary vineyard protection has made such situations extremely rare.

From the start of fungal development, the oxidase secretion by *Botrytis cinerea* inside the berry can be considerable whereas the external signs are barely visible. This situation can be observed in the case of red grapes. The first brown blemishes are more difficult to observe on red grapes than on white grapes. During cold weather, the external vegetation of *Botrytis cinerea* is less developed. These factors must be taken into account when choosing the corresponding sulfiting concentration.

8.7.3 Inhibition, Activation and Selection of Yeasts

Sulfur dioxide is a general antiseptic with a multifaceted activity on different wine microorganisms. Its mode of action has been described in previous sections.

With respect to yeasts, sulfiting is used first and foremost to ensure a delay in the initiation of fermentation, allowing a limited cooling of the grapes. The fermentation is also spread out over a longer period in this manner, avoiding excessive temperatures. More and more often, natural tank cooling is complemented by controlled refrigeration systems.

In the case of white winemaking, the delay in the start of fermentation permits the settling and racking of suspended particles in must.

Sulfiting also makes use of the stimulating effect of sulfur dioxide when used in low concentrations. Consequently, the fermentation speed accelerates, as shown by the curves in Figure 8.8. After an initial slowing of the fermentation at the start, the last grams of sugar are depleted more rapidly. Finally, the fermentation is completed more rapidly in the lightly sulfited must.

During the running off of a tank of red wine that still contains sugar, a light sulfiting (2–3 g/hl) does not block the completion of the fermentation; on the contrary it is known to facilitate it more often then not.

This long-proven effect of sulfiting has been confirmed time and time again. It has been interpreted as the destruction of fungicidal substances by sulfur dioxide. These substances are toxic for the yeast and could come from the grape, *Botrytis cinerea* or even the fermentation itself. An increase in the must proteasic activity has also been considered. This activity would put assimilable amino acids at the disposal of the yeast (Section 9.6.1). Sulfiting probably acts by maintaining dissolved oxygen in the must. Not being tied up in oxidation phenomena, it is available for yeast growth (Section 8.7.2).

Sulfiting has also been considered to affect yeast selection. Apiculated yeasts (*Kloeckera* and *Hanseniaspora*), developing before the others, produce lower quality wines with lower alcohol strength. These yeasts are more sensitive to sulfur dioxide. Therefore, a moderated sulfiting blocks their development. This result has been confirmed by numerous experiments (Romano and Suzzi, 1992), but the research of Heard and Fleet (1988) cast doubts on this generalization. In spite of sulfiting, these strains attained an initial population of 10^6 to 10^7 cells/ml in a few days before disappearing. Moreover, the advisability of eliminating apiculated yeasts and the interest of the successive participation of different yeast species for the production of quality wines are still being considered.

The problem of sterilizing musts by the total destruction of indigenous yeasts through massive sulfiting, or other processes such as heat treatments, followed by an inoculation using selected yeasts will be covered elsewhere.

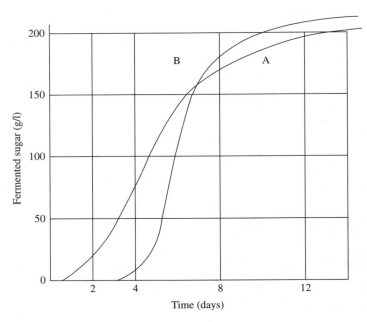

Fig. 8.8. Effect of moderate sulfiting (5–10 g/hl) on alcoholic fermentation kinetics of grape must. (A) Control must. (B) Sulfited must (5 g/hl)

8.7.4 Selection between Yeasts and Bacteria

Sulfur dioxide acts more on wine bacteria than on yeasts. Lower concentrations are consequently sufficient for hindering their growth or suppressing their activity. No systematic studies have been carried out on this subject but this fact is well known and is often demonstrated in practice. For example, in the case of a red wine still containing sugar (a site for simultaneous alcoholic and lactic fermentation), a moderated sulfiting (3–5 g/hl) can initially block the two fermentations. Afterwards, a pure alcoholic refermentation can take place without the absolute necessity of an yeast inoculation.

One of the principal roles of sulfiting in winemaking is to obtain musts much less susceptible to bacterial development, while undergoing a normal alcoholic fermentation. This protection is most necessary in the case of musts that are rich in sugar, low in acidity and high in temperature. The risks of stuck fermentations are highest in these cases.

In summary, sulfur dioxide delays, without blocking, yeast multiplication and alcoholic fermentation. The bacteria, supplied by grapes at the same time as the yeasts, are killed or at least sufficiently paralyzed to protect the medium from their development while the yeasts transform the totality of the sugar into alcohol. The serious danger of bacterial spoilage in the presence of sugar is an important factor in wine microbiology (Section 3.8.2).

In white winemaking and for wines in which malolactic fermentation is not sought, sulfiting can be adopted to inhibit bacteria completely. Incidentally, the light sulfiting of white musts undergoing malolactic fermentation can be insufficient to protect effectively against oxidation.

In red winemaking today, malolactic fermentation has become common practice. Generally speaking, the sulfiting of red grape musts favors wine quality. However, sulfiting must not compromise malolactic fermentation due to its conditions of use and the concentrations employed. To ensure the successful completion of alcoholic fermentation the amount of the sulfurous solution added to grape must should be regulated according to the pH, temperature, sanitary conditions and

Table 8.18. Influence of must sulfiting on the time necessary (expressed in days) for malolactic fermentation initiation in wine after running off (Ribéreau-Gayon et al., 1977)

Sulfiting	Wine No. 1	Wine No. 2
Control 0	40	30
+2.5 g/hl	45	40
+5 g/hl	70	60
+10 g/l	100	100

other factors. Bacterial development must initiate rapidly after the depletion of sugar for exclusive malic acid degradation. The exact SO_2 concentration is difficult to determine, and it varies depending on the region. For red winemaking in the Bordeaux region, 7–10 g/hl seems to be an effective range: below this, the malolactic fermentation is not compromised; above this, it can be considerably delayed (Table 8.18).

8.7.5 Dissolving Power and General Effects on Taste

In red winemaking, sulfiting favors the dissolution of minerals, organic acids and especially phenolic compounds (anthocyanins and tannins) which constitute the colored substances of red wines. The dissolvent activity is due to the destruction of grape skin cells, which yield their soluble constituents more easily in this manner. In fact, the dissolvent effect of sulfur dioxide seems to have been exaggerated in the case of healthy grapes. The better color of wines derived from sulfited must is probably due to a better protection against the oxidasic casse in slightly rotten grapes.

The effectiveness of sulfitic maceration for extracting grape pigments is indisputable, and this process is used for the industrial preparation of commercial colorants. Yet when rigorous experiments are carried out on healthy red grapes, using classic winemaking techniques, no significant color improvement (anthocyanin and tannin concentration and color intensity value) is observed in the presence of a normal sulfiting. Since only the free SO_2 is active, and since this form rapidly disappears in crushed grapes, this effect of sulfiting appears to be exerted for only a brief moment. At the end of fermentation, the effects of maceration time, temperature and pumping-over are more significant.

Nevertheless, the dissolvent effect of sulfiting, with respect to phenolic compounds, is obvious in the case of limited maceration. This operation is not recommended for crushed white grapes before must extraction by pressing. The sulfiting of grapes also has an impact on the color of rosé wines.

Sulfiting also has certain effects on wine quality which still remain poorly defined. The general properties of sulfur dioxide may possibly have indirect consequences (protection against oxidations and the binding of ethanal). In this way, sulfiting often improves the taste of wine—notably in the case of rotten grapes or mediocre varieties. It also protects certain aromas of 'new' wines. Moreover, grape sulfiting does not have an obvious impact on the subsequent development of the bouquet of mature wines.

Certain conditions, such as fermentations in strict anaerobiosis and especially prolonged aging on yeast lees, can lead to the formation of hydrogen sulfide and mercaptans from the added SO_2. The odors of these compounds are disagreeable and can persist in wine.

8.8 THE USE OF SULFUR DIOXIDE IN THE WINERY

8.8.1 Winemaking Concentrations

Considering the rapidity of oxidative phenomena, grape and must sulfiting is only effective if the sulfur dioxide is intimately and rapidly incorporated into the total volume before the start of fermentation. If a fraction of the grape must ferments before being sulfited, it is definitively shielded from the action of the SO_2, because it immediately combines with the ethanal produced by the fermenting yeasts.

In fact, a homogeneous distribution before the start of fermentation is not sufficient. Considering the rapidity of the oxygen consumption by grape must, each fraction of the grape harvest or the must should receive the necessary quantity of sulfur dioxide in the minutes that follow the crushing of the grape or the pressing of the harvest. This is the

only truly effective method of protecting against oxidations. It can be more effective to add 5 g of sulfur dioxide per hectoliter correctly to the harvest than to add 10 g/hl added in poor conditions. A poor sulfiting technique is certainly one of the reasons in the past that led to the use of excessive concentrations.

Based on these principles, the only rational sulfiting method for winemaking consists of regularly incorporating a sulfurous solution into the white grape must as it is being extracted, or for red grapes as soon as they are crushed. A few successive additions of SO_2 into the tank as it is being filled are not truly effective, even after a homogenization at the end of filling. During homogenization, part of the added sulfur dioxide is already in the combined form and thus inactive.

It is therefore also necessary to use a sufficiently diluted sulfur dioxide solution, capable of being correctly incorporated and blended into the must. The direct usage of metabisulfite powder or sulfurous gas in the tank should be avoided. When a tank of red grapes is sulfited by a few additions of a concentrated product during filling, the complete decoloration of certain fractions of the pomace is sometimes observed during the running off. In these cases, the sulfur dioxide was not properly blended, but was instead fixated on certain parts of the grapes, leaving the other parts unprotected.

When choosing the SO_2 concentration to add to the grapes or the must, grape maturity, sanitary state, acidity (pH), temperature and eventual contamination risks must all be taken into account. The choice can sometimes be difficult. Table 8.19 gives a few values for vineyards in temperate climates. The generalization of tank cooling systems and increased hygiene in the wineries, combined with a better understanding of the properties of sulfur dioxide, permit the lowering of the concentrations used in winemaking. Today, sanitary practices in the vineyard avoid grape rot, which once justified the intense sulfitings indicated in Table 8.19.

During the harvest, progressive increases in the sulfiting concentrations can compensate the increasingly significant inoculation (notably bacterial) resulting from the development of microorganisms on the equipment—the inner surface of

Table 8.19. Sulfur dioxide doses for winemaking in temperate climate zones

Status	Sulfur dioxide dose
Red winemaking:	
Healthy grapes, average maturity, high acidity	5 g/hl of wine
Healthy grapes, high maturity, low acidity	5–8 g/hl of wine
Rotten grapes	8–10 g/hl of wine
White winemaking:	
Healthy grapes, average maturity, high acidity	5 g/hl of must
Healthy grapes, high maturity, low acidity	6–8 g/hl of must
Rotten grapes	8–10 g/hl of must

the tanks and the walls of the winery. Problems of difficult final stages of fermentation and microbial deviations are frequently observed in the last tanks filled. Sufficient must sulfiting should avoid these contaminations.

In white winemaking, excessive concentrations (>15 g/hl) followed by a significant sulfiting at the end of fermentation (>4 g/hl) can be a source of reduction odors and should be avoided. Press wines, however, should be more intensely sulfited—especially in the case of continuous presses which cannot be disinfected regularly.

Concerning the sulfiting technique for red winemaking, the solution should be added after grape crushing to facilitate the blending and to avoid evaporation losses and attacks on metallic equipment. Taking into account the transfer of the crushed grapes by a pump with a constant delivery, the sulfurous solution should be injected into the tube immediately after the pump outlet. The sulfiting is suitably distributed and homogenized in this manner. Of course, the injection pump for the sulfurous solution must be properly adjusted and perfectly synchronized with the grape-pump.

The addition of the sulfurous solution after each grape load, by regularly spraying the surface, can only be practiced in small tanks and must also be sufficient in number. Even if it is not completely effective, a homogenization pumping-over is necessary after filling.

In the case of white winemaking, sulfiting must take place after must separation. Sulfiting of the crushed grapes is not recommended since it entails the risk of increasing the maceration phenomena and a fraction of the SO_2 is fixated on the solid parts of the grape.

Considering the oxidation speed of white grape must, sulfiting (which ensures the appropriate protection) should be carried out as quickly as possible. Must extraction equipment (the press cage, mechanical drainer and continuous press) does not supply a constant delivery. Consequently, SO_2 cannot be injected with a pump adapted directly to these outlets. In order to sulfite in this manner, the must has to pass via a small tank through a constant delivery pump. The corresponding manipulation of the must, in particular the pumping, does not protect the must from a slight oxidation before sulfiting.

The sulfiting of white grape musts can also be calculated from the volume of the juice tray at the outlet of the press. During filling, a homogeneous distribution should be ensured.

The necessary volume of a sulfurous solution for sulfiting an entire tank during its filling at the chosen concentration should be prepared in advance. If the system is correctly adjusted, the entire volume of the sulfurous solution should have been injected in to the tank by the time the tank is full.

8.8.2 Storage and Bottling Concentrations

During storage, sulfiting is, first of all, thought to protect wine from oxidation. As an approximation, oxidative risks are present during prolonged storage below 5–10 mg/l for red wines, 20 mg/l for

Table 8.20. Recommended free sulfur dioxide concentrations (mg/l) in wines

Dose type	Red wines	Dry white wines	Sweet white wines
Conservation	20–30	30–40	40–80
Bottling	10–30	20–30	30–50
Expedition doses (cask or container)	25–35	35–45	80–100[a]

[a]This type of wine should be bottled at the production site; bulk expedition should be avoided.

white wines made from healthy grapes and 30 mg/l for white wines made from more or less rotten grapes.

At the microbiological level, sulfiting dry wines must avoid yeast and bacterial development during storage. In dry white wines and red wines having undergone malolactic fermentation, the concentrations used for protection against oxidations are generally sufficient to avoid microbial developments. In red wines that have not undergone malolactic fermentation, the habitual free and total SO_2 concentrations can be insufficient to shield the wine completely from a malolactic fermentation—at least a partial one—during storage.

Of course, the sulfiting rules do not apply to certain kinds of wines (red or white, dry or sweet) with qualities derived from a certain oxidation state or containing ethanal.

Sulfiting also hinders the refermentation of sweet wines, generally provoked by SO_2-resistant yeast strains. The refermentation risks are independent of sugar concentrations, but are influenced by alcohol strength. In satisfactory storage conditions, 50 mg of free SO_2 per liter is required to ensure the storage of a sweet wine with a relatively low alcoholic strength (11%) and 30 mg/l for wines with a high alcohol content (13%).

In practice, carefully adjusted sufficient concentrations must be used to avoid accidental risks. The refermentation of a sweet wine can start in the lees of a tank containing a sufficiently high yeast population to ensure the combination of the SO_2. Simultaneously, at least for a certain amount of time, all of the liquid remains limpid without a refermentation, with 60 mg of free SO_2 per liter. If the fermentative process begins from the lees, the refermentation seems possible in spite of the high concentration of free sulfur dioxide.

The size of the yeast population should always be taken into account to evaluate the effectiveness of a sulfiting. All operations (fining and filtration) that eliminate a fraction of the yeasts permit the lowering of the free SO_2 concentration necessary for conserving sweet wines.

The possibility of lowering free SO_2 concentrations for stabilizing sweet wines results from steps taken in storing wine. Clean (if not sterile) conditions have diminished contaminating populations. These criteria for cleanliness should be applied not only to the product but also to the building, the containers and the material—all contamination sources. Microbiological controls that indicate the number of viable yeast cells are useful tools for adjusting sulfiting.

Table 8.20 indicates free sulfur dioxide concentrations that can be recommended in different situations.

8.8.3 Diminution of Sulfur Dioxide by Oxidation during Storage

The free sulfur dioxide concentration does not remain constant in wines stored in barrels or tanks. There is a continuous loss month after month. Over the years, its concentration decreases even in bottled wine.

The decrease in barrels or tanks results from an oxidation catalyzed by iron and copper ions. Although it is very volatile, a negligible quantity of free sulfur dioxide evaporates during storage in wooden barrels. Nor is it combined. A fairly common error is to consider that any decrease in free sulfur dioxide is the result of a combination with wine constituents. In reality, after the four or five days following the addition of SO_2, the wine constituents no longer bind. An equilibrium is attained and decreases occurring afterwards are due to oxidation. For a new combination to occur, the chemical composition of the wine must be modified. For example, new binding molecules must be formed, such as ethanal, during a limited yeast development or by the oxidation of ethanol when a poorly clarified wine is racked.

The oxidation affecting sulfurous acid forms sulfuric acid. At the pH of wine, it is almost entirely in the form of sulfate. In botrytized and non-botrytized sweet wines with elevated free SO_2 concentrations, a considerable amount of sulfate can be formed (0.5 g/l). Less is formed in dry white and red wines, especially those stored in tanks. In the case of barrel-aged wines, the formation of sulfate by the oxidation of free SO_2 accumulates with the amount resulting from the combustion of sulfur in the empty barrels. This formation lowers the pH and harshens the wine. This phenomenon contributes to the decrease in quality of wines stored in barrels for an excessively long time.

When sulfiting is effected without a measurement beforehand, the wine can be excessively sulfited and its taste affected. In general, the characteristic odor appears at or above 2 mg of active molecular SO_2 per liter. Table 8.15 indicates the corresponding free SO_2 concentrations. To lower the concentration of free SO_2 of a wine, the most effective solution, when possible, is to use this wine to increase an insufficient concentration of free SO_2 of a similar wine.

If such an operation is not possible, the most generally recommended method is to aerate the wine. The effectiveness of this method is based on the slow oxidation of sulfur dioxide. During the days that follow, the higher the temperature, the more rapidly the concentration decreases. Aeration has a limited effectiveness, and 16 mg of oxygen per liter is required to oxidize 64 mg of total sulfur dioxide per liter. This approximately corresponds to a decrease of 42 mg of free SO_2 per liter, taking into account the dissociation of combinations.

The use of hydrogen peroxide is a radical means of eliminating an excess of free SO_2. This method is too severe and is therefore prohibited; it compromises wine quality for a long time.

8.8.4 The Forms of Sulfur Dioxide Used

This antiseptic has the advantage of being available in various forms capable of responding to different situations: gaseous state (resulting from the combustion of sulfur), liquefied gas, liquid solution and crystallized solid.

Sulfurous gas SO_2 liquefies at a temperature of $-15°C$ at normal atmospheric pressure or under a pressure of 3 bars at normal ambient temperature. It is a colorless liquid with a density of 1.396 at 15°C. Placed in 10–50 kg metallic bottles, this form is used for large-quantity additions that can be measured by weighing the bottle, which is placed directly on a scale. A 'sulfidoseur' is used to treat smaller volumes of wine. The graduated container can be precisely filled from the metallic bottle by regulating a pair of small faucets—permitting the addition of precisely measured quantities of the gas.

Liquefied sulfur dioxide is still delivered in vials containing 25, 50 or 75 g of sulfur dioxide for example adapted for sulfiting wine in barrels with capacities of several hundred liters. A special tool perforates these small metal cap-stoppered bottles when they are inside the barrel to be treated.

For SO_2 additions to small volumes of wine, or to have a better incorporation, 5–8% solutions prepared in water or must (to avoid dilution) from liquefied sulfurous gas are used. The quantity needed is weighed. The concentration of the solution is regularly verified by measuring its density (Table 8.21) or by chemical analysis. It tends to decrease in contact with air.

Handling these solutions is disagreeable, since they give off a strong SO_2 odor. Prepared on the premises, they are well adapted to large winemaking facilities such as bulk wineries.

Concentrated 10% solutions, or 18–20% potassium bisulfite solutions, are also used. They are more easily handled than the preceding since they are less odorous. Being more concentrated, however, they are less easily incorporated into wine and must. Legislation limits their use to a single addition of 10 g of SO_2 per hectoliter. They acidify less than the preceding since the acidity in these solutions is partially neutralized. Potassium

Table 8.21. Density (at 15°C) of sulfur dioxide solutions prepared by the dissolution of sulfur dioxide gas in water

Sulfur dioxide (g/100 ml)	Density	Sulfur dioxide (g/100 ml)	Density
2.0	1.0103	6.5	1.0352
2.5	1.0135	7.0	1.0377
3.0	1.0168	7.5	1.0401
3.5	1.0194	8.0	1.0426
4.0	1.0221	8.5	1.0450
4.5	1.0248	9.0	1.0474
5.0	1.0275	9.5	1.0497
5.5	1.0301	10.0	1.0520
6.0	1.0328		

metabisulfite ($K_2S_2O_5$) solutions at 10% diluted in water can also be used. These solutions contain approximately 50 g of sulfur dioxide per liter (5%) and are suitable for limited-volume winemaking. The metasulfite powder should be diluted in water before use. When added directly, it is difficult to blend into the must.

8.8.5 Sulfiting Wines by Sulfuring Barrels

Sulfuring barrels, or small wooden tanks or containers, consists of burning a certain quantity of sulfur in these containers. It is probably the oldest form of using sulfur dioxide in enology. It is used for adjusting the free SO_2 concentration of wines at the moment of racking and also for avoiding microbial contamination when storing empty containers. It has a double sterilizing effect. It is exerted at once on the wine and the internal surface of the container. This practice is part of normal winery operations and could not be replaced by the simple addition of sulfurous solution to the wine. Due to the unpleasant odor imparted to the wine cellar by burning sulfur, its usage can be prohibited by the safety legislation of certain countries. Instead of coming from the combustion of sulfur, sulfurous gas can also be delivered as a bottle of compressed gas.

In any case, sulfur combustion is only applicable to wooden containers. In fact, sulfurous gas, coming from the combustion of sulfur, attacks the internal surface of cement tanks and the coating of metallic tanks. It also accelerates the deterioration of stainless steel.

The sulfur is generally supplied in the form of a wick or ring. It may be coated on a cellulosic weave or mixed with a mineral base (aluminum or calcium silicate). The units most often used are 2.5, 5 and 10 g of sulfur. Chatonnet *et al.* (1993) demonstrated a certain heterogeneity in the quantity of SO_2 produced by the combustion of the same weight wick or ring according to their preparation conditions or storage (fixation of humidity).

From the equation:

$$S + O_2 \longrightarrow SO_2$$

$$32 + 32 = 64 \qquad \textbf{(8.17)}$$

the burning sulfur combines with its weight in oxygen to give double the weight of SO_2. In reality, 10 g of sulfur burned in a 225 l barrel produces only about 13–14 g of SO_2—a 30% loss. One part of the difference is accounted for by the portion of the sulfur that falls to the bottom of the barrel without burning, and the other part by the production of sulfuric acid—a strong acid without antiseptic activity. The sulfiting loss and the acidification of wine (by repeated sulfurings) are explained in this manner.

The combustion of sulfur does not exert its effect by eliminating all of the oxygen from the barrel. The maximum quantity of sulfur that can burn in a 225 l barrel is 20 g, for the maximum production of 30 g of sulfurous gas. At this stage, the combustion stops because the sulfurous gas has the property of hindering its own combustion. It has been determined that approximately 32.5 liters of oxygen are present in the barrel at the moment when the combustion stops, compared with 45 liters beforehand.

These observations lead to the conclusion that the combustion of sulfur is limited. When a 40 g sulfur wick is burned, not all of the sulfur is consumed, even if the wick is burnt to a cinder. About half of it falls to the bottom of the barrel without burning.

The production of SO_2 by the combustion of sulfur in a barrel is therefore irregular. It is especially hindered in humid barrels; for instance 10 g sulfur burned in dry barrel give 12 g SO_2 and only 5 g in humid barrel (Ribéreau-Gayon *et al.*, 1977). In addition, the dissolution of the SO_2 formed is generally irregular during the filling of the barrel. Depending on the filling speed and conditions (by the top or bottom, for example), a more or less significant part of the sulfurous gas is driven out of the barrel. Moreover, the distribution of the sulfiting in the wine mass is not homogeneous. The first wine that flows into the barrel receives more SO_2 than the last. In one example, the free SO_2 increased by 45 mg/l at the bottom of the barrel, by 16 mg/l in the middle and not at all in the upper portion. Consequently, the wine should be homogenized after racking—by rolling the barrel, for example. This sulfiting method should only be used for wines stored in small-capacity containers—say up to 6 hl.

As Chatonnet *et al.* (1993) stated, the combustion of 5 g of sulfur in a 225-liter wooden barrel increases the SO_2 in wine from 10 to 20 mg/l. Sulfur wicks are less efficient (10 mg/l) but more consistent than rings (10–20 mg/l). The latter are more sensitive to their external environment, i.e. moisture.

The combustion of sulfur for the storage of empty barrels will be covered in Volume 2, Section 13.6.2.

REFERENCES

Barbe J.C. (2000) *Les combinaisons du dioxyde de soutro dans les moûts et les vins issus de raisins botrytisés. Rôle des bactéries acétiques*, Thèse de Doctorat, Université Victor Segalen Bordeaux II.

Barbe J.C., de Revel G. and Bertrand A. (2002) *J. Agric. Food Chem.*, 50 (22), 6408.

Barbe J.C., de Revel G., Joyeux A., Lonvaud-Funel A. and Bertrand A. (2000) *J. Agric. Food Chem.*, 48 (8), 3413.

Barbe J.C., de Revel G., Joyeux A., Bertrand A. and Lonvaud-Funel A. (2001a) *J. Appl. Microbiol.*, 90, 34.

Barbe J.C., de Revel G., Joyeux A., Lonvaud-Funel A. and Bertrand A. (2001b) 1ère partie, *Revue F. d'Œnologie*, 189, 26.

Barbe J.C., de Revel G., Perello M.C., Lonvaud-Funel A. and Bertrand A. (2001c) 2ère partie, *Revue F. d'Œnologie*, 190, 16.

Blouin J. (1965) Contribution á l'étude des combinaisons de l'anhydride sulfureux dans les moûts et les vins. Thèse Docteur-Ingénieur, Université de Bordeaux.

Blouin J. (1995) *CR Journée Technique du CIVB*, Bordeaux, France.

Burroughs L.F. and Sparks A.H. (1964) *J. Sci. Food Agri.*, 3, 176

Burroughs L.F. and Sparks A.H. (1973) *J. Sci. Food Agri.*, 24, 187, 199 and 207.

Chatonnet P., Boidron J.N. and Dubourdieu D. (1993) *J. Int. Sci. Vigne Vin*, 27 (4), 277.

Dubernet M. and Ribéreau-Gayon P. (1974) *Vitis* 13, 233.

Guillou-Largeteau I. (1996) Etude de substances de faible poids moléculaire combinant le dioxyde de soufre dans les vins blancs issus de vendanges botrytisées. Mise en évidence et importance du rôle de l'hydroxypropanedial. Thèse Doctorat de l'Université de Bordeaux II (option OEnologie-Ampélologie).

Heard H. and Fleet G.H. (1988) *Austr. NZ Wine Ind. J.*, 3, 57.

Kielhöfer E. and Würdig G. (1960) *Weinberg U. Keller*, 7, 313.

Ribéreau-Gayon J., Peynaud E., Ribéreau-Gayon P. and Sudraud P. (1977) *Sciences et Techniques du Vin*. Vol. 4: *Clarification et Stabilisation. Matériels et Installation*. Dunod, Paris.

Romano P. and Suzzi G. (1992) In *Wine Microbiology and Biotechnology* (ed. G.H. Fleet). Harwood Academic Publishers, Chur, Switzerland.

Sudraud P. (1963) Etude expérimentale de la vinification en rouge. Thèse de Docteur-Ingénieur, Faculté des Sciences de Bordeaux.

Sudraud P. and Chauvet S. (1985) *Conn. Vigne Vin*, 19 (1), 31.

Suzzi G. and Romano P. (1982) *Vini d'Italia*, 24, 138.

Til H.P., Feron U.I. and Degrout A.P. (1972) *Fd. Cosmt. Toxicol.*, 10, 291.

Usseglio-Tomasset L. (1995) Chimie oenologique. Tec and Doc Lavoisier, Paris.

Vaquer J.M. (1988) *Thèse de Doctorat en Pharmacie*, Université de Montpellier, France.

9

Products and Methods Complementing the Effect of Sulfur Dioxide

9.1 INTRODUCTION

Considering the legitimate desire to lower sulfur dioxide concentrations, it is normal to search for adjuvants that complement its action by reinforcing the effectiveness of one of its properties. This chapter covers such chemical products and physical processes that have been or are likely to be authorized by the legislation of different countries. Others are likely to be proposed in coming years.

Sorbic acid, which can be used to increase the antimicrobial properties of sulfur dioxide, is now well known and authorized in many countries. The possibility of using octanoic and decanoic acids will also be covered, though they are not currently authorized. They do not seem to pose any hygiene problems, and they exist naturally in wine; this treatment only reinforces the existing concentrations. Lysozyme from egg white has similar properties. This enzyme is capable of destroying certain bacteria, especially lactic bacteria in wine. Its capacities have been known for a long time but its practical application in the winery has been developed in recent years.

Numerous antibiotics and antiseptics known to act on wine yeasts have not yet been authorized, for reasons of hygiene. One those most recently proposed, 5-nitrofurylacrylic acid, is a powerful fungicide but is highly carcinogenic and induces ethyl intoxication. Another, pimaricine, is autodegradable and has a fungistatic effect, without known secondary effects; it is already authorized in the food industry. A proposal requesting the official approval of this product has been filed in France.

In the 1970s, the use of ethyl pyrocarbonate (Baycovin) in wine was permitted in some countries. This fungicide is very effective for cold-sterilizing wine at the time of bottling. It disappears rapidly, breaking down, mainly, into ethyl alcohol and CO_2, but also releasing tiny quantities of ethyl carbonate that were, nevertheless, significant enough that its use was soon given up altogether. The Bayer pharmaceutical company later marketed a dimethyldicarbonate product, Velcorin, that was just as effective without any of the health risks.

Among the physical processes capable of complementing the antimicrobial properties of sulfur dioxide, the destruction of germs by heat (pasteurization) can be used. Recently, the possibility of destroying germs by high pressure has been demonstrated. This method seems to be effective and affects quality less than heat treatment. The practical conditions of its use remain to be defined and the appropriate equipment to be designed. Of course, all operations that work towards eliminating microorganisms, even partially (racking, centrifugation, filtration and pasteurization), facilitate microbial stabilization and permit the winemaker to lower the SO_2 concentration used.

Ascorbic acid is the most used adjuvant, contributing to the antioxidant properties of sulfur dioxide. The storage of wine with an inert gas is another effective means of avoiding oxidations.

9.2 SORBIC ACID

9.2.1 Physical and Chemical Properties

The formula for sorbic acid contains two double bonds:

$$CH_3-CH=CH-CH=CH-COOH \qquad (9.1)$$

Four isomers exist but only the *trans-trans* isomer is used. Due to its effectiveness and lack of toxicity, its use is authorized in many countries, in particular in the EC, at a maximum concentration of 200 mg/l. It remains prohibited in a few countries (e.g. Austria and Switzerland).

It exists in the form of a white crystallized powder. In a water-based solution, it can be entrained by steam; for this reason, it is found in wine distillate and falsely increases volatile acidity. It has a slightly acidic flavor. Its dissociation constant corresponds to a pK of 4.76. In other words, in wine, it is essentially in the form of a free acid.

Sorbic acid is not very soluble in water (1.6 g/l at 20°C; 5.0 g/l at 50°C) but it is soluble in ethanol (112 g/l at 20°C). Its sodium and potassium salts are very water soluble. In this form, concentrated solutions are prepared for treating wine. Potassium sorbate contains 75% sorbic acid. A solution at 200 g of sorbic acid per liter is prepared by dissolving 270 g of potassium sorbate per liter in water. One liter of this solution can treat 10 hl of wine at 20 g of sorbic acid per hectoliter. Sorbic acid can also be dissolved in alkaline solutions, and 200 g will dissolve in a liter of cold water containing 100 g of KOH. These concentrated solutions must be prepared immediately before use. They become yellow with time.

Certain precautions must be taken when incorporating the concentrated solution into the wine being treated. Due to the pH of wine and the pK of sorbic acid, the latter will be liberated from its salt as soon as the concentrated solution is introduced in the wine. However, this acid is not very soluble: if its concentration at a given moment is too high, it will precipitate. The concentrated solution must therefore be added slowly while being constantly mixed. The use of a dosing pump is recommended.

9.2.2 Antimicrobial Properties

The fungicidal activity of sorbic acid has been tested on yeasts in different circumstances (Ribéreau-Gayon *et al.*, 1977).

The fungicidal concentration, stopping a fermentation (mutage), is relatively high (5 g/l). From 0.5 g/l on, the fermentation is observed to slow and stop before its successful completion. These elevated concentrations do not permit the use of sorbic acid in sweet wine vinification.

The fungistatic concentration, which hinders fermentation in grape must, varies according to must composition (in particular pH), the size of the inoculum and the nature of the strain. The concentration limits cited in the literature are between 100 and 1000 mg/l, with an average value of 300–500 mg/l.

The inhibiting concentration for yeasts in wines containing sugar is lower and depends on the alcohol strength and the pH. Concerning the alcohol content, numerous tests are cited in the literature, taking into account the size of the inoculum and the nature of the yeast strains (Table 9.1).

The pH also has a strong influence on the fungicidal activity of sorbic acid, which increases as the pH decreases. In laboratory experiments (Splittstoesser *et al.*, 1975; Devèze and Ribéreau-Gayon, 1977, 1978), a concentration of 150 mg/l of sorbic acid was needed to hinder the refermentation of a sweet wine at pH 3.1. All other factors being equal, 300 mg/l was necessary to obtain the same results at pH 3.5. At a pH $\geq$3.5, the maximum concentration authorized, (200 mg/l) may be insufficient to ensure the proper stabilization of a wine.

The effect of the pH on the state of sorbic acid has been analyzed. The non-dissociated free acid molecule is known to possess the antiseptic character. Between pH 3.0 and 3.8, the proportion of sorbic acid (pK = 4.76) in its non-dissociated free acid state passes from 98 to 90%. This difference is too small to explain the much more significant impact of the pH. Cell permeability and penetration phenomena of sorbic acid regulated by the pH of the medium may also be involved.

In addition to its action on classic fermentative yeasts, sorbic acid acts against flor yeasts developing on the surface of wine (*Candida*). In capped bottles of red wine containing 10% vol.alcohol with a relatively significant head space and stored upright, 150 mg of sorbic acid per liter ensures their storage for three weeks. Higher concentrations may be needed for lower alcohol strengths, longer storage periods, higher temperatures or wines containing residual yeasts.

The antibacterial properties of sorbic acid are less significant. It exerts practically no activity against acetic acid and lactic acid bacteria. Concentrations of 0.5–1 g/l would be necessary to have a significant effect.

Sorbic acid therefore exerts a selective effect on wine microorganisms and opposes yeast development without blocking bacterial growth. It has the opposite effect to that of sulfur dioxide (which favors yeasts at the expense of bacteria). Consequently, sorbic acid must never be used alone but always associated with an antibacterial product (sulfur dioxide). In wines exposed to air, the amount of volatile acidity formed by acetic acid bacteria can be greater in the presence of sorbic acid, due to the absence of an antagonism with yeasts.

In conclusion, sorbic acid only presents a sufficient effectiveness in practice when associated with a certain concentration of ethanol and free sulfur dioxide. Sorbic acid is an effective adjuvant to sulfur dioxide, since it reinforces its action, but it is not a replacement. Most problems encountered in employing sorbic acid come from its incorrect

Table 9.1. Sorbic acid doses (mg/l) necessary for sweet-wine conservation (laboratory tests with *Sacch. bayanus*) (Ribéreau-Gayon *et al.*, 1977)

Alcoholic strength (% vol.)	Inoculation population		
	5×10^3 cells/ml	50×10^3 cells/ml	500×10^3 cells/ml
10	150	175	200
11	125	150	200
12	100	150	150
13	75	100	150
14	50	75	125

use. A lack of effectiveness and the appearance of strange tastes are generally reported.

9.2.3 Stability and Gustative Impact

Fresh solutions of sorbic acid have no odor and this product does not influence wine aromas. Up to a concentration of 200 mg/l, sorbic acid does not modify gustatory characters of correctly stored wines, either immediately after its addition or after several years of bottle-aging. For certain wines, its impact on taste is perceptible above this value, but it is distinct only for concentrations above 400–500 mg/l. It does not increase the apparent acidity of wine or thin it but does accentuate impressions of astringency, bitterness and harshness, which are perceived in particular in the aftertaste.

In the 1960s, as early as the first treatments, strange odors and tastes were sometimes noted— especially in red wines treated with sorbic acid. For this reason, its authorization was reconsidered. However, when it is correctly used, experience has shown that in normal storage conditions the evolution and development of bottled wine is not affected. These observations do, however, lead to the problem of the stability of this product in wine.

The sorbic acid molecule, like unsaturated fatty acid molecules, possesses two double bonds. As explained in organic chemistry, these bonds can be oxidized by air to form molecules with aldehydic functions. This reaction explains the unpleasant tastes imparted to fatty substances by oxidation. Concentrated aqueous solutions of sorbate effectively become yellow and take on a pungent odor. This observation is proof of a certain chemical instability of sorbic acid. Yet diluted solutions are noted to be significantly more stable: in wine, in particular, the same quantity initially added is found after three years of bottle-aging. Consequently, the chemical instability cannot be presented as an explanation of the organoleptical deviations attributed to sorbic acid.

The appearance of a disagreeable, intense and persistent odor, similar to the odor of geraniums, in wines treated by sorbic acid was quickly determined to be related to bacterial development. In fact, this olfactive deviation appears at the same time as an increase in volatile acidity or simply malolactic fermentation. It can also occur in poorly stabilized bottled wines. Lactic acid bacteria are responsible for this spoilage, and numerous isolated wine strains are capable of metabolizing sorbic acid. The molecule responsible for the geranium-like odor is a derivative of the corresponding alcohol of sorbic acid (Volume 2, Section 8.7.1). This strong-smelling molecule has a perception threshold of less than 1 μg/l.

It is almost impossible to remove the geranium odor from a wine. Since it is still perceptible after significant dilution, blending is not recommended. The most drastic deodorizing treatments fail (fixation on active charcoal, extraction by oil, etc.). The odor passes during distillation and is concentrated in spirits—only a severe oxidation with potassium permanganate eliminates it. Fortunately, the necessary conditions to avoid this serious spoilage (the rational use of SO_2) are now well known and this problem has practically disappeared.

9.2.4 Use of Sorbic Acid

Sorbic acid must be used exclusively for the conservation of wines containing reducing sugars. It serves no purpose in dry wines. In the case of red wines in particular, the development of lactic acid bacteria in the presence of sorbic acid can result in the appearance of an extremely serious olfactive flaw.

Sorbic acid is not a winemaking tool. It does not affect the rules of *mutage* for sweet wines. It is incapable of stopping fermentations that are underway. Sorbic acid is exclusively used for the conservation of sweet wines to avoid their refermentation. It can be added to wine after the elimination of yeasts by racking, centrifugation or filtration.

The concentration used is generally 20 g/hl. It can be decreased in wines with little residual yeast, high alcohol content and/or a low pH. Due to the low solubility of this acid in water, a concentrated solution of the much more soluble potassium sorbate is prepared immediately before treatment. The solution must be introduced slowly into wine and mixed quickly to avoid the insolubilization of the acid.

In wines treated by sorbic acid, the free SO_2 concentration must be maintained between 30 and 40 mg/l to protect against oxidations, bacteria and the gustatory neutralization of aldehydic substances. This SO_2 concentration alone would be insufficient to protect wine against refermentations.

9.3 OCTANOIC AND DECANOIC ACIDS (SATURATED SHORT-CHAIN FATTY ACIDS)

Certain long-chain fatty acids (C_{16} and C_{18}) activate fermentations. Conversely, other shorter-chain fatty acids, in particular C_8 and C_{10} acids, possess a significant fungicidal action (Geneix *et al.*, 1983). They are formed by yeasts during alcoholic fermentation and can contribute to difficult final fermentation stages. This property, combined with their complete innocuousness, led to them being proposed as an adjuvant to sulfur dioxide to ensure sweet-wine stability (Larue *et al.*, 1986). Their use as a wine stabilizer should be approved by official regulation.

Different options are possible. Their total concentration added to wine should not exceed 10 mg/l; for example, 3 mg of octanoic acid plus 6 mg of decanoic acid per liter. Octanoic and decanoic acids are prepared for use by being solubilized in ethanol at 60% volume. The concentrations are calculated so that the addition of the solution does not exceed 1‰ (1 ml/l).

At the indicated concentration, these acids possess a fungicidal effect complementing sulfur dioxide. For the *mutage* of sweet wines, 150 mg of $SO_2 + 9$ mg of fatty acids per liter has the same effectiveness as 250 mg of SO_2 per liter (Table 9.2). The SO_2 saving is significant. The fatty acids should be added 24 hours before sulfiting. In these conditions, the SO_2 addition acts while yeasts are predominantly (if not completely) inactivated. Also, a fraction of the fatty acids is eliminated by fixation on the yeast cells. After this kind of treatment, sweet wines can be conserved with a free SO_2 concentration of around 40 mg/l.

It should be noted that fatty acids are more effective than sorbic acid, which does not permit a decrease of the SO_2 concentration used for *mutage*.

Due to the aromatic intensity of these acids and their esters, the organoleptic effect of such an addition had to be determined. An increase of the four principal wine fatty acids and their ethylic esters has been observed (Table 9.3), but the increase did not represent the totality of the fatty acids that were added. A large portion of them were fixed to the yeast cells during *mutage* and consequently eliminated during clarification. This treatment leads to an increase of a few milligrams (per liter) of the constituents naturally existing in wine. The variations observed are well within the limits caused by the action of certain classic winemaking methods and operations: aerobiosis, the addition of ammonium salts, temperature variations, etc.

Numerous sensory analysis tests were carried out. This addition is not claimed to be completely without effects but, provided that the addition does not exceed 9–10 mg/l, the effects on the

Table 9.2. Sweet-wine sterilization at the end of alcoholic fermentation by the use of octanoic and decanoic acids to complement the action of sulfur dioxide (Larue *et al.*, 1986)

Yeasts assuring fermentation	Sterilization treatment		Viable yeasts 24 h after treatment (cells/ml)
	SO_2 (mg/l)	Octanoic acid (3 mg/l) Decanoic acid (6 mg/l)	
S. cerevisiae	0	0	10^7
	250	0	4×10^2
	150	9	2×10^2
S. bayanus	0	0	7.5×10^6
	250	0	<10
	150	9	<10

Table 9.3. Effect of sterilization conditions at the end of alcoholic fermentation (SO_2 200 mg/l, SO_2 100 mg/l + fatty acids 9 mg/l) on wine composition (fatty acids and their ethyl esters, mg/l) (Larue *et al.* 1986)

Wine number	Total fatty acids in wine[a]		Total ethyl esters in wine[b]	
	SO_2^c	SO_2 + fatty acids[d]	SO_2^c	SO_2 + fatty acids[d]
n°1	2.6	4.1	0.20	0.56
n°2	4.7	6.2	0.30	0.90
n°3	4.6	8.0	0.34	0.95
n°4	3.3	4.9	0.38	0.58
n°5	3.0	4.7		
n°6	12.4	14.3	0.81	1.58
n°7	4.8	6.2	0.26	0.94
n°8	2.6	3.2	0.17	0.36
Average	4.95	5.86	0.35	0.76

[a]Total fatty acids = hexanoic acid, octanoic acid, decanoic acid, dodecanoic acid.
[b]Total ethyl esters of fatty acids = ethyl hexanoate + ethyl octanoate + ethyl decanoate.
[c]Sterilization with 200 mg SO_2/L.
[d]Sterilization with 100 mg SO_2/L + 9 mg fatty acids/L.

organoleptical character are judged to be slight and on the whole positive.

9.4 DIMETHYLDICARBONATE (DMDC)

The use of diethyldicarbonate (DEDC), or ethyl pyrocarbonate (Baycovin), was authorized in the United States and Germany for a few years in the early 1970s. This fungicide is very effective on yeasts; after acting it is hydrolyzed in a few hours after application, at a rate that varies with temperature, according to the following reaction:

$$C_2H_5-O-\underset{\underset{O}{\|}}{C}-O-\underset{\underset{O}{\|}}{C}-O-C_2H_5 + H_2O$$

$$\longrightarrow 2CH_3-CH_2OH + 2CO_2 \qquad (9.2)$$

This product, at doses of a few hundred milligrams per liter, was very effective for sterilizing wine during bottling (Ribéreau-Gayon *et al.*, 1977). It was less useful during wine storage as it disappeared rapidly and ceased to provide protection from repeat contamination.

It rapidly became apparent that the reaction of ethyl pyrocarbonate in wine was more complex than indicated by the reaction shown above. Ethyl alcohol and carbon dioxide were certainly the main degradation by-products, but small quantities of ethyl carbonate were also formed, and its fruit aroma was perceptible above a certain threshold. Most importantly, ethyl pyrocarbonate is a highly reactive molecule and combines with certain substances in wine (organic acids, polyphenols, and nitrogen-based compounds) to produce urethanes, e.g. ethyl carbamate, which is toxic and carcinogenic. Quantities never exceeded 2–4 mg/l, significantly below the official threshold of 30 mg/l in Canada. However, this risk was sufficient for the product to be completely abandoned.

Bayer, the company that produced Baycovin, replaced it with dimethyldicarbonate (DMDC, or Velcorin), considered to have the same sterilizing properties in wine without any of its problems (Bertrand, 1999). DMDC breaks down to form methyl carbamate, which is considered to have practically no toxic effects. Ough *et al.* (1988) demonstrated that 100 mg/l DMDC sterilized wine completely at pH below 3.8 in the absence of SO_2, even if the initial yeast population was greater than 10^7 cells/ml.

On the basis of these findings, DMDC was initially authorized in the United States, then in other countries. Like ethyl pyrocarbonate, DMDC is most effective at the time of bottling, although it has also been suggested for use in stopping the fermentation of sweet (botrytized) wines (Bertrand and Guillou, 1999), thus reducing the amount of SO_2 required. In any case, a certain quantity of free SO_2 is always necessary to protect the wine from oxidation.

In the European Union, this product is currently authorized for use in unfermented beverages at doses below 250 mg/l. In view of its properties, especially the possibility of reducing the use of SO_2, DMDC is currently being tested with a view to registration in the OIV International Code of Winemaking Practices.

DMDC has proved effective not only on fermentation yeasts, but also on those responsible for contamination (Brettanomyces, Volume 2,

Section 8.4.5), as well as, to a lesser extent, bacteria (Ough *et al.*, 1988).

DMDC decomposes to form mainly methanol and carbon dioxide.

$$CH_3-O-\underset{\underset{O}{\|}}{C}-O-\underset{\underset{O}{\|}}{C}-O-CH_3 + H_2O \qquad (9.3)$$

$$\longrightarrow 2CH_3-OH + 2CO_2$$

Methanol is the most important issue in this case, as it is highly toxic. Theoretically, the breakdown of 200 mg/l of DMDC produces 96 mg/l of methanol. There is no regulated limit for this compound (excepted in the U.S.) which is present in all wines (Volume 2, Section 2.2.1). The OIV has set two limits, 300 mg/l for red wines and 150 mg/l for whites. In the United States, the maximum permitted methanol content is 1000 mg/l. Consequently, the use of DMDC may be considered not to have any serious impact on the methanol content of wine.

The breakdown of DMDC also produces nontoxic methyl carbamate, as well as several methyl, ethyl, and methyl-ethyl carbonates. The latter are odoriferous molecules but are present in insufficient quantities to modify wine aroma.

Finally, it should be taken into account that handling this undiluted product involves some risk as it is dangerous if inhaled or allowed to come into contact with the skin. It is advisable to use proper equipment to ensure safe handling.

9.5 LYSOZYME

9.5.1 Nature and Properties of Lysozyme

Ribéreau-Gayon *et al.* (1975 and 1977) were apparently the first to describe the winemaking properties of lysozyme. The authors described a tour of the Médoc wineries in the 1950s with Alexander Fleming, winner of the 1945 Nobel Prize for Medicine for his discovery of penicillin. During a fining operation using egg white, he wondered whether lysozyme, which he had discovered in egg white a few years earlier, could play a role in the microbiological stabilization of wine.

Lysozyme is an enzyme capable of destroying Gram-positive bacteria (Section 4.3.2), such as lactic bacteria in wine. This natural, crystallized substance is capable, at low doses, of causing lysis of the bacteria, i.e. dissolving their cell walls. Egg white contains approximately 9 g/l lysozyme and the standard method of fining may introduce 5–8 g/l into the wine. Ribéreau-Gayon (1975 and 1977) observed that this agent had no effect on acetic bacteria in wine, which is not surprising, as they are Gram-negative. However, they confirmed that the use of crystallized sample of lysozyme very well purified at doses above 4 mg/l achieved its maximum effect within 24 hours and was capable of destroying almost all the lactic bacteria in a wine (Table 9.4).

However, it was not possible to detect any effect of lysozyme during fining with egg white, even at very high doses. There was no difference compared to fining with gelatin or bentonite. The lysozyme in egg white is probably not released into the wine during fining and is precipitated, together with the albumin, by contact with tannins.

At that time, the use of lysozyme for the microbiological stabilization of lactic bacteria was not envisaged, probably due to the fact that the crystallized product was not widely available and the cost of treatment seemed unacceptable.

Table 9.4. Effect of purified lysozyme on lactic bacteria in wine (Number of living bacteria per cm³) (Ribéreau-Gayon *et al.*, 1977)

Doses of lysozyme (mg/l)	Red wine 18 000 000		Red wine 520 000		White wine 20 000
	After 4 h	After 24 h	After 4 h	After 24 h	After 24 h
4	2 600 000	2 300	490 000	12 000	30
8	1 430 000	2 250	200 000	8 000	10
12	19 700	1 750	4 200	2 400	40
16	15 800	2 000	4 700	2 700	10

The use of lysozyme in the dairy and cheese industries has gradually become widespread and it has been demonstrated to have no toxic effects on humans. The resulting increase in availability has led to new interest in the use of this product in winemaking since 1990.

Further research has established that increasing the dose of lysozyme accelerates lysis of the bacteria but has little impact on the number of resistant cells. The bacteria are not all destroyed, irrespective of the dose, so it is impossible to achieve perfect stabilization of the wine (Gerbaux *et al.*, 1997; Gerland *et al.*, 1999). Crystallized lysozyme is produced by Fordas (Lugano, Switzerland) and marketed, at least for winemaking purposes, by Martin-Vialatte (Epernay, France), under the "Bactolyse" brand. Lysozyme is a protein consisting of 125 amino acids (molecular weight: 14.6 Kda) that acts almost immediately, but, at the conditions prevailing in wine (pH), it is rapidly precipitated out or deactivated (e.g. following bentonite treatment). In contrast to SO_2, the activity of lysozyme increases with pH.

9.5.2 Applications of Lysozyme in Winemaking

Several situations in which lysozyme has useful effects during fermentation and wine storage have been described in the literature (Gerbaux *et al.*, 1999; Gerland *et al.*, 1999), even if stabilization in terms of lactic bacteria is not perfect.

1) Inhibiting malolactic fermentation in white wines

In view of the fact that the high dose of SO_2 required in any case to prevent oxidation also has antimicrobial effects, preventing malolactic fermentation in white wines is not usually a problem, except in wines with a high bacteria content and high pH, such as press wines. However, the use of lysozyme makes it possible to achieve the same protection with lower doses of SO_2 (4–5 g/hl). It is necessary to add lysozyme either once (500 mg/l in the must), or twice (250 mg/l in the must and the same in the new wine), as one 250 mg/l dose is not sufficiently effective.

Adding lysozyme does not affect fermentation kinetics. The wines have the same analytical parameters and are unchanged on tasting, provided they are adequately protected from oxidation, but only 30 mg/l of total SO_2 are required, for example, instead of 50.

As lysozyme is deactivated by bentonite, this treatment should be delayed. For the same reason, wines treated with lysozyme react to the heat test, indicating protein instability. However, protein casse develops at high temperatures (50°C), which do not normally occur during wine shipment and storage.

2) Delaying the development of lactic bacteria and malolactic fermentation in red wines

One case where this applies is the fermentation of whole grape bunches, i.e. carbonic maceration and Beaujolais winemaking methods. It is not unusual for malolactic fermentation to start in this type of medium favorable to the development of lactic bacteria before the end of the alcoholic fermentation. Lysozyme (10 g/hl) added to the crushed grapes is at least as effective as sulfiting (5–7 g/hl) and minimizes the risk of an unwanted increase in volatile acidity.

Another situation is that of wines vatted for long periods, when bacterial development increases the risk of malolactic fermentation on the skins. This should always be avoided in view of the risk that there may be trace amounts of residual sugar present, especially if the grapes were not completely crushed prior to fermentation.

Lysozyme's effectiveness at this stage is less clear, in view of its gradual elimination by precipitation with the phenolic compounds. Adding lysozyme earlier in fermentation is immediately effective, but there is always a risk of recontamination, e.g. during pumping-over, while delaying its addition raises the risk of premature malolactic fermentation.

3) Use in cases of difficult alcoholic fermentation

If conditions are unfavorable to yeasts, a decrease in their activity causes fermentation to slow down. This may, in turn, promote the proliferation of lactic bacteria, which are likely to stop alcoholic fermentation completely (Section 12.7.4), making it extremely difficult to restart (Section 3.8.1). It is well known that bacterial growth in a sugar-containing medium provides the most favorable circumstances for lactic spoilage to occur, together with an increase in volatile acidity.

Sulfiting is the most common technique (Section 8.7.4) for controlling unwanted bacterial development. A 200–300 mg/l dose of lysozyme provides a useful complementary treatment, especially in white wines, where lysozyme is more stable than in red wines. Lysozyme not only prevents lactic spoilage, but is also effective just after spoilage starts: the addition of 25 g/hl reduces the bacterial population very rapidly from several million bacteria to fewer than 100.

4) Microbiological stabilization after malolactic fermentation

Microbiological stabilization of wines is necessary to avoid the many problems that may be caused by microorganisms. Lysozyme does not protect sweet wines from fermenting again or eliminate contaminant yeasts (Brettanomyces), nor is it effective against acetic bacteria, which cause volatile acidity in aerated wines.

Lysozyme is, however, effective in eliminating lactic bacteria. A dose of 200 mg/l has a similar effect to sulfiting at 50 mg/l. As this level of SO_2 is generally required in any case to benefit from the other properties of this product (preventing oxidation, as well as eliminating yeast and acetic bacteria), it seems unnecessary to add lysozyme as well.

It has been observed that delaying sulfiting in red wines by adding lysozyme resulted in a more intense color, which also remained more stable (Gerland et al., 1999).

9.6 DESTRUCTION OF YEASTS BY HEAT (PASTEURIZATION)

9.6.1 Introduction

The destruction of microbial germs by heat has been known for a long time, but in enology there have not been as many applications of pasteurization as in other food industries. The reasons are easy to understand.

Bottled wine conserves relatively well due to it alcohol content and acidity, provided that it is conditioned with a sufficient SO_2 (and eventually sorbic acid) concentration after the satisfactory elimination of microbial germs. 'Hot' bottling can contribute to wine stabilization but, unlike beer, the pasteurization of wine was never widespread.

'Hot' bottling ensures the stability of bottled red wines with respect to bacterial development, and sweet white wines with respect to refermentations. It is generally used with wines of average quality that have microbial stabilization problems. Heating to 45 or 48°C sterilizes the wine and the bottle. The presence of free SO_2 avoids an excessive oxidation. Of course, a space appears below the cork after cooling (Volume 2, Section 12.2.4).

Sweet wines are difficult to store during their maturation phase between the completion of fermentation and bottling. For this more or less long period, the wine is stored in bulk, in tanks or in barrels; it normally acquires its stability and limpidity and improves qualitatively, but refermentation risks certainly exist. The destruction of germs by heat should be able to contribute to the required stabilization, while at the same time limiting the sulfur dioxide concentration. The heat conditions necessary for wine stabilization are easy to satisfy without compromising quality. However, the equipment available in a winery makes it difficult to ensure satisfactory sterile conditions for wine handling and storage.

Despite these difficulties, the thermoresistance of wine yeasts is well understood and the practical conditions for microbial stabilization of bulk sweet wine by heat treatment are also clearly defined (Splittstoesser et al., 1975; Devèze and Ribéreau-Gayon, 1977 and 1978). Therefore, the constraints and advantages of these techniques with respect to the desired decrease in sulfur dioxide concentrations can be correctly evaluated.

9.6.2 Theoretical Data on the Heat Resistance of Wine Yeasts

The thermoresistance of microorganisms can be characterized by two criteria: the decimal reduction time (D) which represents the duration of heating, at a constant temperature, required to reduce the population to one-tenth of its initial value; and the temperature variation (Z) which permits the multiplication or division of D by 10. The value of D depends on the microorganism, the culture conditions and the heating environment; the value of z depends almost exclusively on the microorganism.

The following four tested yeast strains are classed according to an increasing resistance to temperature in terms of z and D at 40°C:

Saccharomycodes ludwigi $z = 3.23°C$ $D_{40} = 10.8\,min$

Saccharomyces bayanus $z = 3.94°C$ $D_{40} = 8.45\,min$

Zygosaccharomyces baillii $z = 4.26°C$ $D_{40} = 46.1\,min$

Saccharomyces cerevisiae $z = 4.34°C$ $D_{40} = 65.7\,min$

Studies of the influence of different factors on the thermoresistance of a *Sacch. bayanus* strain have demonstrated the following.

- Yeasts are half as resistant in a wine at 13% as in a wine at 11%.

- Yeasts are about three times more resistant in a wine containing 100 g of sugar per liter than in the corresponding dry wine.

- A wine at pH 3.8 must be heated for a period three times greater than the same wine at pH 3.0.

- Yeast resistance is 10 times greater during the final phase of fermentation than during the log phase of the cells multiplication.

The effectiveness of a heat treatment is expressed in pasteurization units. These units represent the duration of a treatment at 60°C having the same effectiveness on the given microorganism as the treatment being considered. The graph in Figure 9.1 indicates, for wine microorganisms ($z = 4.5°C$), the number of pasteurization units corresponding with a given time/temperature combination. Heating for 3 min at 55.5°C and for 1.4 min at 57°C are equivalent and correspond with 0.3 pasteurization units.

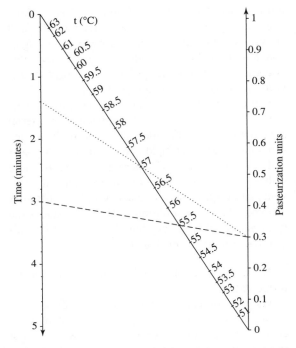

Fig. 9.1. Pasteurization units supplied by a constant temperature, according to time and temperature (Devèze and Ribéreau-Gayon, 1977)

The number of pasteurization units required to obtain a certain level of destruction can be calculated from yeast thermoresistance criteria in a given medium. Laboratory studies anticipated that 0.05 pasteurization units would be sufficient for the destruction of yeasts in a dry wine at 12% vol.ethanol. In practice, 0.5 pasteurization units are required to sterilize such a wine. The uneven heat supplied by industrial heating equipment and the existence of particularly resistant yeast strains at the final stages of fermentation can explain this difference. Table 9.5 shows the necessary conditions for the destruction of germs according to the constitution of the wine.

In addition, a moderate sulfiting is required to protect sweet wines from oxidation. This sulfiting increases yeast sensitivity to heat. The effectiveness of a heat treatment can therefore be improved by injecting sulfur dioxide at the inlet of the temperature exchanger; at elevated temperatures, the proportion of free SO_2 increases.

Table 9.5. Number of pasteurization units (PU) necessary to sterilize different wines (Devèze and Ribéreau-Gayon, 1978)

Alcoholic strength (% vol.)	Concentration in sugar (g/l)		
	0	50	100
10	1.08 PU	1.74 PU	2.83 PU
11	0.73 PU	1.19 PU	1.93 PU
12	0.50 PU	0.81 PU	1.31 PU
13	0.34 PU	0.55 PU	0.90 PU
14	0.23 PU	0.38 PU	0.61 PU

These results show the relative ease of destroying yeasts by heat. They permit the establishment of operating conditions adapted to each particular case. In practice, the principal difficulty arises after the treatment: that is, avoiding subsequent contaminations.

9.6.3 Practical Applications

In practice, stability is satisfactory if the viable yeast population is less than 1 cell/ml. It might be preferable to set a lower limit (for example less than 1 cell per 100 ml) but in this case the sample would have to be filtered for the germ count. This operation can be very difficult, if not impossible—for example, with new sweet botrytized wines. The number of pasteurization units required to sterilize a wine in terms of its constitution is given in Table 9.5. The heating time directly depends on the industrial pasteurization flow rate. From the graph in Figure 9.1, the required temperature can be predicted.

After pasteurization, the wine must be stored in sterile conditions. The tanks and all of the material in contact with the wine must therefore be sterilized—preferably with steam, and if necessary with a chemical disinfectant. In the first case, sterile air must be introduced into the tanks during cooling. In the second case, they must be energetically rinsed with sterilized water.

The results obtained from applying these techniques to a large volume of wine (several thousand hectoliters) led to the proposal of a sweet-wine elaboration method comprising several steps (Devèze and Ribéreau-Gayon, 1978):

1. A well-adapted heat sterilization at the time of *mutage*, when the sugar–alcohol equilibrium is attained. This heat treatment is preceded by the addition of a sufficient amount of SO_2 to obtain a free SO_2 concentration of around 30 mg/l. The sterilized wine is then placed in a sterilized tank.

2. Clarification and stabilization of the wine during the winter and spring following the harvest. The risks of yeast multiplication are limited during this period. After these treatments, the wine is sterilized again and placed in a sterile tank. During these operations, small concentrations of SO_2 are added to maintain the required free SO_2 concentration for avoiding oxidation phenomena.

3. Regular monitoring of the microbiological state of the wine to verify sterility. If the viable yeast population increases exaggeratedly and attains 1000 cells/ml, an additional sterilization is effected. This increase of the yeast population (Table 9.6) can be explained by the presence of an excessive residual population in the wine (>1 cell/ml) preventing a sufficient sterility, or by subsequent contaminations.

4. Sterile bottling, either by filtration or by pasteurization, using established techniques.

If appropriate equipment (which remains relatively expensive) is used, bottled sweet wines can be obtained with the same free SO_2 concentrations as those used for dry white wines. Consequently, the total SO_2 concentration is significantly lowered. Sweet wines stored with only 30 mg of

Table 9.6. Microbiological control, over time, of yeast populations in two sweet-wine tanks, stabilized by pasteurization (Devèze and Ribéreau-Gayon, 1978)

Tank 1		Tank 2	
Date	Viable yeasts/ml	Date	Viable yeasts/ml
Jan. 5	<1	Mar. 11	<1
Feb. 4	<1	Apr. 18	<1
Apr. 18	100	Jun. 23	70
May 16	880	Jul. 6	100
Jun. 1	1540	Jul. 20	200

free SO_2 per liter contain on average 60 mg less total SO_2 per liter than those stored with 50 mg of free SO_2 per liter. The latter concentration is indispensable for avoiding the refermentation of non-pasteurized wines.

9.7 ASCORBIC ACID

9.7.1 Properties and Mode of Action

Ascorbic acid, or vitamin C, exists in fruits and in small quantities in grapes (about 50 mg/l of juice) but it rapidly disappears during fermentation and initial aerations. Wines generally do not contain any.

Ascorbic acid is essentially used in enology (Ribéreau-Gayon et al., 1977) as a reducing agent. Ewart et al. (1987) proposed replacing ascorbic acid with its isomer, erythorbic acid. The latter does not have vitamin properties but possesses the same oxidation–reduction properties. Its industrial production costs are less.

Ascorbic acid was authorized in France as an antioxidant for fruit juices, beers, carbonated beverages and wines, in 1962. Its use does not raise any health-related objections. It is now used in most viticultural countries at a maximum concentration of 150 mg/l, always in association with sulfur dioxide. The recommended concentrations are between 50 and 100 mg/l; higher addition can affect wine taste. As it is completely water soluble (330 g/l), its preparation does not pose a problem. The solution should be prepared at the time of its use. Homogenization should be complete in avoiding all oxygenation, for example by mixing it with sulfur dioxide.

The oxidation mechanism of ascorbic acid has incited much research (Makaga and Maujean, 1994). It functions like an oxidation–reduction system. Its oxidized form is dehydroascorbic acid (Figure 9.2):

ascorbic acid $\longleftrightarrow$

dehydroascorbic acid $+ 2H^+ + 2e^-$ **(9.4)**

This reaction is theoretically reversible but, due to its instability, dehydroascorbic acid disappears.

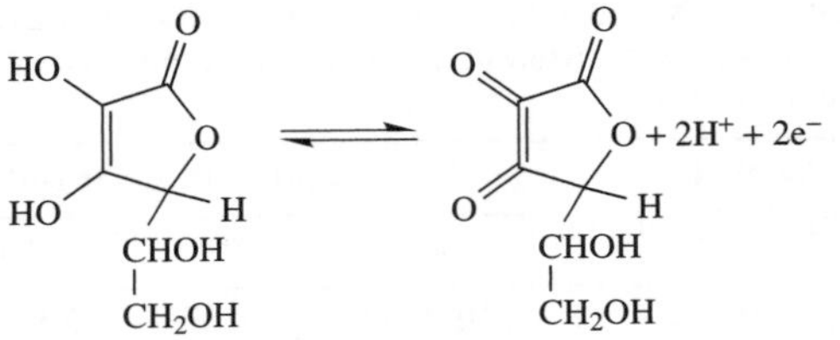

Fig. 9.2. Oxidation of ascorbic acid to dehydroascorbic acid

The two electrons that appear in the course of the reaction reduce certain wine constituents, in particular the ferric ion:

$$2Fe^{3+} + 2e^- \longleftrightarrow 2Fe^{2+} \qquad (9.5)$$

The effectiveness of ascorbic acid in the prevention of iron casse which is exclusively caused by Fe^{3+} ions is explained by the above reactions.

In the presence of oxygen, the oxidation of ascorbic acid leads to the formation of hydrogen peroxide—a powerful oxidant that can profoundly alter wine composition. The presence of a sufficient amount of free sulfur dioxide protects wine from the action of this molecule. It is preferentially oxidized by hydrogen peroxide (Figure 9.3):

$$SO_2 + H_2O_2 \longrightarrow H_2SO_4 \qquad (9.6)$$

The oxidation reaction of ascorbic acid is catalyzed by iron and copper. But, contrary to the direct oxidation of SO_2 by molecular oxygen, the reaction is rapid. It constitutes a simple means of almost instantaneously eliminating dissolved oxygen and preventing the corresponding flaws. Of course, to remain effective, the amount of dissolved oxygen must not be too considerable.

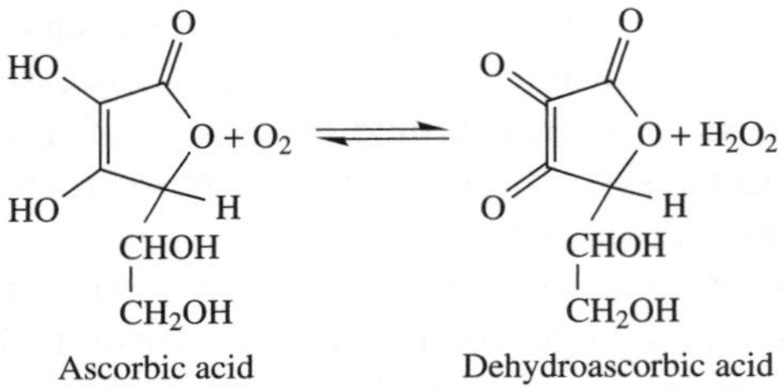

Fig. 9.3. Oxidation of ascorbic acid and formation of hydrogen peroxide

Sulfur dioxide and ascorbic acid therefore have different antioxidant properties. The first has a delayed, but stable, effect which continues over time even in the presence of a subsequent oxygenation. It cannot prevent iron casse, which rapidly appears after an aeration. The second has an immediate effect: it can instantaneously compensate the damage of an abrupt and intense aeration (iron casse), but it acts only as long as the wine is not in permanent contact with air.

Due to the high oxidation sensitivity of ascorbic acid, its effectiveness is only guaranteed when its contact with air is limited. In other words, it protects well against small, brief aerations but not against intense or continued oxidation. Its role is limited to protecting wine from light aerations, following bottling, for example. It is not effective for prolonged storage in tanks or barrels.

The danger of the oxidation of ascorbic acid, especially in the presence of a large amount of oxygen, should also be considered. In these conditions, hydrogen peroxide and sometimes other peroxides are formed. Coupled with the presence of catalyzers, they can cause a thorough oxidation of certain wine constituents, which in the absence of ascorbic acid would not be directly oxidizable by molecular oxygen. The inverse of the desired result can unfortunately be obtained in this manner. This explains some of the problems that can be encountered when ascorbic acid is used incorrectly. For this reason, ascorbic acid should only be used in wines containing a sufficient concentration of free sulfur dioxide, available for the elimination of the hydrogen peroxide formed in the course of oxidations.

9.7.2 Protection Against Enzymatic Oxidations

The addition of ascorbic acid will limit (if not eliminate) must oxidations catalyzed by tyrosinase and laccase. Moreover, it is a substrate of laccase. It does not act by inhibiting the enzymes, as does sulfur dioxide, but rather by monopolizing the oxygen, due to its fast reaction speed. Ascorbic acid is used in this manner in certain countries to complement the protection given by sulfur dioxide against the oxidation of white grape must. Its

use is particularly justified for the protection of mechanically harvested grapes, since it does not act on the maceration, as does sulfur dioxide (Section 13.2.3). This use is not permitted by EU regulations, which only authorize the addition of ascorbic acid in wines.

An effective protection can be obtained in red wines sensitive to oxidasic casse as well as white musts during winemaking. At present, however, the use of ascorbic acid is not widespread in winemaking and not authorized in France probably because the required concentrations are too high to protect musts against oxidations and because sulfur dioxide is more effective.

9.7.3 Protection Against Iron Casse

The aeration of wine oxidizes iron. The amount of ferric iron formed (several milligrams per liter) can be sufficient to induce iron casse. Protection against iron casse can be ensured if the wine receives 50–100 mg of ascorbic acid per liter beforehand (Table 9.7).

Simultaneously, when a wine containing ascorbic acid is aerated, the oxidation–reduction potential slightly increases and then rapidly stabilizes, whereas it continues to increase in the control wine (Figure 9.4). Reciprocally, if ascorbic acid is added to an aerated wine possessing ferric iron, the iron is reduced in the ensuing hours and the

Table 9.7. Protection from iron casse by the addition of ascorbic acid before aeration (Ribéreau-Gayon et al., 1977)

	Fe III 48 h after aeration (mg/l)	Limpidity
White wine (total Fe 18 mg/l)		
Control	8	Cloudy
+25 mg ascorbic acid/l	3	Limpid
+50 mg ascorbic acid/l	1	Limpid
+100 mg ascorbic acid/l	0	Limpid
Red wine (total Fe 15 mg/l)		
Control	6	Slightly cloudy
+25 mg ascorbic acid/l	4	Limpid
+50 mg ascorbic acid/l	0	Limpid
+100 mg ascorbic acid/l	0	Limpid

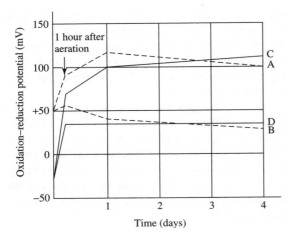

Fig. 9.4. Evolution of oxidation–reduction potential of reduced wines, aerated after the addition of 100 mg of ascorbic acid per liter, compared with control wine (Ribéreau-Gayon *et al*., 1977). (A) Red wine control. (B) Red wine + ascorbic acid. (C) White wine control. (D) White wine + ascorbic acid

oxidation–reduction potential rapidly decreases. The beginnings of an iron casse can be reversed and the corresponding haze eliminated in this manner (Table 9.8).

Ascorbic acid effectively protects against iron casse, which can occur after operations that place wine in contact with air, such as pumping-over, transfers, filtering and especially bottling. In the same conditions, sulfur dioxide acts too slowly to block the oxidation of iron. But, if the wine must be aerated again after a treatment following a first aeration, the ascorbic acid no longer protects the wine. When a wine that has received 100 mg of

ascorbic acid per liter a month earlier is aerated, the evolution of iron III is identical to that in the control. These results lead to the supposition that the added ascorbic acid has disappeared.

9.7.4 Organoleptic Protection of Aerated Wines

In certain cases, ascorbic acid improves the taste of bottled wines. Wines generally taste worse when they contain dissolved oxygen and have an elevated oxidation–reduction potential. Ascorbic acid permits a better conservation of wine freshness and fruitiness—especially in certain types of dry or sparkling white wines. It also decreases the critical phase that follows bottling, known as 'bottle sickness'. The effect is not as considerable or spectacular for all wines but wine quality is never lowered by its use.

The taste improvement due to ascorbic acid depends on several factors. The first is the type of wine. Ascorbic acid is of little interest in the case of wines made from certain varieties or very evolved wines—for example, barrel-aged wines; oxidized white wines, botrytized sweet wine, and fine red wines. On the contrary, it improves the stability of fresh and fruity wines (generally young wines), having conserved their varietal aromas.

Another important factor is the concentration of free sulfur dioxide. It should be situated in an intermediate range between 20 and 30 mg/l, to ensure a refinement of the wine, which in turn presents a fresher aroma with a floral note. For higher or lower free SO_2 concentrations, the qualitative improvement of the wine is less obvious.

Table 9.8. Reduction of iron casse by addition of ascorbic acid, 48 h after the aeration causing the haze (Ribéreau-Gayon *et al*., 1977)

	Fe III (mg/l)		Limpidity 48 h after addition
	At time of addition	48 h after addition	
White wine (total Fe 24 mg/l)			
Control	9	7	Very cloudy
+50 mg ascorbic acid/L	9	2	Limpid
Red wine (total Fe 16 mg/l)			
Control	5	4	Cloudy
+50 mg ascorbic acid/L	5	1	Limpid

Satisfactory results were obtained for sparkling wines produced by the champagne method, by the transfer method, or in pressure tanks. The necessary amounts of sulfur dioxide and ascorbic acid are added to the dosage to ensure concentrations of 20–30 mg/l and 30–50 mg/l, respectively. The coupled addition of the two substances ensures an optimal aroma and improves the finesse and longevity of the wine.

In sparkling wines, ascorbic acid acts not only by its reducing properties but also by its capacity as an oxidation–reduction buffer. Their potential remains stable at 240 mV for several years. In the absence of ascorbic acid, it varies between 200 and 265 mV, according to the effectiveness of corking. This phenomenon clearly affects the organoleptical characters of wine (Makaga and Maujean, 1994).

The use of ascorbic acid insignificantly modifies the use of sulfur dioxide; it permits a slight lowering of its concentration. Yet it possesses other advantages.

9.8 THE USE OF INERT GASES

9.8.1 Wine Storage using Inert Gases

Even before the use of antioxidants (sulfur dioxide and ascorbic acid), the first recommendation for protecting wines against the adverse effects of chemical or microbiological oxidations was to limit their contact with air. Wines were stored in completely filled containers, sometimes equipped with a system permitting dilatation compensation. This recommendation cannot always be followed, if the availability of tanks of a satisfactory size is limited or wine is regularly taken from the same tank for several days. Tanks equipped with a sliding cover which always remains in contact with the surface of the wine were introduced, but the joints between the cover and the inner surface of the tank are rarely satisfactory and their effectiveness is questionable.

Satisfactory results are obtained by storing wine in a partially filled tank with an inert gas, in the total absence of oxygen. Wine storage using inert gas also permits the carbon dioxide concentration

(lowering or increasing) to be adjusted. Although not directly related to protection from oxidation, this subject will nevertheless be covered in this section.

The following gases are authorized for storage: nitrogen, carbon dioxide and argon. Argon is rarely used: it is more expensive than the others and its solubility is limited in wine (4 l/hl). Carbon dioxide is very soluble in wine (107.2 l/hl) and therefore cannot be used alone in partially filled containers. The carbon dioxide concentration of the wine would significantly increase by the dissolution of the gas. It is sometimes used in a mixture with nitrogen (for example, 15% CO_2 + 85% N_2) to avoid the degassing of certain wines that must maintain a moderate CO_2 concentration. Nitrogen is the most commonly used gas. 'R' quality nitrogen is used which contains a little oxygen as an impurity but has no impact on wine. It is less soluble in wine than oxygen (1.8 l/hl compared with 3.6 l/hl) but contrary to oxygen, which reacts with wine constituents by oxidizing them, nitrogen accumulates without reacting. The wine spontaneously becomes saturated in nitrogen during handling in contact with air. Storage in the presence of this gas therefore cannot increase its concentration in wine.

Several principles of inert gas systems for tanks exist but the system adapted to the winery must be well designed. In particular, the installations must be perfectly airtight. Maintaining a slight overpressure is recommended in order to monitor for possible leaks. This method is essentially applicable to perfectly hermetic tanks.

The gases are stored in compressed gas bottles. At the outlet from the bottles, the gas generally undergoes a double expansion. Initially, it is reduced to a pressure of 2 to 8 bars and circulates in copper piping up to the storage tank. A second expansion reduces the pressure to 15–20 mbar or 100–200 mbar. This second solution permits easy identification of piping and tank leaks. Each tank has a separate line and a manometer enabling verification of the pressure and thus the airtightness. Finally, a pressure release valve avoids the unfortunate consequences of operation errors.

Table 9.9. Evolution of the carbon dioxide concentration in wines stored in a carbon dioxide or nitrogen atmosphere, according to fill level of container (Lonvaud-Funel, 1976)

Fill level (%)	CO_2 atmosphere		N_2 atmosphere	
	CO_2 concentration in wine (mg/l)	Increase (%)	CO_2 concentration in wine (mg/l)	Decrease (%)
98	308	7	281	1.5
82	589	106	234	17.8
50	1132	297	144	49.6
18	1708	499	51	82.0

Installations have been specially designed for wine storage using inert gas. Metal tanks are connected together by gas lines, but the tanks can be isolated and individually maintained at a slight overpressure (100–200 mbar). This overpressure attests to the hermeticity of the tanks. It is verified by the manometer reading.

At the beginning of the operation, the tank is completely filled with wine and the hermetic tank vent is secured. A hectoliter of wine is then drained from the faucet at the bottom of the tank. Simultaneously, nitrogen gas is sparged in the upper portion of the tank—replacing the drained wine and creating a nitrogen atmosphere buffer. Next, the internal pressure is adjusted. The tank is then ready for storage. To remove wine, the nitrogen bottle should first be opened, then the gas valve for the tank and finally the wine tank valve. Perfectly clear and stable wines are conserved and protected from oxidations and evaporation for several months in these wine tanks. Nitrogen consumption is extremely limited. The evolution of the taste of these wines is identical to that of wines stored in completely filled tanks.

When there is insufficient wine to fill the tank completely, thus expelling all of the air contained, another solution consists of completely filling the tank with water. It is then emptied under a nitrogen counter-pressure before introducing the wine. In many cases, after partially filling the tank with wine, residual air is expelled by sparging nitrogen in the tank headspace. If these operations are carried out in non-airtight tanks, the inert gas must be constantly renewed—resulting in a considerable consumption of gas. These conditions are not recommended and can lead to a false sense

of security for wine storage. At the end of a few weeks, the wine is oxidized and has lost its CO_2; its odor and taste become insipid.

In any case, this storage system does not release the winemaker from using sulfur dioxide or even permit concentrations to be lowered. This antiseptic remains indispensable for fighting against yeasts and lactic acid bacteria. It must be used at the same concentrations as in full tanks.

Storage under inert gas can cause either an increase or a decrease in the amount of carbon dioxide naturally existing in wine. The data in Table 9.9 show the impact of the wine volume/gaseous atmosphere ratio. The variations are slight, especially with nitrogen, if the tank is practically full. Storage under a carbon dioxide atmosphere easily leads to an excessive increase of its concentration, with corresponding changes in the organoleptic characters of the wine. Conversely, storage under nitrogen causes a considerable decrease in the CO_2 concentration for half-filled tanks.

The study by Lonvaud-Funel and Ribéreau-Gayon (1977) gives the factors permitting the estimation of the CO_2/N_2 mixture which must be used at a given temperature so that a wine conserves its initial dissolved CO_2 concentration during storage, whatever the tank fill level. Inert gases can also be used for ensuring that wine transfers are protected from oxygen, by injecting nitrogen in the lines while pumping wines, for example.

9.8.2 Adjusting the Carbon Dioxide Concentration

Wine tasters are very sensitive to taste modifications caused by the presence of this gas, even below the organoleptic perception threshold. For example, in a red Bordeaux wine, more than 50% of the tasters correctly put in order three samples of a wine containing 620, 365 and 20 mg of carbon dioxide per liter (Ribéreau-Gayon and Lonvaud-Funel, 1976). The characteristic pricking sensation of CO_2 was only perceptible in the first sample. The third sample appeared more insipid than the second, which was judged the best. Yet nothing led the tasters to believe that the difference was related to the CO_2 concentration.

Dry white wines tolerate higher carbon dioxide concentrations. Around 90% of the tasters correctly put in order three samples of the same wine containing, respectively, 250, 730 and 1100 mg of CO_2 per liter. The second sample was preferred overall; the carbon dioxide increased the aroma and the freshness of this wine. Yet the carbon dioxide concentrations should not be exaggerated. Concentration of 1000 mg/l are not as appreciated in dry white wines as one might think.

Due to its organoleptic impact, the carbon dioxide concentration should be correctly adjusted. For each type of wine, there is a corresponding optimal concentration. Red wines tolerate less CO_2 (around 200 mg/l) than dry white wines (around 500–700 mg/l). The more tannic and adapted for aging the wines are, the less they tolerate CO_2.

It can be useful to eliminate excess carbon dioxide rapidly, by agitation, in young red wines intended for early bottling. Racking in the presence of air can decrease the CO_2 concentration by 10%, but this is not always sufficient.

The injection of fine nitrogen bubbles in wine entrains a certain proportion of dissolved gas (carbon dioxide or oxygen) in a wine. The wine delivery rate, with the device in Figure 9.5, can vary from 30 to 120 hl/h. Temperature plays an important role in the effectiveness of this treatment. Below 15°C, the degassing yield is insufficient. The temperature of the wine should preferably be at 18°C. The wine, emulsified with very fine nitrogen bubbles, should then be exposed to air by flowing in a thin film through a shallow tank with a large surface area so that the nitrogen is easily released and entrains the dissolved carbon dioxide.

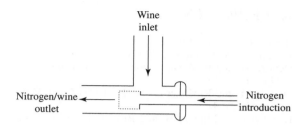

Fig. 9.5. Gas injector, diffusing very fine bubbles in wine circulating through piping (Ribéreau-Gayon *et al.*, 1973)

In tests, simple racking permitted the elimination of 26% of the CO_2. Treating with half a volume of N_2 for a volume of wine eliminated 43% of the CO_2. Four times more nitrogen (2 volumes) only eliminated 54%. It is therefore more reasonable to carry out two consecutive treatments with lower volumes of nitrogen.

In certain cases, the carbon dioxide concentration must be decreased; in others, it must be increased. It can be increased by sparging with carbon dioxide gas; the gas can be injected in the winery piping. The same result can be obtained by placing wine in a partially filled tank, its headspace filled with mixture of a N_2 and CO_2. Lonvaud-Funel (1976) has given the mixture required for obtaining a certain CO_2 concentration according to the respective wine and gas volumes.

These operations are normally carried out at atmospheric pressure. If they were to take place at higher pressures, a gasification would be effected. The operations would no longer be considered as ordinary wine treatments, since gasified wines are subject to special legislation.

The dissolution of carbon dioxide in wine does not differ much from that in water. It depends on the temperature and ranges between 2.43 g/l at 8°C and 1.73 g/l at 18°C. These values correspond with the maximum amount of CO_2 that can be dissolved in wine.

The sparging of wine by carbon dioxide has been suggested. This method can be useful for avoiding oxygen dissolution during transfers and to ensure a protection against oxidations. The wine must be degassed before bottling.

REFERENCES

Bertrand A. (1999) *Viti.*, 249, 25.

Bertrand A. and Guillou I. (1999) *Bull. OIV*, 72, 84.

Devèze M. and Ribéreau-Gayon P. (1977) *Conn. Vigne Vin*, 11 (2), 131.

Devèze M. and Ribéreau-Gayon P. (1978) *Conn. Vigne Vin*, 12 (2), 91.

Ewart A.J.W., Sitters J.H. and Brien C.J. (1987) *Austr. NZ Wine Ind. J.* 1, 59.

Geneix C., Lafon-Lafourcade S. and Ribéreau-Gayon P. (1983) *CR Acad. Sci., Série III*, 296, 943.

Gerbaux V., Meistermann E., Cottereau P., Barrière C., Cuinier C., Berger J.L. and Villa A. (1999) *Bull. OIV*, 72 (819), 349.

Gerbaux V., Villa A., Monamy C. and Bertrand A. (1997) *Am. J. Enol. Vitic.*, 48 (1), 49.

Gerland C., Gerbaux V. and Villa A. (1999) *Revue des Oenologues*, 93S, 44.

Larue F., Murakami Y., Boidron J.N. and Fohr L. (1986) *Conn. Vigne Vin*, 20 (2), 87.

Lonvaud-Funel A. (1976) Recherches sur le gaz carbonique du vin. Thèse de Doctorat Œnologie, Université de Bordeaux II.

Lonvaud-Funel A. and Ribéreau-Gayon P. (1977) *Conn. Vigne Vin*, 11 (2), 165.

Makaga E. and Maujean A. (1994) *Bull. OIV*, 153, 763.

Ough C.S., Kunkee R.E., Vilas M.R., Bordeu E. and Huang M.C. (1988) *Am. J. Enol. Vitic.*, 39, 279.

Ribéreau-Gayon J., Peynaud E., Ribéreau-Gayon P. and Sudraud P. (1975) *Sciences et Techniques du Vin.* Vol. II: *Caractères des vins. Maturation du raisin. Levures et bactéries.* Dunod, Paris.

Ribéreau-Gayon J., Peynaud E., Ribéreau-Gayon P. and Sudraud P. (1977) *Sciences et Techniques du Vin.* Vol. IV: *Clarification et Stabilisation, Matériels et Installations.* Dunod, Paris.

Ribéreau-Gayon P. and Lonvaud-Funel A. (1976) *CR Acad. Agric.*, 62 (7), 491.

Splittstoesser D.F., Lienk L.L., Wilkinson H. and Stamer J.R. (1975) *Appl. Microbiol.*, Sept., 369.

10

The Grape and its Maturation

10.1 INTRODUCTION

The grape constitutes the raw material for producing wines. Its maturity level is the first factor, and certainly one of the most deciding ones, in determining wine quality. It is the result of all of the complex physiological and biochemical phenomena whose proper development and intensity are intricately related to environmental conditions (vine varieties, soils, climate) (Peynaud and Ribéreau-Gayon, 1971; Ribéreau-Gayon *et al.*, 1975; Champagnol, 1984; Huglin, 1986; Kanellis and Roubelakis-Angelakis, 1993; Flanzy, 2000; Roubelakis-Angelakis, 2001).

Compared with other fruits, the study of the grape presents many problems. Berry growth and development are the result of a long and complex reproduction cycle. The ovary, and then the seeds, attract the hormones necessary for their development from the leaves, where they are mainly synthesized. The triggering of the maturation process does not correspond with a true climacteric crisis. It is linked to the drop in growth hormone levels and the appearance of a stress hormone, abscisic acid. In consequence, the behavior of the entire plant strongly influences the development of these processes. Certain studies can be carried out on fruiting microcuttings or potted vines under controlled conditions, but the preponderant influence of environmental parameters on vine behavior requires that a large number of experiments be carried out in the vineyard. The study of maturation therefore comes up against difficulties due to the extreme variability of berry composition, at any given time and for the same variety.

In spite of these difficulties, the observations made each year by researchers at the Faculty of Enology at Bordeaux and by other teams in different wine-producing regions have permitted us to:

- follow and compare the chemical composition modifications of the grape during maturation;

- compare the maturation kinetics over the years, in terms of meteorological conditions;

- compare the evolution of different vineyards, in terms of local environmental conditions;

- forecast maturity dates and thus establish the harvest dates.

These preliminary observations directed subsequent research towards a more thorough study of maturation mechanisms. This chapter will cover the biochemical phenomena characterizing grape maturation and the process of the development of rot. It will also focus on the influence of environmental factors on maturation.

10.2 DESCRIPTION AND COMPOSITION OF THE MATURE GRAPE

10.2.1 The Berry

The grape is a berry, classed in a group of several seeded fleshy fruits. The berries are organized into a cluster. Each berry is attached to the rachis by a small pedicel containing the vessels, which supply the berry with water and nutritive substances (Figure 10.1a). Cluster structure depends on the length of the pedicels: if they are long and thin, the grapes are spread out (Figure 10.1b); if they are short, the bunches are compact and the grapes

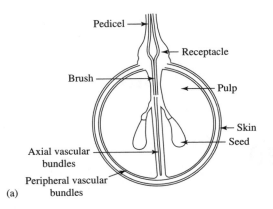

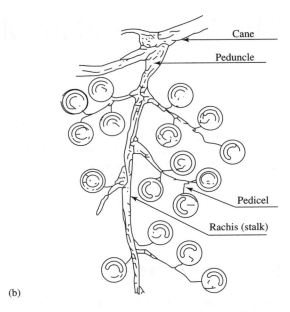

Fig. 10.1. Fruit of the grape vine: (a) grape berry at maturity; (b) structure of grape cluster

are packed together. Varieties used for winemaking often belong to the latter category. Cluster compactness is one of the factors affecting rot sensitivity.

Genetic factors and environmental conditions that characterize berry formation greatly influence its development and its composition at maturity.

10.2.2 Berry Formation

Fruit development is closely related to the modalities of ovule fertilization. Flowering corresponds to the opening of the corolla and the ejection of the calyptra (anthesis). The pollen liberated in this manner can reach the ovary and trigger its growth (*nouaison* or berry setting). The liberation of pollen is facilitated by warm, dry weather. In a cool and humid climate, flowering can be spread out over 10–15 days and sometimes more.

Pollination is normally followed by fertilization, permitting the development of a berry possessing one to four normal seeds. Poor fertilization can lead to the formation of rudimentary seeds (stenospermocarpic seedlessness). The absence of fertilization produces seedless berries (parthenocarpic seedlessness). Seedlessness can be a varietal genetic character, sought after for the production of table grapes (Thompson seedless) or for the preparation of raisins (Corinthe). Non-pollinated, unfertilized ovaries are deficient in growth regulators (polyamines) and form tiny berries that remain green (Colin *et al.*, 2002).

In general, not all of the flowers borne by the cluster are fertilized and become berries. The berry setting ratio decreases as the number of flowers formed on the grape cluster increases. The causes of this phenomenon have been known for a long time. As a general rule, a plant can only supply 100 to 200 berries per bunch with sugar, depending on the variety.

After berry set, a variable proportion of apparently fertilized young berries no longer grow and fall from the plant. This abscission is caused by the hydrolysis of pectins of the middle lamella of the cell walls forming the separation layer at the base of the pedicel. The phenomenon, called shatter (*coulure* in French), is often difficult to distinguish from berry setting in the case of cold weather and overcast skies, which cause an abnormally long flowering—sometimes up to 3 weeks (Figure 10.2). Shatter depends in particular on sugar availability and the effects of climatic parameters on its availability (photosynthesis, sugar migration in the plant). Climatic shatter constitutes the principal cause of yield variability in northern vineyards. In warm climate zones, a water deficiency can bring about the same result.

A varietal-specific sensitivity also exists. Shatter can be complete with Grenache, Merlot, Muscat

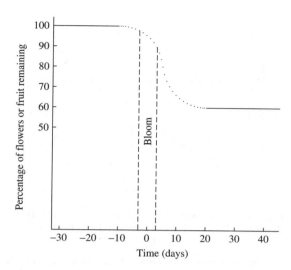

Fig. 10.2. Flower and fruit evolution during bloom and berry setting (Bessis and Fournioux, 1992)

Ottonel or Chardonnay. Other varieties, such as Carignan, Chenin, Sauvignon Blanc, Folle Blanche, Pinot Blanc, Riesling and Cabernet Sauvignon, are much less affected.

Millerandange is related to poor flowering conditions, involving a defective pollination with dead pollen that does not lead to fertilization.

10.2.3 The Developmental Stages of the Grape

In the course of its development, from ovary to ripe fruit, the grape follows an evolution common to all berries. It is generally divided into three phases (Figure 10.3), taking into consideration parameters such as berry diameter, weight and volume:

1. An initial rapid growth or herbaceous growth phase lasting 45 to 65 days; depending on vine variety and environmental conditions. The intensity of cellular multiplication depends on the existence of seeds. Growth hormone concentrations (cytokinins and gibberellins) correspond directly with the number of seeds. The application of gibberellic acid on seedless grapes has become a common viticultural practice (Ito *et al.*, 1969). Cellular growth begins about 2 weeks after fertilization and continues

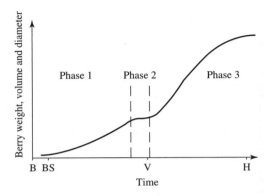

Fig. 10.3. Developmental stages of the grape berry: B, bloom; BS, berry set; V, *véraison;* H, harvest

until the end of the first phase. In the course of this first period, chlorophyll is the predominant pigment. The berries have an intense metabolic activity, characterized by an elevated respiratory intensity and a rapid accumulation of acids.

2. A slowed growth phase during which *véraison* occurs. *Véraison* is characterized by the appearance of color in colored varieties and a translucent skin in white varieties. It is an abrupt phenomenon at the berry level but takes place over several days when different berries of the same bunch are considered. In a vineyard parcel, this phase lasts 8 to 15 day or longer if flowering is very slow. It corresponds with the depletion of growth substance synthesis and an increase in the concentration of abscisic acid.

3. A second growth phase corresponding to maturation. Cellular growth resumes and is accompanied by diverse physiological modifications. The respiratory intensity decreases, whereas certain enzymatic activities sharply increase. This final period lasts 35 to 55 days, during which the grape accumulates free sugars, cations such as potassium, amino acids and phenolic compounds, while concentrations of malic acid and ammonium decrease. Grape size at maturity depends largely on these accumulation processes but also on the number of cells per berry. There is a very close relationship between the dimensions of a ripe grape and the number of seeds it contains (Table 10.1).

Table 10.1. Relationship between number of seeds and berry size at maturity: Merlot variety grapes sampled in 1982 in a Saint-Emilion vineyard (France)

Number of seeds	Berry weight (g)	Juice volume per berry (ml)	Sugar concentration (g/l)
0–1	1.10	0.75	235
2	1.55	1.01	233
3	1.94	1.12	221

10.2.4 Grape Morphology

Each grape comprises a group of tissues (the pericarp) surrounding the seeds. The pericarp is divided into the exocarp (the skin), the mesocarp (the pulp) and the endocarp (the tissue that lines the seed receptacles containing the seeds but is not distinguishable from the rest of the pulp) (Figure 10.1a). The fruit is nourished by a branching vascular network of the rachis, which traverses the pedicels. This vascular bundle then branches out in the pulp. This network can be observed due to the transparency of certain white varieties at maturity.

The skin of the grape forms a heterogeneous region constituted by the cuticle, the epidermis and the hypodermis (Figure 10.4). The cuticle is a continuous layer whose thickness varies depending on the variety: 1.5–4 μm for certain *Vitis vinifera* varieties and up to 10 μm for certain American vines. It begins to develop 3 weeks before anthesis. In the course of berry maturation and development, it becomes increasingly disorganized and its thickness diminishes. The cuticle is generally covered by epicuticular wax (bloom) in the form of stacked platelets, visible by electron microscopy. Wax thickness is relatively constant throughout the course of berry development (about 100 μg of wax/cm^2 of surface).

The epidermis is constituted of one or two layers of tangentially elongated cells whose thickness varies depending on the grape variety. The hypodermis comprises two distinguishable regions: an outer region with rectangular cells and an inner region with polygonal cells.

The pulp is composed of large polygonal cells with very thin, distended cell walls. There are

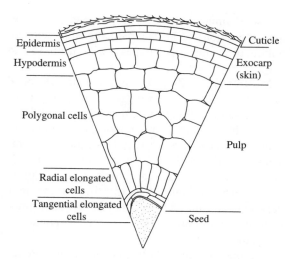

Fig. 10.4. Different grape berry tissues at maturity

25–30 cell layers, organized into three distinct regions.

Each normally constituted seed comprises a cuticle, an epidermis and three envelopes covering the albumen and the embryo.

Grape berry consistency depends on skin and pulp cell wall thickness. Generally, table grape varieties produce plump, thin-skinned grapes (the pulp having thick cell walls), whereas winemaking varieties have tough skins and juicy pulp (pulp with thin cell walls).

On the grape surface, there are between 25 and 40 stomata per berry, depending on the variety. After *véraison*, these stomata no longer function and they necrotize. Rapid fruit enlargement creates tension, resulting in the development of peristomatic microfissures.

10.2.5 Grape Cluster Composition at Maturity

The stalk rachis represents around 3–7% of the weight of a ripe grape cluster. Its chemical composition is similar to the composition of leaves. It contains little sugar (less than 10 g/kg) and an average acid concentration (180–200 mEq/kg). These acids are in the form of salts, due to the large quantity of cations present.

Stalks are rich in phenolic compounds. They can contain up to 20% of the total phenolic compound concentration of the grape cluster, even though they represent a lower proportion of the total weight. These phenolic compounds are more or less polymerized and have a very astringent taste.

The stalk attains its definitive size around the time of *véraison*. Although it loses most of its chlorophyll, it remains green during maturation. It is often completely lignified well after maturity.

The seeds represent 0–6% of the weight of the berry. They contain carbohydrates (35% on average), nitrogen compounds (around 6%) and minerals (4%). An oil can be extracted from the seeds (15–20% of the total weight) which is essentially oleic and linoleic acid. The seeds are an important source of phenolic compounds during red wine-making. Depending on the varieties, they contain between 20 and 55% of the total polyphenols of the berry.

The seeds attain their definitive size before *véraison*. At this time, they have reached physiological maturity. During maturation, the tannin concentration of the seeds decreases whereas their degree of polymerization increases. Conversely, the nitrogen compounds are partially hydrolyzed. The seed can yield up to one-fifth of its nitrogen to the pulp, while still remaining richer in nitrogen than the other solid parts of the grape cluster.

Depending on the grape variety, the skin represents from 8 to over 20% of berry weight. Being a heterogeneous tissue, its importance depends greatly on the extraction method used. Separating the skins by pressing the grapes results in the extraction of the pulp and seeds. This method corresponds best to current enological practices. The sugar concentration of skin cells is very low. For the same weight, the skin is as rich in acids as the pulp but citric acid is predominant. Malic acid, in significant quantities in the skins of green grapes, is actively metabolized in the course of maturation. The majority of tartaric acid is esterified by phenolic acids (cafeic, coumaric). A significant quantity of cations cause the salification of these acids. The contents of the skin cells always have a higher pH than the pulp.

The skin is especially characterized by significant quantities of secondary products of major enological importance (phenolic compounds and aromatic substances). It accumulates these substances during maturation.

The following phenolic compounds are present in the grape skin at maturity: benzoic and cinnamic acid, flavonols and tannins. They are distributed in the cells of the epidermis and the first subepidermal layers in both white and red grapes. In addition, the red grape skin contains anthocyanins, essentially located in the hypodermal cell layers. Exceptionally, in certain years, the cells adjacent to the pulp can be colored. The pulp itself is colored in the case of Tenturier varieties and some American vines or direct producer hybrids. Anthocyanin composition varies from cultivar to cultivar, depending on the anthocyanidin substitution and heterosidic nature of the cultivar (see Volume 2, Section 6.2.3).

The ripe grape skin also contains considerable amounts of aromatic substances and aroma precursors. In certain muscat varieties, the skin can contain more than half of the free terpenols of the berry (Bayonove, 1993). Other chemical families of aromatic substances may also be contained in the skin. Finally, the skin is covered by epicuticular wax, essentially constituted of oleanolic acid.

All of this information is very important from a technological point of view. All methods increasing the solid–liquid contact for color extraction or aroma dissolution should be favored during winemaking.

The pulp represents the most considerable fraction of the berry in weight (from 75 to 85%). The vacuolar contents of the cell contain the grape must—the solid parts (cytoplasm, pectocellulosic cell walls) constituting less than 1% of this tissue. The must is a cloudy liquid, generally slightly colored, having an elevated density due to the many chemical substances that it contains. Sugars are the primary constituents—essentially glucose and fructose. Fructose is always predominant (the glucose/fructose ratio is around 0.9). Saccharose, which is the migratory form of sugar in the plant, exists in only trace amounts in the grape. Other sugars have been identified in the grape: arabinose, xylose, rhamnose, maltose, raffinose, etc. (see Volume 2, Section 3.3.1). The reducing sugar concentration in normal ripe grapes varies from 150 to 240 g/l.

Most of the acids of the metabolism are found in trace amounts in ripe grape pulp (pyruvic, α-ketoglutaric, fumaric, galacturonic, shikimic, etc.). Must acidity, an important element of enological data, is essentially constituted by three acids: tartaric, malic and citric acid (Volume 2, Section 1.2.2). It can vary from 3 to 10 g/l in sulfuric acid or from 4.5 to 15 g/l in tartaric acid, depending on the cultivar, the climate and grape maturity. Phosphoric acid is the preponderant inorganic anion.

The pulp is particularly rich in cations. Potassium, the principal element, is much more abundant than calcium, magnesium and sodium. The other cations are present in much lower concentrations, with iron representing 50% of the remaining cations. Concentrations of metallic trace elements such as lead are infinitesimal, except in the case of accidental pollution. In spite of this concentration in cations, part of the acids remains unsalified. Must pH currently varies between 2.8 and 3.5.

The pulp contains only 20–25% of the total nitrogen content of the berry. The must contains 40–220 mM of nitrogen in its ammoniacal or organic form. The ammonium cation is the most easily assimilable nitrogen source for yeasts and it is often present in sufficient quantities (Volume 2, Section 5.2.2). The amino acid fraction varies from 2 to 13 mM in leucine equivalents (2–8 g/l). Most amino acids are found in grape must at variable concentrations, and a few of them (proline, arginine, threonine and glutamic acid) represent nearly 90% of the total concentration. The relationship between the must amino acid concentration and its organic acid concentration has been known for a long time. The most acidic grapes are always the richest in amino acids. Soluble proteins of the must represent 1.5–100 mg/l.

At maturity, the grape is characterized by a low concentration in pectic substances with respect to other fruits. Pectins represent from 0.02 to 0.6% of fresh grape weight. Differences from cultivar to cultivar and from year to year can be significant.

Only the free pectic fraction, associated to diverse soluble oses, is likely to be found in must. This fraction also contains small amounts of insoluble proteins.

The skin is considered to be the principal source of aromatic substances, but the pulp does contain significant concentrations of these compounds. In certain muscat varieties, the must can contain up to two-thirds of the terpenol heterosides. The pulp is characterized in particular by the accumulation of a diverse variety of alcohols, aldehydes and esters which participate in grape aromas.

There is considerable heterogeneity between different grapes on the same grape cluster. Similarly, the diverse constituents of must are not evenly distributed in the pulp. As a primary technological consequence, the chemical constitution of the juice evolves in the course of grape pressing in white winemaking. The peripheral and central zones (near the seeds) are always richer in sugar than the intermediary zone of the pulp (Figure 10.5). Malic and tartaric acid concentrations increase towards the interior of the berry. Potassium is distributed differently within the grape and often causes the salification of the acids, with the precipitation of potassium bitartrate, in the course of pressing. This heterogeneity seems to apply to all must constituents. Finally, the half of the grape opposite

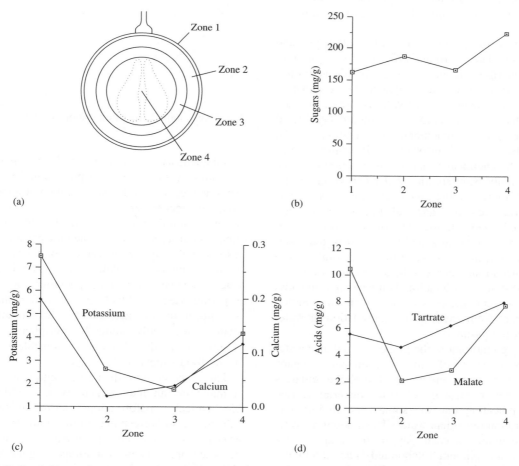

Fig. 10.5. Breakdown of principal constituents inside the grape berry (Possner and Kliewer, 1985) (all results are expressed in mg per g fresh weight). (a) zones; (b) sugars; (c) cations; (d) acids

the pedicel is generally richer in sugars and poorer in acids than the proximal half.

10.3 CHANGES IN THE GRAPE DURING MATURATION

10.3.1 General Characteristics of Maturation

As early as 1897, during his studies on grape respiration, Gerber discovered a respiratory substrate change in berry pulp at *véraison*. These observations were later confirmed by the use of ^{14}C-marked molecules (Ribéreau-Gayon, 1959; 1960). At present, the dominant role of malic acid in the metabolism of ripening fruit is fully established (Ruffner, 1982a).

Most of the primary metabolic pathways have been elucidated through progress in the extraction and study of numerous enzymatic activities. High-performance analytical methods, capable of determining nanograms of volatile substances, are currently being developed and should be able to provide much supplemental information on the secondary metabolism of grapes in coming years.

The biochemical processes of maturation have traditionally been summarized by the transformation of a hard, acidic green grape into a soft, colored fruit rich in sugar and aromas. As already indicated, these transformation can only occur when the grape is attached to the rest of the plant. In this case, the increase in the concentration of a substance in the berry can be due to importation of this substance, on-location synthesis or water loss in the vegetal tissue. Conversely, its diminution can result from exportation, degradation or water gain in the tissue.

During maturation, the grape accumulates a significant quantity of solutes, principally sugars. In spite of berry enlargement (cellular enlargement), the percentage of solid material increases—indicating that the solutes are imported in greater quantities than water. The amount of water that accumulates each day in the grape is the sum of the phloem (elaborated) and xylem sap flux minus the water loss due to transpiration (Figure 10.6). At the start of maturation, the berries

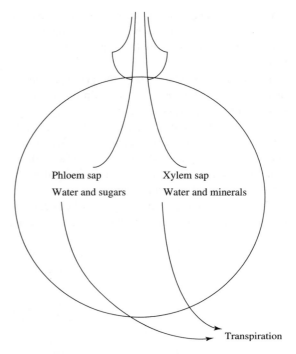

Fig. 10.6. Grape berry alimentation (Coombe, 1989)

simultaneously import water with the sugars, but the amount of water transpired rapidly diminishes as the stomata degenerate; then, transpiration uniquely occurs across the cuticular wax. Sugar accumulation then occurs against the diffusion gradient, often up to considerable concentrations corresponding to a substantial osmotic pressure. In addition, the xylem solute supply strongly diminishes after *véraison*. This phenomenon, due to a partial vascular blockage (or embolism), has an impact on the accumulation of certain substances, especially minerals. Peripheral vessels (Figure 10.1a) then become responsible for most of the food supply to the grape.

The grape is more than an accumulation organ: it maintains an intense activity (respiration and biochemical transformations) during maturation. *Véraison* also corresponds to the synthesis of new enzyme activities and the release of inhibition of other ones. These variations in gene expression cause profound changes in grape metabolism (Robinson and Davies, 2000).

10.3.2 Sugar Accumulation

The most spectacular biological phenomenon of maturation is certainly the rapid accumulation of sugars in the grape from *véraison* onwards. From the start, the inflorescences, due to their growth hormone concentrations, have a strong demand for the products of photosynthesis. However, during the entire herbaceous growth phase, the sugar concentration of green grapes does not exceed 10–20 g/kg in fresh weight (around the same as leaves). The sugars imported daily are metabolized at a high intensity for fruit development but in particular for seed growth and maturation. The nutritive substance demand towards the grape is even more considerable in the days that precede *véraison*.

The depletion of growth hormones, notably auxins, and the increase in abscisic acid concentrations correspond with the lifting of the inhibition of the principal enzymatic activities involved in the accumulation of sugars in pulp cell vacuoles. Saccharose phosphate synthetase, saccharose synthetase and hexokinase are no longer blocked (Ruffner *et al.*, 1995). This accumulation occurs against the diffusion gradient. The transport requires energy to counter the growing osmotic pressure as the sugar concentration increases. Up to 30 bars of pressure can be attained towards the end of maturation. An enzymatic complex associated with the tonoplast of the pericarp cells ensures this transport (Figure 10.7). The sugars, synthesized in the leaves, migrate exclusively in the form of saccharose through the phloem to the grapes (Lavee and

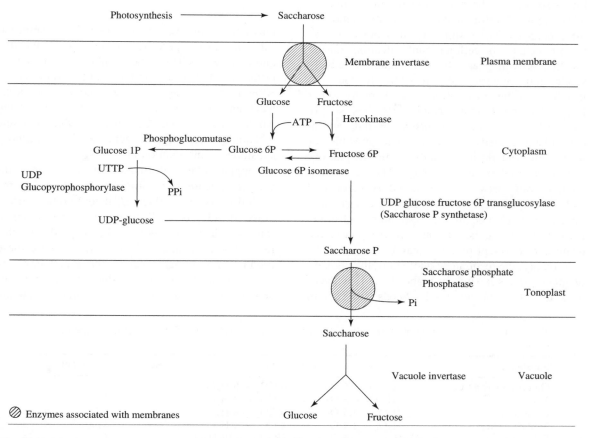

Fig. 10.7. Biochemical mechanism of sugar penetration and accumulation in grape pulp cell vacuoles

Nir, 1986). A first invertase, linked to the plasmic membrane, hydrolyzes the imported saccharose into glucose and fructose. The free sugars are then phosphorylated by the cytoplasmic hexokinase. After the formation of UDP-glucose, the sugars combine again to form saccharose phosphate with the help of a saccharose phosphate synthetase. The energy accumulated in this molecule and liberated by a saccharose phosphate phosphatase linked to the tonoplast permits the accumulation of sugars in the vacuole. These enzymes maintain a sufficient activity during the entire maturation process to accumulate a maximum of 2.5 mmol of sugars per hour per berry. However, the direct transfer of cytoplasmic hexoses via tonoplastic transporters cannot be excluded (Robinson and Davies, 2000).

Profound changes in the metabolic pathways also occur at *véraison*, facilitating the storage of imported sugars. The study of respiration evolution during grape development provides information on these changes. The respiratory intensity increases in proportion to cellular multiplication during the first growth phase. It then remains relatively stable until maturity (Figure 10.8). It does not increase during maturation, as in many other fruits. The most active respiratory sites simply change location. Before *véraison*, the pulp and in particular the seeds are primarily responsible for respiration, but during maturation the respiratory activity is highest in the skins. The respiratory quotient (ratio between the carbon dioxide released and the oxygen consumed) changes at *véraison*, indicating a change in the respiratory substrate. During the entire herbaceous growth phase, the respiratory quotient remains near 1.

In reality, the respiratory quotient of the pericarp of green grapes is slightly higher than 1, whereas it is near 0.7 for the seeds. Seeds are rich in

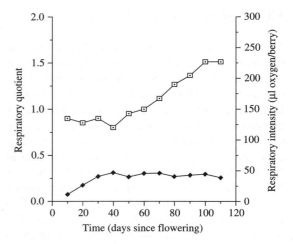

Fig. 10.8. Evolution of respiration during grape development (Harris *et al.*, 1971). □, respiratory quotient; ♦, respiratory intensity

fatty acids, which are mostly likely their respiration substrate. In the pericarp, on the other hand, this quotient results from the combustion of sugars, primarily, but also organic acids (Table 10.2). After *véraison*, the respiratory quotient increases, reaching 1.5 towards the end of maturation. On the whole, it can logically be considered that the grape essentially uses organic acids as its respiratory substrate during maturation.

Supplementary information on metabolic pathway modifications is provided by observing the evolution of the glucose/fructose ratio during grape development. Saccharose is the principal transport form of photosynthesis products; this ratio should be near 1 in the grape where these products accumulate. Yet in the green grape at the start of development, glucose predominates and represents up to 85% of grape reducing

Table 10.2. Respiratory quotient resulting from the complete oxidation of the principal components of grapes

Substance	Overall oxidation reaction	Respiratory quotient value
Glucose or fructose	$C_6H_{12}O_6 + 6\ O_2 \longrightarrow 6\ CO_2 + 6\ H_2O$	1
Malic acid	$C_4H_6O_5 + 3\ O_2 \longrightarrow 4\ CO_2 + 3\ H_2O$	1.33
Citric acid	$C_6H_8O_7 + 4.5\ O_2 \longrightarrow 6\ CO_2 + 4\ H_2O$	1.33
Tartaric acid	$C_4H_6O_6 + 2.5\ O_2 \longrightarrow 4\ CO_2 + 3\ H_2O$	1.60
Fatty acid	$C_{18}H_{36}O_2 + 26\ O_2 \longrightarrow 18\ CO_2 + 18\ H_2O$	0.70

sugars. This ratio, near 5, decreases to 2 at *véraison* and then to 1 at the beginning of maturation. It then remains relatively constant until maturity (between 1.0 and 0.9). Since glucose is more likely to enter into cellular respiration than fructose, the latter preferentially enters into cellular synthesis reactions. This phenomenon explains the elevated glucose/fructose ratio during the herbaceous growth phase of the grape and its decrease after *véraison*, related to a slowing of biosynthetic activity.

10.3.3 Evolution of Organic Acids

From the start to the end of its development, the grape contains most of the acids involved in the glycolytic and shikimic acid pathways as well as in the Krebs and glyoxylic acid cycles. This attests to the functioning of these different pathways. However, their concentrations are generally very low. Tartaric and malic acid represent on average 90% of the sum of the acids. These two acids are synthesized in the leaves and grapes, with a majority produced in the grapes prior to *véraison*. There is no formal proof of the transport of these diacids from the leaves to the grapes (Ruffner, 1982b).

In spite of their chemical similarity, these two acids have very different metabolic pathways. Their evolution is not identical during grape development and maturation. The malic acid/tartaric acid proportion varies considerably according to the grape cultivar and the maturation conditions.

The grape is the only cultivated fruit of European origin that accumulates significant quantities of tartaric acid. Specifically, the L-(+) tartaric acid stereoisomer accumulates in the grape, attaining 150 mM in the must at *véraison* and from 25 to 75 mM in the must at maturity (3.8–11.3 g/l).

This acid is a secondary product of the metabolism of sugars. In fact, there is a significant lag time before obtaining radioactive tartaric acid from the incorporation of $^{14}CO_2$ in the leaves. This phenomenon only occurs in the presence of light. Ascorbic acid is considered to be the main intermediate in the biogenesis of tartaric acid and small quantities are still present in ripe grapes. Even though the ascorbic–tartaric acid transformation is well understood at present, the origin of ascorbic acid is not known with certainty—in spite of all the research carried out over the last 30 years. Two biosynthetic pathways of ascorbic acid appear to exist: one is dependent on plant growth; the other is not. The kinetics of tartaric acid during grape development and maturation are consistent with this dual pathway hypothesis. The herbaceous growth phase is characterized by a rapid accumulation of tartaric acid, related to intense cellular multiplication. During maturation, the tartaric acid concentration remains relatively constant in spite of the increase in berry volume. A small amount of this acid is therefore synthesized during this period. Conversely, there is no formal proof of its catabolism during maturation. The small variation in levels seems rather to be related to the plant's water supply.

Malic acid is a very active intermediary product of grape metabolism. The vine contains the L-(−) malic isomer. The vine assimilates carbon dioxide in the air by a C_3 mechanism (Ruffner *et al.*, 1983). In this manner, during the dark phase of photosynthesis, the leaves and young green grapes fix CO_2 on ribulose 1,5-diphosphate to produce phosphoglyceric acid, which condenses to form hexoses and may also become dehydrated into phosphoenol pyruvic acid. CO_2, catalyzed by PEP carboxylase, is fixed on this acid to form oxaloacetic acid, which is, in turn, reduced into malic acid.

The significant malic acid accumulation during the herbaceous growth phase of the grape (up to 15 mg/g fresh weight–about 95 μmol/berry) is due in part to this mechanism, but a non-negligible proportion results from its direct synthesis by the carboxylation of pyruvic acid. This reaction is catalyzed by the malic enzyme, whose activity is very high before *véraison*.

In any case, the imported sugars are the precursors of the malic acid found in grapes. The malic acid is produced by either catabolic pathways (glycolysis, pentose phosphate pathway) or by β-carboxylation.

Grape maturation is marked by an increase in the respiratory quotient, which suggests the use of this acid for energy production in the grape (Harris *et al.*, 1971). In fact, during maturation, malic acid takes on the role of an energy vector (Figure 10.9). During the herbaceous growth phase, the sugars coming from photosynthesis are transformed into malic acid, which accumulates in the pericarp cell vacuoles (the grape being incapable of stocking significant amounts of starch, as many other fruits do). At *véraison*, due to the severe inhibition of

the glycolytic pathway, malic acid importation from the vacuole permits energy production to be maintained. The activation of a specific permease ensures this transport. The *de novo* synthesis of different malate dehydrogenase isoenzymes supports this hypothesis.

In order to maintain a normal cytoplasmic pH value when energy needs drop (at night, or at a low temperature), the excess imported malic acid

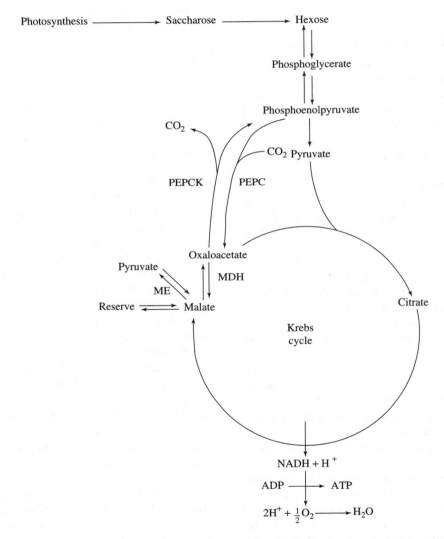

Fig. 10.9. Role of malic acid in the production of energy (ATP) and the formation of different substrates in the grape (Ruffner, 1982b). MDH, malate dehydrogenase; ME, malic enzyme; PEPC, phosphoenolpyruvate carboxylase; PEPCK, phosphoenolpyruvate carboxykinase

is eliminated and transformed into glucose by glu-coneogenesis. PEP carboxykinase decarboxylates part of the oxaloacetic acid formed. Glucose is then formed by the inverse glycolytic pathway. This gluconeogenesis is particularly elevated during *véraison*, but the amount of malic acid transformed into glucose does not exceed 5% of the stocked malic acid, i.e. less than 10 g/l glucose. The presence of abscisic acid in the grape increases the enzymatic activity of gluconeogenesis (glucose 6-phosphatase, fructose 1,6-diphosphate, malate dehydrogenase) (Palejwala *et al.*, 1985).

Malic acid can also be decarboxylated by the malic enzyme. Its affinity constants, which are different after *véraison*, invert the activity (Ruffner *et al.*, 1984). The pyruvic acid formed also contributes to energy production.

During the herbaceous growth phase, tartaric and malic acid essentially play an ionic role. Cation importation and proton consumption during metabolic reactions impose organic acid production from sugars. This ionic regulation seems to occur indifferently with the help of either malic or tartaric acid. Consequently, the sum of these two acids is relatively constant at *véraison* from one year to another for a given cultivar. In spite of their close chemical similarity, these two acids behave very differently in the course of maturation. In spite of their chemical similarity, these two acids behave very differently during ripening: the tartaric acid content of grapes varies very little while that of malic acid follows the decrease in total acidity. At maturity, the sum of these two acids is highly variable, depending on vintage conditions.

10.3.4 Accumulation of Minerals

Potassium is one of the rare minerals translocated by the phloem sap. In the phloem, it permits the translocation of sugars derived from photosynthesis. Consequently, during maturation, the potassium concentration in the grape increases with respect to sugar accumulation kinetics (Schaller *et al.*, 1992).

The xylem sap translocates most other cations in relationship to the amount of water transpired by the grape. Yet transpiration intensity strongly diminishes after *véraison* because of grape skin modifications and stomata degeneration. Most often, calcium accumulation ceases at the start of maturation because of the above grape modifications (Donèche and Chardonnet, 1992). This phenomenon is identical for magnesium, but to a lesser degree. Consequently, the calcium and magnesium concentrations per liter of juice decrease most of the time during maturation.

Being a natrophic plant, the vine accumulates little sodium. This permits a certain level of resistance in salty soils. The concentration of metal trace elements (Zn, Cu, Mn, etc.) is likely to decrease during maturation. The inorganic anion concentration (sulfates, phosphates, chlorides, etc.) continues to increase with the cation concentration, but the incorporation of phosphates, as with magnesium, has often been observed to slow during *véraison* (Schaller and Lohnertz, 1992).

The distribution of minerals in the grape berry is not insignificant and it has an impact on the composition of must at maturity. Potassium is essentially located in the pulp cell vacuoles, but the skin cells also sometimes contain significant amounts.

In theory, the sum of the acids and cations determines must pH. However, in hot years, it depends mainly on the tartaric acid and potassium concentrations, according to the following relationship (Champagnol, 1986):

$$pH = f\frac{[\text{tartaric acid}]}{[\text{potassium}]} \qquad (10.1)$$

10.3.5 Evolution of Nitrogen Compounds

The grape nitrogen supply depends on both the phloem and xylem saps. In these two cases, nitrates are rarely involved. They are only present in small quantities because of their reduction in the roots and leaves.

Nitrogen transport to the grape essentially occurs in the form of ammonium cations or amino acids. Glutamine represents about 50% of the organic nitrogen imported.

There are two intense nitrogen incorporation phases during grape development: the first following berry set, and the second starting at *véraison* and finishing at mid-maturation. Towards the end of maturity, the total nitrogen concentration may increase again. As a result, at harvest, half of the nitrogen in the vegetative part of the plant is stocked in the grapes (Roubelakis-Angelakis and Kliewer, 1992). In unripe fruit, the ammonium cation represents more than half of the total nitrogen. From *véraison* onwards, the ammonium concentration decreases whereas the organic fraction increases. The free amino acids increase by a factor of 2 to 5 during maturation, attaining 2–8 g/l in leucine equivalents. At maturity, the amino acid fraction represents 50–90% of the total nitrogen in grape juice.

The incorporation of the ammonium cation on α-ketoglutaric acid appears to be the principal nitrogen assimilation pathway by the grape. It is catalyzed by glutamine synthetase (GS) and glutamate dehydrogenase (GDH) enzymes. Other amino acids are synthesized by the transfer of nitrogen incorporated on glutamic acid.

Research carried out by numerous authors show that even though the amino acid composition varies greatly, depending on conditions, a small number of amino acids predominate: alanine, γ-aminobutyric acid, arginine, glutamic acid, proline and threonine.

At maturity, arginine is often the predominant amino acid and can represent from 6 to 44% of the total nitrogen of grape juice. In fact, this amino acid plays a very important role in grape berry nitrogen metabolism (Figure 10.10). A close relationship exists between arginine and diverse amino acids (ornithine, aspartic and glutamic acid, proline). As a result, the proline concentration can increase during maturation by a factor of 25–30 through the transformation of arginine. Moreover, aspartic acid constitutes an oxaloacetic acid reserve which, depending on the demand, can be transformed into malic acid or into sugars during maturation.

Maturation is also accompanied by an active proteosynthesis. The soluble protein concentration reaches its maximum before complete maturity and

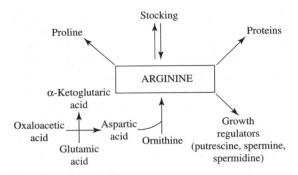

Fig. 10.10. Role of arginine in the nitrogen metabolism of grapes (Roubelakis-Angelakis, 1991)

then diminishes towards the end of maturation. The concentration of grape juice protein can thus vary from 1.5 to 100 mg/l. The concentration of high molecular weight insoluble proteins, often attached to the cell wall, is high from the start of development and continues to increase during maturation.

The juice from mature grapes contains barely 20% of the total berry nitrogen. The remainder is retained in the skins and seeds, even though the latter are likely to liberate soluble forms of nitrogen (ammonium cations and amino acids) in the pulp towards the end of maturation.

10.3.6 Changes in the Cell Wall

The softening of the grape during maturation is the result of significant changes in parietal constituent composition—notably at the cellular level of the pulp. Cellular multiplication and enlargement during grape development and maturation are not accompanied by a proportional increase in the parietal polyosides (Chardonnet *et al.*, 1994). Depending on the varieties, either cell wall deterioration or a relatively constant parietal polyoside concentration results, until the approach of maturity. The pulp texture differences between varieties are explained in this manner.

At the beginning of grape development, the cell walls are primarily composed of cellulose. The *véraison* period is characterized by considerable pectin synthesis to such an extent that it becomes the majority polyoside in some varieties (Silacci and Morrison, 1990). Like a cement, pectins ensure cellulose fiber cohesion. They are formed by the polymerization of galacturonic acid and diverse neutral -oses (rhamnose, galactose and arabinose). A high percentage of the acid functions of galacturonic acid units are methylated.

Maturation is accompanied by a solubilization of these pectins under the influence of several factors. First, pectin methyl esterases (PME) liberate the acid functions of galacturonic acid, resulting in the augmentation of the grape methanol concentration. Cell wall hydration, characterized by swelling, is thus facilitated by increasing the K^+/Ca^{2+} ratio (Possner and Kliewer, 1985). As a result, the pectins are less chelated by calcium; the free acid functions of the galacturonic residues are the site of attack by other enzymatic activities—polygalacturonases and pectin-lyases. Although pectin methyl esterases are present in majority in the grape skin, all of these enzymes are also active in the pulp. This explains the diminution of total pectic substances during grape maturation. This phenomenon is accompanied by an increase of the soluble pectic fraction which is later found in must. The pulp cells are solubilized first. At the end of ripening, variable proportions of pectinolytic enzymes are located in the grape skins.

At maturity, the grape is characterized by a low pectin concentration with respect to other fruits.

10.3.7 Production of Phenolic Compounds

One of the most remarkable characteristics of maturation is the rapid accumulation of phenolic pigments, which give the red grape its enological importance. These phenolic pigments are secondary products of sugar catabolism. Their biosynthetic pathways are present and partially active right at the start of grape development.

Phenolic compounds derived from a simple unit to a single benzene ring are created from the condensation of erythrose 4-phosphate, an intermediary product of the pentose phosphate cycle, with phosphoenol-pyruvic acid. This biosynthetic pathway, known as the shikimic acid pathway (Figure 10.11), leads to the production of benzoic

and cinnamic acid, as well as aromatic amino acids (PHE, TYR). The condensation of three acetyl coenzyme A molecules, derived from Krebs cycle reactions, also leads to the formation of a benzene ring. The condensation of this second ring with a cinnamic acid molecule produces a molecule group known as the flavonoids. These molecules possess two benzene rings joined by a C_3 carbon chain, most often in an oxygenated heterocyclic form. Various transformations (hydroxylation, methoxylation, esterification and glucosidification) explain the presence of many substances from this family in the grape (see Volume 2, Section 6.2).

In these metabolic pathways, phenylalanine ammonialyase (PAL) is the enzyme, which, by eliminating the NH_3 radical, diverts phenylalanine from protein synthesis (primary metabolism) towards the production of trans-cinnamic acid and other phenolic compounds. PAL is located in grape epidermal cells as well as in the seeds. Its maximum activity in the seeds occurs during the herbaceous growth phase; its activity then decreases after véraison to become very low during maturation. PAL activity contained in the grape skin is very high at the start of development, then decreases up to véraison. In colored grapes, PAL activity in the skins increases again at the start of véraison. There is close relationship between its activity and the color intensity of the grape (Hrazdina et al., 1984). Chalcone synthetase is the first specific enzyme of the flavonoid synthesis pathway (condensation of the two rings): its activity strongly increases at the beginning of véraison and then rapidly decreases.

The biosynthetic pathways are active as early as the start of grape development. Consequently, the total phenolic compound concentration continues to increase during this period. The rapid increase in tannin concentration at the beginning of development, however, is followed by a slower accumulation during maturation. The biosynthesis may therefore be less active than the increase in berry volume.

The procyanidinic tannins, derived from flavanol polymerization, attain a maximum concentration in the seeds before véraison. This then strongly decreases to a lower and relatively stable value

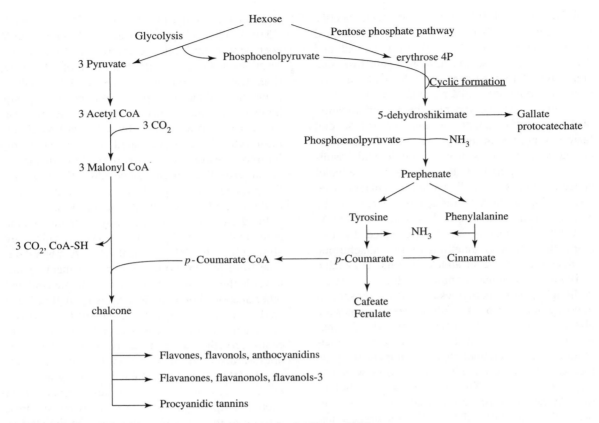

Fig. 10.11. Biosynthesis pathways of phenolic compounds (Conn, 1986)

when the seeds are mature. At *véraison*, the skin tannin concentration is already high—sometimes corresponding to over half of the concentration at maturity (Figure 10.12).

In white grapes, the concentrations of phenolic acids esterified by tartaric acid, flavan-3-ols and oligomeric procyanidins are high at the beginning of development. They then diminish to minimal concentrations at maturity.

In colored varieties, the anthocyanins begin to accumulate in the skins about two weeks before the color is visible. The concentration increases during maturation, but, as with tannins, it attains a maximum and generally diminishes at the time of maturity.

This appearance of anthocyanins is linked to sugar accumulation in the grape but no direct relationship has yet been established. Diverse

parameters, such as sunlight, increase the anthocyanin accumulation speed without affecting the skin sugar concentration (Wicks and Kliewer, 1983).

10.3.8 Evolution of Aromatic Substances

Several hundred different chemical substances participate in grape aroma. In this complex mixture, hydrocarbides, alcohols, esters, aldehydes and other carbon-based compounds can be distinguished (Schreier *et al.*, 1976).

Nearly all of the compounds identified at present are found in numerous varieties that do not possess a particularly specific varietal aroma. For example, a trace of terpenic alcohols is found in neutral-tasting varieties, yet its concentration can attain

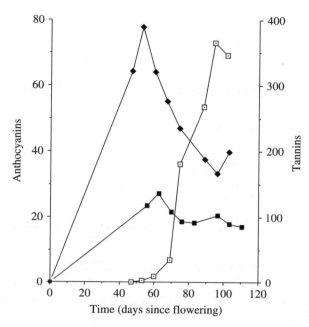

Fig. 10.12. Evolution of phenolic compounds (Darné, 1991) (results expressed in mg/g dry weight): □, skin anthocyanins; ◆, skin tannins; ■, seed tannins

3 mg/l in certain aromatic varieties (Gewürztraminer, Muscat).

For certain varieties, however, the characteristic aroma is the result of a limited number of specific compounds in low concentrations (from nanograms to micrograms). The following compounds and their varietal origins fall into this category: ethylic and methylic esters of anthranilic acid in varieties issued from *Vitis labrusca*, and in particular the Concord grape (Stern *et al.*, 1967); 2-methoxy-3-isobutyl pyrazine in the Cabernet Sauvignon (Bayonove *et al.*, 1976); and 4-mercapto-4-methyl pentan-2-one present in the Chenin variety (Du Plessis and Augustyn, 1981) and identified in Sauvignon Blanc (Darriet, 1993).

The grape aromatic potential is divided into:

- free and volatile odorous substances;

- non-volatile and non-odorous precursors (glycosides, phenolic acids and fatty acids);

- odorous or non-odorous volatile compounds which by their instability are transformed into

other odorous compounds (terpenols, terpenic diols, C_{13} norisoprenoids, etc.).

Terpenic compounds have been studied in particular. Their biosynthetic pathway is schematized in Figure 10.13. The first step produces mevalonic acid from glucose by the acetyl coenzyme A pathway. This principal pathway is generally recognized although another exists by the intermediary of amino acids such as leucine or valine. The second step produces isopentenyl pyrophosphate (IPP) from mevalonic acid. All of the terpenoids are built from this C_5 isoprenic base unit. With the help of the isopentenyl pyrophosphate isomerase, IPP is isomerized into dimethylallyl pyrophosphate (DMAPP). These two isoprenic units play an active role in terpenoid synthesis. One IPP unit condenses with a DMAPP molecule with the help of a prenyl transferase (head–tail condensation of the two molecules) to produce a C_{10} molecule, geranylpyrophosphate (GPP), which constitutes an important junction in terpenoid synthesis. From this compound, the synthetic pathways can form

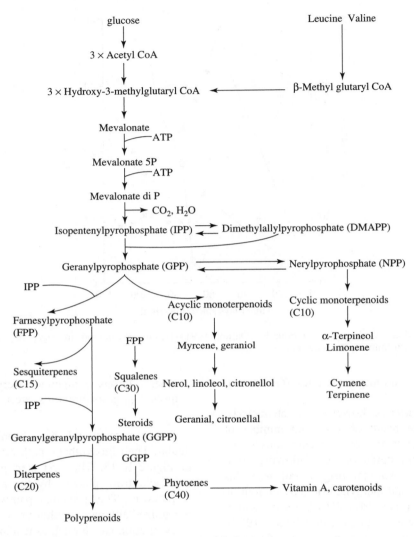

Fig. 10.13. Terpenoid biosynthesis pathway (Bayonove, 1993)

either acyclic or cyclic monoterpenoids or more condensed terpens.

The grape contains many terpenic-based compounds (see Volume 2, Section 7.2). These monoterpenoids exist in a free state and in a bound form of a heterosidic nature. The bound and free terpenol concentration increases during berry development (Figure 10.14). The terpenic heterosides are abundant very early, when the berry is still green (250–500 µg/kg in fresh weight), whereas the free terpenols exist in

only small quantities (30–90 µg/kg in fresh weight). Some are not present at this stage (α-terpineol and citronellol) but begin to appear in significant amounts from *véraison* onwards (linalol, for example). The bound fractions outnumber the free fractions during the entire maturation phase and even increase beyond maturity, whereas the increase in the free fraction slows and its concentration can even decrease. The concentration of some terpenols, such as free linalol and α-terpineol, diminish in this way during

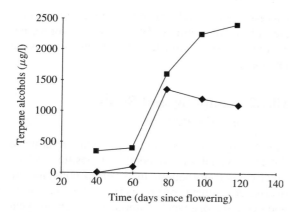

Fig. 10.14. Average evolution of terpene alcohols during Muscat grape maturation (Bayonove, 1993): ◆, free terpene alcohols; ■, bound terpene alcohols

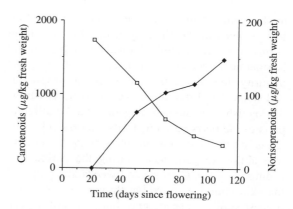

Fig. 10.15. Average evolution of carotenoids and C_{13} norisoprenoids during maturation of Muscat grapes (Bayonove, 1993): □, carotenoids, ◆, norisoprenoids

overripening. This evolution seems to indicate that the stocking of terpenol occurs for the most part in a bound form. All of the terpenols behave in this way with the exception of linalol, whose free fraction sometimes remains greater than the bound fraction throughout maturation.

Other aroma precursor compounds—carotenoids—are well known today (see also Volume 2, Section 7.3.1). These substances share the same origin as terpenols but have a higher molecular weight. The carotenoid concentration in the grape berry varies from 15 to nearly 2500 µg/kg in fresh weight (Razungles, 1985). The most important, in decreasing order, are lutein, β-carotene, neoxanthyn and lutein-5,6-epoxide. These molecules, generally enclosed in cellular organites, are essentially located in the solid parts of the berry: the skin is two to three times richer in carotenoids than the pulp. The carotenoids are found in different proportions in the different parts of the berry, depending on their structure. During maturation, a decrease in the carotenoid concentration and an increase in certain carotenoid-derived molecules such as norisoprenoids are observed (Figure 10.15). The metabolic pathways in the grape leading to the production of odorous substances such as norisoprenoids from carotenoids are not yet known, but carotenoids are known to be sensitive to biochemical oxidation—resulting in the production of ionone-type molecules. Some

norisoprenoids are also found as glycosylated precursors (Volume 2, Section 7.3.2).

Information on other aromatic substances, specific to varietal aromas, is at present very limited. According to Darriet (1993), the 4-mercapto-4-methyl pentan-2-one seems to evolve similarly to free terpenols, with a slight decrease in its concentration towards the end of maturation. Conversely, the unripe grape contains a high concentration of methoxypyrazines (a few dozen nanograms per liter) in certain varieties, such as Cabernet Sauvignon (see Volume 2, Section 7.4). The concentration of these compounds drops significantly in the course of maturation. The highest concentrations are found in the coldest maturation conditions (Lacey *et al.*, 1991). They develop in a very similar way to malic acid (Roujou de Boubée, 2000).

10.4 DEFINITION OF MATURITY—NOTION OF VINTAGE

10.4.1 State of Maturity

The various biochemical processes just described are not necessarily simultaneous phenomena with identical kinetics. Environmental conditions can modify certain transformation speeds, sometimes to the point of upsetting the order of physiological changes in the ripening fruit. Differing from *véraison*, which is a fully defined physiological and biochemical incident (Abbal *et al.*, 1992), grape maturity does not constitute a precise physiological stage. Yet different degrees of maturity can be distinguished. Biologists consider that the different parts of the berry reach maturity successively. The seeds are the first to attain physiological maturity (the ability to germinate) during the period preceding *véraison*. Over several weeks, the pulp and the skin continue to evolve through a maturation process similar to senescence (alteration of the cell wall, accumulation of secondary metabolites).

In enology, pulp maturity corresponds to an optimal sugar/acid ratio; skin maturity is the stage at which the phenolic compounds and aromatic substances attain a maximum concentration. These two kinds of maturity can be distinguished, but the dissociation of the cell wall from the skin must be sufficiently advanced to permit easy extraction of these essential constituents.

Consequently, the definition of maturity varies, depending on the objective. For example, the production of dry white wines requires grapes whose aromatic substances are at a maximal concentration and whose acidity is still sufficient. In certain situations, an early harvest can be interesting. Conversely, when the elaboration of a quality red wine is desired, grape development must be left to continue to obtain the most easily extractable phenolic compounds.

In general, grape maturation results from several biochemical transformations that are not necessarily related to each other. To simplify matters, the increase in sugar concentration and the decrease in acidity are monitored. The accumulation and refinement of white grape aromas and phenolic compounds in red grapes should also be taken into account. The essential property of a quality wine-producing area is to permit a favorable maturation. This corresponds with a harmonious evolution of the various transformations to reach the optimum point simultaneously at the time of the harvest.

In too cold of a climate, the maturation cannot be satisfactory, but in very warm climate the increase in sugar concentration can impose a premature harvest even though the other grape constituents are not at full maturity. Of course, environmental conditions (soil, climate) are involved in these phenomena.

10.4.2 Sampling and Study of Maturation

Monitoring maturation poses problems relating to the large variability of berry composition. When precise data are sought in order to compare the diverse constituents of grapes from one vineyard to another, from one week to another or even one year to another, grape sampling methods are of prime importance. Nothing is more heterogeneous than grapes from the same vineyard at a given moment, even if the same variety is considered.

On a grape cluster, the grapes are formed, change color and ripen one after another over a period of up to 2 weeks, or more in certain difficult conditions. On the same vine, the different grape clusters are never at the same maturity level. The clusters closest to the trunk contain more sugar than those at the extremity of the branches. The ripest grapes are in general the furthest from the ground, as the sap is preferentially conveyed towards the highest and into the longest branches. These differences are even greater when various vinestocks are considered—some vines always develop more quickly than others. It is therefore risky to determine the harvest date from a single vine sampled at random.

Due to this great heterogeneity, a proper monitoring of the maturation of the same parcel requires regular sampling of a sufficient quantity of grapes: 1.5–2 kg, or about 1000–2000 grapes. A larger number of samples are required to ensure that the results are representative of the plot (Blouin and Guimberteau, 2000). The most common method consists of gathering, with shears, three or four grape cluster fragments from 100 vines. Grape clusters under the leaves as well as those directly exposed to sunlight should be gathered, taking them alternately from each side of the row at different heights on the vine. When sampling varieties with compact clusters, this method does not generally take into account the berries located at the interior of the cluster. These berries are often less ripe than the others. In this case, whole grape clusters should be sampled, to obtain a precise idea of the maturation level of the parcel.

In the laboratory, the berries are separated, counted and weighed. The juice is extracted with the help of a small manual press or a centrifugal fruit juice separator. The juice volume is measured and the results are expressed per liter of must. The juice sugar and acid concentrations are then determined.

The study of red grape phenolic compounds requires the manual separation of the skins from the seeds of about 200 berries taken at random in the sampling. Once separated, the skins and the seeds are dried and lyophilized to facilitate the extraction and the determination of their phenolic content.

Aromatic substance monitoring, notably of white grapes, requires the maceration of the solid grape parts with the must beforehand. After a light crushing of the grapes, this maceration is usually carried out for 16 or 24 hours at a low temperature under a carbon dioxide atmosphere. These techniques require adapted equipment and cannot yet be routinely monitored.

10.4.3 Evaluation of the State of Maturity—Maturation Index

Grape monitoring during maturation helps vineyard managers to set the harvest date and maximize the efficiency of their harvest teams according to the ripeness of different cultivars and diverse parcels.

Determining the grape sugar concentration is essential. It is most often effected by an indirect physical measure such as hydrometry or refractometry. If the temperature is not at 20°C, a correction is theoretically necessary, but has little effect on the sugar concentration. The results are expressed in various units, depending on the instruments used. This does not facilitate the interpretation of data originating from different wine-producing countries (Blouin, 1992; Boulton et al., 1995).

These assorted measurement scales are compared in Table 10.3. The degree Oechslé corresponds to the third decimal of the relative apparent

density (D). The relative apparent density permits the evaluation of the sugar concentration. The degree Baumé is approximately converted to relative apparent density by the following formula: $°\text{Baumé} \simeq 144.32(1 - 1/D)$. The degree Baumé of a must corresponds fairly well with the percentage alcohol, at least for values between 10 and 12. The degree Brix (or degree Balling) gives the weight of must sugars, in grams, per 100 g of must. In reality, it is a percentage of the dry matter in must, measured by refractometry or densimetry. This measure is only valid from a certain maturity level onwards (15° Brix). Before this maturity level, organic acids, amino acids and certain precursors of parietal polyosides can have similar refraction indexes to sugar and interfere with the measurement.

In the same way, the relationship between must density and alcohol content is always approximate, since sugar is not the only chemical must constituent that affects density. This measurement is more accurate in white winemaking with non-mucilaginous musts having few suspended particles. The values obtained for must from rotten grapes are inaccurately high. Moreover, the estimation of potential alcohol should take into account the sugar/alcohol transformation ratio. The figures in Table 10.3 use the relationship of 16.83 g/l of sugar per liter for 1% alcohol—the official value retained by the EEC.

Empirical observation of the inverse variation of sugars and acidity during maturation led to the development of a sugar/acidity ratio, called the maturation index. This index is very simple but it should be used with precaution, since there is no direct biochemical relationship between sugar accumulation and acidity loss. More specifically, a given gain in sugar does not always correspond with the same drop in acidity. This ratio is not suitable for comparing different varieties, since varieties exist that are rich both in sugar and in acids. In France, this ratio is calculated from the must sugar concentration (g/l) and the titration acidity expressed in grams of sulfuric or tartaric acid equivalents per liter. Other modes of expression are used in other countries according to the measurement unit used to express the

Table 10.3. Conversion table for various scales used to measure must sugar concentration

Relative apparent density (20°C)	Degree Œchslé	Degree Baumé	Degree Brix	Refractometric measure (in percentage weight of saccharose)	Sugar concentration (g/l)	Potential alcohol (16.83 g of sugar/l for 1% alcohol)
1.0371	37.1	5.2	9.1	10	82.3	4.9
1.0412	41.2	5.7	10.1	11	92.9	5.5
1.0454	45.4	6.3	11.1	12	103.6	6.2
1.0495	49.5	6.8	12.0	13	114.3	6.8
1.0538	53.8	7.4	13.0	14	125.1	7.4
1.0580	58.0	7.9	14.0	15	136.0	8.1
1.0623	62.3	8.5	15.0	16	147.0	8.7
1.0666	66.6	9.0	16.0	17	158.1	9.4
1.0710	71.0	9.6	17.0	18	169.3	10.1
1.0754	75.4	10.1	18.0	19	180.5	10.7
1.0798	79.8	10.7	19.0	20	191.9	11.4
1.0842	84.2	11.2	20.1	21	203.3	12.1
1.0886	88.6	11.8	21.1	22	214.8	12.8
1.0932	93.2	12.3	22.1	23	226.4	13.5
1.0978	97.8	12.9	23.2	24	238.2	14.2
1.1029	102.9	13.5	24.4	25	249.7	14.8
1.1075	107.5	14.0	25.5	26	261.1	15.5
1.1124	112.4	14.6	26.6	27	273.2	16.2
1.1170	117.0	15.1	27.7	28	284.6	16.9
1.1219	121.9	15.7	28.8	29	296.7	17.6
1.1268	126.8	16.2	29.9	30	308.8	18.4
1.1316	131.6	16.8	31.1	31	320.8	19.1
1.1365	136.5	17.3	32.2	32	332.9	19.8
1.1416	141.6	17.9	33.4	33	345.7	20.5
1.1465	146.5	18.4	34.5	34	357.7	21.3

sugar concentration. In Germany, for example, the ratio obtained by dividing the °Oechslé of must by the acidity, expressed in tartaric acid, is currently used.

Attempts have been made in the past to describe the state of maturity, taking into account the respective variations of malic and tartaric acid or the accumulation of cations, but none of the indices developed have significantly improved the evaluation of the maturity level. It seems sensible to take into account the individual variations of each berry constituent separately.

More recently, researchers have focused on the evolution of phenolic compounds during maturation, but the technique of separating the skins from the seeds is awkward and exacting limiting its practical applicability. There is now a rapid whole-berry grinding technique. The grape grinding is followed by a differential phenolic compound extraction, in either a pH 3.2 buffer (compounds easily extractable) or a pH 1 buffer (total potential in phenolic compounds). The density of the solutions obtained is then measured at 280 nm. Information on the total phenolic compound concentration and their extractability is thus obtained.

Unfortunately, no simple methods currently exist that permit an aromatic substance maturation index. Tasting the grape remains, in this respect, the only available criterion for judgments but this does not estimate the subsequent revelation of other aromas.

Micro-imagery by nuclear magnetic resonance has recently been shown to give detailed information on the chemical composition and degradation level of grape cell walls (Pope *et al.*, 1993) but this technique will remain reserved for scientific experimentation for a long-time.

Fourier transform infrared spectrometry, which has recently been developed, should make it possible to assess grape quality more accurately (Dubernet *et al.*, 2000). This method is easy to implement and does only require prior filtration of the samples. It provides a satisfactory evaluation of the potential alcohol, total acidity, pH, and nitrogen content, as well as the color index for black grapes, in a single operation. In addition to this general analysis of the grapes, it is possible to detect the presence of rot (gluconic acid, laccase activity, etc.) or fermentation activity (lactic acid, pyruvic acid, etc.). This new technique, however, only gives reliable results after a long, laborious calibration process using samples analyzed by standard methods.

10.4.4 Effect of Light on the Biochemical Maturation Process

Three factors have major roles in maturation dynamics: light, heat and water availability. In general, they affect vine growth and metabolic activity; their action is well known. Yet these also act directly on grapes, and their effects on metabolic pathways translate into changes in grape chemical composition.

In established grape-growing zones, the availability of natural light does not, in general, limit photosynthetic activity and thus the overall functioning of the plant. In fact, photosynthesis is optimum at a sun radiance (expressed in einsteins, E) of about 700 E/m^2/s. Below 30 E/m^2/s, leaf energy consumption is greater than net photosynthetic production (Smart, 1973). In the absence of clouds, sun radiance is greater than 2500 E/m^2/s. On cloudy days, the radiance varies from 300 to 1000 E/m^2/s. A reduction in photosynthetic activity can thus occur, resulting in a nutrient deficiency in the grape. However, in practice, certain vine trellising methods still cause radiant energy loss. For this reason, wine-growers should ensure that the spacing between vine rows is in proportion to the height of the foliage (0.6–0.8) and should avoid leaf crowding by thinning unwanted shoots in the center of the canopy.

Light has a direct effect on floral induction. Grape cultivar fertility depends greatly on bud light exposure during this induction period.

The effects of sunlight on grape composition are even more numerous and complex. In addition to furnishing the energy for photosynthesis and stimulating certain light-dependent metabolic processes, its radiant effect heats not only surfaces but also the air surrounding vegetal tissue. Grape clusters grown with little light exposure ('shade grapes') always contain less sugar and have a lower pH and a higher total acidity and malic acid concentration than grape clusters directly exposed to sunlight. Light is also essential for phenolic compound accumulation, and phenylalanine ammonialyase (Section 10.3.7) is a photoinductive enzymatic system. In normal conditions, this photoactivation does not seem to be a factor that limits coloration or phenolic compound concentrations in most varieties. Crippen and Morrison (1986) showed that the phenolic composition of shaded and light-exposed grape clusters remained the same in Cabernet Sauvignon. Only certain sensitive red varieties (Ahmeur Bou Ahmeur, Cardinal or Emperor) may exhibit color deficiencies when their grape clusters are not exposed to light. In certain northern vineyards, wines made in climatically unfavorable years are always poorly colored.

The amount of light reaching the grapes also has an impact on the composition and aromatic qualities of the grapes. Exposure to the sun accentuates the decrease in methoxypyrazine content during the ripening of Cabernet Sauvignon grapes. Conversely, partial shade preserves the floral aromas in Muscat grapes.

10.4.5 Influence of Temperature on the Biochemical Processes of Maturation

Temperature is one of the most important parameters of grape maturation and one of the essential factors that triggers it. Temperature affects photosynthetic activity, metabolism and migration intensity in the vine. Its action is not limited to the period of grape development. Its influence on bud burst and flowering dates also has important indirect consequences on grape quality. It is easy to understand that the later the grape develops, the greater the risk that the accompanying maturation conditions will be unfavorable.

Grape growth and development are directly affected by temperature. High temperatures are unfavorable to cellular multiplication. During the herbaceous growth phase, the optimum temperature is between 20 and 25°C. During maturation, temperature affects migration intensity and thus, indirectly, cell growth. Vine temperature requirements during this period are around 20°C (Calo *et al.*, 1992). Too high of a temperature, even for a short time, can irreversibly alter sugar accumulation. Sepulveda and Kliewer (1986) found that temperatures of 40°C during the day and 20°C at night favored sugar accumulation in other parts of the vine to the detriment of the grapes, which received only a small percentage (about 2.5%), with respect to the control (25°C day/15°C night).

As vines have difficulty growing and producing grapes below 10°C, temperatures above this threshold are known as "active temperatures". A strong correlation exists between the sum of the active temperatures during grape development and the grape sugar concentration in a given location. This measurement permits the evaluation of the climatic potential of a given location to ensure suitable grape maturation. Various bioclimatic indexes have been developed to evaluate this potential.

Growing degree-days (Winkler, 1962) are the sum of the average daily temperatures above 10°C from April 1st to October 30th, a 7-month period. This sum is often calculated using monthly averages. Initially established for classifying California into different viticultural zones, this index has become widely used in other countries. The climatic data for the month of October are not useful: in warm zones, the grape has already been harvested; in cool zones, the average temperature in October is often below 10°C. Furthermore, this index does not take the duration of light exposure into account.

The Branas Heliothermic Product (Branas *et al.*, 1946) corresponds to the formula $X \times H \times 10^{-6}$. where X is the sum of the average active temperatures above 10°C for the entire year, and H represents the sum of the length of the days for the corresponding period. Vine-growing is practically impossible when the product is below 2.6. This index gives the most precise results for vineyards established in cool temperate climates where the end of the period containing active temperatures more or less corresponds with harvest time. In extreme cases of warm climates, this period covers the entire year.

In order to obtain a better correlation between bioclimatic data and final grape sugar concentrations, Huglin (1978) proposed a heliothermic index (HI). This index takes into account the maximum daily temperatures over a 6-month period from April 1st to September 30th. In this relationship:

$$HI = \sum_{April\ 1}^{Sept\ 30} \{[(ADT-10) + (MDT-10)] \times K\}/2,$$

$$(10.2)$$

where ADT represents the average daily temperature, MDT the maximum daily temperature, and K is the day-length coefficient—varying from 1.02 to 1.06 between latitudes of 40 to 50 degrees. An HI of around 1400 is the lowest limit for vine-growing. This index has permitted the specification of the needs of different varieties for attaining a given sugar concentration.

The comparison of these different indexes (Table 10.4) shows the difficulty of evaluating the viticultural potential of an area based solely on a temperature criterion, even when corrected for light exposure time. These indices are, however, useful in choosing early- or late-ripening grape varieties to plant in a new vineyard.

In most of the European viticultural zones, cultivars are chosen that reach maturity just before the average monthly temperature drops below 10°C. In warmer climates, this drop occurs later. Consequently, the maturation takes place during a warmer period. Viticultural zones can thus be classified into two categories: Alpha and Beta, depending on whether the average temperature during grape maturity of a given variety is below or above 15°C (Jackson, 1987).

Temperature also strongly influences many biochemical mechanisms involved in grape maturation. For example, malic acid degradation is considerably accelerated during hot weather: malic

Table 10.4. Comparison of different methods of evaluating climates

Viticultural zones	Sum of degree-days (Winkler, 1962)	Heliothermic product (Branas, 1946)	Heliothermic index (Huglin, 1978)
Zone 1 = less than 1390°C			
Geisenheim	995°C	2.6	—
Geneva	1030°C	2.5	—
Dijon	1133°C	—	1710
Conawara	1205°C	—	—
Bordeaux	1328°C	4.0	2100
Zone 2 = 1390°C to 1670°C			
Odessa	1433°C	—	1850
Santiago	1506°C	—	2290
Napa	1600°C	—	2130
Buharest	1640°C	—	—
Zone 3 = 1670°C to 1950°C			
Montpellier	1785°C	5.24	2256
Milan	1839°C	—	—
Zone 4 = 1950°C to 2220°C			
Venice/Verona	1960°C	—	2250
Mendoza	2022°C	2.7 at 7.8[a]	2600
The Cape/Stellenbosch	2066°C		2350
Zone 5 = more than 2220°C			
Split	2272°C	—	—
Palermo	2278°C	—	(Bari) 2410
Fresno	2600°C	—	3170
Algiers	2889°C	—	2600

[a]according to altitude.

enzyme activity (Section 10.2.3) steadily increases between 10°C and 46°C. Temperature does not directly influence tartaric acid concentrations. Elevated respiratory quotients, witnessed at temperatures greater than 35°C, were initially interpreted as the respiratory oxidation of tartaric acid, but at such a temperature this activity corresponds more to the initiation of fermentative phenomena in grape pulp—essentially acting on malic acid (Romieu et al., 1989).

Temperature also has an influence on the composition of grape phenolic compounds. Intensely colored wines are known to be difficult to obtain in extreme temperature conditions (too low or high) though the phenomenon involved can at first appear paradoxical. High temperatures stimulate metabolic reactions, whereas low temperatures curb migration. In either case, however, this corresponds with poor grape sugar alimentation and thus increased competition between primary metabolism (growth) and secondary metabolism (accumulation). The concentration of phenolic compounds is also affected by thermoperiod (Kliewer and Torres, 1972). Raising the nighttime temperature from 15 to 30°C while maintaining a daily temperature of 25°C results in a decrease in grape coloration. The anthocyanins are therefore not a blocked metabolic product but, on the contrary, are reversible. Thus, temperature and sun exposure determine phenolic compound accumulation.

Temperature also exerts a considerable effect on aromatic substances. The aromatic potential of certain white cultivars (Gewürztraminer, Riesling, Sauvignon) are known to be fully expressed only in cool climates, where the maturation period is slow and long. By comparing a cool viticultural zone with a warmer zone in South Australia, Ewart (1987) showed that the total volatile terpene quantity increased more slowly in the cool zone but was higher at maturity. In a cool climate, and especially with shaded grapes, methoxypyrazine concentrations can attain unfavorable organoleptic thresholds (Lacey *et al.*, 1991). Conversely, warm climates can lead to high concentrations of certain phenolic compounds in white cultivars such as Riesling. These compounds confer an excessively astringent character to the wine and lead to the development of a diesel-like odor during aging (Herrick and Nagel, 1985).

Despite the lack of specific experiments, excessive temperatures are known not to be the most favorable conditions for aroma quality.

10.4.6 Impact of the Vine's Water Supply on Grape Ripening

(a) The Effect of Water Availability on the Biochemical Processes Involved in Grape Ripening

Unlike most plants, particularly annual crops, vines are generally grown under less than optimum conditions. Various types of environmental constraints are considered to reduce vine vigor and yields, while maximizing the winemaking potential of the grapes. Among these constraints, a limited water supply plays a major role in vine behavior and grape composition. A moderately restricted water supply, known as "water deficit", generally has a beneficial effect on wine quality. The expression "water stress" should only be used in situations where an excessive lack of water has a negative impact on grape quality or threatens to kill the vines.

Most high-quality wines are produced in areas where annual precipitation does not exceed 700–800 mm. Evidence indicates that high rainfall and excessive irrigation are detrimental to grape quality.

Before *véraison* (color change), water is mainly transferred to the grapes via the xylem and there are close hydraulic relationships between the grapes and the rest of the vine. Any change in the vine's water supply affects sap circulation and, consequently, grape development. The resulting irreversible reduction in grape size is positive from a qualitative standpoint but also reduces yields.

In some countries, the climate may necessitate controlled irrigation of the vines to compensate water losses via transpiration. After *véraison*, the deterioration of xylem circulation leads to a concomitant increase in flows via the phloem. At that stage, the phloem provides the main water supply to the grapes. As phloem sap circulation is not directly related to the vine's water supply, grape growth becomes much less dependent on this factor. A minimum water supply is still necessary, however, for the biochemical ripening processes to proceed normally.

Matthews and Anderson (1989), Duteau *et al.* (1981), and Van Leeuwen and Seguin (1996) showed that water stress caused an increase in the phenolic content of grape juice and skins, with a higher concentration of proline and a lower malic acid content. Inadequate water supply also leads to higher concentrations of terpenic compounds (MacCarthy and Coombe, 1984). Conversely, an abundant water supply leads to an increase in grape volume, with a concomitant decrease in phenolic content. Although the acid concentration is often higher, the juice still has a higher pH (Smart and Coombe, 1983). This is due to an increase in imports of tartaric acid and minerals, especially potassium. The aromatic compounds are also modified, e.g. excess water gives Sémillon grapes a strong herbaceous aroma (Ureta and Yavar, 1982).

While water deficit does not prevent grapes from ripening satisfactorily in terms of their sugar and acid content, excessive water delays the ripening process and alters the chemical composition of the grapes to a considerable extent. In vineyards where irrigation is used, it should be reduced to a minimum after *véraison* to maintain a moderate water deficit.

Finally, heavy rain when the grapes are close to ripening is likely to cause them to burst due to a sudden absorption of water directly through the skins. This phenomenon is less marked at lower temperatures and depends on respiratory intensity.

(b) Monitoring Vine Water Levels

Studying the vine's response to different levels of water supply requires reliable, easily used indicators of water availability in the soil or the water status of the vine.

The first studies of vine reactions to water supply in the late 1960s were based on water balances, carried out using a neutron moisture tester (Seguin, 1970). A probe emitting fast neutrons is inserted in an access tube that stays permanently in the soil. The neutrons are slowed down to a state of thermal agitation when they meet hydrogen atoms. The vast majority of hydrogen atoms in soil are in water molecules; the number of thermal neutrons counted per unit time is thus proportional to the dampness of the soil (humidity by volume). The vines' water consumption between two measurements is calculated by subtracting the second reading from the first and correcting, if necessary, for any precipitation during the interval. Neutron moisture tester studies were used to obtain a detailed view of the water supply in gravel soils in the Haut-Médoc (Seguin, 1975) clay soils in Pomerol, and asteriated limestone in Saint-Emilion (Duteau *et al.*, 1981). Although this was a highly innovative technique at the time, it had several disadvantages. The water balance calculated using this method does not take into account any horizontal inflows of water through the soil or runoff, which may be significant on slope vineyards. After a period of time, roots develop around the access tube and distort the results (Van Leeuwen *et al.*, 2001a). Finally, the vine root systems are often very deep and vineyard geology (gravel, rocky soil, etc.) may make it particularly difficult to install the access tube. Even if neutron moisture testers are used in some New World countries to control vineyard irrigation, the complexity of this technique prevents it from being used more widely. Using Time Domain Reflectometry (TDR) to establish the vineyard water balance is subject to the same difficulties.

Producing a theoretical water balance by modeling is another approach to determine the vines' water supply. The aim is to simulate the water reserves remaining in the soil during the summer on the basis of data on the water available at the start of the season, plus any precipitation, minus losses via evapotranspiration. The most advanced model was developed by Riou and Lebon (2000). In this formula, precipitation could

be determined accurately and evapotranspiration estimated correctly. The main difficulty with this approach is estimating the water reserves at the beginning of the season, which is particularly complex due to the specific conditions in which vines are grown (deep root systems, rocky soil, etc.).

In view of the difficulty in assessing the vines' water balance on the basis of measurements in the soil or modeling, it seemed more practical to measure water levels in the plants themselves. A water deficit causes several measurable alterations in the vine's physiological functions: variations in xylem sap pressure, closing of the stoma, slowdown in the photosynthesis process, etc. When a plant is used as an indicator of its own water status, we refer to "physiological indicators". Among these indicators, leaf water potential is undoubtedly the most widely used because it is reliable and easy to implement. Water potential is measured by placing a freshly picked vine sample (usually a leaf) in a pressure chamber, connected to a bottle of pressurized nitrogen. Only the leaf stalk remains outside the chamber, via a small hole. Pressure in the chamber is gradually increased and the pressure required to produce a sap meniscus on the cut end of the stem is noted. This pressure corresponds to the inverse of the water potential: the higher the pressure required to produce the meniscus on the leaf stem, the more negative the water potential and the greater the water deficit to which the vine has been subjected.

There are three applications for water potential measurements using a pressure chamber: leaf potential, basic leaf potential, and stem potential (Chone *et al.*, 2000).

1. Leaf potential is measured on a leaf that has been left uncovered on a sunny day. This value only represents the water potential of a single leaf. Even if this potential depends on the water supply to the vine, the considerable variability from one leaf to another on the same vine (e.g. due to different sun exposure) leads to a large standard deviation on this measurement, making the value less significant as an indicator.

2. Basic leaf potential is measured in the same way as leaf potential, except that the leaf is picked

just before sunrise. The stoma close in the dark and the water potential in the vine comes back into balance with that in the soil matrix. Basic leaf potential reflects water availability in the most humid layer of soil in contact with the root system, providing, therefore, a more stable value that is easier to interpret than leaf potential measured during the day. It is, however, more difficult to apply, as it requires specific conditions.

3. Stem potential is measured during the day, on a leaf that has been covered by an opaque, airtight bag for at least one hour before the measurement is made. The leaf stoma close in the dark and the leaf potential balances with that of the xylem in the stem. This measurement gives a close approximation of the water supply of the whole plant during the day. Provided certain conditions are observed (measuring time and weather conditions), stem potential is the most accurate of the three pressure chamber applications (Chone et al., 2001a,b).

Carbon 13 isotope discrimination is another physiological indicator of water balance. This isotope represents approximately 1% of the carbon in atmospheric CO_2 and the lighter isotope, ^{12}C, is preferentially involved in photosynthesis. Water deficit causes the stoma to close for part of the day, which slows down CO_2 exchanges between the leaves and the atmosphere and reduces isotopic discrimination. Under these conditions, the $^{13}C/^{12}C$ ratio (known as $\Delta C13$) becomes closer to the ratio in atmospheric CO_2. Measuring $\Delta C13$ in the sugars in must made from ripe grapes (analyzed by a specialized enology laboratory) provides an indicator of the global water deficit to which the vines have been subjected during ripening. $\Delta C13$ is expressed in ‰ in relation to a standard. Values range from -21 to $-26‰$, where $-21‰$ indicates a considerable water deficit and $-26‰$, the absence of water deficit. The advantage of this indicator is that it does not require any field operations other than taking a sample of ripe grapes (Van Leeuwen et al., 2001b; Gaudillère et al., 2002). There is a good correlation between the $\Delta C13$

value measured in must made from ripe grapes and the stem potential.

(c) Impact of Water Balance on Vine Growth and the Composition of Ripe Grapes

A water deficit during the growing season causes profound changes in the physiological functions of the vine. It may progress at varying rates, as shown by the changes in stem potential measured in the same plot of Saint-Emilion vines in 2000 and 2002 (Figure 10.15). When there is a water deficit, the stoma remains closed for part of the day, increasingly restricting photosynthesis as the deficit becomes more severe. A reduction in water supply tends to stop vine shoot and grape growth, affecting the grapes especially before véraison (Becker and Zimmermann, 1984). When the soil dries out around the roots, the tips produce abscisic acid, a hormone that promotes grape ripening. Restricting the water supply to the vine has both negative (restricting photosynthesis) and positive (abscisic acid production, less competition for carbon compounds from the shoot tips, and smaller fruit) effects on grape ripening. If the water deficit is moderate, the positive effects are more marked than the negative factors: the grapes contain higher concentrations of reducing sugars, anthocyanins, and tannins, while the malic acid content is lower (Van Leeuwen and Seguin, 1994). For example, Saint-Emilion wines from the 2000 vintage, when there was an early drop in stem potential, are better than those from the 2002 vintage (Figure 10.16). In cases of severe water stress, photosynthesis is too severely restricted and ripening may stop completely.

In viticulture, it is essential to know to what extent a water deficit has a positive effect on quality and locate the threshold of harmful water stress. The answer to this question depends on the type of production, the types of substances considered, and vine yields.

Most studies concerning the link between the vines' water balance and grape composition have dealt with red wine grapes. It is generally accepted that the red wine grapes can benefit from more severe water deficits than white grapes. On an estate producing both types of wine, it is, therefore,

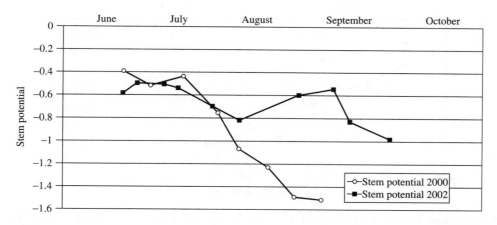

Fig. 10.16. Comparison of variations in stem potential in a Saint-Emilion vineyard in 2000 and 2002 (gravelly soil and Merlot grapes). The more negative the values, the more severe the water deficit

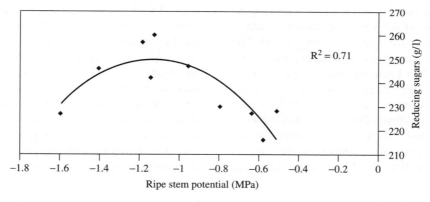

Fig. 10.17. Correlation between the intensity of water deficit (assessed by the stem potential when the grapes were ripe) and the concentration of reducing sugars

logical to plant the red varieties on soils with less plentiful water reserves.

Among the substances that promote red wine quality, sugar accumulation reaches maximum levels when the water balance is moderately restrictive. Grape sugar content is lower both when there is an unlimited water supply and in cases of severe water stress (Figure 10.17). The anthocyanin content increases in a linear manner over the same range of water deficits, reaching a maximum when the water stress is greatest (Figure 10.18). The quality of a red wine depends more on its phenolic content than on the sugar content of the ripe grapes, so red wine grapes may have the potential

to make excellent wine, even if severe water stress has penalized the sugar level of the must.

The issue of the effect of water deficits on quality cannot be settled without discussing yields. The same water deficit may have a positive effect on quality in a vineyard with yields of 30 hl/hectare and lead to blocked ripening with disastrous results at 60 hl/hectare.

(d) Impact of Water Deficit on Early Ripening

The date when grapes ripen depends both on the phenological cycle, which may be assessed by the date of mid-*véraison*, and the rate at which they

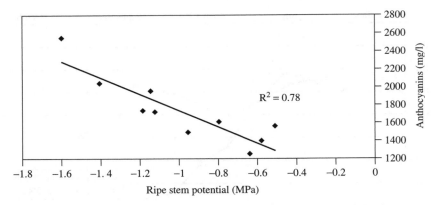

Fig. 10.18. Correlation between the intensity of water deficit (assessed by stem potential when the grapes were ripe) and the anthocyanin concentration

mature, calculated according to Duteau (1990). The earliness or lateness of the phenological cycle depends mainly on the soil temperature, which is related to its moisture content (Morlat, 1989). The ripening rate is largely determined by the vine's water balance (Van Leeuwen and Seguin, 1994). A water deficit promotes rapid ripening by keeping the grapes small (thus making them easier to fill with sugar) and reduces the competition between grapes and shoots for the carbohydrate supply. Figure 10.19 shows an example of the impact of water availability on the ripening rate and

early/late maturity in three plots with very different soils (Van Leeuwen and Rabusseau, unpublished results). To eliminate the impact of temperature on the ripening rate, dates are indicated on the abscissa by the sum of active temperatures starting on August 1st-each day is represented by the average temperature minus ten degrees. The vines on gravel soils and planosol were subject to water deficit and the sugar-acid ratio evolved rapidly towards ripeness. The water supply on the luvisol was not restrictive and the pulp ripened slowly. Although the mid-*véraison* dates were very close

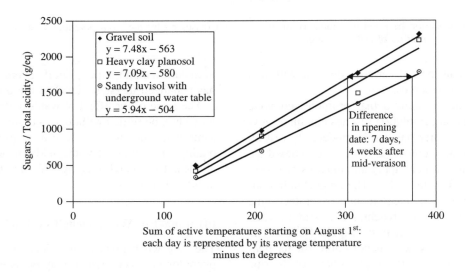

Fig. 10.19. Ripening rates on three soils in Saint-Emilion (Merlot Noir, 2001)

on all the plots, the difference in ripening date was as much as 70° days in the sum of active temperatures, or nearly 7 days after 4 weeks.

The vast majority of damp soils are cool and provide a non-restrictive water supply. Grapes ripen late on these soils as the phenological cycle is delayed and ripening is slow. By the same reasoning, most dry soils are conducive to early ripening. There are a number of highly reputed estates in Bordeaux, especially in Pomerol, but also a few localized cases in Saint-Emilion and the Haut-Médoc, planted on soils with a high clay content. They are unusual as they have high water contents (and are thus cool) but still cause an early water deficit in the vines. This type of clay (smectite) is unusual in that, although it contains large amounts of water, it is unavailable for use by the vines. These soils are conducive to early ripening and although, for historical reasons, Merlot has been planted on them in Pomerol, Cabernet Sauvignon ripens perfectly on the same type of soil in the Haut-Médoc. This example shows that water deficits play an essential role in the early ripening of grapes and have a greater impact than soil temperature. The choice of a grape variety to suit a particular type of soil should depend mainly on its conduciveness to early ripening and, thus, on its water balance (Van Leeuwen, 2001).

(e) Water Balance and Vintage Variation

The water status of a given vintage can be assessed by calculating the water balance. Table 10.5 shows the water status of several vintages in Bordeaux, calculated using the method developed by Riou and Lebon (2000). To eliminate the effect of the soil, we introduced a value, 0, for the water reserves at the beginning of the season, which explains the negative values of the water balance. These values indicate a theoretical water deficit, corresponding to the difference between precipitation and real evapotranspiration (in the absence of stomatal regulation). All the lesser quality vintages without exception had only a slightly negative water balance at the end of September (corresponding approximately to the harvest period). Seasons in which the vines were subjected to a significant

Table 10.5. Correlation between the theoretical water balance from April 1 to September 30 and the quality of the vintage. The more marked the negative water balance, the drier the vintage (based on unpublished work by van Leeuwen and Jaeck)

Vintage	Theoretical water balance on September 30	Vintage quality (Marks out of 20)
1990	−306	19
2000	−290	19
1986	−271	18
1998	−256	18
1995	−241	17
1962	−231	17
1964	−220	17
1997	−211	15
1988	−211	17
1970	−210	18
1961	−207	20
1991	−206	13
1989	−204	19
1985	−198	18
1977	−87	11
1993	−80	14
1954	−80	9
1971	−40	17
1956	−39	9
1968	−33	6
1958	−31	12
1969	−14	12
1973	−12	12
1965	−11	3
1963	−8	3
1992	−4	12
1960	−1	12

water deficit were all great vintages. Even if ripening may be halted on some plots (especially those with young vines, i.e. shallow root systems) in a very dry summer, which has a detrimental effect on the wine, it is interesting to note that, since 1950, there has been no overall quality loss due to water stress in Bordeaux, at least in red-wine producing vineyards.

(f) Ways of Modifying Water Supply in a Vineyard

The ideal water status for producing grapes to make high-quality wine consists of a moderate water deficit, starting early in the season (before *véraison*). The grapes will show less winemaking potential if the vines are not subject to water deficit at all, as well as in cases of severe water stress.

Loss of quality is much more commonly due to a plentiful water supply than due to excessive water stress, even if it is generally unnoticed. When summer rains and water reserves in the soil are such that the vines do not regularly suffer a moderate water deficit, leaf surface must be increased to promote evapotranspiration and vines should be planted on rootstocks that do not take advantage of the plentiful water supply (e.g. Riparia Goire de Montpellier). Quality can also be maximized by selecting an appropriate grape variety (early red and white grape varieties, Van Leeuwen, 2001).

In situations where excessive water stress causes a drop in quality in certain vintages (very dry climate and lack of water reserves in the soil), it is possible to minimize the negative impact on the vines by adapting the vine training system and vegetative growth (Chone *et al.*, 2001b). The best way to protect vines from the negative effects of water stress is by restricting yields. When yields are low, a relatively small leaf surface does not penalize the leaf/fruit ratio. The most widespread form of adaptation to dry conditions is the use of a drought-resistant rootstock (e.g. 110 Richter). It should also be noted that reducing the vines' nitrogen supply reduces their water requirement, by reducing vigor and restricting the leaf surface.

Under extreme conditions, vine-growers may need to irrigate, if permitted by local law. It is considered difficult to grow vines producing viable yields if annual rainfall is under 400 mm. This value may vary, however, depending on the distribution of rainfall throughout the year and the soil's capacity to retain water. In very dry climates, rational irrigation may be a quality factor, while poorly controlled irrigation may also lead to a reduction in winemaking potential. Irrigation should be gradually reduced, so as to produce a moderate water deficit in the vineyard before *véraison*, while avoiding severe water stress. Monitoring the vines' water status by testing stem potential is essential to ensure that irrigation is

perfectly controlled (Chone *et al.*, 2001b). Other promising monitoring methods are currently in the experimental stage.

For many years, the concept of growing vines under restrictive conditions was purely European, mainly in AOC (controlled appellation of origin) vineyards. It is interesting to observe that this idea is being introduced in some New World vineyards. The Australians have successfully tested two irrigation systems that deliberately restrict the vines' water supply. In "Regulated Deficit Irrigation" (RDI), a water deficit is deliberately caused after flowering by stopping irrigation for a period of time (Dry *et al.*, 2001). This is particularly aimed at reducing grape size. "Partial Rootzone Drying" (PRD) involves irrigating both sides of each row separately, alternating at two-week intervals. Thus, part of the root system is always in soil that is drying out. This has been observed to have very clear impact on the grapes' potential to produce high-quality wine, probably partly due to synthesis of larger amounts of abscisic acid than in vines not subjected to any water deficit (Stoll *et al.*, 2001).

10.4.7 Meteorological Conditions of the Year—the Idea of Vintage

The three principal climatic parameters (light, heat and humidity) vary considerably from year to year. Their respective influence on maturation processes is consequently of varying importance and leads to a given grape composition at maturity. The enological notion of vintage can thus be examined.

Variations in meteorological conditions do not have the same influence in all climates. The principal European viticultural regions have been classified into different zones (Figure 10.20). Examining only the sugar concentration in the northern continental zone (Alsacian, Champagne and Burgundian vineyards in France, and Swiss and German vineyards for the most part), the length of sun exposure seems to be the principal limiting factor during grape development (Calo *et al.*, 1992).

This factor is also important during maturation in the North Atlantic zone (Loire and southwestern France vineyards), but is less important

Fig. 10.20. European viticultural climatic zones

in the southern zone (Mediterranean vineyards in Spain, France and Italy). In the latter zone, the hydric factor interferes with the relative consistency of temperature and sun exposure.

High temperatures in this case do not positively affect sugar accumulation, if a considerable hydric stress exists. In the opposite case, they can limit this accumulation by favoring vegetative vine growth when the water supply is not limited.

In the Rioja vineyards of northern Spain, the respective importance (varying from year to year) of the opposing influences of the Atlantic and Mediterranean climate determines wine quality.

Thus the climate/quality relationship can only be represented approximately. The sum of the temperatures, rainfall or length of light exposure does not have the most influence on grape quality, rather, it is their distribution in the course of the vine growth cycle.

Table 10.6. Recent phenological observations on red grape development and maturation (Merlot and Cabernet Sauvignon) in Bordeaux (France) (vintages classified in order of forwardness)

Vintage	Half-bloom (A)	Half-véraison (B)	Harvest (C)	Duration in days		
				A–B	B–C	A–C
1997	23 May	31 July	15 September	69	46	115
1990	27 May	6 August	24 September	69	55	124
1989	29 May	4 August	16 September	67	47	114
1999	31 May	4 August	20 September	65	47	112
2000	3 June	6 August	24 September	64	49	113
1976	4 June	7 August	18 September	64	42	106
1998	4 June	7 August	25 September	64	49	113
1994	4 June	6 August	19 September	63	44	107
1995	4 June	10 August	23 September	67	44	111
1996	4 June	10 August	28 September	67	49	116
1982	5 June	9 August	23 September	65	45	110
1993	6 June	9 August	26 September	64	48	112
1992	6 June	14 August	28 September	69	45	114
2001	7 June	12 August	30 September	66	49	115
2002	7 June	12 August	1 October	66	50	116
1981	12 June	20 August	5 October	69	46	115
1988	12 June	17 August	6 October	66	50	116
1983	13 June	19 August	28 September	67	40	107
1975	14 June	20 August	1 October	67	42	109
1985	15 June	16 August	1 October	62	46	108
1991	15 June	20 August	4 October	66	45	111
1974	15 June	19 August	6 October	65	48	113
1987	15 June	16 August	8 October	62	53	115
1984	18 June	20 August	6 October	63	47	110
1986	20 June	19 August	3 October	60	45	105
1979	21 June	25 August	8 October	65	44	109
1980	25 June	3 September	13 October	70	40	110
1978	26 June	2 September	12 October	68	40	108
1977	27 June	2 September	12 October	67	40	107

In northern vineyards, climatic conditions favoring a forward growth cycle permit grape maturation during a warmer and sunnier period, thus benefiting grape quality. Recent years have permitted the verification of this simple observation in the Bordeaux region (France).

Among the most forward years for grape development, 1982, 1989, 1990, and 2000 produced wines of outstanding quality (Tables 10.6 and 10.7). The climatic conditions of these years are particularly favorable, with warm and sunny days and very little rainfall. At the harvest, Carbernet Sauvignon grapes had high sugar and low malic acid concentrations (Table 10.8). A high cation concentration, as shown by an elevated ash alkalinity, indicated a suitable circulation of water

in the plant and led to relatively high pHs. The long length of maturation in 1990 (55 days on average) resulted in one of the lowest malic acid concentrations in recent years.

Conversely, during late years such as 1980 and in particular 1977, grape development and maturation occurred in unfavorable climatic conditions. The grapes obtained in the same parcels studied were poor in sugar and rich in acids—especially malic acid. The importance of an early growing season for grape quality has also been demonstrated by wine-growing regions with similar climatic conditions, like the Loire Valley in France and New Zealand.

But in a temperate climate, like that of Bordeaux, the moment at which the best or worst

Table 10.7. Comparison between recent vintage quality and climatic conditions from April to September in Bordeaux (France) (vintages classified in order of forwardness)

Vintage	Sum of average temperatures (°C)	Duration of sun exposure (h)	Number of exceptionally warm days $t \geq 30°C$	Rainfall (mm)	Wine quality
1997	3494	1216	24	500	Good
1990	3472	1496	38	319	Exceptional
1989	3463	1463	35	364	Exceptional
1999	3498	1426	17	523	Very good
2000	3447	1454	25	477	Exceptional
1976	3384	1430	29	278	Very good
1998	3373	1226	25	537	Very good
1994	3344	1143	26	620	Very good
1995	3390	1149	36	303	Exceptional
1996	3267	1207	24	531	Exceptional
1982	3331	1262	18	289	Exceptional
1993	3231	1086	17	498	Good
1992	3325	1219	22	557	Mediocre
2001	3357	1505	32	438	—
2002	3309	1414	18	405	—
1981	3223	1144	17	289	Very good
1988	3288	1249	15	362	Very good
1983	3354	1182	24	437	Very good
1975	3250	1256	16	362	Very good
1985	3185	1326	16	311	Exceptional
1991	3419	1370	28	319	Good
1974	3129	1279	17	301	Good
1987	3360	1200	28	368	Good
1984	3111	1308	15	423	Good
1986	3129	1300	21	438	Very good
1979	2938	1183	3	366	Good
1980	3057	1020	9	343	Good
1978	3029	1153	12	320	Good
1977	3044	1135	2	407	Fairly good

Table 10.8. Average composition of Cabernet Sauvignon grapes (sampled at reference vineyards) at harvest according to vintage in Bordeaux (France) (vintages classified in order of forwardness)

Vintage	Weight of 100 berries (g)	Sugar concentration (g/l)	pH	Alkalinity of ash (mEq/l)	Total acidity (mEq/l)	Tartaric acid (mEq/l)	Malic acid (mEq/l)
1997	162	196	—	—	82	—	—
1990	113	199	3.38	50	77	92	31
1989	118	208	3.33	48	93	96	45
1999	134	203	3.57	52	80	87	42
2000	148	213	3.63	50	82	83	39
1976	110	196	3.33	48	97	98	44
1998	149	200	3.55	53	79	82	41
1994	140	193	3.31	45	102	84	56
1995	116	194	3.45	47	86	84	42
1996	138	220	3.38	48	102	88	60
1982	116	200	3.41	48	96	94	48
1993	124	181	3.16	43	94	78	51
1992	134	177	3.26	48	103	83	63
2001	146	204	3.59	53	90	86	54
2002	142	205	3.59	—	90	—	—
1981	110	180	3.36	44	103	91	53
1988	120	191	3.31	47	97	83	57
1983	115	195	3.37	48	107	95	59
1975	119	209	3.30	44	96	91	41
1985	117	196	3.48	48	92	99	35
1991	133	185	3.31	46	96	83	58
1974	107	184	3.27	47	98	92	43
1987	143	176	3.35	46	99	88	55
1984	122	185	3.23	44	119	94	66
1986	115	201	3.36	51	86	92	40
1979	116	174	3.18	44	119	98	60
1980	110	181	3.28	49	112	93	71
1978	119	193	3.26	48	120	91	68
1977	118	170	3.29	44	137	90	85

climatic conditions occur has a greater influence on grape quality than the absolute temperature and the total rainfall during the entire growth cycle (Figure 10.21). Thus grape quality depends on a favorable climatic period towards the end of maturation. The 1978 vintage at Bordeaux is a paradoxical example of one of the latest years (Table 10.6): unfavorable climatic conditions at the beginning of the growth cycle retarded flowering and grape development, but from *véraison* onwards, although the temperature was slightly lower than the seasonal average, a lack of rainfall and considerable sun exposure permitted the grapes to ripen

correctly and attain suitable sugar concentrations (Table 10.8). Among the late vintages of the last 30 years, 1978 is the only year when the grapes reached a satisfactory maturity.

The quality acquired at the beginning of development can be compromised by severe bad weather during maturation. In 1992, grape development was initially precocious thanks to high temperatures and moderate rainfall in the months before flowering. After a July with normal climatic conditions, August was very hot but suffered from an extremely high rainfall (about three times the normal rainfall for August at Bordeaux)

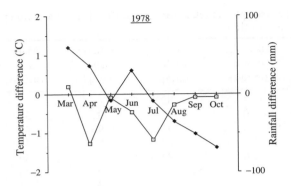

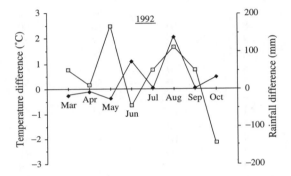

Fig. 10.21. Monthly temperature and rainfall differences with respect to 30-year average for the period from March to October, 1978 and 1992, at Bordeaux (France): □, temperature difference (°C); ♦ rainfall difference (mm)

Table 10.9. Phenolic composition of Cabernet Sauvignon grapes at maturity for three different vintages[a] (Bordeaux France) (Augustin, 1986)

Vintage	Anthocyanins (mg/100 berries)	Tannins (g/100 berries)	
		Skin	Seeds
1985	148	0.35	0.39
1983	132	0.25	0.44
1984	129	0.33	0.52

[a]Vintages are classed by decreasing order of wine quality.

Climatic conditions evidently have an influence on all grape constituents—in particular, secondary metabolites such as phenolic compounds and aromatic substances. Studies on these substances are incomplete and have often been carried out with very different techniques—especially extraction. Table 10.9 gives an example of the phenolic composition of Cabernet Sauvignon grapes at the time of harvest for the 1983, 1984 and 1985 vintages. The skin anthocyanin content is higher in quality vintages. This relationship is not valid for tannins.

10.5 IMPACT OF VARIOUS OTHER FACTORS ON MATURATION

The variability of the maturation process, in terms of vintage climatic conditions, is also regulated, if not controlled, by other parameters.

Some of these parameters are fixed and exert a constant and permanent action: the nature of the soil, the variety and possibly the rootstock as well as plant density and trellising methods. All of these factors are established during the creation of the vineyard. Vine age, to a certain extent, can also be placed in this first category.

Other parameters can be continuously changed. Their modification most often corresponds with a desire to adapt plant reactions to vintage climatic conditions: viticultural practices such as pruning, cluster thinning, hedging, leaf thinning, phytosanitary treatments, etc. Fertilizing is also often placed in this second category. These practices can modify the nature of a soil for a long time.

accompanied by a lack of sunlight. Despite more clement climatic conditions in September, sugar concentrations remained low (Table 10.8). In fact, sugar accumulation is generally rapid in the weeks following *véraison*, and migration becomes slower afterwards. Sugar concentrations rarely increase rapidly during the days before the harvest.

Yet, the harvest should occur in favorable climatic conditions. In 1993, this period was characterized in Bordeaux by heavy rains (more than 180 mm in September). The quality of the vintage dropped substantially during the last days of maturation. Similarly, 1976, a forward vintage benefiting from a warm, dry summer, did not attain the exceptional quality hoped for, because of rains at the end of maturation.

Finally, vintage climatic conditions can produce accidental factors—both meteorological (frost, hale) and sanitary (cryptogamic diseases).

10.5.1 Variety and Rootstock

Rootstocks are used in vine-growing when the chemical composition of the soil or the presence of pests (such as phylloxera) prevents the variety from developing on its own roots. The rootstock develops a different root system than the graft and this results in changes in the water and mineral supply. Grapes of a given variety, grown on the same soil, are known to have a different ionic composition according to their rootstock, but these differences are not sufficiently important to cause a significant variation in must acidity (Carbonneau, 1985).

They are, however, capable of influencing graft photosynthetic activity. The vigor may either increase or decrease, depending on the type of soil. In the richest soils, rootstocks such as 110 R, 140 Ru, 1103 P, SO4 and 41 B confer an excessive vigor to the graft. This heightened vegetative growth slows and limits the maturation process (Pouget and Delas, 1989). In contrast, the Riparia Gloire and 161-49 C rootstocks create a relatively short vegetative cycle, favoring maturation and respecting the general specificity of the variety.

Grape composition at maturity differs when a variety has developed on its own roots as opposed to on a rootstock. These differences essentially affect maturity: they concern the concentrations of sugars, acids and phenolic compounds. Yet if the rootstock is judiciously chosen, the differences that result from divers rootstocks for the same cultivars are always slight (Guilloux, 1981). Only the nitrogen concentration appears to vary significantly (Roubelakis-Angelakis and Kliewer, 1992).

Choosing the variety to suit the climate is a deciding factor for obtaining a good maturation and quality wines. In general, early ripening varieties are cultivated in cold climates (Chasselas, Gewürztraminer, Pinot) and relatively late-ripening varieties in warm zones (Aramon, Carignan, Grenache). In both cases, maturity should occur just before the average monthly temperature drops below 10°C. The maturation process should not take place too rapidly or abruptly in excessively favorable conditions.

Quality cultivars such as Cabernet Sauvignon and Pinot Noir lose much of their aromatic substance and phenolic compound finesse in warm climates. Figure 10.22 indicates the phenological behavior diversity of these two varieties in the different viticultural regions of the world. In a warm climate, characterized by average monthly temperature always above 10°C (for example, Perth, Australia), the duration of development is particularly short—notably at maturation. Conversely, the cycle grows longer in cool and humid temperate climates (French vineyards and in Christchurch, New Zealand).

Choosing a variety for a given area depends greatly on its ability to reach a sugar concentration of 180–200 g/l during maturation, but ripe grape quality is also affected by other chemical constituents.

The tartaric/malic acid ratio varies considerably from one variety to another. At maturity, the grapes of most varieties contain more tartaric than malic acid. Some varieties, however, always have a higher concentration of malic acid than of tartaric acid: Chenin, Pinot and Carignan. In a warm climate, varieties having a high tartaric/malic acid ratio are preferably chosen.

Peynaud and Maurié (1953) had already noticed that variability in organic acid concentration causes a very variable nitrogen concentration from one variety to another. Even more than the total nitrogen concentration, amino acid composition varies greatly—to such an extent that it is used by certain authors as a means of varietal discrimination. Arginine and proline concentrations can vary by a factor of 10 to 15, depending on the variety—for example, from 300 to 4600 mg/l for proline. The proline/arginine ratio is relatively constant from one vintage to another in the same grape variety.

The different varieties also seem to have a large diversity in phenolic composition. A study carried out on the principal French red grape varieties cultivated in the Mediterranean or Atlantic climate showed that concentration variations according to

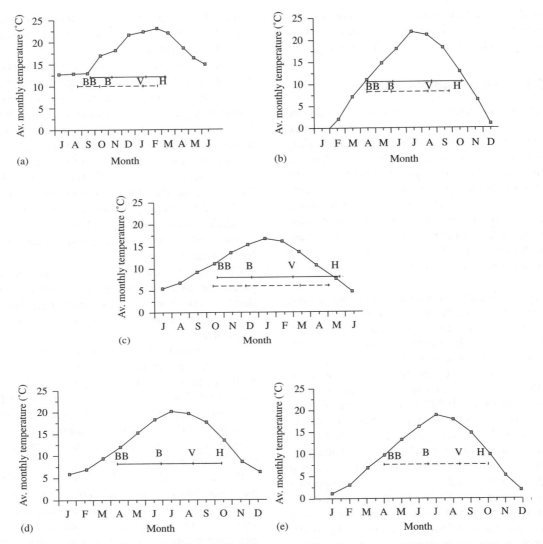

Fig. 10.22. Phenological behavior of Cabernet Sauvignon and Pinot Noir according to climate: (a) Perth, Australia; (b) Prosser, WA, USA; (c) Christchurch, New Zealand; (d) Bordeaux, France; (e) Beaune, France. (a)–(c) after Jackson and Lombard (1993). BB, budburst; B, bloom; V, *véraison;* H, harvest; □—□, temperature curve; ├———┤, vegetative cycle of Cabernet Sauvignon; |- - -|, vegetative cycle of Pinot Noir

the climate are significantly less than according to the variety (Bisson and Ribéreau-Gayon, 1978).

Similarly, fluctuations in grape phenolic content from one vintage to another and for a given variety are less than the variations between varieties. The genotypic effect of the variety is thus preponderant on grape phenolic compound richness. Anthocyanidic and procyanidic profiles vary greatly with respect to the variety and can therefore be used in varietal discrimination (Calo *et al.*, 1994).

The variability of grape aromatic content is even greater. Some varieties possess characteristic aromas. At present, not all of the molecules responsible for these aromas have been identified. In certain varieties, such as the Concord, descendant of native American vines (*Vitis labrusca, Vitis rotundifolia*), the grapes always exhibit a foxy odor due to methyl anthranilate (Bailey, 1988).

Similarly, disease resistant hybrids such as Castor and Pollux, obtained by crossing *Vitis vinifera* and native American vines, are often characterized by a strawberry odor resulting from the presence of a few nanograms of 2,5-dimethyl-4-hydroxy-2,3-dihydro-3-furanone (furaneol) (Rapp, 1993).

The Cabernet family of varieties possesses notes of vegetal aromas due to the presence of pyrazine derivatives (see also Volume 2, Section 7.4) (Bayonove *et al.*, 1976).

Finally, within the muscat group, the free and glycosidic terpenol profile and concentrations vary greatly with respect to the variety (see also Volume 2, Section 7.2) (Bayonove, 1993). Sulfurous compounds and volatile phenols responsible for diverse aromas (medication, blackcurrant, etc.) most likely vary in the same way (Rapp, 1993).

Varieties differ considerably with respect to each other, and choosing a variety suitable for local environmental conditions is a deciding factor in wine quality. When such choices are well established by viticultural tradition and scientific observation, quality improvement depends on clonal selection within the variety. Clonal selection has been successfully developed for Cabernet Sauvignon, Pinot Noir, Chardonnay, Riesling and Gewürztraminer (Schaeffer, 1985). It focuses on limiting varietal shatter sensitivity, increasing membrane selective permeability (higher sugar concentrations) and modifying berry volume and grape cluster morphology.

Although many empirical observations note the influence of vine age on wine and grape quality, little scientific work has been devoted to this subject. According to Dring (1994), the young vine develops a root system adapted to its environment during its first years. At the end of its fourth year, a functional equilibrium between the roots and the metabolic activity of the aerial parts of the vine is established. A mycorrhizum most often facilitates the mineral nutrition of the young vine (Possingham and Grogt Obbink, 1971).

If the vine-grower then succeeds in respecting the vine–soil–climate equilibrium and in

regulating the harvest volume, notably by pruning, the plant can develop sufficient reserves in the old wood to ensure the prerequisites for proper maturation each year. Old vines are thus less sensitive to yearly climatic variations and most often produce grapes rich in sugar and secondary products favorable to wine quality.

10.5.2 Soil Constitution and Fertilization

The influence of soil on grape composition and wine quality is definitely the most difficult to describe. The soil, by its physical structure and chemical composition, directly affects root system development and consequently the vine water and mineral supplies. It exerts an equally important effect on the microclimate. Soil color and its stone content profoundly modify the minimum and maximum temperatures as well as the light intensity in the lower atmosphere surrounding the grape clusters. Whether with schist plates, limestone pebbles or siliceous gravel, vine-growers have long made the most of this 'second sun' to improve grape maturation.

As mentioned previously (see Section 10.4.6), a regular water supply is needed for grape development and maturation.

This water from the soil transports the minerals that are necessary for growth in the plant. The ionic concentration of this solution is related to the nature of the soil and the fertilizers added, but a large amount of the available minerals is the result of biological activity in the soil. A potential disequilibrium can seriously affect vine growth. The best-known example is the increase in the exchangeable phytotoxic copper concentration in old, traditional vineyards that have received many sulfur and copper-based treatments to ensure the sanitary protection of the vine. Under the influence of bacteria in the soil, the sulfur is oxidized into sulfates which accumulate in the soil. The resulting soil acidification causes copper solubilization (Donèche, 1976).

Many synthetic pesticide residues can similarly disrupt certain soil reactions, notably the biological mineralization of nitrogen, but there has been no research on the consequences of these phenomena on grape maturation and composition.

Many studies have focused on the influence of different levels of nitrogen and potassium fertilization. The removal of these minerals by the harvest is relatively low, compared with other crops. Since the roots exploit a large volume of soil, vine mineral needs are relatively low.

For example, annual nitrogen fertilization should not exceed 30 kg/ha, which is largely sufficient for meeting the plants' needs. Above this value, nitrogen exerts a considerable effect on vine vigor, and excessive vegetative growth blocks the maturation process. In this case, the grape crop is abundant, but sugar and phenolic compound concentrations are low and the grapes are rich in acids and nitrogen compounds. Excessive addition of nitrogen also increases the concentrations of ethyl carbamate precursor and histamine, which are likely to lower the hygienic quality of wine (Ough *et al.*, 1989). The effect of nitrogen on vigor can be limited by water supply deficiencies in warm climates. Temporary or permanent cover crop between vine rows may lead to a deficit in the vines' nitrogen supply due to competition, but also as a result of mineral nitrogen fixation (or denitrification). Nitrogen deficiency in the grapes may lead to fermentation problems in the must and may also, above all, have a detrimental effect on their synthesis of phenolic compounds and a large number of aromatic substances.

The problem of potassium is more complex. This cation predominantly participates in must and wine pH and acidity. Facing the fairly general increase of wine pH during recent years, much research tends to show that the soil is responsible for this high potassium supply, due to excessive soil richness or fertilization. However, a direct relationship between excess potassium fertilization and decreased grape acidity has not been demonstrated definitively in all cases.

Potassium actively participates in grape sugar accumulation. In years with favorable climatic conditions, the ripe grape imports large amounts of potassium. Due to the high malic acid degradation characterizing such a maturation, must acidity is principally the result of the tartaric acid concentration. Insolubilization of tartaric acid salt in the course of winemaking greatly lowers the acidity.

Some vineyards are established on salty soils. High sodium chloride concentrations increase the osmotic potential of the soil solution. As a result, the plant must strongly increase its respiratory intensity to ensure the necessary energy for its mineral nutrition. In hydroponics, this lowers vine vigor and results in a more forward maturity. The sugar concentration increases but not the amount of phenolic compounds. In vineyards, these effects are modified by using specific rootstocks (saltcreek, dodridge and 1613). On salty soils, potassium, magnesium and organic acid concentrations decrease whereas calcium and chloride concentrations increase.

In conclusion, the primordial influence of soil has been recognized for a long time in the form of viticultural *terroirs*. The soil must create favorable conditions for grape development and maturation (mineral and water supply and microclimate). The temperature above the soil and its water content also have an impact on the earliness or lateness of the vines' growing season (Barbeau *et al.*, 1998; Tesic *et al.*, 2001). These two parameters give an initial indication of the quality of a *terroir*. But a quality *terroir* must also limit the consequences of weather variations from one year to another. Soil study is difficult since all of the factors likely to influence the biochemical processes of maturation should be taken into account.

Our understanding of the role of soil in the intrinsic quality of wine still rests essentially on empirical data. Each grape variety does, however, excel in particular soil types. Thus Cabernet Sauvignon predominates in the Médoc *appellation* (France) where this variety ripens on sandy, gravely hilltops and produces rich and complex wines, but tradition shows that the best results on clay-rich parcels in the Médoc flat land and dales are obtained with Merlot.

10.5.3 Management of Vine Growth

Grape maturation is also influenced by other permanent factors, established during the creation of the vineyard. Planting density, row spacing and canopy placement (existence and positioning of wire trellising) condition plant physiology through root development with respect to the soil and use of sunlight by the leaves. These factors directly affect vine vigor. Their action on the grape can only be indirect, notably acting on the microclimate surrounding the grape clusters (temperature and sun exposure).

Rigorous experiments in this domain are difficult to carry out. The existing criteria for establishing a vineyard are primarily empirical but vine vigor has been shown to increase when plant density decreases, with the risk of a retarded maturation.

In northern and temperate regions, tradition (verified by research) recommends relatively high planting densities of around 10 000 vines/ha. In a drier, Mediterranean climate, optimum quality is often obtained with a density of between 3000 and 5000 vines/ha. In spite of the water deficit, the high density restricts potassium imports and maintains a good acidity level in wines made from these grapes.

In conditions favoring vegetative growth (irrigation in warm and sunny climates), excessive leaf crowding should be limited by low plant densities (1000–2000 vines/ha), and adapted training and pruning methods. Canopy management is of major importance in this case (Carbonneau, 1982).

As well as the climate, soil fertility imposes its own rules. According to Petit-Lafitte (1868), 'the poorer the soil, the higher the plant density'. A very compact root system thus permits the maximum exploitation of the soil potential. The same reasoning should be used in dry soils.

10.5.4 Vineyard Practices for Vigor Control

Vine management and growing are characterized by severe measures limiting vegetative development and the amount of fruit. A certain canopy surface is required for grape alimentation and a relationship exists between this surface and grape quality.

The development of the canopy surface to fruit weight ratio can be used to evaluate grape quality. In an example with Tokay, Kliewer and Weaver

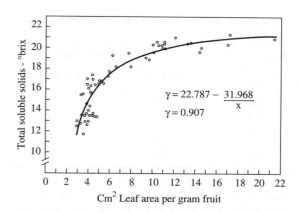

Fig. 10.23. Regression of total soluble solids (°Brix) of 'Tokay' berry juice at harvest (September 24) on leaf area per unit crop weight (cm^2/g). (Kliewer and Weaver, 1971)

(1971) showed that grape sugar concentration diminished sharply when this ratio was lower than 10 cm^2/g (Figure 10.23). Proline and phenolic compound concentrations were similarly affected. Di Stefano *et al.* (1983) obtained identical results for terpenic compound concentrations in white muscat. In general, the necessary canopy surface for the maturation of 1 g of fruit varies from 7 to 15 cm^2/g for most *Vitis vinifera* varieties, but increasing this ratio above these values has little effect on grape composition, as shown by the sugar concentration curve in Figure 10.23.

Winter pruning is the first operation carried out in the vineyard. It consists of controlling vine production by leaving only a certain number of buds capable of producing inflorescences. Vines respond very differently to pruning. Some varieties have such fertile buds at the base of the shoots that pruning is often ineffective for yield control. Bud fertility varies greatly at the same position on the shoot, depending on the variety. Cultivar productivity can also vary with respect to the climate. Many factors must thus be considered. Yet most specialists agree that low yields are needed to obtain proper grape maturation.

Increases in yield have long been known to affect grape sugar concentration negatively (Table 10.10). This problem is especially troubling in vineyards that use varieties close to

Table 10.10. Relationship between crop yield and sugar concentration

Grape weight per wine (kg)	Sugar concentration (g)	
	Per vine	Per liter of must
0.4	63	235
0.8	126	225
1.2	168	200
1.6	184	165
2.0	182	130

their cultivation limits. In very warm regions, the harvest can be delayed to attenuate this phenomenon. In northern vineyards, the high variability of annual climatic conditions determines the ratio of grape yield to sugar concentration. Experience shows that great vintages, obtained in particularly favorable climatic conditions, most often correspond with abundant grape crops.

In a warm and sunny climate, increasing canopy surface and improving sun exposure combined with controlled irrigation will often increase yields without lowering grape quality (Bravdo *et al.*, 1985), but several consecutive abundant crops can lead to depletion of vine reserves.

After berry set, excess grapes can be removed by thinning. Manual thinning is expensive. Thinning 30% of the crop before *véraison* results in a 15% increase in the sugar concentration and a 5% drop in acidity. Chemical berry thinning currently remains experimental and is an extremely delicate operation, undertaken at berry set.

Other techniques are available to the vine-grower to modify the physiological behavior of the vine. Depending on soil fertility and climatic conditions, trimming or hedging can slow vine vigor and limit leaf crowding. A single topping (removing recent shoot growth), at the end of flowering, diminishes the risk of shatter by limiting sugar competition between young grape clusters and apical shoots.

According to Koblet (1975), considering that a shoot bears about 200 g of grapes, 10–14 leaves per shoot are necessary to ensure their maturation, assuming a canopy surface to grape weight ratio of 8–10 cm^2/g. Trimming, which leaves a maximum of 14–16 leaves per shoot,

improves the maturation process and increases sugar and secondary metabolite concentrations. Trimming too severely produces opposite results. Trimming also presents the advantage of lowering the evaporation surface, thus limiting hydric stress risks in certain environmental conditions (lack of rainfall, soil water deficiencies).

Older, less active leaves can also be removed from the base of the shoot. Correct leaf thinning essentially exposes grape clusters to sunlight, improving grape maturation and limiting the risk of rot. Leaf thinning around grape clusters thus reduces malic acid concentrations and the 'pyrazine' character while augmenting grape anthocyanin concentrations for Cabernet Sauvignon. The vegetal aromatic character of Sauvignon Blanc can also be lowered in this manner (Arnold and Bledsoe, 1990).

Leaf thinning seems to be very effective, especially when practiced just after *véraison*. The result of this operation depends greatly on climate, variety and canopy placement. A partial leaf thinning before berry setting lowers the future crop volume, due to flower-fall. But this delicate technique can lower fertility during the following vegetative cycle (Candolfi-Vasconcelos and Koblet, 1990).

Finally, chemical substances are now available to the vine-grower to slow vine vigor and accelerate the maturation process. These are usually growth hormone biosynthesis inhibitors (Reynolds, 1988).

10.5.5 Effects of Disease and Adverse Weather

Late frosts and hailstorms occurring in the spring often produce the same effects as shoot removal by topping, but latent bud development is disorganized, resulting in bushy vegetation. Flowering is considerably extended and grape cluster maturation is uneven. The latest grape clusters have difficulty in reaching maturity. Damage caused by summertime hail alters grape cluster alimentation: affected grapes wither or are attacked by parasites, and a rapid harvest may be necessary.

Diverse causes can result in more or less severe vine defoliation. Maturation is difficult due to insufficient grape alimentation. A late downy mildew attack can cause total leaf loss in certain very sensitive varieties, such as Grenache. Similarly, leaves infected with powdery mildew always lower grape quality. Parasite development leads to significantly reduced crop yields, and very late attacks hinder grape maturation.

A potassium deficiency can cause leaf-scorch flavescence and premature leaf-drop—and in consequence a decrease in grape sugar and phenolic compound concentrations. Black leaf, encouraged by overproduction and soil dryness, often accompanies potassium deficiencies.

Bunch stem necrosis also lowers crop quality and can cause crop loss. It is often linked to excessive yields and magnesium deficiencies. Certain varieties are particularly sensitive: Gewürztraminer, Sauvignon Blanc, Ugni Blanc and, notably, Cabernet Sauvignon. This necrosis results in a decrease in sugar, anthocyanin, fatty acid and amino acid concentrations whereas the grapes remain rich in organic acids (Ureta *et al.*, 1981).

10.6 BOTRYTIS CINEREA

10.6.1 Gray Rot and Noble Rot

In addition to the diseases already mentioned (downy and powdery mildew), one of the principal causes of crop quality degradation is grape rot due to the development of various microorganisms (bacteria, yeasts or other fungi).

The principal microorganism responsible is usually *Botrytis cinerea* which is 'a ubiquitous fungus except perhaps in desert zones' (Galet, 1977). Endowed with a great polyphagy, this saprophyte can exist on senescent or dead tissue such as vine wood. It is also capable of waiting for favorable conditions in diverse resistant forms (sclerotia or conidia, with a high dissemination capacity). The presence of water on the surface of vegetal tissue and an optimal temperature of 18°C are ideal conditions for the germination of the resistant forms and mycelial growth. Conidial germination is possible at temperatures between 10°C

and 25°C. In these conditions, the contamination area of this fungus is very large and covers a great number of the world's viticultural regions. Grape gray rot thus remains one of the major concerns of vine-growers.

In a few areas of the world, particular conditions permit *Botrytis cinerea* to develop on mature grapes. This process results in an overripening that increases the sugar concentration while improving quality. The parasitized grape dehydrates and the sugars are more concentrated than the acids. Most importantly, the grape acquires the characteristic aromas that permit the production of renowned sweet white wines such as Sauternes-Barsac, Côteaux du Layon (France), Tokay (Hungary) and Trockenbeeren auslese (Germany and Austria).

Noble rot requires specific environmental conditions. The many studies undertaken have not yet been able to define these conditions, precisely but, in general, *B. cinerea* development in the form of noble rot is thought to be favored by alternating dry and humid periods. Nighttime humidity, dew and frequent morning fogs in the valleys of certain rivers stimulate fungal development, whereas warm and sunny windy afternoons facilitate water evaporation—limiting fungal growth.

Many factors participate in this phenomenon:

- The soil, by its nature and possibly its drainage, should permit the rapid elimination of rain water.

- The canopy placement and surface should permit a maximum number of grape clusters to be aerated and exposed to sunlight.

- The grape cluster structure should be fairly dispersed.

The nature of the variety also greatly affects grape sensitivity to *B. cinerea*, but no direct relationship seems to exist between variety type and noble rot quality. Differences essentially originate from the level of maturation precocity. Pucheu-Planté and Leclair (1990) showed the importance of the nature of the clone on noble rot quality.

10.6.2 Grape Sensitivity to *Botrytis cinerea*

Vine inflorescences can suffer from rot attacks if the climatic conditions are favorable to *B. cinerea* development. Peduncular rot causes flowers to fall and consequently a sharp drop in future crop volume. During the entire herbaceous growth period, the grape is resistant to this parasite.

Gray rot rarely occurs between fruit set and *véraison*. In 1983 and 1987, northern European vineyards suffered early attacks, sometimes affecting up to 30% of the berries but the reasons for the loss of resistance in green grapes are still not known.

In certain cases, of compact grape cluster varieties with elevated grape setting rates, a few berries can become detached from their pedicel and remain imprisoned inside the grape cluster. These damaged grapes constitute a direct penetration path for the fungus, circumventing the natural resistance of the grape. More often, the contamination is the consequence of another phenomenon, such as hail or other parasites. After *véraison*, the grape rapidly becomes more or less sensitive to *B. cinerea*.

These behavioral differences of the grape are due to multiple causes that will now be examined briefly without an in-depth study of the pathology of the grape–*B. cinerea* relationship.

In the first place, the green grape skin, covered by a thick cuticle, constitutes an effective barrier against parasites. Since Bonnet's (1903) initial research, a resistance scale of the principal *Vitis* species has been established based on the cuticle thickness of their respective berries. American varieties whose cuticle thickness varies from 4 μm (*Vitis rupestris*) to 10 μm (*Vitis coriacea*) have better protected berries than European species (*Vitis vinifera*), whose cuticle thickness is from 1.5 to 3.8 μm. This observation led to the production of *V. vinifera* and American species hybrids that are effectively more resistant to gray rot, but these hybrids do not usually produce quality grapes on the best *terroirs*.

The same relationship between cuticle thickness and *B. cinerea* resistance was encountered in *V. vinifera* varieties but on a smaller scale. The sensitive varieties all have a cuticle thickness of less than 2 μm (Karadimtcheva, 1982).

During maturation, the cutin and wax quantity per surface unit increases. This accumulation is more intense when the grapes are exposed to sunlight in an environment relatively low in humidity. Contact between berries is always characterized by a lower cuticle thickness.

Although the fungus possesses a cutinolytic activity, it is very low. In fact, direct penetration of the grape cuticle by *B. cinerea* enzymatic digestion has not been proven. Only a developed mycelium produces sufficient amounts of cutinase to attack a neighboring berry cuticle (often less thick if the grape cluster is very compact). In the surface of the cuticle there are perforations that are a potential point of entry for mycelial filaments (Blaich *et al.*, 1984). Resistant hybrids have fewer perforations than sensitive varieties. The number of these cuticle perforations increases in the course of maturation.

Under the cuticle, the exocarp also participates in the resistance to *B. cinerea*. According to Karadimtcheva (1982), the external layer of the hypodermis in certain varieties resistant to gray rot comprises more than seven rows of thin, elongated cells, with a total thickness exceeding 100 μm. In sensitive varieties, it contains only four to six cell rows, with a total thickness of 50–60 μm.

The extent of the thickening of the epidermal cell walls, occurring in the course of maturation, varies depending on the variety. The most sensitive varieties have the thickest cell walls. This phenomenon is caused by the partial hydrolysis of pectic compounds by endogenous grape enzymes (Section 10.2.5). The increase in soluble pectins varies greatly, depending on the variety. In consequence, grape skins exhibit varying degrees of sensitivity to enzymatic digestion by the exocellular enzymes of *B. cinerea* (Chardonnet and Donèche, 1995).

In addition to this mechanical resistance, the grape skin contains preformed fungal development inhibitors. All epidermal cells possess tannin vacuoles. These phenolic compounds exert a weak fungistatic effect on the pathogen.

As in many fruits, an inhibitor of the endopoly-galacturonase of *B. cinerea* is also contained in the grape skin cell walls. This glucoproteic substance is liberated during cell wall degradation by the pathogen. The concentration and persistence of this inhibitor vary according to the stage of development and the variety considered.

The green grape skin is also capable of synthesizing phytoalexins in response to an infection (Langcake and Pryce, 1977). These stilbenic derivatives (trans-resveratrol and its glycoside, trans-piceid, dimer ε-viniferin, α-viniferin trimer, β-viniferin tetramer) have fungicidal properties (Figure 10.24).

Resveratrol is obtained by the condensation of *p*-coumaroyl CoA with three malonyl CoA units in the presence of the stilbene synthase (Figure 10.25). Viniferine formation is then ensured by grape peroxidases.

At the time of a parasite infection, the normal flavonoid metabolism (Section 10.3.6) is diverted towards stilbenic derivative production by the action of the stilbene synthase. Remarkably, after *véraison*, this capacity is very rapidly lost even in pathogen-tolerant American vines (Figure 10.26).

Thus during maturation, the grape loses most of its physical and chemical defenses. The parasite sensitivity differences observed for diverse varieties and clones essentially result from differences in their grape development time. Microperforations of the cuticle and stomatic fissures (Section 10.2.4) also constitute a passageway for the efflux of grape exudate, which is indispensable for conidial germination and the proliferation of *B. cinerea* (Donèche, 1986). Chemical modifications of the grape during maturation—notably an increase in the sugar, amino acid and soluble pectin concentrations—furnish the fungus with essential nutrients for mycelial growth.

10.6.3 Noble Rot Infection Process

Some observations have led to the conclusion that *Botrytis cinerea* is sometimes present inside grapes as soon as they set in (Pezet and Pont, 1986). When it comes out of the latency phase, it develops

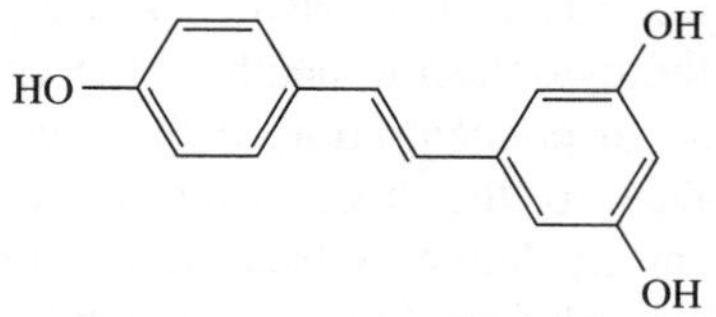

trans-Resveratrol

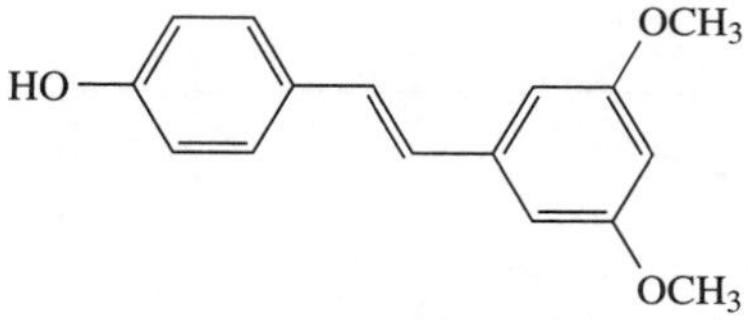

trans-Pterostilbene

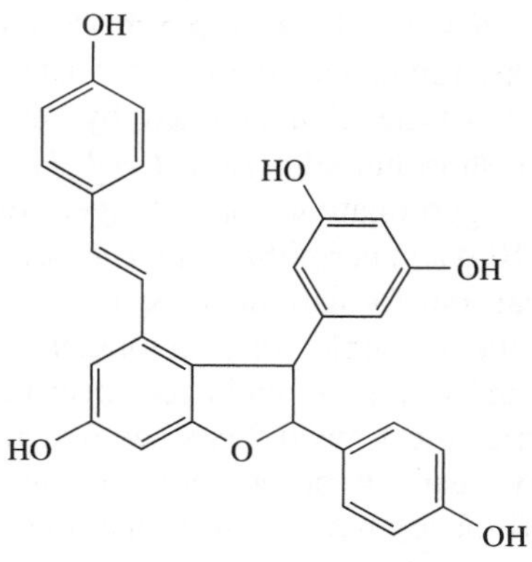

ε Viniferine

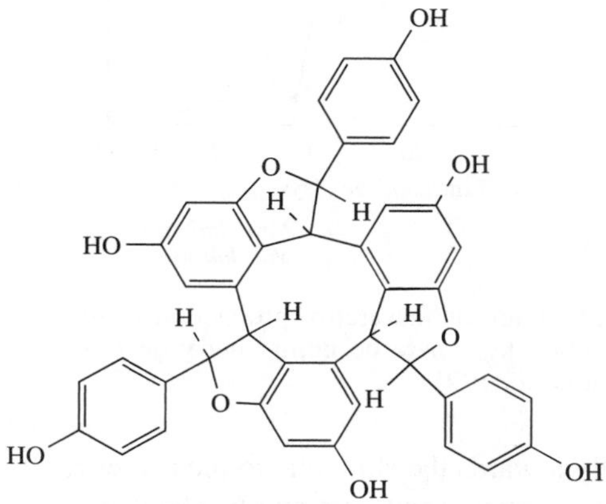

α Viniferine

Fig. 10.24. Principal stilbenic derivatives identified in the vine (Langcake and Pryce, 1977)

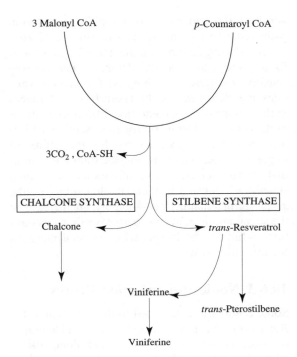

Fig. 10.25. Biosynthesis pathways of stilbenic phyto-alexins in the vine (Jeandet and Bessis, 1989)

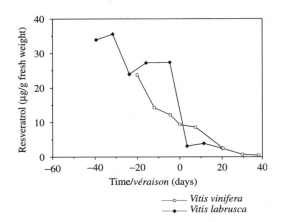

Fig. 10.26. Evolution of resveratrol production in response to infection by a parasite during berry development (Jeandet *et al.*, 1991)

preferentially towards the skin, due to the presence of antagonist enzymes (chitinase and *B. glucanase*) in the flesh cells. An exogenous nutrient supply facilitates the elongation of the germinative tube.

When the mycelial hypha reaches a microfissure, it penetrates the grape. Thus *B. cinerea* development occurs mainly in the grape's superficial cell walls. More precisely, the mycelial filaments are located in the middle lamella of the pectocellulosic cell walls. The latter are degraded by the enzymes of the fungus (pectinolytic, cellulasic complex, protease and phospholipase enzymes).

When the mycelium has totally overrun the pectocellulosic cell walls of the epidermal and immediate sub-epidermal cells, white grapes take on a chocolate-brown color that is characteristic of the *pourri plein* stage. The mycelium then produces filaments that emerge from the skin surface by piercing the cuticle or by taking advantage of the diverse fissures used for the initial penetration. The filament extremities differentiate by producing conidiophores, whose conidia later detach and contaminate nearby berries.

The cell walls of vegetal tissue are so greatly modified that they can no longer ensure their functions. In particular, berry cell hydration is no longer regulated. It can vary with respect to climatic conditions and, in ideal conditions, should lead to a characteristic desiccation accompanied by the cytoplasmic death of the epidermal cells. The sugar concentration of these cells is considerable. Due to the high osmotic pressure, the fungus can no longer subsist and stops developing. This shriveled *confit* stage, known as *pourri rôti*, is used for elaborating sweet white wines (Table 10.11).

The infection process, from healthy grape to *pourri rôti* grape, lasts from 5 to 15 days, depending on environmental conditions. This overripening period must be characterized by an alternation between short humid periods (3–4 days), favoring conidial germination, and longer dry periods (about 10 days) permitting grape concentration and chemical transformations to occur.

For quality noble rot development, the phenomenon must be rapid and occur near maturity, but the berries must reach this stage intact.

Many years of observation in Sauternes vineyards (France) have shown that the first symptoms of attack appear 15–20 days before maturity. Regardless of climatic conditions, *B. cinerea* development is slow until maturity. At this stage,

Table 10.11. Principal modifications of chemical grape constitution by a noble rot attack

Constituent	Per liter of must		Per 1000 berries	
	Healthy berry	Noble rotted berry	Healthy berry	Noble rotted berry
Weight (g)	—	—	2020	980
Volume of must (ml)	—	—	1190	450
pH	3.33	3.62	3.33	3.62
Sugar concentration (g)	247	317	294	143
Glycerol (g)	0	7.4	0	3.4
Alkalinity ash (mEq)	33	81	39	36
Total acidity (mEq)	123	112	146	50
Tartaric acid (mEq)	71	33	85	15
Malic acid (mEq)	81	117	96	54
Citric acid (mEq)	2.7	3.5	3.2	1.6
Acetic acid (mEq)	5.4	6.9	6.4	2.4
Gluconic acid (mEq)[a]	0	10.6	0	4.8
Ammonium (mg)	85	56	101	25
Amino acids (mg)	1282	1417	1526	638
Proteins (mg)	2815	3795	3350	1708

[a]Musts from infected grapes contain from 1 to 2.5 g gluconic acid/l.

the parasite spreads rapidly and its growth is explosive. At a certain moment, a high percentage of grapes simultaneously reach the *pourri plein* stage (Figure 10.27).

Grape maturation at a vineyard, in a parcel or even on the same grape cluster, is never absolutely synchronous (Section 10.4.1). As soon as a berry approaches maturity, it is contaminated by conidia from a nearby rotten berry. This asynchronism makes successive sorting necessary during the harvest ('successive sorting' is a local term which means that successive handpicking is used to ensure that only noble rotted grapes are harvested). The use of different varieties with varying degrees of precocity is beneficial in practice: for example Muscadelle, Semillon and Sauvignon Blanc in Sauternes (France) or Fürmint, Harslevelü and Muskat Ottonel in Tokay (Hungary).

10.6.4 Changes in the Chemical Composition of Noble Rot Grapes

Physical data (juice volume, berry weight) effectively express the grape desiccation phenomenon. Grapes can be concentrated by a factor of 2 to 5, depending on climatic conditions.

The enological profile of must obtained from botrytized grapes is specific (Table 10.11). This juice is very rich in sugars but its acidity is similar to that of juice obtained from healthy grapes. The tartaric acid concentration is often even lower and the pH higher (from 3.5 to 4.0), attesting to the

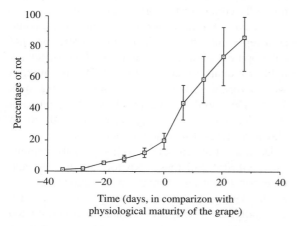

Fig. 10.27. Evolution of a noble rot attack in a Sauternes vineyard (France) (average of 10 years of experimentation on different parcels)

concentration of other substances such as potassium ions. Compounds not present or in negligible concentrations in healthy grapes are encountered in considerable quantities, especially in *pourris rôtis* grapes. For example, glycerol and gluconic acid can reach concentrations of 7 g/l and over 3 g/l in botrytized must, respectively.

But this concentration phenomenon masks profound chemical constitution changes resulting from the biological activity of *B. cinerea*. As Müller-Thurgau demonstrated as early as 1888, these changes affect sugars and organic acids.

B. cinerea uses little pectocellulosic cell wall residue for its development. As a result, the contaminated grape becomes rich in galacturonic acid derived from the degradation of pectic compounds. It prefers to assimilate glucose and fructose accumulated in the pulp cells, and 50% of the sugars are lost in the production of these noble rot wines.

Metabolic studies *in vitro* have shown that the young mycelium of *B. cinerea* possesses the enzymes of the Embden–Meyerhof pathway, the hexosemonophosphate shunt and the tricarboxylic acid cycle (Donèche, 1989). It directly oxidizes glucose into gluconic acid. The latter, according to a process identical to the Entner–Doudouroff pathway, permits the young mycelium to synthesize substantial quantities of cellular material. However, when the fungus is partially deprived of oxygen, mycelial growth is low and the complete oxidation of glucose is accompanied by the liberation of glycerol in the environment.

Thus the initial fungal development under the grape skin is marked by considerable glycerol accumulation (Figure 10.28). When *B. cinerea* emerges on the outside of the grape and reaches its stationary phase, remarkably, it can no longer assimilate gluconic acid. This acid, which accumulates in the grape, is a characteristic secondary product of significant sugar degradation. The gluconic acid concentration depends on the duration of the external development of the fungus, varying from 5 to 10 mEq/berry.

The glycerol concentration of contaminated grapes also varies according to the duration of the respective internal and external fungal development phases. The glycerol concentration

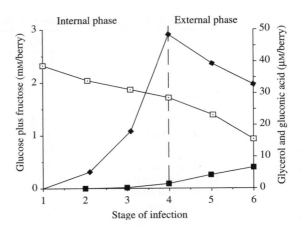

Fig. 10.28. Sugar assimilation and secondary product formation during a noble rot attack (Donèche, 1987): □, glucose + fructose; ◆, glycerol; ■, gluconic acid. Stages of infection: (1) healthy berry; (2) spotted berry (spot diameter less than 5 mm²); (3) spotted berry (spot diameter greater than 5 mm²); (4) fully rotted berry (completely spotted); (5) appearance of mycelial hyphae on the berry surface; (6) *pourri rôti*

is between 50 and 60 μmol per berry at the *pourri plein* stage. Despite the concentration phenomenon, only 10–40 μmol exist per berry at the *pourri rôti* stage. Part of the accumulated glycerol is oxidized by a glycerol dehydrogenase in the course of the external development phase of *B. cinerea*. Musts obtained from botrytized grapes normally contain 5 to 7 g/l of glycerol.

The concentration ratio of glycerol to gluconic acid represents the length of internal and external development phases of the parasite. It constitutes a noble rot quality index (Figure 10.29). In vintages with favorable climatic conditions (for example, 1981, 1982 and 1985), rapid grape desiccation from the *pourri plein* stage onwards leads to an elevated glycerol to gluconic acid ratio.

During a botrytis attack, other polyols (mannitol, erythritol and meso-inositol) are formed and their concentrations increase in the grape (Bertrand *et al.*, 1976). *B. cinerea* also produces a glucose polymer, which accumulates in contaminated grapes. Its concentration can attain 200 mg/l in must. This glucan is often at the root of subsequent wine clarification difficulties (Dubourdieu, 1978).

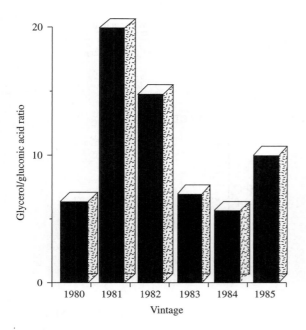

Fig. 10.29. Influence of vintage on noble rot quality, expressed by the glycerol/gluconic acid ratio (Donèche, 1987)

B. cinerea development is always accompanied by the degradation of the principal acids of the grape. This biological deacidification lowers initial grape acidity by 70% on average. Tartaric acid degradation is progressive in the course of the infection process, and it stimulates sugar assimilation. Malic acid is generally less degraded: it occurs especially at the end of a botrytis attack and corresponds to a strong energy demand in order to accumulate reserve substances in the developing conidia.

The other acids are less degraded. This sometimes leads to their increased concentrations in *pourris rôtis* grapes. Citric acid is an example and can be synthesized by certain *B. cinerea* strains but, in spite of the concentration phenomenon, its concentration rarely exceeds 8 mEq/l. Acetic acid behaves similarly.

Mucic acid has also been observed to accumulate. It is a product of the oxidation of galacturonic acid (Wurdig, 1976). This acid is capable of precipitating in wine in the form of calcium salts, but this phenomenon is rare and seems limited to northern vineyard wines. More generally, it also affects wines made from grapes that were insufficiently ripe when Botrytis set in.

B. cinerea development is also to the detriment of grape nitrogen compounds. The fungus degrades grape proteins and liberates the nitrogen in amino acids with the help of proteases and amino oxidases diffused in the grape. *B. cinerea* then assimilates this nitrogen and synthesizes the metabolic proteins necessary for its growth. The grape thus becomes rich in exocellular fungal proteins. Musts obtained from *pourris rôtis* grapes contain less ammonium and more complex forms of nitrogen than musts from healthy grapes.

Like many other fungi, *B. cinerea* produces an exocellular laccase: *p*-diphenol oxygen oxidoreductase (Dubernet *et al.*, 1977). This enzyme oxidizes numerous phenolic compounds. It is involved in the pathogenetic process and its synthesis is induced by two groups of substances. The first group comprises phenolic compounds (gallic and hydroxycinnamic acid), most likely toxic to the fungus. The second group consists of pectic cell wall substance degradation products (Marbach *et al.*, 1985). The fungus adapts the molecular structure of this exocellular laccase to the pH of the host tissue and the nature of the phenolic compounds present. The quantity of the enzyme produced is also regulated.

Laccase transforms the principal white grape phenolic compounds (cafeic and p-coumaric acids—both free forms and forms esterified by tartaric acid) into quinones (Salgues *et al.*, 1986). These quinones tend to polymerize, forming brown compounds. These compounds are most likely responsible for the characteristic chocolate color of *pourris pleins* grapes.

Towards the end of development, the fungus produces less laccase and this enzymatic activity tends to decrease at the *pourri rôti* stage. Botrytized grape musts are less sensitive to oxidation than supposed.

Aromatic substances are greatly modified during a botrytis attack. Glycosidases produced by *B. cinerea* hydrolyze the terpenic glycosides. The fungus oxidizes the free terpenic compounds, which seem to have fungicidal properties, into less

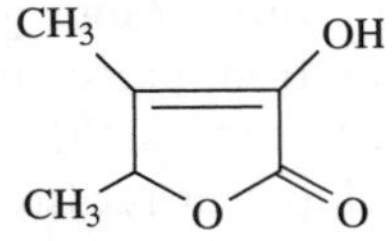

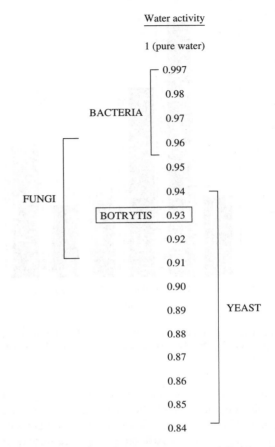

Fig. 10.30. Sotolon

odorous products (Bock *et al.*, 1986). Even though aldehydes are reduced to their corresponding alcohols, *B. cinerea* development is more characterized by the accumulation of furfural, benzaldehyde and phenylacetaldehyde (Kikushi *et al.*, 1983). According to Masuda *et al.* (1984), sotolon (hydroxy-3-dimethyl-4,5-2(5H) furanone) (Figure 10.30) is one of the principal compounds involved in the characteristic *rôti* aroma of botrytized grapes, but much research is still needed to discover the exact components of this specific aroma.

After a botrytis attack, grapes and must contain polyosides with phytotoxic and fungistatic activities. *B. cinerea* also produces divers antibiotic substances: botrytidial, norbotryal acetate and botrylactone. Some of these substances can be the source of fermentation difficulties.

10.6.5 Gray Rot and Other Kinds of Rot

Large quantities of a variety of epiphytic microflora (bacteria, yeasts, and fungal spores) are present on the grape skin surface. The development or reproduction of a given microorganism is, above all, determined by environmental conditions (temperature and free water).

When healthy grapes ripen under dry conditions, the low level of water activity on the skin surface promotes the proliferation of osmophilic microorganisms, especially yeasts (Rousseau and Donèche, 2001). Average water activity (from condensation or fog) is required for fungi to develop, while bacteria need large quantities of free water, generally from heavy rainfall, before they can multiply (Figure 10.31).

Furthermore, the various microorganisms interact (antagonism and competition for nutrients). Indeed, a number of yeast strains capable of

Fig. 10.31. Minimal values of water activity for the growth of epiphytic microorganisms in the grape

restricting the development of *B. cinerea* are used in organic disease control.

Noble rot is a regular phenomenon developing uniformly throughout the vineyard. Gray rot attacks, however, are usually very heterogeneous. Partially or totally infected grape clusters are often encountered on one plant whereas the grapes of the neighboring vinestock are totally untouched.

The gray rot infection process by *B. cinerea* is identical to the noble rot process previously described, but early fungal development is difficult to detect on red grapes. The external development of the fungus is certainly the most characteristic trait of gray rot. The conditions leading to the death of the fungus in the case of noble rot do not occur. A mycelial 'felt' forms on the surface of grapes. The contamination of neighboring grapes is facilitated by the intense biological activity of this mycelium. All viticultural factors increasing grape cluster compactness and maintaining a high amount of moisture on the grapes thus favor the spreading of the disease.

The chemical composition of grapes is greatly modified in the course of a gray rot attack. All of the intermediary products between noble rot and gray rot can be encountered.

B. cinerea consumes grape sugars while accumulating glycerol and gluconic acid. Contrary to noble rot, sugar concentration by grape dehydration remains low in comparison with sugar degradation (Figure 10.32). Consequently, the sugar concentrations of musts obtained from grapes infected by gray rot rarely exceed 230 g/l. The fungus also accumulates large quantities of gluconic acid during its external development phase (more than 10 μEq/berry or more than 3 g/l of must).

Malic and tartaric acid degradation is more significant than in the case of noble rot. Up to 90% of the initial concentrations present in healthy grapes can be degraded. The fungus also accumulates higher amounts of citric and acetic acid in the contaminated grape, but the acetic acid concentration rarely exceeds 5 μEq per berry.

The differences between the two kinds of rot are even more pronounced when considering phenolic

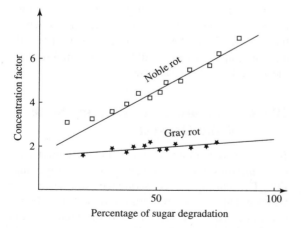

Fig. 10.32. Relationship between sugar degradation and berry dehydration according to the type of rot (Donèche, 1992)

compounds. These are much more oxidized by laccase, especially in red grapes whose skin is rich in phenolic substrates. Laccase activity increases as mycelium grows and indicates age of gray rot. The risk of color breakdown, known as oxidasic casse, is considerable when the must is exposed to air after crushing (Chapter 11).

In contrast to noble rot, which gives sweet white wines their specific qualitative aromas, gray rot often causes aromatic flaws. The grapes and wines obtained often are marked by characteristic mold or undergrowth odors. The responsible compounds are cuticular fatty acid (1-octen-3-one, 1-octen-3-ol) or terpenic compound (unidentified) derivatives formed during pellicular maceration by the mycelial biomass (Bock *et al.*, 1988).

Other fungi are often simultaneously present with *B. cinerea*. As a result, the rot takes on varied colors: black (*Aspergillus niger*), white, blue or green (*Penicillium* sp., *Cladosporium* sp.). These fungi develop less mycelial biomass than *B. cinerea*; glycerol and gluconic acid accumulation is less substantial. The contaminated grapes are often extremely bitter and possess aromatic flaws originating from amino acid and skin phenolic compound transformations. These compounds give to wines phenol and iodine odors (Ribéreau-Gayon, 1982). *Cladosporium* possesses a much higher laccase activity than *B. cinerea*; in addition, laccase synthesis by *B. cinerea* increases in the presence of *Aspergillus* or *Penicillium* (Kovac, 1983). All fungi have similar water requirements and the most decisive parameter in their selection is, certainly, temperature. For this reason, the toxinogenic strains of *Aspergillus* are most widespread in hot-climate vineyards.

Some grapes have a strong "damp earth" smell due to the accumulation of geosmin. This compound, derived from the biosynthesis of terpenoids, is formed by several strains of *Penicillium*, in the presence of *B. cinerea*. Research is in progress to identify the factors behind this phenomenon, which affects some vineyards on a regular basis.

A second category of microorganisms exists on the surface of grapes. Differing from fungi, this category generally does not possess cell wall hydrolysis enzymes and therefore cannot penetrate grapes with intact skins. This group is essentially made up of oxidative yeasts and acetic acid bacteria.

B. cinerea frequently exerts a powerful antagonistic influence and hinders the multiplication of these microorganisms, but in warm condition acetic acid bacteria proliferate, utilizing sweet juice that escapes from the fissures created by the emergence of *B. cinerea* at exterior of the grape. The evolution of the grape from the *pourri plein* stage onwards is thus different and leads to sour rot (*pourriture aigre*) (Figure 10.33) (Section 12.2.1).

These bacteria transform the glycerol, formed beforehand by *B. cinerea*, into dihydroxyacetone. Among these acetic acid bacteria, *Gluconobacter* species oxidize glucose with the help of membrane dehydrogenases. Sugar degradation is thus substantial and is accompanied by the accumulation of gluconic, keto-2- and keto-5-gluconic and diketo-2,5-gluconic acid in the grape. The production of these ketonic compounds substantially increases the combining potential of the must with sulfur dioxide.

The development of these acetic acid bacteria is also characterized by acetic acid production. The musts obtained from grapes infected with sour rot can contain more than 40 g of acetic acid and up to 25 g of gluconic acid per liter. Since these bacteria only slightly degrade grape acids, the musts obtained have extremely low pHs. This is the worst form of rot.

Yeasts may also be involved in grape contamination, either alone or associated with acetic acid bacteria. In fact, yeasts have been identified as responsible for certain acid rot attacks in Mediterranean vineyards. This disease is caused by oxidative yeast development (*Candida, Kloeckera, Hanseniaspora*). It is known that certain phytopathogenic strains are likely to cause lesions in plant tissue. Tartaric acid of the grape is not attacked. The formation of gluconic, acetic and galacturonic acid greatly increases acidity. These yeasts produce a small amount of ethanol and the must possesses high concentrations of ethyl acetate and ethanol.

Damaged grapes cannot avoid this alteration in development. Grape skin lesions can occur at any stage of development. The causes are diverse: bursting due to a rapid water flux, insect bites or skin degeneration due to overripeness.

Exceptionally, *B. cinerea* can be observed to develop exclusively, forming a crescent-shaped mycelial mass to obstruct the fissure (Donèche, 1992). Changes in the chemical constitution of the grape are thus also characteristic of a gray rot attack.

Usually, however, the sweet juice seeping out of the damaged grape favors the multiplication of oxidative yeast and acetic acid bacteria. These microorganisms are generally transported by the insects responsible for the lesions. The grape thus inevitably evolves towards vulgar rot.

Grapes infected with vulgar rot (sour or acid) cannot be used to make wine. Their presence in the vineyard must be detected as soon as possible and the grape clusters should be eliminated to limit the spreading of the disease.

10.6.6 Evaluating the Sanitary State of the Harvest

B. cinerea development, alone or associated with other microorganisms, lowers potential grape quality. The enological consequences are serious in wines made from altered grapes: oxidations, degradation of color and aromas, and fermentation and clarification difficulties. The objective measurement of the sanitary state of the harvest therefore presents an obvious interest.

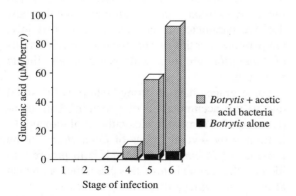

Fig. 10.33. Evolution of gluconic acid concentrations in grapes during development of *B. cinerea*, or of *B. cinerea* followed by development of acetic acid bacteria

For a long time, only visual evaluation methods were available to the vine-grower for judging the extent of a contamination. This technique takes only external fungal development on the berries into account. Yet, *B. cinerea* has already partially altered the grape before emerging on the surface of the grape. To complicate the matter further, the infection spots are difficult to see on the surface of a red grape.

At present, a grape crop selection criterion is based on monitoring the laccase activity within the grape, secreted early on by *B. cinerea*. Due to the natural presence of another oxidoreductase (tyrosinase) in healthy grapes, a specific substrate must be used for measuring the laccase activity in must. Two measurement methods exist.

The first method is based on a polarographic measure of the must oxygen consumption in the presence of a laccase-specific substrate. This method is not very sensitive and the elimination of must phenolic compounds is not indispensable. It has the advantage of taking into account all of the oxidasic activities likely to exist in must obtained from contaminated grapes, but this method may not be able to differentiate between slightly contaminated and perfectly healthy grapes. A machine based on this method has been developed which automates this analysis (Salgues *et al.*, 1984).

The other process includes a colorimetric measurement that makes use of syringaldazine, a laccase specific substrate (Harkin and Obst, 1973). This colorless orthodiphenol is stable with respect to chemical oxidation as well as the presence of tyrosinase and polyphenoloxidase, in healthy grapes. The quinone formed in the presence of laccase has an intense rose-mauve color (Figure 10.34). The speed at which it appears is measured by spectrophotometry (Dubourdieu *et al.*, 1984). The reaction must be carried out on a non-sulfited must. The phenolic compounds must also be eliminated by percolation on a polyvinylpolypyrolidone (PVPP) column to avoid their interference in the analysis. The results are expressed in laccase activity units per milliliter of must. A laccase activity unit is defined as the quantity of the enzyme capable of oxidizing a nanomole of syringaldazine per minute in analysis conditions. Manual analysis by this method is relatively quick (5–10 minutes, depending on the percolation time for the sample). It is simple to use since ready-to-use kits exist, including PVPP cartridges and ready-to-use reactants as well as a colored chart indicating the corresponding laccase units. A semi-quantitative determination is thus possible in the winery without using colorimetry. An automated analyzer also exists based on the same

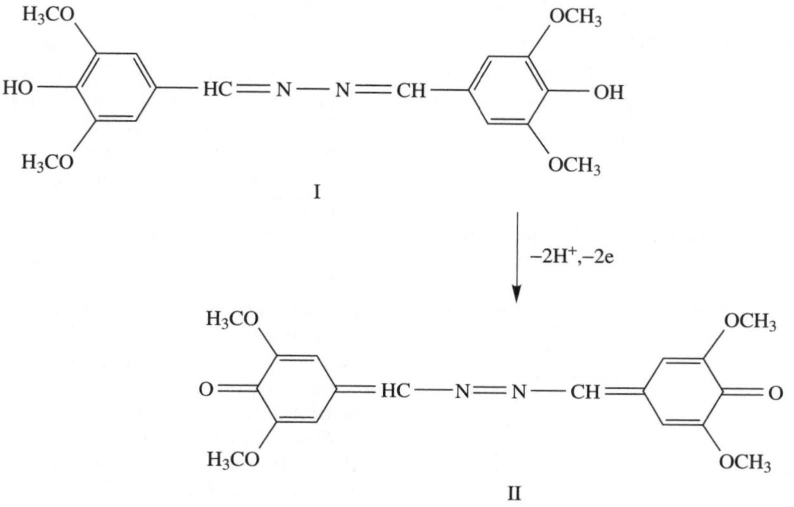

Fig. 10.34. Oxidation reaction of syringaldazine by laccase

principle and adapted to large-volume operations. The results, obtained in 2 minutes, are given in laccase units, as with the manual colorimetric method.

A strong correlation has been demonstrated between visually determined grape contamination levels, and laccase activity (Table 10.12), but Redl and Kobler (1992) emphasized that the laccase concentration does not permit the estimation of the total phenolic compound decomposition level of rot contaminated grapes. Notably, there is no correlation between the laccase activity and the total phenolic compound index, determined by spectrophotometry at 280 nm.

Cagnieul and Majarian (1991) developed an immunological method for detecting *B. cinerea*. Polyclonal antibodies are used which recognize the presence of specific polyosides, secreted by the fungus. Other microorganisms present on the grape, such as *Aspergillus, Penicillium, Cladosporium*, acetic acid bacteria and oxidative yeasts, do not interfere with this immunological test. Thanks to its sensitivity, the fungus can be detected 20–30 days before the harvest, providing adequate time for applying fungicides. This method is also capable of differentiating between wines made from botrytized grapes and healthy grapes at any given stage of the winemaking process (Fregoni *et al.*, 1993).

In future, Fourier transform infrared spectrophotometry will be an invaluable tool for assessing the condition of harvested grapes. This method is already capable of detecting the presence of rotten grapes, but there is no close correlation with the measurement of laccase activity.

Table 10.12. Relationship between the level of rot determined by visual inspection and laccase activity measured by the oxidation of syringaldazine (Redl and Kobler, 1992)

Level of rot (%)	Laccase activity (units/ml)
<1	0.39
1–5	0.78
6–10	2.25
11–25	6.56
26–50	8.12
51–100	15.86

10.7 CONCLUSION

In conclusion, progress in vine-growing and disease prevention has greatly improved wine quality, not only by diminishing wine flaws but also through permitting the harvest to be delayed until optimal maturity.

New risks have arisen from this progress. Improved vineyard practices can result in excessive plant vigor. Above a certain level, increased vegetative growth is always detrimental to the biochemical maturation processes. The start of these processes is at least delayed and thus may occur in unfavorable climatic conditions. Excess production is another risk: diluted grapes are obtained, producing wines with little structure, color and aroma.

The grape is more than a reserve stocking organ. At harvest time, it still possesses an intense metabolic activity. Particular attention should therefore be given to grape handling, especially when a percentage of the grapes are rot infected. In this case, the grapes contain many additional enzymes of fungal origin.

REFERENCES

Abbal P., Boulet J.C. and Moutounet M. (1992). *J. Intern. Sci. Vigne Vin*, 26, 231.

Arnold R.A. and Bledsoe A.M. (1990). *Am. J. Enol. Vitic.*, 41, 74.

Augustin M.M. (1986). Etude de l'influence de certains facteurs sur les composés phénoliques du raisin et du vin. Doctorat d'Université, Bordeaux.

Bailey L.H. (1988). *The Evolution of Our Native Fruits*. MacMillan Co., New York.

Barbeau G., Asselin C. and Morlat R. (1998). *Bull. OIV*, 805–806, p. 247.

Bayonove C. (1993). In *Les acquisitions Récentes en Chromatographie du Vin* (ed. B. Donèche). Tec & Doc Lavoisier, Paris.

Bayonove C., Richard H. and Cordonnier R. (1976). *CR Acad. Sci., Série C*, 283, 549.

Becker N. and Zimmermann A. (1984). *Bull. OIV*, 641–642, p. 584.

Bertrand A., Pissard R., Sarre C. and Sapis J.C. (1976). *Conn. Vigne Vin*, 10, 427.

Bessis R. and Fournioux J.C. (1992). *Vitis*, 31, 9.

Bisson J. and Ribéreau-Gayon P. (1978). *Ann. Technol. Agric.*, 27, 827.

Blaich R., Stein U. and Wind R. (1984). *Vitis*, 23, 242.

Blouin J. and Guimberteau G. (2000), *Maturation et maturité du raisin*, Féret Ed., Bordeaux.

Blouin J. (1992). *Techniques d'analyses des moûts et des vins*. Dujardin-Salleron, Paris.

Bock G., Benda I. and Schreier P. (1986). *J. Food Sci.*, 15, 659.

Bock G., Benda I. and Schreier P. (1988). *Z. Lebensmittel*, 186, 33.

Bonnet A. (1903). *Ann. Ec. Nat. Agri. Montpellier*, 3, 58.

Boulton R.B., Singleton V.L., Bisson L.F. and Kun-Kee R.E. (eds) (1995). Chapman & Hall Enology Library. Chapman & Hall, New York.

Branas J., Bernon G. and Levadoux L. (1946). *Eléments de viticulture générale*. Published by the authors, printed by Delmas, Bordeaux.

Bravdo B., Hepner Y., Loinger C., Cohen S. and Tabacaman H. (1985). *Am. J. Enol. Vitic.*, 36, 125.

Cagnieul P. and Majarian W. (1991). In *Comptes Rendus de la 3ème Conférence Internationale sur les maladies des plantes* (Bordeaux, 1991), ANPP Ed., Paris.

Calo A., Costacurta A., Tomasi D., Becker N., Bourquin H.D., de Villiers F.S., Garcia de Lujan A., Huglin P., Jaquinet L. and Lemaitre C. (1992) *Riv. Vitic. Enol.*, 3, 3.

Calo A., Tomasi D., Cravero M.C. and Di Stefano R. (1994). *Riv. Vitic. Enol.*, 3, 13.

Candolfi-Vasconcelos M.C. and Koblet W. (1990). *Vitis*, 29, 199.

Carbonneau A. (1982). *Prog. Agric. Vitic.*, 99, 290.

Carbonneau A. (1985). In *Proceedings of the International Symposium on cool climate Viticulture and Enology* (eds D.A. Heatherbell, P.B. Lombard, F.W. Bodyfelt and S.F. Pryce) Oregon University Experiment Station Technical Publication No. 7628, Eugene.

Champagnol F. (1984). *Eléments de physiologie de la Vigne et de Viticulture Générale*. Published by the author, Montpellier, France.

Champagnol F. (1986). *Prog. Agric. Vitic.*, 103, 361.

Chardonnet C. and Donèche B. (1995). *Vitis*, 34, 95.

Chardonnet C., Gomez H. and Donèche B. (1994). *Vitis*, 33, 69.

Chone X., Tregoat O. and Van Leeuwen C. (2001a), *J. Int. Sci. Vigne Vin*, H.S., 47.

Chone X., Van Leeuwen C., Dubourdieu D. and Gaudillère J.P. (2001b), *Ann. Botany*, 87 (4), 447.

Chone X., Tregoat O., Van Leeuwen C. and Dubourdieu D. (2000), *J. Int. Sci. Vigne Vin*, 34 (4), 169.

Colin L., Cholet C. and Geny L. (2002), *Aust. J. Grape and Wine Res.*, 8, 101.

Conn E.E. (1986). *Recent Advances in Phytochemistry*, Vol. 20: *The Shikimic Acid Pathway*. Plenum Press, New York.

Coombe B.G. (1989). *Acta Hortic.*, 239, 149.

Crippen D.D. and Morrison J.C. (1986). *Am. J. Enol. Vitic.*, 37, 1986.

Darne G. (1991). Recherches sur la composition en anthocyanes des grappes et des feuilles de vignes. Thèse Doctorat ès Sciences, Bordeaux.

Darriet Ph. (1993). Recherches sur l'arôme et les précurseurs d'arôme du Sauvignon. Applications technologiques. Thèse Doctorat Université, Bordeaux.

Di Stefano R., Corina L. and Bosia P.D. (1983). *Riv. Vitic. Enol.*, 36, 263.

Donèche B. (1976). Effets du mancozèbe sur la microflore des sols de vignobles et participation des microorganismes ... sa dégradation. Thèse Doctorat de 3ème cycle, Université de Bordeaux II.

Donèche B. (1986). *Agronomie*, 6, 67.

Donèche B. (1987). Etude biochimique de la relation hôte–parasite dans le cas du raisin et de *Botrytis cinerea*. Thèse Doctorat ès Sciences, Bordeaux.

Donèche B. (1989). *Can. J. Bot.*, 67, 2888.

Donèche B. (1992). In *Wine Microbiology and Biotechnology* (ed. G.H. Fleet). Harwood Academic Publishers, Chur, Switzerland.

Donèche B. and Chardonnet C. (1992). *Vitis*, 31, 175.

Dubernet M., Dubernet M., Coulomb S., Lerch M. and Traineau I. (2000), *Rev. Fr. Oenol.*, 185, 18.

Dubernet M., Ribéreau-Gayon P., Lerner H.R., Harel E. and Mayer A.M. (1977). *Phytochem.*, 16, 191.

Dubourdieu D. (1978). Etude des polysaccharides sécrétés par *Botrytis cinerea* dans la baie de raisin. Incidence sur les difficultés de clarification des vins de vendanges pourries. Thèse Doctorat de 3ème cycle, Bordeaux.

Dubourdieu D., Grassin C., Deruche C. and Ribéreau-Gayon P. (1984). *Conn. Vigne Vin*, 18, 4.

Du Plessis C.S. and Augustyn O.P.H. (1981). *S. Afr. J. Enol. Vitic.*, 2, 101.

Duteau J. (1990), In *Actualités œnologiques 89*, Dunod Ed., Paris.

Duteau J., Guilloux M. and Seguin G. (1981), *Conn. Vigne Vin*, 15 (3), 1.

Düring H. (1994). *Am. J. Enol. Vitic.*, 45, 297.

Dry P., Loveys B., McCarthy M. and Stoll M. (2001), *J. Int. Sci. Vigne Vin*, 35 (3), 129.

Ewart A.J.W. (1987). In *Proceedings of Sixth Australian Wine Industry Conference*, Adelaide (ed. T.H. Lee).

Flanzy C. (2000), *Œnologie: fondements scientifiques et technologiques*, Tec et Doc Ed., Lavoisier, Paris.

Fregoni M., Perino A. and Vercesi A. (1993). *Bull. OIV*, 745–746, 169.

Galet P. (1977). Les maladies et les parasites de la vigne. Paysan du Midi, Montpellier, France.

Gaudillère J.P., Van Leeuwen C. and Ollat N. (2002), *J. Exp. Bot.*, 53 (369), 757.

Gerber C. (1898). *Ann. Sci. Nat. Bot.*, 4, 1.

Guilloux M. (1981). Evolution des composés phénoliques de la grappe pendant la maturation du raisin. Influence des facteurs naturels. Thèse Doctorat Université, Bordeaux.

Harkin J.K. and Obst J.R. (1973). *Experientia*, 29, 381.

Harris J.M., Kriedemann P.E. and Possingham J.V. (1971). *Vitis*, 9, 291.

Herrick I.W. and Nagel C.W. (1985). *Am. J. Enol. Vitic.*, 36, 95.

Hrazdina G., Parsons G.F. and Mattick L.R. (1984). *Am. J. Enol. Vitic.*, 35, 220.

Huglin P. (1978). *CR Acad. Agr. Fr.*, 64, 1117.

Huglin P. (1986). *Biologie et Écologie de la Vigne.* Payot, Lausanne.

Ito H., Motomura Y., Konno Y. and Hatayama T. (1969). *Tohoku J. Agric. Res.*, 20, 1.

Jackson D.I. (1987). *Vinifera Winegrowers J.*, 14, 144.

Jeandet P. and Bessis R. (1989). *Bull. OIV*, 703–704, 637.

Jeandet P., Bessis R. and Gautheron B. (1991). *Am. J. Enol. Vitic.*, 42, 41.

Kanellis A.K. and Roubelakis-Angelakis K.A. (1993). in *Biochemistry of Fruit Ripening* (eds G.B. Seymour, J.E. Taylor and G.A. Tucker). Chapman & Hall, London.

Karadimtcheva B. (1982). *Bull. OIV*, 613, 246.

Kikushi T., Kadota S., Suehara H., Nishi A., Tsubaki K., Yano H. and Harimaya K. (1983). *Chem. Pharm. Bull.*, 31, 659.

Kliewer W.M. and Torres R.E. (1972). *Am. J. Enol. Vitic.*, 23, 71.

Kliewer W.M. and Weaver R.J. (1971). *Am. J. Enol. Vitic.*, 22, 172.

Koblet W. (1975). *Wein-Wiss*, 30, 241.

Kovac V. (1983). *Bull. OIV*, 628, 420.

Lacey M.J., Allen M.S., Harris R.L.N. and Brown W.V. (1991). *Am. J. Enol. Vitic.*, 42, 103.

Langcake P. and Pryce R.J. (1977). *Phytochem.*, 16, 1193.

Lavee S. and Nir G. (1986). in *CRC Handbook of Fruit Set and Development* (ed. S.P. Monselix). CRC Press, Boca Raton, FL.

MacCarthy M.G. and Coombe B.G. (1984). *Acta Horticulture*, 171, 447.

Marbach I., Harel E. and Mayer A.M. (1985). *Phytochem.*, 24, 2559.

Masuda M., Okawa E., Nishimura K. and Yunome H. (1984). *Biological Chem.*, 48, 2707.

Matthews M.A. and Anderson M.M. (1989). *Am. J. Enol. Vitic.*, 40, 52.

Morlat R. (1989). *Le terroir viticole: contribution à l'étude de sa caractérisation et de son influence sur les vins. Application aux vignobles rouges de la moyenne vallée de la Loire*, Thèse Doctorat ès Sciences, Université Bordeaux 2.

Müller-Thurgau H. (1888). *Landwirt. Jarbücher*, 17, 83.

Ough C.S., Stevens S.D. and Almy J. (1989). *Am. J. Enol. Vitic.*, 40, 219.

Palejwala V.A., Parikh H.R. and Modi V.V. (1985). *Physiol. Plant*, 65, 498.

Petit-Lafitte A. (1868). *La vigne dans le Bordelais* (ed. J. Rothschild) Paris.

Peynaud E. and Maurié A. (1953). *Ann. Technol. Agr.*, 2, 12.

Peynaud E. and Ribéreau-Gayon P. (1971). in *The Biochemistry of Fruits and their Products*, Vol. 2 (ed. A.C. Hulme) Academic Press, London.

Pezet R. and Pont V. (1986). *Rev. Suisse Vitic. Arboric. Hortic.*, 18, 317.

Pope J.M., Jonas D. and Walker R.R. (1993). *Protoplasma*, 173, 177.

Possingham J.V. and Grogt Obbink J. (1971). *Vitis*, 10, 120.

Possner D.R.E. and Kliewer W.M. (1985). *Vitis*, 24, 229.

Pouget R. and Delas J. (1989). *Conn. Vigne Vin*, HS, 27.

Pucheu-Planté B. and Leclair P. (1990). in *Comptes Rendus du 4ème Symposium International d'Œnologie (Bordeaux, 1989)*, Dunod, Bordas, Paris.

Rapp A. (1993). in *Les Acquisitions récentes en chromatographie du vin* (ed. B. Donèche) Tec & Doc Lavoisier, Paris.

Razungles A. (1985). Contribution á l'étude des caroténodes du raisin: teneur et localisation dans la baie, évolution au cours de la maturation. Thèse Ecole Nationale Supérieure Agronomique, Montpellier, France.

Redl H. and Kobler A. (1992). *Mitteilungen Klosterneuburg*, 42, 25.

Reynolds A.G. (1988). *Hort. Science*, 23, 728.

Ribéreau-Gayon G. (1960). *Vitis*, 2, 113.

Ribéreau-Gayon J., Peynaud E., Ribéreau-Gayon P. and Sudraud P. (1975). in *Traité d'Œnologie: Sciences et Techniques du vin*. Vol. 2: *Caractères des vins, maturation du raisin, levures et bactéries*. Dunod, Paris.

Ribéreau-Gayon P. (1953). *CR Acad. Agri.*, 39, 800–807.

Ribéreau-Gayon P. (1982). *Bull. OEPP*, 12, 201.

Riou C. and Lebon E. (2000), *Bull. OIV*, 73, 837–838, 755.

Robinson S.P. and Davies C. (2000), *Aust. J. Grape and Wine Res.*, 6, 175.

Romieu Ch., Robin J.P., Nicol M.Z. and Flanzy Cl. (1989). *Conn. Vigne Vin*, 23, 165.

Roubelakis-Angelakis K.A. (1991). in *Proceedings of International Symposium Nitrogen in Grape and Wine*, Seattle, WA.

Roubelakis-Angelakis K.A. (2001), *Molecular Biology and Biotechnology of the Grapevine*, Kluwer Academic Publishers, Dordrecht.

Roubelakis-Angelakis K.A. and Kliewer W.M. (1992). *Hort. Rev.*, 14, 407.

Roujou de Boubée D. (2000), *Recherches sur la 2-méthoxy-3-isobutylpyrazine dans les raisins et les vins. Approches analytique, biologique et agronomique*. Thèse de Doctorat, Université de Bordeaux 2.

Rousseau S. and Donèche B. (2001), *Vitis*, 40 (2), 75.

Ruffner H.P. (1982a). *Vitis*, 21, 247.

Ruffner H.P. (1982b). *Vitis*, 21, 346.

Ruffner H.P., Brem S. and Rast D.M. (1983). *Plant Physiol.*, 73, 582.

Ruffner H.P., Possner D., Brem S. and Rast D.M. (1984). *Planta*, 160, 444.

Ruffner H.P., Hürlimann M. and Skrivan R. (1995). *Plant Physiol. Biochem.*, 33, 25.

Salgues M., Olivieri C., Chabas M. and Pineau J. (1984). *Bull. OIV*, 638, 308.

Salgues M., Cheynier V., Gunata Z. and Wylde R. (1986). *J. Food Sci.*, 55, 1191.

Schaeffer A. (1985). in *Proceedings of the International Symposium on Cool Climate Viticulture and Enology* (eds D.A. Heatherbell, P.B. Lombard, F.W. Bodyfelt and S.F. Price). Oregon State University Experiment Station Technical Publication No. 7628, Eugene.

Schaller K.O. and Löhnertz O. (1992). *Vitic. Enol. Sci.*, 47, 202.

Schaller K.O., Löhnertz O. and Chikkasubbanna V. (1992). *Vitic. Enol. Sci.*, 47, 36.

Schreier P., Drawert F. and Junker A. (1976). *J. Agric. Food Chem.*, 24, 331.

Seguin G. (1970), *Les sols de vignobles du Haut-Médoc. Influence sur l'alimentation en eau de la vigne et sur la maturation du raisin*, Thèse Doctorat ès Sciences, Université Bordeaux 2.

Seguin G. (1971). *Conn. Vigne Vin.*, 3, 293.

Seguin G. (1975), *Conn. Vigne Vin.*, 9 (1), 23.

Seguin G. (1983). *Bull. OIV*, 623, 3.

Sepulveda G. and Kliewer W.M. (1986). *Am. J. Enol. Vitic.*, 37, 20.

Silacci M.W. and Morrison J.C. (1990). *Am. J. Enol. Vitic.*, 41, 111.

Smart R.E. (1973). *Am. J. Enol. Vitic.*, 24, 141.

Smart R.E. and Coombe B.G. (1983). in *Water Deficits and Plant Growth*, Vol. VII: *Additional woody crop plants* (ed. T.T. Kozlowski). Academic Press, New York.

Stern D.J., Lee A., McFadden W.H. and Stevens K.L. (1967). *J. Agric. Food Chem.*, 15, 1100.

Stoll M., Loveys B. and Dry P. (2001), *J. Exp. Bot.*, 51, 1627.

Tesic D., Wolley D.J., Hewett E.W. and Martin D.J. (2001), *Aus. J. Grape Wine Res.*, 8, 15.

Ureta C.F. and Yavar O.L. (1982). *Conn. Vigne Vin*, 16, 187.

Ureta C.F., Boidron J.N. and Bouard J. (1981). *Am. J. Enol. Vitic.*, 32, 90.

Van Leeuwen C. (2001), *J. Int. Sci. Vigne Vin*, H.S., 97.

Van Leeuwen C. and Seguin G. (1994), *J. Int. Sci. Vigne Vin*, 28 (2), 81.

Van Leeuwen C., Chone X., Tregoat O. and Gaudillere J.P. (2001a), *The Australian Grapegrower and Winemaker*, 449, 18.

Van Leeuwen C., Gaudillere J.P. and Tregoat O. (2001b), *J. Int. Sci. Vigne Vin*, 35 (4), 195.

Wicks A.S. and Kliewer W.M. (1983). *Am. J. Enol. Vitic.*, 34, 114.

Winkler A.J. (1962). *General Viticulture*. University of California Press, Berkeley, CA.

Wurdig G. (1976). *Weinwissenschaft*, 112, 16.

11

Harvest and Pre-Fermentation Treatments

11.1 INTRODUCTION

The definition of maturity, the biochemical transformations of grapes during maturation and related subjects have been described in Chapter 10. Grape maturity varies as a result of many parameters and is not a precise physiological state.

In certain conditions (for example, dry white winemaking in warm climates), grapes are sometimes harvested before complete maturity. In other conditions (in temperate climates, for example), the natural biochemical phenomena may be prolonged when unfavorable climatic conditions have

disrupted normal maturation kinetics, and this has become a tradition in certain northern vineyards ('late harvest', *beerenauslese*, etc.). In other regions, *Botrytis cinerea* in the noble rot form causes overripening (Sections 10.5 and 14.2.2).

All on-vine overripening methods increase the ratio of sugar to acid. Grapes accumulate sugar while breaking down malic and/or tartaric acid. In all cases, this natural drying process lowers crop volume due to water loss.

Similar results are sought by exposing picked grapes to sunlight or storing them in ventilated buildings. These techniques are used in grape-growing regions as different from one another as Jerez (Spain) and Jura (France).

Must quality can also be improved after the harvest but these wine adjustments, whether physical or chemical, should not be made simply to compensate for basic viticultural inadequacies. These various processes are strictly regulated, to avoid potential abuse and to avoid their becoming standard practices with the objective of replacing the work of nature.

The grape remains a living organism after it is picked. Many enzymes maintain sufficient activity to ensure various biochemical processes. Enzymatic activity is regulated by cellular compartmentation, which continuously limits available substrates. Enzymatic activity is higher in rot infected grapes, in which case grapes should be maintained intact for as long as possible. Great care should be given to their harvest and transport.

Pre-fermentation practices at the winery destroy the structure of grape cells. Enzymes are placed in direct contact with abundant substrates, resulting in explosive enzymatic reactions. Laccase activity is the best-known reaction: secreted by *B. cinerea*, it alters phenolic compounds in the presence of a sufficient amount of oxygen. Not all of the enzymes present are harmful to quality—pectinolytic enzymes, for example, favor must clarification by parietal constituent hydrolysis. For several years, enologists have sought to amplify these favorable reactions in using more active, industrially produced enzymes.

Thanks to an increased fundamental understanding of grape constituents and their enzymatic transformation mechanisms, manufacturers have greatly improved equipment design, treatments and technological processes. Many methods for maintaining, increasing and, if necessary, correcting grape quality are currently available to the enologist. Enology will likely bring about other improvements in future years, notably with respect to phenolic compounds and aromatic substances.

Nevertheless, two significant constraints persist. International trade imposes increasingly strict regulations, encouraged by the consumer's desire for natural products. Also, certain technological methods require costly equipment and this level of investment is restricted to large wineries.

A bridge between viticultural practices and winemaking methods, pre-fermentation treatments demand great care from the enologist.

11.2 IMPROVING GRAPE QUALITY BY OVERRIPENING

Overripening is a natural prolongation of the maturation process, but differs from maturation on a physiological level. The maturing of stalk vascular tissues progressively isolates the grapes from the rest of the plant. As a result, crop volume generally diminishes since evaporative water loss is no longer compensated for by an influx from the roots. Overripening is also characterized by an increase in fermentative metabolism and alcohol dehydrogenase activity (Terrier *et al.*, 1996). Noble rot is also a process which ameliorates grapes by overripeness; it is described in Sections 10.6 and 14.2.2.

11.2.1 On-Vine Grape Drying

The grapes are left on the vine for as long as possible with this natural drying method—sometimes even after grape cluster peduncle twist. The berries progressively shrivel, losing their water composition. They produce a naturally concentrated must, richer in sugar and aromatic substances. Acidity does not increase in the same proportions and can even decrease by malic acid oxidation. Other biochemical maturation phenomena also occur; notably, the skin cell walls deteriorate. This methods should therefore be used only with relatively thick-skinned varieties to limit the risk of *B. cinerea* development.

11.2.2 Off-Vine Grape Drying

In certain regions, this method can be limited to simply exposing grapes to sunlight for a variable length of time. In the Jerez region, Pedro Ximenez variety grapes are exposed to the sun on straw mats for 10–20 days before pressing (Reader and Dominguez, 1995). The grapes are turned over regularly and covered at night to protect them from moisture. During this process, locally called *soleo*, must density regularly attains 1.190 to 1.210 but sometimes exceeds 1.235. The juice yield is very low (250–300 l/tonne) and only vertical hydraulic presses are capable of extraction. The resulting juice is extremely viscous and dark, with pronounced grape aromas. The Pedro Ximenez variety is particularly rich in organic acids. The heat of the Andalusian sun provokes the formation of a significant quantity (50–75 mg/l) of hydroxymethylfurfural from fructose.

In this same region of Spain, as well as in many other Mediterranean vineyards (Greece, Cyprus, Italy, Turkey, etc.), this sun-drying method is applied to muscat grape varieties (Alexandria Muscat, for example). More than simply concentrating grape sugar, sun-drying in particular increases the typical aroma of the must. These musts attain high free and odorous terpenic alcohol concentrations.

In the Jura region, healthy grapes are sorted on the vine and the different varieties are judiciously gathered (in particular, Savagnin). The selected grape clusters are hung to dry, spread out on wooden grids covered with straw or suspended on wires in strongly ventilated storage rooms. This drying method results in extensive waste, not only by desiccation but also by rot. Contaminated grapes are removed regularly. This operation lasts 2–4 months. The grapes are generally pressed after Christmas, producing musts containing 310–350 g of sugar per liter (legal minimum = 306 g/l) with a higher than normal volatile acidity. The yield is approximately 250 l/tonne.

11.2.3 Artificial Overripening

Natural grape drying is a difficult operation to master, due especially to the risks of rot-induced grape alteration. Since the beginning of the 20th century, enologists have been trying to replace this natural process with an adapted technology. The principles of an industrial overripener are simple. The equipment consists of circulating hot and dry air over the grapes, which are placed in small boxes inside the heated compartment. The ventilation system circulates 2500–5000 m^3 of dry air (below

15% relative humidity) per hour, at a temperature varying from 25 to 35°C.

According to recent experiments which confirm earlier tests (Ribéreau-Gayon *et al.*, 1976), the apparatus reduces the grape crop mass by 10–15% in 8–15 hours and increases the alcohol strength by 1.5% volume in potential alcohol. The decrease in acidity, by the oxidative degradation of malic acid, varies according to air temperature.

Other biochemical phenomena probably accompany this artificial overripening. Wines obtained from these treated grapes are richer in color and tannins and are always preferred at tastings. This treatment is exclusively for red winemaking, since the resulting increased phenolic compound concentrations are detrimental to quality white winemaking.

Equipment costs and utilization constraints limit the use of this technique, but its effectiveness is proven.

11.3 HARVEST DATE AND OPERATIONS

First and foremost, the grapes should be protected from attacks and contaminations such as Eudemis and rot, right up to the harvest.

Optimal enological maturity depends on grape variety, environmental conditions and wine type (Section 10.4.1). Thus a perfect knowledge of *vèraison* conditions and half-*vèraison* dates will permit the vinegrower to organize the harvest according to the various maturity periods. Maturity analysis monitoring complements this information (Sections 10.4.2 and 10.4.3).

The grape crop should be harvested under favorable climatic conditions. After rainfall, the grape clusters retain water that is likely to dilute the must. The grapes should therefore be allowed to drain, at least partially. The most recent research measuring water activity on the surface of grapes has shown that it needs to remain there for at least two hours. Morning fog can also cause must dilution. Harvesting should begin after the sun has dried the vines, but at the same time the prolonged maceration of harvested grapes in the juice of the inevitably burst grapes during the warmest hours of the day should be avoided.

Mechanical harvesting facilitates the realization of the above recommendations. Progress in harvest machine technology has helped to avoid berry alteration and excessive vegetal debris. Thanks to its speed and ease of use, the harvester permits a rapid harvest of grapes at their optimal quality level and at the most favorable moment. Manual grape-picking can be even more selective and qualitative, but its cost is not justifiable in all wineries.

Whatever the harvest method, the vinegrower's principal concern should be the maintenance of grape quality.

11.3.1 Grape Harvest

From ancient times to recent years, harvest methods have barely evolved—other than slight improvements in tools for grape cutting and gathering.

In certain appellations (Champagne, for example) and vineyards, quality concerns prohibit mechanized harvesting. In noble rot (Section 10.5) vineyards, it cannot be implemented because the harvester is not capable of selecting grapes that have reached the proper stage of noble rot. Everywhere else, since the beginning of the 1970s, mechanical harvesting has undergone spectacular development as a result of increased production costs and the disappearance of manual labor (Vromandt, 1989).

Lateral or horizontal strike harvesting techniques are easily adapted to traditional vine training methods. The grapes are shaken loose by two banks of flexible rods which straddle the vine row. The banks of rods transmit an alternating transversal oscillation to the vines (Figure 11.1). This movement transmits a succession of accelerations and decelerations to the grape clusters which results in individual grapes, partial grape clusters or entire grape clusters falling. The adjustment of these machines is complicated and requires a complete mastery of the technique. The number, position and angle of the rods or rails in the banks must be chosen in accordance with the training and pruning

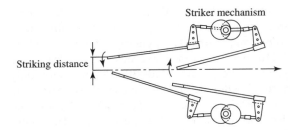

Fig. 11.1. Principle of lateral strike harvesters (Jacquet, 1983)

system used. Finally, the striking frequency must be adapted to the forward speed of the harvester. The frequency is adjustable from 0 to 600 strikes per minute on conventional harvesters; some more recent models attain up to 1400 strikes per minute.

Harvesters should be adjusted to allow for not only the variety (the ease of grape dislodgement depends on the variety) but also the pruning method employed and the canopy density at the time of the harvest. For example, a harvester with an insufficient striking frequency in thick foliage does not harvest the entire grape crop. On the contrary, an excessive oscillation amplitude and striking speed will transmit a lot of kinetic energy to the rod banks, so that the rod-strikes burst the poorly protected berries.

New striking methods are responsible for most of the recent improvements in mechanical harvesting. On conventional harvesters, the rod ends are not attached; thus their inertia is not controlled. In more recent models, manufacturers have eliminated this free rod end. Their solutions vary with respect to the various machines (Figure 11.2). In some cases, two rod ends are connected by an articulated link and these flexible rails strike by bending. In other cases, striker shafts at the front and back of the machine oscillate semi-rigid rods at different amplitudes.

The dislodged grapes are gathered on an impermeable mobile surface made up of a series of overlapping plastic elements: in general, two lines of plastic elements shaped like fish-scales. These elements yield to vinestocks and trellis posts by moving on their rotation axis. Lateral grill or perforated belt transporters then drain the grapes as

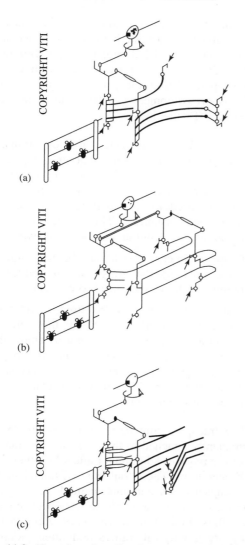

Fig. 11.2. New lateral strike harvester models (Vromandt, 1989): (a) flexible striker; (b) dual-drive rigid striker; (c) differential rod bank striker

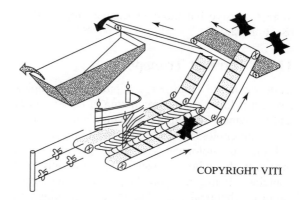

Fig. 11.3. Lateral strike harvester (Vromandt, 1989)

The MOG should be eliminated as quickly as possible, preferably before becoming covered in juice. It is usually removed by blowers. Extractor fans placed above the conveyor belt are effective. When these extractors are properly adjusted and combined with destemming screens, they are capable of reducing the rate of miscellaneous rubbish to 0.5%. The crop is generally stored temporarily in one or two hoppers with a capacity of 8–20 hl. These hoppers are capable of dumping the crop directly into the transport containers.

Ribéreau-Gayon *et al.* (1976) had already indicated that rigorous comparative studies between manual and mechanical harvest quality were practically impossible. The two methods would have to be examined on a sufficiently large and homogeneous parcel, and the harvest reception and winemaking equipment would have to be capable of identically handling a large immediate grape supply from mechanical harvesting or a progressive grape supply from manual harvesting.

An experienced enologist is capable of examining the grape crop visually to compare mechanical harvest and manual harvest quality, including the proportion of bursted or rotten grapes and the presence of leaves and leaf-stalks (intact or lacerated).

Recent research has confirmed those initial observations (Clary *et al.*, 1990). Prudent harvesting with a correctly adjusted harvester produces similar results to classic manual harvesting. Manual harvesting, however, continues to permit

they are conveyed. Upon reaching the extremity of the machine, the grapes are transferred to a shelf or bucket elevator (Figure 11.3). This type of machine permits high harvest speeds, to the detriment of crop quality. Juice losses may represent up to 10% of total weight.

Striker mechanisms inevitably entrain leaves, leaf fragments and other vine parts (MOG— material other than grapes) with the grape clusters.

more extensive but more expensive grape sorting (Section 12.2.3).

11.3.2 Harvest Transport

Choosing harvest transport equipment is a complex issue. It is linked to the organizing of harvest work and the winery reception installation and is subject to certain enological and economical constraints.

From an enological viewpoint, grapes should arrive at the winery intact. More precisely, the container should transport the grapes in the physical or biochemical state obtained after picking and transfer them to the reception bin. The exaggerated bruising and crushing of grapes can be avoided by:

- using shallow transport containers (not exceeding a depth of 0.8 m);

- using easily cleaned material to ensure proper hygiene;

- limiting the number of grape transfers and the load and dumping height.

Vine-growers in certain viticultural regions (Champagne, for example) are required to follow these strict rules. Small perforated containers, preferably plastic, are used for grape-picking to ensure grape quality from the first step of the harvest. These containers are stacked on an open trailer and gently emptied at the winery.

Mechanical harvesting produces a different grape supply rate compared with manual picking:

- Hourly crop volume is considerable (from 4 to 10 tonnes per hour) and the daily duration is often 12 hours.

- The harvest is partially destemmed and crushed—sometimes with a lot of juice (10–30% of total juice).

- The harvest is full of MOG (leaves, leaf-stalks, shoot fragments) and sometimes small animals.

Mechanical harvest transport does not follow the same rules. The harvested grapes should be brought rapidly to the winery after the juice has been separated from the solid parts to the extent possible. Sulfur dioxide should not be added to the unseparated harvest; it favors the maceration of the solid material during transport. Similarly, carbon dioxide only protects must from oxygen when separated from the grapes (Jacquet, 1995).

Transport equipment can be grouped into two categories (Figure 11.4):

1. Removable containers are placed on a transport chassis—sometimes several at once, depending on their size (0.2–10 hl). In some cases, the containers have large capacities (10–150 hl) and are used for transporting the grape crop over long distances. This operation is not recommended from an enological viewpoint.

2. Grape recipients should not be too large, to avoid crushing the grapes and to reduce the number of times they need to be transferred.

3. Fixed containers may be used, corresponding to a transport unit. Within this category, dumping containers instantaneously empty their entire crop into the reception bin, and containers equipped with screw-pumps progressively empty the crop. Thus they can adapt to any reception installation (Figure 11.4).

The transport container for harvest trailers is permanently attached to the chassis. The vineyard therefore requires additional trailers for other vineyard operations. These trailers are distinguished by their capacity and dumping method. The smallest (from 20 to 30 hl) are able to pass between vine rows, thus eliminating intermediary crop transfers. They are gravity-tilt trailers and often require a costly recessed reception area. High-capacity bin trailers are shallower, which limits grape bruising and crushing, but they are too large to pass between vine rows. These containers empty the grapes into the reception bin using a hydraulic lift system. Elevator bin trailers also exist and are capable of lifting their containers up to 1.5–2 m before emptying, depending on the model. This system does not require a recessed reception area. The harvest can be directly fed to the first step of the winemaking process (destemmer—crusher or press). Similarly, screw-bin trailers equipped with pumps (Figure 11.4) eliminate the need for a reception

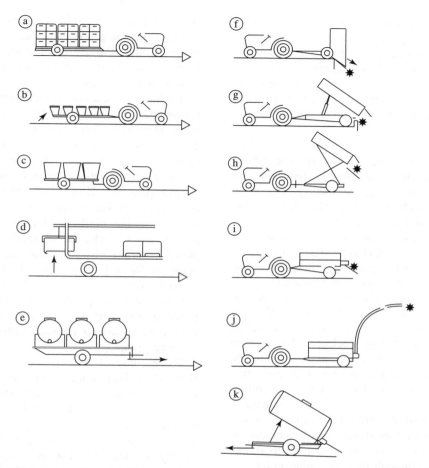

Fig. 11.4. Different harvest transport containers (Ribéreau-Gayon *et al.*, 1977). Portable containers: (a) 20–90 l stackable boxes; (b) 60–100 l plastic or wooden containers; (c) 600–800 l harvesting bin; (d) 1000–2000 l high-capacity containers, transported by truck; (e) 15–20 hl portable tanks. Trailers: (f) 15–25 hl gravity dumping bin; (g) 20–30 hl mechanical dumping bin; (h) elevator dumping bin; (i) screw-driven bin; (j) screw-and pump-driven bin; (k) pumper tank

installation. Although these systems are very practical, the mechanism of some screw-pumps is too brutal and can decrease crop quality.

Mechanically harvested grape crops, especially white, require rapid draining. This is most often effected with grills forming a double floor in the bottom of bin trailers.

11.3.3 Cleaning and Sorting the Grape Crop

These operations include eliminating MOG (leaves and stalks) and damaged, unripe or rotten grapes.

During manual harvesting, MOG is eliminated (at least for red grape crops) at the same time as the crushing and stemming process. The same equipment is used for cleaning mechanically harvested grape crops, but the different aspect of these grape crops would justify adapted machinery, which will certainly be developed in the future.

Sorting the harvest to eliminate bad grapes is only possible with intact grapes. This operation is difficult with machine-harvested grapes. Cutters, however, may precede the harvester to eliminate most of the damaged, spoiled or unripe grapes.

However, better results have been obtained using automatic sorting systems with vibrating screen.

Manually harvested grapes have the undeniable advantage of being able to undergo an effective sorting. This can be carried out as the transport bin is being filled. At present, sorting tables are used and workers, placed around these tables, remove bad grapes.

Among the various sorting tables proposed by manufacturers, there are models adapted to the back of the tractor that directly feed the bin trailer. Usually, however, the sorting table is an independent, detachable unit installed between the reception area and the first piece of winemaking equipment. A sorting table essentially consists of a conveyor belt on a metal chassis. The belt is driven by rollers powered by an electric motor. The belt speed should be slow (less than 5 m/min) to limit workers' eye fatigue. The belt is often made of food-quality rubber and is sometimes perforated. An articulated plastic belt is also used. The sorting tables are often slightly inclined (5–10%) to facilitate draining. Only intact grapes should be supplied to the sorting table. Screw-pumps often damage grapes and therefore compromise sorting effectiveness. Ideally, small-capacity containers should be emptied directly on the table, in which case the grapes are spread out as the containers are emptied. Vibrating tables are now used to spread the grapes out, although they are rather noisy.

11.3.4 Grape Selection and Selective Must Extraction by Low Temperature Pressing

Generally, in a homogeneous parcel containing only one grape variety, maturation intensity can vary from one grape cluster to another and even from one grape to another on the same cluster (Section 10.3). During manual grape-picking, and even more so during machine harvesting, the level of maturity of grapes is difficult to distinguish. A best, the incoming grape crop can be sorted according to its sugar concentration and sanitary state.

In noble rot regions, grapes and grape cluster fragments are selected by successive (sorting) harvesting. Only the grapes having reached the *rôti*

stage are picked (Sections 10.5 and 14.2.2). Even in this case, the grape-picker cannot always precisely evaluate the degree of grape concentration. Furthermore, the state of the harvest does not necessarily make the appropriate choice possible.

In white winemaking, a method called controlled temperature pressing currently permits selective must extraction from grapes richest in sugar. These grapes freeze at a lower temperature than those less rich in sugar. This modern technique, which is authorized as a harvest selection method by European legislation, originates from a traditional grape-picking method in certain viticultural regions. In the northern vineyards of Germany, Austria and Canada, the winemaker benefits from the severe climate by harvesting and pressing white grapes while they are partially frozen. A must particularly rich in sugar is thus obtained and is used to make the highly prized ice wines (*eiswein*). This pressing method strongly accentuates the concentrations of sugar and aromatic substances of these overripe grapes.

The method of 'selective cryoextraction' consists of cooling grapes until only those richest in sugar remain normal. The others are frozen solid and are not compressible. The grapes are then immediately pressed. Only the must from the grapes richest in sugar and therefore highest in quality is extracted (Chauvet *et al.*, 1986).

The method is carried out differently depending on the harvest method. With manual harvesting, grapes in small containers are placed in a freezing chamber. After cooling, the boxes are manually emptied into the press. When grapes are harvested mechanically, this method is less effective. The grapes must be drained before cooling and are then frozen in bulk by a freezing device. Liquid nitrogen is often the cooling source.

Noble rot grape winemaking has found particularly interesting applications for this technique (Section 14.2.4b) but it can also be applied to healthy grapes for making dry white wines. In addition to sugars, other elements of grape chemical constitution are modified. Even though the total acidity and malic acid concentrations are higher in the selected must, the tartaric acid concentration generally decreases. The increased potassium concentration resulting from cryoextraction makes a portion of this acid insoluble. Phenolic compound concentrations remain stable, indicating that the cryoextraction acts essentially on the pulp without altering the skin. The wines produced are richer and more complex.

Supra-extraction (Section 13.3.6) is derived directly from cryoextraction. It consists of subjecting the grapes to freeze-defrost cycles, then pressing them at room temperature. The ice crystals tear the cell walls, so pressing extracts even more compounds from the grapes.

11.4 ACIDITY ADJUSTMENTS OF THE HARVEST

Generally, in temperate climates and in traditional winemaking regions, regular grape sampling and maturity assessment and the correct choice of grape varieties ensure a proper level of harvest acidity. In the Bordeaux region (France), the low acidity of Merlot is often compensated by the higher acidity of Cabernet and possibly Malbec, planted together in the same vineyard. But acidity may need to be corrected when making single-variety wines or in extreme climatic conditions.

Acidity adjustments consist of either increasing (acid additions) or decreasing (acidity reduction) total must acidity. Various products are used for this purpose. Contrary to overripening techniques, acidity adjustments are strictly regulated in the European Community and in other countries that respect OIV (Office International de la Vigne et du Vin) recommendations.

11.4.1 Acidification

In hot winegrowing regions and during exceptional ripening years in temperate zones, a considerable amount of malic acid is degraded during maturation. In order to maintain the freshness and firmness of wines desired by consumers, especially in whites, the acidity should be increased by adding an acid. High acidity also enhances the protective effects of sulfur dioxide.

Strong inorganic acids, such as hydrochloric, phosphoric and sulfuric acid, may not be added to wine and are prohibited in all countries. European legislation and the OIV only recommend tartaric acid, but this acid tends to harden wines and should be added with caution.

For a total acidity comprised between 3.0 and 3.5 g/l as H_2SO_4 (4.6–5.4 g/l as tartaric acid), 50 g of tartaric acid should be added per hectoliter. If the total acidity of the grapes is below 3 g/l as H_2SO_4, then 100 g of tartaric acid should be added per hectoliter. In any case, legislation limits this acidification to 1.5 g of tartaric acid per liter (Table 11.1). The need for acid additions can also be determined with respect to must pH. For example, they are necessary for pH above or equal to 3.6. In low acidity musts, the production of succinic and lactic acid during fermentation tends to increase acidity in greater proportions, and this should be considered when determining the need for an acid addition.

In both red and white winemaking, tartrate should be added before, or, preferably, toward the end of fermentation. Allowing for precipitation, the addition of 100 g of tartaric acid per hectoliter increases the acidity by 1 g/l expressed as H_2SO_4 (1.5 g/l as tartaric acid). Acid additions, however, should never be calculated to bring acidity up to normal levels. As adding tartaric acid in solubilizes potassium, it tends to have more effect on pH and flavor than on total acidity levels. Harvesting a portion of the crop before full maturity or using grapes from secondary flowering can provide natural acidification, but these methods are not recommended as they are detrimental to quality.

The addition of citric acid is not a useful solution, since the total citric acid concentration limit is set at 1 g/l as citric acid. This limit, imposed by European legislation, does not noticeably increase the total acidity. Furthermore, lactic bacteria can break down this acid during winemaking, increasing volatile acidity.

Calcium sulfate (plaster or gypsum) was traditionally used in certain viticultural regions (Jerez in Spain). Its addition at doses of 1.25–2.25 g/l lowered must pH by precipitating calcium tartrate. This method has now been practically abandoned. Although effective, it is not advisable; moreover,

Table 11.1. Grape acidity correction regulations[a] in the European Community (except Portugal)

Schematic delimitation of viticultural zones	Acidification[b,c]		Deacidification[e]
	Normal years	Exceptional years[d]	
(a) Vineyards of Belgium, Luxemburg, Holland, Great Britain, Austria and Germany (except the Baden region)	—	—	Authorized
(b) Vineyards of Baden and the northern half of France (Loire Valley included)	—	—	Authorized
(C Ia) Vineyards of the southern half of France, except meridional, and north-west of Spain	—	Authorized max. 1.5 g/l	Authorized
(C Ib) Alpine vineyards of northern Italy	—	Authorized max. 1.5 g/l	Authorized
(C II) Meridional vineyards of France except zone (C III); vineyards of the northern half of Spain; Italian vineyards except zones (C I) and (C III)	Authorized max. 1.5 g/l	—	Authorized
(C III) Vineyards of the eastern Pyrénées, Var and Corsica in France; vineyards of the southern half of Spain; Italian vineyards of Basilicate, Pouilles, Calabria, Sicily and Sardinia; the vineyards of Greece	Authorized max. 1.5 g/l	—	Authorized

[a]Grape musts, partially fermented grape must and new wines that are still fermenting.
[b]By addition of tartaric acid exclusively.
[c]In finished wines, acidification permitted up to 2.5 g/l.
[d]EC declares exceptional years.
[e]No limit.

its use must be mentioned on the wine label (Benitez *et al.*, 1993) (Volume 2, Section 12.4).

Cation exchange treatments in must and wine are authorized in some countries. When used for stabilizing tartaric precipitation, these treatments acidify the resulting product; but in normal treatment conditions, the pH is not lowered by more than 0.2. Eliminating potassium by electrolysis (Volume 2, Section 12.5) also causes a slight decrease in pH.

11.4.2 Deacidification

When grapes do not reach complete maturity in northern vineyards, grape acidity can be considerable. In these conditions, malic acid concentrations are almost always greater than those of tartaric acid. When the biological degradation of malic acid is not desired due to the organoleptical changes that it causes, the juice must be chemically

deacidified. This practice is authorized in many countries and various products are available.

The two main compounds, calcium carbonate and potassium bicarbonate, react according to the same mechanism (Blouin and Peynaud, 2001). When they combine with tartaric acid, H_2T, the carbonate is broken down into carbonic acid, released as CO_2, and the calcium or potassium forms an insoluble salt with the tartrate, which then precipitates out.

— with calcium carbonate

$$H_2T + CaCO_3 \longrightarrow CaT \downarrow + CO_2 \uparrow + H_2O$$
$$150\ g \quad\ 100\ g$$

— with potassium bicarbonate

$$H_2T + KHCO_3 \longrightarrow KHT \downarrow + CO_2 \uparrow + H_2O$$
$$150\ g \quad\ 100\ g$$

Theoretically, in both cases, 1 g/l of deacidifying agent neutralizes 1.5 g/l of H_2T, giving a decrease in acidity of 20 meq/l or 1 g/l expressed in H_2SO_4 (or, of course, 1.5 g/l expressed in tartaric acid). In practice, the reaction is less efficient, especially in the case of $KHCO_3$, where higher doses are recommended to achieve the same level of deacidification (Table 11.2), so this product should be reserved for minor acidity corrections. In any case, it should always be borne in mind that these deacidifying agents act exclusively on tartaric acid, so they should not be used to try to adjust acidity to normal levels, which would necessitate the elimination of excessive amounts of tartaric acid and cause an unacceptable increase in pH.

In the case of calcium carbonate, deacidification results directly from the salt's chemical reaction, so maximum deacidification is rapidly achieved and is relatively predictable. However, if, for any reason, the wine has a high calcium content, there is a risk of further precipitation at a later date, even in bottle.

However, the deacidification mechanism involving potassium bicarbonate is more complex. Following an initial decrease in acidity due to this salt's reaction with tartaric acid, the formation of potassium bitartrate upsets the ion equilibrium and precipitates, producing a secondary deacidification that is not, theoretically, predictable. It is advisable to conduct laboratory trials to determine the appropriate dosage (Flanzy, 1998).

For all these reasons, it is advisable to use calcium carbonate to deacidify musts with excessive acidity levels before or during fermentation. Potassium bicarbonate should be reserved for slight corrections (after laboratory trials), during the final preparation phase. It is, in many cases, preferable, to implement deacidification in two stages.

Table 11.2. Product doses for the deacidification of musts and wines

Initial total acidity (g H_2SO_4/l)	$CaCO_3$ (g/hl)	$KHCO_3$ (g/hl)
Less than 7.0	50	75
From 7.0 to 7.5	75	110
From 7.5 to 8.0	100	—
Greater than 8.0	125	—

Another authorized product, neutral potassium tartrate, is rarely used, due to its cost and its low deacidifying power. To lower the total acidity by 1 g/l as sulfuric acid, requires 2.5–3 g of neutral potassium tartrate per liter. This product deacidifies by precipitating potassium hydrogentartrate, which possesses an acid function.

All of these products act uniquely by precipitating tartaric acid, since the potassium and calcium salts of malic acid are soluble. Yet insufficiently ripe grapes contain excess malic acid.

Wucherpfennig (1967) recommended a deacidification technique in Germany based on the precipitation of a double calcium malate and tartrate salt, insoluble above pH 4.5. With this method, a fraction of the must to be treated is completely neutralized by calcium carbonate, containing a small amount of calcium malate and tartrate seed crystals. After precipitation of the double salt, this strongly deacidified volume is filtered before being blended back into the untreated fraction. This procedure was developed and perfected by Haushofer (1972). Where M (in hl) is the total volume of must to deacidify, the fraction to be treated (M_d) by calcium carbonate is calculated by using the following formula:

$$M_d = \frac{(T_1 - T_2)}{T_1} \times M \qquad (11.1)$$

T_1 is the titratable must acidity in grams of tartaric acid per liter; T_2 is the desired acidity after the treatment. At a must volume M_d, a quantity of calcium carbonate is added, given by the formula: $0.67 (T_1 - T_2)M$.

When there is too much surplus malic acid in the initial must, part of the calcium carbonate added to the treated volume produces soluble calcium malate. At the time of final blending, the surplus calcium reacts with the tartaric acid of the untreated must. The deacidification thus occurs in two steps and acts essentially on the tartaric acid. The tartaric acid concentration therefore has to be adjusted in order to decrease malic acid concentrations substantially, by double salt formation (Usseglio-Tomasset and Bosia, 1992). This result can be obtained by using a mixture of calcium carbonate and calcium tartrate, also containing a small amount of the double calcium salts of malic and tartaric acid to favor calcium tartromalate crystallization. The deacidification power of this mixture depends on the proportion of calcium tartrate used. In this case, the use of this deacidification process becomes complex. It only reaches final equilibrium after a long time.

In white winemaking, this deacidification should be effected after must clarification but before fermentation. Aromatic ester production by yeasts is facilitated by a moderate pH. Conversely, this treatment permits a more precise deacidification of red wines when performed at the end of alcoholic fermentation, at the time of running off. It can also help to trigger malolactic fermentation.

European legislation does not impose must deacidification limits, but there is a limit of 1 g/l of the total acidity as tartaric acid for wine. Table wines must have a minimum total acidity of 4.5 g/l as tartaric acid. In any case, this treatment must be declared and cannot be combined with an acidification.

11.5 INCREASING SUGAR CONCENTRATIONS

Certain regions have difficulty in producing quality wines. Chapter 10 discussed the climatic and soil conditions which excessively favor vine growth and grape development. These conditions lead to musts with high sugar concentrations, but the resulting wines lack finesse and aromatic complexity.

In the best *terroirs* of northern vineyards, unfavorable climatic conditions during difficult vintages often hinder maturation. Many parameters (reduced photosynthesis, continued vegetative growth, excessive crop yields, etc.) limit grape sugar accumulation; thus, adjusting the natural sugar concentration can be useful. Of course, these adjustments must be limited and are not capable of replacing a complete maturation. In particular, their use should not incite premature harvesting or exaggerated crop yields.

Various subtractive techniques increase sugar concentrations by eliminating part of the water found in grapes—similarly to natural overripening.

The crop yield is consequently lowered. Although some of these techniques are still in the experimental stage, they are largely preferred by international authorities over additive techniques (Tinlot, 1990). Additive techniques, despite their long length of use, always have the inconvenience of increasing crop volumes by the addition of an exogenous product—sugar or concentrated must (Dupuy and de Hoogh, 1991).

11.5.1 Subtractive Techniques

These techniques, similarly to drying and cryoextraction, consist of eliminating part of the water contained in grapes or in must. Two physical processes can be used: water evaporation, or selective separation across a semi-permeable membrane (reverse osmosis). European legislation has set the limits for these treatments: a 20% maximum volume decrease and a 2% volume maximum alcohol potential increase.

Quite apart from installation costs, the considerable crop yield loss resulting from the use of these methods hindered their proliferation, and chaptalization was preferred for a long time. In recent years, a focus on increasing wine quality has renewed an interest in these methods. In red wines in particular, tannin concentrations are simultaneously increased. Of course, these methods should never be used with the intent of correcting excessive crop yields.

Heat concentration is a longstanding method often used in the food industry. For more than 40 years, it has been used to make concentrated must, but potential wine quality must not be compromised during the heating process. The denaturation of thermosensitive must constituents and the appearance of organoleptic flaws (hydroxymethylfurfural, for example) must therefore be avoided. Earlier equipment, operating at atmospheric pressure, required relatively high temperatures, which produced off-aromas. For this reason, certain French appellations prohibited their use. Today, vacuum evaporators have lowered the evaporation temperature to 25–30°C and interesting qualitative results have been obtained. Even lower evaporation temperatures are possible but the must delivery rate becomes too low.

In addition, horizontally grouped tubular exchangers permit continuous high-speed treatment of must. This system limits the risk of heat bands—the prolonged contact between a fraction of the must and the hot exchanger surface. A thermocompressor, acting as a heat pump, extracts part of the must vapor and mixes it with the vapor produced by the steam generator. This system has the twofold advantage of lowering the treatment temperature and saving energy (by favoring must evaporation). Current equipment can treat from 10 to 80 hl of must per hour, with an evaporation capacity of 150–1200 l/h. In controlled appellation zones, this treatment should be effected in a closed circuit directly linked to the fermenter. The evaporation occurs at a low temperature (25–30°C) and in these conditions the concentration factor is always less than 2 (Berger, 1994). The sugar concentration of must treated by the concentrator remains practically constant during its operation cycle. Iron and malic acid are concentrated to the same degree as the sugars, but potassium and tartaric acid concentrations are lower, due to their partial precipitation during the treatment (Table 11.3). This technique, however, is not recommended for concentrating musts made from grape varieties with marked varietal aromas.

Peynaud and Allard (1970) used reverse osmosis to eliminate water from grape must at ambient temperatures. The results obtained were satisfactory from an enological viewpoint, but problems in regenerating the cellulose acetate membranes stopped this technique from being developed further. The usefulness of reverse osmosis for

obtaining wines of superior quality was confirmed by Wucherpfennig (1980) but it was the development of composite membranes that gave rise to new experiments with reverse osmosis (Guimberteau et al., 1989).

Figure 11.5 illustrates the principle of reverse osmosis. An appropriate membrane is used to separate a concentrated saline solution (A) from a more diluted solution (B). The difference in chemical potential tends to make the water pass from the low potential compartment to one with the higher the potential (direct osmosis); the latter

Table 11.3. Must concentration with a vacuum evaporator (Cabernet Sauvignon, Bordeaux, 1992)

Constituent	Initial must	Concentrated must	Condensed vapors
Sugar concentration (g/l)	179	204	0
pH	3.26	3.39	3.80
Total acidity (g/l H_2SO_4)	5.1	5.6	0.05
Malic acid (g/l)	4.2	4.8	0
Tartaric acid (g/l)	6.2	6.8	0
Potassium (g/l)	1.7	1.9	0
Iron (mg/l)	1.8	2.1	0

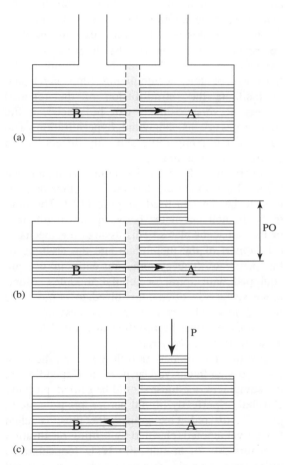

Fig. 11.5. Principle of reverse osmosis (Guimberteau et al., 1989): (a) direct osmosis; (b) osmotic equilibrium (OP = osmotic pressure); (c) reverse osmosis (P = pressure greater than osmotic pressure)

is diluted. The diffusion of water stops when the internal pressure of compartment A (osmotic pressure) counterbalances the pressure that diffuses the water across the membrane. If a pressure greater than this osmotic pressure is exerted on compartment A, the direction of the diffusion of the water is reversed (reverse osmosis) and solution A is concentrated.

A pressure of at least twice the osmotic pressure must be exerted on the must to force the water of the must to cross the membrane. This concentrates various must constituents. During water transfer, molecules and ions retained by the membrane are apt to accumulate on its surface, increasing the real must concentration to be treated and thus the required pressure. To limit this concentration polarization phenomenon, the filtering side of the membrane must be cleaned to minimize the thickness of this accumulating limit layer and to facilitate the retrodiffusion of the retained solutes. Reverse osmosis can be placed in the same category as tangential hyperfiltration. The equipment is comparable and only differs by the nature of the membrane.

Several models exist, depending on membrane layout. Plate modules, derived from filter-presses, used to be the first used. (Figure 11.6). The fluid to be treated circulates between the membranes of two adjacent plates. This assures the mechanical support of the membrane and the draining of the permeate. The systems currently in use are equipped with spiral or tubular modules. Membrane surface is often maximized to compensate for the low delivery rate of these systems. Several modules must be installed in parallel to have a satisfactory delivery rate.

Due to the extremely thin flow stream, the must undergoes an intensive clarification. Depending on the equipment, the must should be settled, partially clarified or perfectly limpid before this process. In practice, for red wine grapes, the must is taken from a vat, cooled, clarified by settling or filtration to a turbidity of around 400 NTU (concentrated by reverse osmosis, and returned to its original vat. In certain cases, to avoid the formation of potassium hydrogentartrate crystals, the use of metatartric acid has been suggested. Depending on

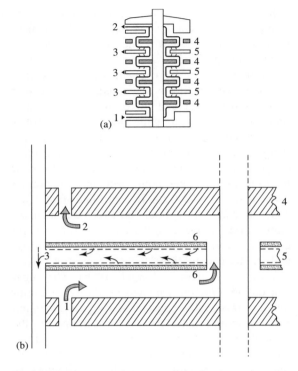

Fig. 11.6. Plate membrane module for reverse osmosis: (a) schematic diagram of entire module; (b) detail of cell. (1) Must to be treated; (2) concentrated must; (3) permeate; (4) intermediary plate; (5) membrane support; (6) membrane

the modules, the working pressure varies from 60 to 120 bars, the temperature is regulated between 15 and 25°C and the permeate delivery rate (the quantity of water removed over a period of time) is from 0.5 to 5 l/h/ m² of filter surface.

Several findings have been confirmed in experiments over the last 10 years (Berger, 1994). The membranes have a solute retention rate of over 99.5%, but losses increase when the number of modules in parallel is increased in order to attain elevated continuous treatment delivery rates. In this case, traces of sugars and minerals are found in the permeate (Table 11.4). Tastings reveal that the permeate sometimes gives off sulfurous odors reminiscent of low tide. The retentate or enriched must is of a good organoleptical quality. Generally, the concentrations of malic acid, metals, phenolic compounds and macromolecules such as proteins

Table 11.4. Concentration by reverse osmosis (Berger, 1994)

Constituent	Initial must	Concentrated must (retentate)	Water eliminated (permeate)
Sugar concentration (g/l)	175	434	1.75
pH	3.10	2.93	3.53
Total acidity (g/l H_2SO_4)	5.9	11.5	0.1
Malic acid (g/l)	4.5	9.7	0.1
Tartaric acid (g/l)	6.8	9.2	0.6
Potassium	1.17	1.43	0.03
Iron (mg/l)	8.3	17	1.3
Absorbance at 280 nm	8.7	20	0.05

and polyosides increase proportionally with the sugar concentration. Salification phenomena cause a limited modification of the concentrations in tartaric acid and potassium and the pH.

Subtractive techniques increase tannin concentrations by decreasing juice volume with respect to pomace volume and are therefore more often applied to red winemaking than white. In this manner, they act like tank-bleeding (Section 12.5.8), but without the loss of sugar. These enrichment methods have the distinct advantage of being self-limiting for technical (concentration of bad tastes and flaws) and especially economic (operation costs and volume losses) reasons. They represent a considerable cost: in equipment investment, operation costs and crop yield loss. The last inconvenience does not exist if the crop yield is above authorized limits. The main interest of these techniques is to treat must made from grapes soaked by heavy rain during the harvest.

11.5.2 Additive Techniques

Contrary to subtractive techniques, increasing the sugar concentration by the addition of an exogenous product is not directly limited by technical constraints. These techniques are therefore very strictly regulated. Such strict regulations have led the technical branches of regulatory organizations to develop sophisticated analytical methods to verify that the regulations are indeed being followed: magnetic resonance of the deuterium of ethanol, the isotopic ratio of ^{13}C by mass spectrometry, etc. (Martin *et al.*, 1986). Must sugar concentration can be increased by directly adding pure sugar, concentrated must or rectified concentrated must. It is generally legally limited to a 2% volume increase of the alcoholic strength—more under certain conditions. In practice, however, this increase should be limited to 1–1.5% volume to avoid disequilibrating the wine by an excessive vinosity which would mask wine fruitiness. Within responsible limits, sugar addition is an effective

Table 11.5. Analytical comparison between concentrated white must and the same concentrated must rectified (Brugirard, 1987)

Constituent	Concentrated must	Rectified concentrated must
Density (20°C)	1.3620	1.3535
Sugar concentration (g/l)	871	852
Potential alcohol (% vol.)	51.23	50.11
Total acidity (g/l H_2SO_4)	12.5	0.25
Iron (mg/l)	20.3	0
Copper (mg/l)	1.1	0
Ashes (mg/kg of sugars)	177	<1.2
Total phenols (mg/kg of sugars)	478	152

means of increasing the gustative quality of wine by augmenting body and harmony.

Sugaring, better known as chaptalization since the end of the 18th century, consists of adding refined white saccharose to must. The saccharose must be at least 99% pure but can be derived from any plant (sugar cane, sugar beet, etc.). The quantity of saccharose required to increase a wine by 1% volume alcohol varies from 16 to 19 g/l, depending on the yeast strain, must oxygenation and the initial sugar concentration. European legislation has established precise doses: 18 g/l in red winemaking and 17 g/l in white winemaking. The saccharose is dissolved in a fraction of must. This operation should be effected during the first one-third of alcoholic fermentation and during a pumping-over. The saccharose dissolves more quickly in warm must and the simultaneous aeration stimulates the fermentative activity of the yeasts. The augmentation of alcoholic strength by chaptalization modifies some of the constituents of the corresponding wine. The total acidity decreases by 0.1–0.2 g/l as H_2SO_4 for 1% volume alcohol added. This diminution is caused by an increased potassium bitartrate precipitation. In red winemaking, phenolic compound extraction increases by approximately 5% for each additional 1% volume of ethanol. The increase in volatile acidity is negligible as long as the chaptalization is not exaggerated. For alcoholic strengths greater than 13% volume, it increases by approximately 0.05 g/l as H_2SO_4. Glycerol and dry residue increase in lesser proportions than ethanol. For a long time, the differences in concentrations of the various wine constituents, expressed in ratios, were the only methods available for enforcing chaptalization regulations. Today, isotopic methods can distinguish the sugar and thus the ethanol formed in terms of its botanical origin (grape, sugar beet or sugar cane).

Various atoms have isotopes: deuterium ^2H for hydrogen ^1H and ^{13}C for carbon ^{12}C, for example. In a multiple atom molecule of natural origin, a very small but variable quantity of these atoms can be replaced by the corresponding isotopes. The ethanol molecule possesses several hydrogens. An isotopomer is formed if one or more of its

hydrogens, for example, are replaced by deuterium. The proportion of these different isotopomers depends on the origin of the fermented sugar. Indeed, the vine's photosynthesis mechanism is different from that of other crops (beets, sugar cane) and the deuterium content varies according to latitude. Isotopomers are determined by nuclear magnetic resonance (NMR). This method is capable of detecting the addition of sugar beet sugar (chaptalization) by comparing a sample wine with control wines made in the same geographical region during the same vintage. In order to evaluate a mixed addition of sugar beet and sugar cane sugar during fermentation, this first analysis must be complemented by a carbon isotope determination using mass spectrometry.

The addition of rectified concentrated must (RCM) is similar to chaptalization. This colorless liquid contains an equimolar mixture of glucose and fructose. The rectified concentrated must is obtained by dehydration. All compounds other than sugar are eliminated by ion exchange resin treatment. (Table 11.5). European legislation has precisely defined the characteristics of this product. The RCM must have a refractometric index greater than or equal to 61.7%. The RCM, diluted to 25° Brix (Section 10.3.3), must not have a pH greater than 5, an optic density greater than 0.1 at 425 nm or a conductivity greater than 120 µS/cm. The legal limits are set at less than or equal to the following concentrations: titration acidity, 15 mEq/kg (of sugar); sulfur dioxide, 25 mg/kg; total cation concentration, 8 mEq/kg; and hydroxymethylfurfural, 25 mg/kg.

RCM is used in the same manner as pure saccharose and leads to the same chemical constituent modifications in the corresponding wines. It has several inconveniences with respect to saccharose. First of all, RCM is more expensive. Equally important, it is not in crystallized form; therefore its purity and stability over time depend on preparation and storage conditions. Attempts have been made to separate the glucose and fructose from the must in order to crystallize them separately, but its production costs are too high. Otherwise RCM increases the dilution of the product. Finally,

Table 11.6. Sugar additions and limits for quality wines produced in determined regions according to European legislation (values expressed in % vol.)

	European viticultural zones					
	A	B	CI		C II	C III
			(a)	(b)		
Potential alcohol before addition	6.5	7.5	8.5	9	9.5	10
Addition limits in potential alcohol (% vol.)						
Normal year	3.5	2.5	2	2	2	2
Exceptional year	4.5	3.5	2	2	2	2
Authorized volume increase (%)[a]						
Normal year	11	8	6.5	6.5	6.5	6.5
Exceptional year	15	11	6.5	6.5	6.5	6.5
Total alcohol content (% vol.)						
Minimum commercial	9	9	9	9	9.5	10
Maximum commercial	Variable in France according to *appellation controleé*. Not limited in most other countries					

[a]By addition of concentrated or rectified concentrated must.

a similar product prepared by the hydrolysis of saccharose can be used to adulterate and sometimes counterfeit RCM. The only advantage of RCM is that it is derived from grapes. This reference to its origin is secondary, since the purest possible product is desired and obtained through appropriate treatments.

Must constituent modifications are even more pronounced when non-purified concentrated must is used. These concentrated musts are usually obtained by indirect heating. All organic and mineral elements in the musts are concentrated. They have very low pHs (often lower than 3) despite the insolubilization of a portion of the tartaric acid in the form of potassium and calcium salts. In fact, sulfur dioxide is used in high concentrations for conserving the must before concentration. The SO_2 is oxidized into sulfates during the treatment. Concentrated musts are also very rich in iron and are apt to cause iron casse. They are always very dark in color, due to the decomposition of sugars and nitrogen compounds in this hot and acidic environment. The resulting products intensify the color produced by phenolic compounds. The minimum concentrated must sugar concentration is 582 g/l (51° Brix). Concentrated must greatly modifies the composition of the wine obtained—especially white wines: a slight increase in total and volatile

acidity and a more significant increase in glycerol, dry extract and phenolic compounds in red wines. Concentrated must often comes from the same grape varieties as the must to be enriched and sometimes from the same geographical zone.

Due to the diversity of its viticultural zones, the European Community has developed very elaborate legislation on this subject (Table 11.6). In many appellation zones, notably in France, the legislation is ever more restrictive.

11.6 ENZYMATIC TRANSFORMATIONS OF THE GRAPE AFTER ITS HARVEST

When conserved intact after its harvest, the grape still maintains an intense physiological respiratory activity—which is utilized during drying. The lack of oxygen and the depletion of available respiratory substrates rapidly provoke the initiation of fermentative processes in the berry.

The cellular destruction of grapes during pre-fermentation treatments results in oxygen dissolution, despite the precautions taken. Two enzyme categories, oxidoreductases and oxygenases, are responsible for many grape constituent transformations. They often harm grape quality.

Depending on winemaking techniques, the duration of grape solid maceration in must varies. During this period, hydrolase-type enzymes act on grapes and must. These enzymes are responsible for the hydrolysis of diverse macromolecules such as proteins, polyosides, heterosidic derivatives and various esters. Their action often improves the grape/must mixture. This maceration phase should therefore sometimes be prolonged.

11.6.1 Hydrolysis Enzymes

The active proteosynthesis that characterizes maturation is responsible for the high protein concentration in mature grapes. In must, proteins often represent 50% of the total nitrogen. In white winemaking, part of these proteins are insoluble and are eliminated during clarification. Endogenous grape proteases are apt to hydrolyze these proteins into soluble forms. These forms are more easily assimilated by yeasts during fermentation. The enzymes catalyze the hydrolysis of the peptidic bond between two amino acids (Figure 11.7).

The grape possesses a low and constant proteasic activity during its herbaceous growth phase. From *véraison* onwards, this activity strongly increases. In a ripe grape, the proteasic activity is essentially located in the pulp (Table 11.7). But the proteases are generally bound to cell structures. Healthy grape juice thus has relatively few proteases (30% of total proteasic activity).

Table 11.7. Distribution of proteasic activity in the different areas of a healthy grape (expressed as % of total activity per berry) (Cordonnier and Dugal, 1968)

Grape area	Proteasic activity
Pulp + juice	81.6
Skin	17
Seeds	1.4

All physical treatments of grapes (mechanical harvesting, stemming, crushing) increase the proportion of soluble proteases. The higher free amino acid concentration of these musts attest to this (Cantagrel *et al.*, 1982).

Grape proteases are acidic, with an optimum pH near 2.0. In the pH range of must, 40–60% of the potential proteasic activity exists. Protein hydrolysis activity during the pre-fermentation phase varies greatly, depending on grape maturity and harvest treatments. This certainly affects fermentation kinetics but the relationship has never been established. A slight sulfur dioxide addition (around 25 mg/l), however, has been confirmed to stimulate proteasic activity. This explains, at least partially, its activation effect on fermentation (Section 8.7.3).

Finally, botrytized grapes also contain fungal proteases. Contrary to grape proteases, these are soluble and pass entirely into the must. *Botrytis cinerea* aspartate proteinase has an optimum pH in the vicinity of 3.5. Whatever their origin, proteases are thermostable; they increase soluble nitrogen, even during thermovinification.

Among fruits, the grape is one of the least rich in pectic substances. These substances are predominately located in skin cell walls (Section 10.2.6). Must consequently contains a small amount of these compounds in the soluble form (0.5–1 g/l expressed in galacturonic acid). Depending on harvest treatments, some insoluble skin compounds may be extracted. Must pectic substance concentrations can thus attain 2.5 g/l. They are principally associated with cellular debris and must sediment. Most are rapidly hydrolyzed by pectolytic enzymes of the grape. Ethanol subsequently precipitates the rest of them at the end of fermentation. Wine is therefore practically devoid of must pectic substances (Usseglio-Tomasset, 1978).

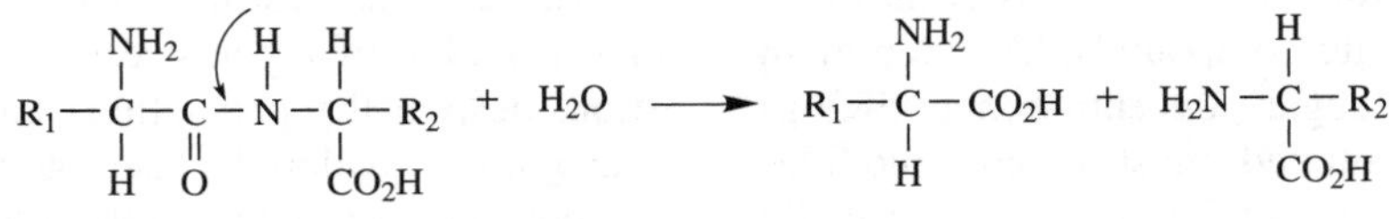

$$R_1 - \overset{\displaystyle NH_2}{\underset{\displaystyle H}{C}} - \overset{\displaystyle H}{\underset{\displaystyle O}{C}} - N - \overset{\displaystyle H}{\underset{\displaystyle CO_2H}{C}} - R_2 + H_2O \longrightarrow R_1 - \overset{\displaystyle NH_2}{\underset{\displaystyle H}{C}} - CO_2H + H_2N - \overset{\displaystyle H}{\underset{\displaystyle CO_2H}{C}} - R_2$$

Fig. 11.7. Protease mode of action by hydrolysis of peptide bonds

The ripe grape is rich in pectin methyl esterase (PME) and polygalacturonases. Ripe grapes have high pectin methyl esterase and polygalacturonase contents, but contain no pectin lyase. PME is not a hydrolysis enzyme but rather a saponification enzyme. It liberates the acid functions of galacturonic units, resulting in the accumulation of methanol in the must (Figure 11.8). Grape PME is thermostable and has an optimum pH of 7 to 8. Its activity is reduced at the pH of must, but its action beforehand in the grape results in a significant decrease in the degree of esterification of liberated pectic compounds in the must. This action is essential, because the polygalacturonase can only act on the free carboxylic functions of the galacturonic units. Two types of polygalacturonase exist in the grape: exo-polygalacturonases exert their hydrolytic action sequentially, beginning at one end of the poly-galacturonic chain; endo-polygalacturonases act at random on the interior of the chains. In the latter case, although the pectin chain hydrolysis is very limited, the endo-polygalacturonic activity leads to a rapid and significant decrease in must

viscosity. The viscosity is reduced by one-half when fewer than 5% of the glycolytic bonds are broken. Grape polygalacturonases retain a significant activity at the pH of must (optimum activity pH is between 4 and 5). The homogalactur-onane zones are susceptible to rapid hydrolysis. The rhamnogalacturonane zones are more resistant, due to the presence of side-chains of arabinose and galactose.

When the *Botrytis cinerea* fungus infects grapes, it synthesizes cellulase, pectinase, and protease enzymes that break down the cell walls. (Section 10.5.2). In addition to PME and polygalactur-onases, *Botrytis cinerea* produces a lyase that cuts pectic chains by β elimination (Figure 11.8). This endolyase activity is not influenced by the ester-ification level of the carboxylic functions of the galacturonic units. All pectolytic enzymes of this fungus have an optimum pH near 5 and are therefore very effective during a botrytis attack. Must made from contaminated grapes consequently contains very few pectic substances. A polyoside, glu-can, secreted by *Botrytis cinerea* is mostly responsible for their viscosity (Volume 2, Section 3.7.2).

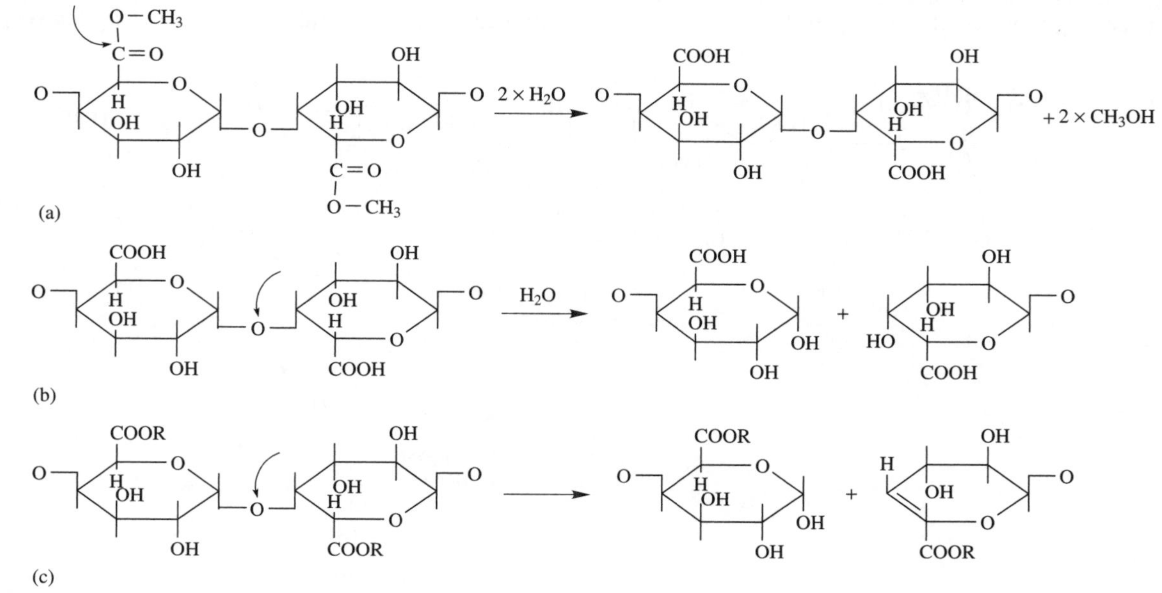

Fig. 11.8. Mode of action of different pectolytic enzymes: (a) pectinesterases; (b) polygalacturonases; (c) pectinlyases (R = H or CH$_3$)

The pectolytic enzymes produced by the grape or the fungus are fairly resistant to sulfur dioxide. Their activity is, however, reduced at temperatures below 15°C and above 60°C.

In muscat-type aromatic varieties, a considerable proportion of their aromatic potential is in the form of terpenic heterosides—non-odorous in ripe grapes (Section 10.2.8). During pre-fermentation treatments, enzymatic hydrolysis of these compounds increases must aromatic intensity (Figure 11.9). This phenomenon is enhanced by maceration of grape solids because of the high concentration of bound terpenic compounds in skins.

These compounds, called diglucosides, are composed of a glucose associated with another sugar, i.e. rhamnose, arabinose or apiose. The hydrolysis of these heterosides requires two sequential enzymatic activities. A β-L-rhamnosidase, an α-L-arabinosidase or a β-D-apiosidase must act on the molecule before the β-D-glucosidase is able to exert its action (Figure 11.10). In practice, this hydrolysis is relatively limited. Grape glycosidases have an optimal activity at a pH between 5 and 6, and they only retain part of this activity at the pH of must. These glycosidases are very specific and are not active on certain terpenic heterosides, notably tertiary alcohol derivatives, such as linalol. Moreover, β-glucosidase is strongly inhibited by free glucose (Bayonove, 1993). In contaminated grapes, the glycosidases secreted by *Botrytis cinerea* are more active. But the fungus totally degrades the aromatic potential of the grape (Section 10.5.3). Furthermore, the β-glucosidase of *Botrytis cinerea* is inhibited by the gluconolactone it produces.

Fig. 11.9. Demonstration of the enzymatic hydrolysis of terpenic glycosides (Bayonove, 1993)

11.6.2 Oxidation Enzymes

A green leaf-type odor or herbaceous note is produced when vegetal tissue (especially leaves,

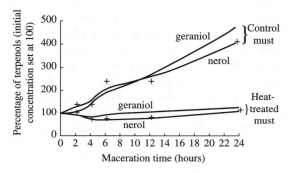
Terpenyl glycosides

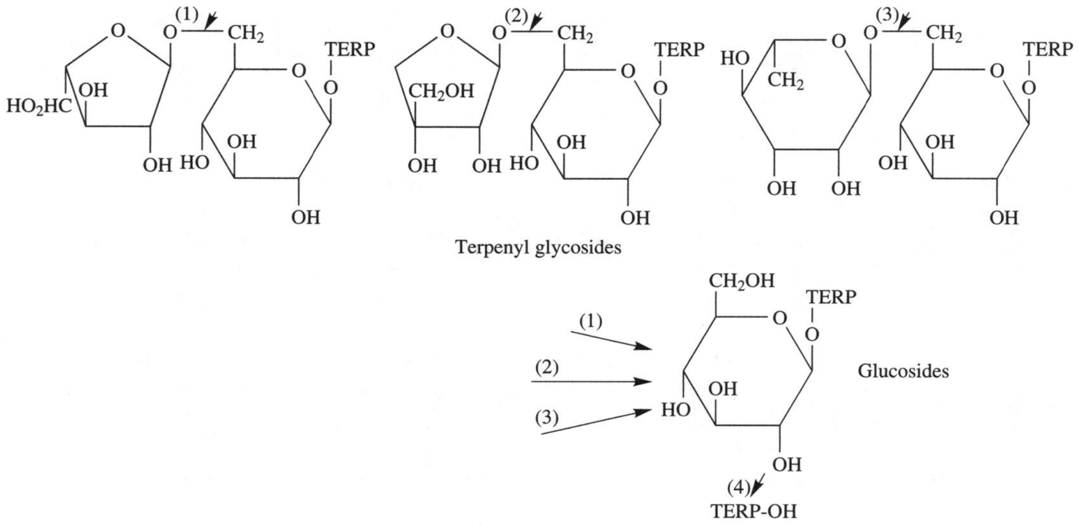

Glucosides

Fig. 11.10. Enzymatic hydrolysis mechanism of terpenic glycosides (Bayonove, 1993). (1) α L-Arabinofuranosidase; (2) β D-apiosidase; (3) α L-rhamnosidase; (4) β D-glucosidase

but also fruit) is crushed. This phenomenon was shown to exist in grapes by Rapp *et al.* (1976). Four enzymatic activities are sequentially involved (Figure 11.11). First, an acylhydrolase frees the fatty acids from membrane lipids. Next, the lipoxigenase catalyzes the fixation of oxygen on these C_{18} unsaturated fatty acids. This enzyme preferentially forms hydroperoxides in C_{13} from linoleic and linolenic acids. The peroxides obtained are then cleaved into C_6 aldehydes. Some of them are reduced to their corresponding alcohols by the alcohol dehydrogenase of the grape (Crouzet, 1986). These alcohols are responsible for their corresponding odors. Since the cleavage enzymes are linked to membrane fractions, the aldehyde concentrations are proportional to the intensity of solids maceration. To limit their concentration

during white winemaking, a sufficiently clear must should be obtained as quickly as possible (less than 200 NTU) (Dubourdieu *et al.*, 1986).

Grape cellular structure breakdown during pre-fermentation treatments is also accompanied by other enzymatic oxidations. Oxygen consumption speed thus varies from 0.5 to 5 mg/l/min, depending on must origin. This variation is caused for the most part by the oxidation of phenolic compounds.

Ripe grapes contain an orthophenol oxygen oxidoreductase, also known as cresolase, catechol oxidase and tyrosinase. Its activity is extremely variable, depending on the grape variety and degree of ripeness. (Dubernet, 1974). Tyrosinase consists of a group of isoenzymes differing in inductor nature and catalyzed activities (Mayer and Harel, 1979).

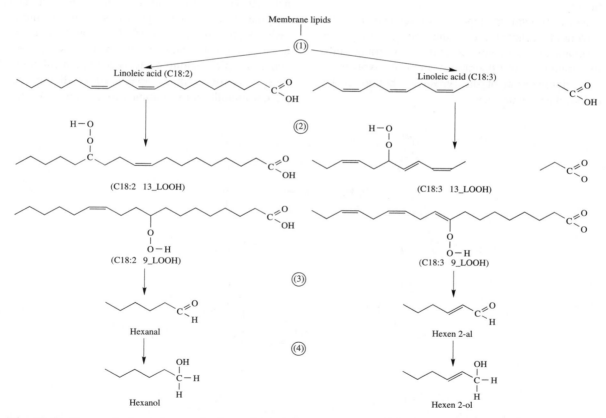

Fig. 11.11. Enzymatic formation mechanism of aldehydes and C_6 alcohols, responsible for grassy flavors (Crouzet, 1986). (1) [Acyl]hydrolase; (2) lipoxygenase in the presence of oxygen; (3) peroxide cleavage enzyme; (4) alcohol dehydrogenase

In white grape must, this enzymatic activity preferentially oxidizes tartaric derivatives of hydroxycinnamic acids (1), majority phenolic compounds in grape pulp (Figure 11.12). The quinones produced (2) are unstable and likely to enter into two different reactions (Figure 11.13). First, these very reactive quinones can condense with other phenolic compounds (flavonoids), forming polymerized products. Their color evolves from yellow to brown according to the degree of condensation (Singleton, 1987). The quinones are also apt to react with a strongly reductive molecule such as glutathion. This reaction produces a colorless derivative, S-glutathionyl-2-trans-cafeoyltartaric acid, known as the Grape Reaction Product or GRP (3) (Salgues et al., 1986). This derivative is not oxidizable by tyrosinase and thus does not modify the color of the must.

Must browning depends of course on the flavonoid concentration and consequently on mechanical treatments of that favor grape stalk maceration. These operations are also involved in the solubilization of the tyrosinase bound to the chloroplast membranes.

Yet the trapping of quinones by glutathion limits oxidation phenomena. Must browning is therefore also dependent on the glutathion concentration.

The tyrosinase of grapes is active but unstable at the pH of must (optimum activity at pH 4.75). Temperatures above 55°C or the addition of more than 50 mg of sulfur dioxide per liter are necessary to denature this enzymatic activity. Lower sulfur dioxide concentrations only modify oxidation rates. In fact, the bisulfite ions regenerate the potential enzyme substrates by reducing the quinones formed. Finally, treating must with bentonite reduces the soluble fraction of tyrosinase.

Phenolic compound oxidation is much more dangerous when the grapes have been attacked by Botrytis. Botrytized grapes contain a p-phenol oxygen oxidoreductase known as laccase (Dubernet, 1974). Contrary to tyrosinase, this fungal enzyme is stable at the pH of must and is more resistant to sulfur dioxide. It is also able to oxidize a greater number of phenolic substrates and molecules belonging to other chemical families. Laccase is thus capable of oxidizing the phenol-glutathion (3) complex to quinone (4). The

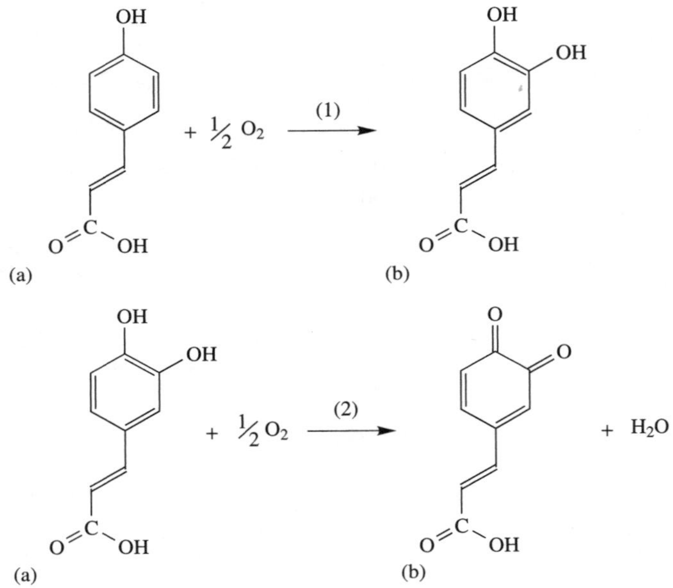

Fig. 11.12. Mode of action of grape tyrosinase on hydroxycinnamic acids (Mayer and Harel, 1979). (1) Cresolase activity (a: coumaric acid, b: cafeic acid) (2) Catecholase activity: (a) cafeic acid; (b) quinone

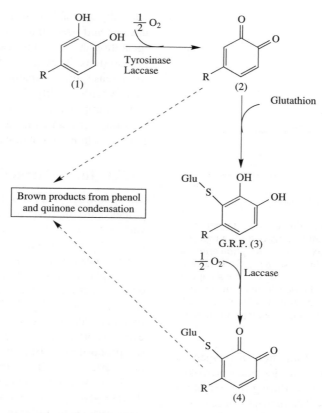

Fig. 11.13. Oxidation mechanisms of healthy grape must by tyrosinase and botrytized grape must by laccase (Salgues, 1986)

glutathion, therefore, can no longer trap quinone (Salgues *et al.*, 1986). More brown condensation products are formed from the same initial phenolic compounds by laccase than during oxidation by tyrosinase.

If the glutathion concentration is elevated, the quinone (4) can be partially reduced to phenol with the fixation of a second glutathion molecule. This new derivative is no longer oxidizable by laccase. Oxidation phenomena and the corresponding browning are thus limited. This second reaction is not likely in botrytized grape musts (Salgues *et al.*, 1986). The oxygen consumption rate is not higher than in healthy grape must (Section 8.7.2), but the action of sulfur dioxide is slower (Figure 11.14). The contaminated grape contains many other oxidases that also consume oxygen (glucose oxidase, amino oxidase, etc.). A temperature of 50°C

destroys laccase more quickly than tyrosinase. This thermal denaturation is the only possible treatment as adding bentonite only very slightly decreases laccase activity.

Peroxidases have long been proven to exist in grapes (Poux and Ournac, 1972). This enzymatic activity is essentially located in grape cell vacuoles. It most likely plays an important role in the oxidative metabolism of phenolic compounds during maturation (Calderon *et al.*, 1992). During pre-fermentation treatments, the activity of this enzyme seems to be limited by a peroxide deficiency. A low sulfur dioxide concentration is sufficient to destroy these peroxidases.

An increased understanding of these oxidation phenomena has spurred the development of a pre-fermentation technology called white must hyper-oxygenation (Müller-Späth, 1990) (Section 13.4.1).

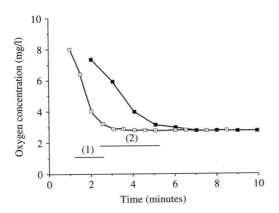

Fig. 11.14. Effect of sulfur dioxide on oxygen consumption in musts made from healthy and rotten grapes. (1) Time necessary to stop oxygen consumption after the addition of sulfur dioxide in must made from healthy grapes. (2) Time necessary to stop oxygen consumption after the addition of sulfur dioxide in must made from rotten grapes. □, oxygen consumption in must made from healthy grapes; ■, oxygen consumption in must made from rotten grapes

A sufficient and controlled addition of oxygen as soon as cellular structures are degraded provokes the denaturation of tyrosinase during the oxidation reactions that it catalyzes. The disappearance of the enzyme and the depletion of oxidizable phenolic substrates thus make the must stable with respect to oxidation. The condensation products responsible for browning should be eliminated before fermentation. Due to its possible impact on aromatic elements, this technique seems better adapted to certain cultivars. It is not applicable to botrytized grapes, due to the resistance of laccase.

11.7 USE OF COMMERCIAL ENZYMES IN WINEMAKING

The beneficial action of diverse hydrolysis enzymes from grapes is often limited by must pH or an insufficient activity due to the limited duration of pre-fermentation treatments. Manufacturers have developed better adapted enzymatic preparations, essentially from diverse species of fungi (*Aspergillus, Rhizopus* and *Trichoderma*). Research in this field is very active (van Rensburg and Pretorius, 2000). The enzymatic profile of

currently available commercial preparations is still unclear and users must develop their own experimentation. Many countries permit the use of these preparations. They are added as early as crushing to increase juice extraction, or to finished wine to improve filterability. These methods can also be used to improve color extraction and must quality (settling, fermentability and aromatic intensity) in red and white winemaking, respectively.

11.7.1 Juice Extraction

The addition of pectolytic enzymes in crushed grapes can improve juice extraction for certain varieties very rich in pectic substances (Muscat, Sylvaner, etc.). Commercial preparations contain diverse enzymatic activities which are active at a low pH: pectin methyl esterases, polygalacturonases, pectin lyases and hemicellulases. At a concentration of 2–4 g/hl, 15% more juice can be obtained during a settling period of 4–10 hours; even a shorter settling period (1–2 hours) increases the proportion of free run (Table 11.8). Effectiveness varies according to the nature of the grapes.

These pectolytic preparations can also contain diverse glycosidases (Cordonnier *et al.*, 1989) and proteases (Schmitt *et al.*, 1989), responsible for secondary transformations. Their degree of purity must therefore be assured.

11.7.2 Must Clarification

Pectolytic preparations lower white must viscosity and thus accelerate sedimentation (Figure 11.15). In less than an hour, the colloidal equilibrium is destabilized, resulting in a rapid sedimentation and increased must limpidity. A more compact must deposit, facilitating static settling, is rare. This treatment can lead to excessive juice clarification. Its use should be determined according to must

Table 11.8. Enzymatic treatment of crushed grapes (Kadarka variety, Hungary; enzymatic preparation: Vinozyme 2 g/hl) (Canal-Llaubéres, 1989)

Juice	Control	With enzyme
Free run	63%	93%
Press	37%	7%

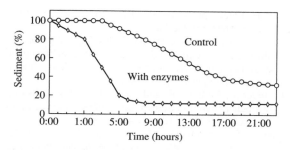

Fig. 11.15. Effect of pectolytic enzymes on the sedimentation speed of white must lees (Canal-Llaubères, 1989)

composition. The enzymatic degradation of pectic compounds is subsequently demonstrated by a distinct improvement in the filterability of the musts and wines obtained. These wines are often better prepared for tangential filtration (Volume 2, 11.9).

In red winemaking, these preparations are used in particular for press wines and heat-treated grapes and must. In the latter case, the must is very rich in pectic compounds and devoid of endogenous grape enzymes. These are destroyed by heat (Martinière and Ribéreau-Gayon, 1973). Pectolytic enzymes can also be used at the time of running off after a traditional maceration.

In botrytized grapes, the pectic compounds are degraded for the most part and replaced by a fungal polymer, glucan (Section 10.5.3). A glucanase is industrially prepared from *Trichoderma* sp. fungus cultures (Dubourdieu *et al.*, 1981). The enzyme is preferably added (1–3 g/hl) after fermentation. Its action takes from 7 to 10 days and must occur at a temperature equal to or greater than 10°C. Higher doses are required in red winemaking since phenolic compounds partially inhibit the glucanase. Industrial glucanase also affects the yeast cell walls and improves the wine's colloidal stability.

11.7.3 Color Extraction and Stabilization

Red wine color results from maceration of grape solids (skins, pips and sometimes stalks) during alcoholic fermentation. Phenolic compound extraction thus depends on many factors: grape variety,

grape maturity, length of maceration, number of pumping-overs, temperature, etc. (Section 12.5). Adding pectolytic enzymes at the start of maceration can facilitate this extraction (Table 11.9). The resulting wine is richer in tannins and anthocyanins with a higher color intensity and redder tint.

This treatment also improves the organoleptical characters (notably structure) of the wine (Canal-Llaubères, 1992). It apparently favors color stabilization by forming polymerized pigments (Parley *et al.*, 2001). Further research is needed to evaluate the stability of these changes during aging. These preparations also contain β-D-glucosidase, likely to hydrolyze anthocyanin glycosides (van Rensburg and Pretorius, 2000).

11.7.4 Freeing of Aromas

The glycosidases contained in commercial pectolytic enzymes are capable of partially hydrolyzing terpenic glycosides (Table 11.10). The first

Table 11.9. Influence of pectolytic enzymes on color extraction in red winemaking (Merlot in Bordeaux, France, 1988; Vinozyme 3 g/hl at filling) (Canal-Llaubères, 1990)

Wine (20 days of maceration)	Control tank	Enzymed tank
Absorbance at 280 nm	64	66
Tannins (g/l)	3.5	3.8
Anthocyans (mg/l)	768	895
Color intensity	1.58	1.68
Tint	0.44	0.40
Absorbance at 420 nm (%)	27.8	26.2
Absorbance at 520 nm (%)	63.0	65.3
Absorbance at 620 nm (%)	9.2	8.5

Table 11.10. Liberation of terpenols by enzymatic hydrolysis (Gewürtraminer 1985; Novoferm 12 = 15 ml/hl, 1 month incubation at 18°C) (Canal-Llaubéres, 1990)

Terpenols (µg/l)	Control wine	Enzyme wine
Linalol	141	151
Terpineol	74	75
Citronellol	45	52
Nerol	53	104
Geraniol	216	358
Total	529	740

tests of these enzymes were conducted on dry wines because of the inhibiting effect exerted by glucose on the β-glucosidase. These enzymes may also act on other aromatic compounds present in the form of odorless precursors in certain grapes.

This treatment is intended to complete the terpenic compound transformations effected by yeasts during fermentation. However, it releases all the terpenic alcohols too rapidly. The pleasant-smelling monoterpenes, such as linalol, nerol, and geraniol, may be converted into more stable forms during aging, including terpineol, which has a less attractive aroma (Park, 1996). (Volume 2, Section 7.2)

In any case, care should be taken to avoid enzyme preparations containing cinnamate decarboxylase as it may lead to the development of ethyl-phenols with a highly unpleasant musky odor (Volume 2, Section 8.4.3).

Enzymatic preparations should never contain cinnamate decarboxylase. This enzyme can lead to the formation of ethyl-phenols with a very disagreeable animal odor (Chapter 2).

REFERENCES

Bayonove C. (1993) Les composés terpéniques. In *Les Acquisitions récentes en Chromatographie du Vin* (ed. B. Donèche), pp. 99–120. Tec & Doc Lavoisier, Paris.

Benitez J.G., Grandal-Delgado M.M. and Martin J.D. (1993) Study of the acidification of sherry musts with gypsum and tartaric acid. *Am. J. Enol. Vitic.*, 44, 400–404.

Berger J.L. (1994) Les applications de l'osmose inverse. In *Les Acquisitions récentes dans les Traitements Physiques du vin* (ed. B. Donèche), pp. 65–80. Tec Doc Lavoisier, Paris.

Blouin J. and Peynaud E. (2001), *Connaissance et de Vin*, 3rd Edition, Dunod, Paris.

Brugirard A. (1987) Aspect œnologique des moûts concentrés rectifiés et utilisation pratique en vinification. Rev. Œnol. 44, 9–12.

Calderon A.A., Garcia-Florneciano E., Munoz R. and Ros Barcelo A. (1992) Gamay grapevine peroxidase: its role in vacuolar anthocyani(di)n degradation. *Vitis*, 31, 139–147.

Canal-Llaubères R.M. (1989) Les enzymes industrielles dans la biotechnologie du vin. *Rev. Œnol.*, 53, 17–22.

Canal-Llaubères R.M. (1992) Enzymes in winemaking. In *Wine Microbiology and Biotechnology* (ed.

G.H. Fleet), pp. 477–506. Harwood Academic Publishers, (hur, Switzerland).

Cantagrel R., Symonos P. and Carles J. (1982) Composition en acides aminés du moût en fonction du cépage et de la technologie, et son influence sur la qualité du vin. *Sci. Aliments*, 2(HS 1), 109–142.

Chauvet S., Sudraud P. and Jouan T. (1986) Sélection des baies et enrichissement des moûts par cryoextraction sélective. *Viti*, 101, 11–38.

Clary C.D., Steinhauer R.E., Frisinger J.E and Peffer T.E. (1990) Evaluation of machine vs hand-harvested Chardonnay. *Am. J. Enol. Vitic.*, 41, 176–181.

Cordonnier R. and Dugal A. (1968) Les activités protéolytiques du raisin. *Ann. Technol. Agric.*, 17 (3), 189–206.

Cordonnier R.E., Gunata Y.Z., Baumes R.L. and Bayonove C.L. (1989) Recherche d'un matériel enzymatique adapté à l'hydrolyse des précur-seurs d'arômes de nature glycosidique du raisin. *Conn. Vigne Vin* 23, 7–23.

Crouzet J. (1986) Les enzymes et l'arôme des vins. *Rev. Fr Œnol.*, 102, 42–49.

Dubernet M. (1974) Recherches sur la tyrosinase de *Vitis vinifera* et la laccase de *Botrytis cinerea*. Applications technologiques. Thèse de doctorat, Université de Bordeaux II.

Dubourdieu D., Villetaz J.C., Desplanques C. and Ribé-reau-Gayon P. (1981) Dégradation enzymatique du glucane de *Botrytis cinerea*. Application à l'amélioration de la clarification des vins issus de raisins pourris. *Conn. Vigne Vin*, 15, 161–177.

Dubourdieu D. Ollivier C. and Boidron J.N. (1986) Incidences des opérations préfermentaires sur la composition chimique et les qualités organoleptiques des vins blancs secs. *Conn. Vigne Vin*, 20, 53–76.

Dupuy P. and De Hoogh J. (1991) *The enrichment of wine in the European Community—Wageningen Agricultural University*. Report EUR 13239 EN, 1–151, Commission of the European Communities.

Flanzy C. (1998), *Oenologie. Fondement scientifique et technologique*. Lavoisier Tec. Doc. Paris.

Guimberteau G., Gaillard M. and Wajsfelner R. (1989) *Conn. Vigne Vin*, 23 (2), 95–118.

Haushofer H. 1972, La désacidification des moûts et des vins par formation d'un sel double. *Conn. Vigne Vin*, 6, 373.

Jacquer P. (1990) Le tri manuel de la vendange à la loupe. *Viti*, 148, 121–124.

Jacquet P. (1995) Les systèmes de transport. *J. Int. Sci. Vigne Vin*, H.S., 7–19.

Jacquet P. (1995) Les tables de tri. *J. Int. Sci. Vigne Vin*, H.S., 27–32.

Martin G.J., Guillou C., Naulet N., Brun S., Tep Y., Cabanis J.C., Cabanis M.T. and Sudraud, P. (1986) *Sci. Alim.*, 6, 385–405.

Martinière P. and Ribéreau-Gayon J. (1973) Etude expérimentale de l'influence du chauffage des raisins sur la vinification. *Ann. Technol. Agric.*, 22, 1–20.

Mayer A.M. and Harel E. (1979) Polyphenol oxidases in plants. *Phytochem.*, 18, 193–211.

Müller-Späth H. (1990) Historique des expérimentations de vinification sans SO₂ et par oxygénation. *Rev. Fr. Œnol.* 124, 5–12.

Park S.K. (1996), *Food Biotechno.*, 5, 280–286.

Parley A., Vanhanen L. and Heatherbell D. (2001), *J. Grape and Wine Rese.*, 7, 146–152.

Peynaud E. and Allard J.J. (1970) Concentration des moûts de raisin par osmose inverse. *CR Acad. Agric.*, 56 (18), 1476–1478.

Poux C. and Ournac A. (1972) Détermination de la peroxydase dans le raisin. *Ann. Technol. Agric.*, 21 (1), 47–67.

Rapp A., Hastrich H. and Engel L. (1976) *Vitis*, 15, 29–36.

Reader H.P. and Dominguez M. (1995) Fortified wines: Sherry, Port and Madeira. In *Fermented Beverage Production* (eds A.G.H. Lea and J.R. Piggott) pp. 159–207. Blackie Academic and Professional, Glasgow.

Ribéreau-Gayon J., Peynaud E., Ribéreau-Gayon P. and Sudraud P. (1976) Amélioration de la vendange. In *Sciences et Techniques du vin* Vol. 3: *Vinifications, transformations du vin*, pp. 3–28. Dunod, Bordas, Paris.

Rigaud J., Cheynier V., Souquet J.M. and Moutonet M. (1990) Mécanismes d'oxydation des polyphénols dans les moûts blancs. *Rev. Fr. Œnol.*, 124, 27–31.

Salgues M., Cheynier V., Gunata Z. and Wylde R. (1986) Oxydation of grape juice 2-S-glutathionyl caffeoyl tartaric acid by *Botrytis cinerea* laccase and characterization of a new substance: 2,5-di-*S*-glutathio-nyl caffeoyl tartaric acid. *J. Food Sci.*, 51, 1191–1194.

Schmitt A., Köhler H., Miltenberger A. and Curschmann K. (1989) Versuch zum einsatz pektolytischer enzyme mit proteolytischer nebenaktivitat. *Deutsche Weinbau*, 11, 408–414.

Singleton V.L. (1987) Oxygen with phenols and related reactions in must, wines and model systems: observations and practical implications. *Am. J. Enol. Vitic.*, 38, 69–77.

Terrier N., Sauvage F.X. and Romieu C. (1996) Absence de crise respiratoire, induction de l'activité alcool deshydrogénase et diminution de l'acidité vacuolaire lors de la maturation du raisin In *Œnologie 95*, Compte Rendu 5ème Symposium International d'Œnologie, pp. 24–28. Tec & Doc Lavoisier, Paris.

Tinlot R. (1990) La situation mondiale de l'enrichissement. *Vignes Vin*, 5, 51–54.

Usseglio-Tomasset L. (1978) Acquisitions récentes sur les phénomènes colloïdaux dans les moûts et les vins. *Ann. Technol. Agric.*, 27, 261–274.

Usseglio-Tomasset L. and Bosia P.D. (1992) La désacidification des moûts selon la méthode allemande. *Bull. OIV*, 731–732, 5–13.

Van Rensburg P. and Pretorius J.S. (2000), *S. Agri. J. Enol. Vitic.*, 21, 52–73.

Vromandt G. (1989) Dossier vendangeuses. *Viti* 137, 83–103.

Wucherpfennig K. (1967) Tendances dans l'évolution du travail des vins en Allemagne. In *Compte Rendu 2ème Symposium International d'Œnologie* pp. 411–433, INRA.

Wucherpfennig K. (1980) Possibilités d'utilisation des processus membranaires dans l'industrie des boissons. *Bull. OIV*, 583, 186–205.

12

Red Winemaking

12.1 GENERALITIES

Red wine is a macerated wine. The extraction of solids from grape clusters (specifically from skins, seeds and possibly stems) accompanies the alcoholic fermentation of the juice. In conventional red winemaking, extraction of grape solids is by means of maceration, which occurs during must fermentation. Other methods exist that dissociate fermentation and maceration, such as thermovinification.

The localization of red pigment exclusively in skins, at least in the principal varieties, permits a slightly tinted or white wine to be made from the colorless juice obtained from a delicate pressing of red grapes. Wines for the elaboration of champagne are a good example. The designation *blanc de blanc* was created to distinguish white wines derived from white varieties and those from red. Finally, varietal nature is not sufficient for characterizing the origin of a red wine. Maceration intensity is of prime importance.

The length and intensity of maceration are adjusted according to grape variety and the type of wine desired. In fact, maceration is a means by which the winemaker can personalize the wine. *Primeur* wines are made to be drunk young: their aromas and fruitiness greatly outweigh phenolic compound concentrations, but premium wines require a sufficient tannin concentration to develop properly during aging.

Grape quality directly influences grape skin maceration quality in red winemaking and is thus of the greatest importance. In fact, the grape skin is more affected than the juice by cultivation techniques, maturation conditions and sanitary state. Vintage and growth rankings are therefore much more clearly defined with red wines than whites. In the Bordeaux region, anthocyanin and tannin concentrations in the same parcel can vary by as much as a factor of two, from one year to another, according to maturation conditions. Must acidity and sugar concentrations can fluctuate by 50% and 15%, respectively. These numbers are not surprising, since the plant requires a lot of energy to synthesize anthocyanins. For this reason, the northernmost vineyards produce only white wines. In any case, when phenolic compound concentrations are examined in relation to environmental conditions, their nature, properties and localization in the tissues must also be considered. Enologists readily define 'good' tannins as those that give wines a dense structure without aggressiveness, and 'bad' tannins as those characterized by vegetal and astringent herbaceous savors. The nature and chemical properties of these various phenolic compounds are covered in Chapter 6 of the second volume of this series. This highlights the need to wait until the grapes reach full phenolic maturity, which may occur later than physiological ripeness. Similarly, high levels of methoxypyrazines in insufficiently ripe grapes of certain varieties (especially Cabernet Sauvignon) are responsible for a herbaceous, green bell pepper character in must and wine that is considered a defect above certain levels (Volume 2, Section 7.4)

Grape composition and quality variability result in heterogeneous grape crops. Grape selection can compensate for this heterogeneity and tanks should be filled with a homogeneous single-variety grape crop that has the same sanitary state and level of maturity. *Terroir*, quality, vine age, rootstock, fruit loads, and a number of other factors should be taken into consideration. Appropriate vineyard management methods are increasingly being applied to

achieve the low yields essential to ensure perfect grape ripeness and high quality. This batch selection, effected at filling time, must be maintained during the entire winemaking process, until the definitive stabilization after malolactic fermentation. The best batches are then blended together to make a wine of superior quality. The complementary characteristics of the various batches often produce a blended wine that is superior in quality to each of the batches before blending.

The grape crop should also be carefully sorted to eliminate damaged or unripe grapes. This operation can be effected in the vineyard during picking or in the winery at harvest reception. At the winery, the grapes are spread out on sorting tables. A conveyor belt advances the crop, while workers eliminate bad grapes. A concern for perfection in modern winemaking has led to the generalization of such practices. Their effectiveness is even more pronounced when they are applied to grape crops of superior quality.

Red grape crop heterogeneity requires specific winemaking techniques to be adapted according to the crop. Much remains to be learned in optimizing the various grape specifications.

The generalization of malolactic fermentation is another characteristic of red winemaking. This phenomenon has been recognized since the end of the last century but, until the last few decades, it was not a consistent component of red winemaking. For a long time, a slightly elevated acidity was considered to be an essential factor in microbial stability and thus contributed to wine quality. Moreover, red wine must acidification was a widespread practice. It has currently disappeared for the most part, since it is only justified in particular situations. Today, on the contrary, malolactic fermentation is known to produce a more stable wine by eliminating malic acid, a molecule easily biodegraded.

It was in temperate regions that malolactic fermentation (MLF) first became widespread. These wines, which are rich in malic acid, are distinctly improved, becoming more round and supple. MLF was then progressively applied to all red wines, even those produced in warm regions already having a low acidity. This type of fermentation may not be advisable in all regions and another method of stabilizing red wines containing malic acid should be sought.

The classic steps in red winemaking are:

- mechanical harvest treatments (crushing, destemming and tank filling);

- vatting (primary alcoholic fermentation and maceration);

- draining (separation of wine and pomace by dejuicing and pressing);

- final fermentations (exhaustion of the last grams of sugar by alcoholic fermentation and malolactic fermentation).

There are currently many variations on each stage in traditional winemaking, but the operations described in this chapter constitute the basic method for producing high quality red wines. It does, however, require considerable tank volume capacity and many constraining manipulations. In consequence, other techniques have been developed. The standard order of certain operations has been changed to make a certain level of automation possible—for example, in continuous vinification and heat extraction (Section 12.8).

Finally, fermentation with carbonic maceration takes advantage of the special aromatic qualities produced by fermenting whole grapes under anaerobic conditions (Section 12.9). This special fermentation gives these wines specific organoleptic characters.

12.2 MECHANICAL HARVEST TREATMENTS

12.2.1 Harvest Reception

Diverse methods, adapted to each winery, are used to transport the harvest from grapevine to winery. In world-renowned vineyards, small-capacity containers are carefully manipulated by hand. In most vineyards, the harvest is transported in shallow bed trailers or trucks. Whatever the container capacity, the grapes should be transported intact without being crushed. Transport containers should also be kept clean. If the transport time is long the grapes should be transported during the cooler hours of the night.

Red grapes are certainly less sensitive to maceration and oxidation phenomena than white grapes, but microbial contamination is likely to occur in a partially crushed harvest, left in the vineyard, especially in the presence of sunlight. These risks must be avoided.

During mechanical harvesting, the grapes are transported in high-capacity containers. Speed and hygiene are even more important in this case, since the grapes are inevitably partially crushed with this method.

Small-volume containers are emptied manually. More generally, a dumping trailer is used, which empties its load into the receiving hopper (Figure 12.1). In high-capacity installations, the bins are placed on a platform which dumps the grapes sideways—thus avoiding excessive truck and tractor maneuvering. If the winery is equipped, the grape crop may pass on a sorting table (Section 11.3.3) before reaching the receiving hopper. Manual sorting is only effective if the grapes are whole. It is almost impossible to combine with mechanical harvesting: at best, obviously damaged grapes can be removed from the vines before the harvester arrives. Otherwise, sorting may take place in the vineyard, immediately after the grapes have been picked, or when the grapes arrive at the winery. In the latter case, the grapes should be transferred to the sorting table manually to spread them evenly. Transfer screws should not be used, as they crush the grapes and make sorting impossible.

Two sorting tables may be necessary at wineries producing high quality wine. The grapes are initially sorted when they arrive from the vineyard. The second sorting operation, after destemming, removes any small fragments of stems and leaves, etc. that were missed during the first sorting operation, and is followed by crushing. There are now increasing numbers of machines, based on various techniques, available to do this operation automatically.

Receiving hoppers are available in various designs. In small wineries, they may be installed directly above the crusher-stemmer and filled

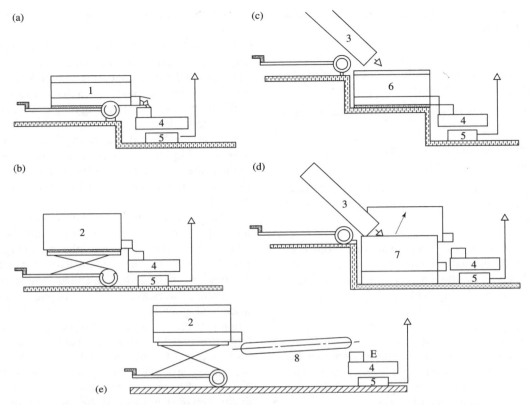

Fig. 12.1. Examples of harvest receiving equipment for red winemaking (Source: P. Jacquet, Bordeaux, personal communication). (a) Gondola with screw conveyor, gravity dumping. (b) Elevator gondola with screw conveyor, adjustable dump height. (c) Gravity dump gondola. (d) Gravity dump gondola, elevator hopper designed to gravity feed destemmer—crusher. (e) Sorting table between trailer and destemmer—crusher. Key: (1) trailer with a screw conveyor; (2) same trailer with an elevator system; (3) dumping trailer; (4) destemmer—crusher; (5) grape pump; (6) stationary hopper feeding the destemmer—crusher by gravity; (7) elevator hopper with adjustable height to feed the crusher by gravity; (8) sorting table

directly from the transfer vehicle. In general, a perpetual screw in the bottom of the hopper regulates throughput and it should turn slowly to avoid excessive crushing of the grapes. Throughput may be increased by using a larger-diameter hopper.

When buying grapes according to weight and sugar concentration, these values must be determined at the time of reception. Grape crop heterogeneity complicates the determination of the sugar concentration. The sample should therefore be taken after crushing and homogenization. The sanitary condition of the grapes may also be assessed at this stage by analyzing their laccase activity (Section 10.6.6).

At the outlet of the crusher—destemmer, a pump distributes the grapes to a given tank. Sulfur dioxide is added at this time (Sections 8.7; 8.8.1) and any necessary addition.

Grape handling should be minimized, limiting transfer distances and maximizing the use of gravity. Rough handling is likely to shred or lacerate stem tissues, so that sap is liberated from vegetal tissue and later found in wine. The suspended solids concentration simultaneously increases; in fact, this measurement may be used to evaluate equipment quality. The most quality-oriented solution consists of sorting and destemming the grapes by hand, then crushing

Red Winemaking

them, if necessary, through a wooden screen, thus eliminating the need to crush them mechanically. Finally, the must is transferred without pumping. Of course, only the most prestigious estates can afford the high cost of these techniques.

In partially botrytized grapes, a brutal mechanical action on grapes disperses a glucidic colloid (glucan) in the must. Glucan is produced by *Botrytis cinerea* and is located between the pulp and the skin inside the berry. The wine obtained is difficult to clarify. When the same grapes are carefully handled, the wine is clarified much more easily. Wine clarification difficulties with botrytized grapes are always observed in the same wineries. The type of equipment used is often responsible.

12.2.2 Crushing

Grapes are traditionally crushed to break the skin in order to release the pulp and the juice. This operation is probably one of the most ancient harvest treatments. Partial crushing can be obtained by the traditional technique of treading the grapes. High-speed centrifugal crusher—destemmers assure an energetic crushing. There are also many other systems between these two extremes.

The consequences of crushing are as follows:

1. The juice is aerated and it is inoculated by yeasts. The fermentation is quicker and the temperature higher. In certain circumstances, a slower fermentation speed and lower temperatures can be obtained through not crushing (carbonic maceration, Section 12.9.4).

2. Aeration can be harmful. In partially rotted grapes, it can provoke an oxidasic casse.

3. Crushed grapes can be pumped, and sulfiting is more homogeneous.

4. All of the juice is fermented; at the time of running-off, the press wine does not contain sugar.

5. Crushing has a significant effect in facilitating maceration and accentuating anthocyanin and tannin dissolution. An energetic crushing intensifies this effect. Tannin concentrations proportionally increase more rapidly than the color.

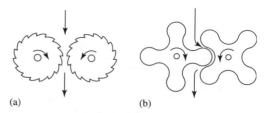

Fig. 12.2. Crusher roller design: (a) spiral ribbed rollers; (b) grooved rollers with interconnecting profiles

This increased maceration can be an advantage in certain cases but it tends to increase the herbaceous astringency and disagreeable tastes of average varieties.

Premium wine grapes are traditionally lightly crushed to burst the berries without lacerating the solid parts. Crushing is used to facilitate fermentation and avoid residual sugar in press wines. Methods other than crushing should be used to increase maceration (vatting time, pumping-over operations, temperature). They better respect wine quality. Even when carbonic maceration is not strictly used (Section 12.9), winemakers may wish to avoid crushing the grapes for great wines with long vatting periods, to avoid brutal damage to the plant tissues.

Two kinds of crushers exist. Roller crushers (Figure 12.2) are coated with plastic; the opposing rollers turn in opposite directions and their spacing is easily adjustable. This system works well but delivery rates are limited. High-speed perforated wall crushers (Figure 12.3) can be either horizontal or vertical. A beater projects the grape clusters against a perforated wall, and the burst grapes pass through the perforations. These machines simultaneously destem the grapes. As they are rough on the grapes, they are not recommended, especially in making high quality wines.

12.2.3 Destemming

This operation, also known as destalking, is now considered indispensable (after much discussion of its advantages and disadvantages for a long time).

Destemming has a number of consequences:

1. A primary and financially important advantage of this operation is the reduction of the required

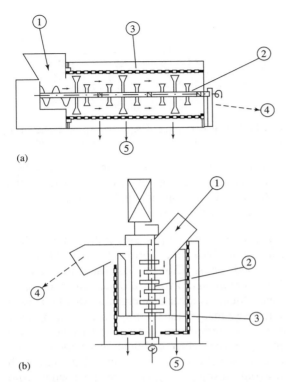

(a)

(b)

Fig. 12.3. Operation principle of (a) a horizontal destemmer and (b) a vertical centrifugal destemmer. Key: (1) hopper; (2) shaft with arm and paddles; (3) perforated cylinder; (4) stem outlet; (5) destemmed grape outlet

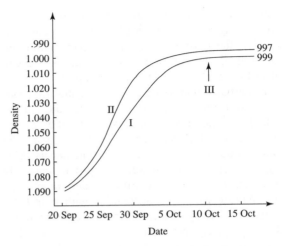

Fig. 12.4. Influence of stem on alcoholic fermentation (Ribéreau-Gayon *et al.*, 1976). I: destemmed grapes. II: non-destemmed grapes. III: running-off date

tank capacity by 30%. In addition, the pomace volume to be pressed is greater with a stemmed grape crop. Although the stem facilitates juice extraction during pressing, a higher-capacity press is required.

2. Fermentations in the presence of stems are always quicker and more complete (Figure 12.4). The stem facilitates fermentation not only by ensuring the presence of air but also by absorbing calories, limiting temperature increases. Fermentation difficulties are rarely encountered with stemmed grapes.

3. The stems modify wine composition. They contain water and very little sugar, thus lowering alcohol content. Moreover, stem sap is rich in potassium and not very acidic. Destemming therefore increases must acidity and alcohol content.

4. With botrytized grapes, stems protect wine color from oxidasic casse. The laccase activity of *Botrytis cinerea* is most likely fixated or inhibited.

5. Destemming most significantly affects tannin concentrations. Table 12.1 indicates the approximate proportion of phenolic compounds supplied by the various parts of the grape cluster. In this experiment, 54% of the total tannins come from grape skins, 25% from seeds and 21% from stems. Results may vary according to grape quality and grape variety. More precise details on the nature and concentrations of phenolic compounds from the various parts of the grape cluster are given in Chapter 6 of the second volume.

Table 12.2 recapitulates the principal modifications of wine constitution caused by destemming. Despite the increase in total phenolic compounds in the presence of stems, color intensity diminishes. This long-observed fact is interpreted as the adsorption of grape skin anthocyanins on the ligneous surface of the stems. This interpretation has been confirmed in a model solution containing anthocyanins and tannins; either a stem extract or the stems themselves is added. In the first case, the tannin concentration increases considerably, while the

Table 12.1. Influence of different parts of the grape cluster on phenolic compounds and wine color (Malbec, fermentation at 25°C, vatting time 10 days) (Ribéreau-Gayon et al., 1976)

Component	Juice	Juice + skins	Juice + skins + seeds	Juice + skins + seeds + stem
Color[a]:				
Intensity	—	1.81	1.40	1.17
Tint	—	0.39	0.43	0.48
Anthocyanins (g/l)	—	0.98	0.94	0.85
Tannins (g/l)	—	1.75	2.55	3.25
Total phenolic compounds (permanganate index)	5	32	47	56

[a]Intensity = OD 420 + OD 520
Tint = OD 420/OD 520
(OD 420 and OD 520 = optical density, under 1 mm thickness, at 420 nm and 520 nm).

color intensity slightly increases. In the second case, the tannins increase but the color intensity decreases. Tannins play an important role in the color of mature wines. Although wines made from stemmed harvests have less color when young, they become more colored than their destemmed counterparts in the course of aging.

6. The increased tannin and phenolic compound concentration of wines made from stemmed harvests can increase wine quality in certain cases, e.g. for young vines and wines with insufficient structure without the stems. Yet grape stems are likely to give vegetal and disagreeable herbaceous tastes to wines. In general, when finesse is favored, destemming is indispensable. In any case, the decision of

Table 12.2. Influence of stem on wine composition (Ribéreau-Gayon et al., 1976)

Component	Destemmed	Not destemmed
Alcoholic strength (% vol.)	13.2	12.7
Total acidity (mEq/l)	86	78
Volatile acidity (mEq/l)	11.4	11.4
Total phenolic compounds (permanganate index)	38	58
Color[a]:		
Intensity	1.28	1.18
Tint	0.51	0.57

[a]See Table 12.1.

a total or partial destemming must take into account stem quality, which is related to variety and maturity level.

In the past, the grape crop was destemmed by hand directly in the vineyard or, more generally, in the winery, by rubbing the grape clusters with rakes against a wooden hurdle. Today, this operation is carried out mechanically (Figure 12.3). Destemmers comprise a perforated cylinder, with a shaft equipped with paddle-like arms running through its center. When the shaft turns, it draws in the grape clusters and expels the stems out the other end. The juice, pulp and grape skins pass through the perforations. The continual quest for higher output has lead to increased rotation speeds and replacing the paddle-shaped rods with beaters. The beaters apply sufficient force to burst the grapes without the need of a crusher. Vertical shaft destemmers (Figure 12.3) can treat 20–45 metric tons per hour. They operate at 500 rpm and the centrifugal force evacuates the stems by the top of the machine. These machines have a brutal action on the grapes and produce fine suspended solids, imparting vegetal and herbaceous tastes to wines. Their use should be avoided, at least for the production of premium wines (Section 12.2.2).

Crushing and destemming are generally effected by the same piece of equipment, but in certain cases it would be desirable to have the option of not destemming. For a long time, with conventional crusher–destemmers, crushing preceded destemming. Today, there is an increasing number of machines that eliminate the stems before crushing the grapes. The stems do not pass between the crusher rollers. In this manner, the risk of shredding the stems is lowered. This order of operation increases must quality, since stem shredding liberates vegetal vacuolar sap, which is bitter and astringent.

A quality destemmer should not leave any berries attached to the stem. Reciprocally, the stem should not be impregnated with juice. The stem should also be entirely eliminated, with no broken fragments remaining. The laceration of ligneous stem tissue by the machine can seriously affect quality and in these instances destemming should be avoided.

Attempts have been made to eliminate residual stem waste after the destemmer and before the crusher. Quite large quantities are removed in this way.

12.3 FILLING VATS

12.3.1 Filling Vats and Related Operations

In the case of fermentation with carbonic maceration (Section 12.9), vats must be filled directly from the top with uncrushed grapes, which obviously requires a very complex system. Otherwise, grapes are usually received at a single winery location and transferred to the fermentation vats after destemming and crushing. Transfer pumps must do as little damage as possible to grape tissues and distances should be kept to a minimum, with as few bends as possible in the hoses. This operation can be carried out manually, without pumping. As the must increases in volume during fermentation, about 20% empty space should be left in each vat. If anti-foaming agents (Section 3.2.5) are used, less headspace is required.

A considerable volume of gas is released during fermentation, approximately 50 l of carbon dioxide per liter of must fermented. Ensuring that a flame stays alight inside the fermentation vessel before going inside helps check for oxygen, in view of the danger of asphyxiation from carbon dioxide.

The grapes must be sulfited adequately and homogeneously during transfer to the vat (Section 8.8.1).

Several operations may be carried out during transfer of the grapes/must, or in the following few hours. Firstly, they may be inoculated with a fermenting must (a few percent corresponding to 10^6 cells/m) or dried active yeast (LSA), S. cerevisiae, chosen from among the various commercial strains (over 100). The main qualities required are the aptitude to complete fermentation successfully and heat resistance. The impact of yeast strains on the character of red wines is less marked than in the case of white wines (Section 13.7.2). Winemakers must still ensure, however, that the strain selected is suitable for the type of wine being made. Recommended doses of 10–25 g/hl correspond to inoculation with $2.10^6 - 10.10^6$ cells/m. Indigenous yeasts must be inhibited by appropriate doses of SO_2 to ensure effective seeding. Dried yeasts must be reconstituted prior to use, by mixing them into a mixture of must and water (1:1) at 40°C. The reconstituted yeast must be spread evenly through each vat.

Acidity can be corrected (Section 11.4) during the initial transfer into vat or at a later time. If sugar levels need to be increased (chaptalization, Section 11.5.2), this is best done when the must is warm at the beginning of fermentation. The sugar dissolves more easily and the subsequent aeration stimulates fermentation, while relatively low sugar levels promote multiplication of the yeast cells during the growth phase.

If an assay (Section 3.4.2) indicates a nitrogen deficiency, ammonium sulfate (10–30 g/hl) may be added as soon as the vat has been filled, or, preferably, once fermentation has started.

Adding tannin during fermentation had been abandoned for a long time, but the quality of the products now available, particularly those made from white grape skins or fresh grape seeds, has revived interest in this procedure. These products are not only considered capable of improving body and tannic structure, but also of stabilizing color by promoting condensation of anthocyanins and tannins (Volume 2, Section 6.3.10).

It is thus useful to add tannin early in the fermentation process, when the tannins have not yet been extracted from the grape seeds, so that they can react with the anthocyanins released early in vatting. According to some authors, the results are uneven due to the low solubility of the tannins and the difficulty of mixing them into the must (Blouin and Peynaud, 2001), so it is preferable to add the product after running-off. High doses (20–50 g/hl) are required to raise the initial tannin levels by approximately 10%.

Another operation currently attracting some interest is the addition of pectolytic enzymes to promote extraction of phenolic compounds (Sections 11.7.3; 12.5.1), for the purpose of obtaining wines with a higher tannin content, but less astringency and bitterness (Blouin and Peynaud, 2001).

Glycosidases may also be used to promote extraction of terpenic aromas—particularly useful in making Muscat wines. Care must be taken in traditional red winemaking to avoid producing off-aromas. The use of enzymes in winemaking requires further scientific research.

Some juice can also be bled off at this stage (Section 12.5.8), mainly to eliminate rainwater and juice that has not yet absorbed compounds from the skins. Decreasing the quantity of must facilitates concentration of the phenolic compounds during vatting. This operation is generally carried out after the vat has been filled and the juice has been separated from the pomace. Water is eliminated (e.g. by reverse osmosis or vacuum evaporation) (Section 11.5.1) at the same time. The results are very similar but these methods maintain the natural grape sugars. These techniques are capable of concentrating the must by 5–10%, or even as much as 20%. Excessive concentration of the must changes the flavor balance of the wine completely and it is preferable to adapt vineyard management methods and reduce yields on the vine to achieve similar results.

12.3.2 Principal Vatting Systems

Various types of fermentor exist. They are distinguished by the aeration level supplied to yeasts and the modulation of skin contact. Aeration helps to ensure a complete fermentation, and skin contact modulation influences maceration and phenolic compounds extraction.

Fermentation releases gas within the must. The bubbles rising toward the surface of the fermentor entrain solid particles, which unite and agglomerate, forming the cap. The skin cap is maintained at the top of the fermentor by the pressure of the released gas.

Pomace plays an important role. First and foremost, during maceration, it yields its constituents (anthocyanins and tannins). These compounds are indispensable components of the character of red wine. Yeast multiplication is also particularly intense within the pomace: $10–50 \times 10^6$ cells/ml have been observed in the juice at the bottom of the fermentor and $150–200 \times 10^6$ cells/ml in the juice impregnating the pomace.

Although no longer recommended, open floating-cap fermentors are still used in small-scale installations. They were used in the past because the extended contact with air permitted successful fermentations, even in musts containing high concentrations of sugar. Moreover, temperature increases are less significant. Yet, the inconveniences are undeniable. Alcohol losses can attain and sometimes exceed 0.5% (Section 12.6.1; see Table 12.10). The risk of oxidasic casse with botrytized grapes is also certain. Additionally, as soon as the active fermentation stops, the pomace cap surface is no longer protected from aerobic germs development. Bacterial growth is facilitated and contamination risks are high due to the large surface area of this spongy surface. As soon as the fermentation slows, the pomace cap should be regularly immersed to drown the aerobic germs. This operation, known as cap punching (pigeage), can only be carried out manually in small-capacity fermentors. If necessary, it can be mechanically effected with a jack or another piece of equipment. Submerging the pomace cap also contributes to the extraction of its constituents. It also aerates the must and homogenizes the temperature. But this type of fermentor does not permit a long maceration. The tanks must be run off before the carbon dioxide stops being released. Afterwards, spoilage risks in the pomace cap are certain and the resulting press wine would have an elevated volatile acidity. Manually removing the upper layer of the most contaminated part of the pomace cap is not sufficient, nor is covering the tank with a tarpaulin after fermentation.

To avoid pomace cap spoilage and to eliminate the laborious work of regularly punching down the cap, systems have been developed that maintain the cap immersed in the must—for example, under a wooden hurdle fitted to the tank after filling. The must in contact with air is permanently renewed by the released gas. Acetic acid bacteria have more difficulty developing in this environment. The compacting of the pomace against the wooden hurdle does not facilitate the diffusion of its constituents, and several pumping-overs are therefore recommended to improve maceration.

Today, most red wines are fermented in tanks that can be closed when the carbon dioxide release rate falls below a certain level. The complete protection from air permits maceration times to be prolonged, almost as long as desired. The tank can be hermetically sealed by a water-filled tank vent (Figure 12.5) or simply closed by placing a cover on the tank hatch. In the latter case, the CO_2 which covers the upper part of the tank disappears over time and the protection is not permanent. The tank should therefore be completely filled with wine or a slight pumping-over operation should be carried out twice a day to immerse the aerobic germs.

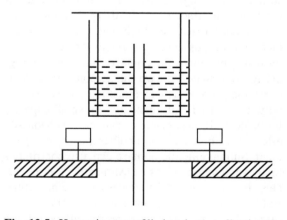

Fig. 12.5. Hermetic water-filled tank vent allowing the release of CO_2 from tank during fermentation without air entering

For a long time, the major inconveniences of the closed fermentor were a considerable temperature increase and the absence of oxygen. As a result, fermentations were often long and difficult, and stuck fermentation occurred frequently. Today, these two inconveniences are mitigated by temperature control systems and pumping-over operations with aeration, permitting the dissolution of the necessary oxygen for a successful fermentation.

In conclusion, this fermentor design avoids alcohol loss by evaporation. Press wine quality is greatly increased, while the laborious work of cap punching is eliminated. This kind of tank has also been empirically observed to facilitate malolactic fermentation.

12.3.3 Fermentor Construction

Red wine fermentors have been successively made of wood, concrete and steel, and on occasion plastic.

Wood is a noble material and wooden tanks have long been part of the tradition in great winemaking regions. New wood releases aromatic compounds into the wine during fermentation but this property is attenuated after a few years and this phenomenon no longer occurs during fermentation. Disadvantages are that wooden tank maintenance is difficult, that old wooden tanks are a source of contamination and bad tastes, and that wooden tanks are not completely hermetic. They must sometimes be expanded with water before use, with all of the corresponding risks of microbial contamination. In addition, the flat ceiling of a truncated tank is rarely hermetic—this kind of tank is not suitable for prolonged wine conservation. Wood is also a poor heat conductor. Wooden tanks are subject to considerable temperature increases that must be compensated by appropriate refrigeration systems; yet when the fermentation is completed, they retain the heat generated for a long time, favoring a post-fermentation maceration.

Concrete permits the effective use of available space, since the tanks are manufactured on site, but the acids in wine attack concrete. The inner tank walls must therefore be protected. The tanks can be coated with a 10% solution of tartaric acid applied three times at intervals of several days. In these conditions, the inner tank walls are coated with calcium tartrate. The wine in the tank contributes to maintaining the coating. It is, however, preferable to coat the inner tank walls with an innocuous and chemically inert lining such as epoxy resins or araldite. Whatever lining is used, the coating of these tanks requires continuous maintenance. Concrete is a better heat conductor than wood, but refrigeration systems are still indispensable. These tanks are completely hermetic and can be used for wine storage.

Steel, particularly stainless steel, is the material most often used today for manufacturing fermentors. Two categories of stainless steel exist: one contains molybdenum; the other does not. Chrome−nickel−molybdenum steel is more resistant to corrosion and it is necessary for the long-term conservation of sulfited white wines, especially in partially filled tanks: in the humid atmosphere above the wine, sulfur dioxide gas is concentrated and the condensation formed on the tank walls is corrosive. For red winemaking and storage in completely filled tanks, the less expensive chrome−nickel steel is sufficient.

Stainless steel tanks have the significant advantage of being hermetic and easily fitted with various types of equipment. Their internal and external maintenance is also facilitated; their inner walls are impregnable. Stainless steel also has a good thermal exchange, avoiding excessive temperature increases. In certain cases, red wine fermentations can occur at 30°C without cooling. In any case, cooling is simplified: a cool liquid is circulated within the double wall of the tank or in an integrated thermal exchanger. When a sufficient amount of cool water is available, running water over the exterior of the tank can be sufficient. In the 1960s and 1970s, stainless steel tanks represented a considerable advance in temperature control compared with wooden and concrete tanks and this superior control was much appreciated by winemakers. Today, however, it has been observed that these tanks insufficiently warm the fermenting must when the ambient temperature is too low. This phenomenon is accentuated in cases where the tanks have been placed outside to lower the cost of investment. As soon as fermentation stops, the tank

temperature rapidly decreases to the ambient temperature; as a result, maceration phenomena, which are influenced by temperature, are slowed. To master red winemaking, temperature-controlled (heating and cooling) stainless steel tanks are necessary. However, in any cases, as heat inertia is limited, it is difficult to obtain homogeneous temperatures in the post-fermentation phase. Recent developments in cooling equipment have led to renewed interest in wooden or concrete fermentation vats, where homogeneous temperatures are easier to maintain.

Tank capacity must also be considered when designing a winery. High-capacity tanks are of course economical, but tank size should not be exaggerated and should be adapted to the winery (50−350 hl). It is difficult to control the various steps in winemaking in vats containing over 350 hl. Tanks of limited capacity permit superior batch selection and skin extraction due to increased skin contact. The tank should be filled before the start of fermentation and for this reason the filling time should not exceed 12 hours.

Tank shape is important for red winemaking. The exchange surface between the pomace and the juice should be sufficient. The dimensions of sheet steel used during manufacturing sometimes result in tanks that are too high with respect to their diameter. Reciprocally, tanks should not be too wide. In this case, pomace leaching is greatly reduced and pumping-over operations lose their effectiveness; air contact can also be excessive. Tank height should slightly exceed tank diameter. High-performance pumping-over systems can compensate for disproportionately high tanks to a certain extent.

12.3.4 Equipping Fermentors

Basic red winemaking tanks should be equipped with the following:

- Two juice evacuation taps on the lower part of the tank, placed at different heights to facilitate racking (elimination of sediment), with an orifice at the lowest point of the tank for emptying and cleaning.

- One or two doors: one a bit higher permits the tank to be emptied after draining; the second, lower door is less essential but can be useful for cleaning the tank.

- A gauge to indicate the filling height.

- A tasting spigot for taking samples.

- A thermometer.

- A hermetic lid at the top of the tank. A water-filled tank vent (Figure 12.5) makes the tank completely airtight, while allowing the liquid to expand.

This basic configuration has often been complemented with additional, more specific add-ons. Steel tanks lend themselves particularly well to additional equipment. Nowadays, several manufacturers offer vats specifically designed for fermenting red wines. They are equipped with complex attachments that turn them into complete systems for monitoring and controlling fermentation. For example, Selector System (Gimar Tecno 15 040, Occimiano, A.I., Italy) has an automated vat cleaning system, programmable pumping-over with or without aeration and/or spraying the pomace cap, temperature control, and management of fermentation kinetics according to changes in specific gravity, so that the entire fermentation cycle can be programmed and controlled directly by the system.

Temperature control systems are generally the first add-on. Automatic temperature monitoring, sometimes continuous, is standard. Temperature probes, which must be properly placed to ensure correct measurements, are often part of a more elaborate temperature control system. Initially, these systems simply consisted of flowing cool water over the exterior of the tank. Today, cooling fluids (water or a dilute glycol solutions) are often circulated through a double wall in the tank or an internal thermal exchanger. The latter is more efficient but makes tank cleaning more difficult.

Tanks not only need to be cooled, but also heated on occasion. An identical system is therefore used to circulate a heated liquid (hot water). The liquid can be sent through the same thermal exchanger, or preferably another pipeline. Since the liquid (the juice) and the solid (the pomace) are separated in the tank, a pumping-over operation is required, at the same time as cooling or heating, to homogenize the temperature.

In high-performance installations, when a tank reaches a maximum preset temperature, a pumping-over operation is automatically performed to homogenize the temperature. If the temperature remains too high after this first operation, the tank is cooled. Automated temperature control systems have been designed that regulate the temperature throughout the entire fermentation process.

Automatic pumping-over systems have also been sought, to facilitate skin extraction (Section 12.5.5) and permit the aeration of fermenting must (Section 12.4.2). In the past, the carbon dioxide released during fermentation and the resulting pressure were used to pump the fermenting must to an upper tank, after cooling if necessary. Opening a valve releases the pressure, causing the must to cascade on the pomace cap. Due to the complexity of this system, the pump should be specifically adapted to the tank. This pumping-over operation can be done with or without aeration in the tank located below the fermentor. Aeration could be better controlled by injecting a predetermined quantity of oxygen in the lines. Pumping-over frequency and duration should of course be adapted to the must. Pumping-over too often may make the wine excessively hard and astringent. Various systems are available to improve pomace leaching and intensify skin contact.

Injections of pressurized gas at 3 bars (nitrogen, CO_2, or even air) can replace conventional pumping-over operations. A specially adapted pipe injects the gas through the piping of the lower part of the tank. Results that would normally take over an hour with a traditional pumping-over operation (rotating irrigator, stream breaker, etc.) are obtained in a few minutes. This system has been combined with standard pumping-over at the beginning and end of maceration, using food-grade gas.

Other methods complementing or replacing pumping-over operations can be employed to improve pomace extraction. Hydraulically controlled pistons have been developed which immerse the pomace cap, replacing the traditional cap punching method (*pigeage*). Various systems break up the pomace cap inside the tank, particularly

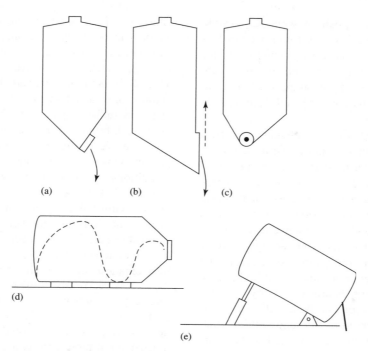

Fig. 12.6. Self-emptying tanks: (a, b) self-emptying by gravity; (c) screw-conveyor self-emptying tank (screw conveyor is incorporated into the chute at the bottom of the tank); (d) rotating tank cake break-up and pomace evacuation; (e) hydraulic dump tank

rotating cylindrical tanks. Due to the continuing evolution of these systems, a more detailed description is difficult (Blouin and Peynaud, 2001), but their use is covered in Section 12.5.8.

Another piece of equipment in high demand is an automatic pomace removal system which evacuates the skins from the tank toward the press, replacing the arduous task of manually emptying the tank.

A number of self-emptying tanks have been proposed. Models with inclined floors (20° slope) improve pomace evacuation. The worker removes the pomace with an adapted rake without having to enter the tank. Automatic self-emptying tanks are also available and several models have been proposed (Figure 12.6). The most simple are cylindrical with an extremely inclined floor (45° slope) ending in a large door. The slope of the tank floor and the door dimensions should be appropriate to the nature and viscosity of the grape crop; for example, long vatting times dry pomace, making evacuation more difficult. With this type

of evacuation system, the entire tank contents should be emptied in one go; a sufficiently large receiving system linked to the press must therefore be placed below the tank. Progressively emptying the tank may cause the pomace to get stuck: it forms an increasingly compacted arch, which is very difficult to break.

Hydraulic systems evacuate pomace by inclining the tank. Screw conveyor systems inside tanks permit a controlled and regular pomace evacuation (Figure 12.6); they are used in rotating tanks.

12.4 CONTROLLING ALCOHOLIC FERMENTATION

12.4.1 Effect of Ambient Conditions

In the past, red winemaking methods in warm and cool climates were differentiated. In certain variable-climate vineyards, warm and cool vintage winemaking techniques were also distinguished. Problems linked to fermentation temperature control and grape composition were responsible for these distinctions. These differences are less important today. The necessary conditions for successful winemaking are known: they are adapted to the nature of the grape crop and are not difficult to carry out, as long as the appropriate equipment is available.

A cool year or cool climate is characterized by a late and often insufficient maturity. Grape acidity is elevated and the musts are thus relatively protected against bacterial attack. However, there is a risk of botrytis attacks and the formation of oxidasic casse, since cool climates often correspond to rainy climates. In addition, grape crops arriving at the winery are often characterized by relatively low temperatures in cool years. As a result, the initiation of fermentation can be difficult, even more so when the grapes are washed by rain; the natural yeast inoculation can be insufficient.

Ferré (1958) observed in Burgundy region vineyards that fermentation was activated in 12 hours at 25°C, in 24 hours at 17–18°C and in 5–6 days at 15°C; it was nearly impossible at 10°C. These numbers are of course approximate and depend on many other factors, in particular the yeast inoculation concentration. Tanks should not be left at insufficient temperatures. The resulting fermentations are often slow and incomplete, with a risk of mold development beforehand. Tanks are also immobilized for prolonged periods, which can create problems in vineyards that use each tank several times during the harvest.

The must should therefore be warmed as quickly as possible to 20°C. If the fermentation does not begin shortly after warming, the temperature rapidly drops down to its initial value. A simultaneous yeast inoculation is required to avoid this problem; it also accelerates the fermentation and thus provokes a more considerable temperature increase. If the temperature becomes too elevated, cooling may be required after these operations which accelerate the fermentation. Aeration also remains useful, as long as the harvest is not susceptible to oxidasic casse.

In contrast with a cool year, a warm year or warm region produces a forward harvest. The resulting

must is rich in sugar and so a complete fermentation can be difficult to obtain. The low acidity also increases bacterial risks and requires adapted sulfiting (Section 8.8.1). All of the harvest conditions combine to produce an elevated fermentation temperature, and temperature control systems are therefore indispensable. In such a situation, the risks of a stuck fermentation and consequently bacterial spoilage are maximal. Paradoxically, the highest quality wines can be made in these winemaking conditions. In temperate climates, the greatest vintages have long been known as the most difficult ones to vinify, but winemaking methods adapted to these conditions are relatively recent. Temperature control in particular has been essential. Although a moderate temperature (20°) is necessary to initiate fermentation correctly, the temperature should not be excessive. Yeasts in their growth phase are particularly heat sensitive: when the initial temperature is between 26 and 28°C, the increase in temperature during the yeast growth phase makes stuck fermentation more common and increases the risk of producing excessive volatile acidity. Initial cooling of the grape crop is therefore recommended.

Establishing the temperature during fermentation is dependent on many factors concerning fermentation kinetics and skin extraction by maceration. Stuck fermentations are likely to occur when the temperature exceed 30°C. Slightly lower and relatively constant temperatures (25–28°C) are advised for musts with elevated sugar concentrations and in difficult fermentation conditions. Premium quality wines capable of aging require a maceration permitting considerable phenolic compounds extraction (Section 12.5.5). Elevated temperatures play an essential role in this phenomenon. After a successful fermentation, the temperature can be raised to above 30°C to increase this extraction. Grape quality should of course also be considered before prolonging skin contact (Section 12.5.8). *Primeur* wines, however, are made to be drunk young and respect the fruity character of the grape; lower fermentation temperatures are recommended for these wines (25°C).

In difficult fermentation conditions linked to excessive temperatures, several palliatives were formerly recommended. Limited crushing slowed the fermentation process and thus produced less heat. Another method consisted of simultaneously filling several fermentors over several days in the hope that the regular addition of fresh grapes would moderate the fermentation process. In the latter case, sulfiting is not sufficient to avoid the increased risks and resulting consequences of stuck fermentation and hydrogen sulfide production. Finally, elevated temperatures justify early draining. This process, which separates the juice from the pomace, lowering bacterial risks, is employed in warm regions. Long vatting times, however, are practiced in more temperate zones.

Today, must aeration or, more specifically, aeration of yeasts during their growth phase (Section 3.7.2), along with temperature control, is the most effective way of helping difficult fermentations. It is carried out during pumping-over operations, or, possibly, by means of microoxygenation. Other processes capable of facilitating completion of fermentation (e.g. adding nitrogen or cell hulls, Section 3.6.2.) are described in Chapter 3.

12.4.2 Pumping-over Operations and Must Aeration

The disadvantages of open fermentors have already been covered (Section 12.3.1): alcohol evaporation, bacterial spoilage risks, etc. These fermentors do, however, permit air contact and therefore a better fermentation. Fortunately, the same effect can be obtained with a closed fermentor. Pumping-over can assure sufficient air contact with the must, supplying the needed oxygen. This operation consists of letting fermenting must flow in contact with air and then pumping it back into the upper part of the tank. The effectiveness of this method has been known in the Bordeaux region since the end of the 19th century. In-depth research was carried out in the 1950s. Due to its simplicity, its use has been widespread.

The numbers in Table 12.3 (Ribéreau-Gayon *et al.*, 1951) demonstrate the effectiveness of pumping-over operations for improving the fermentation process. They also show that yeasts better resist elevated temperatures, when aerated, but there is a certain amount of confusion as to the

Table 12.3. Effect of aeration by pumping-over on fermentation kinetics (Ribéreau-Gayon *et al.*, 1951)

Time	Tank aerated by pumping-over		Non-aerated tank (without pumping-over)	
	Temperature (°C)	Density	Temperature (°C)	Density
Day 1	22	1.088	23	1.088
Day 2	26	1.084	26	1.084
Day 4	32	1.047	29	1.073
Day 6	20	0.996	27	1.045
Day 10	—	—	27	1.020
Day 20	—	—	20	1.002

most opportune time to aerate. The same authors demonstrated that correct timing of oxygenation is essential. Early aerations at the beginning of fermentation help to prevent stuck fermentations: the yeasts are in their growth phase, and oxygen is utilized to improve their growth and produce survival factors (Section 3.7.2). Early pumping-over operations have the additional advantage of avoiding alcohol loss by evaporation.

An aeration carried out on the second day of fermentation is the most effective. The effectiveness of later aerations, in the presence of fermentation difficulties, is greatly diminished (Figure 12.7), sometimes to the point of being non-existent. In the final stages of fermentation, the yeast does not make use of oxygen, since ethanol and other toxic metabolites hinder its nitrogen assimilation. A nitrogen addition in the final stages of fermentation, therefore, does not help to re-establish fermentation activity, even after aeration.

Pumping-over with aeration is only beneficial at certain moments, but the pumping-over operation in general has other effects. It homogenizes the temperature, sugar concentrations and yeast population of the fermentor, compensating the effects of the more active fermentation in and just below the pomace cap (Section 12.3.1). Above all, this operation facilitates extraction of compounds from the pomace (anthocyanins and tannins)and enhances maceration (Section 12.5.4).

The pumping-over process is schematized in Figure 12.8. The fermenting must flows from a faucet located at the lower part of the fermentor. It should be equipped with a filtering system inside

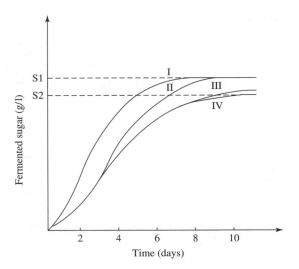

Fig. 12.7. Effect of momentary aeration (by pumping-over operations at different times) on fermentation kinetics. I: open tank, permanent aerobiosis—all sugar in must (S1) is fermented. II: closed tank, aeration on 2nd day by pumping-over—the fermentation is accelerated, with respect to total anaerobiosis, and complete. III: closed tank; aeration on 6th day—the acceleration of fermentation is insignificant. IV: closed tank in total anaerobiosis—the fermentation stops at a sugar concentration (S2) lower than S1

the tank to stop seeds and skins from blocking the orifice, since the obstruction of this orifice presents a serious problem in early pumping-overs before skin cap formation. The must flows from a certain height into a container with a capacity of several hundred liters. The pressure of the falling juice produces an emulsion which facilitates oxygen dissolution. Running the must over a flat surface is also recommended, to increase air contact. Specially equipped faucets intensify the emulsion. The aerated must is then pumped back to the upper part of the fermentor, soaking the pomace cap. Aeration may be eliminated by pumping must in a closed system, directly from faucet to pump to the upper portion of the fermentor. Pomace leaching may also be avoided by placing the pipe, at the upper part of the fermentor, below the pomace cap into the liquid.

Admittedly, the system in Figure 12.8 and its use are based on empirical data. The quantity of

oxygen dissolved in must with this system cannot be even approximated. However, the quantity of oxygen dissolved in must exposed to air is of the order of 6–8 mg/l and varies according to temperature. The quantities necessary to avoid stuck fermentation are approximately 10–20 mg/l, which can be obtained by pumping-over with aeration twice, 24 hours apart. Actual dissolved quantities during pumping-over operations are probably lower. Experience shows that this amount is sufficient. Nevertheless, a system permitting controlled oxygen addition (from a compressed gas bottle, for example) would be preferable. Indeed, this is the aim of process known as microoxygenation. The precise amount of oxygen necessary must also be determined. Of course, oxygen is added simply to assure yeast growth and survival, and a quantity greater than the optimum dose has no adverse effects on yeasts. Nevertheless, enzymatic oxidations in must may occur, despite the protection of carbon dioxide during fermentation. Tannins protect healthy red grape juice from excessive oxidation. For this reason, they better tolerate aeration than white grape musts, which are not generally pumped over (Section 13.7.3). With more or less rotten red grapes, oxidasic casses are easily triggered and the amount of oxygen added should be limited, if not nil.

Certain technical requirements must be met for pomace extraction to be effective (Section 12.5.5). Unfortunately, they are not always satisfied in practice. This process does not circulate must to assure the direct dissolution of pomace constituents by leaching, but it does replace the saturated must impregnating the pomace cap with must taken from the bottom of the fermentor. Approximately two-thirds of the pomace cap is immersed in the fermenting must and one-third floats above the liquid. All of the must should be pumped over and the entire pomace cap should be soaked to obtain satisfactory results. These conditions can be difficult to realize in narrow parallelepiped fermentors, especially if the lid is not located in the center of the tank. The same limited must fraction participates in this pumping-over operation. Draining (possibly with aeration) a third to a half of the tank volume and then brutally

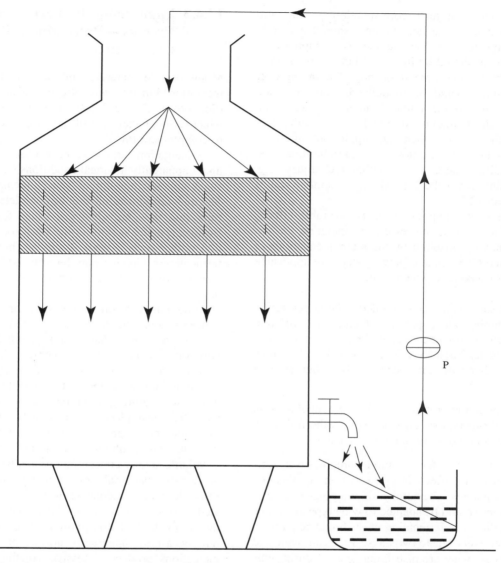

Fig. 12.8. Pumping-over operation, showing must aeration below and pomace leaching above (P = pump)

releasing the must from the top of the tank permits the immersion of the entire pomace cap (rack-and-return). During this process, the pomace cap descends in the tank. Various systems (cables) permit the cap to be broken up and reformed.

Various types of rotating irrigators exist. They must be placed at the center of the fermentor and assure the thorough soaking of the entire surface. To be fully effective, their pump delivery rate must be sufficient. Increasing the flow rate during a pumping-over operation can suffice for modifying tannin concentrations and consequently wine style. Even in ideal conditions, the liquid may pass through preferred passages in the pomace cap. Depending on operating conditions, pumping-over effectiveness with respect to pomace extraction is extremely variable. Close monitoring is indispensable.

A volume of juice corresponding to one-third to one-half of the tank volume should generally be pumped over. The number of pumping-over operations should be increased but not their duration. In any case, the frequency of pumping-over operations should be modulated. This operation contributes to the tannic structure of wine and favors the extraction of the highest quality tannins, making wine rich and supple, but an excessive tannin concentration can lead to hard, aggressive, disagreeable wines. Other techniques (e.g. punching down the cap) also give similar results (Section 12.5.5).

Due to its simplicity and its favorable effects, pumping-over is an essential operation in red winemaking. Inspired by the recommendations of Peynaud (1981), the following steps are applicable to Bordeaux-style winemaking:

1. As soon as the tank is filled, a homogenization pumping-over operation blends the different grape crops and evenly distributes the sulfur dioxide. Yeast may be inoculated at that time, together with any other additives, but aeration is unnecessary.

2. Pumping-over with aeration is essential as soon as fermentation starts, as well as the following day and, possibly, the day after that.

3. During active fermentation, pumping-over operations are effected for extraction of phenolic compounds. The number of pumping-over operations should be adapted to wine type and grape quality. The fermenting must should be pumped over every one to two days. 'Free-run' wine and 'press' wine are also homogenized during this operation.

4. After fermentation, pumping-over operations should be discontinued in hermetic, closed tanks to avoid oxygen exposure. In open tanks, pumping-over operations are continued. Extended macerations in partially filled tanks require a short pumping-over operation twice per day to immerse aerobic germs.

5. Pumping-overs rarely suffice to restart slow or stuck fermentations.

12.4.3 Monitoring the Fermentation Process—Determining its Completion

Monitoring temperature and density during fermentation kinetics has already been described (Sections 3.2.2 and 3.2.3). It is indispensable in winemaking. Other controls are also recommended to complement this data.

Fermentation rates have been observed to vary under apparently identical conditions (temperature, sugar content, amount of yeast inoculated, etc.). Sluggish fermentations may be completed successfully, but they are always a cause for concern. Besides specific factors in the must, one explanation is that several yeast strains are involved and fermentation kinetics may be affected by antagonism between them (Killer effect, Sections 1.7 and 3.8.1).

At the end of alcoholic fermentation, malic acid concentrations should be determined and monitored if necessary. Malolactic fermentation (MLF) normally occurs after the complete depletion of sugars. An early initiation of MLF is generally linked to alcoholic fermentation difficulties and insufficient sulfiting. In certain cases, the two fermentations take place simultaneously, even though the antagonistic phenomena between yeasts and bacteria tend to inhibit alcoholic fermentation.

Volatile acidity concentrations can be monitored to identify bacterial contamination. It is indispensable when malolactic fermentation is initiated during slow alcoholic fermentations, before the complete depletion of sugar (Section 3.8). On rare occasions, apparently normal alcoholic fermentations produce excessive volatile acid concentrations. In this case, the fermenting must has most likely been contaminated by the fermentor or poorly maintained equipment. Spoilage bacteria can also produce volatile acidity before the start of alcoholic fermentation. After fermentation, its origin is more difficult to identify. Renovating the winery and replacing old equipment generally eliminate this contamination risk.

Yet the production of volatile acidity does not always indicate the presence of bacteria. Yeasts may also produce volatile acidity. In certain, as yet poorly defined, cases, excessive amounts of volatile acidity are produced (0.4–0.6 g/l in H_2SO_4, or 0.5–0.7 g/l in acetic acid). By determining the concentrations of the two lactic acid isomers (L(+)-lactic acid and D(−)-lactic acid), the origin of the acetic acid can be identified. During fermentation, yeasts produce a few dozen mg/l of the former and less than 200 mg/l of the latter (Lafon-Lafourcade and Ribéreau-Gayon, 1977). Higher values indicate the involvement of lactic acid bacteria in the production of volatile acidity, but the possibility of high volatile acidity production levels by yeasts should always be considered. Moreover, the standard methods used to protect against bacterial spoilage have no effect on yeasts. Additionally, wine acidification tends to increase yeast-based volatile acidity production. Certain yeasts have an increased capability for volatile acidity production which attains a maximum in the course of fermentation, tending to decrease toward the end. In red winemaking, an excessive temperature (28°C) at the initiation of fermentation contributes to elevated volatile acidity levels.

The final stages of fermentation should be closely monitored. When the density drops below 1.000, this measurement is no longer sufficient to measure precisely the evolution of the fermentation. Moreover, the relationship between possible residual sugar and density is complex. When fermentation is complete, wine density can vary between 0.991 and 0.996, according to alcohol content. In addition, free-run wines always have a lower density than press wines, which are rich in extracted constituents.

The completion of fermentation is verified by chemically measuring the sugar concentration. For a long time, the reducing property of sugars was exclusively used to determine their concentration, but methods based on this characteristic also measured other substances in addition to fermentable sugars (glucose and fructose). Due to this 'interference', fermentations were considered complete when these methods indicated less than 2 g of sugar per liter. This value actually signified the presence of less than 2 g of reducing agents per liter, including, among other substances, glucose and fructose. Due to an increased quantity

of reducing agents in press wines and wines made from rotten grapes, fermentations are considered complete at approximately 3 g/l in these cases. Today, more and more glucose-and fructose-specific analysis methods are available. Their overall concentration should not exceed several hundred milligrams per liter when the fermentation is complete. This is necessary to avoid spoilage due to the development of contaminant yeasts (*Brettanomyces*) during barrel-aging (Volume 2, Sections 8.4.5; 8.9.6).

12.5 MACERATION

12.5.1 The Role of Maceration

Red wines are macerated wines. Maceration is responsible for all of the specific characteristics of sight, smell and taste that differentiate red wines from white wines. Phenolic compounds (anthocyanins and tannins) are primarily extracted, participating in the color and overall structure of wine. Yet aromas and aroma precursors, nitrogen compounds, polysaccharides (in particular, pectins) and minerals are also liberated in the must or wine during maceration.

The corresponding chemical elements come from the skins, seeds and sometimes the stems. Each of these organs supplies chemically and gustatorily different phenolic compounds. The gustatory differences are confirmed by tasting wines made in the presence of one or more of these organs. Stems give wine herbaceous flavors and seeds contribute to harshness. Skins contact alone produces a supple but incomplete wine that is too fluid in structure. Skins and seeds contact makes a more balanced wine. The phenolic compounds of each organ also vary according to variety, maturation conditions and other factors. Furthermore, in the same organ (for example, in the grape skin), herbaceous, vegetal and bitter substances along with leafy and grassy substances are located alongside phenolic compounds favorable to wine quality. Fortunately, the latter substances are extracted before the others.

Consequently, the maceration should be modulated and fractionated. Only useful grape constituents should be dissolved—those positively contributing to wine flavor and aroma. The extraction of these desirable substances should be maximal, if not total.

The concentration of substances in grape tissues detrimental to wine quality increases as grape quality diminishes. This phenomenon can be verified by chewing a grape skin after the pulp and seeds have been removed by pressing the berry between the thumb and index finger. Initially, mild savors evolve toward mellow tannins. Afterwards, vegetal sensations become increasingly bitter and aggressive. The transition rate from pleasant to disagreeable sensations varies according to grape quality. The evolution of tannin quality can be evaluated during maturation in this manner. The same experience effected on seeds leads to similar results. Harshness and astringency diminish during maturation, while sensations of body and harmony increase.

An abundance of pleasant-tasting substances useful for winemaking and a lack of unpleasant ones characterize the grapes of top-ranked growths. These characteristics typify mature years, i.e. great vintages. Such wines are capable of undergoing the most intense extractions and prolonged vatting times. The resulting high tannin concentrations are necessary to ensure their long-term aging. Lesser quality red wines, made for immediate consumption, have relatively short macerations—more flaws than qualities would result from longer macerations.

The extraction of pomace constituents during maceration should therefore be modulated according to grape variety and quality and also the style of wine desired (see Volume 2, Section 6.6). Yet each grape crop is capable of producing a given type of wine, depending on natural factors (the *terroir*).

Premium wines require a tannic structure which should not compromise finesse and elegance. These wines are difficult to produce and require grapes of superior quality benefiting from great *terroirs* and great growths. Light, fruity red wines are relatively easy to obtain—grape quality is not essential, but if grape quality (variety, maturity, sanitary state, etc.) is insufficient, tannic red wines rapidly become heavy, coarse and without charm. Short vatting times and limited maceration lessen the occurrence of disagreeable characteristics.

A number of methods are available to the winemaker to adjust extraction levels during maceration. They essentially influence tissue destruction and favor the dissolution of phenolic compounds. Techniques are continually evolving and engineers regularly propose new solutions. Current methods will be described later in this chapter, each one probably having preferential effects on one or more groups of extracted substances. For example, brutal crushing promotes the extraction of bitter and herbaceous substances. Percolation of must, on the contrary, favors supple and full-bodied tannins. Constituent extractability of various organs varies with several factors (variety, maturity level, etc.).

Enzymatic reactions, activated by grape enzymes, are involved in cell wall degradation. They favor the dissolution of their vacuolar contents (Section 11.7.3). Commercial enzymatic preparations have recently been developed to activate these phenomena; they have pectinase, cellulase, hemicellulase and protease activities of diverse origins (Amrani-Joutei, 1993). These enzymes seem to favor the extraction of skin tannins over skin anthocyanins. They act on the tannins linked to the polysaccharides of the cell wall, giving the enzymatic wine a more full-bodied character than the control wine.

Touzani *et al.* (1994) obtained encouraging results from an enzymatic pool produced from *Botrytis cinerea* cultures, not containing laccase. This preparation attacks cell walls and favors the anthocyanin extraction over tannin extraction. It could therefore be interesting for making *primeur*-style wines, to be drunk young, fruity and rich in color but not very tannic.

Future research will be required to determine the selective effect of this enzymatic extraction on the various phenolic compounds, according to conditions. Its effect, with respect to standard practices for regulating maceration, also needs to be explored. Regardless of the mechanisms involved, tasting has confirmed the interest of using enzymatic preparations for maceration during red wine-making.

These various results demonstrate the utility of a better chemical and gustatory understanding of the molecules involved in maceration phenomena.

The extraction of a specific combination of these molecules could thus be obtained according to the maceration technique used. Certain practices may be beneficial to wine quality in some situations but not in others. Only fundamental research on the chemistry of phenolic compounds will be capable of giving definitive answers. These studies are complicated by the extreme complexity and the high reactivity of the molecules involved (Volume 2, Section 6.3).

In traditional winemaking, maceration occurs during vatting (*cuvaison*), while the pomace soaks in the juice. Alcoholic fermentation occurs in the juice, producing ethanol and raising the temperature. Both ethanol and temperature participate in the dissolution of pomace constituents.

12.5.2 Different Types of Maceration

There is a current trend to distinguish between the various types of maceration, other than standard extraction during fermentation:

1) High-temperature extraction prior to fermentation used in thermovinification (Section 12.8.3), either followed by normal fermentation, or separate fermentation of the juice.

2) Cool-temperature extraction prior to fermentation, aimed at enhancing aromatic complexity. The start of fermentation is postponed by maintaining low temperatures and an appropriate level of SO_2, as well as by delaying inoculation with active yeasts.

   A more elaborate form of this technique consists of cooling the grapes to around 5°C, by injecting liquid CO_2 or dry ice, and maintaining this temperature for 5–15 days. The temperature shock bursts the grape skin cells and releases intensely colored juice (Blouin and Peynaud, 2001). Once the must has been heated to normal temperature, fermentation proceeds as usual. The purpose of this technique is to obtain wines with high concentrations of phenolic and aromatic compounds. The results of this rather laborious method are not universally appreciated. Further research is required to identify the conditions required to produce

deep-colored, aromatic wines without any rustic or herbaceous character. Satisfactory results have been achieved with Pinot Noir (Flanzy, 1998), producing finer, fruitier wines. In any case, the results are better than those obtained with cold maceration, following stabilization with considerable doses of SO_2, which gave deep-colored wines that were lacking in varietal character and tended to dry out on the end of the palate (Feuillat, 1977).

3) Post-fermentation vatting is required by the best premium quality red wines to prolong skin contact after the end of fermentation, sometimes combined with an increase in temperature (final, high-temperature maceration, 12.5.5).

12.5.3 Principles of Maceration

The passage of pomace constituents, particularly phenolic compounds (anthocyanins and tannins), into fermenting juice depends on various elemental factors. The results constitute overall maceration kinetics. The phenomena involved are complex and do not cause a regular increase in extracted substances. In fact, among these various factors, some tend to increase phenolic compounds, while others lower concentrations. Moreover, they do not necessarily always act in the same manner on the various constituents of this group.

Maceration is controlled by several mechanisms (see also Volume 2, Section 6.6.1):

1. The extraction and dissolution of different substances. Dissolution is the passage of cell vacuole contents from the solids phase into the liquid phase. This dissolution depends first of all on vine variety and grape maturity levels. This is especially important for anthocyanins. In certain cases, strongly colored musts are obtained immediately after crushing. In other cases, a period of 24–48 hours is required. Tissue destruction through enzymatic pathways or crushing facilitates dissolution. The more intense the crushing, the more dissolution is favored. Finally, dissolution depends on the various operations that participate in tissue destruction: sulfiting, anaerobiosis, ethanol, elevated temperatures, contact time.

2. Diffusion of extracted substances. Dissolution occurs in the pomace, and the impregnating liquid rapidly becomes saturated with extracted substances; exchanges therefore stop. Further dissolution is dependent on the diffusion of the extracted substances throughout the mass. Pumping-over or punching down the pomace cap renews the juice impregnating the pomace cap. This diffusion is necessary for suitable pomace extraction. It homogenizes the fermentor and reduces the difference between the phenolic compound concentrations of free-run wine and press wine.

3. Refixation of extracted substances on certain substances in the medium: stems, pomace, yeasts. This phenomenon has been known since Ferré's (1958) observations (Section 12.2.3).

4. Modification of extracted substances. This hypothesis still requires further theoretical interpretations. Anthocyanins may temporarily be reduced to colorless derivatives (Ribéreau-Gayon, 1973). The reaction appears to be reversible, since the color of new wines exposed to air for 24 hours increases, with the exception of those made from rotten grapes. Anthocyanin–Fe^{3+} ion complex formation may be involved in this color increase in the presence of oxygen. Ethanol may destroy tannin–anthocyanin associations extracted from the grape (Somers, 1979). In the same environmental conditions, free anthocyanins are less colored than tannin–anthocyanin combinations, which are formed again during aging and assure color stability.

The quantity of anthocyanins and tannins found in wine depends first of all on their concentration in the grape crop. Ripe grapes are the first condition for obtaining rich and colored wines. However, only a fraction of the phenolic compound potential of the grape is found in wine. Their concentration depends not only on the ease of phenolic compound extraction but also on the extraction methods used. The phenolic compound concentrations of various components of grape clusters and wine have been compared. Approximately 20–30% of the phenolic potential of grapes is transferred to

wine. The loss is significant and efforts have been made to improve this yield but, due to the complexity of this phenomenon and the molecules involved, a simple solution is difficult to find. Finally, "bleeding" a vat (Section 12.5.9) is a way of raising tannin levels by reducing volume. Eliminating water by other techniques (Section 11.5.1) achieves similar results, but keeps all the sugar in the must.

In red winemaking, maceration must be adapted to suit the grapes constitution (Volume 2, Section 6.6.2).

12.5.4 Influence of Maceration Time (Vatting Time)

The dissolution of phenolic compounds from solids into fermenting must varies according to maceration time, but no proportional relationship between maceration time and phenolic compound concentration exists. Color intensity has even been observed to diminish after an initial increase during the first 8–10 days (Ferré, 1958; Sudraud,

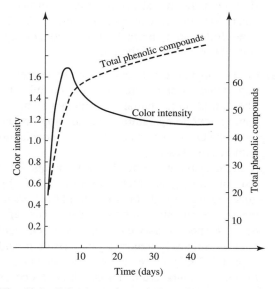

Fig. 12.9. Color intensity and phenolic compound concentration evolution of red wines according to maceration time (Ribéreau-Gayon *et al.*, 1970). Color intensity is defined as the sum of the optical densities at 420 and 520 nm at 1 mm thickness, (CI = OD 420 + OD 520). Total phenolic compounds are determined by the permanganate index

1963). The graphs in Figure 12.9 depict the evolution of a maceration extended well beyond normal conventions. In this laboratory experiment, the color intensity passes through a maximum on the eighth day and then diminishes. The evolution of total phenolic compounds is different: during an initial phase lasting a few days, their concentration increase is rapid and then slows afterwards. This behavioral difference is due to tannin concentrations (skins and seeds) in the grape crop being 10 times greater than anthocyanin concentrations. In both cases, the concentrations increase during the first few days. Afterwards, tannin losses are proportionally less significant and an overall increase is always observed. Certain varieties have very low tannin concentrations. In this case, tannins evolve similarly to color (Figure 12.9). Other experiments have shown that the nature and properties of tannins vary in function of maceration time.

Similarly, the various grape organs (skins, seeds and stems) contain specific phenolic compounds. Their extraction varies according to diverse conditions. Skin anthocyanins are extracted first; ethanol is not required for their dissolution. Skin tannin extraction begins soon after, facilitated by the increasing presence of ethanol during fermentation. A relatively long maceration is necessary for seed tannin extraction. The presence of ethanol is required to eliminate lipids. The skins contain the most supple tannins, but they can become bitter if grape maturity is insufficient. Seed tannins are harsher but less bitter.

Evidently, these notions pertaining to the evolution of color and anthocyanins in terms of maceration time primarily concern new wines—anthocyanins are in fact the essential elements of their color. As wine matures, the role of tannins becomes increasingly important. Extended vatting times produce more colored wines, even if the resulting new wines initially appear to confirm the contrary.

The causes of this drop in color intensity, after several days of vatting, have been examined and interpreted. Stems have long been known to decrease the intensity of wine color (Section 12.2.3); this phenomenon is the result of their adsorbing anthocyanins. White grape skins have also been shown to adsorb anthocyanins when placed in red grape must. A yeast biomass in a fermenting medium adsorbs both anthocyanins and tannins.

Chemical reactions also diminish color intensity. Grape tannin–anthocyanin combinations are destroyed and anthocyanins are reduced to colorless forms during these reactions (Section 12.5.1).

These facts lead to an important conclusion. On approximately the eighth day of maceration, wine color intensity is at its maximum and tannin concentrations are limited, permitting fruity sensations to be conserved. This vatting method is best adapted to wines for early drinking. In contrast, long vatting times produce rich tannic wines capable of extended aging, but these elevated tannin concentrations require grapes of high quality.

12.5.5 Influence of Pumping-over and Cap Punching (*Pigeage*)

Pumping-over is important in red winemaking, at least for certain varieties such as Merlot and Cabernet Sauvignon. Section 12.4.2 describes the objectives and steps of this operation. In addition to introducing oxygen, it plays a major role in extracting compounds from the pomace and homogenizing the contents of the vat. It plays a major role in pomace extraction and tank homogenization. At skin and juice separation, the drawn-off wine and press wine have more similar tannin concentrations when pumping-over operations are carried out. The numbers in Table 12.4 show the dual effect of pumping-over operations. Increasing the number of pumping-over operations accentuates these effects. An effective pumping-over should thoroughly leach the pomace cap. Precise experiments have shown that phenolic compound concentrations and color intensity vary significantly according to pumping-over conditions. "Rack-and-return" is recommended for this purpose and consists of running-off part of the fermenting must, then reintroducing it all at once over the broken-up pomace.

Pumping-over has an additional advantage that is more difficult to evaluate. This operation does not affect tissue integrity and promotes the

Table 12.4. Influence of pumping-over on color and tannin dissolution in open fermenters (Sudraud, 1963)

Time	No pumping-over		Two pumping-overs, on days 1 and 3	
	Tannin (permanganate index)	Color intensity[a]	Tanin (permanganate index)	Color intensity[a]
Maceration 3 days	39	0.83	46	0.93
Maceration 6 days	43	0.87	48	0.98
Maceration 10 days	45	0.89	52	1.04
Run-off wine at the time of running-off	48	0.93	56	1.16
Press wine at the time of running-off	102	1.35	95	1.30

[a]See Table 12.1.

extraction of higher-quality tannins, at least in some grape varieties. It imparts a rich and full-bodied structure to wines, without bitterness and vegetal characteristics.

The timing of pumping-over operations influences the selective extraction of skin and seed tannins. Skin tannins are more easily released and are sufficient for making *primeur* wines, but seed tannins are necessary to obtain a premium wine. Pumping-overs in the final stage of fermentation are used to extract these compounds. The advantages of this technique have led to its general use. However, in certain cases it is abused, producing aggressive and disagreeable tannic wines; the resulting press wines are thin and unusable.

Gas injection systems have been proposed to replace traditional pumpings (Section 12.4.2, Figure 12.8) for the extraction of pomace constituents. These systems consist of introducing a specially adapted pipe into the lower part of the tank. The pipe injects carbon dioxide, nitrogen or filtered air into the fermentor. The tank contents are churned up briskly. This operation is much more rapid than traditional methods, and a duration of 1 minute per 100 hl of grape crop at a pressure of 3 bars is recommended. The technique should not be used during the final stage of fermentation (when the density falls below 1.010). Continuing injections beyond this value would lead to the definitive disintegration of the pomace cap.

In certain cases, punching down can replace pumping-over. As its name suggests, punching

down consists of plunging the skin cap completely into the liquid. This immersion results in the disintegration of the cap and increases maceration phenomena. The process promotes seed tannin extraction and thus increases the tannic structure of wine. For certain varieties (e.g. Pinot Noir), punching down the cap produces better results than pumping over, but there is some concern that it may give a slightly "rustic" character to Cabernet Sauvignon and Merlot wines, especially if used to excess. Various types of tanks can be equipped with automatic punch-down systems (Blouin and Peynaud, 2001). This equipment is better adapted to small-capacity tanks.

12.5.6 Influence of Temperature

Heat is a means of degrading tissues. It increases the dissolution of pomace constituents and accelerates maceration. The technique of heating crushed red grapes has been used for a long time and is particularly important in thermovinification (Section 12.8.3).

In addition to this extreme case, temperature is an essential factor in standard maceration. It should be sufficiently high to assure satisfactory extraction of phenolic compounds. The experiment (Sudraud, 1963) in Table 12.5 clearly illustrates its impact.

Both the average and maximum temperature affect extraction. The results of a laboratory experiment in Table 12.6 indicate the simultaneous influence of maceration time and temperature. When maceration is prolonged, an elevated temperature

Table 12.5. Influence of fermentation temperature on dissolution of phenolic compounds (Sudraud, 1963)

Temperature	Total phenolic compounds (permanganate index)	Color intensity[a]
20°C	44	0.71
25°C	48	0.87
30°C	52	0.96
20–37°C (average 29.5)	52	1.21
25–37°C (average 32.6)	60	1.43

[a]See Table 12.1.

can exacerbate the drop in anthocyanin concentrations and color intensity. An elevated maceration temperature also favors the extraction of yeast mannoproteins, which participate in the production of soft and full-bodied wines.

Moderate maceration temperatures (25°C) are preferred for the production of *primeur*-style wines. These are made to be drunk young and this approach gives them a good color while conserving

their fruity aroma characteristics. Moderate temperatures are also recommended when there is a risk of fermentation difficulties (elevated sugar concentration). An elevated temperature (30°C) extracts the tannins required to produce a premium wine capable of long aging; higher temperatures would promote further extraction but would also compromise yeast activity—they should thus be used with caution.

In stainless-steel vats, especially if they are installed in the open air, excessively low fermentation temperatures may cause problems in certain years in some climates. These temperatures do not permit a sufficient maceration. Moreover, as soon as the fermentation stops, such tanks can rapidly drop in temperature, no longer producing calories. The temperature in wooden tanks develops differently.

Tank temperature control systems have permitted an almost perfect regulation of the maceration temperature during fermentation. Cool grape crops and excessively cool fermenting juice can be

Table 12.6. Influence of maceration temperature on dissolution of phenolic compounds (Ribéreau-Gayon *et al.*, 1970)

Time and temperature	Tint[a]	Color intensity[a]	Anthocyanins (g/l)	Tannins (g/l)	Total phenolic compounds (permanganate index)
Maceration 4 days Temperature					
20°C	0.54	1.04	0.54	2.2	39
25°C	0.52	1.52	0.63	2.4	45
30°C	0.58	1.46	0.64	3.3	55
Maceration 8 days Temperature					
20°C	0.45	1.14	0.59	3.0	43
25°C	0.56	1.62	0.61	3.2	48
30°C	0.56	1.54	0.62	3.6	55
Maceration 14 days Temperature					
20°C	0.53	1.16	0.49	2.5	48
25°C	0.51	1.36	0.59	3.5	58
30°C	0.56	1.44	0.58	3.8	59
Maceration 30 days Temperature					
20°C	0.56	1.45	0.38	3.5	63
25°C	0.67	1.20	0.39	3.7	67
30°C	0.80	1.47	0.21	4.3	72

[a]See Table 12.1.

warmed to a suitable temperature, but since an integrated heat exchanger warms only a portion of the tank contents, a pumping-over operation is required to homogenize the tank. Temperature control must of course be set according to fermentation kinetics. For example, the initiation of fermentation requires a moderate temperature (20°C), since the yeast is relatively heat sensitive during its growth phase (Section 3.7.1 and 3.8.1) and there is a greater risk of stuck fermentation at that stage.

Glories *et al.* (1981) developed a technique called heated post-fermentation maceration. The objective of this method is to separate the fermentation phase at a moderate temperature from the maceration phase. After fermentation, the tank contents are warmed to between 35 and 40°C for several days. In this process, only the wine is directly heated, the pomace cap being heated indirectly. The wine is heated to between 50 and 60°C during the entire maceration period. This process has not been observed to alter the taste of wine. In addition, malolactic fermentation takes place normally. As long as this method is effected in the absence of residual sugar, there is no risk of bacterial spoilage, resulting in increased volatile acidity. The method most significantly affects phenolic compound concentrations and color intensity, which are greatly increased by hot maceration after fermentation.

In the experiment in Table 12.7, tasting reveals wine A to be diluted because of an elevated crop yield and insufficient maturity. Wine B is more full-bodied and has more taste but no vegetal character. Wine B is greatly improved when compared with wine A, which is clearly lower in quality. Applying heated post-fermentation maceration to wines that are already naturally rich greatly intensifies their concentration—which could be interpreted as an improvement, except that this concentration is often at the expense of gustatory finesse: the tannins rapidly become hard and rustic, and this tannic astringency tends to increase during aging. The repeated pumping over required to maintain the temperature during this process can magnify these flaws. Suspended solids are often produced and press wines often become unusable. The above observations emphasize the need to consider wine quality before using this method.

Temperature drops in steel tanks should be avoided after the completion of fermentation. A post-fermentation maceration at 30°C during several days often favors wine quality. This approach emulates the thermal conditions in wooden tanks following fermentation. "High-temperature, post-fermentation vatting" must not be confused with "thermovinification" (Section 12.8.3).

Temperature control, in addition to pumping-over operations and vatting time, is another means of modulating extraction during maceration. Temperature regulation can profoundly modify the tannic structure of wine. But there is a risk of also increasing the rustic character of wine. The nature of the grape crop and the type of wine desired are criteria which should be considered when determining optimum temperature regulation.

Table 12.7. Effect of 'heated post-fermentation maceration' on wine composition

Component	Traditional winemaking (Wine A)	Heated post-fermentation maceration (4 days at 45°C) (Wine B)
Color intensity[a]	0.45	0.67
Tint[a]	0.82	0.75
Total phenolic compounds (Folin index)	29	38
Anthocyanins (mg/l)	273	329

[a]See Table 12.1.

12.5.7 Effect of Grape Sulfiting and Alcohol Produced by Fermentation

The impact of must sulfiting on pigment extraction is covered in Section 8.7.5. Sulfur dioxide destroys cell tissue and promotes the dissolving of pomace constituents, but in traditional winemaking with ripe, healthy grapes, pumping over, temperature and vatting time have a greater impact. All things considered, SO_2 has little influence on the color intensity and phenolic compound concentration of normal red wines.

However, the dissolvent effect of SO_2 is manifest in rosé winemaking, since the phenolic compound concentration is low in this case. In fact, SO_2 can be detrimental to white winemaking. This dissolvent effect may also affect red grapes if they are insufficiently ripe, and pigment extractability is poor. In that case, sulfiting facilitates anthocyanin extraction in the early stages, especially during cool-temperature maceration (Section 12.5.2).

With rotten grapes, sulfiting does not improve the extraction of pigments; instead, it prevents the laccase activity of *Botrytis cinerea* from destroying them. The numbers in Table 12.8 show that high sulfur dioxide concentrations increase the total phenolic compound concentration and color intensity in the case of highly contaminated grapes. The elevated tint value in the control sample is due to a yellow component, characteristic of oxidasic casse.

The impact of the ethanol produced by fermentation seems to be complex. According to Somers (1979), it is involved in the decrease

Table 12.8. Influence of sulfiting botrytized grapes on phenolic compounds of the resulting wines (Sudraud, 1963)

Sulfiting	Total phenolic compounds (permanganate index)	Color[a] intensity	Tint[a]
Control	32	0.53	0.76
Sulfited at 10 g/hl	41	0.63	0.42
Sulfited at 20 g/hl	55	0.83	0.43

[a]See Table 12.1.

in color intensity observed during vatting. The mechanism seems to correspond to the destruction of tannin–anthocyanin combinations, resulting in the liberation of free anthocyanins, which are less colored. At the same time, alcohol is considered to participate in tissue destruction and, as a result, in the dissolution of pomace constituents (Table 12.9). In large wineries containing many fermentors, with relatively homogeneous grape crops, the wines with the highest tannin concentrations and color intensities are often observed to have the highest alcohol strength.

12.5.8 Impact of Various Mechanical and Physical Processes Acting Directly on the Pomace (Flash-détente)

The impact of crushing grapes has already been covered in Section 12.2.2. Energetic crushing increases the diffusion of solid tissue components, but, according to a general rule, the corresponding tissue destruction promotes the extraction

Table 12.9. Influence of alcohol on extraction of pomace phenolic compounds in a model solution (10 days of maceration at $20°C$, pH 3.2) (Canbas, 1971)

Alcohol	Tannins (g/l)	Total phenolic compounds (permanganate index)	Anthocyanins (mg/l)	Color intensity[a]
0% vol.	0.66	12	169	1.95
4% vol.	0.96	16	214	3.60
10% vol.	1.32	20	227	6.35

[a]See Table 12.1.

of inferior-quality tannins. These tannins impart vegetal and herbaceous tastes to the must and resulting wine. Unfortunately, no measures are currently available to confirm this fact. Yet when compared with wines made from energetically crushed grapes, pumping-over operations have been observed to favor the extraction of soft and more agreeable tannins. The extent of the negative effect caused by excessive crushing is according to the grape variety and its degree of vegetal character. However, all things being equal, quality grape crops are much more sensitive to crushing intensity than ordinary grapes.

Similarly, the breaking up and punching down of the pomace cap has long been used to increase maceration while protecting the wine from bacterial development. In the past, these operations were common and carried out manually in small-capacity open fermentors. They are no longer possible in today's large fermentors, but tanks can now be equipped with mechanical devices (screw, helix, jack and piston-based) which assure the breaking up, reshaping and punching down of the pomace cap (Blouin and Peynaud, 2001). The pomace cap can also be broken up by a mechanical claw, which is also used for devatting. When the claw is not functioning, the tank is closed by a removable cover. The operation of these various systems is fairly complex but they are effective in terms of extracted substances. They can be applied to certain varieties (Pinot, Section 12.5.4), but with other varieties, these methods can rapidly lead to excessively hard and disagreeable tannins.

Rotating cylindrical fermentors have similar constraints. A fixed internal device inside the tank breaks up the cap and is also used for devatting the fermentor. This kind of fermentor is relatively expensive. It has the same advantages and inconveniences as the equipment described in the above paragraph. Grape crops with low concentrations of phenolic compounds are extracted rapidly. In other situations, a satisfactory extraction is also obtained rapidly, permitting the fermentor to be used several times during the same harvest. This system does not necessarily give satisfactory results for producing premium wines.

Another recently-developed process, (IMECA DIF, 34 800 Clermont–L'Héraut, France) aimed at intensifying maceration of pomace in grape must, is known as *Flash-détente* (Boulet and Escudier, 1998). It consists of bringing the crushed grapes rapidly to a high temperature, then chilling them almost instantaneously in a high vacuum. It may be considered a variation on thermovinification (Section 12.8.3). Under these conditions, the skin cell tissue structure is completely degraded and components essential to wine quality (phenolic compounds, polysaccharides, aromas, etc.) are rapidly released during later vatting.

The grapes are destemmed and part of the juice separated out. As soon as the crushed grapes come into the heating chamber, it is brought up to a high temperature ($70°C–90°C$) very rapidly by direct injection of saturating vapor at $100°C$, produced using the reserved juice. The must is then transferred by means of a positive-displacement pump into a high-vacuum chamber connected to a vacuum pump coupled with a condenser. When the hot must enters the vacuum chamber, the water it contains is vaporized, cooling it rapidly to the boiling point of water under the vacuum conditions used, i.e. $30°C–35°C$.

According to Boulet and Escudier (1998), this sequence has three consequences: the crushed grapes are cooled in less than one second, the grape cell walls are broken, and there is only a very low concentration of residual oxygen in contact with the must.

The water recovered from the vacuum chamber is condensed, and represents 7–12% of the total volume of the must. Part of this is used to produce the hot steam used in the system, in addition to that made from the separated juice. The rest may be added back into the must, making it more concentrated. This is only possible if permitted by legislation. This technique does not dilute the must as the steam used to heat the must is made from grape juice, but may concentrate it, as part of the water is eliminated.

If the treated grapes are pressed immediately, the resulting wine is similar to that produced by fermentation combined with standard techniques for heating the must (Section 12.8.3). If, however, the must is left on the skins at a suitable temperature after *Flash-détente* treatment, the pomace extraction kinetics are much faster than in a normal winemaking situation, reaching maximum anthocyanin and polyphenols levels after 3–4 days.

This high-temperature treatment destroys laccase, so it is extremely suitable for grapes affected by *Botrytis cinerea*. Pectolytic enzymes are also partially destroyed, so the must becomes viscous after *Flash-détente* and is difficult to drain. Pressing is facilitated by mixing treated must with a sufficient quantity of untreated juice.

Generally speaking, according to Boulet and Escudier (1998), wines treated by *Flash-détente* contain 30–60% more polyphenols than controls, even after the traditional 3-week vatting time on the skins.

The resulting wines have more intense color and a better tannic structure. They have different aromas, but do not lose their varietal character. This winemaking technique is obviously better suited to some grape varieties than others.

These mechanical and physical extraction techniques will become more widely used when the substances extracted from different grapes under various conditions and their properties are better known, so that the processes most likely to enhance quality can be applied in each case.

A better understanding of the nature and properties of the substances extracted in different maceration conditions will lead to the development and use of mechanical techniques enhancing quality maceration phenomena. In this sector, as in many others, empiricism has preceded research. There is currently no theoretical knowledge of these phenomena that permits the explanation and prediction of the results observed.

12.5.9 The Maceration Process: Grape Quality and Tannin Concentrations in Wines

(a) Grape Quality

In Section 12.1, the importance of grape quality was emphasized. It directly influences the consequences of maceration. The quantity and quality of phenolic compounds are directly related to grape variety, *terroir*, maturity level, disease status, etc.

Proper maturation conditions are essential to the accumulation of phenolic compounds. The climate plays a major role in phenolic compound production, since this requires a considerable amount of energy. Vine culture techniques also affect maturation. Moreover, phenolic compound accumulation is limited in young vines; therefore, relatively old vines are necessary for premium wine production.

Crop yields also greatly affect the accumulation of tannins and anthocyanins, but this factor must be interpreted carefully. In some cases, the same climatic criteria that favor quality can also favor quantity. In vintage-dependent, temperate vineyards, the best quality years are sometimes also the most abundant. Reciprocally, low-yield vintages do not necessarily produce the best quality grapes. When considering crop yields, plant density should also be taken into account: must sugar concentrations have long been known to diminish when per vine production increases. Vineyards with a long tradition of quality choose to maintain their plant density at 10 000 vines /ha in poor soils. In this manner, a satisfactory production is assured while respecting grape quality. In richer soils, lower plant densities decrease cultivation costs (as low as 2000 vines /h, and sometimes even lower). As a result, to have the same production per hectare, higher yields per vine are required. In satisfactory climatic conditions, the grapes on these vines ripen normally, producing a relatively large harvest, but in less satisfactory maturation conditions, crop volume is more apt to delay maturity, with low plant densities as opposed to high plant densities.

The relationship between vine production and maturation conditions (in particular, phenolic compound concentrations) is complex and difficult to interpret. Practices that increase vine vigor (fertilizing, rootstock, pruning, etc.) are known to delay maturation. Phenolic compounds are the first substances affected. When production is excessive, wines rapidly become diluted and lack color. Some grape varieties (e.g. Cabernet-Franc) are more susceptible to flavor dilution due to excessively high yields than others (e.g. Cabernet Sauvignon). A carefully measured equilibrium between an acceptable selling price and optimal wine quality will determine crop yields and consequently the future of premium red wines. The future of great red wines no doubt depends on maintaining a balance between producing enough wine to achieve reasonable profitability and keeping yields low to optimize quality. A few Bordeaux winegrowers have started keeping their yields extraordinarily low—about half the normal level (20–30 hl/hectare)—to produce extremely concentrated wines that find a market at unusually high prices.

Excessively vigorous vines and excess rain, leading to berry swelling, also cause abundant harvests. Various techniques are available to mitigate the resulting defects. The first of there is cluster thinning, which consists of eliminating a portion of the grape clusters between setting and *véraison*. Cluster thinning should preferably be carried out near *véraison*. At this time, grapes manifesting physiological retardation can be eliminated; the vegetation is also less affected. This green harvest, however, is difficult work and its effectiveness is limited. The retained grapes swell to compensate for the thinning. When 30% of the grapes are removed, crop yields generally decrease by only 15%. Vine vigor and pruning should preferably be regulated to assure a crop yield corresponding to quality grapes.

Eliminating a fraction of the must can also increase tannin concentrations of dilute grape must. This method increases the ratio of skins and seeds (pomace) to juice. A few hours after filling the fermentor, as soon as the juice and solids can be separated, some of the juice is drawn off (approximately 10–20% of the total juice volume). This operation significantly affects tannin concentrations and color intensity. The method should be used with caution, as excessive concentration of the must can lead to exaggeratedly aggressive tannins. The volume of must to be drawn off depends on skin quality, maturity, the absence of vegetal character and on grape disease status. The drawn-off juice can be used to make rosé wine. It is not advisable to throw away the

excess must or rosé wine to avoid pollution. In any case, it constitutes a loss of production, which must be compensated by producing a better quality wine capable of fetching a higher price.

Siégrist and Léglise (1981) have obtained data illustrating the importance of the solid part of the harvest. In their study, a Pinot Noir must containing 60% juice and 40% pomace is higher in quality than a similar must containing 80% juice and 20% pomace.

Instead of drawing off juice, it is now possible to eliminate water directly from the grape must (Section 11.5.1). Two methods currently exist: the first circulates the must across membranes which retain water by reverse osmosis (Degremont, Inc.); the second evaporates water in a low-temperature (20–24°C) forced vacuum (Entropie, Inc.). These techniques have the additional advantage of increasing the sugar concentration, thus eliminating the need for chaptalization in some cases.

In addition to phenolic compound concentrations, the nature and properties of these substances also play an essential role in the maceration process and its consequences.

The potential dissolution of skin pigment varies, especially according to maturity level. Phenolic maturity corresponds to a maximum accumulation of phenolic compounds in the berry. Cellular maturity is defined with respect to the level of cell wall degradation (Volume 2, Sections 6.5.3; 6.5.4). Extraction of phenolic compounds increases with this degradation level (Amrani Joutei, 1993). Augustin (1986) defined the anthocyanin extraction coefficient (A_E) as follows:

$$A_E = \frac{\text{wine anthocyanins}}{\text{mature grape anthocyanins}} \times 100$$

This coefficient varies from one year to another. For example, the following values were obtained for Merlot and Cabernet Sauvignon: 46.5 in 1983, 26.5 in 1984 and 39.0 in 1985. Moreover, these values correspond fairly well with the maturity level, expressed by the ratio: (sugar concentration)/(total acidity). The same coefficient varies less for tannins (26.8 in 1983, 23.2 in 1984 and

30.7 in 1985). It does not correspond to juice maturity.

The organoleptic quality of tannins is directly related to maturation conditions. Enologists have defined this quality in terms of 'good' tannins and 'bad' tannins. The chemical understanding of these phenolic compounds has made it possible to make a better choice of winemaking techniques that optimize the quality of various kinds of grapes. A perfect state of phenolic maturity not only supposes a maximum tannin concentration; it also corresponds to soft, non-aggressive, non-bitter tannins.

Environmental conditions (*terroir* and climate) and grape variety determine this phenolic maturity, which can be illustrated by Cabernet Sauvignon. In cool climates, its insufficiently ripe tannins take on a characteristic vegetal note. The same flaw can occur in excessively hot climates: the rapid sugar accumulation forces harvesting before the tannins reach their optimum maturity. A harmonious maturation of the various constituents of the grape characterizes great *terroirs* and great vintages. When conditions permit, grapes should never be harvested before complete phenolic maturity. Harvest dates based on sugar/acid ratios should be delayed, when necessary, so that tannins may soften. To ensure this maturation, several more days are sometimes needed before harvesting. During this period, grapes should be protected against *Botrytis* attacks in certain situations; in other situations, excessively high sugar concentrations should be avoided by close monitoring.

(b) Wine Tannin Concentration

By taking into account the previously mentioned notions, general red winemaking principals can be improved for the better control of maceration time and intensity.

If grapes have low anthocyanin and tannin concentrations, only light red wines should be made. These wines, however, should be fresh and fruity. A limited concentration of grape phenolic compounds nevertheless merits an explanation. It can be a varietal characteristic, which must be taken into account. Vine cultivation conditions, favoring crop yields over quality, can also be responsible.

Adapted winemaking techniques are necessary in these cases. Techniques for compensating a phenolic deficiency are palliative and are not a substitute for perfect grape maturity.

Grapes rich in phenolic compounds are capable of making premium wines. Tannins play at least as important a role in wine aging potential as alcohol or acidity. Their role is at least as important as that of alcohol and acidity. However, tannin quality also contributes to aging potential. For example, common varieties, incomplete maturity and poor sanitary conditions contribute aggressive phenolic elements. Their addition in wine should be limited, if not totally avoided. Viticultural traditions have led to the establishment of the longest vatting times in the best *terroirs*. Reciprocally, rosé wines should be made from grape crops whose quality does not improve with maceration. Intermediate techniques can also be used.

In the 1970s, the great Bordeaux wines were considered to have insufficient tannic structure. The young wines did not taste well and there was concern that they would not age as gracefully as older vintages. There were certainly significant changes in vineyard management practices during this period, leading to higher yields and less concentrated must. However, at the same time, progress in winemaking improved management of the fermentation process. The resulting clean, fruity wines no longer needed many years' aging for certain defects to be attenuated. Nowadays, thanks to recent developments in vineyard management and winemaking techniques, Bordeaux wines have good structure and are already enjoyable immediately after vinification.

The extraction of phenolic compounds should also be modulated according to the anticipated aging potential of wine. Some experts believe that recent premium red Bordeaux wines lack tannic aggressiveness; they are thought to be too easy to drink when young and not capable of long aging. According to such expects, the tannic aggressiveness of past vintages has contributed to their present quality and extended aging potential, but this line of reasoning is highly debatable. First of all, in past vintages (for example, from the beginning of this century), the fermentations were less pure and the grapes were less healthy, even though crop yields were low and the wine concentrated. As a result, these wines were aggressive when young. The harshness of the tannins was reinforced by the elevated acidity (less ripe grapes and no malolactic fermentation). Many years were required to soften the tannins. In certain limited cases, great vintage wines resulted. Today, wines are more pleasant to drink at the end of fermentation because of improved winemaking and viticultural techniques. It is possible to judge these wines and evaluate their quality when they are still young. The commercial value of these wines is often established within a few years of their production, when offered to the market. A disagreeable-tasting young wine would be difficult to sell in today's market by simply arguing that it should improve with considerable aging.

Despite the agreeable taste of present-day premium wines immediately following fermentation, they are still capable of long-term aging. Additionally, the number of well-made wines is much higher than in the past. Yet not all wines lend themselves to long-term aging. *Terroir* and vintage also participate in a wine's aging potential. Vine cultivation conditions leading to high crop yields also limit the proper development of grape constituents. Truly great vintage wines, however, are fruity and enjoyable when young, although they have sufficiently high levels of good-quality tannins to age well for a remarkably long time. Although as pleasant as lighter new wines, these wines are capable of long-term aging. They are made from the grapes that best support extended maceration, resulting in a harmonious tannic structure.

Thus, in Bordeaux in the 1990s, winegrowers have reverted to more quality-oriented vineyard management practices. In particular, yields have been reduced to produce wines that are both more complex and more intense, as well as fruity and well-balanced. In certain cases, winemakers aim for extreme concentration, by keeping vine yields very low (20–30 hl/hectare) and emphasizing on extraction (bleeding off, pumping over, and long vatting times). The resulting rich flavors are reinforced by marked oakiness. Of course, these wines must be sold for sufficiently high prices to justify these expensive techniques. A number of these wines have been commercially successful, indicating that their quality has been recognized.

These wines are appreciated for their deep color, their rich aromas, featuring oak as an essential element, and their powerful structure and complex flavors. They stand out from other wines in blind tastings and are real "competition wines". As accompaniments to a meal, however, they are less enjoyable due to their aggressiveness, which may dominate to the point of being barely acceptable. It is easy to understand the variable appreciation of these wines.

Another consequence of this type of production is a standardization of quality that is more due to winemaking techniques than natural factors. In general, these wines are made with noble grape varieties from well-known winegrowing areas. However, successful wines have been produced by these methods from *terroirs* that had never been recognized as top quality, as well as from others that certainly had been recognized. Finally, there is no information available as yet on their aging potential. It is understandable that there should be some doubt concerning the long-term future of this type of production and its attendant prestige.

In conclusion, only the best grape varieties grown on the best *terroirs* produce wines that combine the high tannin content indicative of aging potential with aromatic finesse and complexity. On tasting, these wines are not only superbly concentrated, but also well-balanced and elegant. Winemakers today are aware that excessive tannin extraction tends to mask a wine's fruit and that perfect balance is the sign of a well-made wine.

12.6 RUNNING OFF AND PRESSING

12.6.1 Choosing the Moment for Running Off

Choosing the optimal vatting time is a complicated decision with many possible solutions. It depends on the type of wine desired, the characteristics preferred (tannin intensity and harmonious structure are not always compatible) and the nature of the grape. This decision also depends on winemaking conditions. For example, only closed fermentors permit extended vatting times. In open fermentors, the must, in full contact with air, ferments easily, but the risks of bacterial spoilage and alcohol loss make short vatting times necessary (Section 12.3.1).

In the 1950s in France, vatting times tended to be shortened from the traditional 3 or even 4 weeks. The goal of this approach was to produce more supple and less tannic wines, but the major reason was the preoccupation with avoiding bacterial spoilage. Ferré (1958) was a principal advocate of short vatting times for quality wines: 'Vatting times can be reduced to 5 or 6 days without affecting wine quality; vatting times longer than 8 days should be avoided, if only to reduce the alcohol loss occurring in open fermentors.' The data in Table 12.10 are important; they indicate the amount of alcohol loss that can occur.

More recently, new techniques (aeration through pumping over, temperature control, etc.) have made it possible to prolong vatting in closed vats without risking spoilage. Winemakers also aim to achieve greater concentration in many types of wine. Today, premium wines often have vatting times of 2–3 weeks. Extended vatting times are chosen to increase tannin concentrations but, according to analysis, the third week does not significantly increase this concentration. The prolonged vatting time nevertheless has a 'maturing effect' on the tannins. This maturation softens the tannins and improves the gustative quality of wines. The chemical transformations during this phenomenon are not known precisely, but they can be appreciated by tasting macerating wines between their 8th and 20th day of vatting. The oxidation of tannins is a possible explanation of these transformations. The oxygen introduced during pumping-overs would be responsible for this oxidation. Controlling this phenomenon would represent a considerable advance in winemaking.

Certain vineyards macerate their wines for only 2–4 days. The wines produced are ordinary. This technique is often used in hot climates, because short vatting times eliminate the risk of significant bacterial spoilage. Additionally, longer vatting times (and thus greater extraction) risk increasing gustatory flaws to the detriment of finesse. In fact, maceration intensity should be established in accordance with grape quality. Maceration is shortened for ordinary varieties and in poor quality *terroirs*; improved grape varieties in quality viticultural regions allows extended maceration.

Adjusting the vatting time is a simple method for modifying the maceration and it is therefore one of the most variable characters of red winemaking from one region to another. Its duration should be chosen by the winemaker according to grape quality and cannot be generalized: it varies from one vineyard to another, one year to another and even one fermentor to another, since grape quality is never homogeneous. This quality depends on the maturity level of the grapes (resulting from vine exposition and age) and their disease state. Winemaking equipment should never be the determining factor for deciding vatting times, but unfortunately too many wineries do not have sufficient tank capacity. Winemakers are therefore

Table 12.10. Alcohol loss according to vatting time in an open tank (Ferré, 1958)

Vatting time (days)	Density at devatting	Temperature at devatting (°C)	Alcoholic strength of wine (% vol.)	Alcohol loss (% vol.)
0	1.082	15	11.1	—
1	1.074	17	11.1	0
2	1.071	19	11.0	0.1
3	1.030	23	10.9	0.2
4	1.017	28	10.6	0.5
5	0.999	24	10.5	0.6
6	0.997	23	10.4	0.7
7	0.997	21	10.3	0.8

sometimes forced to run off wine prematurely in order to free up tank space. In such cases, vatting times can be too short.

Three types of vatting techniques are summarized below.

1. Running off before the end of fermentation—the wine still contains sugar, and the must density is between 1.020 and 1.010. This short vatting time of 3–4 days is generally recommended for average-quality wines coming from hot climates. This method is adopted for producing supple, light, fruity wines for early drinking, but it can also be used to attenuate excessive tannin aggressiveness due to variety or *terroir*.

2. Running off immediately after fermentation, as soon as the wine no longer contains sugar—approximately the 8th day of maceration. In these conditions, a maximum color intensity with a moderate tannin concentration is expected (Section 12.5.3, Figure 12.9). The gustatory equilibrium of new wines is optimized. Their aromas and fruitiness are not masked by an excessive polyphenol concentration. This vatting method is recommended for premium wines which are to be rapidly commercialized. The resulting wines are not harsh or astringent and can be drunk relatively young. When the grape crop is exceptionally ripe and thus very concentrated, premium wines may also be made in this manner. Finally, open fermentors must be run off immediately following the end of fermentation.

3. Running off several days after alcoholic fermentation. Vatting times may exceed 2–3 weeks. This method is often used to produce premium wines. The tannins assuring the evolution of the wine are supplied during this extended maceration (Figure 12.9). After several years, free anthocyanins have all but disappeared. Wine color is essentially due to combinations between anthocyanins and tannins. When making premium wines, successful winemaking requires a compromise. On the one hand, the tannin concentration must be sufficient to ensure long-term aging. On the other,

the wine should remain fairly soft and fruity. These criteria are important, since wines are often judged young.

In fact, vatting times do not follow precise rules. They depend on the kind of wine desired and on grape quality.

12.6.2 Premature Fermentor Draining due to External Factors

Sometimes fermentors must be drawn off before the ideal tannin concentration has been attained. This operation is recommended for stuck fermentations (Section 3.8.1). For reasons already mentioned (Section 3.8.3), these is the risk of development of lactic acid bacteria in sugar-containing musts with inactive yeasts. The volatile acidity would consequently increase dramatically. Drawing off the juice is a means of eliminating the majority of the bacterial population located in the pomace. Sulfiting can be effected at the same time (3 g/hl). This operation may, of course, delay malolactic fermentation, but the sulfur dioxide concentration should be calculated to allow the alcoholic fermentation to restart while blocking bacterial activity.

Various vine diseases alter grape crops. As a result, disagreeable tastes often appear in wine. Early draining may help to lessen the severity of these alterations. Gray rot (*Botrytis cinerea*) is a typical example. Certain vineyards are susceptible to gray rot, since the maturation period coincides with the rainy season. Fortunately, the pesticides currently available have greatly reduced the frequency of this disease. Botrytis has multiple effects on grape constitution and wine character (Chapter 10). Their impact influences maceration decisions.

First of all, the various forms of rot impart mushroom-like, iodine-like and moldy odors to wine. A short vatting time avoids their concentration.

Moreover, *Botrytis* secretes laccase. This enzyme has a very high oxidative activity (Sectin 11.6.2), and it can rapidly alter a red wine exposed even briefly to air. In this case, laccase

Table 12.11. Laccase destruction (expressed in arbitrary units) and oxidasic casse protection according to sulfiting (free-run wine) (Dubernet, 1974)

SO₂ added (mg/l)	Corresponding free SO₂ (mg/l)	Initial laccase activity	Laccase activity after	
			6 days	15 days
Wine no. 1				
0	0	0.16	0.16	0.16
50	16	0.16	0.02	0.00
100	28	0.16	0.00	0.00
Wine no. 2				
0	0	0.13	0.13	0.12
50	18	0.13	0.10	0.07
100	34	0.13	0.03	0.01
Wine no. 3				
0	0	0.16	0.16	0.16
50	28	0.16	0.11	0.09
100	56	0.16	0.08	0.03

analysis or, more simply, an oxidative casse test is advised before running-off. This test consists of filling a wine-glass halfway and leaving it in contact with air for 12 hours. The wine is considered to risk a casse if: (i) it changes color; (ii) it is turbid; (iii) there is sediment in the glass; (iv) its tint is less brilliant; (v) there is an iridescent film on the surface; (vi) the color becomes a brownish yellow.

If the results of these tests are positive, an extended maceration is not necessary. A prolonged vatting time would in fact intensify the flaws due to grape rot. In this case, the wine should also be sulfited at the time of running-off the wine. Malolactic fermentation will of course be more difficult, but all oxidative risks are avoided during draining.

Certain measures should be taken when wines made from rotten grapes are run off. First of all, sulfur dioxide has a high combination rate in these wines. The sulfur dioxide concentrations must therefore be relatively high (5 g/hl, or more). In the presence of SO_2, enzymatic activity is instantly inhibited, but the complete destruction of laccase activity is slow. At concentrations of 20–30 mg of free SO_2 per liter, several days are required to destroy this enzyme completely. Fortunately, during this time, the sulfur dioxide protects the wine against oxidasic casse. After the complete

destruction of laccase, the protection is definitive and independent of the presence of free SO_2.

The data in Table 12.11 demonstrate the effectiveness of SO_2 in destroying laccase. Three different wines receive 0, 50 and 100 mg of sulfur dioxide per liter. The second column indicates the combination rate, which is particularly elevated in wine 1. This wine also has the lowest residual free SO_2 concentrations. The following columns indicate the decrease in laccase activity (expressed in arbitrary units) after sulfiting. In wine 3, the enzyme has not totally disappeared within 15 days of sulfiting at a high concentration (10 /g/hl).

Figure 12.10 indicates the role of sulfiting in protecting against oxidasic casse in wine. The wine contains laccase and is exposed to air. Figure 12.10a corresponds to the aeration of a non-sulfited sample. The laccase activity diminishes according to time but does not disappear. In the first phase, the red color component (OD 520) and the yellow color component (OD 420) increase. In the second phase, the oxidasic casse appears with an increase in the yellow color component and a decrease in the red color component. In phase 3, the oxidasic casse causes a precipitation of colored matter.

Figure 12.10b corresponds to the evolution of a sample of the same wine, exposed to air, after being sulfited at 36 mg/l. The free SO_2

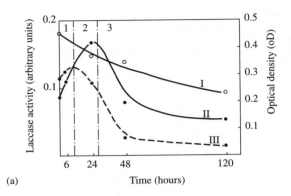

(a)

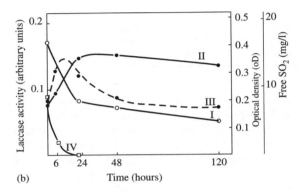

(b)

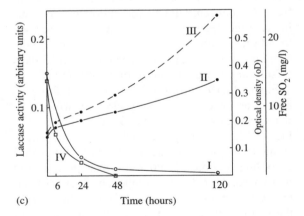

(c)

Fig. 12.10. Evolution of laccase activity, red color (OD 520), yellow color (OD 420) and free SO₂ concentration upon air contact (Dubernet, 1974): (a) non-sulfited sample; (b) sample sulfited at 36 mg/l; (c) sample sulfited at 55 mg/l. I: laccase activity. II: OD 420 (optical density at 420 nm, 1 mm thickness), yellow. III: OD 520 (optical density at 520 nm, 1 mm thickness), red. IV: free SO₂

disappears after 24 hours, but the laccase activity is not entirely destroyed. As long as the wine contains free sulfur dioxide, it is protected against oxidasic casse (the red color component and the yellow color component increase). The oxidasic casse occurs afterwards, lowering the red color component.

Figure 12.10c indicates the evolution of the same wine exposed to air, after sulfiting at 55 mg/l. When the free SO₂ concentration falls to zero after 48 hours, the laccase activity has been completely destroyed. The wine is thus definitively protected from oxidasic casse. The yellow color component and especially the red color component increase with exposure to air.

12.6.3 Running Off

The running-off operation consists of recovering the wine which spontaneously flows out of the fermentor by gravity. The wine is then placed in a recipient where alcoholic and malolactic fermentations are completed.

In the traditional, quality-orientated European vineyards, the drawn-off wine was collected in small wooden barrels. The wooden fermentors were not hermetic enough to protect wine from contact with air. Concrete and stainless steel tanks have been recommended since their development, for wine storage during the completion phase of fermentation. This completion phase precedes barrel aging. The tanks must of course be completely full and perfectly airtight.

When wines are barreled down directly, without blending beforehand, the wine batches may be heterogeneous. Yeasts and bacteria participate in these differences and they govern the completion of the fermentations. As a result, wine composition (residual sugar, alcohol and tannin concentrations) may be affected. The less the grapes are crushed and the fewer the pumping-over operations, the greater the difference between barrels of wine.

Temporarily putting the wine in vats, a technique that came into general use in Bordeaux in the 1960s, offers four advantages. First, it presents an opportunity to blend the wines. Second, yeast and bacteria cells may be evenly distributed; fermentations are thus more easily completed. Third, abrupt temperature drops occurring in small containers are avoided (they can hinder the completion of fermentation). Fourth, the daily analysis of fermentation kinetics, including the completion of alcoholic and malolactic fermentation, is easier and more rigorous with a limited number of large tanks than with a great number of small barrels.

New wines nevertheless evolve differently according to the storage method used. Slow final fermentation stages (up to several months) accentuate evolution differences with respect to storage conditions. Wine clarification in tanks occurs more slowly and is more difficult to obtain than in barrels. Carbon dioxide concentrations are also maintained over a longer time, negatively impacting wine taste. Tanks are also known to generate reduction odors from lees, such as hydrogen sulfide or mercaptans.

Since the late 1990s, it has become increasingly popular to run the wine off into barrel immediately (Section 12.7.2), as malolactic fermentation in wood has been shown to enhance aromatic complexity as well as the finesse of oak character. In fact, it is not known whether this undisputed improvement is due to the effect of bacteria on molecules released by the oak, or the fact that the new wine is still warm when it is put into barrel. The fact remains that, if red wines are to be barrel-aged, they should be run off into barrel as soon as possible. We now have all the necessary techniques to avoid the problems that led to the abandonment of barrel-aging in the past: blending, temperature control, analytical monitoring of fermentation in individual barrels, etc.

12.6.4 Pressing

After the wine is run off from the fermentor, the drained pomace is emptied from the tank and pressed. Self-emptying and automatic devatting fermentors are capable of executing this operation automatically (Section 12.3.3). Devatting can also be carried out manually, but this is laborious.

These automatic alternatives do not always respect quality criteria. In fact, fermented skins are more sensitive to the shredding and sometimes grinding effect of mechanical solutions than fresh grape crops. As a result, suspended solids are formed and press wines are turbid, bitter and sometimes colorless. Furthermore, pressing must be rapidly effected, due to pomace sensitivity to oxidation phenomena. Finally, in some cases, the addition of a fraction of press wine to run-off wine can improve overall wine quality. The goal of obtaining quality press wines is therefore completely justifiable.

To simplify devatting, a method was developed consisting of energetically mixing fermentor contents to disperse the pomace and homogenize the tank. A pump transfers the mixture to the press; the juice and skins are then separated in the press cage. This method, however, is detrimental to wine quality: the brutal mechanical action on the pomace induces vegetal and herbaceous tastes; furthermore, these flaws are not limited to the press wine—they are distributed to all of the wine.

To assure wine quality, devatting should be carried out manually and the pomace extracted carefully. A screw or, even better, a conveyor-belt system is used to transport the pomace out of the tank. Clearly, a worker must be inside the tank to feed the pomace transport system (the absence of carbon dioxide must be verified before a worker enters the tank). Ideally, the extracted pomace should fall directly from the transport system into the press. The press should therefore be mobile and capable of being placed in front of each tank. This, however, is not always possible. Moreover, this kind of pressing operation affects winery cleanliness. For this reason, pomace pumps are used to transfer drained grape skins to the press through a pipeline. The press is immobile and generally located outside of the tank room, favoring winery cleanliness, but this set-up jeopardizes wine quality. Since the appearance of these pumps on the market 20 years ago, their operation has been much improved. Current models have less of an impact on tissue integrity, especially with short pipelines, but the high pressure required to displace the pomace through the pipeline, especially through its bends, is detrimental to wine quality. This system therefore always affects the quality of press wine. The addition of wine to the pomace to facilitate its transport further diminishes overall wine quality.

When the press cage cannot be placed in front of the tank door, a conveyor belt system can be used as long as the tank and the press are not too far from each other. Another possibility is to fill several 100/l containers directly in front of the tank. These containers can then be transported to the fixed location press and emptied into it.

Oxidation should be avoided during all pomace handling (devatting, transport and pressing). All material and receiving tanks should also be perfectly clean. Good hygiene avoids the possible

development of acetic acid bacteria. In fact, these bacteria may already be present in the pomace, if the vatting time was long and the fermentor not completely hermetic.

Presses currently used are illustrated in Figure 12.11. They are also used in white winemaking. But fermented skins are pressed more easily than fresh skins. In fact, a smaller press capacity is needed for red winemaking. When the pomace is pressed, the solids must be broken up between each pressure increase-decrease cycle so that more juice

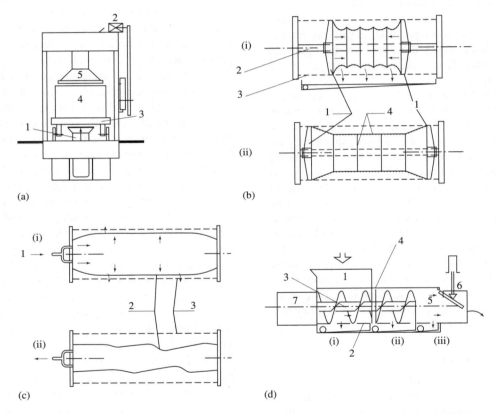

Fig. 12.11. Different types of press (Source: P. Jacquet, Bordeaux, personal communication). (a) Vertical hydraulic press. The shaft (1), driven by an electric motor (2), raises the mobile press bottom (3) and presses the pomace in the basket (4) against the fixed head (5). (b) Moving-head press (two heads) (1). (i) In the pressing phase, the heads advancing toward each other with the movement of the threaded axle (2), inside the cage (3), press the pomace. (ii) During head retraction, the chains and the hoops (4) break up the cake, while the cage is being rotated. (c) Bladder press. (i) The injection of compressed air (1) is responsible for pressing by inflating the bladder (2) against the press cage (3). (ii) After decompression, the rotation of the cage (crumbling) breaks up the cake. (d) Continuous press: (1) hopper; (2) perforated cylinder, filtering the juice; (3) screw; (4) anti-rotation system preventing the pomace from rotating with the screw; (5) compression chamber; (6) restriction door; (7) motor. Two or three levels of juice selection (i, ii, iii) are possible along the length of the screw

can be obtained. Each of these operations has an impact on the quality of the press wine, so it should be kept in separate batches.

Vertical hydraulic presses were of the oldest design. They produced good quality press wine, but loading, unloading, and breaking up the pomace between pressure cycles were awkward, labor-intensive operations. There has recently been renewed interest in this type of press, as it can easily be moved in front of the vat door for filling. Breaking up the pomace has been simplified, or even eliminated altogether, by inserting efficient drains through the pomace, which makes it possible to extract a large volume of good-quality press wine in a single operation without applying excessive pressure.

Moving head presses have the significant advantage of being automatic and sometimes programmable. Chains and press rotation are used to break up the cake after decompression. This operation produces suspended solids and can lead to olfactory defects.

Pneumatic presses comprise a horizontal press cage and an inflatable membrane. Air forced into the membrane crushes the pomace against the cage. After decompression, cage rotation breaks up the cake. The lack of a central shaft, as opposed to moving head presses, increases the press capacity of a cage of the same size. This type of press produces the highest quality results. Another option is a pneumatic press continually fed by an axial pomace pump, but this practice decreases wine quality and is not recommended. Part of the advantages of a pneumatic press are lost with this technique. In any case, the pomace should be transported as short a distance as possible with a minimum number of bends in the pipeline.

Quite a few years ago, continuous screw presses were fairly popular, due to their ease of use and high pressing speed. Yet, even with a large-diameter screw turning slowly, these presses have a brutal action on pomace; press wine quality is affected. Screw presses always produce lower quality wines than other presses. Due to pressing variations along the length of the screw press, the press wine receiving tank should be divided for separate collection of the batches corresponding

with the first pressing, second pressing, etc. of discontinuous presses. The small volume of the last batch is generally very low in quality: it should be eliminated and distilled.

12.6.5 Composition and Use of Press Wines

The wine impregnating the pomace constitutes the press wine. Its volume during winemaking depends on the level of pulpiness of the grapes. In the Bordeaux region, it represents approximately 15% of the finished wine on average. Press wine contains an interstitial wine. This wine is easy to separate from the skins and relatively similar to free-run wine, when the fermentor has been well homogenized by pumping-over operations and the grape correctly crushed. It is, however, also made up of a wine which saturates the pomace tissues. This wine is very different from free-run wine and much more difficult to extract. Following this principle, two kinds of press wine are generally separated. The first press wine (approximately 10% of the finished wine or two-thirds of the press wine) is obtained through a direct pressing. When pomace handling and pressing are correctly effected, the first press is of a good quality. The second press wine (approximately 5% of the total quantity of wine and one-third of the press wine) is not as good in quality, as it is obtained at high pressure after the press-cake has been broken up. This damages grape tissues that have become more fragile during fermentation, releasing substances with bitter, herbaceous overtones and accentuating the characteristic astringency of press wines, due to their high tannin content.

Grape quality primarily affects press wine quality. Ordinary quality varieties and grapes ripened in hot climates can produce press wines containing a high concentration of aggressive and vegetal tannins. Press filling conditions and pressing methods also affect press wine quality, i.e. both the number of pressings with cake break-up and the maximum pressure. Finally, a slow and regular pressure increase, even between two pressings, is beneficial to press wine quality. A single pressing, without pomace cake break-up, is recommended for premium wines, of course, while assuring a slow and regular pressure increase up to the maximum. This method produces less press wine but of a superior quality.

All elements, except for alcohol, are concentrated in press wine. Table 12.12 gives an example of this phenomenon. The alcohol content decreases by 40%, and in some cases by even more. The presence of reducing agents is most likely responsible for the higher sugar concentration. Uncrushed grapes may also liberate unfermented sugar during pressing. The volatile acidity of press wine is always higher than in free-run wine—indicating an increased bacterial risk in the pomace. Total wine acidity is generally also a little higher, but the higher mineral concentration also increases the pH of press wine. More phenolic compounds (anthocyanins and tannins) are present, reflected in the extract values. Press wines also contain more nitrogen compounds. In certain hot climates, high maturity levels lead to extremely concentrated grapes: the resulting press wines are so rich in tannins that their tasting can be too bitter and astringent to market. These wines should be distilled. Although there is little analytical data on this, press wines also contain polysaccharides and other colloids that add body to the overall flavor.

Press wine quality also depends on winemaking conditions. Repeated pumping-overs, elevated maceration temperatures and other techniques that increase maceration will deplete the pomace of qualitative phenolic compounds. The resulting press wines lack body and color and are dominated by astringent and vegetal savors. These inferior-quality wines cannot be blended with free-run wines to improve overall wine structure and quality, and must sometimes be distilled or eliminated, representing a considerable loss in wine volume. Press wine quality must therefore be ensured by avoiding excessive maceration and extraction techniques.

The decision to blend press wines with free-run wines is complicated. It not only depends on both free-run and press wine quality but also on the type of wine desired. In general, press wines are not added when making primeur-style wines for early drinking, except when the press wines are excessively light. Moreover, press wines should not be incorporated into wines made from ordinary and rustic varieties. Premium wines made from very concentrated grapes are often very tannic, even after short vatting times; the addition of press wines does not improve their overall quality. When possible, press wines that are not used for blending should be distilled.

Press wines are, however, required most of the time for making premium wines in temperate climates. In this case, press wines generally have higher tannin concentrations than free-run wines and are often excessively astringent. Due to their colloidal structure, adding a small percentage of press wine makes a fuller and more homogeneous finished wine. But even press wines without flaws generally have a heavy odor which masks the fruity character of new wines. Immediately following the addition of press wines, wine aroma is less

Table 12.12. Composition of free-run and press wine (Ribéreau-Gayon *et al.*, 1976)

Component	Free run wine	Press wine
Alcoholic strength (% vol.)	12.0	11.6
Reducing sugars (g/l)	1.9	2.6
Extract (g/l)	21.2	24.3
Total acidity (g/l H_2SO_4)	3.23	3.57
Volatile acidity (g/l H_2SO_4)	0.35	0.45
Total nitrogen (g/l)	0.28	0.37
Total phenolic compounds (permanganate index)	35	68
Anthocyanins (g/l)	0.33 .	0.40
Tannins (g/l)	1.75	3.20

refined and fruity, but this flaw tends to disappear with aging. The wine is, however, fuller and more balanced and harmonious—capable of long-term aging. In certain quality wine regions, press wines are generally considered indispensable to wine quality.

Delaying the addition of press wines so that the clarification process may take place can be beneficial to wine quality. The press wines may undergo fining or pectolytic enzyme treatments (0.5 g/hl) at the time of draining, before malolactic fermentation. Too long of a delay (for example, until the spring following the harvest) causes the free-run wine to evolve. As a result, it may not blend well with the typical savors of press wine. The best solution is to decide whether to add a certain proportion of press wine soon after the completion of malolactic fermentation. The press wine percentage (5–10%) must be determined according to the anticipated aging potential of the wine. When the ideal blend ratio is obtained after laboratory trials, these proportions are used for blending the various batches in the winery. At this stage, a certain level of tannic aggressiveness should be sought. These tannins improve the barrel and bottle aging potential of the wine. The press wine may also be progressively blended during the months following fermentation, to compensate for thinning (which always accompanies the first stage of wine maturation). With this process, the wine is at its optimum quality during the period when it is judged and sometimes sold.

12.7 MALOLACTIC FERMENTATION

12.7.1 History

Research on malolactic fermentation of red wines, its role and its importance have greatly influenced the evolution of post-Pasteurian enology. Various concepts have been developed leading to contradictory winemaking methods. Several decades were necessary for the establishment of a general doctrine in all viticultural regions.

Ribéreau-Gayon and Peynaud (1961) wrote:

For twenty or so years, a better understanding of the malolactic fermentation phenomenon, its agents, its mechanism and its factors has permitted considerable progress in malolactic fermentation research. An ever increasing number of observations and studies in concerned winemaking regions have also participated in a better understanding of this process. Yet viticultural regions are slow to apply this information. Its practice seems to spread slowly from one region to another and is difficult to establish. The complication of winemaking methods by these notions with respect to certain simplistic theories has created a certain amount of resistance. The need to modify outdated, but generally accepted, doctrines has also slowed progress. It is surprising that solidly established and widely confirmed notions have encountered so many obstacles.

Malolactic fermentation is both relatively simple and extremely important in practice, and all sensible winemaking and red wine storage techniques take its existence and laws into account. It is an important element in premium wines, even in complete maturity years. In addition, it regulates wine quality from year to year. The less ripe the grapes and therefore the higher the malic acid concentration, the more malolactic fermentation lowers wine acidity. The differences in acidity of wines from the same region are much smaller than those of the corresponding musts.

Another less readily accepted consequence of malolactic fermentation is an improvement in biological stability caused by bacteria that eliminate highly unstable malic acid, which results in an increase in pH.

The existence and importance of malolactic fermentation were not easily recognized. It occurs in variable conditions which make proving its existence difficult. If it takes place during or immediately following alcoholic fermentation, it can be completed without being noticed, but it can also occur several weeks or months after alcoholic fermentation. Since little carbon dioxide is released, the phenomenon is sometimes almost imperceptible. The decrease in total acidity observed can also be interpreted as a potassium hydrogeno tartrate precipitation. Additionally, the chemical analysis of malic acid, especially in the presence of tartaric acid, had always been difficult. The determination of malic acid concentrations by paper chromatography was the first simple and significant method (Ribéreau-Gayon, 1953); it could be used in the winery and permitted the diminution of malic acid to be monitored. It greatly contributed to the establishment of the notion of malolactic fermentation.

Malolactic fermentation is nevertheless a winemaking tradition. It occurred irregularly but did exist in past red wines. The data in Table 12.13 are significant in this respect. It was not until the decade from 1963 to 1972 that malolactic fermentation became systematic. A better control of microbial spoilage simultaneously permitted the lowering of volatile acidity concentrations, essentially affecting maximum values. Bordeaux was the forerunner with this systematic control of malolactic fermentation, which occurred much later in many viticultural regions throughout the world. Although not pertaining directly to this chapter, the figures in Table 12.13 concerning the alcohol content are interesting: they show that chaptalization has permitted the alcohol content of recent vintage Bordeaux wines to be regulated in comparison with past vintages, but maximum values have remained similar over the years.

The first observations of malolactic fermentation date back to the end of the 19th century in Switzerland and Germany and to the beginning of the 20th century in France. The data in Table 12.14, pertaining to winemaking in 1896, give characteristic examples of malolactic fermentation. Researchers at that time were not capable of correctly interpreting the informations: they focused in particular on the volatile acidity increase, following the bacteria population increase observed under the microscope. They attributed the lowering of total acidity to potassium hydrogentartrate precipitation—the disappearance of malic acid was not even considered. This situation was thought to be the beginning of a serious microbial contamination that should absolutely be avoided.

Past researchers also noticed that sulfiting of must resulted in higher acidity wines. This phenomenon was interpreted as a greater dissolution of the acids of pomace in the presence of sulfur dioxide. The idea that bacteria were inhibited and that malic acid was not degraded was not even considered.

Table 12.13. Analyses of different vintages of red wines, Medoc and Graves vineyards (analyses performed in 1976; (Ribéreau-Gayon, 1977)

Period	Number of samples	Levels	Alcoholic strength (% vol.)	Total acidity (g/l H_2SO_4)	Volatile acidity (g/l H_2SO_4)	Total SO_2 (mg/l)	Malic acid (g/l)	Sugar (g/l)
1906–31	12	Min	10.0	3.72	0.53	27	0	2.0
		Max	12.3	5.59	0.90	83	3.5	4.2
		Av.	11.1	4.40	0.71	41	0.8	2.5
1934–42	11	Min	9.9	3.92	0.46	27	0	2.0
		Max	13.1	5.39	1.10	105	3.5	3.8
		Av.	11.2	4.32	0.75	53	0.6	2.4
1943–52	17	Min	10.3	3.43	0.45	28	0	2.0
		Max	12.7	5.29	1.05	70	2.0	3.8
		Av.	11.3	4.03	0.70	45	0.2	2.3
1953–62	18	Min	11.1	3.19	0.45	29	0	2.0
		Max	13.3	5.29	0.72	80	1.5	2.5
		Av.	11.9	3.55	0.57	52	0.1	2.1
1963–72	18	Min	11.8	3.09	0.37	51	0	2.0
		Max	12.7	3.53	0.50	89	0	2.5
		Av.	12.0	3.38	0.42	67	0	2.0

Table 12.14. Alcoholic and malolactic fermentations in tank; results obtained in 1896 (Gayon, 1905)

Tank no.	Duration of fermentation (days)	Remaining sugar (g/l)	Total acidity (g/l H_2SO_4)	Volatile acidity (g/l H_4SO_4)	Number of bacteria under microscope
1	2	116	4.35	0.12	0
	5	45	4.25	0.12	0
	6	23	4.25	0.13	1–2
	7	6	3.19	0.21	25–30
	10	3	3.15	0.35	30–35
	13	1	3.15	0.36	35–40
2	2	104	4.25	0.10	0
	5	29	4.25	0.10	0
	6	4	4.47	0.12	0
	7	1	4.20	0.17	4–5
	8	1	3.35	0.23	20–25
	13	1	3.40	0.45	50–60

The general existence of this phenomenon was established from 1922 to 1928 in Burgundy by the research of L. Ferré (1922) and from 1936 to 1938 by the studies of J. Ribéreau-Gayon in Bordeaux (both authors cited in Ribéreau-Gayon *et al.*, 1976). The importance of this second fermentation was demonstrated to be an essential step in making premium red wines. Yet the apparently simple corresponding notions were difficult to accept. For a long time in enological works, malolactic fermentation was described in the chapters covering diseases and spoilage. Certain enology schools contested both the existence and especially the value of this second fermentation.

The importance and utility of malolactic fermentation were slow to be established because of the involvement of lactic acid bacteria, considered to be contaminating agents (Table 12.14). Their frequent presence in red winemaking was thought to correspond with the beginning of spoilage that should be avoided at all costs. Pasteur once said: 'Yeasts make wine, bacteria destroy it'. It seemed pretentious at the time to go against the beliefs of a great scientist. Finally, the idea that the same bacteria could be beneficial when they degrade malic acid and detrimental when they attack other constituents was difficult to accept.

Furthermore, a slightly elevated acidity was considered in the past to be a sign of quality. A low pH effectively opposes bacterial development and thus can limit the production of volatile acidity. However, malic acid is a highly biodegradable molecule and its disappearance results in a biological stabilization of the wine, even though the pH increases. When a red wine containing malic acid is bottled, there is always a risk that malolactic fermentation will start in the bottle after a few months, resulting in spoilage due to gassiness and an increase in volatile acidity.

The diagram in Figure 3.9 (Chapter 3) summarizes the principles of correct present-day red winemaking. Lactic acid bacteria should only be active when all of the sugar has been fermented. Interference between the two fermentations should be avoided (Section 3.8.1). It compromises the completion of alcoholic fermentation and can result in a considerable increase in volatile acidity, if the bacteria decompose the remaining sugar. When there is no longer any sugar, the bacteria develop mainly by degrading malic acid, the most easily biodegradable molecule. In this case, the bacteria have a beneficial effect. The high biodegradability of malic acid requires its elimination. The bacteria are thus beneficial during this process.

As soon as the malolactic fermentation is completed, the same bacteria can rapidly become detrimental and certain precautions are necessary to avoid this unwanted evolution. The bacteria are apt to decompose pentoses, glycerol, tartaric acid, etc. These transformations cause common wine diseases (lactic disease, *amertume*, *tourne*, etc.), which increase volatile acidity and lactic acid concentrations to a variable degree. Total acidity is thus also increased. The second fermentor in Table 12.14 gives an example of this phenomenon. Between the 8th and 13th day, the bacteria population, total acidity and volatile acidity increase considerably. These increases indicate that the transformation is no longer a pure malolactic fermentation.

Fortunately, lactic acid bacteria have a preference for malic acid—otherwise, present-day premium wines would not exist—but care must be taken to assure useful microbial transformations while avoiding harmful ones. Despite its acidity and alcohol content, wine is alterable, but luckily not too alterable.

The errors committed in certain French wineries during the 1950s due to winemaking principles at the time are understandable. Wine must was massively sulfited to be absolutely sure of avoiding bacterial contamination. On the one hand, the wine did not benefit from the advantages of malolactic fermentation. On the other hand, since the wine was not stored in sterile conditions, it remained susceptible to subsequent contamination. An untimely and uncontrolled malolactic fermentation could therefore occur at any moment.

In view of the gradual decrease in total acidity observed in many vineyards today, there may be some doubt as to the absolute need for malolactic fermentation. In future, steps may be taken to prevent it in certain, specific cases. Of course, for the moment, that is only a hypothesis as, according to our present understanding of these phenomena, malolactic fermentation is still an indispensable stage in red winemaking.

Current technology should lead to the development of stabilization methods preventing uncontrolled malolactic fermentations. The first step is to avoid excessive contamination, even though absolute sterility is difficult (if not impossible) to obtain. Physical methods such as heat treatments are the most effective methods for eliminating lactic acid bacteria. Various sterile bottling techniques exist that make use of either filtration or heat treatments. Among chemical methods, sulfiting is effective due to the antibacterial effect of bound sulfur dioxide. Lactic acid bacteria inhibitors also exist: egg white lysozyme (Section 9.5.2), fumaric acid and nisin. Their use need to be authorized. These substances are not always completely effective, nor are they perfectly stable. In any case, the high resistance of certain strains in wine should be taken into account, especially when wine pH is high.

12.7.2 Wine Transformations by Malolactic Fermentation

This section provides further details on the chemical and flavor changes that occur in wine during malolactic fermentation (Section 6.3.3). The mechanism reactions involved are described in Chapter 5 and the overall reaction of this phenomenon is shown in Figure 12.12. This reaction is a simple decarboxylation, explaining the loss of an acid function. In practice, at the pH of wine, malic acid is partially neutralized in the form of dissociated salts, thus in an ionic form, but the overall phenomenon described remains the same. Each time that a molecule of malic acid, in the free acid or ionic form, is degraded, a free acid function is eliminated. Only a limited amount of

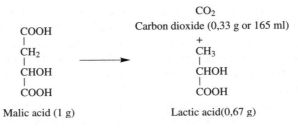

Fig. 12.12. Malolactic fermentation: overall reaction

carbon dioxide is released, but it is perceptible if the cellar is quiet. It can, in fact, be the first sign of the initiation of malolactic fermentation.

The decrease in acidity following malolactic fermentation varies according to the malic acid concentration and thus grape maturity. This decrease in acidity can be from 2 g/l in H_2SO_4 to sometimes 3 g/l (3–4.5 g/l in tartaric acid). Total acidity decrease from 4.5–6.5 g/l in H_2SO_4 (6.75–9.75 g/l in tartaric acid) to 3–4 g/l H_2SO_4 (4.5–6 g/l in tartaric acid). The fermentation of 1 g of malic acid per liter lowers the total acidity by approximately 0.4 g/l in H_2SO_4 (0.6 g/l in tartaric acid).

The preceding reaction does not explain acetic acid production, but volatile acidity always increases during malolactic fermentation. This production is due, at least in part, to citric acid degradation (Section 4.3.3). Although a molecule of citric acid produces two acetic acid molecules, this degradation is always limited because grapes do not contain large quantities of this acid.

Bacteria also produce volatile acidity from the degradation of pentoses. In fact, these sugars might be used as energy sources. Malic acid degradation does not seem sufficient to ensure cell energy needs (Henick-Kling, 1992).

Observations show that volatile acidity increases at the end of this phenomenon, when malic acid is almost entirely depleted. Moreover, this increase is even greater when malolactic fermentation is facilitated (low acidity musts, for example).

Table 12.15 shows the main chemical transformations in wine during malolactic fermentation. In this case, it is incomplete, as the wine still contains 0.5 g/l malic acid that has not been degraded.

The results (mEq/l) assess the consequences. The lactic acid formed corresponds to half of the malic acid transformed. The diminution in fixed acidity corresponds approximately to the difference between the loss in malic acid and the gain in lactic acid.

The chemical transformations of wine by malolactic fermentation are much more complex in reality. Malolactic fermentation also produces ethyl lactate, the formation of which contributes to the sensation of body in wine (Henick-Kling, 1992). Additionally, other secondary products have been identified, the most important being diacetyl, produced by bacteria (a few milligrams per liter), that belongs to a complex pool of production and degradation mechanisms. At moderate concentrations, this secondary product contributes to aromatic complexity, but above 4 mg/l the characteristic butter aroma of this substance dominates.

Another transformation attributed to lactic acid bacteria is the decarboxylation of histidine into histamine, a toxic substance. This reaction does not occur often and is carried out by certain bacterial strains in specific conditions. It is responsible for elevated histamine concentrations (10 mg/l or higher) sometimes found in certain wines.

Wine color modification always accompanies malolactic fermentation. Color intensity decreases and the brilliant red tint diminishes. This modification is due to the decolorization of anthocyanins when the pH increases, but condensation reactions between anthocyanins and tannins are probably also involved. These reactions modify and stabilize wine color.

Table 12.15. Analysis of a wine before and after malolactic fermentation (Ribéreau-Gayon et al., 1976)

Acids	Concentrations (g/l)[a]		Concentration (mEq/l)		
	Before	After	Before	After	Difference
Total acidity	4.9	3.8	100	78	−22
Volatile acidity	0.21	0.28	4.3	5.6	+1.3
Fixed acidity	4.7	3.6	96	73	−23
Malic acid	3.2	0.5	48	8	−40
Lactic acid	0.12	1.8	1.4	20	+19

[a]Total acidity, volatile acidity and fixed acidity are expressed in H_2SO_4.

The organoleptic character of the wine is also greatly improved. First, wine aromas are more complex. Wine bouquet is intensified and the character and firmness of the wine are improved, as long as the lactic notes are not excessive. Malolactic fermentation conditions (bacterial strains and environmental and physical factors) certainly influence results, and this fact is illustrated by effecting malolactic fermentations on white wines, which are, of course, simpler and thus more sensitive to changes brought about by malolactic fermentation. These transformations merit further study. Harmful aromatic flaws may occur, especially with difficult malolactic fermentations and toward the end of this phenomenon.

The taste of the wine is also considerably improved. The role of deacidification becomes more important when the initial malic acid concentration of wine increases. The softening of wine is due first of all to a decrease in acidity. The substitution of the malic ion by the lactic ion also contributes. In fact, malic acid corresponds to the aggressive, green acid of unripe apples. Lactic acid is the acid found in milk; it has a much less aggressive taste. Additionally, the association of the flavor of malic acid with the astringency of tannins is not harmonious. This phenomenon permits red wines to lose their acid and hard character. They become softer, fuller and fatter—essential elements for a quality wine.

Attempts have been made to determine the influence of bacterial strains and operating conditions on organoleptical changes in red wines brought about by malolactic fermentation. At present, no definitive results have been found.

According to certain recent theories, red wine quality is even more greatly improved when malolactic fermentation takes place in barrels (Section 12.6.3). Wine aroma is more complex and fine and the oak character more integrated; tannins are fuller and more velvety. These differences are already present at the end of malolactic fermentation, but they are often less flagrant at the time of bottling. When this technique is applied correctly, it has no detrimental effects, but requires a lot of extra work in the cellar, especially in handling and inspecting large numbers of barrels.

Finally, concerning the transformations in wine, the degradation of the malic acid molecule leads to a biological stabilization even though the pH increases (Section 12.7.1).

12.7.3 Monitoring Malolactic Fermentation

It is essential to determine the initiation of malolactic fermentation and to monitor the diminution and complete depletion of malic acid in each tank. For a long time, the malic acid concentration was difficult to determine chemically—it could only be extrapolated through comparing total acidity before and after malolactic fermentation. In this case, simultaneous potassium hydrogentartrate precipitation could also lower total acidity, falsifying the estimate of malic acid concentrations. In due course, paper chromatography appeared (Ribéreau-Gayon, 1953). This analytical tool, which permitted a simple, visual method for monitoring the diminution of malic acid (Figure 12.13), represented a considerable advance and greatly contributed to the general use

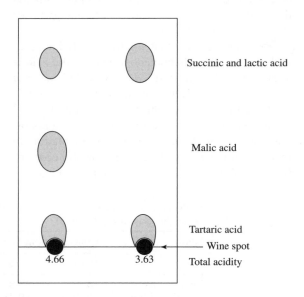

Fig. 12.13. Separation of organic acids found in wine by paper chromatography (Ribéreau-Gayon, 1953). Left: wine that has not undergone malolactic fermentation. Right: the same wine after malolactic fermentation. Total acidity is expressed in g/l H_2SO_4

of malolactic fermentation in wineries. Although not a very precise method for analysis of malic acid, it also permits the gustatory characteristics of a wine to be compared according to the stage of malolactic fermentation.

Paper chromatography is easy to use and is always widely employed in wineries to monitor this second fermentation, though the method is sometimes slightly varied. The Bordeaux region, for example, produces 4–6 million hl red wine per year, representing some 40,000–60,000 vats that have to be tested at least twice, requiring a considerable number of analyses. Today, there is also a more accurate enzymatic method for assaying malic acid. The reagents are expensive but the analysis may be automated and is especially suitable for checking the completion of malolactic fermentation. They are also well adapted to verifying the completion of malolactic fermentation, but the reactives are expensive.

Malolactic fermentation is normally monitored after the wine has been run off—in other words, after the completion of alcoholic fermentation. Yet malolactic fermentation may begin prematurely, when musts are insufficiently sulfited or inoculated with bacteria before alcoholic fermentation. In this case, the two fermentations may be slow but complete. The overlapping of the two fermentations can lead to a stuck alcoholic fermentation and in this situation the bacteria are also apt to produce volatile acidity from sugars (Sections 3.8.1 and 3.8.2). This must be closely controlled, to avoid a serious accident, and the monitoring of both the malic acid concentration and volatile acidity is recommended. When the alcoholic fermentation is not complete, the sugar concentration should also be closely followed. If bacterial spoilage begins to occur, the wine should be immediately sulfited (3 g/hl).

Experience has shown that, even in sugar-containing media, lactic acid bacteria during their growth phase do not produce acetic acid and decompose only malic acid (Section 3.8.3). The complete depletion of malic acid, however, greatly increases the risk of serious alterations when the wine still contains sugar. For example, malolactic fermentation sometimes occurs before the wine has

been run off, during the post-fermentation phase. In this case, the wine may still contain some residual sugars, especially in the case of slightly crushed grapes.

After skin and juice separation and pressing, the tanks are completely filled with wine. The malolactic fermentation process is then monitored daily, using paper chromatography. This second fermentation generally takes a few days to a few weeks (maximum). An excessive delay of its initiation is generally due to winemaking errors: too much sulfiting or too low a pH or wine temperature. In the past, certain vineyards and sometimes entire viticultural regions claimed that malolactic fermentation was impossible. A winemaking error, most often excessive sulfiting, was generally responsible.

If malolactic fermentation does not commence, the fear of an increase in volatile acidity does not justify prematurely sulfiting the wine. In fact, in the absence of sugar, lactic acid bacteria begin to develop by degrading malic acid into lactic acid. Yet an excessive delay of the initiation of the fermentation does require other precautions to be taken. To avoid excessive oxidation, the wine should be sulfited. Although the wine sediment contains bacteria, it may be necessary to remove it to prevent reduction odors. Malolactic fermentations have then been observed (especially in the past) to occur normally in the spring or summer following the harvest.

The diminution of malic acid is also monitored in order to choose the ideal moment for definitively stabilizing the wine. Sulfiting at 3–8 g/hl (press wine) permits this stabilization. Before sulfiting, the wine is racked to eliminate a fraction of the bacteria with the coarse sediment. This operation should follow the complete depletion of malic acid. After its initiation, malic acid degradation is normally completed in several days. In rare cases, bacteriophages destroy bacteria (Section 6.5): they stop malolactic fermentation and several hundred milligrams of malic acid per liter may remain undegraded. In this situation, the wine must be reinoculated to restart the fermentation.

The residual malic acid content should be under 100 mg/l when the wine is stabilized by sulfiting and it would be dangerous to bottle a wine containing over 200 mg/l. At the very minimum, lactic acid bacteria should be allowed to continue until concentrations have fallen below 200 mg/l, to prevent future bacterial alterations. This stabilization should not be excessively delayed: bacterial spoilage, such as increased volatile acidity, is most likely to occur in the final stages of malolactic fermentation. The possibility of strains highly resistant to sulfur dioxide, especially at high pHs, should also be considered, but they are not a significant problem if the wine no longer contains sugar, malic acid and citric acid and if its conservation temperature is relatively low. The evolution of these wines should nevertheless be closely monitored.

12.7.4 Conditions Required for Malolactic Fermentation: Influence of Acidity, Temperature, Aeration and Sulfiting

Bacterial development conditions are described in Chapter 6. During the first hours after the fermentor is filled, the bacteria originating on the grape develop rapidly. As soon as alcoholic fermentation is initiated and ethanol is formed, the bacteria population greatly decreases and environmental conditions become increasingly hostile—only the most resistant strains are capable of surviving. During separation and pressing, the contact of the wine with contaminated equipment may increase the population of these resistant strains. These resistant strains, remain latent for a variable period, then start malolactic fermentation when the population reaches levels of the order of 10^6 cells/ml.

In good winemaking practice, the latent phase should be sufficiently long to avoid the undesirable overlapping of the two fermentations (Section 12.7.3). This phase should also be short enough so that the malic acid may be degraded within a reasonable amount of time.

Techniques may be employed to influence malolactic fermentation but the ideal conditions for this phenomenon remain ill-defined. Malolactic fermentability of wine varies according to the region, vineyard and year. Wine also tends to ferment better in large containers than in small ones. These facts are difficult to interpret but environmental and nutritive conditions of bacteria appear to play an important role. Bacterial growth is limited by alcohol and acidity in wine; in addition, bacteria are incapable of synthesizing certain essential substances (nitrogen compounds, amino acids and growth factors). Specific deficiencies may therefore make certain malolactic fermentations difficult. Yet, in practice, modifying wine composition does not improve malolactic fermentability to any significant extent, except by increasing its pH. Each bacterial strain probably has an optimum nutritive medium.

Optimum growth conditions for grape origin bacteria are described in this section. Inoculation will be covered in Section 12.7.5.

Alcohol is the first limiting factor of malolactic fermentation. Malic acid concentrations often decrease fastest in tanks containing the lowest alcohol concentrations. *Leuconostoc oenos* (now known as *Oenococcus oeni*) is predominantly responsible for malolactic fermentation in red wines and it cannot grow in alcohol concentrations exceeding 14% volume. Some lactobacilli can resist 18–20% volume alcohol and are apt to cause spoilage in fortified wines. Besides alcohol production, the wine yeast strain responsible for alcoholic fermentation affects bacterial growth and malolactic fermentation. It yields macromolecules (polysaccharides and proteins) to the medium. The enzymatic systems of the bacterial cell wall hydrolyze these substances.

The following factors participate in the control of malolactic fermentation: acidity (Section 6.2.1), temperature (Section 6.2.4), aeration (Section 6.2.5), vatting time and sulfiting (Section 6.2.2).

(a) Influence of Acidity

As acidity increases, a growing number of bacterial species is inhibited. Malolactic fermentation becomes increasingly difficult but, simultaneously, it is increasingly pure. Malic acid is predominantly degraded. The degradation of other wine components is slight, thus limiting the increase in volatile acidity.

Lactic acid bacteria growth is optimum at a pH between 4.2 and 4.5. In the pH range of wine (3.0–4.0), malolactic fermentation speed increases with the pH. The pH limit for growth is 2.9 but even at 3.2 bacterial growth is very limited. Malolactic fermentation becomes possible at a pH of 3.3 or higher.

Malolactic fermentation is necessary with insufficiently ripe grapes but the high malic acid concentrations in these grapes hinder this fermentation. With very ripe grapes having a low acidity, the impact on wine taste is less significant but malolactic fermentation occurs easily and the risk of bacterial spoilage is much higher.

When the pH is excessively low, wine can be deacidified to facilitate the initiation of malolactic fermentation: for example, 50 g of $CaCO_3$ per hectoliter can be added to the wine (Chapter 11). The role of this deacidification is to rectify the pH without removing an excessive amount of tartaric acid. This operation must take into account the decrease in total acidity brought about by malolactic fermentation, according to the malic acid concentration in the wine. This operation should be effected on a fraction of the total volume (20–30%, for example). This deacidified fraction is used to initiate the natural deacidification reactions (malolactic fermentation followed by potassium hydrogentartrate precipitation).

(b) Influence of Temperature

The effect of temperature is twofold. First, an elevated temperature (above 30°C) during vatting can affect bacteria. Second, bacterial growth and the initiation of malolactic fermentation require a certain temperature range. The impact of temperature on bacterial growth depends on the alcohol content of the wine. For 0–4% volume of ethanol, the optimal growth temperature is 30°C, as opposed to 18–25°C for an alcohol strength of 10–14% volume (Henick-Kling, 1992). In practice, the optimal temperature for malolactic fermentation is between 20 and 25°C. The fermentation is slowed outside these ranges.

New wines should therefore be maintained at a temperature of at least 18°C. Temperature is the simplest means of influencing malolactic fermentation. In the past, the wine cellars were not temperature-controlled and the cold autumn air was the principal factor blocking malolactic fermentations.

The fermentation of malic acid is slow at 15°C, whereas it is complete in a few days at 20°C. When initiated at a suitable temperature, malolactic fermentation is generally completed, even when the temperature drops to 10°C. Its initiation in winter is unlikely if the temperatures are unfavorable: fermentation will most likely occur the following spring, when the temperature rises naturally. Sulfiting should be carefully timed to avoid blocking this phenomenon.

The fermentation should be conducted at as low of a temperature as possible (18–20°C, for example). The low temperature makes malolactic fermentation slower but limits the risk of bacterial spoilage—in particular, excessive volatile acidity production due to the transformation of substances other than malic acid.

(c) Influence of Aeration

Each bacterial species has its specific needs. In practice, these different needs are not known. Malolactic fermentation is possible for a large variation of aeration. This factor does not seems to be preponderant. Wine oxygenation by contact with air generally accelerates the initiation of malolactic fermentation, but saturating wine with pure oxygen delays or completely blocks it. In practice, a moderate aeration is often beneficial to malolactic fermentation (Peynaud, 1981).

(d) Influence of Sulfiting

Bacteria are known to be highly sensitive to sulfur dioxide (Section 8.6.3). They are much more sensitive to it than yeasts (Section 8.7.4). Moderate concentrations of sulfur dioxide assure a pure alcoholic fermentation without bacterial contamination—always dangerous in the presence of sugar. Both free and bound sulfur dioxide have an effect on bacteria. Over several months of storage, wine is regularly sulfited; this operation increases the total SO_2 concentration and malolactic fermentation becomes difficult, if not impossible, even with a low concentration of free SO_2.

Sulfiting particularly affects malolactic fermentation in two circumstances: sulfiting the crushed grapes during tank filling and sulfiting wine at skin and juice separation.

In normal winemaking, the wine should not be sulfited at running off to avoid compromising malolactic fermentation. There are two exceptions, corresponding to accidental factors: contaminated harvests and stuck fermentations (Section 12.6.2). In the first case, a light sulfiting (2–5 g/hl) protects against oxidasic casse; in the second case, it avoids lactic disease. In both cases, the situation is serious enough to justify making malolactic fermentation more difficult. After sulfiting at 2.5 g/hl at running off malolactic fermentation has been reported to be delayed until the following summer. Sulfiting at 5 g/hl can definitely block fermentation. In these exceptional cases, the wine should be massively inoculated with wine that was not sulfited at running off with a normal malolactic fermentation.

New red wines should not be sulfited immediately. This can pose a problem if the initiation of malolactic fermentation is slow. Bacterial spoilage is, of course, unlikely because lactic acid bacteria initially degrade malic acid, but oxidation can be detrimental to wine quality. In general, premature sulfiting can make malolactic fermentation impossible.

Adding sulfur dioxide when the must is put into vat also has an impact on malolactic fermentation. The fermentation may be delayed to a variable extent, depending on the concentration of sulfur dioxide used and the way it is mixed into the must (Section 8.8.1) and may, in extreme cases, even be permanently inhibited. The concentration chosen must be sufficient to retard malolactic fermentation, to avoid its interference with alcoholic fermentation and the associated risks, but not so excessive that the malolactic fermentation cannot be completed within a reasonable time period.

The action of sulfur dioxide depends not only on the concentration chosen but also on grape composition. The pH and disease state of grapes, in particular, influence the SO_2 binding rate. The ambient temperature is also a factor. In Bordeaux region conditions, 5 g/hl has little effect on delaying malolactic fermentation, 10 g/hl clearly slows it and at 15 g/hl or higher it becomes impossible. In northern-climate vineyards, 5 g/hl can be sufficient to stop it, but in hot regions, malolactic fermentation may still occur at 20 g/hl.

Determining harvest sulfiting levels is difficult. Sulfiting, however, remains a particularly sensitive method for modulating the malolactic fermentation process. Proper sulfiting permits a complete alcoholic fermentation, while avoiding spoilage phenomena, without compromising or excessively delaying the fermentation. The use of lysozyme (Sections 9.5.1 and 9.5.2) has been recommended to supplement the effect of SO_2 in delaying the development of indigenous bacteria and, thus, the start of malolactic fermentation.

12.7.5 Malolactic Fermentation Inoculation

The optimum conditions for obtaining malolactic fermentation in new wines were recommended in the last section. Most often, this fermentation begins within a reasonable amount of time, but it is not always initiated spontaneously. At optimum conditions in a winery where alcoholic fermentation has already occurred, malolactic fermentation often occurs in at least a few of the fermentors. It can therefore be propagated throughout the winery by massively inoculating the other fermentors. For example, a third of the volume of a fermentor with a completed malolactic fermentation can be mixed with two-thirds of the volume of a fermentor with a difficult malolactic fermentation. Fermentors are sometimes inoculated with the lees from a nearly completed malolactic fermentation vat. The fermentation is thus (hopefully) completed within a few weeks instead of a few months. The elimination of malic acid no longer poses significant practical constraints. Better control of winemaking conditions, especially temperature control and sulfiting, have led to the progressive resolution of past difficulties.

Mixing wines to inoculate other fermentors, however, may oppose the legitimate desire of selecting batches according to grape quality. Definitive blending is generally carried out several weeks after malolactic fermentation, when tasting permits a more exact judgment of wine quality.

The temperature must be maintained while waiting for spontaneous malolactic fermentation and this can become costly. Finally, indigenous bacteria are not necessarily higher in quality than commercial strains: an inoculation with selected strains could be preferable.

As a result, research has been focused for a long time on developing commercial, selected bacterial strains which can be inoculated into wine to ferment malic acid. The possibility of implanting genuine malolactic fermentation starters in wine is definitely interesting but it has also posed many difficulties, which have been progressively resolved, though these solutions are not definitive. Such implanting is general practice in many wineries in the world but is not used systematically in all of them. In general, *Oenococcus oeni* strains are used and this bacterial species is best adapted to malolactic fermentation.

For a long time, direct inoculation of new wines after alcoholic fermentation constantly failed. The bacteria introduced could not develop, due to the environmental conditions (pH and alcohol content) encountered in wine. Bacterial populations were observed to regress rapidly, resulting from cell death. The interpretation of this situation can be summarized simply. To have a sufficient biomass, commercial bacteria are initially cultivated in an environment promoting their growth and are therefore adapted to these specific environmental conditions. When placed in wine, a much less favorable environment, they must adapt in order to multiply and initiate malolactic fermentation. This adaptation becomes increasingly difficult, the more the composition of the two media differs. Indigenous bacteria from the grape, however, undergo a progressive selection according to their ability to adapt to changing environmental conditions. They more easily ensure malolactic fermentation than commercial strains. The difficulty in using commercial strains led to the experimental development of the techniques described below, even if they are no longer in use.

- inoculating must before alcoholic fermentation, when the alcohol-free environment is most favorable to bacterial growth;
- using a sufficiently large non-proliferating bacterial biomass to degrade malic acid without cellular multiplication;
- inoculating wine after alcoholic fermentation with a commercial biomass which has undergone a reactivation phase just before use;
- inoculating wine with a commercially prepared biomass which is already adapted to wine.

(a) Inoculating Must before Alcoholic Fermentation

In traditional winemaking, bacteria from the harvest multiply in the sugar-containing must before the initiation of alcoholic fermentation (Section 6.3.1). From this initial population, a progressive selection, during alcoholic fermentation, results in a reduced population, which is, however, relatively well adapted to environmental conditions. This reduced population capable of carrying out malolactic fermentation can be increased by inoculating the must with *Oenococcus oeni* before the initiation of alcoholic fermentation.

To avoid the risk of inhibiting yeasts by a bacterial inoculation, it is advised to inoculate simultaneously with yeasts and bacteria. Current commercial preparations, freeze-dried or frozen, contain $10^{11}-10^{12}$ viable cells. An inoculation of 1 g/hl corresponds to 10^6-10^7 cells/ml. Bacteria can be directly added to the must without preparation beforehand.

Proposed since the 1960s, this method seems to be a satisfactory solution to the problem of inoculation for malolactic fermentation. It has even been used to obtain malolactic fermentations in harvests sulfited at 15 g/hl, as long as the bacterium starter is added at the time of the initiation of alcoholic fermentation—for example, during the first pumping-over when the free sulfur dioxide has disappeared.

The results, however, are not always as satisfactory as supposed. In practice, three situations can occur:

1. After a significant population decline, bacterial growth occurs toward the end of alcoholic fermentation; malolactic fermentation initiates simultaneously and completes rapidly. This is the ideal situation.

2. The population decline leads to their complete disappearance. Malolactic fermentation kinetics are not improved. The inoculation has no effect.

3. A difficult alcoholic fermentation, accentuated by antagonistic phenomena between yeasts and the high bacterial population, leads to a stuck alcoholic fermentation and premature growth of lactic acid bacteria in a sugar-containing medium. Volatile acidity is produced. *Oenococcus oeni* is the best adapted bacterium for malolactic fermentation. It is, however, a heterofermentative coccus which forms acetic acid from sugars. The production of volatile acidity is a serious accident.

It is almost impossible to establish the ideal conditions for consistently obtaining the first situation for every wine, in terms of its composition (alcohol content, pH). These conditions are influenced by the selection of an adapted homofermentative strain and the respective yeast and bacterium inoculation concentrations.

Considering the serious dangers of this technique, inoculating with *Oenococcus oeni* before the initiation of alcoholic fermentation is not advised. Even when simultaneously inoculating with active yeasts, the risk of slow and sometimes stuck fermentations is too great. The sugar-containing medium would be left to lactic acid bacteria. This technique is, nevertheless, still used regularly in some wineries. The risk of an increase in volatile acidity has probably not been accurately assessed.

More recently, another attempt to inoculate must before alcoholic fermentation was made using a *Lactobacillus plantarum* starter (Prahl *et al.*, 1988), making use of non-proliferating cells.

(b) Inoculating with Non-proliferating Bacteria

Having witnessed the difficulty of obtaining lactic acid bacteria growth in wine, Lafon-Lafourcade (1970) studied the possibility of obtaining malic acid degradation by using a biomass sufficiently abundant and rich in malolactic enzyme so that the reaction can occur without cellular multiplication.

When the evolution of an inoculated bacteria population in wine is studied, an abrupt drop in the number of viable cells is observed in the first hours. Afterwards, the decline is slower. After several days, bacterial growth may occur, but this growth is too uncertain to be used as a technique for initiating malolactic fermentation. Yet, during the decline of the population, the malolactic enzyme supplied by the bacteria induces the partial degradation of malic acid. In this case, the bacteria do not act as a fermentation starter but rather as a potential enzymatic support.

Despite efforts to establish the necessary conditions, a *Oenococcus oeni* biomass inoculated in wine is not capable of degrading all of the malic acid present. The complete reaction can only be obtained by massive inoculations (1–5 g/l), which are not feasible in practice. In general, when the population has completely disappeared, the reaction stops, leaving malic acid. In addition, the malolactic activity of commercial preparations rapidly diminishes during conservation, even at low temperatures.

The kinetics of the reaction could possibly be improved by fixing cells or even enzymatic preparations on solid supports. The resulting protection with respect to the medium could increase the average duration of the enzymatic activity. Wine would circulate in these reactors to be 'demalicated'. At present, this research has not lead to practical applications.

The reaction would of course be easier in the must before alcoholic fermentation, but this technique is not feasible with heterofermentative *Oenococcus* strains. The risk of these bacteria developing in a sugar-containing medium cannot be taken, since volatile acidity production would be significant (see below).

However, a *Lactobacillus plantarum* biomass could be introduced in the must. This homofermentative strain uniquely produces lactate from sugars. Prahl *et al.* (1988) demonstrated that a *Lactobacillus plantarum* preparation could be inoculated into the must at the time of filling the fermentor to degrade malic acid. The preparation contains 5×10^{11} viable cells/g. A concentration of 10 g/hl is used, corresponding to 5×10^7 cells/ml. Malic acid degradation is initiated rapidly; it then continues slowly and is completed during alcoholic fermentation. These bacteria are not resistant to ethanol. As a result, their activity progressively diminishes; sugar assimilation is negligible and no volatile acidity production is observed. This method is simple and has no adverse organoleptic effects but its use is limited, due to the risk of the bacteria population completely disappearing before the end of the reaction. Furthermore, the bacteria are sensitive to free sulfur dioxide. For these various reasons, the general application of this technique is not possible for the moment.

(c) Inoculating with Commercial *Oenococcus oeni* Preparations after Reactivation

Oenococcus oeni is the best-adapted strain for malolactic fermentation in wine. It is involved in practically all spontaneous fermentations. Due to the presence of ethanol, adding this strain in wine after alcoholic fermentation results in a significant decline in its population. Part of the malic acid may be degraded but the cellular multiplication necessary for assuring a complete malolactic fermentation does not consistently occur.

Lafon-Lafourcade *et al.* (1983) were the first to show that bacteria survival could be improved during their transfer to wine, as long as the population is brought to a suitable physiological state beforehand. These authors proposed using the expression 'reactivation' to designate this operation. In fact, this is not a simple precultivation. The population increase that accompanies this operation is a beneficial side effect, but is not the primary objective sought.

Many authors have used this idea of reactivation. Although many different procedures have been proposed, that of Lafon-Lafourcade *et al.* (1983) is the most used. Non-sulfited grape juice is diluted to half its original concentration (80 g/l of sugar per liter); a commercial yeast autolysate is added (5 g/l); and the pH is adjusted to 4.5 with $CaCO_3$. After several hours, commercial biomasses inoculated at 10^6 cells/ml at 25°C produce fermentation starters rich in malolactic enzymes. These starters are also more resistant in wine than non-reactivated starters. Populations increase to 10^6, 10^7 and 10^9 cells/ml after 2 hours, 24 hours and 6 days of reactions, respectively.

The starters prepared in this manner are inoculated into wine after alcoholic fermentation. Table 12.16 attests to the effectiveness of this operation. In all cases, wine is inoculated at 10^6 cells/ml. By the end of 12 days, cellular multiplication has occurred and malolactic fermentation is nearly complete, if the starter has undergone a reactivation of 24 hours or 6 days. A 2-hour reactivation is insufficient. Without reactivation, the population declines and malolactic fermentation is still not initiated after 12 days.

Table 12.16. Effect of bacteria reactivation conditions on malolactic fermentation (Lafon-Lafourcade *et al.*, 1983)

Measurement on 12th day	Non-reactivated biomass inoculation	Inoculation by reactivated biomass: duration of reactivation		
		2 hours	24 hours	6 days
Population (cell/ml)	10^5	3×10^7	4.4×10^7	9.4×10^7
Malic acid degraded (g/l)	0	2.3	3.7	3.6

Initial malic acid concentration: 4.5 g/l.
Bacteria inoculation: 10^6 cells/ml.
Temperature: 19°C.

In practice, the reactivated starter preparation added to wine should not exceed a concentration of 1/1000, since the yeast autolysate is highly odorous. To obtain a cellular concentration of 10^6-10^7 in wine, its concentration must be between 10^8 and 10^9 cells/ml in the reactivation medium with a reactivation time of 48–72 hours. Commercial starter preparations contain $10^{10}-10^{11}$ viable cells per gram. The reactivation medium must therefore be inoculated at 10 g/l.

This method is effective, but it does require a certain knowledge of microbiological methods—not always possible in wineries. This constraint limits its development. Many wineries prefer spontaneous malolactic fermentation, even though it requires more time.

(d) Inoculating with Commercial *Oenococcus oeni* Preparations not Requiring a Reactivation Phase

For a long time, attempts to inoculate commercial biomasses directly into wine after alcoholic fermentation failed. Bacteria populations had difficulty adapting to the physicochemical conditions of wine.

The reactivation procedure previously described could be assumed to confer an indispensable characteristic to bacteria. It would therefore be very difficult (if not impossible) to obtain commercial preparations ready for use in wine. However, since 1993, Chr Hansen's Laboratory Danmark A/S has marketed a starter, under the name Viniflora Oenos, that can be inoculated directly into wine immediately after alcoholic fermentation. Experimental results obtained with this preparation in the laboratory and in the winery have shown that bacterial growth and malolactic fermentation can be obtained 15 days in advance, with respect to a control (Figure 12.14). No organoleptic flaws are observed.

The effectiveness of this preparation is based on selecting a suitable strain, in terms of its resistance to alcohol, pH, SO_2 and other various limiting factors in wine. It also depends on the particular preparation conditions of the commercial biomasses. This preparation includes a progressive

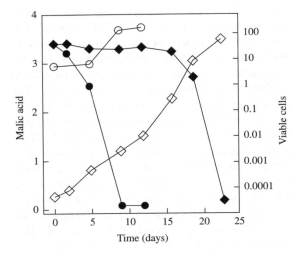

Fig. 12.14. Initiation of malolactic fermentation in red wine (Cabernet Sauvignon, Graves, 1992) by direct inoculation with a freeze-dried preparation (*Viniflora oenos*, Chr. Hansen's Laboratory, Danmark H/S). Temperature = 20°C; pH = 3.5; alcohol = 11.8% vol.; total SO_2 = 5 mg/l; glucose = 0.30 g/l; fructose = 0.45 g/l. Malic acid (g/l): —●— (inoculated medium); —◆— (control). Viable cells (10^6/ml): —○— (inoculated medium); (10^6/ml) —◇— (control)

adaptation to the limiting environmental conditions of wine.

These preparations have not yet been proven to initiate recalcitrant malolactic fermentations. Accelerating this phenomenon by several days is an advantage, but is not essential.

12.8 AUTOMATED RED WINEMAKING METHODS

12.8.1 Introduction

In red winemaking, the complexity of the operations linked to controlling skin extraction lends itself to the development of manufacturing processes permitting the automation of winemaking steps. Equipping fermentors to provide a certain level of automation has already been discussed (Section 12.3.3). In the 1960s, the development of two particular winemaking techniques focused on automation: continuous winemaking and thermovinification (heating the harvest). The *Traité*

Sciences et Techniques du Vin (Ribéreau-Gayon *et al.*, 1976) was edited while these techniques were being developed. It gave a detailed description of these methods (approximately 100 pages). They were subject to the same popularity that all innovations were causing at this time. Some techniques were being generalized that, in reality, were best suited to specific applications. Today, the use of these techniques is on the decline; they are still worth mentioning but no longer justify a detailed description.

12.8.2 Continuous Winemaking

Initially, the development of continuous winemaking was based on the advantages of continuous fermentation. This method was adopted in certain industries using fermentation because of its rapidity and regularity. Continuous fermentation is generally conducted in a communicating fermentor battery. Fresh must enters at one extremity and the fermented product flows out from the other. In these conditions, the multiplication of yeasts is controlled and their population and activity are at their maximum. The same conditions may be reproduced by regularly supplying a single fermentor with must at the bottom and extracting the fermented product from the top at the same rate as the supply.

In continuous red winemaking, fermentation and maceration are sought simultaneously. For this reason, continuous fermentation cannot provide the full benefits of traditional winemaking techniques.

Continuous winemaking permits rigorous operation control and good work organization. It is best applied to high-volume winemaking of the same quality and style wines.

Continuous fermentors (Figure 12.15) comprise a 400 to 4000 hl stainless steel tower. A 4000 hl system can handle 130 metric tons of harvest per day and it can produce approximately 23 000 hl of wine in 3 weeks. An annual wine production of 40 000 hl is necessary to justify the costs of such a system. These fermentors permit the daily reception of fresh grapes and the evacuation of an equivalent amount of partially fermented wine and skins. In the upper part, a rotating rake removes the

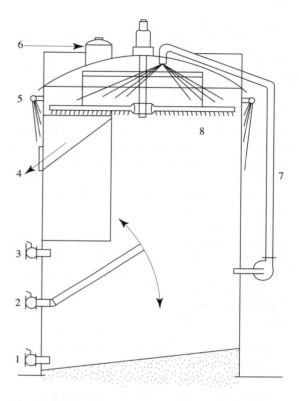

Fig. 12.15. Continuous fermenter: (1) seed evacuation; (2) adjustable-speed wine outlet valve; (3) grape must inlet valve; (4) pomace evacuation; (5) cooling nozzles; (6) expansion dome for wine storage; (7) pumping-over conduit; (8) pomace extraction rake. (Ribéreau-Gayon *et al.*, 1976.)

skins toward a continuous press. A quarter of the total volume of the tower is renewed each day. This corresponds with a 4-day average maceration time. The seeds which accumulate at the bottom of the tank are regularly eliminated; seed maceration for long periods in the presence of alcohol can confer herbaceous flavors and excessive astringency to wine. The weight of the seeds thus eliminated depends on tank dimensions and sometimes attains 1 metric ton per day.

The winemaker has all of the equipment required for controlling operations at his disposal. Before being transferred to the fermentor, the harvest is automatically sulfited by a dosing pump. Pre-fermentation adjustments such as modifying acidity and chaptalization can be performed.

Temperature control and pumping-over operations are automated. The daily supply of fresh grapes minimizes temperature increases: in similar conditions, the temperatures in continuous fermentors are 5–7°C lower than in traditional batch fermentors.

In continuous winemaking, environmental conditions are favorable to yeast growth. The yeast population is approximately two times greater than in traditional winemaking, sometimes reaching 2×10^8 cells/ml. For this reason, fermentation is rapid. It is further accelerated by the introduction of oxygen. Wines flowing out of the fermentor still contain sugar but are saturated with yeasts. The completion of alcoholic fermentation is facilitated. The principle of continuous winemaking favors the most ethanol-tolerant yeasts; apiculated yeasts are eliminated. The alcohol yield is consequently slightly higher (0.1–0.2% volume). The glycerol concentration simultaneously decreases by 1 g/l on average. Finally, the decrease in pectolytic enzyme activity in an alcoholic medium decreases methanol concentrations.

Maceration is regulated by the daily supply of fresh grapes. Its conditions must be perfectly controlled. The maceration starts in an alcoholic environment and at an elevated temperature—conditions that promote extraction of phenolic compounds. The maceration is relatively short but it can be increased by pumping-over operations. The concentration of phenolic compounds in the wine is related to the frequency of pumping over and addition of fresh grapes.

When this method is correctly applied, the resulting wines have no significant organoleptical differences with respect to traditionally made wines.

Continuous fermentors present a particularly high risk of bacterial contamination. Their operating conditions lend themselves to lactic acid bacteria development, and malolactic fermentations can be initiated since the fermentor is continuously supplied with fresh grapes. In this sugar-containing environment, lactic disease may occur inside the fermentor. To avoid this dangerous contamination, a homogeneous sulfiting is recommended. The SO_2 concentrations should be slightly higher than in traditional winemaking. Lactic acid isomer analysis is particularly effective for detecting bacterial contamination in continuous fermentors (Section 12.4.3) (Peynaud et al., 1966). Contamination by lactic acid bacteria can thus be detected (well before the detection of bacteria under the microscope) through monitoring acetic acid production and using paper chromatography to observe the evolution of the concentration of malic acid. Excessive microbial contamination can require the immediate stopping and draining off continuous fermentors.

There are several advantages to this method. The quality of the products is at least identical, if not superior, to that from traditional winemaking; space, labor and material are saved; temperature increases are less significant; malolactic fermentation is facilitated; and the control of the operations is grouped together and therefore more efficient.

The first inconvenience of this method is the risk of bacterial contamination, to which the winemaker should be alert. These fermentors also need a continuous supply of grapes, even during weekends, regardless of the frequency and speed of the harvest. For this reason, continuous and traditional winemaking methods should be employed simultaneously to adapt to varying conditions.

The primary disadvantage is the need to mix grapes of different origins and quality. Grapes cannot be selected, nor can their diversity be expressed in the wine: a single type is produced. This approach is contrary to current winemaking concepts—the diversity of grape origins is now emphasized. For this reason (at least in France), after a period of development, this technique lost popularity.

In the first half of the 20th century, various winemaking methods using continuous fermentation were studied, in particular in the Soviet Union. The first industrial continuous fermentors appeared in Argentina in 1948 and were later developed in Algeria and in the south of France (Midi). The largest expansion of this method was in the 1960s and 1970s, when about 100 of these plants were built. Today, their use is on the decline.

12.8.3 Thermovinification: Heating the Harvest

Heating whole or crushed grapes promotes the diffusion of phenolic compounds from the skins. Colored musts are thus obtained. This phenomenon has been known for a long time; it was referred to even in the 18th century. Attempts have long been made to increase red wine color by heating.

Until fairly recently, heating methods remained very empirical. Only part of the harvest was heated; it was then blended with the rest of the tank and underwent traditional winemaking methods.

The idea is not new but, during the last 30 years, industrial heating processes have developed. They permit large volumes of grapes to be heated rapidly to high temperatures (65–75°C). Various techniques are used, although heating the grapes directly with steam has been almost entirely abandoned (Peynaud, 2001). Destemmed, crushed grapes may be heated directly in a tubular heat-exchanger, heated by steam or, preferably, hot water, or plunged into juice that has been separated from the solids and heated.

The pressed juice may be cooled before fermentation, but if the must is to be fermented on the skins, the solid and liquid components must be cooled together, which is a much more complex operation, requiring special equipment.

Products based on this method were developed with two distinct objectives. In one application, the method was integrated into traditional winemaking to increase concentrations of phenolic compounds, especially anthocyanins (color). In the other, it was used to automate red winemaking, thus decreasing the cost of labor.

Heating the grapes to extract more color is not currently in favor, at least in *appellation d'origine contrôlée* vineyards. First of all, fermentors are now preferably equipped with temperature control systems, which permit a more flexible use of heat to promote the extraction of phenolic compounds. Excessive heating of the entire crushed grape crop, combined with a traditional maceration, might cause excessive tannic bitterness—the wine is consequently without finesse. The increased must color obtained through heating the crushed grapes has also been shown to be unstable, disappearing during fermentation (Table 12.17).

In addition, even if new thermovinification wines are more colored than traditionally made wines, they progressively lose this advantage during maturation.

Thermovinification lines were developed with the goal of automating winemaking. The destemmed, crushed grapes are heated to between 65 and 75°C, then transferred to a vat for up to an hour. The results depend on the temperature used and on the length of time the heat is applied. They are then cooled and pressed. The highly colored juice is then fermented. During this time, it loses part of

Table 12.17. Evolution of anthocyanin and color intensity in a heated and pressed must, compared with traditional winemaking, during alcoholic fermentation (Ribéreau-Gayon *et al.*, 1976)

Duration of fermentation (hours)	Traditional winemaking		Thermovinification of red grape must	
	Anthocyanins (mg/l)	Color intensity[a]	Anthocyanins (mg/l)	Color intensity[a]
0	252	0.37	816	3.08
3	248	0.45	810	2.98
6	244	0.47	824	3.35
10	200	0.48	936	3.36
24	260	0.51	596	1.23
72	302	0.81	508	1.00
96	400	1.16	540	1.20
End of fermentation	468	0.75	476	0.92

[a]See Table 12.1.

its color. All of the operations can be automated, which results in substantial savings in labor costs. Moreover, this system significantly decreases the amount of fermentor volume needed. This alone can justify the installation of a thermovinification line.

Whatever the heating method used, it is recommended that the must or crushed grapes should be cooled before the initiation of fermentation, which must take place at approximately 20°C. Excessive production of volatile acidity by yeasts can hopefully be avoided.

Tasting results are not always homogeneous and depend on grape composition, and on heating and maceration conditions. The participation of these factors is poorly understood. In certain cases, the wines obtained have more color and are better than the traditionally made control wines. They can be rounder and fuller bodied, while still having a fruitiness giving them personality. In other cases, they have abnormal tastes, an amylic dominant vegetal aroma, a loss of their freshness and a bitter aftertaste.

Figure 12.16 shows that the temperature should be higher than 40°C for 15 minutes to obtain significant color extraction, but the extraction is not increased for temperatures above 80°C. Identical results are observed for the tannins. For this reason, a temperature of 70°C for 10 minutes corresponds to a standard thermovinification treatment.

Heating grapes destroys the natural pectolytic enzymes of the grape and so spontaneous clarification of new wines is difficult. This circumstance intensifies potential gustatory flaws. Adding commercial pectolytic enzymes can resolve this problem, but their effectiveness varies.

Destruction of oxidases and protection against oxidations are favorable consequences of thermovinification. Rotten grapes benefit the most from this treatment as they contain laccase, which has a significant oxidizing activity. However, enzymes are only destroyed at temperatures over 60°C, while their activity increases with temperature up to that point, so the must has to be heated very rapidly. Enzymatic activity actually increases at temperatures below 60°C. The increase in temperature during this process must therefore be rapid. Finally, it is accepted that heating Cabernet

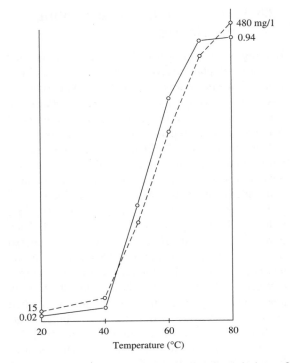

Fig. 12.16. Anthocyanin extraction and evolution of color intensity according to temperature (Ribéreau-Gayon *et al.*, 1976). _____ Color intensity (OD 520 + OD 420) - - - - Anthocyanins (mg/l)

Sauvignon must attenuates the green bell pepper character produced by methoxypyrazines in insufficiently ripe grapes.

Heating also affects fermentation kinetics. Yeast activity continues at temperatures that yeasts generally do not support. At temperatures well above those that kill yeasts, heated musts ferment easily. However, this heating destroys nearly all of the yeasts originating on the grapes. A second natural inoculation occurs during the subsequent handling of juice and skins and this new population rapidly becomes significant. Thus manual inoculation the harvest is unnecessary. Heating is therefore not a viable method for killing the indigenous yeast population, which should be eliminated when using selected yeast strains. This activation of the fermentation is not due to a natural selection of thermoresistant yeasts; it is most likely caused by the dissolution, or at least dispersal, of activators in the grape must belonging to the steroid family.

These activators come from the grape skins. Flash-pasteurization, rapid heating to a high temperature, has also been suggested as a means of restarting a stuck fermentation (Section 3.8.3).

Nitrogen compounds may also be involved in the improvement of fermentation kinetics. Heating the crushed grapes increases not only the total nitrogen and amino compound concentrations, but also the consumption of nitrogen during fermentation.

Heating grapes puts many complex chemical and microbiological phenomena into play. Yet until these phenomena are better understood, separating the maceration and fermentation phase has no distinct advantage. In addition, the performance of temperature control systems used in traditional batch fermentors is continually improving. These systems often produce higher quality wines. For this reason, thermovinification techniques no longer present the same interest as they did not so long ago.

12.9 CARBONIC MACERATION

12.9.1 Principles

Like all vegetal organs, the grape berry has an aerobic metabolism. Respiration produces the necessary energy to ensure its vital functions. In this complex chain of reactions, the grape makes use of oxygen from air to decompose sugar into water and carbon dioxide. Yet when many plants are deprived of air, they adopt an anaerobic metabolism and produce ethanol from sugars. *Saccharomyces cerevisiae* is the classic example of this phenomenon. The anaerobic metabolism is significant because this yeast has a good tolerance to ethanol.

The whole, uncrushed berry also develops an anaerobic metabolism when placed in a carbon dioxide atmosphere. During this phenomenon, various chemical and physicochemical processes occur, especially ethanol production. They are linked to the functioning of the cells in the whole berry but, in contrast to yeasts, grape berry cells are not very tolerant of ethanol. Ethanol production is therefore limited: it varies from 1.21 to 1.89% volume for the Carignan variety, regardless of the must sugar concentration, when between 184 and 212 g/l (Flanzy *et al.*, 1987). The intensity of anaerobic metabolism is in accordance with the variety, the vintage, and maceration temperature and duration.

Enzyme systems in the grape cells, particularly alcohol dehydrogenase, cause the phenomena that give carbonic maceration wines their specific character.

Anaerobic metabolism occurs whenever the oxygen concentration is low, in either a gaseous or liquid environment, but in a liquid environment the intensity of the phenomenon diminishes (Flanzy *et al.*, 1987). Whole grapes immersed in must undergo a less intense anaerobic metabolism than the same grapes placed in a carbon dioxide atmosphere. This diminution is due to exchanges between the berry and the ambient environment, which are greater in the liquid phase than in a gaseous atmosphere. The diffusion of sugars, phenolic compounds and malic acid across the grape skin toward the solution lowers the concentration of anaerobic metabolism substrates in the berry. In addition, the diffusion occurs in both directions. When intact berries are placed in a medium containing alcohol, their ethanol concentration is increased, thus inhibiting anaerobic metabolism. This observation demonstrates the importance of the condition of the grape crop for carbonic maceration. The higher the proportion of uncrushed grapes, the more effective is carbonic maceration.

Pasteur is credited with first noting the taste modification of whole berries during fermentation. He confirmed his observations by placing grapes in a bell jar filled with carbon dioxide. These grapes took on a vinous odor and taste reminiscent of fermented grapes. He concluded that crushing grapes has a dominant impact on red winemaking. This is the basic principle behind fermentation with carbonic maceration, initiated by M. Flanzy in 1935 and studied in detail by C. Flanzy (1998).

Before mechanized crushing, when grapes were still crushed by foot, many berries remained whole. A certain degree of carbonic maceration occurred in the fermentor. At the same time, the juice of uncrushed grapes was progressively released by the weight of the harvest, thus fermenting more slowly. Consequently, the fermentor temperature was moderate. In warm climates, winemakers directly benefited from this phenomenon in the past.

Carbonic maceration comprises two steps:

- The fermentor is filled with whole grapes under a blanket of carbon dioxide and kept at a moderate temperature (20–30°C) for 1–2 weeks. The atmosphere of the fermentor is then saturated with CO_2 for 8–15 days. This is the pure carbonic maceration phase. Anaerobic metabolism reactions modify grape composition. Substances from the solid tissue disintegrated by anaerobiosis are also diffused in the juice and the pulp.

- The fermentor is emptied and the pomace is pressed. The juice is then run off, the pomace pressed, and the free-run and press wines are usually assembled prior to normal alcoholic and malolactic fermentation.

In fact, it is impossible to fill a fermentor with only whole berries. Some are crushed and their juice undergoes a normal alcoholic fermentation. During maceration, additional grapes continue to be crushed as the processes during anaerobiosis weaken cell tissue. The fermentation of completely crushed grapes and pure carbonic maceration occur simultaneously to varying degrees. The condition of the grapes influences the amount and intensity of carbonic maceration. In practice, during the first step of winemaking, yeast-based fermentation always accompanies the anaerobic metabolism of the berry. The winemaker should take steps to minimize this interference.

12.9.2 Gaseous Exchanges

During the first hours of anaerobiosis, the berry tissues to absorb carbon dioxide. Metabolic pathways make use of this dissolved CO_2. Using CO_2 marked with ^{14}C, it has been demonstrated that the gas is integrated not only into various substrates, malic acid and amino acids, but also into sugar and alcohol. The volume of carbon dioxide dissolved into the berry in this manner is temperature dependent. It represents 10% of berry volume at 35°C, 30% at 20°C and 50% at 15°C (Flanzy *et al.*, 1987).

The berry metabolism simultaneously releases CO_2 which eventually attains an equilibrium with the amount absorbed. In a closed system, the equilibrium is established in 6 hours at 35°C, 24 hours at 25°C and 3 days at 15°C (Flanzy *et al.*, 1987).

The initial CO_2 concentration controls the intensity of the anaerobic maceration phenomena, reflected by ethanol production. In certain experimental conditions, for a given time and temperature, this production can vary by a factor of two, depending on the CO_2 concentration in the atmosphere (20–100%).

12.9.3 Anaerobic Metabolism

It has long been known that the grape berry is capable of producing ethanol. This production is always low and depends on the variety. According to different authors, it varies from 1.2 to 2.5% volume or from 0.44 to 2.20% volume. The speed and limits of ethanol production are governed by temperature (Figure 12.17). Maximum production is obtained earlier at higher temperatures than

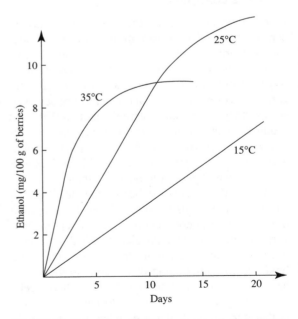

Fig. 12.17. Ethanol formation in grapes during anaerobic metabolism, according to temperature (Flanzy *et al.* 1987)

at lower temperatures, but a higher maximum is obtained at lower temperatures.

Temperature is consequently a major factor in the intensity of anaerobic metabolism. Raising the temperature of excessively cool grape crops is therefore recommended. Heating conditions also have an impact.

The yield of the transformation of sugar to alcohol is difficult to determine. It seems to be similar to the alcoholic yield of yeasts—18.5 g of sugar per 1% percent volume of ethanol. Various secondary products are simultaneously formed: 1.45–2.42 g of glycerol, 21–46 mg of ethanal, approximately 300 mg of succinic acid and 40–60 mg of acetic acid per liter. The presence of all of these compounds indicates the existence of a mechanism similar to yeast-based alcoholic fermentation. Yet in this case, the glyceropyruvic fermentation portion would be greater, since the average glycerol/ethanol ratio × 100 is 18–20% instead of 8%.

During anaerobic metabolism, total berry acidity diminishes. Peynaud and Guimberteau (1962) demonstrated in rigorous laboratory experiments that tartaric and citric acid concentrations remained constant while malic acid concentrations dropped sharply. The degree of this decrease depends on the variety: 32% for Petit Verdot, 42% for Cabernet Franc, 15% for Grenache Gris and 57% for Grenache Noir. As with ethanol production, temperature affects malic acid degradation (Figure 12.18). It regulates the speed and limit of the phenomenon.

Malic acid diminution is a major effect of carbonic maceration. Ethanol is produced after

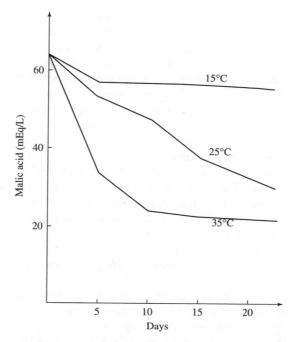

Fig. 12.18. Malic acid degradation during anaerobic metabolism, according to temperature (Flanzy *et al.*, 1987)

a double decarboxylation. Yeasts use an identical mechanism (Figure 12.19). Two enzymes have been confirmed as being involved in these reactions—they are even considered as markers of an anaerobic metabolism. The specific activity of the malic enzyme reaches its maximum between the 3rd and 4th day of anaerobiosis at 35°C. During the same period, the alcohol dehydrogenase is progressively inactivated. This is probably linked to the accumulation of ethanol (Flanzy *et al.*, 1987).

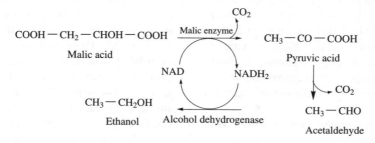

Fig. 12.19. Malic acid degradation by the grape berry in anaerobiosis

Traces of succinic, fumaric, and shikimic acid, but not lactic acid, are generally considered to be formed by this anaerobic metabolism, while the ascorbic acid content decreases. Ascorbic acid concentrations decrease. There are also significant changes in the nitrogen compounds, including an increase in the amino acid content, probably dissolved from the solids in the must, as well as a decrease in protein nitrogen.

The anaerobic metabolism also causes a breakdown of the cell walls, with hydrolysis of the pectins leading to an increase in the methanol content, which may reach levels up to 80 mg/l, corresponding to the hydrolysis of approximately 500 mg pectin.

Finally, within 30 minutes, the development of an anaerobic metabolism in the grapes leads to a significant decrease in the ATP and ADP molecules responsible for energy transfers in biological systems. After an initial decrease of approximately 20% at the time of passing into anaerobiosis, the energy charge (EC = $(ATP + \frac{1}{2}ADP)/(ATP + ADP + AMP)$) stabilizes for 8–10 days before decreasing again. In anaerobiosis, the regeneration ability of energy-rich bonds (ATP, ADP) is limited (Flanzy et al., 1987).

An important result of carbonic maceration during red winemaking is the characteristic aroma produced. The nature and origin of the molecules involved in this aroma remain unknown. According to Flanzy et al. (1987), the formation of aspartic acid from malic acid, along with succinic and shikimic acid, may be the source of aroma precursors. These researchers also noticed differences in higher alcohol and fermentation ester concentrations with respect to wines that did not undergo carbonic maceration. The principal difference is the increase in various aromatic derivative concentrations: vinyl-benzene, phenyl-2-ethyl acetate, benzaldehyde, vinyl-4-gaïacol, vinyl-4-phenol, ethyl-4-gaïacol, ethyl-4-phenol, eugenol, methyl and ethyl vanillate. Ethyl cinnamate, in particular, was proposed as an indicator of carbonic maceration wines.

Peynaud and Guimberteau (1962) pointed out that the simultaneous action of the intracellular berry and yeast cell metabolisms were responsible for the agreeable aroma produced in fermentors. They conducted controlled laboratory experiments to elaborate on these findings. After 8 days of anaerobiosis at 25°C in the total absence of yeasts, whole grapes release weak aromas which are not always agreeable. Reduction aromas are even produced in a nitrogen atmosphere. The researchers concluded:

> If there is a biochemical transformation of essential substances in anaerobiosis, it does not appear to be in the right direction. These observations do not concur with the development of agreeable aromas noted during winemaking with carbonic maceration. The aroma improvement may be due in particular to the action of yeasts.

The carbonic maceration aroma is probably due to the successive action of the anaerobic metabolism of berry, yeast and perhaps bacteria, but the mechanisms of these transformations remain to be determined. In 1987, Flanzy et al. again took up this hypothesis.

12.9.4 Grape Transformations by Carbonic Maceration

In a fermentor undergoing carbonic maceration, the grape berry is transformed by anaerobic metabolism reactions of its own cells. These reactions are independent of any yeast involvement and have been covered in the preceding section. Tissue degradation favors the maceration phenomena involved. Phenolic compounds, anthocyanins, nitrogen compounds and other components of the solid parts of the berry are diffused in the juice of the pulp.

The data in Table 12.18 express the consequences of these phenomena. A slight increase in nitrogen and possibly mineral concentrations is observed. There is also a systematic increase in total polyphenol concentrations. The dissolution of anthocyanins also results in an increase in color intensity. This increase is considerable for certain varieties, but in general the juice obtained is simply pink. In this case, temperature also plays an essential role (Figure 12.20). At elevated temperatures, phenolic compound

Table 12.18. Modification of composition of Cabernet-Franc and Petit Verdot grape juice in anaerobiosis (8 days at 25°C in a CO_2 or nitrogen atmosphere) (Peynaud and Guimberteau, 1962)

Component	Cabernet Franc			Petit Verdot		
	Control	CO_2	Nitrogen	Control	CO_2	Nitrogen
Reducing sugars (g/l)	200	162	162	145	109	104
Ethyl alcohol (g/l)	0	15.9	15.1	0	17.5	18.3
Glycerol (g/l)	0.23	2.65		0.60	2.05	
Acetaldehyde (mg/l)	12	47	54	16	58	37
Methyl alcohol (mg/l)	0	50	50	0	70	50
Total nitrogen (mg/l)	532	588	574	490	588	588
Permanganate index[a]	9	12	10	9	19	18
Color intensity[b]	0.20	0.30	0.21	0.09	0.72	0.65
Tint[b]	0.90	0.71	0.79	1.00	0.56	0.60
pH	3.25	3.40	3.40	3.00	3.35	3.30
Total acidity (mEq/l)	96	80	84	134	98	102
Ash alkalinity (mEq/l)	52	52	50	49	55	52
NH_4^+	8.4	5.2	7.2	6.4	5.2	5.6
Sum of cations (mEq/l)	**156**	**137**	**141**	**189**	**158**	**160**
Tartaric acid (mEq/l)	92	92	94	110	96	98
Malic acid (mEq/l)	50	29	34	65	44	42
Citric acid (mEq/l)	2.5	2.3	2.1	3.0	2.5	2.9
Phosphoric acid (mEq/l)	2.1	2.1	2.1	3.1	3.1	3.1
Acetic acid (mEq/l)	0.6	1.8	1.8	0.6	2.4	1.8
Succinic acid (mEq/l)	0	5.0		0	5.0	
Sum of anions (mEq/l)	**147**	**132**	**134**	**182**	**153**	**148**

[a]Total phenolic compound index.
[b]See Table 12.1.

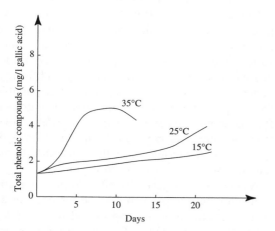

Fig. 12.20. Influence of temperature on phenolic compound diffusion in juice pulp during anaerobic metabolism (Flanzy *et al.*, 1987)

concentrations increase over 8–10 days and then diminish.

According to Bourzeix (1971, cited in Ribéreau-Gayon *et al.*, 1976), only 0.7 g of phenolic compounds per liter are found in juice which has undergone carbonic maceration from grapes originally containing a 4 g/l potential concentration. Approximately 150 mg of anthocyanins pass into the juice out of the 1650 mg contained in a kilogram of fresh grapes.

Tint decreases during carbonic maceration. It is expressed as the degree of yellow coloration with respect to the degree of red coloration (optical density at 420 nm divided by optical density at 520 nm). This ratio diminishes during carbonic maceration. The anthocyanins diffuse more rapidly in the juice than the colorless phenolic compounds. In general, tannin and anthocyanin extraction is more limited in carbonic maceration than in traditional winemaking. This can be an advantage or an inconvenience, depending on the type of wine.

Aroma evolution should of course be taken into account when considering grape transformations by carbonic maceration, but no experimental results currently exist.

12.9.5 Microbiology of Carbonic Maceration

Carbonic maceration creates particular developmental conditions for yeasts and bacteria. These conditions are different in the two stages of the process. They are also different with respect to traditional winemaking.

In the first stage, the yeasts originate from the grape or are added as a yeast starter. They develop in juice produced by the progressive crushing of a portion of the grapes in a carbon dioxide atmosphere (containing no oxygen) and a non- or slightly sulfited environment. At the time of running off and pressing, yeast populations attain 10^8 and sometimes 2×10^8 cells/ml. This significant yeast development is explained by the low ethanol concentration and the presence of bloom constituents (oleic and oleanolic acid). These unsaturated fatty acids, like sterols, activate the fermentation and compensate for the absence of oxygen to a certain degree. This large and active population ensures a rapid fermentation of sugar during the second stage of this process. The increase in nitrogen, assimilable by yeasts, during anaerobic metabolism certainly favors fermentation. Yet if elevated temperatures (35°C) are attained during the carbonic maceration phase, all or part of the yeast population may be destroyed and alcoholic fermentation may become stuck, creating an opportunity for bacterial growth. In this case, both volatile acidity and inhibitors such as ornithine are produced. These inhibitors reinforce fermentation difficulties and increase the risks of stuck fermentations (Flanzy et al., 1987).

Carbonic maceration consequently facilitates the development of lactic acid bacteria and the malolactic fermentation process. The risk of bacterial spoilage is also greater, especially during difficult alcoholic fermentations.

Many factors promote bacterial development:

- the absence of sulfiting or at least the irregular sulfiting of a heterogeneous medium;

- the presence of carbon dioxide, promoting their growth;

- the involvement of the latent phase population in a slightly alcoholic environment;

- the presence of steroids and fatty acids from the bloom.

During the second stage, malolactic fermentation occurs in favorable conditions due to the increased pH and the improved nitrogen supply. Bacteria may develop in the presence of residual fermentable sugars, with the consequent risk of spoilage. Alcoholic and malolactic fermentation should never overlap (Section 3.8.2). Freshly-picked grapes must be sulfited prior to carbonic maceration, even if it is difficult to distribute the sulfur dioxide evenly among the berries. The various phases of the microbiological processes should also be rigorously monitored.

12.9.6 Using Carbonic Maceration

Different systems can be developed to make use of carbonic maceration. Its success depends on anaerobic metabolism intensity, which itself depends on fruit integrity, degree of anaerobiosis, possible traces of oxygen, duration and temperature.

Harvesting, transport and vatting methods must take account of the integrity of the grapes and grape clusters. Flanzy et al. (1987) have described various methods, including placing picked grapes in small containers to avoid crushing and using a system to fill the fermentors gently, after possibly weighing the grapes, to limit the bursting of grapes. Pumps should never be used to transfer grapes from the receiving area to a fermentor. Conveyor systems are preferred since they maintain tissue integrity better than worm-screw pumps.

Anaerobiosis is generally effected in a hermetic fermentor, but grapes can also be wrapped in an airtight plastic tarpaulin and placed in a wooden crate. The crates filled with grapes at the vineyard are transported to the winery and stored.

The grapes are not destemmed before carbonic maceration. Berry integrity may sometimes be compromised by necessary mechanical operations but with certain varieties, in certain regions, the presence of stems may introduce herbaceous notes and a degree of bitterness during carbonic maceration. Stem elimination should therefore be considered in some cases. Destemmers without rollers that do not crush the grapes should be used. Laboratory experiments have shown that the metabolism is less intense for berries that have been detached from their peduncle than for whole grape clusters. Similarly, current mechanical harvesting methods do not permit carbonic maceration to be effected in satisfactory conditions. In the future, new equipment combined with adapted vine growing methods will perhaps permit a mechanical harvest better suited to the needs of carbonic maceration.

Whatever the precautions taken while filling the fermentor, some grapes are inevitably crushed and release juice. During the anaerobic phase, other grapes are progressively crushed, increasing juice volume. In an experiment at 25°C with the Carignan variety, 15% of the total free-run juice is released in 24 hours, 60% on the 5th day and 80% on the 7th day of maceration. Variety, maturity and fermentor height are factors influencing the formation of this free-run juice. The homogenization pumping-over operation, when effected, is also a factor. It is sometimes used for even distribution of the sulfur dioxide (3–8 g/hl), which is necessary to avoid microbial spoilage. The use of lysozyme has been envisaged (Section 9.5.2) to prevent the premature development of lactic bacteria.

A fermentor undergoing carbonic maceration therefore contains the following:

1. Whole grapes immersed in a carbon dioxide atmosphere, poor in oxygen. They are the most affected by anaerobic metabolism. In addition, they are located in an environment with an increasing ethanol concentration. At a certain partial pressure, ethanol diffuses into the berries in anaerobiosis. At pressing, the press juice has a higher alcohol content than the juice from a solely anaerobic metabolism.

2. Whole grapes immersed in must from crushed grapes. They undergo a less intense anaerobic metabolism than (1).

3. Must from certain crushed grapes undergoing yeast-based alcoholic fermentation. Crushed

grapes macerate in this juice, the fermentation of which occurs at the base of the fermentor. It must be carefully monitored to avoid bacterial spoilage. Acetic acid bacteria may develop when the fermentation develops slowly. The addition of a yeast starter in full activity helps to avoid this problem. It also gives protection from the inopportune development of lactic acid bacteria in the case of slowed yeast activity. When the pH is excessively high (pH 3.8), tartaric acid may be added to the juice at the bottom of the fermentor (up to 150 g/hl, taking into account the total must volume anticipated at the end of the anaerobic maceration phase). Sulfiting is also indispensable for inhibiting lactic acid bacteria (3–8 g/hl). Homogenization pump-overs must be minimized when making these additions, otherwise, free-run juice volume is increased. During this fermentation phase, microbial activity must be monitored through the disappearance of sugar and malic acid, the increase in volatile acidity and possibly the analysis of L(+)-lactic acid, whose presence indicates bacterial activity.

When the addition of sugar (chaptalization) is judged necessary, it is effected after devatting, at the start of the second fermentation stage.

Anaerobiosis is obtained by filling an empty fermentor with carbon dioxide from an industrial gas cylinder or a fermenting tank. After filling, the carbon dioxide supply must be continued for 24–48 hours to compensate for possible losses and dissolution in the grape. After this period, fermentation emissions compensate for losses. The extinguishing of a candle flame when placed in the tank verifies anaerobiosis.

The temperature and duration of the anaerobic phase are essential parameters of carbonic maceration. The elevation of the temperature is less important with carbonic maceration than with crushed grapes, which have more active fermentations. In hot climates, this fact was used to the winemaker's advantage, when controlling temperatures was more difficult than today. The anaerobic metabolism, however, must take place at a relatively high temperature (30–35°C) for this method to be fully effective. Yet the temperature must not exceed 35°C, above which this metabolism is affected. Maceration for 6–8 days at 30–32°C is recommended. An insufficient temperature can be compensated for by prolonging maceration time—for example, 10 days at 25°C or 15 days at 15°C—but the result is not necessarily identical. In some regions, excessively low temperatures (15–20°C) restrict the use of carbonic maceration, as the reactions are slowed down. Systems have been devised to warm the grape crop; this operation is always complex. Immersing the grape crop in warm must or wine is laborious and several days are needed to obtain a perfect homogeneity of heat. Attempts at heating the grapes directly on the conveyor belt, using microwave technology, have met with limited success.

The moment of devatting should be chosen according to the style of wine desired. This difficult decision is based on experience but takes into account the evolution of density and temperature, color and tannic structure, aroma and juice flavor along with the degree of grape degradation and pulp color. Pumping the must over once or twice before devatting enhances the must's aromatic intensity and tannic structure.

During devatting, the grapes should be carefully pressed using a horizontal moving-head press or a pneumatic press. These presses do not affect tissue structure. Due to the presence of whole grapes, pressing capacity must be considerable (one-third to one-half higher than in traditional winemaking). Pressing is also slower. Grapes may be crushed just before pressing to simplify this operation. Since the press wine is potentially organoleptically richer than the free-run juice, the press/free-run ratio should be as high as possible.

At the time of devatting, the free run juice has a density between 1.000 and 1.010 and the press juice between 1.020 and 1.050.

Table 12.19 compares free run and press juice composition for traditional (crushed grapes) and carbonic maceration winemaking. In carbonic maceration press wines, the alcohol content is higher (caused by ethanol fixation) and the acidity lower (due to malic acid degradation). These wines also have lower concentration of phenolic compounds and other extracted components; their dissolution is diminished.

Due to their complementary composition, free run wines and press wines should be blended immediately after pressing, before the completion of alcoholic and malolactic fermentation. Bacterial contamination in the free-run wine, leading to premature malolactic fermentation and the risk of an increase in volatile acidity, is the only reason for fermenting the free-run and press wine separately. Microbiological analysis should be systematic at this stage, followed by sulfiting and reseeding the

Table 12.19. Run-off and press wine analysis comparing traditional winemaking with carbonic maceration (Flanzy *et al.*, 1987)

Component	Winemaking using crushed grapes		Carbonic maceration	
	Free-run	Press	Free-run	Press
Alcoholic strength (% vol.)	12.05	10.96	11.15	13.00
Density at 15°C	0.9949	0.9991	0.9966	0.9920
Glycerol (g/l)	9.29	9.75	9.10	7.91
Dry extract (g/l)	23.8	32.0	25.5	19.2
Total acidity (g/l H_2SO_4)	3.30	3.50	3.50	2.80
pH	3.90	4.05	3.93	3.90
Total nitrogen (mg/l)	154	425	144	123
Color intensity[a]	388	912	510	487
Tannic matter[b]	1342	2550	1582	1440

[a]Sum of optical densities at 420 and 520 nm.
[b]Sum of optical densities at 260 and 280 nm.

must with fresh yeast to complete the alcoholic fermentation, if necessary.

During the second fermentation phase, the complete transformation of sugar into alcohol is generally very quick. It is carried out at 18–20°C to preserve aroma components. Afterwards, the favorable conditions permit the easy initiation of malolactic fermentation. Despite the existence of two distinct phases, carbonic maceration requires less time than traditional winemaking. This method is therefore well adapted for wines that are quickly put on the market.

12.9.7 Characteristics of Wines Made by Carbonic Maceration

Table 12.20 (Flanzy *et al.*, 1987) compares the composition of traditionally made wines (crushed grapes) and wines having undergone carbonic maceration (effected at 25 and 35°C). The importance of temperature in anaerobic metabolism is shown. At 35°C, this technique permits the same tannic structure as traditional winemaking. In general, density and dry extract, fixed acidity and phenolic compound concentrations are lower with carbonic

maceration than with traditional winemaking. This winemaking technique produces a lighter wine, containing less substances from the solid parts of the grape. The method has advantages when used with rustic grape varieties—it avoids the excessive extraction of aggressive olfactive and gustatory elements lacking finesse. In other cases, carbonic maceration may result in insufficient structure and an impression of thinness, or results anywhere in between.

In Table 12.20, volatile acidities were observed to be relatively high—often greater than 0.5 g/l in H_2SO_4 sometimes attaining 0.69 g/l (0.65 and 0.84 g/l in acetic acid). These numbers indicate the inherent bacterial risk associated with this winemaking method.

The structural difference of a wine having undergone carbonic maceration, with respect to a traditionally made wine, as pointed out by laboratory analysis, is reflected in its organoleptic characters. Carbonic maceration produces supple, round, smooth and full wines. For this reason, they are often used in blends to improve wine quality. However, this positive characteristic in certain situations can be negative in others: wines

Table 12.20. Analytical comparison of the composition of wines having undergone carbonic maceration at 25°C (CM25) and 35°C (CM35) and of the same wines made from crushed grapes (CG) (analysis carried out four months after the end of malolactic fermentation, 1983 vintage) (Flanzy *et al.*, 1987)

Component	Carignan			Mourvèdre		
	CG	CM25	CM35	CG	CM25	CM35
Alcoholic strength (% vol.)	11.6	11.4	11.4	12.25	12.25	12.35
Ash (g/l)	3.0	2.7	3.0	3.4	2.6	2.8
Ash alkalinity (mEq/l)	34.5	32.5	34.5	41.0	33.0	33.3
Glycerol (g/l)	8.0	7.0	7.3	9.0	7.7	8.5
Total nitrogen (mg/l)	196	146	238	179	120	129
Total acidity (g/l H_2SO_4)	3.10	3.10	3.00	3.00	3.30	3.30
Volatile acidity (g/l H_2SO_4)	0.41	0.34	0.51	0.43	0.54	0.69
Tartaric acid (mEq/l)	24.9	24.0	25.7	22.2	19.3	19.9
Malic acid (mEq/l)	0	0	0	0	0	0
pH	3.71	3.61	3.74	3.85	3.71	3.80
Potassium (mEq/l)	35.3	32.8	34.8	42.5	34.8	38.4
Total SO_2 (mg/l)	38	57	27	44	45	57
Optical density 520×10^3 (red)	394	393	384	570	410	505
Optical density 420×10^3 (yellow)	101	94	108	146	109	142
Optical density 280×10^3 (tannins)	810	770	903	1250	980	1240
Total polyphenols (g/l gallic acid)	1.573	1.436	1.755	2.690	2.120	2.736
Anthocyanins (g/l)	0.509	0.518	0.474	0.738	0.527	0.624

can become thinner and more fluid and, depending on the variety and maturity level, the less abundant tannins can also be more bitter, probably due to the presence of the stems.

Carbonic maceration is certainly most interesting from an aromatic viewpoint. It produces wines with a unique aroma. Some have accused this technique of producing uniform wines and of masking the aromas of quality varieties (Ribéreau-Gayon et al., 1976). Other authors (Flanzy et al., 1987) find that the aromas of certain varieties (Muscat and Syrah are intensified. This technique has also been observed to increase the aromatic intensity of relatively neutral varietal wines (Aramon, Carignan).

Changes in the concentrations of secondary products of alcoholic fermentation have been reported. In particular, aromatic substances specific to this winemaking method seem to be produced. Yet the nature and origin of the corresponding molecules are not always clear, in spite of the considerable research that this technique has incited. The typical aroma seems to be acquired during the anaerobic metabolism phase, but the yeasts seems to be involved in its expression.

The description of the specific aromas of carbonic maceration wines is confronted by the well-known difficulties of tasting vocabulary. According to experts, carbonic maceration wines have a dominant fruitiness with notes of cherry, plum and fruit pit, whereas traditionally made wines have a dominant vinosity with notes of wood, resin and licorice. In addition, the various aromatic components are more harmoniously blended in carbonic maceration wines.

Carbonic maceration is best applied in making *primeur* wines for early drinking. Experts, however, do not agree on the aging potential of these wines. For some, carbonic maceration wines lose their specificity after two years of aging, but are always better than the equivalent traditionally made wines. For others, these wines evolve poorly after a year of aging: they lose their characteristic aromas and do not undergo the harmonious gustatory development of traditional wines. When evaluating the differences of opinion regarding this technique, the variety should be considered and carbonic maceration conditions should also be taken into account. The temperature is particularly important, since it determines the intensity of the anaerobic metabolism.

Carbonic maceration is essentially used for red winemaking. It is best adapted to certain varieties, such as Gamay. Reservations have always been expressed about using this technique in regions known for their fine wines with aging potential, due to concerns that varietal character may be lost. The technique is also used for rosé (Section 14.1.1) and fortified wines, and has been used experimentally to produce white wines and base wines for sparkling wines and spirits, but has not been further developed. Contaminated harvests (more than 15% rotten grapes) and mechanically harvested grapes should not undergo carbonic maceration.

REFERENCES

Amerine M.A., Berg H.W., Kunkee R.E, Ough G.S. and Singleton V.L. 1980 *The Technology of Wine Making*. Avi Publishing Co., Westport, CT.

Amrani-Joutei K. 1993 Localisation des anthocyanes et des tanins dans le raisin. Etude de leur extrabilité. Thèse Doctorat de l'Université de Bordeaux II. Mention Œnologie-Ampélologie.

Augustin M. 1986 Etude de l'influence de certains facteurs sur les composés phénoliques du raisin et du vin. Thèse Doctorat de l'Université de Bordeaux II.

Blouin J. and Peynaud E. 2001, *Connaissance et travail du Vin*, 3ème Edition, Dunod, Paris.

Boulet J.C. and Escudier J.L. 1998, In *OEnologie. Fondements scientifiques et œnologiques*, Flanzy C. editeur, Lavoisier, Tec et Doc, Paris.

Boulton R.B., Singleton U.V., Bisson L.F. and Kunkee R.E. 1995 *Principles and Practices of Winemaking*. Chapman & Hall Enology Library, New York.

Bourzeix M. 1971, *cité par J. Ribéreau-Gayon et al.*, 1976.

Canbas A. 1971 Les facteurs de dissolution des composés phénoliques au cours de la vinification. Thèse Doctorat de 3ème cycle, Université de Bordeaux II.

Dubernet M. 1974 Recherches sur la tyrosinase de *Vitis vinifera* et la laccase de Botrytis cinerea. Thèse Doctorat de 3ème cyle, Université de Bordeaux II.

Ferré L. 1922, *cité par J. Ribéreau-Gayon et al.*, 1976.

Ferré L. 1958 *Traité d'Œnologie Bourguignon*. INAO, Paris.

Feuillat M. 1997, *Revue des Œnologues*, 82, 29.

Flanzy C. 1998, *Œnologie. Fondements scientifiques et technologiques*, Lavoisier, Tech et Doc, Paris.

Flanzy M. 1935, *C.R. Acad. Agric.*, 21, 935.

Flanzy C., Flanzy M. and Benard P. 1987 *La vinification par macération carbonique*. INRA, Paris.

Gayon U. 1905 *Préparation et conservation des vins*. Pech éditeur, Bordeaux.

Glories Y., Ribéreau-Gayon P. and Ribéreau-Gayon J. 1981 CR Acad. Agric., 623

Henick-Kling T. 1992 In *Wine Microbiology and Biotechnology* (ed. G.H. Fleet). Hartwood Academic Publishers, Chur, Switzerland.

Lafon-Lafourcade S. 1970 *Ann. Technol. Agric.* 19 (2), 141–154.

Lafon-Lafourcade S. and Ribéreau-Gayon P. 1977 *CR Acad. Agric.*, 551.

Lafon-Lafourcade S., Carre E., Lonvaud-Funel A. and Ribéreau-Gayon P. 1983 *Conn. Vigne Vin*, 17, 55–71.

Peynaud E. 1981 *Connaissance et Travail du Vin*. Dunod, Paris.

Peynaud E. and Guimberteau G. 1962 *Ann. Physiol. Vég.*, 4 (2), 161–167.

Peynaud E., Lafon-Lafourcade S. and Guimberteau G. 1966 *Am. J. Enol. Vitic.*, 4, 302.

Prahl C. and Nielsen J.C. 1993 *The development of Leuconostoc oenos malolactic cultures for direct inoculation*. Chr. Hansen's Laboratory Danemark A/S, Horsholm, Denmark.

Prahl C., Lonvaud-Funel A., Korsgaard S., Morrison E. and Joyeux A. 1988 *Conn. Vigne Vin*, 22, 197–207.

Ribéreau-Gayon J. and Peynaud E. 1961 *Traité d'Œnologie*, Vol. II. Béranger, Paris.

Ribéreau-Gayon J., Peynaud E. and Lafourcade S. 1951 *Ind. Agric. Alim.*, 68, 141.

Ribéreau-Gayon J., Peynaud E., Ribéreau-Gayon P. and Sudraud P. 1976 *Sciences et Techniques du Vin*, Vol 3: *Vinification. Transformations du Vin*. Dunod, Paris.

Ribéreau-Gayon J., Peynaud E., Ribéreau-Gayon P. and Sudraud P. 1977 *Sciences et Techniques du Vin*, Vol. 4: *Clarification et Stabilisation. Matériels et Installations*. Dunod. Paris.

Ribéreau-Gayon P. 1953 *CR Acad. Agric.*, 39, 807.

Ribéreau-Gayon P. 1973 *Vitis*, 12, 144.

Ribéreau-Gayon P. 1977 *CR Acad. Agric.* 63, 120.

Ribéreau-Gayon P., Sudraud P., Milhe J.C. and Canbas A. 1970 *Conn. Vigne Vin*, 2, 133.

Siégrist J. and Léglise M. 1981 *CR Acad. Agric.*, 67, 300.

Somers T.C. 1979 *J. Sci. Food Agric.*, 30, 623–633.

Sudraud P. 1963 Etude expérimentale de la vinification en rouge. Thèse Docteur-Ingénieur, Faculté des Sciences de Bordeaux.

Touzani A., Muna J.P. and Donèche B. 1994 *J. Int. Sci. Vigne et Vin*, 28 (1), 19.

13

White Winemaking

13.1 GENERAL NOTIONS AND DISTINCTIVE CHARACTERISTICS OF WHITE WINEMAKING

13.1.1 The Essential Role of Pre-fermentation Operations in Dry White Winemaking

Although red wines are obtained by the alcoholic fermentation of musts in the presence of the solid parts of the berry (skins and seeds), white wines are exclusively produced by the fermentation of grape juice. Thus, in the production of white wines, juice extraction and varying degrees of clarification always precede alcoholic fermentation. It is the absence of skin contact in the alcoholic phase, and not the color of the grape, that distinguishes white winemaking from red winemaking. White wines can therefore be made from red grapes having white juice, if the grapes are pressed in conditions that prevent grape skin anthocyanins from coloring the must. This is the case of *blancs de noirs* from Champagne, made from Pinot grapes.

That is not to say that white winemaking does not include any maceration. If this term designates solubilization of solid components in juice, a certain degree of maceration is inevitably associated with white winemaking. It occurs in the absence of alcohol during the pre-fermentation phase, at the time of juice extraction and clarification.

Varietal aromas and aroma precursors are located in the grape skin or in the underlying cell layers in most quality cultivars (Volume 2, Chapter 7). Yet these zones are also the richest in grassy-smelling and bitter-tasting substances, especially when the grapes are not completely ripe, are stricken with rot or are from a *terroir* less favorable for producing quality wines. The taste of a dry white wine, made from a given grape, therefore depends greatly on the conditions of various pre-fermentation operations: harvest, crushing, pressing and clarification.

All winemaking includes a selective extraction of grape components; white winemaking does not escape from this general principle. Winemaking not only consists of carrying out the alcoholic fermentation of must or grapes but also, and especially, extracting the best part of the grape berry while limiting the diffusion of substances in the liquid phase capable of generating olfactory and gustatory flaws.

In red winemaking, fractional extraction of grapes occurs primarily during alcoholic fermentation and maceration. The winemaker influences the future taste of a red wine by adjusting vatting times (Section 12.5). By adjusting various operations during vatting, the winemaker approaches day by day, over a period of 2–3 weeks, the desired tannin, color and aroma concentrations for the wine. During vatting, time is the winemaker's ally.

In white winemaking, conditions for the extraction of berry components are radically different, since maceration phenomena occur before alcoholic fermentation. In this case, pre-fermentation treatment conditions control the passage of compounds responsible for the qualities and flaws of grapes into must. The quality of a dry white wine, made from given grapes, depends above all on grape and must manipulation during production. In other words, the art of making dry white wines lies in knowing how to press the grapes and clarify the musts in a manner that simultaneously extracts and preserves potential grape quality. For certain varieties (Sauvignon, Muscat, etc.), a limited skin contact (pre-fermentation skin maceration) before pressing can be useful in facilitating the diffusion of varietal aromas and their precursors in the juice. The winemaker has only a limited amount of time to extract components from the grape skins before the juice begins to ferment—generally a few hours to a few days maximum. In addition, the choices that are made during the pre-fermentation phase are definitive: pressing time and program, juice selection, possible skin maceration, blending free run and press juice and degree of must clarification.

Therefore, in dry white winemaking, the important choices are made before alcoholic fermentation. Afterwards, corrections and adjustments are practically impossible. When alcoholic fermentation is initiated, the taste of the dry white wine is already largely determined. Even the decision to tank or barrel ferment is made fairly early. In fact, barrel-fermented wines should be barreled at the initiation of fermentation, to avoid the wood dominating the wine later (Section 8.8). This decision must be made as early as possible so that barrel purchases can be planned properly. It is too late to barrel wine after a tank fermentation, even if wine quality would have justified barrel aging. Juice clarification is another example of definitive prefermentation decisions: it is impossible to mitigate its consequences subsequently, during fermentation. In fact, no satisfactory methods exist to stimulate lagging fermentations of overclarified juice or to eliminate vegetal and reduced odors that appear during the fermentation of poorly clarified juice.

Botrytized sweet winemaking constitutes an extreme case of the fundamental decisions involved concerning fractional berry component extraction. The most important decisions made by the winemaker essentially concern picking grapes at the ideal noble rot stage (Section 14.2.2). Noble rot is not only an overripening by water loss, like raisining, but also and especially an intense enzymatic skin maceration paired to a *Botrytis*-specific metabolism. The decisive part of botrytized sweet winemaking occurs in the grape on the vine.

In conclusion, each type of winemaking contains a key phase during which the decisions of the winemaker have a determining and almost irreparable effect on wine taste: vatting for red wines; pre-fermentation operations for dry white wines; and noble rot development and picking conditions for botrytized sweet wines.

13.1.2 White Wine Diversity and Current Styles

White wines are generally thought to present a greater diversity of styles than red wines (Ribéreau-Gayon *et al.*, 1976). In fact, apart from still and sparkling wines, white wines can be dry or contain a varying amount of residual sugars (from several grams to several dozen or, sometimes, even more than 100 grams per liter). These differences can occur in wines made from the same varieties and parcels—the grapes being picked at different maturity levels. This is the case of great German, Austrian and Alsacian Rieslings and Gewürztraminers, Hungarian Furmints, Bordeaux Semillons and Sauvignons, Loire Valley Chenins, etc. The fixed acidity of dry and sweet white wines can also vary greatly (from 3 to 6 g/l expressed as H_2SO_4). Moreover, dry wines may undergo malolactic fermentation, universally used for red wines. Intense or discrete, predominantly marked by the variety or only by secondary products of alcoholic fermentation, white wines also seem to have more diverse aromas than red wines. This relatively rich typology characterizing white wines may be divided into two general categories which also concern red wines:

1. Premium wines improve during bottle aging by developing a bouquet.

2. *Primeur* wines, incapable of aging, are to be drunk young.

In addition, barrel matured white wines are partially or totally made in new barrels. In other wines, the organoleptic character supplied by the oak is not sought—these wines are made in neutral vessels (tanks or used barrels). Finally, like certain red wines, some white wines are distinguishable by their oxidative characteristic (sherry and yellow wines). Yet most are made in the virtual absence of oxygen and under the protection of antioxidants such as sulfur dioxide and ascorbic acid to preserve their fruity aroma.

The diversity of white wine types and winemaking methods has strongly diminished during the last 20 years, due to the trend towards a world market, a standardization of consumer tastes and a general trend of producers to imitate a few universally appreciated models. The growing influence of the wine critic on the market has certainly amplified and accelerated this convergence of white wines towards a few widely recognized types. Four categories currently distinguish international dry white wines: neutral, Chardonnay, Sauvignon and aromatic white wines.

(a) Neutral White Wines

Neutral white wines do not possess a particular varietal aroma. They only contain young wine fermentation aromas—essentially due to ethyl esters of fatty acids and acetates of higher alcohols produced by yeasts (Section 2.3), when the fermentation of clarified juice occurs at relatively low temperatures (16–18°C). These wines are appreciated especially for their thirst-quenching character, due to their refreshing acidity possibly reinforced by the presence of carbon dioxide (0.6–1 g/l). They should be low in alcohol and without bitterness. Their fleeting aroma rarely lasts for more than a year of storage; these white wines are generally bottled a few months after the completion of fermentation and should be drunk within the year that follows the harvest. A particular varietal aroma is not sought nor is an expression of *terroir* expected in these white wine 'beverages'. They are generally made from high-yielding, slightly or non-aromatic varieties such as Ugni Blanc (Italian Trebbiano), Maccabeu (Spanish Viura), Airen (also of Spanish origin and having the highest planted varietal surface area in the world), White Grenache, Clairette, etc. Unfortunately, neutral white wines are sometimes produced with noble varieties, due to excessively high crop yields and unfavorable soil and climate conditions. This is often the case of Semillon at yields greater than 60 hl/ha, or

Sauvignon grown in hot climates regardless of yields. In the 1970s, these simple and inexpensive white wines sold well, especially when promoted by a strong brand name. Today, the demand for them has dropped, as the market orientates itself towards more expressive white wines, particularly in Anglo-Saxon countries. Oxidized wines possibly containing several grams of sugar and non-oxidized wines containing several grams of sugar have all but disappeared.

(b) Chardonnays

Chardonnay is the principal current international white wine standard. White Burgundy wines supplied the original model (Meursault, Chassagne-Montrachet, Chablis, etc.). The top estates from this region are among the best dry white wines in the world. Their wines are powerful, firm, aromatically intense and 'sweet', although they do not contain residual sugar. The great white Burgundies are distinguished by their aging potential. During the aging process, they develop a remarkable reduction bouquet. In its zone of origin, the Chardonnay variety produces grapes rich in both sugar and acid, often reaching 13% potential alcohol for an acidity between 6 and 7 g/l (expressed as sulfuric acid) and remarkably low pH (3.1–3.3). The traditional Burgundy winemaking method, with barrel fermentation and on-lees aging, has profoundly influenced current white winemaking methods. Today, enological research has shed light on and justified these Burgundy-origin empirical practices, put to use world-wide and not just for Chardonnay.

During the last 20 years, Chardonnay was largely planted in European Mediterranean climates and New World vineyards. Along with Cabernet Sauvignon for red wines, it is certainly the variety best adapted to climatic conditions warmer than its original *terroir*. All Chardonnay producers try to attain the Burgundy archetype, like Cabernet Sauvignon wines strive to attain the top-ranked growth model of the Medoc. Excellent Chardonnays are found in many viticultural regions throughout the world, but the diversity of expression of this variety in different Burgundy

terroirs or climates still remains more fascinating for the wine buff.

(c) Sauvignons

Inspired by the wines of Central France (Sancerre and Pouilly-Fumè), Sauvignons constitute another important world standard for dry white wines. Their often intense and complex typical aroma is easily recognized. Certain volatile substances responsible for this aroma as well as their precursors in the grape have recently been identified (Volume 2, Chapter 7). The Sauvignon aroma is more sensitive to climatic conditions during maturation than the Chardonnay aroma. It is therefore less constant and stable and more difficult to reproduce. The aromatic expression of Sauvignon is often disappointing in Mediterranean climates. It has consequently been less universally successful than Chardonnay. Due to its cool climate, New Zealand without a doubt produces one of the most aromatic Sauvignons in the New World. On average, Sauvignon wines have a lesser aging potential than Chardonnay wines, except in very particular situations. Sauvignon also originated in the Bordeaux region and is nearly always blended with Semillon in this area. The Sauvignon contributes the fruitiness, the firmness and the acidity, while the Semillon gives the wine body, richness and bouquet during aging. These two varieties are particularly complementary. During recent years, Sauvignon winemaking methods have undergone many changes—including a return to barrel fermentation of musts originating in the best *terroirs* as well as on-lees aging of new wines, whatever the fermentation method (barrel or tank). Current Chardonnay and Sauvignon winemaking methods are very similar, but malolactic fermentation is rarely practiced on Sauvignon wines (Section 13.7.6).

(d) Aromatic White Wines

Various aromatic white wines compose the fourth group. Sometimes, these wines are famous and made from premium varieties. Their geographical territory has remained limited to their original regions. An exhaustive list of these wines is not included in this text, but a few examples will be given.

Within this group, the dry white, premium quality German and Alsacian wines are worth mentioning. These wine styles are also made in Austria and continental Europe. The most notable varieties are Riesling, Pinot Gris and Gewürztraminer. Late harvesting of these varieties produces premium sweet wines capable of considerable aging. They have a characteristic aroma reminiscent (at least in part) of their grape or juice aroma. These floral or Muscat varieties are distinguishable from simple savor varieties such as Sauvignon, Chardonnay, Chenin, etc. The juice of simple savor varieties is not very fragrant, but their wines have a characteristic varietal aroma essentially derived from odorless precursors located in the grape. The role of volatile terpene alcohols and certain norisoprenoids in the aroma of Muscat varieties has been largely studied and proven (Volume 2, Chapter 7). The specific aroma of these different varieties, however, is for from being totally elucidated.

Several regional varieties also exist which, for diverse reasons, have until now only produced typical wines in relatively limited zones. Some of them have never been planted outside their region of origin, while others lose their character in warmer climates. French varieties include Chenin Blanc (Savenières, Loire Valley), Viognier (Condrieu) and Petit and Gros Manseng (Jurançon). Albarino is in the north of Spain and the remarkable and rare Petite Arvine is in the Swiss Valais.

13.2 WHITE GRAPE QUALITY AND PICKING CRITERIA

Varietal aroma finesse, complexity and intensity are the primary qualities sought after in a dry white wine. Its personality is due to varietal expression or, more precisely, its particular aromatic profile on a given *terroir*. Fermentation aroma components are present in all wines and are not very stable over time. These esters and higher alcohols produced by yeasts are not sufficient to give a white wine an aromatic specificity, but they were the first to be measured by gas phase chromatography because of their relatively high concentrations in wines.

Consequently, in the past, the importance of their contribution to the aromatic quality of dry white wine was exaggerated. It has been widely accepted that aromatic quality is mainly due to the primary aroma—the aroma originating in the grape—even though the volatile compounds responsible are far from being identified and the production mechanisms from grape to wine remain unknown. The handicap of neutral varieties is thus explained: no winemaking method can compensate. For all that, the varietal aroma is not the only character of a dry white wine. The balance of acidity and softness, sensations of volume, structure and persistence and the impression of density and concentration also play an important role in quality appreciation. Healthy ripe grapes must be used to obtain a wine with all of these characteristics. The grape disease state and maturity level, in particular aromatic, are the essential harvest selection criteria for making quality dry white wines. Harvest time and methods, (mechanical or manual) influence these two essential parameters and are therefore very important.

13.2.1 Disease State

White grape varieties are susceptible to gray rot due to *Botrytis cinerea* development on the grape (Section 10.6). In a given region, the more forward white varieties are more subject to this disease than red varieties. Obtaining healthy grapes with Sauvignon, Semillon and Muscadelle grapes is much more difficult than with Merlot and Cabernets. Muscats in Mediterranean climates, Chardonnay in Champagne and Chenin in the Loire Valley are also affected.

From early contamination of the grape cluster (latent since bloom), *Botrytis* can develop explosively near harvest time. Feared by winegrowers, this pathogen is triggered by severe rains near *vèraison* and during maturation. The fungus contaminates both green and burst berries, degrading the skin—the site of aromas and aroma precursors.

Even a relatively small percentage of botrytized grapes in the crop always seriously compromises the aromatic quality of dry white wines. Gray rot on white grapes results in a decrease in varietal aroma, a greater instability of fermentation aromas and the appearance of olfactory flaws. These consequences of gray rot on the aroma of white wines are much more serious than oxidasic casse. This casse is a direct manifestation of the laccase activity on wine color, especially with red and rosè wines (Section 12.6.2), but it can be observed in certain white wines—in particular, bottled sparkling wines—even several years after bottle fermentation.

Although empirically witnessed with all aromatic varieties, the harmful effects of gray rot on the primary grape aroma has only been quantified with the muscat variety by measuring monoterpene alcohol concentrations in musts (Boidron, 1978). When *Botrytis* contaminates 20% or more of a grape crop, the total terpenic alcohol concentration of Frontignan Muscat or Alexandria Muscat drops by nearly 50% with respect to healthy grape must concentrations (approximately 1.5–3 mg/l). The most fragrant terpene alcohols (linalol, geraniol and nerol) are the most affected. These alcohols are partially transformed into less fragrant compounds, such as linalol oxides, α-terpene alcohol and other compounds (Rapp *et al.*, 1986), themselves original components of healthy juice. This rapid degradation of terpenes by *Botrytis cinerea* can be observed in the laboratory in a fungus culture on a medium supplemented with monoterpene alcohols.

Gray rot is also observed to affect the specific aromas of other varieties. A relatively small percentage of gray rot (less than 10%) diminishes the Sauvignon varietal aroma in wine. This aroma is due at least in part to very fragrant volatile thiols, existing in trace amounts (a few nanograms or a few dozen ng/l) in wines (Volume 2, Chapter 7). These aromas are essentially found in the grape in the form of odorless precursors linked to cysteine (Tominaga *et al.*, 1996). *Botrytis cinerea* may directly degrade the free and bound aromas of Sauvignon, but this has not been clearly demonstrated. This type of degradation would only explain an aroma loss corresponding to the percentage of botrytized grapes. Yet the reaction of fragrant thiols with quinones formed by the oxidation of grape phenolic compounds has been clearly proven to exist. *Botrytis* laccase activity in a must

containing phenolic compounds inevitably leads to the formation of quinones. The quinones trap Sauvignon varietal aroma as it is formed during alcoholic fermentation. When Sauvignon must is insufficiently sulfited during the pre-fermentation phase, it is oxidized. The resulting combination of thiols and quinones produces wine with a slight or non-existent varietal aroma (Section 13.4.1).

Paradoxically, noble rot does not destroy the specific aroma of white varieties used to make great botrytized sweet wines (Section 14.2.3). In the Sauternes region, the lemon and orange fragrances of Semillon and Sauvignon are even enhanced, as is the mineral character of Riesling or the lychee aroma of Gewürztraminer in the Alsacian or German noble rot wines. The bouquet of dry wines made from healthy grapes and sweet wines made from botrytized grapes of the same variety and from the same *terroir* has even been observed to converge during bottle aging. In the ideal noble rot case, the intense skin maceration of the ripe grape under the action of fungal enzymes promotes the diffusion of free and bound varietal aromas in the must. These aromas are concentrated without being degraded. This process is different from raisining, in which the grapes are concentrated by the sun which burns the grape skin. Most of the varietal specific aromas are lost and a character peculiar to raisins is acquired, varying little from one white variety to another. Theoretically, a small proportion of noble rotted grapes could be added to a grape crop intended for dry white winemaking, but in practice this is difficult. At the time of the healthy white grape harvest, however, most of the rot-infected grapes on the vine correspond to early *Botrytis* sites developed on the unripe grape and thus gray rot.

Gray rot also greatly diminishes the intensity of fermentation aromas of dry white wines. Among the exocellular enzymes liberated by *Botrytis* in the infected grape, esterases exist whose activity persists in juice (Dubourdieu *et al.*, 1983). They are capable of catalyzing the rapid hydrolysis of esters produced by yeasts during alcoholic fermentation. Figure 13.1 shows the hydrolysis kinetics of these different fragrant compounds in a dilute alcohol medium in the presence of a

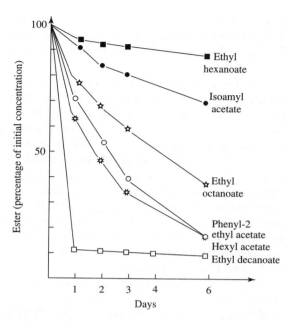

Fig. 13.1. Enzymatic hydrolysis (pH 3.4 at 18°C 11% ethanol) of different esters by an exocellular protein extract from *Botrytis cinerea* (Dubourdieu *et al.*, 1983)

Botrytis enzymatic extract. The occurrence of gray rot is more detrimental to neutral variety wines, containing essentially fermentation aromas.

Finally, gray rot seriously affects the aromatic distinctness of dry white wines. Varietal aromas are masked while dusty, dirty and moldy aromas appear. It also promotes the development of rancid, camphorated and waxy odors, appearing later during maturation and especially in the bottle. This type of olfactory flaw is comparable to a premature oxidative aging of white wines (Volume 2, Section 8.2.3). Gray rot is not solely responsible and white wines made from healthy grapes can also contain the flaw. The responsible compounds and their formation mechanisms remain to be discovered.

The level of *Botrytis cinerea* contamination of a grape, in the form of gray rot, therefore constitutes a determining criterion for evaluating grape quality, whether red or white. But white grape crops, and consequently dry white wine quality, are more affected at lower levels of gray rot contamination than red grapes. The visual examination of

the percentage of botrytized berries, despite its insufficiencies, was the only method available to winegrowers. A dozen or so years after the pioneering research of Dubernet (1974) on *Botrytis cinerea* laccase, new methods of analyzing this enzymatic activity in juice (Section 10.6.6) were developed. This new way of quantifying *Botrytis* development on the grape appeared promising. Two methods were proposed: the first, a polarographic method, measured the oxygen consumption in a must sample with a Clark electrode (Salgues *et al.*, 1984); the second, a more sensitive colorimetric method, used syringaldazine as a specific reactive that produces a pink quinone in the presence of laccase (Dubourdieu *et al.*, 1984; Grassin and Dubourdieu, 1986). Results are expressed in laccase units (Section 10.6.6).

Juice from healthy grapes is evidently devoid of laccase. Coming from infected grapes, it can contain from one to several dozen units per milliliter, depending on the fungal development stage, the variety and the climatic conditions influencing the concentration contained in the grapes. Before the appearance of *Botrytis* conidiophores, the infected berries contain little laccase activity (1–2 units/ml). Activity considerably increases with sporulation (15–20 units/ml) and continues to grow, due to concentration, during the shriveling of the grape (Table 13.1). Universal enological tolerance thresholds are always difficult to establish for a grape defect: they depend on the level of quality or perfection desired for a wine. Ideally, a white grape crop should not contain any botrytized berries; the laccase activity should be zero,

Table 13.1. Development of *Botrytis cinerea* on the grape berry and laccase activity of juice (Grassin and Dubourdieu, 1986)

Development of *Botrytis*	Laccase activity units/ml[a]
Healthy grape	0
Full rot without conidiophores	1–2
Appearance of conidiophores	15–20
Shriveled rotted grapes	20–70

[a] 1 laccase activity unit corresponds to the quantity of enzymes capable of oxidizing 1 nmol of syringaldazine per minute in laboratory conditions.

or at least less than 1 unit/ml. The oxidasic casse threshold for red wines is greater than 3 units/ml and the sensitivity limit of the laccase measure by the colorimetric method is 0.5 units/ml. In the event of a gray rot attack, the grapes must be manually sorted in the vineyard; this is the sole means of maintaining the quality of the healthy portion of the harvest (Section 13.2.3).

Although less widespread than gray rot, sour rot (Section 10.6.5) can seriously affect the disease state of grape crops in localized areas, when the maturation occurs in a warm and humid climate. In the Bordeaux region, white grapes, in particular Sauvignon, are more sensitive to this disease than red grapes. It has not been extensively studied and is poorly known, despite its seriousness. The grapes take on a brick-red color within a few days, while letting some of the juice flow out, and they simultaneously give off a strong acetic acid odor. The microbial agents responsible for this acetic fermentation are a combination of aerobic yeasts (*Hanseniaspora uvarum*) and acetic bacteria. Fruit flies are known to be the contamination vector (Bisiach *et al.*, 1982: Guerzoni and Marchetti, 1987), but the exact causes of the berry contamination by the microorganisms which habitually make up the microfauna of the grape surface have not been elucidated. The development of sour rot is encouraged (like gray rot) by excessive swelling of the berries following heavy precipitations during maturation. The pressure of surrounding grapes can often detach certain grapes from the pedicel. Contamination can occur, beginning at this rupture zone. Microscopic epidermal fissures permitting juice flow, invisible to the naked eye, may also be a cause. The evolution of the grape crop towards gray rot or sour rot from these situations depends on environmental conditions. When excessive temperatures (higher than 30°C) block the development of *Botrytis cinerea*, sour rot quickly appears and is capable of destroying the entire harvest in a few days. While the development of gray rot in humid climates ceases with the return of hot and dry conditions, sour rot continues its growth inexorably—whatever the meteorological conditions. Young, vigorous vines with a superficial root structure planted in well drained soils are the most sensitive to sour rot. This phenomenon is aggravated by bird and insect damage in vines located in urban areas and zones well lit at night.

Sour rot obviously damages dry white wine quality more than gray rot. Musts made from partially sour grapes can contain more than 1 g of acetic acid and several grams of gluconic acid per liter (Section 10.6.5). They have very high sulfur dioxide combination rates caused by ketonic substances formed by the acetic bacteria metabolism. Their propensity for premature fermentations makes natural settling particularly difficult to effect. Finally, to combat the spreading of sour rot, the harvest must often be started before complete maturity and the grapes must be rigorously sorted in the vineyard.

The presence in the harvest of even a small proportion of grapes infected by powdery or downy mildew leads to the appearance of characteristic olfactory flaws, having an earthy and moldy smell. These odors can adversely affect the aroma of dry white wine. Fortunately, these types of grape spoilage have become rare.

13.2.2 Maturity and Setting the Harvest Date

The need for picking ripe grapes to make good wine is well understood. Yet optimum grape crop maturity (whether red or white) is difficult to define and there is no universal notion of grape maturity. It depends on the latitude of the vineyard, the climate, the vintage, the variety, and the parcel as well as the type of wine desired.

Must sugar concentration and acidity do not solely define the maturity of grapes destined to produce dry aromatic wines. Aroma and aroma precursor concentrations are also determining factors. No systematic relationship, however, exists between optimum grape aroma concentrations and maximum grape sugar concentrations—no more than between the latter and optimum grape acidity levels for a given type of wine. The characteristics of Chardonnay maturity are not the same in Meursault and Champagne or on a ranked growth *terroir* and a generic appellation. It is therefore impossible to establish a general rule for this subject. The notion of aromatic maturity is often used by certain enologists and winemakers, this language can be misleading. Optimum maturity can only correspond to a level of grape maturity that produces the best wine from a grape crop of a given parcel. Furthermore, the optimal aroma composition of a grape is not easy to define. In fact, the grape, like all fruits, progressively loses its vegetal and herbaceous aromas during maturation, to acquire fruity aromas which are more or less stable towards the end of maturation.

The formation of these different aromas in wine is relatively complex. Some exist in a free state in the grape; others are formed from precursors located in the must, during the pre-fermentation phase under the action of grape enzymes, or during alcoholic fermentation through yeast metabolism. The grape has a potential for both undesirable herbaceous flavors and sought-after fruity aromas. These two potentials evolve in the opposite direction during maturation. Theoretically, such changes should be measurable. According to the theoretical representation in Figure 13.2, the grape has an optimum composition in the 5 week following *véraison*, but it is not yet possible to monitor

analytically the aromatic evolution of the principal varieties—with the exception of Muscat, whose free terpene alcohol concentration gives an indication of the intensity of the characteristic floral aroma. Until enology makes further progress in this area, standard maturity assessments, half-bloom and half-*véraison* dates and grape-tasting, which helps to evaluate the aromatic maturity of the harvest, must be used to determine the harvest date.

Standard maturity assessment, from *véraison* until harvest, follows the evolution of three principal parameters: berry weight, sugar concentration and total acidity. It is also useful to measure the malic acid concentration and the pH, but these analyses are not often carried out by winemakers. A maximum sugar concentration without loss of berry weight indicates the completion of maturation. Overripening, which begins when berry weight diminishes, is generally depicted by an additional increase in sugar and possibly acid concentrations. Overripening is rarely sought for white grapes used for the production of dry white wines, due to the accompanying aroma losses.

A minimum concentration of must sugar has been empirically determined for each variety, within a particular region, for producing dry white wines of satisfactory quality. For example, in the Bordeaux region, Sauvignon and Semillon must have at least 190 g and 176 g of sugar per liter, respectively. Below these limits, regardless of winemaking methods, the wines obtained have a vegetal aroma. They lack finesse and rarely express the personality of the *terroir*. Similarly, the optimum acidity of ripe white grape musts is specific to both the location of the vineyard and the variety used. In Bordeaux, the optimum acidity at the time of the harvest is between 5 and 6 g/l (expressed as H_2SO_4) for Sauvignon and 4–5 g/l for Semillon. These value limits correspond to average acidities and sugar concentrations of samples taken over several years at the time of ideal maturity (Table 13.2).

When grapes have reached their minimum sugar concentration, harvest is possible but several other conditions must also be satisfied: grape-tasting indicates the disappearance of herbaceous aromas whereas fruity aromas, characteristic of the variety,

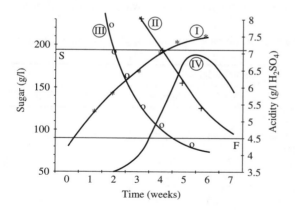

Fig. 13.2. Theoretical schema of the evolution of Sauvignon grapes during the 7 weeks following *véraison*. I: evolution of sugar concentration (S = minimum concentration in France). II: evolution of acidity (optimum: between 5 and 6 g/l H_2SO_4). III: evolution of vegetal flavor (arbitrary units) (F = minimum perception threshold). IV: evolution of fruit character potential (arbitrary units)

Table 13.2. Average Sauvignon and Semillon must composition at harvest in representative parcels in Graves-classed growth (Bordeaux)

Year	Sauvignon		Sémillon	
	Sugar (g/l)	Acidity (g/l H_2SO_4)	Sugar (g/l)	Acidity (g/l H_2SO_4)
1987	212	5.0	193	4.3
1988	217	5.2	198	4.0
1989	213	5.9	195	4.7
1990	210	4.3	193	3.2
1991	188	6.8	160	6.3
1992	161	6.1	155	4.7
1993	200	5.2	168	4.2
1994	201	5.8	183	4.5
1995	192	5.2	186	3.6
1996	193	7.2	180	5.2
Average	198	5.7	180	4.5
Potential alcohol (%)	12.0		10.9	

are present; approximately 40 days have passed since half-*véraison* (plus or minus 10%); and the acidity is within in the optimum range for the given variety and vineyard. In general, the slower the rate of decrease in acidity during maturation, the longer the harvest can be delayed without fear of losing fruity varietal aromas. In the best *terroirs* for making aromatic dry white wines capable of considerable aging, the grape remains fruity and sufficiently acidic in the final stages of maturation.

These *terroirs* have slow and complete maturation conditions. Conversely, excessively hot climates, early harvests and excessive water stress in the summer are unfavorable to the aromatic evolution of white grapes.

Table 13.3 gives an example of Sauvignon maturation on two different Bordeaux soils: a sandy-gravely soil (G) and a sandy-clay soil on compact limestone (C). Soil G filters well and has a low water reserve. The water supply is limited,

Table 13.3. Maturation characteristics of 1993 Bordeaux Sauvignon on sandy-gravel soil (G) and sandy clay on calcareous soil (C) (Sergent, 1996)

Characteristic	Sample dates						
	Aug. 11	Aug. 19	Aug. 25	Sep. 1	Sep. 9	Sep. 16	Sep. 20
Soil (G)							
Berry weight (g)	0.98	1.17	1.34	1.44	1.47		
Véraison percentage	81	97	100				
Sugar concentration (g/l)	112	139	149	174	197		
Total acidity (meq/l)	290	200	150	128	106		
Malic acid (meq/l)	212	156	102	73	55		
pH	2.57	2.84	3.06	3.15	3.16		
Soil (C)							
Berry weight (g)	1.03	1.25	1.33	1.62	1.54	1.67	1.55
Véraison percentage	66	97	100				
Sugar concentration (g/l)	81	121	138	160	184	193	193
Total acidity (meq/l)	424	266	202	172	142	124	110
Malic acid (meq/l)	301	173	120	85	62	54	32
pH	2.43	2.64	2.82	2.86	2.89	3.01	3.05

making the vine more forward than in soil C, which has fewer hydric constraints. The grapes on soil C may be picked up to 2 weeks later than those on soil G and are slightly overripe. Soil C grapes undergo a slower maturation. At the same maturity level in the two soils, soil C grapes are slightly more acidic with less malic acid and a lower pH. Over the years, soil C grapes are observed to remain fruitier during the maturation process than soil G grapes. The harvest date can consequently be set to correspond with practically the maximum sugar concentration desired. On the gravely-sandy soil G, the characteristic Sauvignon aroma can be almost totally lost in the course of a week. Early harvesting on this type of soil is necessary, not only due to the forwardness of the *terroir* but also due especially to the instability of the varietal aroma. In a given vineyard, an understanding of the forwardness and behavior of parcels greatly influences the reasoning behind determining harvest dates.

13.2.3 Harvest

White grape harvesting for quality wine production has long been known to be more difficult and require more precautions than red grape harvesting. More sensitive to oxidation, easily masked by olfactory flaws, white wines have a more fragile aroma than red wines. The aroma can be partially lost or altered as early as the harvest, if certain rules are not followed. Harvest conditions must be such that the grapes picked are healthy and their enological maturity (sugar, acidity and aroma concentrations) is as uniform as possible. Leaves, petioles, dirt and assorted debris should be avoided in the harvest. From harvest to their arrival at the winery, the grapes must be as intact as possible to limit must oxidation and stem maceration.

The grapes should be harvested at a temperature below 20°C. In warm climates, harvesting may have to occur at night or in the early morning; but moisture on the grape clusters should be avoided, as it can be a significant cause of dilution.

The choice of harvesting method depends on grape maturity and disease status on one hand and on economic constraints on the other. White grapes can be harvested manually or mechanically, all at once or in several stages, with or without sorting in the vineyard or on sorting tables at the winery.

In temperate-climate vineyards sensitive to gray rot, multiple selective manual harvesting optimizes dry white wine quality. Only healthy grapes reaching the desired maturity level are picked. Infected grapes are eliminated in the vineyard. This is the most effective sorting method. Spoiled grapes and grape clusters are left at the vine, unripe grapes are picked at a later date. Leaf removal and cluster thinning, carried out during the year, combine to avoid cluster crowding and promote sun exposure. In this manner, ripe grape clusters are more easily identified by the grape-picker. Well planned pruning, the early elimination of base-bud shoots and laterals, leaf removal near *véraison* and grape-cluster thinning should all be carried out with the objectives of promoting a healthy sanitary state and homogenous grape maturity. These efforts not only improve wine quality but also facilitate harvesting.

Multiple selective harvesting has long been recognized to increase grape quality. Chaptal (1801), restating established principles, wrote:

> Only healthy and ripe grapes should be picked; all infected grapes should be discarded with care and unripe grapes should be left on the vine. The harvest is carried out two to three times in places where wine quality is a great concern. In general, the first cuvée is the best. Some countries nevertheless harvest all grapes at the same time. The characteristics of the good and bad are expressed at the same time. A much lower quality wine is thus produced compared with the potential of the grapes, if more precautions were taken during the harvest.

In certain years, grapes may have a homogenous maturity and perfect sanitary state. In these situations, all of the grapes can be harvested at the same time—multiple harvesting is not necessary.

Due to its lower cost, its speed and its simplicity, mechanical harvesting has been increasingly adopted over the last 20 years. Its effect on dry white wine quality can be negligible in optimum sanitary and maturity conditions but mechanical

harvesting of a heterogeneous crop always sacri-
fices wine quality. In this case, it is of economic
and enological interest to have the infected grapes
removed by a picking team before harvesting the
healthy portion of the crop.

Mechanically harvested white grapes must be
protected against oxidation. Sulfiting, however,
must be avoided since it promotes the extraction
of phenolic compounds. The addition of dry ice to
the crop is a preferable alternative. Some countries
use ascorbic acid, but this antioxidant is only
authorized for treating wines in the EC.

Whether manually or mechanically harvested,
the grapes should be transported rapidly to the
winery in containers that minimize berry crushing.

13.3 JUICE EXTRACTION

13.3.1 General Principles

In dry white winemaking, pre-fermentation operations (grape and juice handling and treatments)
are deciding factors in final product quality
(Section 13.1.1). Their role is multiple. They must
extract and clarify juice in a relatively limited
amount time while minimizing juice loss. In addition, the diffusion of certain grape skin substances
in the juice, in particular fruity aromas and aroma
precursors, must be promoted during these operations. The dissolution of herbaceous-odor and
bitter-tasting compounds, associated with the solid
parts of the berry, must simultaneously be limited.
The formation of substances capable of decreasing
the stability of extracted fruit aromas must also
be avoided. Oxidized or oxidizable phenolic compounds in particular are able to trap certain aromas
(Section 13.2.1).

Before describing the different techniques used
and their consequences for juice and wine composition, juice extraction principles should be
discussed.

The fermentation of juice containing too many
suspended solids (resulting from juice extraction) does not produce quality dry white wine.
In fact, high concentrations of suspended solids
in juice are known to have detrimental effects

on wine quality (Section 13.5). The first criterion of a juice extraction method, therefore, is
its ability to produce clear juice with a turbidity
as near as possible to desired levels (200 NTU).
All winemaking is a series of elementary operations and each one must be conceived with
the idea of facilitating the others that follow.
The lower the concentration of suspended solids
in the juice after draining or pressing, the easier it is to accomplish juice clarification; conversely, clarification becomes impossible after a
poorly adapted pressing that produces excessive
suspended solids. When designing winery equipment, this criterion is not always sufficiently taken
into account. The production of suspended solids
during juice extraction has other disadvantages:
it indicates that the grape has undergone severe
mechanical treatment and consequently a greater
amount of herbaceous character substances are diffused in the juice.

Proper juice extraction should also limit oxidation phenomena, the dissolution of phenolic
compounds from the skins, seeds and stalks, and
pH increases linked to potassium extraction from
the solid parts of the grape. Resulting oxidation
phenomena and juice browning can be evaluated
by measuring the absorbance of filtered juice at
420 nm. The dissolution of phenolic compounds
is measured by the phenolic compound index (the
optical density of the juice at 280 nm is subtracted
from the optical density at the same wavelength
of the same juice percolated on PVPP). The pH
increase during pressing merits being followed
more systematically in practice. The evolution of
juice electrical conductivity during pressing can
also provide interesting information on pressing
kinetics.

The objectives described above are better
attained when the following conditions are satisfied:

• low pressing pressure;

• limited mechanical action capable of triturating
grape skins;

• slow and progressive pressure increases;

• high volume of juice extracted at low pressure;

- juice extraction at a temperature lower than 20°C;

- limited crumbling and press-cake breaking during pressing;

- minimum air contact—rapidly protected from air exposure and sulfited.

The transformation of grapes into juice can be obtained by different methods. Juice extraction can be immediate or preceded by a skin maceration phase. It can be continuous or in batches, with or without crushing and destemming. Continuous and immediate juice extraction processes (very widespread until recently in high-volume wineries, despite its disastrous consequences on juice quality) are fortunately being abandoned; they will therefore be covered only briefly. Immediate whole or crushed grape batch pressing and skin maceration will be described in more detail.

13.3.2 Immediate Continuous Extraction

In this process (Figure 13.3), the grapes, crushed by rollers beforehand, fall by gravity into a continuous inclined dejuicer containing a helical screw and are transferred into the continuous press placed

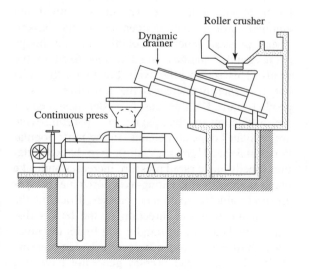

Fig. 13.3. Continuous juice extraction process

below. Continuous dejuicers are capable of treating large amounts of grape crops (several hundred kilograms per minute), liberating a high proportion of free run (70%). They have the disadvantage of producing juice with a high concentration of suspended solids and elevated turbidity (1000–10 000 NTU). The percentage of lees after natural settling is between 30 and 50%. Clarifying this volume of suspended solids is problematic: it requires costly large-scale filtration or centrifugation equipment which has insufficient treatment rates. Moreover, the high speed of continuous dejuicers limits the diffusion of aroma components from the skins into the juice. Sometimes even at the same turbidity after clarification, juice having undergone continuous dejuicing has more difficult fermentations than juice from batch pressing—more apt to extract compounds indispensable to yeasts.

Continuous presses extract the remaining 30% of juice contained in the skins after continuous dejuicing. The skins are pushed into a cylinder by a large helical screw against a restriction to compact the skins and form a plug. Due to the tearing and grinding of the grapes caused by the screw, the juice obtained with this equipment, regardless of its performance, is bitter, vegetal, colored and high in tannins and has an elevated pH (Peynaud, 1971; Maurer and Meidinger, 1976). The wines obtained could never make up a quality blend (*appellation*). Speed is the only advantage of this press, which is capable of throughputs of up to 100 metric tons/hour. The use of this extraction method is rare today, due to the development of high-capacity pneumatic presses that are capable of high throughputs while maintaining the quality of batch pressing.

13.3.3 Immediate Batch Extraction without Crushing

Also called whole cluster pressing, this extraction process is based on pressing methods used in Champagne. In this famous sparkling-wine region, the objective is to obtain white juice even from red grapes. The grape skin must not be triturated during handling or pressing.

Intact grapes are placed into the press. To maintain skin tissue integrity, they are not crushed, pumped or destemmed. The juice is extracted in batches: the filling, pressing and emptying operations are carried out successively and make up a cycle. In small installations treating premium quality grapes, the small containers (wooden tubs or crates) used to transport the whole grapes are also used to fill the presses. They are easily maneuvered on pallets by means of a forklift.

For larger-scale production, if trailers are used to transport grapes from the vineyard to the winery, they must be capable of emptying their contents into the press without the need for a must pump, which inevitably crushes the grapes. Various techniques are effective. Trailers can dump by gravity into a screw-driven hopper feeding a conveyor belt leading to the press. Hydraulic-lift trailers capable of elevating themselves to the level of the press may also be used. The grapes are then transferred into the press with a helical screw. A helical screw-based system does not result in a significant amount of burst berries and does not adversely affect wine quality as long as: (i) the transfer distance does not exceed 4–5 m; (ii) the height of grapes above the screw is kept to a minimum (a few dozen centimeters); (iii) the screw diameter is sufficient (30–40 cm); (iv) its rotation speed is slow. Belt-driven hoppers have recently appeared, containing a conveyor belt: they transfer whole grapes in ideal conditions, but this system can be difficult to clean.

Three principal types of batch presses are used: vertical presses, moving head presses and pneumatic presses. These same presses are used for red winemaking (Section 12.6.4, Figure 12.11). Press operating conditions have a greater influence on the quality of fresh grapes than fermented skins, which contain only approximately 15% of the total wine produced.

Vertical screw presses are the oldest, since their operating principle was invented by the Greeks. They are the archetypal press (Hiéret, 1986). At the beginning of the 20th century, vertical screw presses were progressively replaced hydraulic presses that made use of hydraulic pressure to compress the berries. In vertical presses with a mobile basket, the hydraulic jack lifts the basket and compresses the berries from top-to-bottom in the direction of the fixed pressure plate. In fixed basket presses (traditional Champagne presses), the top to bottom compression is produced by a mobile pressure plate equipped with a hydraulic jack that lowers itself.

The quality of juice extracted by vertical hydraulic presses is indisputable, since the pressure is exerted without triturating the grapes. The juice has a low concentration of suspended solids due to the filtration resulting from the cake thickness. This type of press requires elevated pressures, from 4 to 5 bars during the first pressing to 14 bars for the last. The extended pressing time and the percolation of juice across the skins increases the concentration of fragrant compounds from the skins in the must. The primary disadvantages of these vertical hydraulic presses are their slow throughput and the labor-intensive operation of breaking the press cake. In most installations, they have been replaced by rotating horizontal presses, either moving-head or pneumatic, permitting the cake to be broken up mechanically.

In threaded-axle moving-head presses, depending on the direction of basket rotation, the plates (heads) approach from each end (compression phase) or separate from each other (decompression phase). The separation of the heads provokes the break-up of the cake. The press is filled and emptied through central openings in the basket. These presses generally contain internal hoops connected by stainless steel chains fixed to the heads. This set-up effectively breaks up the cake but also sharply increases the formation of suspended solids. For this reason, models designed for the production of champagne and sparkling wine are without hoops and chains. In the most popular horizontal head press (Vaslin), two basket rotation speeds make rapid pressure increases and cake break-up possible. Rotating-axle horizontal presses are preferable to fixed-axle presses. Rotating the axle in the opposite direction of the basket displaces the heads more rapidly and limits the number of basket rotations necessary between pressing cycles. Moving-head presses generally have five to six pressure levels (up to 9 bars). Pressing can be controlled manually or automatically—the program is modulated according to the nature of the grape. The pressing quality obtained with this type of press depends a lot on the choice of pressing cycle. Increasing the pressure too quickly and excessive, too rapid and ill-timed cake break-up lead to vegetal and oxidized juice with suspended solids. Slow manual pressing, while monitoring throughput and juice turbidity, obtains the best results.

Figure 13.4 gives an example of a pressing cycle with a horizontal moving-head press (Vaslin, 22 VT) in manual mode using whole, healthy and ripe Sauvignon. All pressings, as well as the first three cake break-ups, are executed at a slow basket

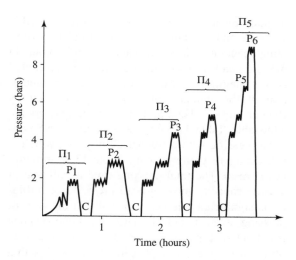

Fig. 13.4. Whole-cluster pressing cycle in manual mode with a moving-head press (Vaslin 22 VT). ρ_1 to ρ_6 = pressure level; Π_1 to Π_5 = pressure step; C = crumbling, drying or cake break-up

rotation speed. Each time after retracting the heads, must extraction is normally resumed at a pressure lower than before breaking up the cake. Table 13.4 states the various pressing times and juice volume and turbidity for this diagram. The drained juice and the juice from the first pressing are the most turbid. As soon as the pomace cake is formed, it acts as a filter and 50% of the juice is extracted without breaking up the cake. The juice from the first two pressings (Π_1 and Π_2) constitutes the free run juice. The press juice is composed of the last three pressings and represents approximately 15% of the total extracted volume. Total pressing time exceeds 3 hours, but the overall turbidity of the pressings (about 500 NTU) is satisfactory with respect to the 200 NTU preferred for a juice before fermentation. Consequently, the percentage of lees obtained through natural settling is generally less than 10%.

Correctly carried out to obtain quality juice, pressing with a horizontal moving-head press is necessarily slow. Additionally, due to mechanical constraints imposed by the basket, presses larger than 60 hl cannot be manufactured; thus their use for quality white wines and small installations is to limited.

In pneumatic presses, the pressure exerted to extract the juice is applied to the grape clusters by an internal membrane which is inflated by an on-board air compressor. The maximum pressure attained by a pneumatic press is 2 bars. Different models exist: perforated basket or closed tank, equipped with drains, with an axial bladder or side-mounted membrane and filled axially or through doors. They can function manually or automatically with more or less sophisticated programs.

Table 13.4. Evolution of must turbidity during whole-cluster pressing of Sauvignon grapes in a moving-head press (Vaslin 22 VT)

Pressing	Time (min)	Volume (hl)	Juice %	Turbidity (NTU)
Draining	15	0.6	4	630
P1	60	6.9	46	690
P1 + P2	50	5.0	34	290
P1 + P2 + P3	40	1.3	8	580
P2 + P3 + P4	30	0.9	2	350
P3 + P4 + P5	15	0.3	2	310
Total	210	15.0	100	550

Closed tank presses are preferable to perforated basket presses, since the grapes can be more easily protected with a blanket of carbon dioxide during filling. These presses can also be used for skin maceration in the correct conditions (Section 13.3.5). Additionally, the tank press has a greater mechanical resistance than the perforated basket press for the same metal thickness. Juice collection by the drains in tank presses also limits juice oxidation. In basket presses, the juice flows out in a thin layer, increasing oxidation risks. Membrane tank presses are currently the most popular, especially for high-capacity presses. The membrane is located on the half of the tank opposite the drains. The largest membrane presses currently have a 350 hl capacity. Filling, pressing, crumbling and emptying is depicted in Figure 13.5. During the pressing phase, the tank is immobile with the drains facing the bottom. During crumbling, the membrane is deflated and the basket rotates.

Axial filling is only an option when the press is filled with crushed grapes (Section 13.3.4). This method leads to an increase in suspended solid concentrations, obtained after pressing.

As with horizontal moving-head presses, the pressing quality of pneumatic presses depends on the chosen pressing method and cycle. The general rules are the same. The maximum volume

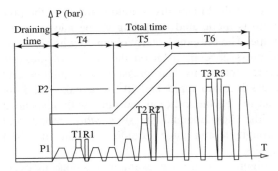

Fig. 13.6. Standard pressing program of a Bucher pneumatic press. T1, T2, T3, duration of sustained pressure; T4, low pressure period; T5, increasing pressure period; T6, maximum pressure period. R1, R2, R3, number of rotations during crumbling in different phases of the cycle

of juice must be extracted at the lowest possible pressure and crumbling must be limited. Crumbling generates less suspended solids than in moving-head presses but the oxidation promoted is not negligible. Figure 13.6 is a representation of the standard pressing program used by Bucher presses. Time and pressure parameters are adjustable. Table 13.5 gives an example of a Sauvignon pressing with a Bucher 22 hl press, which is interesting to compare with Table 13.4 (moving-head press). The total pressing time is half that of a same capacity moving-head press. Juice turbidity is significantly lower, especially at the start of pressing, even though the largest volume of juice is extracted at this time. Turbidity varies relatively little during pneumatic pressing. Compared with a moving-head press, a pneumatic press can liberate a significantly clearer juice more rapidly using much lower pressures. Suspended sediment deposit is often less than 5%.

The most recent pneumatic presses have fully programmable systems, permitting the operator to

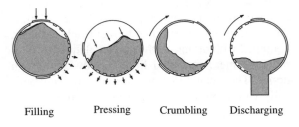

Filling   Pressing   Crumbling   Discharging

Fig. 13.5. Operation of a closed-tank membrane press

Table 13.5. Evolution of must turbidity during whole-cluster pressing of Sauvignon grapes with a pneumatic Bucher (22 hl) according to an automatic program (Figure 13.5)

Pressure	Time (min)	Volume (hl)	Juice %	Turbidity (NTU)
Low (0.2 bars)	55	13.2	88	400
Increasing from 0.2 to 2 bars	27	1.6	10	350
Maximum (2 bars)	8	0.2	2	300
Total	90	15	100	463

carry out several dozen sequences within a pressing cycle. Within each sequence, the pressure, duration and number of tank rotations during crumbling are defined. In this manner, the pressure can be increased by adjustable increments. The juice produced has a low turbidity and sometimes does not require clarification. The pressing programs that limit the number of crumblings and thus juice oxidation are by far the best. By measuring juice flow rate, the system determines the pressure level at which the press must operate and the amount of time that this pressure should be maintained. The press incrementally increases the pressure and determines the time and intensity of crumbling.

The most important characteristic of pneumatic pressing is the low increase in the concentration of phenolic compounds in juice during the pressing cycle (Maurer and Meidinger, 1976). The juices of the final pressings, or at least a greater percentage of final press juices, may be blended into the finished product.

In conclusion, different batch pressing systems have been successively used in the making of quality white wines. Clear juice extraction and slow total pressing times were two inherent characteristics of vertical presses. With this kind of press, even if the winemaker had wanted to press more quickly and less carefully, this was impossible. With a moving-head press, clear juice extraction is dependent on pressing slowly and correctly. Yet these presses have often been incorrectly used to work more quickly, producing juices of inferior quality with suspended solids. The creation of pneumatic presses radically changed the constraints on white grape pressing. Clear juices are obtained with much reduced pressing times and a lower proportion of inferior quality press juice. The total pressing time, however, must be sufficiently long to permit the diffusion of characteristic aromas of the grape into the juice.

13.3.4 Advisability of Crushing and Destemming with Immediate Extraction

Crushing consists of breaking the grape skin for immediate liberation of pulp and part of the juice. Grooved rollers with adjustable spacing have long been used for crushing. Turning in opposite directions, they seize the grape clusters and crush the berries. The machinery should be adjusted so that the stalk and seeds are maintained intact during this operation. The crusher should be placed above the press to permit gravity filling. This set-up eliminates the need for a must pump, which always generates suspended solids. This principle is not always respected. In high-volume wineries, pneumatic presses are frequently filled axially—using a must pump. Draining is facilitated by the periodic rotation of the basket. In this case, the presses are used as dynamic drainers before pressing. These operations always increase the formation of suspended solids.

Crushing before pressing has the advantage of permitting a more significant draining of the grape crop during the filling of the press. Press capacity can be increased by 30–50%. Additionally, crushed grapes are pressed slightly more quickly than whole-clusters. On the negative side, the juice obtained is at least two times more turbid than after whole-cluster pressing. Lees volume after natural settling is approximately 20% if a moving-head press is used, and slightly lower with a pneumatic press. The suspended solids are liberated in the drained juice before the press cake is capable of playing its role as a filter. Consequently, when a closed tank press is filled with crushed grapes, instead of draining immediately, the drains should be closed during the first half of filling to limit the formation of suspended solids.

The mechanical action exerted on the skins by crushing seems to promote the diffusion of aroma components into the juice. Nevertheless, by rapidly liberating the juice, crushing curtails skin contact. The two phenomena work against each other. The role of crushing on primary aromas located in the skins is uncertain.

Crushing is generally thought to increase herbaceous flavors (hexanol, *cis*-3-hexenol and *trans*-2-hexenol) in juice and wine, especially in the case of insufficient grape maturity (Table 13.6). Even after adjusting juice to suitable turbidity levels before fermentation, the wines obtained have considerably

Table 13.6. Influence of Semillon grape maturation and crushing on concentrations (mg/l) of C_6 compounds (hexanol + hexenols) in Semillon wines

State of grapes	Harvest A[a]	Harvest B[a]
Crushed	2.1	1.5
Not crushed	1.4	1.0

[a]Harvests A and B are separated by 10 days.

higher C_6 alcohol concentrations with early harvested and crushed grapes.

Destemming white grapes intended for immediate pressing also presents certain disadvantages. The stalks act as a drain during pressing. Removing them increases draining time and the number of crumblings required. To facilitate the pressing of mechanically harvested grapes, certain manufacturers (Bucher) have equipped their pneumatic presses with complementary drainage systems. The presence of stalks during pressing also limits the concentration in juice of thermolabile proteins which cause protein casse in white wines. Wines are therefore stabilized with lower bentonite concentrations when produced from juice extracted from non-destemmed grapes (Volume 2, Section 6.6.2).

If press capacity permits, crushing and destemming should be avoided and the grapes should be hand-harvested and pressed immediately. These operations are only necessary when the grape undergoes skin maceration before pressing.

13.3.5 Skin Maceration

Experience and cautiousness have led to the creation of general white winemaking principles recommending as little maceration as possible with the solid parts of the grape cluster. The diffusion of substances from these solids into the juice leads to various flaws in the wine: vegetal aromas of unripe grapes, astringency and bitterness of phenolic compounds from seeds, skins and stems and moldy, earthy and fungal odors from spoiled grapes. It should therefore be avoided. With grapes of heterogeneous maturity levels and disease status, immediate and rapid juice extraction followed by rigorous press selection is indispensable. Draining rooms, used in the 1950s

for stocking crushed grapes before pressing, were abandoned for this reason (Ribéreau-Gayon et al., 1976). They provoke oxidation and an uncontrolled maceration in the presence of stalks.

Yet with certain varieties, when soil and climatic conditions combine to produce perfectly ripe and healthy grapes, skin maceration can be sought for the better extraction of grape skin components that participate in the aroma, body and aging potential of dry white wines. In this case, the positive elements largely outweigh the negative elements linked to insufficient maturity or a high state of disease.

Slow pressing contributes to the extraction of aromatic elements from the grape skin. In fact, press juice is the result of a certain degree of maceration. In certain cases, adding it to wine is desirable; in other cases, it should be avoided. The higher the sugar concentration and aromatic intensity and the lower the pH, the more adding press juice improves wine quality. In vineyards in the Bordeaux region, incorporation of Sauvignon or Semillon press juice before fermentation is systematic when the grapes come from old vines and the best parcels. It is avoided with juice from young vines, insufficiently ripe grapes and excessively high yielding vines Section (13.2.2).

Skin maceration consists of voluntarily permitting a contact phase between the skins and the juice in controlled conditions. An adapted tank is filled with moderately crushed, destemmed grapes. Several hours later, the drained juice is collected and the drained pomace is pressed.

Results reported, as well as winemakers' opinions on skin maceration and the quality of wine obtained by this method, are sometimes contradictory. This is not surprising, since the nature of the grape (variety, disease status and maturity) and maceration conditions (temperature, tank and grape handling) greatly influence its effect. According to certain authors (Ough, 1969; Ough and Berg, 1971; Singleton et al., 1975), skin contact lasting for more than 12 hours results in coarse, phenolic wines of inferior quality. Others (Arnold and Noble, 1979) find that the skin maceration of Chardonnay significantly improves aroma quality and wine structure without increasing bitterness and astringency. The best results, in this particular case, have been obtained with relatively long maceration (16 hours). Shorter maceration has been recommended for Austrian varieties (Haushoffer, 1978).

Skin maceration grew increasingly popular in France during the mid 1980s (Dubourdieu *et al.*, 1986; Ollivier, 1987). This operation produces satisfactory results with white Bordeaux varieties (Sauvignon, Semillon and Muscadelle) as well as with Muscats, Chardonnay and Gros-manseng—as long as it is carried out with the proper material on healthy grapes (negative response to laccase activity test) with homogeneous maturity.

The grapes, completely destemmed, are transferred to the maceration tank with a must pump, the tank having been filled beforehand with a layer of carbon dioxide to avoid oxidation. Sulfiting is avoided, to limit the extraction of phenolic compounds. Different installations are possible.

The first solution consists of carrying out the skin maceration in a pneumatic press, if it is airtight. When maceration is complete, the juice is drained and the skins are pressed. This system has enological advantages and is simple. The grapes are only transferred once, thus eliminating oxidation. The primary disadvantage is the immobilization of the press.

Skin maceration is generally carried out in a tank equipped with a system permitting the drained juice (70%) to be removed and the drained skins to be transferred to the press by gravity. The volume of this tank must be triple that of the press.

Skin maceration in a membrane tank (Elite-Pera) is a process situated between maceration in a pneumatic press and tank maceration. At the end of maceration, the juice is collected first by natural draining and then by inflating the tank membrane, incrementally increasing pressure (0.1–0.25–0.4 bar). Ninety per cent of the juice can be collected by this method. The drained skins are transferred by gravity into the press with the help of a screw conveyer. The juices obtained by this method do not contain many suspended solids (200–300 NTU) and are particularly well protected from oxidation.

Grape temperature must be maintained below 15°C during maceration. Circulating cold fluid through the jacket or cooling-coil to refrigerate the tank directly is not possible without agitating the grapes, but this operation is not recommended because it promotes the formation of suspended solids and the extraction of phenolic compounds. The crushed grapes may also be cooled with a tube heat exchanger. This process requires considerable cooling capacity and draws the grapes through small-diameter piping with many bends. Increased production of suspended solids may result. Another method consists of incorporating liquid carbon dioxide into the grape crop during filling at the outlet of the must pump. The grapes are cooled without a supplemental mechanical treatment. In addition, the oxygen dissolved in the must during crushing is eliminated by the flow of CO_2. The grapes are also transferred to the tank under an inert atmosphere. It requires 0.8 kg of CO_2 to cool 120 kg of destemmed grapes by 1°C.

Maceration times vary from 12 to 20 hours, depending on the winery. At controlled temperatures (10–15°C) and in the absence of oxygen, this time period seems to permit a suitable extraction of aromatic compounds from the skins without the risk of significant dissolution of phenolic compounds.

Pressing macerated grapes does not pose any particular problems. Due to the destruction of the pectic structure by grape enzymes, the grapes can be pressed at low pressures with only one to two crumblings necessary. The first pressing is immediately reincorporated with the free run juice. The final pressings are left separate. The decision to incorporate them with the free run and other pressings is made after clarification.

Skin maceration results in a decrease in must acidity and an increase in pH (Table 13.7). These changes are linked to the liberation of potassium from the skins and the resulting partial salification of tartaric acid. The acidity can decrease by as much as 1–1.5 g/l (expressed as H_2SO_4), but the degree of these changes depends on the variety and the *terroir*. Acidity and pH often vary less in Chardonnay than in other varieties such as White Grenache (Cheynier *et al.*, 1989).

Table 13.7. Influence of pre-fermentation maceration on total must acidity before clarification (1985 harvest) (Dubourdieu *et al.*, 1986)

Variety	Control[a]		Pre-fermentation maceration		
	Total acidity (g/l H_2SO_4)	pH	Total acidity (g/l H_2SO_4)	pH	Duration (in hours)
Sauvignon 1	5.6	3.05	4.05	3.35	8
Sauvignon 2	5.3	3.15	4.00	3.35	12
Sauvignon 3	6.05	3.43	4.75	3.53	12
Sauvignon 4	6.9	2.98	5.50	3.30	18

[a]Must obtained by whole-cluster pressing

Table 13.8. Influence of an 18-hour pre-fermentation maceration in practical conditions on must phenolic compounds (Dubourdieu *et al.*, 1986)

Variety	Start of maceration		End of maceration	
	OD 280	Phenolic compound index	OD 280	Phenolic compound index
Sauvignon 5	4.4	3.5	6.5	4.9
Semillon 1	4.6	3.1	5.6	4.3
Muscadelle 1	4.3	3.2	6.1	4.4

[a]Maceration for 18 hours at 20°C

Table 13.9. Influence of pre-fermentation maceration in practical conditions on wine phenolic compounds (Dubourdieu *et al.*, 1986)

Variety	Immediate pressing		Pre-fermentation maceration	
	OD 280	Phenolic compound index	OD 280	Phenolic compound index
Sauvignon 1	6.7	3.3	7.5	3.3
Sauvignon 3	6.3	3.3	8.1	4.7
Sauvignon 4	5.6	3.3	5.8	3.0

With respect to whole grape pressing, skin maceration also provokes an increase in optic density at 280 nm and the phenolic compound index (Table 13.8) but the differences observed in wines are less marked (Table 13.9) and the optic density at 280 nm remains well under 10—the upper limit generally accepted for white wines.

Maceration increases the amino acid concentration in juice, resulting in an improved fermentation speed, which is often observed in practice. Macerated grapes also produce juice and wine that is richer in neutral polysaccharides (Table 13.10) and proteins than pressed whole clusters. Wines made from macerated grapes

Table 13.10. Influence of pre-fermentation maceration on total polysaccharide concentrations (mg/l) in wine (Dubourdieu *et al.*, 1986)

Variety	Whole-cluster pressing	Pre-fermentation maceration[a]
Sauvignon 1	389	469
Sauvignon 3	356	547
Sauvignon 4	385	520
Sauvignon 6	266	359
Semillon 1	362	435
Semillon 2	228	442
Muscadelle 1	290	373

[a]Maceration: 12 hours for Sauvignon and 18 hours for other varieties.

require higher bentonite concentrations to be stabilized (Volume 2, Section 6.6.2).

Skin maceration makes the most of the aromatic potential of the grapes and in general it significantly enhances varietal aroma without increasing herbaceous flavors. In Muscat wines, these sensory differences can be analytically interpreted by measuring free and bound terpene alcohols. Baumes *et al.* (1989) observed increases of 576–742 μg/l in free terpenes and 689–1010 μg/l in bound terpenes. Measuring 4-mercapto-4-methyl pentan-2-one in Sauvignon wines also indicates the obvious role of maceration in the varietal aroma of this variety in wines. The 10 ng/l concentration in the control wine almost doubles to 18 ng/l in wine made from macerated grapes.

13.3.6 Cryoselection and Supraextraction

Chauvet *et al.* (1986) initially developed these techniques to improve the quality of juice intended for sweet winemaking (Section 14.2.4b), but they are also of interest for dry white winemaking. The process consists of cooling whole grape clusters in small crates for 20 hours or so in a walk-in freezer at a temperature of −2 to −3°C. Two phenomena—cryoselection and supraextraction—are at the origin of these changes in juice composition observed with respect to traditional pressing.

Cryoselection corresponds with pressing grapes at low temperature. Only the sweetest grapes remain unfrozen and release their juice. A quality juice is obtained, the volume increasingly limited as the temperature is lowered. After thawing, a second pressing releases a lower quality juice, coming from grapes with a lower sugar concentration.

Supraextraction corresponds to pressing whole grapes after they have been thawed. The freezing and thawing of the skins and lower epidermal layers results in modifications in tissue ultrastructure. In certain aspects, it produces an effect comparable to skin maceration. Notably, aromas and aroma precursors are released more easily from the grape. Extraction of skin phenolic compounds is, however, lower than with skin maceration and even immediate whole grape pressing (Ollivier, 1987).

Sugar extraction from the skins is also increased by 0.3–0.6% potential alcohol during supraextraction. Despite its slowness and elevated cost, supraextraction promotes the aromatic expression of certain noble white varieties.

13.4 PROTECTING JUICE FROM OXIDATION

13.4.1 Current Techniques

Oxygen is often said to be the enemy of white wines. In fact, except for rancio wines, whose flavor results from intense oxidation during production, white wines are protected from oxygen (or at least from oxidative phenomena) during the winemaking process and maturation. These precautions are taken to protect the fruity aromas of young wine and to avoid browning. They also promote the later development of a reduction bouquet in premium wines during bottling aging.

The oxidation of substances in white wine can occur at any time during winemaking. While the need to protect white wine from oxidation after fermentation is generally accepted, protecting must from oxidation is not unanimously considered necessary.

Most winemakers prefer limiting air contact with crushed grapes and white juice as much as possible. An adapted sulfur dioxide addition to juice blocks the enzymatic oxidation of phenolic compounds. This philosophy is based on empirical observation: juices of many grape varieties must conserve a green color during the prefermentation phase to be transformed into fruity white wines. Oxidation phenomena must consequently be avoided as much as possible.

Other enologists believe, on the contrary, that musts too well protected from oxygen give rise to wines that are much more sensitive to oxidation. Furthermore, pressing experiments carried out in an air-free environment have shown that these wines brown quicker in contact with air than wines made from traditionally pressed grapes (Martinière and Sapis, 1967). In addition, these wines are more difficult to stabilize with sulfur dioxide.

Muller-Späth (1977) was the first to contest the need to sulfite white juice before alcoholic fermentation. His research clearly showed that adding pure oxygen to non-sulfited juice before clarification improves the stability of white wine color without producing oxidation-type flaws. This process, called hyperoxidation or hyperoxygenation, consists of oxidizing juice polyphenols to precipitate them during clarification and eliminate them during alcoholic fermentation.

Must oxidation results in a varying degree of color stabilization of white wines, depending on variety (Schneider, 1989; Cheynier *et al.*, 1989, 1990; Moutounet *et al.*, 1990). Hyperoxygenation has also been used successfully on an experimental basis to discolor and improve the quality of second pressing Pinot Noir and Meunier juice in Champagne (Blank and Valade, 1989). The impact of this technique on the aromatic quality of the wine varies according to the variety and the tasting panel. The effect on aroma is sometimes judged favorable or neutral for Alsacian and German varieties, Chardonnay and Chasselas (Fabre, 1988; Müller-Späth, 1988; Cheynier *et al.*, 1989), but hyperoxygenation, or simply not protecting musts from oxidation, considerably affects the aroma of Sauvignon Blanc (Dubourdieu and Lavigne, 1990). The 4-methyl-4-mercaptopentan-2-one concentration decreases when the must is less well protected from oxidation (Figure 13.7). The mechanisms of this phenomenon will be discussed in the next Section (13.4.2). Juice oxidation also decreases the aromatic intensity of other varieties such as Semillon and Petit and Gros Manseng, whose aromatic similarity to Sauvignon is due to the participation of sulfur-containing compounds (Volume 2, Chapter 7). The best Chardonnay wines also seem to be made by limiting juice oxidation. Boulton *et al.* (1995) shared this opinion, believing that juice hyperoxidation harms the varietal aroma of wines.

13.4.2 Mechanisms of Juice Oxidation

Oxygen consumption in juice is essentially due to the enzymatic oxidation of phenolic compounds. Two oxidases (Section 11.6.2) are involved (Dubernet and Ribéreau-Gayon, 1973,

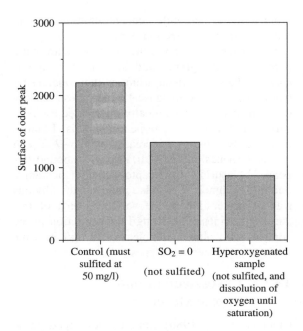

Fig. 13.7. Influence of must oxidation on 4MMP concentration in Sauvignon wines (Dubourdieu and Lavigne, 1990)

1974): tyrosinase in healthy grapes; and laccase from *Botrytis cinerea*, which only exists in juice from botrytized grapes. Laccase activity can be specifically measured to evaluate grape health analytically (Sections 10.6.6 and 13.2.1).

The substrates of tyrosinase are almost exclusively cinnamic acids and their esters with tartaric acid (caftaric and coumaric acids). It transforms caftaric acid into quinones (Section 11.6.2, Figure 11.12). These oxidation reactions are extremely quick. The oxygen consumption speed in juice, when first put into contact with air, can exceed 2 mg/l/min whereas it is around 1–2 mg/l/day in wine. A certain degree of oxidation in juice inevitably results during white winemaking before protection by sulfur dioxide. The decrease in the speed of oxygen consumption during successive oxygen saturations is caused much more by the depletion of the substrate, caftaric acid, than by the inhibitive effect of the oxidation products formed. Adding caftaric acid reestablishes the initial consumption rate (Moutounet *et al.*, 1990) but laccase is capable of catalyzing the

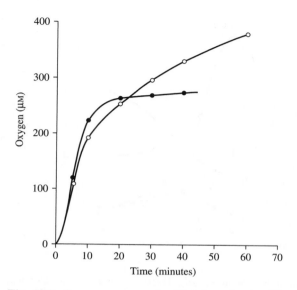

Fig. 13.8. Oxygen consumption kinetics of a healthy Colombard must (—●—) and of one with 10% contamination by *Botrytis cinerea* (—○—). (Moutounet *et al.*, 1990)

oxidation of a large variety of substances. It not only acts rapidly but also continues over a much longer time period (Figure 13.8).

The quinone formed from caftaric acid has several possible destinations:

1. It can combine with quinone traps in juice—in particular, glutathione (Section 11.6.2, Figure 11.13), a highly reductive tripeptide with a free sulfhydryl group, found in concentrations up to 100 mg/kg in certain varieties. The product formed is S-glutathionyl-2-caftaric acid, initially called the Grape Reaction Product (GRP) (Cheynier *et al.*, 1986). This oxidation–reduction reaction regenerates the *o*-diphenol function. Tyrosinase has no action on this glutathione derivative, but it can be oxidized by laccase. The caftaric acid quinone can also combine with other juice-reducing agents, such as ascorbic acid. The coupled oxidation regenerates caftaric acid. As long as glutathione (GSH) and ascorbic acid concentrations are elevated, consumption of must oxygen does not result in quinone accumulation or juice browning.

2. The caftaric acid quinone can also enter into coupled oxidations with flavonoids as well as GRP. Flavonoid and GRP quinones are formed. These then combine with glutathione to form di-*S*-glutathionyl caftaric or GRP2.

3. The caftaric acid quinone is capable of condensing with *o*-diphenols, first with caftaric acid. The color and insolubility of the products formed increases with their degree of condensation. Flavonol quinones also enter into condensation reactions, resulting in strongly colored and later insoluble products.

The oxygen consumption speed of white juice and the nature of the products formed therefore depends on initial concentrations of caftaric acid, glutathione, ascorbic acid and flavonoids in juice. Variety and, most likely, grape maturation conditions influence the proportion of caftaric acid and glutathione found in juice. The differences in reaction of two juices to the presence of oxygen are illustrated in Figure 13.9 (Rigaud *et al.*, 1990). Colombard juice is rich in glutathione and ascorbic acid, and its oxidation frees few caftaric acid quinones. In an initial phase, as caftaric acid is formed, it is reduced by ascorbic acid. When ascorbic acid is depleted, caftaric acid combines with glutathione to form GRP, which accumulates. Juice color changes from green to beige. There is no browning and few reactions coupled with the flavonoids, when quinones are not available. Ugni Blanc juice contains relatively little glutathione and no ascorbic acid: it behaves differently. Oxygen consumption is more rapid and a large quantity of quinones are formed, resulting in a pronounced browning of the juice with a continued increase in the orange nuance. The caftaric acid quinone enters into reactions coupled with flavonoids and GRP, whose concentration decreases.

The oxidation phenomena linked to the properties of tyrosinase and laccase are rapid and are present as early as crushing and pressing. Unsulfited juices exposed to air consume a variable quantity of oxygen according to their caftaric acid and flavonoid concentrations.

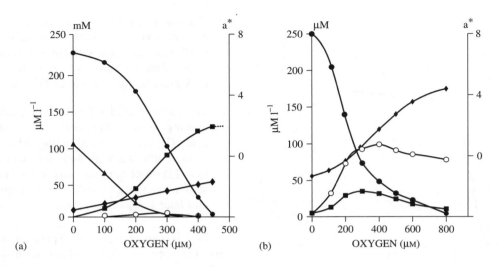

Fig. 13.9. Oxidation of (a) Colombard and (b) Ugni Blanc must (Rigaud *et al.*, 1989). ○ caftaric acid quinone; ● caftaric acid; ■ GRP; ▲ ascorbic acid; ♦ measurement of the red color

The effects of juice hyperoxygenation on the stability of white wine color are variable, due to the existence of several reactional mechanisms.

If color stabilization by oxidation is sought, other juice components must not be adversely affected—especially aroma. The difficulty of obtaining fruity Sauvignon wines from oxidized or brown-orange tinted juices remained unexplained for a long time, but it is now known that various thiols, playing a role in the Sauvignon aroma, are very sensitive to oxidation. They tend to produce disulfur bonds in the presence of oxygen and, more importantly, combine very rapidly with quinones. Sulfiting protects these aromas by blocking quinone formation. This treatment is effective, even if implemented on partially oxidized juice, since it reduces the quinones (Darriet, 1993).

13.4.3 Techniques for Protecting Juice from Oxidation

The winemaker can implement various complementary techniques to limit juice oxidation:

- sulfiting—antioxidant and antioxidasic activity;
- adding ascorbic acid—antioxidant effect;

- cooling grapes and musts to slow oxidation reactions;
- heating musts at 60°C for several minutes to destroy oxidases;
- handling grapes in the absence of air to limit the dissolution of oxygen;
- clarification to eliminate a portion of the tyrosinase activity associated with solids and to limit the oxidasic activity of juice.

Sulfiting is the first, most simple and effective method of protecting juice from oxidation. To destroy tyrosinase, 50 mg/l of sulfur dioxide per liter must be added to the juice. If the grapes are healthy, this addition definitively blocks enzymatic oxidation mechanisms. Sulfiting must be carried out with higher concentrations to inactivate tyrosinase in highly colored press juices, containing quinones.

The entire sulfur dioxide addition should be made at the same time. It should be fully homogenized into the juice. Sulfiting at concentrations below 50 mg/l should be avoided, since this only delays oxidation phenomena and juice browning; in time, all of the oxygen contained in the juice is consumed. The worst method consists of progressively adding small quantities of SO_2. The total amount of oxygen consumed in these conditions by a juice in contact with air is greater than in an unsulfited juice. The final color of the two juices, in terms of oxidation, is practically equal.

Sulfiting grapes promotes the extraction of skin phenolic compounds. Protecting with ascorbic acid (10 g/hl) does not have this disadvantage, but this powerful reducing agent is not antioxidasic. Like low sulfur dioxide concentrations, it only limits browning by reducing quinones but does not limit oxygen consumption. Grapes must be in limited contact with oxygen when using ascorbic acid. For example, they should be handled in the presence of dry ice when filling maceration tanks or pneumatic presses. Juices not protected by sulfur dioxide should not be allowed to stagnate in contact with air in press pans during pressing. Their large surface area promotes rapid oxidation.

Cooling grapes and juices is extremely effective in slowing juice oxidation, and it should be used systematically. Oxygen consumption is three times faster at 30°C than at 12°C (Dubernet and Ribéreau-Gayon, 1974). Cooling with liquid CO_2 while filling tanks and presses associates handling the harvest in an inert atmosphere as soon as juice appears with the effects of cooling.

Cryoextraction or supraextraction (pressing whole grapes at temperatures below 0°C) considerably limits oxidation. This technique enhances the fruity character of dry white wines, compared with pressing at ambient temperatures. Not only are aromas and aroma precursors freed by the freezing and thawing of the skins, but oxidative phenomena are also limited during pressing.

Clarification (Section 13.5) limits oxidasic activity but does not prevent juice from browning. A sufficient soluble tyrosinase activity remains in fresh juice unprotected from oxygen, allowing rapid browning. Clarification is a means of eliminating oxidation products—in particular, condensed flavonoids formed during coupled oxidations.

Heating juice theoretically destroys oxidases but it must occur quickly after extraction. The heating process must also be rapid. It is rarely used in practice.

13.5 CLARIFICATION

13.5.1 Formation and Composition of Suspended Solids and Lees

Freshly extracted grape juice is more or less turbid. It contains suspended solids of diverse origin: earth, skin and stem fragments, cellular debris from grape pulp, insoluble residues from vineyard treatment products, etc. Macromolecules in solution or in the course of precipitating are also involved in juice turbidity. Among them, grape pectic substances play an essential role (Volume 2, Section 3.6). With rotted grapes, juice turbidity is also caused by the presence of polysaccharides, especially $(1-3:1-6)$-β-D-glucane, produced by *Botrytis cinerea* in the berry. A few milligrams of this substances is enough to provoke serious clarification difficulties (Volume 2, Sections 3.7 and 11.5.2). These glucidic macromolecules influence juice turbidity through the Tyndal effect (Volume 2, Section 9.1.2). Acting as protective colloids (Volume 2, Section 9.4), they also hinder clarification by limiting or blocking particle flocculation and sedimentation phenomena as well as clogging filter surfaces. Natural grape pectinases (or those added by the winemaker) acting on the colloidal structure of the juice facilitate natural settling. After several hours, the juice separates into two phases: a more or less opalescent clear juice and a deposit varying in thickness. The latter contains different colored successive strata: greenish brown in the lower portion of the deposit and green to light beige in the upper portion. Some winemakers distinguish between the heavy deposit that forms first during natural settling and the light deposit that accumulates more slowly. Clarification consists of separating (by racking, for example) the clear juice from the lees before alcoholic fermentation.

The quantity of lees formed during juice extraction and the speed of sedimentation speed depend on variety, grape disease status, maturity and especially winemaking methods (crushing, draining, pressing, etc.) (Section 13.3).

In normal conditions, juice turbidity generally decreases during grape maturation (Hadjinicolaou,

1981). This evolution results from the hydrolysis of pectic substances in the berry by pectic enzymes of the grape (endopolygalacturonase and pectin esterase). In dry weather conditions, the grape remains pulpy and the juice is more difficult to extract and clarify, due to a lack of pectic activity. Towards the end of maturity, the soluble acid polysaccharide (pectin) concentration in juice generally evolves in parallel with juice turbidity. This is a good potential indicator of clarification (Robertson, 1979; Dubourdieu et al., 1981; Ollivier, 1987). When the pectin concentration in juice continually lowers during maturation, the juice is generally easy to clarify. In the opposite case, clarification is more difficult and exogenous pectic enzymes must be used.

High levels of rot in the harvest increase juice turbidity and make clarification difficult, due to the protective colloidal effect of glucan produced by *Botrytis*. A low concentration of rot (less than 5%) tends to facilitate juice clarification, due to a pectinase activity in contaminated grapes that is nearly 100 times higher than in healthy grapes.

Juice extraction methods have a prime influence on the formation of suspended solids. Slow batch pressing while minimizing crumbling obtains the clearest juices (Section 13.3.3).

The exact physical structure and chemical composition of lees remain unknown. They are made up of varying sized particles of less than 2 mm. They are generally observed to contain essentially insoluble polysaccharides (cellulose, hemicellulose, pectic matter) and relatively few nitrogen compounds, essentially insoluble proteins not utilizable by yeasts (Table 13.11). They also contain mineral salts and a significant amount of lipids—most likely from cellular membranes. This lipidic fraction contains a slightly higher proportion of unsaturated than saturated fatty acids. The principal fatty acids are linoleic acid (C18:2), palmitic acid (C16:0) and oleic acid (C18:1) (Table 13.12). They enter into the composition of membrane phospholipids of grape cells but a small proportion also exists in a free state, adsorbed to lees particles (Lavigne, 1996). These particles are definitely utilizable by yeasts.

Table 13.11. Lees composition (%) (Alexandre et al., 1994)

Component	%
Total neutral polysaccharides	71.9
Total nitrogen	2.6
Ashes	5.5
Acid polysaccharides	5.2
Lipids	7.8
Total	93

Table 13.12. Total fatty acid composition of lees (%) (Alexandre et al., 1994)

Component	%
Lauric acid C12:0	8.3
Palmitic acid C16:0	25.0
Palmitoleic acid C16:1	5.5
Stearic acid C18:0	22.2
Oleic acid C18:1	22.2
Linoleic acid C18:2	25.0

13.5.2 Influence of Clarification on Dry White Wine Composition

Winemakers have long observed an improvement in dry white wine quality resulting from proper juice clarification. Clarifying juice improves wine quality more dramatically, when compared with the unclarified juice, when there is a high concentration of suspended solids in the initial juice. Wines made from juices containing too many suspended solids have heavy, green aromas and bitter tastes. They are also more colored, richer in phenolic compounds and their color is less stable to oxidation. At the end of fermentation, they often contain reduction odors, more or less difficult to eliminate by aeration and racking. Inversely, the fruity character of the variety is more distinct and stable in wines made from clear juice. Some of these empirical observations based on tasting have been interpreted by analysis.

Since the 1960s, clarification has been known to improve the fermentation aromas of dry white wines (Crowell and Guymon, 1963; Bertrand, 1968; Ribéreau-Gayon et al., 1975). Wines made from clarified juices have lower concentrations of heavy-odor higher alcohols, and higher concentrations of ethyl esters of fatty acids and higher-alcohol acetates, which have more pleasant aromas.

Clarification also limits the concentration of C_6 alcohols in wines (Table 13.13) (Dubourdieu *et al.*, 1980). Before fermentation, juices essentially contain C_6 aldehydes (hexanal, *cis*-3-hexenal and *trans*-2-hexenal) formed by enzymatic oxidation of linolenic and linoleic acid during pressing. The detailed mechanisms of these reactions are described in Section 11.6.2. These compounds are not very soluble in juice and most likely remain partially associated with the must deposit. During alcoholic fermentation, they are systematically reduced into the corresponding alcohols by the yeast and pass into solution in the wine. The elimination of must lees therefore helps to lower vegetal aromas in dry white wines. The influence of clarification increases when pressing and handling of grapes become more brutal and maturity decreases.

Initial research demonstrating the enological value of clarification generally reported the percentage of clear juice obtained by different clarification methods but rarely specified the turbidity of the clarified juices. Yet relatively small turbidity variations have been shown to have a determinant influence on alcoholic fermentation kinetics and wine composition.

More recent research has focused on the influence of the degree of clarification on the production of off-odor sulfur-containing compounds during alcoholic fermentation and the more or less stable reduction off-odors that result (Lavigne-Cruège, 1996; Lavigne and Dubourdieu, 1997) (Volume 2, Section 13.6.2). Some heavy sulfur-containing compounds produced by yeasts increase juice turbidity (Table 13.14). Yet, considering the perception threshold and olfactory descriptors of these various compounds (Volume 2, Section 13.6.2), only methionol (methylthio-3-propanol-1), with a disagreeable odor of cooked cabbage, is significantly involved in the off-odor observed when juice turbidity exceeds 250 NTU. Methionol is stable in wine and it cannot be eliminated by racking and aeration. This troublesome

Table 13.13. Influence of must clarification on C_6 alcohols concentrations (hexanol + hexenols) in wine (Dubourdieu *et al.*, 1980)

Must treatment	Must turbidity (NTU)	C_6 alcohols in wine (mg/l)
Non-clarified must	400	2.0
Clarified must	260	1.0
Lees	ND[a]	2.1
Filtered lees	8	0.9

[a]ND = not determined

Table 13.14. Influence of must turbidity on concentration of heavy sulfur-containing compounds in wines (μg/l) (Lavigne-Cruège, 1996)

Substances	Must turbidity			Perception threshold of substance (model solution)
	120 NTU	250 NTU	500 NTU	
2-Mercapto-ethanol	113	140	179	130
Methyl-2-tetrahydro-thiophenone	102	131	191	70
2-Methylthio-ethanol	61	61	66	250
Ethyl methylthio-3-propanoate	1	2	2	300
Methylthio-3-propanol-1 acetate	5	6	6	50
Methylthio-3-propanol-1 (methionol)	1097	1958	3752	1200
Methylthio-4-butanol-2	35	66	60	80
Dimethyl-sulfoxide	363	728	1448	odorless
Benzothiazole	28	26	29	50
3-Methylthiopropionic acid	85	178	310	50

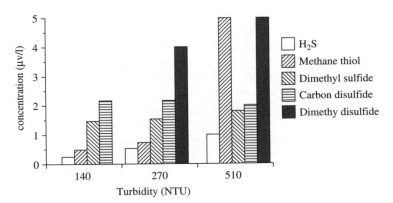

Fig. 13.10. Influence of must turbidity on the formation of light volatile sulfur-containing compounds by yeasts (Lavigne, 1996)

consequence of insufficient clarification is definitive. The degree of juice turbidity must therefore be adjusted with precision to maintain the aromatic finesse of dry white wine.

Increased methionol formation in insufficiently clarified juices cannot be interpreted as a methionine enrichment of juice by lees. In fact, the lees do not contain soluble amino acids and the acid protease is incapable of freeing them by hydrolyzing proteins in the juice. Furthermore, even prolonged contact between the juice and lees does not result in an increased concentration of amino acids.

Some experiments demonstrate the role of the lipidic fraction of the lees in the excessive production of methionol during alcoholic fermentation. This fraction most likely promotes the incorporation of methionine in yeasts which is transformed into methionol according to the Ehrlich reaction (Section 2.4.3). The lipidic fraction of the lees has also been shown to be involved in limiting acetic acid production by yeasts. The practical consequences of these phenomena are obvious. If juice turbidity is too low, an insufficient concentration of long-chain unsaturated fatty acids risks inducing excessive production of acetic acid by yeasts. If the turbidity is too high (greater than 250 NTU), an excess of these same fatty acids promotes excessive methionol formation.

Juice turbidity also influences the production of volatile sulfur-containing compounds by yeasts (Volume 2, Section 8.6.2): H_2S, methanethiol,

dimethyl disulfide, carbon disulfide. Wine made from juice fermented at 510 NTU has a pronounced reduction off-odor (Figure 13.10). In the same wine fermented at 270 NTU, dimethyl disulfide is present in concentrations above the sensory threshold but does not produce a reduction characterized flaw. However, when the methanethiol concentration exceeds its sensory threshold (0.3 µg/l), which is the case with a less clarified must, the quality of wine aroma is immediately lowered. Methanethiol thus plays a major role in reduction flaws in dry white wines following insufficient clarification. Sulfur and certain pesticide residues in the lees explain the effect of juice turbidity on the formation of light sulfur-containing compounds.

At equal turbidity levels, sulfiting also influences the production of heavy and light volatile sulfur-containing compounds by yeasts. Methionol and hydrogen sulfide, for example, increase greatly with the sulfur dioxide concentration used. This concentration must not exceed 5 g/hl, added in totality as soon as the juice is received. The free sulfur dioxide concentration should not be adjusted before alcoholic fermentation, or during or after clarification, since this practice does not provide greater oxidation protection for the juice and systematically promotes the production of sulfur-containing compounds by yeasts.

The role of clarification in fruity varietal aromas is not well known and winemakers' observations in this respect are sometimes contradictory.

Insufficiently clarified juices, especially those containing insoluble products of phenolic compound oxidation, can produce wines with decreased varietal aromas. On the contrary, overclarification (less than 50 NTU) also decreases the fruity aroma of dry white wines. This phenomenon has been observed with Muscat, Chardonnay, Sauvignon, Semillon, Mansengs, etc. It is exacerbated by difficult fermentation conditions, excessively slow fermentations with increased volatile acidity production. The varietal aroma of wines made from excessively clarified juices is sometimes masked by an artificial, banana, amylic or soapy aroma, linked to the presence of a significant quantity of esters.

It is difficult to recommend an optimum turbidity that is valid for all varieties. A range between 100 and 250 NTU is generally used, since it is a suitable compromise between a good alcoholic fermentation and aromatic finesse.

13.5.3 Effect of Clarification on Fermentation Kinetics

Slow and stuck fermentations of dry white wines are well known consequences of clarification. This phenomenon, varying in intensity according to juice composition and clarification methods, has incited much research and been interpreted differently in the past.

Clarification depletes must microflora. Inoculating juice with yeasts after clarification has long been known to hasten the initiation of fermentation but does not noticeably change its duration or the quantity of residual sugar present upon its completion (Ribéreau-Gayon and Ribéreau-Gayon, 1954). Clarification is not simply an 'unyeasting' of the juice.

A variety of physical actions also contribute to the stimulating effect of alcoholic fermentation by suspended solids. By providing nucleation sites for gas bubbles, suspended solid particles have been suggested to promote the elimination of CO_2 from the fermentation medium—thus limiting its inhibitive effect on yeasts. This effect is very limited at tank pressures found in dry white winemaking. Suspended solids are also thought

to promote yeast multiplication by serving as a support. As a matter of fact, the addition of various supports such as infusorial earth (Schanderl, 1959), bentonite (Groat and Ough, 1978) and cellulose (Larue *et al.*, 1985) improve the fermentation speeds of severely clarified musts but, at equal turbidity, does not have the same effect as fresh suspended solids.

Suspended solids also supply yeasts with nutritional elements and adsorb certain metabolic inhibitors. In fact, these two effects are related and significant. The lipid fraction of suspended solids provides the principal nutritional supply (Section 13.5.1)—in particular, long chain unsaturated fatty acids (C_{18}) that the yeast can incorporate into its own membrane phospholipids. Sugar and amino acid transport systems across the yeast membrane are consequently improved. Due to their hydrophobic lipid content, suspended solids are capable of adsorbing toxic inhibitive fatty acids freed in the juice during alcoholic fermentation (C_8, C_{10}, C_{12}). The combination of these two effects (lipidic nutrition and toxic fatty acid adsorption) produces a survival factor effect for yeasts (Section 3.5.2) (Ollivier *et al.*, 1987; Alexandre *et al.*, 1994).

In Table 13.15 and Figure 13.11, seven lots corresponding to different clarification levels were constituted from fresh Muscadelle juice. Lowering juice turbidity prolongs alcoholic fermentation. In the case of the most clarified juice (lot C), this can lead to a stuck fermentation. Supplementing lot C with either colloids or soluble macromolecules re-establishes fermentation conditions similar to lot F, the least clarified. In the final stage of fermentation, the colloids or soluble macromolecules act as survival factors (Section 3.5.2), maintaining a higher viable population (Figure 13.11). The adsorption of C_8 and C_{10} fatty acids by suspended solids is easily demonstrated in a hydroalcoholic medium model in the laboratory (Table 13.16). In the same conditions, they have a fixating capacity similar to a 0.5 g/l commercial preparation of yeast hulls (Section 3.6.2) (Lafon-Lafourcade *et al.*, 1979).

Slow alcoholic fermentations observed in extremely clarified juices are always accompanied by increased acetic acid concentrations in wines

Table 13.15. Influence of the colloidal composition of clarified Muscadelle must in different conditions on the length and completion of alcoholic fermentation (Ollivier *et al.*, 1987)

Component	Treatments[a]					
	CS	CC	CE	CN	CN + HF	CN + SM
Turbidity (NTU)	280	62	1.5	2.6	120	6.8
Proteins (mg/l)	—	506	412	356	382	548
Total polysaccharides (mg/l)	344	323	218	318	323	540
Length of alcoholic fermentation (days)	18	25	33	54	25	31
Residual sugar at the end of fermentation (g/l)	1.2	2.0	2.0	2.7	2.0	1.9

[a]CS, cold settled must; CC, coarse clarification; CE, cold settling + enzymatic clarification; CN, centrifuged must; HF, colloidal haze; SM, soluble macromolecules.

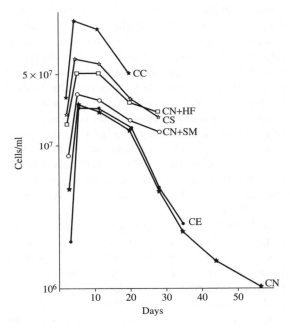

Fig. 13.11. Influence of must turbidity on yeast populations during alcoholic fermentation: CS, cold settled must; CC, coarse clarification; CN + SM, centrifuged must + soluble macromolecules; CE, cold settling + enzymatic clarification; CN: centrifuged must; CN + HF, centrifuged must + colloidal haze

(Section 2.3.4). The degree of this effect varies depending on the yeast strain used (Table 13.17). It is intensified by high sugar concentrations.

13.5.4 Clarification Methods

The most simple and effective juice clarification method is natural settling—the natural settling of

Table 13.16. Influence of the amount of haze formation on fatty acid adsorption (mg/l) after 24 hours of contact (Ollivier *et al.*, 1987)

Turbidity (NTU)	Hexanoic acid	Octanoic acid	Decanoic acid
1	5.6	10.8	4.1
15	5.6	11.0	4.3
31	5.0	10.2	3.3
62	4.4	7.5	1.5
124	5.3	8.7	2.0

Table 13.17. Influence of must turbidity on volatile acidity in wines (g/l H_2SO_4) for two *Saccharomyces cerevisiae* strains (Zymaflore VL1 and Levuline ALS-EC 8)

Turbidity	VL1	EG8
500	0.10	0.14
250	0.12	0.14
100	0.20	0.40
50	0.30	0.55

suspended solids followed by a careful racking. Free run and the first pressing on the one hand and subsequent pressings on the other are collected separately in proportionally wide tanks, preferentially by gravity, and are then sulfited. Soon after pressing, an initial clarification should be carried out to separate the gross lees already formed. The supernatant is pumped from the top of the tank. The hose is progressively lowered into the tank while the surface of the liquid (well illuminated by a hand-held lamp) is observed. The operation is stopped as soon as the hose nears the lees.

The lees from the last pressings of juice are often brown. Even after filtration, the resulting juice should not be blended with the free run: it should be fermented separately. The gross lees from the first pressings are effectively clarified by filtration, which should be carried out as soon as possible since this deposit is very fermentable. The filtrate can be blended with juice which has already undergone an initial racking.

The juice should be cooled to 5–10°C before the second sedimentation to slow the initiation of alcoholic fermentation and limit oxidation. Its duration varies, depending on the juice. With certain pressing methods, the second racking is sometimes not necessary because the juice is already sufficiently clear.

Precisely adjusting clarification levels requires the use of a nephelometer. This device should be standard equipment in every winery that produces dry white wine. A direct nephelometric measurement is much more rapid, convenient and accurate than determining the percentage of particles in conic centrifuge tubes, as recommended in some works. Figure 13.12 gives an example of corresponding values between turbidity and solid percentages. The optimum turbidity range of 100–250 NTU corresponds to 0.3–0.5% of

particles. The nephelometric measurement is much more precise than the solid percentages. In addition, the nephelometric reading is direct, whereas the solid percentage measurement require at least 5 minutes of centrifugation.

Samples are taken from the middle of the decanting tank to monitor juice turbidity evolution during clarification. When optimum juice turbidity is attained, a second racking is carried out as indicated above. Once this operation is accomplished, the turbidity of the clear racked juice must be verified. If it is too high due to error, additional settling is necessary. If, on the contrary, it is too low, fine lees must be added to the clear juice.

To obtain the highest quality juice from final pressings, suitable for blending with juice from first pressings, commercial pectinases should be used to maximize clarification. Very low turbidity, from 10 to 15 NTU, should be obtained. The light lees of these press juices are highly colored by phenolic compounds and should be eliminated. Like heavy lees from press juice, they are not worth filtering. Optimum press juices turbidity (100–200 NTU) is obtained by adding the appropriate quantity of fine lees from the natural settling of the corresponding free run.

When juice clarification is too slow, due to insufficient activity of natural grape pectinase, settling can be accelerated by using commercial pectinases from *Aspergillus niger*. These preparations should be pure and not contain cinnamate esterase activity (Volume 2, Section 8.4.3), to limit vinyl-phenol production by yeasts. Like grape pectinases, commercial pectinases have several activities—notably a pectin esterase activity that demethylizes pectic chains and an endopolygalacturonase activity that hydrolyzes osidic bonds between galacturonic residues. The use of exogenous pectinases can result in excessive juice clarification. Juice turbidity must often be adjusted after settling by adding fine lees. Juice clarity must be fastidiously adjusted to assure a complete fermentation and allow the expression of aromatic quality of dry white wines.

After settling, the juices from the last pressings must be evaluated to determine if they can be blended with first pressing juice. There are no

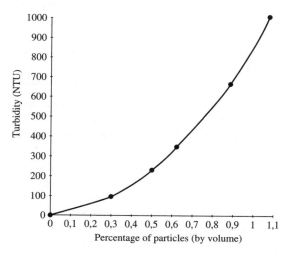

Fig. 13.12. Example of the correlation between turbidity expressed in NTU and percentage of solid particles in must

general rules to guide this decision. Tasting, evaluation of visual color (the least oxidized possible), phenolic compound index, sugar concentration and especially pH are taken into account. For example, for a Bordeaux Sauvignon, press juices with a pH greater or equal to 3.5 are not blended, but the addition of quality press juice to free run intensifies the varietal aroma of certain Sauvignon wines. Press juices should be blended before alcoholic fermentation.

Other juice clarification methods exist using more or less expensive equipment: centrifugation, filtration, tangential microfiltration and carbon dioxide or nitrogen flotation. For various reasons, these techniques do not produce as high quality wine as natural settling. Centrifugation always causes a certain amount of oxidation and filtration systems generally produce wines lacking aromatic intensity. These techniques are even less justified today, since pneumatic presses extract relatively clear juice requiring minimal settling. Replacing continuous extraction systems with pneumatic presses in large installations has made all of the mechanical clarification systems practically obsolete.

13.5.5 Clarification Methods for Lees

The volume of lees obtained after natural settling represents a sizable proportion of the harvest.

Two types of filters are used for lees filtration: diatomaceous earth rotary vacuum filters and plate-and-frame filters (using 1–1.5 kg of perlite/hl of lees filtered). These two methods extract clear juice (less than 20 NTU), without clogging, at a rate of 1–2 hl/h/m². Their recuperation rate is near 90% when the lees to be filtered contain 10% solids. There is practically no juice loss (Dubourdieu *et al.*, 1980; Serrano *et al.*, 1989). The juice obtained can be blended with clear juice from natural settling without a quality difference detectable by tasting or analysis.

Plate-and-frame filters are easier to use than rotary vacuum filters—especially for small wineries. They also have the advantage of not exposing the juice to air.

13.6 JUICE TREATMENTS AND THE ADVISABILITY OF BENTONITE TREATMENTS

Sugar concentration and acidity adjustments are described in Sections 11.4 and 11.5. They should be carried out after clarification.

In the past, bentonite treatments were also recommended to eliminate proteins in the juice—responsible for instability of juice clarity (Milisavljevic, 1963; Ribéreau-Gayon *et al.*, 1976) (Volume 2, Sections 5.5 and 5.6). The use of bentonite in juice before fermentation offers long-known advantages but also presents certain more recently discovered inconveniences.

Treating juice with bentonite is recommended for wines which are to be clarified shortly after the completion of alcoholic fermentation. Additional handling of the wine is avoided in this manner at time when the wine is thought to be more fragile. However, protein stability tests carried out on juice are imprecise and therefore not practical. The same bentonite concentration is generally used on juices from a given winery. In spite of this treatment, white wines are sometimes unstable at bottling and require an additional bentonite treatment.

If white wines are to undergo barrel or tank on lees aging (Sections 13.8 and 13.9), bentonite treatments are not recommended for two reasons:

1. Maintaining wine for several months on lees containing bentonite, with weekly stirring, has been observed to damage organoleptic quality.

2. On-lees aging naturally stabilizes white wines with respect to protein precipitation (Volume 2, Section 5.6.4). The mechanisms of this phenomenon have long occurred unnoticed. Yeast autolysis progressively releases different mannoproteins in wine with a strong stabilizing power with respect to the proteins responsible for proteic casse. On-lees maturation of White wine diminishes the bentonite concentration necessary for stabilization by a factor of 2 to 4.

After barrel or tank maturation, on-lees aged white wines are treated with bentonite at relatively low concentrations, determined by precise and reliable protein stability tests.

13.7 FERMENTATION OPERATIONS

13.7.1 Filling

Fermentors are filled with clarified juice. Approximately 10% of the tank volume is left empty to avoid the overflowing of foam (Section 3.2.5) produced during the tumultuous phase of alcoholic fermentation.

Different clarified juices must often be assembled when filling a high-capacity fermenting tank. This operation requires several elementary precautions. Before blending the clarified juices from different tanks, fine lees which have settled after racking must be reincorporated into the juice. In addition, juice that has not initiated fermentation should not be blended with fermenting juice, since the yeasts fermenting one juice produce H_2S in the presence of free SO_2 from the other juice. The initiation of fermentation must occur after the constitution of the blend.

13.7.2 Yeast Inoculation

Within the last 20 years or so, the use of active dry yeast (ADY) in winemaking has increased considerably. It has replaced the traditional practice of yeast starters in many wineries. In this formerly widespread method, a juice is strongly sulfited (10 g/hl) to eliminate spoilage yeasts and promote the growth of wine yeasts. It is then inoculated into newly filled fermentors at a concentration of 2–5% after several days of spontaneous fermentation.

The kinetics of spontaneous dry white wine fermentation are fairly haphazard. The speed and degree of completion of fermentation vary, depending on the indigenous strain present. In addition, clarification 'un-inoculates' juice to a certain degree. A slow fermentation due to a low yeast population can result. Sometimes quality wild microflora can produce spontaneous fermentations with excellent results. Selection of wine yeast strains began by isolating strains with successful spontaneous fermentations.

Today, approximately 30 active dry yeast strains belonging to *Saccharomyces cerevisiae* are used in white winemaking. They have been selected based on more or less empirical criteria of their enological aptitudes in different winemaking regions of the world.

The active dry yeast strain chosen for white winemaking has significant consequences on fermentation kinetics and the development of varietal aromas. A difficult fermentation always gives rise to dull wines lacking aromatic definition and intensity. The most important quality of a yeast strain intended for dry white winemaking is the ability to ferment completely a juice with a turbidity of between 100 and 200 NTU containing up to 220 g of sugar per liter, without excessive production of volatile acidity. This ability to ferment clarified juice is not widespread among wild yeasts, and wild yeast inoculum of spontaneous fermentations sometimes do not contain such strains.

Some strains like 71B, produce high ester concentrations—in particular, higher alcohol acetates contributing the fermentation aromas of dry white wines. Their use is only recommended for neutral grape varieties—they mask the varietal aroma of noble varieties.

Other strains, like VL1, were selected for their low vinyl–phenol production. These compounds possess rather unpleasant pharmaceutical aromas. Above a certain concentration, they dull the aroma of dry white wines (Volume 2, Sections 8.4.2 and 8.4.3). These strains have low cinnamate decarboxylase activity. During alcoholic fermentation, this enzyme catalyzes the partial transformation of p-coumaric and ferulic acid found in juice into vinyl-4-phenol and vinyl-4-gaiacol. Since this enzyme is inhibited by phenolic compounds, only white wines can contain quantities of vinyl-phenols likely to affect their aroma. The use of strains with low cinnamate decarboxylase activity is recommended—particularly for white juices containing high concentrations of hydroxycinnamic acid.

The role of yeasts in the varietal aroma of wines is poorly understood. With the exception of the terpenic aromas of Muscat varieties and the Sauvignon aroma, little research has been dedicated to the aromatic characteristics of other varieties. Yeasts have been shown to free only small amounts of free terpene alcohols from terpenic glycosides present in juice. The yeast strain used for fermenting Muscat juice, therefore, does not greatly influence the terpene alcohol composition in wine. However, many winemakers have observed the particular aptitude of certain *Sacch. cerevisiae* strains (EG8, 2056, VL3) to intensify the varietal aroma of certain grape varieties (Sauvignon, Gewürztraminer, Mansengs) (Volume 2, Chapter 7).

As a result, the winemaker must understand not only the fermentative behavior of the yeast strain used but also its effect on the specificity of the wine made. The composition of the grape; aroma precursors is responsible for the aromatic intensity of the wine. The role of the yeast is to transform this grape aroma potential into free aromas. *Ipso facto*, the aromatic character revealed varies according to the vintage and *terroir*. A good yeast strain permits the expression of the finesse and complexity of the aromatic character of the grape, but it does not detract from this character by revealing flaws, masking it with excessive fermentation aromas or caricaturing it by revealing only some of the particular nuances. If yeast strains are chosen judiciously, the use of selected yeasts will not lead to a standardization of dry white wine aromas.

White juice should be inoculated with active dry yeast at a concentration of 10–15 g/hl or 10^6 cells/ml of juice, immediately after clarification. The cells are reactivated beforehand for 20 min in a water and must mixture (1:1) at 40°C. If the must was clarified at low temperatures (10–12°C), it is not necessary to wait for the must temperature to rise before inoculating, since early inoculation guarantees the implantation of the starter.

The tank or barrel contents should be homogenized at the time of the inoculation. In this manner, the suspended solids are well blended during yeast growth. In high-capacity tanks without an agitator, the blending operation is difficult and the starter should be pumped into the lees at the bottom of the tank, as opposed to over the possibly overclarified must at the top of the tank.

13.7.3 Addition of Ammonium Salts and Juice Aeration

The general mechanisms that link the nitrogen and oxygen needs of yeasts during the alcoholic fermentation process (Sections 2.4.2, 3.4.2 and 3.5.2), explain the advisability of the addition of ammonium salts and the necessity of aeration during dry white winemaking.

Assimilable nitrogen concentrations (ammonium cation and amino acids except for proline) in white juice from cool-climate vineyards (northern and Atlantic) are generally sufficient to assure normal yeast multiplication but, even in these climates favorable to white varieties, an insufficient nitrogen supply to the vine or excessive summer dryness can sometimes result in juices deficient in assimilable nitrogen. Viticultural conditions favoring this situation are varied but always foreseeable: superficial root systems of young vines; winter root asphyxiation in poorly drained soils; light soils with an insufficient water reserve; and ground cover strongly limiting water and nitrogen supplies. The winemaker should pay close attention to these potential problems, since nitrogen-deficient white juices almost always produce heavy white wines with little fruit and aging potential.

White juices with concentrations of less than 25 mg of the ammonium cation or 160 mg of assimilable nitrogen per liter should be supplemented with ammonium sulfate. Assimilable nitrogen concentrations in suspected nitrogen-deficient white juices should be systematically analyzed using the formol index (Aerny, 1996), a simple method easily performed at the winery with a pH meter. Some white musts can contain less than 40 mg of assimilable nitrogen per liter. In these extreme cases, supplementing the must with 30 g of ammonium sulfate per hectoliter (the maximum allowable dose permitted in the EEC) is not sufficient to re-establish a suitable nitrogen supply to yeasts. Chronic nitrogen deficiencies should be corrected at the vineyard by appropriate viticultural practices.

Ammoniacal nitrogen is added either all at once at the time of inoculation or in two additions, the second occurring at the same time as the aeration on the second or third day of alcoholic fermentation. The second method sometimes permits a more rapid fermentation.

In the past, due to fears of aroma loss through oxidation, white juices were not aerated during alcoholic fermentation in high-capacity tanks. Opinions today are less categorical. The varietal aroma of white wines is affected by oxidation during pressing and draining (Section 13.4.2) but an aeration in the first half of fermentation has no effect on the fruit aroma of aromatic varieties. At this stage, the considerable reducing power of yeasts very effectively protects the aromas from oxidation. If the addition of oxygen sometimes diminishes fermentation aromas (esters and fatty acids) of dry white wines, it is caused by the resulting stimulation of alcoholic fermentation. The risk of slow or stuck fermentations, associated with strictly anaerobic conditions, is more serious than a minimum loss of transient fermentation aromas.

The juice should be aerated during pumping-over operations—maintaining suitable air and liquid contact. Oxygen gas can also be directly injected into fermenting juice by an aerating device. This oxygen addition (2–4 mg/l) should occur in the first days of alcoholic fermentation during the yeast exponential growth phase. Oxygen permits the synthesis of sterols—essential cell membrane components and yeast survival factors during the stationary phase. Aeration becomes necessary for the completion of alcoholic fermentation when juice turbidity is low and sugar concentrations are high. Adjusting juice clarity, measuring and correcting (when necessary) assimilable nitrogen concentrations and aerating at the right moment along with yeast inoculation are the principal factors governing successful fermentations in dry white winemaking.

13.7.4 Temperature Control

In traditional white winemaking, alcoholic fermentation was carried out in small containers (barrels or small tuns) located in cool cellars with temperatures between 12 and 16°C. In these conditions, the fermentation temperature remained close to the cellar temperature from throughout the fermentation. It rarely exceeded 22–25°C during the most active phase of fermentation. These barrel and tun fermentation conditions still exist in historical quality white wine regions (Burgundy, Sauternes, Graves, Loire Valley, Alsace, etc.). In the last 15 years, many wineries, in a desire to improve white wine quality, have reverted back to the age-old technique of barrel fermenting. Temperature control problems are much less of an issue in these small containers.

The temperature of high-capacity fermenting tanks must be controlled to avoid excessive temperatures during fermentation. Today, most wineries have tanks equipped with temperature control systems. A cooling system maintains water at a low temperature (4–6°C) in an insulated tank. A second system distributes the cool water through exchangers, placed inside the fermenting tank or set up as a jacket on the outside. These cooling systems were developed fairly recently. In the 1950s, the migration from small to large fermentors occurred without taking into account the heat exchange consequences of these changes.

As in red winemaking, excessively high temperatures (above 30°C) can be the cause of stuck fermentations, but this problem rarely occurs today, since winemakers have means of cooling juice before and during fermentation. They are also aware of the need to avoid excessive temperatures to limit aroma loss. Fermentation temperatures above 20°C diminish the amount of esters produced by yeasts and increase higher alcohol production (Bertrand, 1968). The effect of temperature on varietal wine aroma is much less clear. At very high temperatures (28–30°C), the rapid release of carbon dioxide entrains certain substances, causing aroma loss, but lower-temperature fermentations at 18°C or at 23–24°C do not necessarily produce wines with a significant varietal aroma difference. Fermentations at 18°C or lower are therefore not a means of enhancing the fruit character of aromatic varieties. Pre-fermentation operations and selection of yeast strains have a much greater influence. The temperature kinetics of high-capacity fermentors should simply be modeled after temperature evolution in barrel fermentations, with a maximum temperature of around 22–23°C at mid fermentation—progressively decreasing to the cellar temperature by the end of fermentation.

Untimely temperature drops should be avoided at all stages of fermentation in both barrels and tanks. For example, a tank temperature should not be lowered from 23–16°C over a few hours, to avoid subsequent temperature control. The thermic shocks that yeasts undergo in these conditions promote slow and even stuck fermentations (Section 3.7.1).

13.7.5 Completion of Alcoholic Fermentation

The duration of dry white wine fermentation depends on several parameters: juice extraction conditions; sugar and assimilable nitrogen concentrations; turbidity; yeast strain; aeration; and fermentation temperature. The winemaker can adjust and control all of them. A slow or stuck fermentation is most often the result of carelessness and always affects wine quality. The alcoholic fermentation of a white wine should not exceed 12 days. Longer fermentations should not be sought after except in the case of exceptionally high sugar concentrations.

Juice density is measured daily to monitor alcoholic fermentation kinetics. When the density drops to approximately 0.994–0.993, sugar concentrations are then measured daily to verify the completion of fermentation. Fermentation is considered complete when less than 2 g of reducing sugars per liter remain. The fermentors are then carefully topped off. Subsequent operations depend on whether malolactic fermentation is carried out.

If malolactic fermentation is not desired, the wine temperature is lowered to around 12°C. The lees are stirred daily by agitation or pumping, avoiding oxygen dissolution. This operation makes use of the reducing power of yeast lees to protect wine from oxidation. The formation of reduction odors in the lees is simultaneously avoided (Section 13.9).

After 1–2 weeks, the wine is sulfited at 4–5 g/hl. Until recently, on-lees aging in high-capacity tanks was not considered possible, due

to the appearance of reduction odors. The lees were rapidly eliminated by racking shortly after sulfiting. Today, by taking certain precautions, white wines can be on-lees aged even in tanks (Section 13.9).

13.7.6 Malolactic Fermentation

Malolactic fermentation is always sought with red wines but is practiced less often for white wines. Its use depends on the variety and wine region.

Chardonnay in Burgundy and Chasselas in Switzerland are two classic examples of on-lees aged white wines which systematically undergo malolactic fermentation after alcoholic fermentation. The primary objective of this transformation is to deacidify the wine. This is especially true of premium quality white Burgundies. Before malolactic fermentation, their total acidity can be as high as 7 g/l expressed as H_2SO_4 with a corresponding low pH. Malolactic fermentation also increases the biological stability of the wine. For example, to avoid an accidental malolactic fermentation in the bottle, champagnes undergo malolactic fermentation after alcoholic fermentation, in controlled conditions. Malolactic fermentation also contributes to the aromatic complexity of Chardonnay wines. A Chardonnay wine that has not undergone malolactic fermentation cannot be considered a great Chardonnay. Malolactic fermentation does not lessen the varietal aroma of chardonnay; on the contrary, it develops and stabilizes certain aromatic and textural nuances, making the wine more complete. Unfortunately, not fully understanding the varietal aroma of chardonnay, enology still cannot provide an explanation of these phenomena at the molecular level.

Today, many Chardonnay wines made throughout the world according to the Burgundy model undergo malolactic fermentation more for the aromatic consequences than for deacidification and stabilization. In these same cases, the juices are often acidified to be capable of malolactic fermentation. These practices may shock a European winemaker but are employed to produce a certain type of wine.

For most other varieties, such as Sauvignon, Semillon, Chenin and all Alsacian, German and

Austrian varieties, malolactic fermentation noticeably lowers the fruity character of white wines. Other methods should be used to lower acidity when this operation is necessary, but the role of lactic acid bacteria on the aroma of wines made from these varieties should be studied, at least to justify the tradition of avoiding malolactic fermentation with these varieties.

When malolactic fermentation is desired, the wines are topped off and maintained on lees after alcoholic fermentation, without sulfiting, at a temperature between 16 and 18°C. The containers must be fully topped off and the lees stirred weekly to avoid oxidation. With proper winemaking methods, in particular moderate sulfiting, malolactic fermentation spontaneously initiates after a latent phase of variable length that can be shortened with the use of commercially prepared malolactic inoculum (Section 13.7.5). Some wineries even keep non-sulfited wines, having completed malolactic fermentation, at low temperatures from one year to another to use as a malolactic starter culture. In regions where malolactic fermentation is systematically practiced, its initiation does not pose any particular problems, since the entire installation (notably the barrels) contains an abundant bacterial inoculum and malolactic fermentation is difficult to avoid. If wine begins to oxidize while waiting for the initiation of malolactic fermentation, it should be lightly sulfited (2 g/hl). Malolactic fermentation is not compromised and the aromatic character of the wine is preserved. Once the malic acid is degraded, the wines are sulfited at 4–5 g/hl and maintained on lees until bottling.

13.8 MAKING DRY WHITE WINES IN BARRELS

13.8.1 Principles

Dry white wines capable of substantial aging are traditionally fermented and matured in small containers. This practice was widespread at the beginning of the century. In France, it continued in certain prestigious Burgundy *appellations*. At the beginning of the 1980s, barrel fermenting and aging of white wines surged in popularity, affecting nearly all wine regions in the world. However, the use of barrels is not suitable for all wines; also, implementing a barrel program is difficult and very costly.

The yeasts play an essential role in the originality of the traditional Burgundy method of barrel-aging white wine. Contrary to red wine, which is barreled after the two fermentations, white juice is barrel fermented and then aged on lees in the same barrel for several months without racking. During this aging process there are interactions between the yeasts, the wood and the wine. Unknown to enology for a long time, these different phenomena are better understood today. They encompass several aspects: the role of exocellular and parietal yeast colloids; oxidation–reduction phenomena linked to the presence of lees; the nature and transformation by yeasts of volatile substances yielded by the wood to the wine; and barrel fermenting and aging techniques.

13.8.2 The Role of Exocellular and Parietal Yeast Colloids

The yeast cell wall is composed of glucidic colloids—essentially β-glucans and mannoproteins. Its detailed molecular structure is now well understood (Section 1.2.2).

The macromolecular components of the yeast cell wall, particularly the mannoproteins, are partially released during alcoholic fermentation and especially during on-lees aging. In the laboratory on a model medium, contact time, temperature and agitation of the yeast biomass promote the release of these substances (Volume 2, Section 3.7) (Llaubères *et al.*, 1987). All of these conditions occur in traditional on-lees barrel aging. A wine barrel fermented and aged on total lees with weekly stirring (bâtonnage) has a higher glucidic yeast colloid concentration than a wine fermented and aged on fine lees in a tank for the same time period (Figure 13.13). The difference in concentration can attain 150–200 mg/l.

The release of mannoproteins is the result of an enzymatic autolysis of the lees. β-Glucanases present in the yeast cell wall (Section 1.2.2) maintain a residual activity several months after

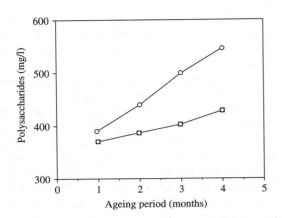

Fig. 13.13. Evolution of the total polysaccharide concentration in white wine during tank aging on fine lees (□), or barrel aging on total lees (○)

cell death. They hydrolyze the parietal glucans—anchor points of the mannoproteins released into the wine.

The direct organoleptic influence of polysaccharides on the body or fullness of on-lees aged white wines has never been clearly established, but the polysaccharides released during on-lees aging (from the yeast cell wall, for example) are capable of combining with phenolic compounds in white wines (Chatonnet *et al.*, 1992). The total polyphenol index and the yellow color thus steadily diminish in the course of on-lees barrel aging. Moreover, after several months of aging, wines that are barrel aged on total lees are less yellow than the same wine aged on fine lees in a tank (Figure 13.14). The lees limit the ellagic tannin concentration, originating from oak in particular. Tannins given off by the wood are fixed on the yeast cell walls and the polysaccharides (mannoproteins) released by the lees. A wine conserved on lees therefore has a lower overall tannin concentration as well as a much lower proportion of free (reactive) tannins (Figure 13.15).

In addition, on-lees aging lowers white wine sensitivity to oxidative pinking. This problem, characterized by a color evolution towards a grayish-pink (Simpson, 1977), occurs when wine is slightly oxidized during stabilization or bottling.

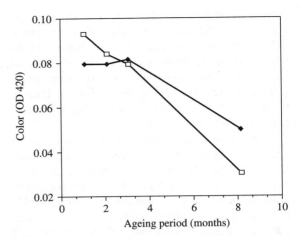

Fig. 13.14. Evolution of the yellow color (OD 420) of a wine tank aged on fine lees (♦) and aged in new barrels on total lees (□) (Chatonnet *et al.*, 1992)

Young white wines, in particular Sauvignon, whose musts were carefully protected from oxidation, are especially sensitive to this color change. The compounds involved in these phenomena are not known. Contrary to anthocyanidins, they are not discolored by varying the pH and sulfiting, but the pink color disappears upon exposure to light. Even if the pink color of these wines generally disappears after several months of bottle maturation, this problem can lead to commercial law suits. The sensitivity of white wine to this oxidation can be evaluated by measuring the difference in absorbance of the wine at 500 nm, 24 hours after adding hydrogen peroxide (Figure 13.16). This value multiplied by 100 is the sensitivity index. If it is greater than 5, there is a definite risk of pinking. Wine sensitivity to pinking remains fairly constant in the course of aging racked wine on fine lees, in barrel or in tank, but diminishes rapidly on total lees (Table 13.18). The yeast lees probably adsorb the precursor molecules responsible for pinking, but neither casein fining nor PVPP treatment is capable of significantly decreasing wine sensitivity to pinking. The addition of ascorbic acid (10 g/hl) at bottling is the only effective preventive treatment.

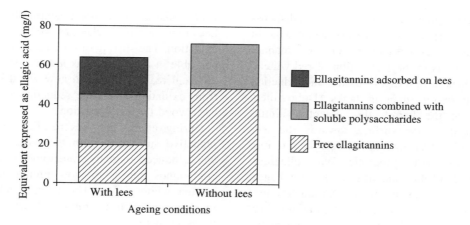

Fig. 13.15. Proportions of ellagic tannins of a barreled wine aged on lees or without lees (Chatonnet *et al.*, 1992)

Table 13.18. Evolution of the pinking sensitivity index in the same wine aged on fine lees and on total lees (Lavigne, unpublished results)

Aged	November	January	March	April
On fine lees	17	13	8	8
On total lees	17	4	0	0

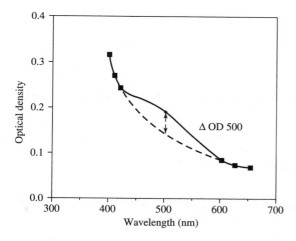

Fig. 13.16. Pinking sensitivity determination of a white wine 24 hours after the addition of hydrogen peroxide (Simpson, 1977). Dashed line = interpolation of the adsorption curve without pinking, of OD 400, 410, 420, 600, 625, 650. Δ OD 500 = difference between the calculated and measured value at OD 500. PSI (pinking sensitivity index) = Δ OD 500 $\times$ 100

The release of mannoproteins during on-lees aging also increases tartaric and protein stability in white wines (Volume 2, Sections 1.7.7, 5.6.3 and 5.6.4).

13.8.3 Oxidation–Reduction Phenomena Linked to the Presence of Lees

Maintaining white wines on total lees in tanks after sulfiting is difficult without taking certain precautions (Section 13.9). Disagreeable sulfurous odors rapidly appear, making racking necessary. With proper must clarification and sulfiting (Section 13.5.2), barrel aging permits prolonged contact with total lees without the development of reduction odor flaws. Inversely, when a dry white wine is separated from its lees and stored in new barrels, it more or less rapidly loses its fruit character and develops oxidative odors. These resinous, waxy and camphorated odors intensify during bottle aging. The lees are thus indispensable to the proper evolution of dry white wine in barrels. They act as a reducing agent, in a manner similar to tannins in the maturation of red wine.

White wines have a higher oxidation–reduction potential in barrels than in tanks (Dubourdieu, 1992). Inside the barrel, this potential diminishes from the wine surface towards the lees (Figure 13.17). Over time, the barrels seem to lose some of their oxidative properties. Wood ellagitannins, released in lesser quantities as the barrel ages, contribute to its oxidizing power. A reduction tendency consequently occurs more often in used barrels than in new barrels. Stirring homogenizes the wine oxidation–reduction potential (Figure 13.18). Lees reduction is blocked, as well as surface wine oxidation. The stirring of on-lees wines is as indispensable in new barrels as in used barrels, but for different reasons. Wine in new wood is protected from oxidation and wine reduction is avoided in used wood by this operation.

During aging, the lees release certain highly reductive substances into the wine, which limit wood-induced oxidative phenomena. These same compounds appear to slow premature aging of bottled white wines. The nature and formation mechanisms of these compounds are described in the next section.

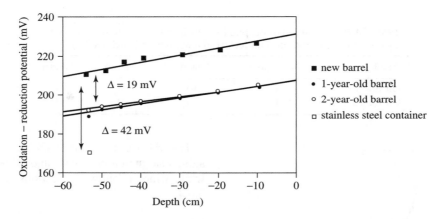

Fig. 13.17. Influence of the aging method on the oxidation–reduction potential of white wine (Dubourdieu, 1992)

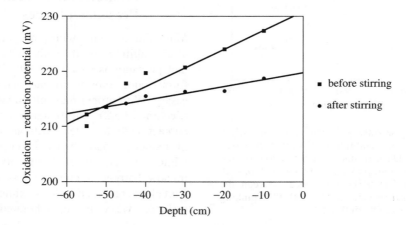

Fig. 13.18. Influence of stirring on the oxidation–reduction potential of barrel-aged white wine (Dubourdieu, 1992)

13.8.4 Nature of Volatile Substances released by Wood and their Transformation by Yeasts

Among the many volatile substances released by wood in wine, volatile phenols, β-methyl-γ-octalactones and phenol aldehydes are the principal compounds responsible for the wood aroma of barrel-aged wines. Volatile phenols, in particular eugenol, give wine smoky and spicy aromas. *Cis*- and *trans*-methyl-γ-octalactones are responsible for the coconut aroma. Volatile phenols, essentially vanillin, produce vanilla notes. Furanic aldehydes have grilled aromas but their perception threshold is much higher than concentrations found in wine. Their olfactory impact is thus negligible.

Barrel-fermented wines overall have less wood aroma than the same wines barreled after alcoholic fermentation (Chatonnet *et al.*, 1992). This phenomenon is essentially linked to the reduction of vanillin by yeasts into vanillic alcohol, which is almost odorless. Furanic aldehydes are also reduced into alcohols.

On-lees aging after alcoholic fermentation also influences the wood aroma of white wine. In terms of aging, the wood character is less pronounced and better incorporated if the wine is maintained on total lees (Chatonnet *et al.*, 1992). Yeasts are capable of fixing and continue to transform certain volatile compounds as they are released from the wood.

Late barreling (after alcoholic fermentation) and prematurely eliminating lees will produce white wines with excessive wood character. These methods have unfortunately been practiced in the past in certain regions, where winemakers have applied red wine aging techniques to white wine.

13.8.5 White Wine Barrel-Aging Techniques

Due to its high cost and the care required, the use of new barrels is only economically feasible for the production of relatively expensive, premium quality wines produced in limited quantities. Barrel aging should be applied to dry white wines capable of aging and slowly developing a bottle-aged bouquet. These wines are the most sought after by connoisseurs and fetch the best prices. After several years of aging, the wood character of these great white wines is perfectly incorporated into the overall bouquet. Their aging potential is improved, but not conferred, by the judicious use of barrels. The use of new barrels is not necessarily suitable for dry white wines intended to be drunk young and made from fruity varieties. Fruit character is less intense, as the wood masks aromatic expression. To satisfy the current (but perhaps temporary) demand for oaked wines in certain markets, it can be tempting to barrel ferment and age ordinary wines. In this case, the barrel is simply a means of compensating for an aromatically deficient wine. The widespread use of these practices may actually lower the consumer appeal of oaked wines over time.

The most popular barrels in France for white winemaking are made from fine-grain oak from forests in central France, notably Allier. In these mature forested areas, sessile oak (*Quercus sessilis*) essentially constitutes the oak population. Fine-grain oak contains much higher concentrations of odorous compounds, in particular β-methyl-γ-octalactones, than coarse-grain oak. The latter—largely pedunculate oak (*Quercus pedunculata*)—comes from isolated trees or brushwood under full-grown sessile oaks. The coarse-grain (Limousin) is less odorous, but contains much higher tannin concentrations. White wines made with Limousin oak therefore have a more pronounced yellow color and tannic character. The technique is rarely used for wines, particularly whites, but spirits are generally aged in coarse-grain wood.

Toasting, carried out during barrel production, considerably influences the aromatic impact of the wood on wine. Barrels are toasted to between medium and high so that the very fragrant fine-grain wood does not dominate the fragile aroma of white wines (Table 13.19).

Intermediate-grain wood from Burgundy may also be used for white winemaking. This wood is not very tannic; it is less fragrant than fine grain and is best adapted to medium toasting.

Fine-grain oak from central and northern Europe, in particular Russia, also produces acceptable barrel

Table 13.19. Influence of wood origin (Allier, Limousin) and toasting intensity[a] of the color and wood character of white wines (Chatonnet, 1995)

Character	Control tank	Allier				Limousin			
		L	M	H	VH	L	M	H	VH
Total polyphenols (OD 280/PVPP)	3	4	3.9	3.9	3.8	5.2	4.3	4.7	4.4
Color (OD 420)	0.1	0.12	0.13	0.13	0.08	0.42	0.47	0.48	0.48
Furfural (mg/l)	0	0.9	3.6	4.9	3.5	1.8	2.55	4.8	4.3
Methyl-5-furfural (mg/l)	0	0.8	1.1	0.75	0.5	0.9	0.95	0.8	0.4
Furylic alcohol (mg/l)	0	0.5	5.1	4.8	4.2	4	3.6	4.3	1.8
Total furans (mg/l)	0	2.2	9.8	10.45	8.2	6.7	7.1	9.9	6.5
trans-methyl-octalactone (mg/l)	0	0.13	0.17	0.053	0.037	0.067	0.051	0.023	0.012
cis methyl-octalactone (mg/l)	0	0.29	0.14	0.089	0.114	0.095	0.095	0.055	0.058
Total Methyl-octalactones (mg/l)	0	0.42	0.31	0.142	0.151	0.162	0.146	0.078	0.07
Gaïacol (μg/l)	2	10	18.5	38	65	6	12	21	33
Methyl-4-gaïacol (μg/l)	0	10	14	24	29	10	11	14	18
Vinyl-4-gaïacol (μg/l)	150	98	114	149	117	104	110	99	74
Ethyl-4-gaïacol (μg/l)	0	9	9	14	15	4	4	4	13
Eugenol (μg/l)	0	27	29	38	28	13	13	19	23
Phenol + o-cresol (μg/l)	8	25	26	47	41	26	27	17	35
p-Cresol (μg/l)	—	1	1	2	—	0	1	1	0
m-Cresol (μg/l)	—	2	3	4	—	2	2	1	0
Vinyl-4-phenol (μg/l)	300	197	206	319	210	187	211	214	131
Vanillin (mg/l)	0	0.29	0.35	0.36	0.2	0.2	0.64	0.43	0.1
Syringaldehyde (mg/l)	0	0.49	0.69	1.4	1.8	0.27	0.4	—	—
Total aldehyde phenols (mg/l)	0	0.88	1.04	1.76	2	0.47	1.04	—	—

[a]L, light toast; M, medium toast; H, high toast; VH, very high toast.

wood. It has a similar composition to that of oak grown in central France. In identical winemaking conditions, white wines made in fine-grain Russian barrels and French barrels are very similar in taste (Chatonnet *et al.*, 1997).

American white oak (*Quercus alba*) is very fragrant. It is rarely used for premium white winemaking, as excessive concentrations of β-methyl-γ-octalactone are apt to be released, totally masking the wine's character. American oak is recommended for rapidly oaking ordinary white wine.

Barrel preparation for white winemaking is relatively simple. New barrels, delivered unsulfited, are simply rinsed with cold water and drained for a few minutes before use. Used barrels, stored empty and regularly sulfured, are apt to release SO_2 into the juice during filling. As a result, abnormally high levels of H_2S are formed during alcoholic fermentation and are capable of generating strong reduction odors (Lavigne, 1996). This phenomenon is particularly pronounced if fermenting juice is barreled. Used barrels must consequently be filled with water 48 hours before use, to eliminate the SO_2 likely to be released in the fermenting juice.

The barrels are placed in a cool cellar (16°C) and are filled either before or at the start of alcoholic fermentation. A 10% headspace should be left, to avoid foam overflow during maximum fermentation intensity. The fine lees and/or the yeasts should be carefully put into suspension to homogenize the juice before barreling. After barreling, the tank dregs (lees and deposit) should be scrupulously distributed in each barrel of the lot. At the start of fermentation, barreling replaces aeration during the yeast multiplication phase. If the barrels are filled with juice before fermentation, aeration (by introducing air or oxygen) is necessary when fermentation is initiated. If these different precautions are not taken, fermentation is irregular from one barrel to another in the same lot. In barrel as in tank, difficult fermentations are often the result of human error or negligence.

As soon as fermentation is nearly complete, the barrels are topped off with juice from the same lot. Sluggish fermentations can often be reactivated by topping off the barrel with a wine lot that has recently completed a successful fermentation. This technique is equivalent to using a starter (10%) composed of a population in the stationary phase—resistant to inhibition factors. At the end of alcoholic fermentation, the barrels are stirred daily until sulfiting (Section 13.7.6). Wines undergoing malolactic fermentation are not sulfited until its completion.

During barrel maturation, stirring and topping off should occur weekly, with free SO_2 concentrations maintained around 30 mg/l.

13.9 CONTROLLING REDUCTION ODOR DEFECTS DURING WHITE WINE AGING

13.9.1 Evolution of Volatile Sulfur Compounds in Dry White Wine During Barrel or Tank Aging

Barrel-fermented and aged dry white wines are most often maintained on yeast lees during the entire aging process. In this type of winemaking, if the olfactory reduction aromas do not appear during alcoholic fermentation, they rarely occur later. Stirring wine frequently to put the lees in suspension and limited oxidation across barrel staves inhibit the formation of off-odor sulfur-containing compounds in the wine.

During the barrel aging of wine, volatile wine thiols, H_2S and methanethiol—normally present at the end of fermentation—decrease progressively (Figure 13.19). This phenomenon occurs more rapidly in new barrels, probably due to greater oxygen dissolution and the oxidizing effect of new wood tannins (Lavigne, 1996).

Despite the relative ease of barrel aging dry white wine on total lees, the winemaker must still pay close attention to the winemaking factors (clarification and sulfiting) that influence the production of sulfur-containing compounds by yeasts. In fact, even if a wine is in a new or used barrel, racking and the definitive separation of the foul-smelling lees from the wine must be carried out, if a reduction flaw exists at the end of alcoholic fermentation. The quality of barrel aging is greatly compromised, particularly in new barrels; in the absence of lees, the dry white wine is not protected from oxidation.

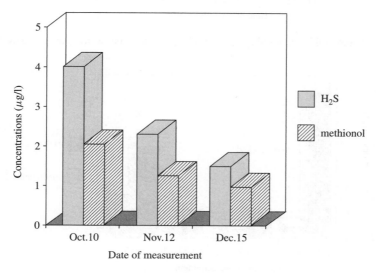

Fig. 13.19. Evolution of light sulfur-containing compounds in barreled white wine (without flaws) on total lees (Lavigne, 1996)

Controlling reduction aroma defects in dry white wines during aging in high-capacity tanks is more difficult. The presence of lees inevitably leads to the development of reduction odors within the first month of aging, whatever the reduction state of the wine after alcoholic fermentation (Figure 13.20). In most cases, tank-aged dry white wines are systematically racked and separated from their lees. In these conditions, if the gross lees are eliminated early enough, before reduction aroma defects occur, the wine can be aged on fine lees without risk. Early racking helps to stabilize light sulfur-containing compound concentrations. This opinion was clearly stated in the last Bordeaux

enological treaty (Ribéreau-Gayon *et al.*, 1976). It was founded simply on observation, without any supporting analytical data:

> The principal danger of storing white wines on yeast lees in high-capacity tanks is the development of hydrogen sulfide and mercaptan odors. But even if the presence of yeasts is not accompanied by these characteristic off-odors and tastes, their rapid elimination results in fresher and more aromatic wines which better conserve their positive characteristics.

Aeratively racking wine in the tank without separating the lees is not sufficient to avoid the

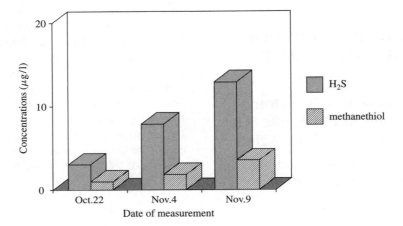

Fig. 13.20. Evolution of light sulfur-containing compounds in white wine on total lees in tank (Lavigne, 1996)

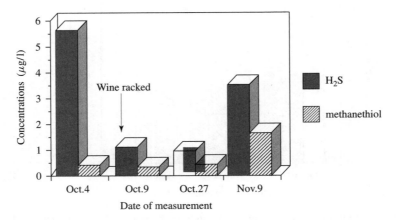

Fig. 13.21. Evolution of light sulfur-containing compounds in white wine in tank racked with its lees (Lavigne, 1996)

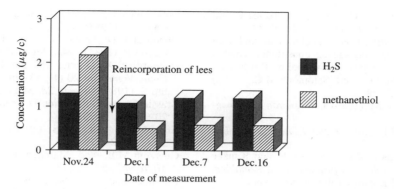

Fig. 13.22. Evolution of H_2S and methanethiol concentration in a tank aged white wine after the reincorporation of its lees (Lavigne, 1995)

development of disagreeable odors. Immediately after racking, H_2S and methanethiol concentrations diminish, but within a month, the time necessary for the lees to settle on the bottom of the tank, the defect reappears (Figure 13.21). The compacting of the lees under the pressure exerted at the bottom of high-capacity tanks seems to promote reduction phenomena in white wines after sulfiting (Lavigne, 1996).

The ability of yeasts to generate foul-smelling sulfur compounds in these conditions progressively diminishes during aging and totally disappears after a few weeks. The loss of the lees sulfitoreductase activity, catalyzing the reduction of SO_2 into H_2S, explains this development.

13.9.2 Aging Dry White Wine in a High-Capacity Tank on Lees

As long as the sulfitoreductase activity remains in yeasts, dry white wines fermented in high-capacity tanks cannot be stored on their lees without the risk of developing reduction off-odors. However, if the lees are temporarily separated from the wine until the reductive activity stops, they can be reincorporated afterwards—there is no longer a risk of generating sulfur-containing compounds (Lavigne, 1995).

In practice, wine is racked several days after sulfiting. The lees are stored separately in barrels. This initial step of the aging method stabilizes sulfur-containing compound concentrations in the

wine on fine lees, and at the same time avoids reduction odors from the gross lees. Simultaneously, H_2S concentrations progressively diminish in the barreled lees. Only one day after separation from the wine, these lees no longer contain methanethiol. After approximately one month, the lees are reincorporated into the wine. At this stage, not only do the lees no longer generate sulfur-containing compounds but their addition also provokes an appreciable decrease in the concentration of methanethiol in the wine (Figure 13.22).

The use of fresh lees is an authorized enological practice, used to correct the color of prematurely oxidized white wine. The ability of lees to adsorb certain volatile wine thiols has been discovered more recently (Lavigne and Dubourdieu, 1996).

Yeasts, taken at the end of fermentation and added to a model solution containing methanethiol and ethanethiol, are capable of adsorbing these volatile thiols. They are fixated by the yeast cell wall mannoproteins. During aeration, a disulfur bond is formed between the cysteine of the cell wall mannoproteins and the thiols from the wine.

REFERENCES

Arnold R.A. and Noble A.C. (1979) *Am. J. Enol. Vitic.*, 30 (3), 179–181.

Aerny J. (1996) *Revue Suisse Arboric. Hortic.*, 28 (3), 161.

Alexandre H., Nguyen Van Long T., Feuillat M. and Charpentier C. (1994) *Rev. Fr. Oenol.*, 145, 11–20.

Baumes R., Bayonove C., Cordonnier R., Torres P. and Seguin A. (1989) *Rev. Fr. Oenol.*, 116, 6–11.

Bertrand A. (1968) Utilisation de la chromatographie en phase gazeuse pour le dosage des constituants volatils du vin. Thèse 3ème Cycle, Université de Bordeaux.

Bisiach M., Minervini G. and Salomone M.C. (1982) *Bull. OEPP*, 12, 5–28.

Blank G. and Valade M. (1989) *Le Vigneron Champenois*, 6, 333–345.

Boidron J.N. (1978) *Ann. Technol. Agric.*, 27 (1), 141–145.

Boulton R.B., Singleton V.L., Bisson L.F. and Kunkee R.E. (1995) *Principles and practices of Winemaking*. Chapman & Hall Enology Library, New York.

Chaptal (1801) *Traité théorique et pratique sur la culture de la vigne avec l'art de faire le vin*. Vol. 2, 41–42. Delalain, Paris, reprinted by Edition d'art et technique (Morannes).

Chatonnet P. (1995) Influence des procédés et de la tonnellerie et des conditions d'élevage sur la composition et la qualité des vins élevés en fûts de chêne. Thèse Doctorat Université de Bordeaux II

Chatonnet P., Dubourdieu D. and Boidron J.N. (1992) *Sci. Aliments* 12 (4), 666–685

Chatonnet P., Sarishivilli N.G., Oganessyants L.A., Dubourdieu D. and Cordier B. (1997) *Rev. Fr. Oenol.*, 167, 46–51.

Chauvet S., Sudraud P. and Jouan T. (1986) *Bull. OIV*, 661–668, 1021–1043.

Cheynier V., Trousdale E.K., Singleton V.L., Salgues M. and Wylde R. (1986) *J. Agric. Food Chem.*, 34, 217–221.

Cheynier V., Rigaud J., Souquet J.M., Barillère J.M. and Moutounet M. (1989) *Am. J. Enol. Vitic.*, 40 (1), 36–42.

Cheynier V., Rigaud J., Souquet J.M., Duprat F. and Moutounet M. (1990) *Am. J. Enol. Vitic.*, 41, 346–349.

Crowell E.A. and Guymon J.F. (1963) *Am. J. Enol. Vitic.*, 14, 214

Dubernet J. and Ribéreau-Gayon P. (1973) *Conn. Vigne Vin*, 4, 283.

Dubernet J. and Ribéreau-Gayon P. (1974) *Vitis*, 13, 233.

Dubernet M. (1974) Recherches sur la tyrosinase de Vitis vinifera et la laccase de Botrytis cinerea. Applications technologiques. Thèse Doctorat 3ème Cycle, Université de Bordeaux II.

Dubourdieu D. (1992) *Le bois et la qualité des vins et eaux-de-vie. J. Int. Sci. Vigne Vin*, Special Issue, 137–145.

Dubourdieu D. and Lavigne V. (1990) *Rev. Fr. Oenol.* 124, 58–61.

Dubourdieu D., Hadjinicolaou D. and Bertrand A. (1980) *Conn. Vigne Vin*, 14 (4), 247–261.

Dubourdieu D., Hadjinicolaou D. and Ribéreau-Gayon P. (1981) *Conn. Vigne Vin*, 18 (4), 237–252

Dubourdieu D., Koh K.H., Bertrand A. and Ribéreau-Gayon P. (1983) *C.R. Acad. Sci.*, 296, 1025–1028.

Dubourdieu D., Grassin C., Deruche C. and Ribereau-Gayon P. (1984) *Conn. Vigne Vin*, 18 (4), 237–252.

Dubourdieu D., Ollivier Ch. and Boidron J.N. (1986) *Conn. Vigne Vin*, 20 (1), 53–76.

Fabre S. (1988) *Objectif 28*, Mars, 21–25.

Grassin C. and Dubourdieu D. (1986) *Conn. Vigne Vin*, 20 (2), 12, 5–28.

Groat M. and Ough C.S. (1978) *Am. J. Enol. Vitic.*, 29 (2), 112–119

Guerzoni M.E. and Marchetti R. (1987) *Appl. Env. Microbiol.*, 53, 35–130.

Hadjinicolaou D. (1981) Incidence des opérations préfermentaires sur la fermentescibilité des moûts et les caractères organoleptiques des vins blancs. Thèse Doctorat, Université de Bordeaux II.

Haushoffer H. (1978) *Ann. Technol. Agric.*, 27 (1), 221–230.

Hiéret J.P. (1986) *L'outillage traditionnel de la vigne et du vin en bordelais*. Presses universitaires de Bordeaux, Université de Bordeaux 3.

Lafon-Lafourcade S., Geneix C. and Ribéreau-Gayon P. (1979) *Appl. Env. Microbiol.*, 47, 1246–1249.

Larue F., Geneix C., Park M.-K., Murakami Y., Lafon-Lafourcade S. and Ribéreau-Gayon P. (1985) *Conn. Vigne Vin*, 19, 41–52.

Lavigne V. (1996) *Rev. Fr. Oenol.*, 155, 36–39.

Lavigne V. and Dubourdieu D. (1996) *J. Int. Sci. Vigne Vin*, 30 (4), 201–206.

Lavigne V. and Dubourdieu D. (1997) *Rev. Oenol.*, 85, 23–30.

Lavigne-Cruège V. (1996) Recherches sur les composés volatils soufrés formés par la levure au cours de la vinification et de l'élevage des vins blancs secs. Thèse Doctorat, Université de Bordeaux II.

Llaubères R.M., Dubourdieu D. and Villetaz J.C. (1987) *J. Sci. Food Agric.,* 41, 277–286.

Martinière P. and Sapis J.C. (1967) *Conn. Vigne Vin*, 2, 64.

Maurer R. and Meidinger F. (1976) *Deu. Weinbau* 31 (11), 372–377.

Milisavljevic D. (1963) In *1er Symposium International d'Œnologie de Bordeaux*, Editions INRA, Paris.

Moutounet M., Rigaud J., Souquet J.M. and Cheynier V. (1990) *Rev. Fr. Oenol.*, 124, 32–38.

Müller-Späth H. (1977) *Die Weinwirtschaft*, 6, 1–12.

Müller-Späth H. (1988) *Objectif 28*, Mars 21-25, 15–19.

Ollivier C. (1987) Recherches sur la vinification des vins blancs secs. Thèse Diplôme d'Etudes et de Recherches de l'Université de Bordeaux II.

Ollivier C., Stonestreet Th., Larue Fr. and Dubourdieu D. (1987) *Conn. Vigne Vin*, 21 (1), 59–70.

Ough C.S. (1969) *Am. J. Enol. Vitic.*, 20 (2), 93–100.

Ough C.S. and Berg H.W. (1971) *Am. J. Enol. Vitic.*, 22 (3), 194–198.

Pacottet P. (1921) *Vinification*, Baillières, Paris.

Peynaud E. (1971) *Connaissance et travail du vin.* Dunod, Paris.

Rapp A., Mandery H. and Niebergall H. (1986) *Vitis,* 25, 79–84.

Ribéreau-Gayon J. and Ribéreau-Gayon P. (1954) *Xème Congrès International des Industries Agricoles et Alimentaires,* Madrid.

Ribéreau-Gayon P., Lafon-Lafourcade S. and Bertrand A. (1975) *Conn. Vigne Vin,* 9 (2), 117–139

Ribéreau-Gayon J., Peynaud E., Ribéreau-Gayon P. and Sudraud P. (1976) *Traité d'Oenologie Sciences et Techniques du Vin,* Vol. 3, pp. 363–364. Dunod, Paris.

Rigaud J., Cheynier V., Souquet J.M. and Moutounet M. (1990) *Rev. Fr. Oenol.,* 124, 27–31.

Robertson G.L. (1979) *Am. J. Enol. Vitic.* 30 (2), 182–186.

Salgues M., Olivieri C., Chabas M. and Pineau J. (1984) *Bull. OIV,* 57 (638), 309–311.

Schanderl H. (1959) *Die Mikrobiologie des Mostes und Weines.* Eugen Ulmer, Stuttgart.

Schneider V. (1989) *Die Weinwirtschaft,* 1, 15–20.

Sergent D. (1996) Adaptation du Sauvignon et du Sémillon à la diversité géologique et agro-pédologique de la région des Graves. Incidences du climat et des sols sur le comportement de la vigne, la maturation des raisins et les vins. Thèse de Doctorat de l'Université Victor Ségalen Bordeaux 2.

Serrano M., Paetzold M. and Dubourdieu D. (1989) *Conn. Vigne Vin,* 23 (4), 251–262.

Simpson R.F. (1977) *Vitis,* 16, 286–294.

Singleton V.L., Sieberhagen H.A., de Wet P. and Van Wyk C.J. (1975) *Am. J. Enol. Vitic.,* 26 (2), 62–69.

Tominaga T., Masneuf I. and Dubourdieu D. (1996) In *Vème Symposium International d'Oenologie,* pp. 44–49. Tech & Doc Lavoisier, Paris.

14

Other Winemaking Methods

14.1 ROSÉ WINES

14.1.1 Definition

In many countries, especially within the EC, laws have been established to define wine. Yet red, white and rosé wines are never specifically defined, although a classification system would be particularly useful for rosé wines. In fact, at least in France, certain treatments (potassium ferrocyanide), authorized for white and rosé wines, are prohibited for red wines. In addition, each of these different kinds of wines may be subject to specific regulations.

Various characteristics give rosé wines their charm. These fruity wines have a light structure and are served chilled; they can accompany an entire meal. Although some have obtained a good reputation, they are generally not premium wines. Several winegrowing areas (e.g. Côtes de Provence) have acquired a reputation for producing fresh, fruity rosé wines. They are certainly not easy to make and are not always given all the necessary attention (Castino, 1988), nor are the best grapes usually set aside for making rosé wine. In some cases, making a rosé may be the best way of attenuating certain defects in red wine grapes, e.g. insufficient ripeness, rot, or off-flavors. Rosé wine may also be a by-product of drawing juice out of vats to enhance the concentration of the remaining red wine. In France, rosé wines are usually dry, while in other countries, notably the United States, many off-dry or sweet rosés may contain 10–20 g/l residual sugar.

Due to the diversity of grape crops and winemaking techniques used, it is practically impossible to establish a technological definition of rosé wines. Further complicating the situation, mixing red and white grapes is authorized in certain cases, but blending red and white wines is prohibited, with very few exceptions. Color is therefore the only criterion for defining rosé wines—falling somewhere between the color of red and white wines. Characteristic value ranges for the different color parameters should be established for the various kinds of rosé wines (see Table 14.1).

To a certain extent, the current trend is towards lighter rosé wines.

Rosé wines have some similarities to red wines; they are often made from red grape varieties and contain a small quantity of anthocyanins and tannins. They are also refreshing, like many white wines, and for this reason white winemaking techniques are also used in their production.

There is a large range of rosé wines in terms of color intensity and attempts have been made to characterize them by analytical parameters (Garcia-Jares *et al.*, 1993a,b). Clarets, at one end of the spectrum, are light red wines having some body; they require a short skin maceration and are softened by malolactic fermentation. After standard rosé wines, the other end of the spectrum comprises lightly colored wines (California blush wines), refreshing and light like white wines. 'Stained white wines' resemble white wines but are made from red grapes or have been in accidental contact with red wines. The expression *blanc de blanc* was created to distinguish white wines made from white grape varieties.

Although appearing slightly yellow in color, white wines made from unmacerated red grapes can contain a small amount of anthocyanins, in a colorless state due to the SO_2. A pink coloration

Table 14.1. Comparison of the composition of rosé and red wines (Blouin and Peynaud, 2001)

	Color intensity /cm	Total polyphenol index	Anthocyanins (mg/l)
Rosé	0.7–2.1	8–18	20–50
Young red wine	2.2–5.1	10–30	90–250
Red wine with aging potential	>6	>40	>350

after the addition of concentrated HCl indicates their existence. This reaction is very sensitive and clearly differentiates 'stained' wines from *blanc de blanc* wines, which do not change color. Slight contamination in white wines can occur even in correctly washed and maintained wooden containers that have contained red wine.

Many red grape varieties are suitable for making rosé wines. The Côtes de Provence vineyards specialized in rosé wine production have developed a blend of grape varieties that provide well-balanced color, bouquet, and body (Flanzy, 1998). Cinsault, Syrah, and Mourvèdre add aromatic depth, finesse, and elegance to the basic varieties, Carignan and Grenache.

Among the variety of methods for making rosé wines, immediate pressing and drawing off are the most common. Carbonic maceration (Section 12.9) is not very widely used, but it produces interesting, complex aromas in full-bodied rosé wines (Adré *et al.*, 1980). However, it frequently results in wines that are too deep in color for a classic rosé, even if the anaerobic phase is short and the temperature is controlled (35°C for 36 hours or 25°C for 48 hours), so they must be blended with lighter-colored wines.

It is important to monitor grape ripeness. Grapes intended for fine, crisp rosé wines should not be overripe (potential alcohol not exceeding 12% by volume and relatively high acidity), while those intended for fuller-bodied, softer rosés need slightly higher potential alcohol levels and lower acidity. Only healthy grapes should be used, as wine quality is likely to suffer if over 15% of the raw material is affected by rot.

It is best to harvest the grapes when the temperature is cool, at night or in the early morning, to preserve their fruitiness. Most grapes for rosé wines are handpicked, but a properly adjusted harvesting machine also gives good results.

14.1.2 Importance of Color in Characterizing the Various Types of Rosé Wines

Characterizing rosé wines by analytical parameters has long been sought. Table 14.2 indicates that

Table 14.2. Phenolic compounds, anthocyanins and color of different types of French rosé wines (Ribéreau-Gayon *et al.*, 1976)

Wine	Total phenolic compound index	Anthocyanins (mg/l)	Tannins (mg/l)	Color intensity[a]	Tint[b]	Tannin/ anthocyanin ratio
Anjou				0.10–2.00	0.50–1.80	
Béarn	7–14	14–74	150–430	0.76–1.18		4.3–10.4
Bordeaux rosé	7–11	35–41	440–850	0.69–1.67		10.0–21.1
Bordeaux clairet	10–14	115–160	720–800	1.05–1.50		5.3–6.3
Côtes de Provence (direct pressing)	7–11	14–55	80–320	0.38–1.19	0.80–1.98	5.6–15.8
Côtes de Provence (free run)	7–15	11–62	63–270	0.51–1.76	0.58–1.62	2.1–7.8
Midi (direct pressing)	10–14	13–35	180–320	0.63–1.19	0.80–1.17	5.6–15.8

[a]OD 420 + OD 520 (for 1 mm thickness).
[b]OD 420/OD 520 (for 1 mm thickness).

anthocyanin concentrations are between 7 and 50 mg/l for rosé wines obtained from directly pressed fresh grapes. For rosé wines obtained by drawing off after a short maceration, the maximum concentration is 100 mg/l.

The tannin/anthocyanin ratio permits the two winemaking methods (direct pressing and drawing off) to be differentiated. This ratio diminishes as the maceration time increases and it is higher when the grapes are directly pressed.

André *et al.* (1970, 1971) strongly insisted that color is important when evaluating rosé wines. Rosé wine color has a very large range of intensity and tint. When the color is intense, it verges on bright red; reciprocally, the clearest wines have a yellow nuance. The color intensity expresses a more or less full-bodied structure.

The same researchers discussed the results of a tasting that demonstrated the influence of color on the evaluation of rosé wines. Six rosé wines were submitted to the evaluation of a tasting group. The wines were classed according to their average score and were ranked according to three different criteria:

- a classification taking only color into account;

- a classification resulting from a traditional tasting (color, odor, taste);

- a classification effected without taking color into account (dark glasses).

The first two tastings produced similar results but the blind tasting led to very different results. With standard rosé wine, without specific character, color was the essential element of its evaluation, as long as the wine had no flaws.

Rosé wine color is directly influenced by grape variety. It depends on the anthocyanin concentration in the skins and their dissolution speed. All else being equal, Carignan produces more colored wines than Grenache, and Cinsault produces the lightest wines. The tint of the color also depends on variety. A prevailing yellow color is the result of a higher extraction of tannins with respect to anthocyanins. The color of rosé wines made from Grenache is predominantly yellow, but Carignan produces dark pink wines, with purple nuances.

14.1.3 Rosé Winemaking by Direct Pressing

This widely used method consists of using white winemaking techniques on red grapes, directly pressing fresh grapes. A certain degree of maceration is necessary to obtain color. It is done directly in the press cage, while the crushed grapes are being drained. This method therefore does not require as quick an extraction as white winemaking methods.

Pressing methods have an important effect on wine quality. Increasing the pressure, of course, increases total phenolic compound extraction. In addition, after each time that the press cake is broken up, extracted tannin concentrations increase more quickly than anthocyanin concentrations, and the yellow tint increases. The different press juice should therefore be selected in the course of extraction and be discerningly blended with the free run juice. The press juice from the last pressing cycle may be eliminated because, besides its vegetal taste, it supplies more tannins than anthocyanins.

Immediately after extraction, the juice should be protected from oxidation by sulfiting (5–8 g/hl). In theory, clarification seems less important in rosé winemaking than in white winemaking, but this practice refines wine aroma and diminishes the iron concentration. Must can be treated with bentonite. Anthocyanin fixation results in a slight color decrease but it is brighter and less sensitive to oxidation. It is not advisable to use bentonite with pectolytic enzymes.

Extreme clarification of the must is not required. As in the treatment of white must (Section 13.5), turbidity levels below 50 NTU may lead to difficulty in fermentation, while levels above 250 NTU may result in herbaceous off-odors. Low doses (0.5–2 g/hl) of pectolytic enzymes may facilitate settling.

Fining with casein, gelatin, or bentonite may be helpful in clarifying the must (Flanzy, 1998), especially when the grapes are botrytized, taking care to use small enough doses not to affect the flavor of the finished wine. Settling residues may be clarified in the same way as those from white must are clarified (Section 13.5.5).

Commercial yeasts should be selected for their fermentation capacity and performance in revealing aromas. Temperature should be maintained at approximately 20°C. The poor fermentability of certain musts may lead to problems in completing fermentation, alleviated by adding nitrogen and, especially, oxygen.

In the past, malolactic fermentation was not customary—the freshness and fruitiness of these wines was considered indispensable. Today, this second fermentation is used to make these wines fuller. It is often difficult to carry out and requires a more moderate sulfiting of the grape crop.

Rosé wines should be kept at relatively low temperatures to preserve their aromas, with an adequate dose of free sulfur dioxide (20 mg/l).

Immediately after sulfiting, wines are slightly discolored and appear more yellow; but in the long term the color is more stable, with a more affirmed pure red nuance. This type of wine is generally intended to be enjoyed young so it is important to avoid loss of color, especially as wine that has been heavily sulfured to prevent malolactic fermentation may take a long time to recover. Color may be stabilized by adding tannin extracted from grape seeds (10 g/hl).

14.1.4 Making Rosé Wines by Skin Contact or Drawing Off

Deeper-colored, fuller-bodied rosé wines are produced by leaving the skins and seeds in contact with the juice for a short time, to extract more anthocyanins and tannins. However, excessive skin contact may result in too much color, accompanied by marked astringency and bitterness.

The juice may be kept in contact with the grape solids in the press for short periods of time (2–20 hours): this technique is known as pre-fermentation skin contact. This process may also take place in a vat for a longer period (10–36 hours), then some of the juice is drawn off and fermented as a rosé wine. Skin contact is primarily aimed at making rosé wine, while drawing off is mainly intended to produce a more concentrated red wine from the remaining must in the vat, with the rosé wine made from the drawn-off juice as a by-product. In the Côtes de Provence vineyards, which specialize in rosé wine production, 40% are made using skin contact, 10% by drawing off juice, and the remaining 50% by pressing the red grapes immediately (Masson, 2001). Pre-fermentation skin contact is known to enhance softness and fruit, while reducing acidity.

In both cases, the crushed, stemmed, sulfited grapes are transferred either directly to a pneumatic press with the drains closed, or to a vat. After a variable period on the skins (2–36 hours), the juice is separated from the solids, either by pressing or by drawing off all or part of the liquid from the vat. The must is then fermented in the same way as juice from grapes that have been pressed immediately.

In certain cases, only part of the juice (10–20%) may be drawn off from the vat. Once a certain

Table 14.3. Comparison of different rosé wines made from the same grapes (Sudraud *et al.*, 1968)

Winemaking method	Total phenolic compound index	Anthocyanins (mg/l)	Tannins (mg/l)	Color intensity[a]	Tannin/ anthocyanin ratio
Direct pressing	6	7	100	0.41	14.3
Maceration for 12 hours:					
Without SO$_2$	11	26	320	0.52	12.3
SO$_2$ at 10 g/hl	16	100	760	1.53	7.6

[a]See Table 14.1.

volume of juice has been drawn, it is refilled with crushed grapes. This technique is not only to produce a rosé wine but also to enhance the phenolic content and color of the remaining red wine, by increasing the solid/liquid ratio in the vat. In some cases, the main purpose of drawing off is to improve the quality of the red wine.

Contact time, temperature and sulfiting are factors that influence phenolic compound dissolution and color in rosé wines (Castino, 1988). Sulfur dioxide is known to have a certain dissolvent power (Section 8.7.5). It is not manifested during traditional red winemaking, due to the preponderant effects of other factors (duration, temperature and pumping-over). Yet when maceration is limited, the effect of sulfiting is obvious. Table 14.3 shows the impact of the winemaking technique on the color intensity and phenolic compound concentrations of rosé wines. Sulfiting promotes anthocyanin dissolution and color enhancement. It is not easy to control the conditions that will produce the required color and phenolic structure, as they depend on the specific characteristics of the wine.

The success of such rosé winemaking is based above all on the use of healthy and perfectly mature quality grape varieties. Malolactic fermentation is a general practice and becomes all the more necessary as the maceration phenomenon increases. A low acidity softens the tastes of the tannins.

14.2 BOTRYTIZED SWEET WINES (SAUTERNES AND TOKAY)

14.2.1 Introduction

Due to their lack of tannins, white wines tolerate a large diversity of structure such as varying acidities

and the presence of carbon dioxide (sparkling wines) or sugar (sweet wines). Most sweet wines are white wines. Only a few special red wines, or fortified red wines, such as port, contain sugar, but they receive additional alcohol.

Sweet wines correspond to an incomplete fermentation, leaving a certain proportion of grape sugar that has not been transformed into alcohol. Wines are arbitrarily judged to be semi-dry, sweet and syrupy sweet (*liquoreux*) according to their sugar concentration: up to 20 g/l, 36 g/l and above 36 g/l, respectively. The sugar concentration is sometimes expressed in potential alcohol: for example, $12 + 2$ signifies a wine containing 12% volume alcohol and 36 g of sugar per liter (2% volume of potential alcohol).

Semi-dry and sweet winemaking are fairly similar to dry winemaking, but the grapes must have a sufficient sugar concentration and the fermentation must be stopped before completion, either naturally or by a physical or chemical process.

Syrupy sweet wine making is different. The required high sugar concentration cannot be attained during maturation. Certain processes must concentrate the juice and certain winemaking steps are unique to these wines.

Drying, freezing and noble rot are used to concentrate juice. Due the importance of noble rot, it will be discussed in more detail later in this section.

Grapes can be dried naturally by the sun, when left on the vine, or by artificial heating (Section 11.2). This overripening process can be used to make different types of wine. The drying of the grapes results in a varying degree of concentration, but the enzymatic systems of the fruit permit a greater concentration of sugars than acids. Grapes can also be dried for up to several months in a closed room which may or may not be heated. This method results in musts containing up to 400 g of sugar per liter, capable of producing syrupy sweet wines. The fermentation of these juices is difficult and the price of these wines is high, due to volume loss and production costs.

Freezing grapes on the vine produces ice wine (Eiswein), well known in Alsace and Germany. The grapes are left on the vine until the winter frosts. The temperatures, -6 to $-7°C$, lead to the partial freezing of the least ripe grapes. By pressing the grapes at low temperatures, only the juice from the ripest grapes, containing the most sugar, is extracted. Cryoextraction (Section 14.2.4c) seeks to reproduce this natural process artificially. Making such wines is to subject to weather conditions and not always possible every year. The method is difficult and expensive, and should only be used to make premium wines.

14.2.2 Noble Rot

The biology of *Botrytis cinerea* and its development in the form of noble or vulgar rot have been described in (Section 10.6). This overripening process, noble rot, permits the production of great botrytized sweet wines. These exceptional wines can only be made in specific conditions. Their production is therefore limited.

The Sauternes-Barsac region is certainly one of the most highly esteemed areas for noble rot sweet wines but other regions exist in France (Loupiac, Sainte-Croix du Mont, Monbazillac, Anjou), in Germany (Moselle) and in Hungary (Tokay).

Noble rot presupposes fungus development on perfectly ripe grapes. Sémillon and Sauvignon grapes must attain 12–13% vol. potential alcohol and have a pH of less than 3.2 before any fungal development. At this time, the berries are golden with slightly brown thick skins. This result can only be attained on certain *terroirs*, with low crop yields (40–45 hl/ha), before berry concentration by noble rot. Mycelial filaments penetrate through microfissures and decompose the grape skin. This decomposition is the result of an intense enzymatic maceration of the grape skin. The grape then attains the *pourri plein* (full rot) stage and has a

brownish color. The berry does not burst: it maintains its shape but the skin no longer maintains the role of a protective barrier from the external environment. The berry acts like a sponge and is concentrated as the water evaporates. The grapes are harvested when they attain the *rôti* stage. The concentration in the berry leads to an increase in internal osmotic pressure that causes the death of the fungus. The second phase of *Botrytis cinerea* development must therefore occur soon after the full rot phase, before subsequent *Botrytis* development can result in gray rot. The distinction between noble rot and gray rot is not always obvious.

Late-season weather and harvesting conditions are essential for noble rot development. Maximum grape concentration should not always be sought, because gray rot may develop when conditions are not ideal. Attempts have been made to harvest at the full rot stage and then concentrate the grapes by artificial means.

Alternating humid and sunny periods are essential for reaching the perfect state of maturation. Gray rot occurs when *Botrytis cinerea* develops in extremely humid conditions. The development of noble rot requires a particular climate, ideally with morning fogs to assure fungal growth, followed by warm afternoon sunshine to concentrate the grapes, for a relatively long period of 2–4 weeks. In Bordeaux vineyards, these meteorological conditions correspond with the establishment of a high pressure ridge extending the anticyclone from the Azores to the north-east. Noble rot can also develop rapidly in the Gironde region after a short period of rain, caused by oceanic depressions, followed by a sunny and dry spell (low humidity, 60%) with winds from the north to north-east. This type of weather is generally associated with the presence of an anticyclone in north-eastern Europe.

Noble rot develops progressively on different grape clusters and even on different grapes on the same grape cluster. The grapes must therefore only be picked when they attain their optimum maturation state. Selective harvesting ensures that grape-pickers only remove the noble-rotted grape clusters or grape cluster fractions during each picking. Climatic conditions dictate the number of selective pickings each year—up to three or four.

Harvesting can continue until november in the north hemisphere. Due to the evolution of grape crop quality from one day to another according to climatic conditions and the evolution of rot, juice should be vinified separately according to harvest date.

Vinegrowing for the purpose of making botrytized sweet wines requires more meticulous care than for making dry white wines. This is particularly the case in oceanic climates, favoring the early implantation of *Botrytis* on sensitive varieties. In the Sauternes region, Sémillon and Sauvignon require shorter cane pruning and the rigorous control of vegetative growth, in particular early deleafing (before *véraison*) in the grape zones.

Variable proportions of bunch rot can coexist with noble rot. The involvement of acetic acid bacteria results in the production of volatile acidity. Gray rot can also form odors and tastes resembling mushrooms, mold, iodine and phenol. In addition, considerable amounts of carbonyl-based compounds may be produced, which bind sulfur dioxide and make these wines difficult to stabilize.

They are produced by acetic bacteria in the *Gluconobacter* genus present on the grapes. 5-Oxofructose is one of the main substances responsible for this phenomenon (Sections 8.4.3 and 8.4.6).

14.2.3 The Composition of Musts Made from Grapes Affected by Noble Rot (Section 10.6.3) and the Resulting Wines

Noble rot can reduce crop volume by up to 50%. Low crop yields of 15–25 hl/ha affect grape quality.

The fungus consumes a large quantity of grape sugar to assure growth. According to Ribéreau-Gayon *et al.* (1976):

> 50% more sugar and twice the juice volume is obtained from healthy grapes with respect to noble-rotted grapes, for the same acreage. Natural concentration and the resulting considerable crop yield losses are responsible for these richer and better quality wines.

This fungus also consumes a considerably higher proportion of acid than sugar. This phenomenon is beneficial to wine quality, since the acidity increases much more slowly than the sugar concentration. It is not a simple concentration by water evaporation, but rather a biological deacidification. *Botrytis cinerea* has the rare property of degrading tartaric acid: its concentration decreases more than the malic acid concentration and the pH consequently increases by 0.2 units.

Table 10.11 takes the modifications caused by noble rot into account. The lower sugar and acidity concentration due to weight and volume loss in the 1000 berries corresponds with a significant increase in the sugar concentration and a slight decrease in acidity of the juice.

Botrytis cinerea forms glycerol and gluconic acid from sugar. These two compounds play an important role in characterizing rot quality. Glycerol is produced at the start of *Botrytis* development. Its concentrations increase as the rot becomes more noble. Gluconic acid is formed much later and corresponds with a poor rot evolution. Wines made from healthy grapes should contain less than 0.5 g of gluconic acid per liter, noble rot wines between 1 and 5 g and gray rot wines more than 5 g. The higher the glycerol/gluconic acid ratio, the better the rot quality is.

The development of *Botrytis cinerea* also results in the production of two polysaccharides (Dubourdieu, 1982). One, with a complex structure, has antifungal properties and inhibits alcoholic fermentation. The other is a β-glucan with colloid protector properties and it impedes clarification of new wine. It is produced inside a viscous gel, located between the skin and the pulp. Its level of diffusion in the juice is related to grape handling and treatment conditions. All brutal mechanical operations, such as crushing, pumping and pressing, diffuse glucan in the juice and make the wine more difficult to clarify.

Botrytis cinerea development corresponds with enzymatic changes in the grape. Laccase, an oxidation enzyme, replaces the grape's tyrosinase and is much more active on phenolic compounds than the latter. Noble-rotted musts are probably relatively well protected from oxidation, since most phenolic substrates of the grape are already oxidized by the time of harvest. Specific varietal aromas seem to be fairly well protected from oxidation during noble rot development (Section 13.2.1). In addition, *Botrytis* synthesizes a number of enzymes (cellulase, polygalacturonase), permitting its penetration into the grape. An esterase that hydrolyzes fermentation esters has been isolated, thus accounting for the differences in aroma between wines made from grapes affected by noble rot and from other white wines.

The molecule sotolon also participates in the aroma (roasted, crystallized fruit, honey) of noble rot wines (Section 10.6.4).

Another characteristic of wines made from grapes affected by noble rot is their relatively high volatile acidity level. Its origin can be accidental, due to the presence of lactic acid bacteria in juice and especially acetic acid bacteria on grapes. Yeasts also form some volatile acidity, due to the corresponding fermentation difficulties of these juices (Section 2.3.4).

Operations likely to reduce the production of volatile acidity include (Section 14.2.5) supplementing the must with ammoniacal nitrogen, which should be adjusted to 190 mg/l at the start of fermentation, combined with seeding with a suitable yeast and aeration. The aim is to increase the cell population to a maximum, as this minimizes the formation of volatile acidity. For this reason, European Union legislation has specified higher limits for volatile acidity in botrytized wines. Lafon-Lafourcade and Ribéreau-Gayon (1977) sought to specify the origin of this volatile acidity. Two lactic acid isomers and ethyl acetate were analyzed in wines. Lactic acid bacteria produce large quantities of the former and acetic acid bacteria the latter. The authors concluded the following:

- If D (−)-lactic concentrations are more than 200 mg/l, lactic disease may exist.

- If L (+)-lactic concentrations are more than 200 mg/l, malolactic fermentation is responsible.

- If ethyl acetate concentrations are more than 160 mg/l, acetic acid bacteria are involved.

If these three parameters are less than the indicated values, the yeasts are entirely responsible for the volatile acidity. Volatile acidities between 0.9 and 1.3 g/l in H_2SO_4 (1.1–1.6 g/l in acetic acid) of many Sauternes wines were found to be produced exclusively by yeasts, with no bacterial involvement.

14.2.4 Noble Rot Juice Extraction

(a) Pressing Grapes

The general rules of white winemaking should be followed when transporting noble-rotted grape crops. The depth of grape crops, in particular, should be kept to a minimum during transport, to avoid spontaneous crushing.

Upon their arrival at the winery, the grapes should be manipulated with care. Rough handling provokes the excessive formation of suspended solids and vegetal tastes. It would also make the wines more difficult to clarify, because of the diffusion of glucan from *Botrytis cinerea* in the juice. Winemaking techniques should make use of gravity, as much as possible. Pumps on self-emptying gondolas are often too brutal.

Noble-rotted grapes are generally crushed and macerated in the liberated juice. This operation helps to extract the sugar. The grapes are not destemmed to facilitate the circulation of juice in the pressed skins, but manual destemming is sometimes practiced after the first pressing with the stems. In this manner, the tannic and vegetal substances of the stems are not conferred to the must during subsequent pressings at high pressure. The extraction of noble-rotted juice is difficult: a high pressing pressure must be used and the press cake must be broken up between pressings.

Due to their high viscosity, these grape crops cannot be drained before pressing. They are directly transferred to the press cage by gravity or a conveyor belt. Pressing is certainly the most difficult and essential operation in botrytized sweet winemaking, and incorrect operation of the presses can sacrifice quality. These operations in particular must be carried out slowly and delicately.

Continuous presses should not be used. They are too brutal even when equipped with a large diameter screw turning very slowly. They shred the grape crop and produce juices with a lot of suspended solids, and glucan-rich wines that are difficult to clarify.

The older vertical cage hydraulic presses are effective and produce few suspended solids in the juice. The press cake is broken up manually with these presses and the pressing cycles are slow. Moving-head presses are therefore preferred because of their ease of operation and satisfactory yield. After two to three pressing cycles, the last cycle can be effected by a hydraulic press, which permits higher pressing pressures. Pneumatic presses, used for other types of grape crops, are not well adapted for pressing noble-rotted grapes. Their pressing pressure is insufficient. Although manufacturers have created special models with up to 3 bars of pressure, these pneumatic presses do not properly extract the juices with the highest sugar concentrations, above 22 or 23% volume of potential alcohol.

Selecting the best juice from the various pressing cycles is a difficult problem. With healthy grapes, the drained juice and the juice from first pressing cycles are the richest in sugar; the last pressing cycles are vinified separately. With noble-rotted grapes, the juice with the highest sugar concentration and the highest iron and tannin concentrations is the most difficult to extract: it is released in the last pressing cycles. The addition of this juice to the blend can therefore improve wine quality but should be done prudently.

(b) Cold Pressing (Cryoextraction)

Cryoextraction permits the selection of the most sugar-rich and therefore ripest grapes. Chauvet et al. (1986) first proposed this technique.

Cryoextraction is based on a law of physics. Raoults's law states that the freezing point of a solution lowers as the solute concentration increases. When the temperature of a white grape crop is lowered to 0°C, only the grapes containing the least amount of sugar are frozen. By pressing at this temperature, a selected juice is obtained which represents only part of the total juice volume. The potential alcohol content of this juice, however, is higher. By further lowering the temperature of the grape crop and the pressing, the number of frozen berries is increased. As a result, the selected must volume is further diminished and the potential alcohol content increased. Table 14.4, which gives the results of an experiment, specifies the increase in potential alcohol content for grape crops pressed at different temperatures. The selected must volume represents between 60 and 80% of the total volume, depending on the temperature.

The selection of the most sugar-rich and thus ripest grapes by cryoextraction is the primary cause of improved wine quality. Freezing may also concentrate grapes diluted by rainy harvest seasons. A phenomenon called supraextraction is also involved. Freezing and thawing grapes results in tissue destruction and a better sugar extraction.

In practice, the grape crop is collected in small containers and frozen in a walk-in freezer at varying temperatures as low as −16°C. The grapes are then pressed in a horizontal moving-head press or in a pneumatic press. The freezing point of the richest grapes is sufficiently low that their juice may be extracted without difficulty. Table 14.5

Table 14.4. Effect of pressing temperature (cryoextraction) on the volume and alcohol potential (% vol.) of selected and residual musts from noble rot grapes (Sauternes) (Chauvet et al., 1986)

Control musts: potential alcohol (% vol.)	Pressing temperature (°C)	Selected musts		Residual musts		Volume ratios selected must/ blended must
		Volume (hl)	Potential alcohol (% vol.)	Volume (hl)	Potential alcohol (% vol.)	
16.7	−5	116	19.4	19	11.3	86
16.0	−6	81.5	19.6	21.5	12.3	79
16.0	−8.5	74.5	20.7	33	9.8	69
16.2	−10	65	22.0	37	13.1	64
16.0	−11	67	23.2	33	10.7	67
17.0	−15	60	23.4	40	12.6	60

Table 14.5. Improvement of Sauternes must quality in 1985 by using cryoextraction (Chauvet *et al.*, 1986)

Vineyard	Grape temperature at pressing (°C)	potential alcohol (% vol.)	
		Control must	Selected must
A	−13	15.9	21.5
B	−9	13.2	19.8
C	−7	13.0	19.0
D	−7	13.6	20.3
E	−9	18.0	23.3

gives the results obtained for five Sauternes vineyards in 1985. Installing a cryoextraction system is expensive but energy costs are reasonable. In any case, this technique can only be applied for limited grape crop volumes.

The improvement in must quality is not only limited to an increase in potential alcohol content. These musts have the same characteristics as their late-harvested counterparts: they are more difficult to ferment; the risk of volatile acidity production is greater; and the sulfur dioxide binding rate is higher. However, these cryoextraction selected wines are clearly preferred over the control wines (Chauvet *et al.*, 1986) due to their increased concentration, finesse and distinctiveness.

This technique is applicable to all types of white grape crops but is especially interesting for noble-rotted grape crops. Excessive humidity during overripening can prevent the grapes from obtaining the desired concentration levels. Prolonging the concentration phase would only increase grape degradation and lower crop quality, but, by picking early, cryoextraction can be used to eliminate excess water and select the grapes capable of producing quality botrytized sweet wines. The quantity is diminished but quality is maintained. This method can complement manual sorting during the harvest, but cannot replace it.

(c) Sulfiting Juice

Sulfiting has several effects during white winemaking:

1. It protects against oxidation by inhibiting laccase, produced by *Botrytis cinerea*. Due to the

considerable oxidation of grape phenolic substrates during overripening, however, the oxidation risks are less significant than one would think.

2. By blocking fermentation for several hours, it permits a coarse clarification by sedimentation.

3. It is also capable of inhibiting the development of acetic bacteria, often present in large quantities on botrytized grapes, as well as lactic bacteria, responsible for malolactic fermentation. Finally, sulfiting restricts the proliferation of some spoilage microorganisms, particularly apiculated yeasts that "waste" sugar and form unwanted by-products in the wine.

4. It destroys antifungal substances and thus facilitates alcoholic fermentation.

Sulfiting juice intended for botrytized sweet wine production has often been criticized. This operation leads to increased concentrations of bound sulfur dioxide, which remains definitively in the wine. Subsequent SO_2 additions must therefore be limited to remain within legal total SO_2 limits, thus compromising the microbiological stabilization of wine. In practice, this inconvenience of sulfiting is attenuated by the fact that only 40–60% of the SO_2 added in juice is found in the bound form in wine. The rest is oxidized into SO_3.

To conclude, a light sulfiting at 3–5 g/hl, for example, is generally recommended at this stage.

(d) Juice Clarification, Bentonite Treatment and Juice Corrections

Clarification is an indispensable operation in white winemaking, but its advantages are more disputed for botrytized sweet wines. It has been blamed for making wines thinner. This criticism seems excessive—in most cases, clarification improves aromatic finesse and gustatory qualities. However, the natural settling of botrytized musts is difficult to effect: the suspended solids and the highly dense juice have similar specific weights. Must viscosity is also high due to the high sugar concentration and the presence of glucidic colloids produced by

Botrytis cinerea. In practice, natural settling for 18–24 hours results in partial clarification, permitting large particles to be eliminated. The addition of pectolytic enzymes to noble-rotted juice presents no clarification advantages. Pectinases are secreted by *Botrytis cinerea* in the grape and are found in abundant quantities in the juice. The average pectinase activity ratio between healthy grapes and noble rotten grapes is estimated to be 1:10. Natural settling for 3–4 days at 0°C permits a more effective clarification. This kind of clarification should only be used when there are doubts about rot quality.

As with dry white winemaking, overclarification can lead to large fermentation problems and increased acetic acid production. Must turbidity should not be as low as in dry white winemaking (100–200 NTU); 500–600 NTU or even a slightly higher turbidity is perfectly acceptable. Moreover, botrytized sweet wines are not subject to the same problems related to insufficient clarification as dry white wines—the development of reduction odors and vegetal tastes, oxidability, etc.

Bentonite has been widely used as a settling aid for juices in large fermentors during white winemaking but is no longer used to make barrel-fermented wines that are aged on lees. The effectiveness of this treatment in the making of noble-rotted sweet wines has long been questioned. The high colloid concentration in these wines appears to diminish the adsorbing power of bentonite with respect to proteins. Moreover, adding bentonite before fermentation as opposed to after racking does not facilitate sedimentation. The presence of bentonite can also make racking, following the completion of alcoholic fermentation, more difficult. For these different reasons, wines are treated with bentonite a few months before bottling, after a proteic instability laboratory test.

Depending on the legislation, various adjustments such as the addition of sugar or modification of acidity can be made to juice before the initiation of fermentation. These adjustments should be limited to avoid disequilibrating the wine.

The addition of ammoniacal nitrogen often promotes the fermentation of these musts, depleted by the development of *Botrytis cinerea*. An addition of 10–15 g of ammonium sulfate per hectoliter, 25–40 mg of ammonium ion (NH_4) per liter, is generally recommended.

The addition of 50 mg of thiamine per hectoliter is often beneficial. It is not only a growth factor but can also limit the combination rate of sulfur dioxide during wine storage by promoting the decarboxylation of ketonic acids (pyruvic acid and α-ketoglutanic acid) (Lafon-Lafourcade *et al.*, 1967).

When dry white wines and noble-rotted wines are made in the same winery, adding a small quantity of lees from healthy grape must, after clarification, can considerably improve the fermentability of noble-rotted musts and diminish the production of volatile acidity (Dubourdieu, unpublished results). Depending on the amount available of the healthy grape must sediment suspension, 2–4 liters of sediment should be added to each barrel (1 to 2%), resulting in a turbidity increase of 200–400 NTU. The sediment used should be sulfited, unfermented and conserved at a low temperature.

14.2.5 Fermentation Process

(a) Fermentation Difficulties

Noble-rotted musts are known to be difficult to ferment. The high sugar concentration is the principal limiting factor but nutritive deficiencies provoked by the growth of *Botrytis cinerea* are also responsible. Among possible deficiencies in the must, nitrogen is a key factor. Masneuf *et al.* (1999) analyzed 32 musts intended for dry white winemaking and reported a mean available nitrogen content of 182 mg/l, measured by formol titration. The same analysis of 20 samples of botrytized must intended for sweet white wine production yielded a mean value of 84 mg/l only (Bely *et al.*, 2003). Polysaccharidic antifungal substances affecting the fermentation process are also involved. For an identical sugar concentration, juice is more difficult to ferment when grapes are deeply attacked by *Botrytis cinerea*.

It is advisable to adjust the ammonium sulfate content to 190 mg/l at the beginning of fermentation to facilitate the process and minimize the production of volatile acidity. Later addition of nitrogen supplements is likely to have a negative impact.

Inoculation of the juice is strongly recommended. The chosen yeast strain should be highly ethanol tolerant and should produce little volatile acidity in difficult fermentation conditions. The dry yeasts should not be introduced directly into the juice to start the fermentation for this type of winemaking. A yeast starter should be prepared in diluted must, supplemented with NH_4^+ and yeast hulls, then seeded with dried active yeast at a dose of 2.5 g/hl of the total volume to be inoculated. The starter is added to the must on the second day of fermentation, at a rate of 2% of the total volume. This increases the maximum yeast population, which controls fermentation rate and volatile acidity production (Section 2.3.4). In one experiment, adding yeast in this way reduced the final volatile acidity content by 20%.

Noble-rotted wines were traditionally fermented in small wooden barrels. This method creates favorable conditions for this kind of winemaking. The juice from each day of harvesting can be separated according to quality. Temperatures are also better controlled, since the fermentation occurs at near ambient temperatures, but the cellar may need to be heated if juice temperatures are too low or if the ambient temperature drops too low. Barrel fermentation also permits a continuous microaeration, promoting yeast activity and a complete fermentation. Barrel-fermented musts with high sugar concentrations produce more alcohol than the same must in a large fermentor, because the fermentation stops earlier in a tank. Furthermore, the presence of carbon dioxide protects fermenting must from oxidation, and phenolic compounds, the main oxidation substrates, are destroyed in the grapes by *Botrytis cinerea* as it develops.

Tank-fermented musts have an even greater need for aeration, since this fermentation occurs in stricter anaerobic conditions. In sweet winemaking, oxygen should be introduced during the stationary phase of the yeast population growth cycle, rather than during the growth phase, as is the case when making other types of wine (Section 3.7.2).

Later aeration has been shown to prevent excessive increases in volatile acidity, which mainly occurs in the early stages of fermentation. Temperature control is also indispensable in these conditions to ensure that the fermentation temperature remains within reasonable limits (20–24°C).

Rapid and vigorous fermentations result in less aromatic wines and should be avoided, but exaggeratedly slow fermentations are not a factor for quality.

(b) Stopping Fermentation (*Mutage*)

Sugar concentration and alcohol content determine the gustatory equilibrium of this kind of wine. The sweetness of sugar must mask the burning characteristic of alcohol. Reciprocally, the latter must balance the heaviness of a high sugar concentration. For this type of wine, the alcohol content and potential alcohol strength should approximately satisfy the following relationship: 13 + 3, 14 + 4, 15 + 5. This equilibrium is generally obtained by blending several wine batches, some containing more sugar, others more alcohol.

Fermentation rarely stops spontaneously at the exact alcohol/sugar ratio desired. In some cases, it can go too far; in other cases, it becomes excessively slow and the increase in volatile acidity is more significant than the decrease in the sugar concentration. At this point, the fermentation must be stopped. This operation generally consisted of adding sulfur dioxide to the wine.

Sulfurous gas was traditionally added directly to warm wine, since the yeasts are more sensitive to SO_2 at higher temperatures. More recently it has been supposed it is better, before adding SO_2, to wait the wines were allowed to cool and a portion of the yeasts was eliminated by racking. For a given antiseptic concentration, the lower the initial population, the smaller is the residual population. The wine should be protected from air while racking, if possible by an inert gas atmosphere, to avoid the oxidation of ethanol and the formation of traces of ethanal, which combines strongly with SO_2.

In any case, a massive concentration of sulfur dioxide (20–30 g/hl) should be added, to block all yeast activity instantaneously and avoid even limited ethanal production by the yeasts. Several days after the addition, the free SO_2 concentration should be verified. A value of 60 mg/l is suitable for storage. Adjustments can be effected at this time if necessary.

14.2.6 Aging and Stabilization

The organoleptical quality of botrytized sweet wines improves considerably after several months of barrels maturation and several years of bottle aging. The bouquet takes on finesse and complexity—reminiscent of confect fruit and toasted almonds. The wine becomes harmonious on the palate. The sweetness is perfectly balanced by the alcohol and a note of acidity gives a refreshing finish. The wines are not heavy and syrupy, in spite of their high sugar concentration. These transformations remain poorly understood even today. They occur in conditions, particularly oxidation–reduction conditions, reminiscent of red wine maturation and aging.

Premium botrytized white wines are barrel-aged for 12–18 months, sometimes even 2 years or more. The bungs are maintained upright. As a result, the barrels must be topped off once per week during this maturation process. The barrels are hermetically closed much earlier than for red wines (1–2 weeks after *mutage*). Due to the risk of refermentation, all wine must be handled with the utmost cleanliness during the topping off operation.

The first racking is generally effected at the beginning of December, after the first cold spell. The objective is to separate the coarsest lees and conserve the finest. Rackings are then carried out every 3 months. At each racking, the barrels are carefully rinsed with cold water and then sterilized with hot water at 80°C or with steam. After being drained, they are sulfited before being refilled.

Microbiological stabilization is difficult with these wines. Refermentations are always possible, despite their low fermentability linked to the involvement of noble rot. First of all, the SO_2 combination rate can be high (Section 8.4). It is not always possible to obtain the necessary concentration of free SO_2 concentration, while remaining within the total SO_2 concentration limits imposed by legislation. The free SO_2 concentration should be maintained at approximately 60 mg/l. This difficulty rarely occurs in the presence of noble rot and is generally linked to at least a partial involvement of gray rot. Due to their antiseptic properties, sorbic acid and fatty acids (C_8 and C_{10}) can be used as chemical adjuvants to sulfur dioxide (Sections 9.2 and 9.3) during the storage of these wines.

Various physical processes can complement the effect of SO_2 and help to stabilize these wines (Section 9.4). Wine conservation at low temperatures (around 0°C) hinders but does not definitively inhibit yeast development, but heating wine at 50–55°C for several minutes can totally destroy the yeast population (Section 4.4). Heat-sterilized wines must be stored in sterile conditions to prevent subsequent contamination. Sterile storage, however, poses practical problems and is not possible in wooden casks (Volume 2, Section 12.2.3).

The difficulty of storing high sugar concentration wines, capable of refermenting, has incited the development of other techniques. A dry wine and a 'sweetening reserve' can be prepared separately. The sweetening reserve is a partially fermented juice containing 2–2.5% vol. alcohol and 150–200 g of sugar per liter. Only the second wine fraction is difficult to store, but its volume is limited (15–20% of the total volume). The wine can therefore be sulfited at a high concentration (100 mg/l) and stored at a low temperature. The two wines are blended just before bottling in a sterile environment. Wine aroma conservation is maximized with this method but the technique is only applicable to wines containing a maximum of 30 g of sugar per liter. Botrytized sweet wines, however, cannot be made by this method.

In addition, the presence of glucan makes botrytized sweet wines difficult to clarify by filtering or fining. This substance has colloid protector characteristics (Volume 2, Section 11.5.2). Yet when correctly barrel matured, a bentonite treatment is sufficient to fine these wines and is sometimes not even necessary. In this case, the majority of clarification problems are due to poor rot quality and poorly adapted working conditions.

14.2.7 Tokay Wine

Tokay is a famous botrytized sweet wine produced in Hungary. Several types of wine exist under the same name. The most renowned is the Tokay Azsu, which is a perfectly balanced, perfumed and delicate sweet wine. As with Sauternes wines, *Botrytis cinerea* transforms the grape to form noble rot, but in different conditions.

These wines are prepared with Azsu grapes which are concentrated on the vine, by both drying and noble rot. The grapes are ground (mechanically today) to obtain a type of paste. A high-quality new wine in the final stage of fermentation, concentrated in alcohol, acidity and extract, is then poured over this paste. The wine and paste are then macerated for 24–36 hours, permitting the sugar and different aromatic elements of the Azsu grapes to be diffused in the wine. This mixture is then pressed to separate the pomace and the wine. The wine is then aged in a cask. The amount of paste added to the wine corresponds to the different types of Tokay Azsu. The wine is classified by the number of back-baskets (*puttonyos*) of paste (20–25 kg) per 136-liter cask. The following types of Tokay can thus be distinguished:

- Three back-baskets per cask Tokay Azsu: this contains at least 60 g of natural sugar per liter and it must be aged for at least 3 years in barrel and bottle.

- Four back-baskets per cask Tokay Azsu: this contains at least 90 g of natural sugar per liter and is aged for at least 4 years.

- Five back-baskets per cask Tokay Azsu: this contains at least 120 g of natural sugar per liter and is aged for at least 5 years.

- Six back-baskets per cask Tokay Azsu: this contains at least 150 g of natural sugar per liter and is aged for at least 6 years.

The quality of Azsu *eszencia* (essence of Azsu) is even higher. It is produced in certain specific vineyards with vine-dried and noble-rotted grapes, ideal for creating this wine. The corresponding must has a high sugar concentration and is difficult to ferment. The wine contains about 250 g of sugar per liter and its alcohol content is between 6 and 8% vol.

Tokay wines are improved by air-exposed aging in cool cellars. These cellars are galleries dug into calcareous rock which maintain the wine at a constant 10°C temperatures. Due to their high sugar and alcohol concentration, these wines are not very sensitive to microbial spoilage and can be conserved in partially filled casks.

14.3 CHAMPAGNE AND SPARKLING WINES

14.3.1 Introduction

There are many methods for making "sparkling wines", but this term refers exclusively to wines that have undergone alcoholic fermentation in a closed vessel. Artificial carbonation by saturation with carbon dioxide gas (CO_2) does not produce the same quality bead, and wines made effervescent by this method are known as "artificially carbonated wines".

Champagne is the most prestigious sparkling wine. It is produced according to *Appellation d'Origine Contrôlée* regulations, in a delimited area, using strictly defined grape varieties and winemaking techniques.

Sparkling wine is generally made in a two-stage process. The first consists of making a base wine with particular characteristics. The wine is blended and cold-stabilized, then fermented for the second time. In the case of Champagne, this second fermentation (called "prise de mousse") takes place in the bottle that will ultimately be delivered to the consumer, once the yeast deposit has been disgorged. The use of the expression "*méthode champenoise*" for sparkling wines produced, with the same method, outside the delimited Champagne appellation is now prohibited and has generally been replaced by "*méthode traditionnelle*". The advantages of second fermentation in bottle have also been combined with clarification by filtration, achieved by emptying the wine into a vat after its second fermentation, then filtering and treating it, if necessary, prior to bottling in the final bottles for shipment (Section 14.3.5).

In the case of sparkling wines made by the "Charmat" or "cuve close" method, the second fermentation takes place in airtight vats (Section 14.3.5). The wine is then filtered, bottling liqueur is added, and the blend is bottled under pressure, ready for delivery.

In the past, some wines became sparkling spontaneously, when they were bottled with some residual sugar before fermentation was completed. They were unreliable to make, as the second fermentation was uncontrolled and could produce insufficient pressure, or, on the contrary, result in excess pressure that caused the bottle to explode. Several variations on this technique are still used today, but are now much better controlled. Fermentation is stopped by refrigerating the wine once it has reached a specific residual sugar level. Once it has been bottled, it is allowed to warm up naturally and fermentation is completed. This technique is relatively simple to use as the chilled must after stopped fermentation may be stored as long as required, then bottled at an appropriate time.

In the later 17th century, problems in controlling the second fermentation led the Champagne winegrowers to devise their present system of separating the complete fermentation of a dry white base wine from the second fermentation in bottle, with the addition of a controlled amount of sugar corresponding exactly to the carbon dioxide pressure required.

14.3.2 Fermenting Base Wines

(a) Principles

The aim is to make a base wine with a moderate alcohol content, generally a maximum of 11% vol., as more ethanol will be formed during the second fermentation in bottle and the total overall alcohol content is to remain below 13% vol. The base wine should also have a certain level of fresh acidity, to ensure the right balance in the finished product. The grapes are thus harvested at an earlier stage of ripening than in other *Appellation d'Origine Contrôlée* vineyards. Consequently, the grapes must be pressed very carefully to avoid skin contact and the resulting bitterness and herbaceous character. In the case of Champagne, this is especially vital when pressing the black Pinot Noir and Pinot Meunier grapes that are blended with the white Chardonnay. Picking and pressing conditions must, therefore, be carefully controlled.

Handpicking is followed by sorting to eliminate any defective, damaged, or rotten grapes, as a relatively low percentage of spoiled grapes have a negative impact on quality. The whole grapes are transported in recipients containing 45–50 kg, with holes to drain off any juice, thus avoiding skin contact and accidental fermentation, as well as maintaining the grapes in aerobic conditions. These containers are carefully washed after each use.

(b) Pressing and Extracting the Must

The grapes must be pressed soon after picking, without crushing, to avoid contact between skin and juice, and the various fractions of must are kept separately. Each pressing operation takes approximately four hours. Two types of presses are used. The hydraulic presses traditionally used in Champagne are round or square, with a large surface area to ensure that the grape layer is no more than 80–90 cm thick. Pressure is kept low to avoid crushing the skins. Horizontal presses have also been used for a number of years now, but only platen presses without chains or, preferably, pneumatic presses (Valade and Blanck, 1989). The different fractions of the must reflect the uneven ripeness of the grapes and the varying composition of the vacuolar sap. The *cuvée* corresponds to the juice from the middle part of the grape flesh, which is both the sweetest and the most acidic. The outer part of the flesh is sweet, but less acidic, due to the salification of the organic acids in the vicinity of the skin. The flesh closest to the seeds has the highest acidity and the least sugar.

Pressing methods are standardized and the juice is collected in small vats, known as "*belons*". Traditional Champagne presses used to hold 4000 kg grapes. Two or three pressings in quick succession and loosening the pomace after each one (Valade and Pernot, 1994) produced 2050 l of top-quality must (enough to fill ten of the 205 l barrels used in Champagne). This was known as the *cuvée*. The next two pressings produced 410 l (2 barrels) of "premières taille" and a third gave a further barrel of "deuxième taille". The final pressing in a standard hydraulic press produced 200–300 l of press wine ("rebêche"), intended for distillation rather than for making Champagne. In presses with a large surface area, the edges are subjected to less pressure than the center, so the pomace is brought from the edges towards the center between each pressing. These presses may be operated automatically (Valade and Pernot, 1994).

The must is separated in a similar way in the case of horizontal presses, which are, however, easier to use.

A 1993 regulation specifies that 4000 kg grapes must produce 25.50 hl of clarified must, allowing for 2–4% sediment. Only the *cuvée* (20.50 hl) is used to make prestigious Champagnes, while the "taille" (5 hl) produces fruitier, faster-maturing wines that are generally included in non-vintage *brut* blends.

Analysis of the different pressings, as they come out of the press, shows significant variations in composition (Table 14.6). As pressing continues, total acidity decreases, as do both the tartaric and malic acid levels. Mineral content and pH increase, as do the phenolic content and color intensity, while the sugar level remains relatively constant. The aromatic intensity and finesse of wines made from successive pressings also diminishes, but we do not have the means to analyze these changes.

It is essential to take great care harvesting, pressing, clarifying, and separating the must to maximize the quality of the finished Champagne (Moncomble *et al.*, 1991). Not only do these precautions make it possible to produce practically colorless must with very little sediment from black grapes, but they are also vital to preserve the finesse and quality of the end product.

(c) Must Clarification and Fermentation

For the reasons outlined above, the must should be perfectly clarified prior to fermentation (Pernot, 1999). Generally, the must is clarified immediately after pressing at the pressing room in the vineyards. Some Champagne producers clarify the must again on arrival in the fermentation cellar. Sulfite is added (5–8 g/hl) and the sediment is left to settle down naturally. In some years, depending on the condition of the harvested grapes, pectinases may be added to facilitate flocculation and settling.

Champagne musts contain large concentrations of nitrogen compounds, especially proteins, which contribute to the quality of the finished Champagne, especially, persistence of the bead. However, proteins are also involved in instability problems, leading to turbidity. Some producers add tannins (5 g/hl) to flocculate any unstable proteins. Bentonite may also be used for this purpose, at doses not exceeding 30–50 g/hl.

Traditionally, the initial alcoholic fermentation ("bouillage") took place in oak barrels, in cellars where the temperature was never higher than 15–20°C. Some producers still barrel-ferment their wines to enhance aromatic complexity and flavor. However, most base wines are now fermented in coated steel or, mainly, stainless-steel vats, as they are easy to maintain at temperatures below 20°C. The aim is for the fermentation to continue

Table 14.6. Physicochemical characteristics of Champagne must (Valade and Blanck, 1989)

	Cuvée			Second pressing			Third pressing		
	1982	1985	1986	1982	1985	1986	1982	1985	1986
Density	1072	1075	1073	1071	1075	1072	1071	1075	1073
Sugar (g/l)	163	171	164	161	171	163	161	170	165
Total acidity (g/l H_2SO_4)	8.1	8.5	9.2	6.8	6.9	8.0	5.9	6.0	6.4
Tartaric acid (g/l)	7.4	8.2	8.6	6.5	6.9	7.8	5.9	6.5	7.1
Malic acid (g/l)	6.6	7.7	8.6	5.7	6.8	7.8	5.1	6.5	7.1
pH	3.05	3.10	3.05	3.18	3.27	3.17	3.33	3.43	3.43
Potassium (mg/l)	1790	1650	1965	1965	1995	2385	2160	2320	3175
Total nitrogen (mgN/l)	575	650	790	570	675	850	585	760	955
Available nitrogen (mgN/l)	202	327	420	192	327	460	185	345	490

smoothly and uninterruptedly until the residual reducing sugar content is below 2 g/l. Although the must is usually chaptalized, this is not always necessary as the aim is to achieve an alcohol content no higher than 10–11% vol. Completing fermentation is not usually a problem, especially as the must is systematically seeded with selected yeasts.

Winemaking methods in other sparkling wine-producing areas are generally based on those used in Champagne, but may be simplified to reduce costs. Although other grape varieties do not have the same qualities as the Chardonnay, Pinot Noir, and Pinot Meunier used in Champagne, they make suitable base wines, provided they are picked with a sufficiently high acidity level, e.g. Chenin Blanc in the Loire Valley, Ugni Blanc in Bordeaux, Mauzac in Limoux, Maccabeo in Catalonia, etc.

(d) Malolactic Fermentation

Champagne is an elegant, fruity wine that requires a certain level of acidity (around 6 g/l H_2SO_4). The drop in acidity that occurs during the second fermentation must be taken into account in making the base wine (Section 14.3.3).

According to E. Peynaud (cited by Ribéreau-Gayon et al., 1976), when he analyzed Champagnes in the 1950s, he found that malolactic fermentation was uncommon and there was a considerable difference in flavor between wines with and without malic acid. The same author added that there had been many changes since that time (increase in pH, decrease in sulfiting, etc.), resulting in a greater vulnerability to bacterial activity and increasingly frequent uncontrolled malolactic fermentations that were unknown in the past.

There is still some discussion concerning the beneficial effect of malolactic fermentation on Champagne aromas. If it is properly controlled, it improves the quality of acidic wines, especially Chardonnay, as the bacterial activity enhances their aromas (Section 13.7.6). In other cases, it may result in wines lacking freshness that age too rapidly and may even necessitate the addition of tartaric acid to raise the acidity level (Volume 2, Section 1.4.3).

To ensure microbiological stability and avoid the serious consequences of malolactic fermentation

during second fermentation ("prise de mousse") or bottle aging ("conservation sur lattes"), malic acid must be eliminated from the base wine prior to bottling. This solution has been found most effective and is the most widely used, although other methods are under investigation.

The factor with the greatest impact on malolactic fermentation is the sulfur dioxide content, as it is inhibited by even small doses of free SO_2. It is also affected by the combined SO_2 content and becomes very difficult, or even impossible, at total SO_2 levels above 80–100 mg/l. Careful observation (Ribéreau-Gayon et al., 1976) showed a correlation between the total SO_2 content and malolactic fermentation in Champagne during the second fermentation in bottle. Current filtration technology makes it possible to sterile-bottle the base wine and inoculate it with pure yeast, thus avoiding malolactic fermentation in bottle without excessive use of SO_2. Lysozyme may also assist in stabilizing the base wine (Gerbaud et al., 1997; Pilatte et al., 2000).

However, it is not always easy to start malolactic fermentation at the right time in wines with very high acidity (Section 13.7.6) and it is essential to adjust the SO_2 content and temperature for that purpose. It is also possible to inoculate a properly prepared starter, but this is a rather laborious operation. Seeding with reactivated bacterial biomass, initially developed for red wines (Section 12.7.5), has considerably improved malolactic fermentation conditions in Champagne base wines (Laurent and Valade, 1993) and suitable products are now commercially available.

14.3.3 Second Fermentation in Bottle: The Champagne Method

(a) Preparing and Bottling a Cuvée

The new base wines are clarified, racked, filtered, and fined in the usual way. Base wines may be fined using isinglass (1.5–2.5 g/hl) or gelatin (4–7 g/hl), with or without tannins (2–4 g/hl) (Marchal et al., 1993). Tannin may be required to deal with the instability resulting from these wines' relatively high protein content, while it is essential to maintain a sufficient level of protein to give a high-quality bead. Moderate doses of bentonite may also assist in protein stabilization.

A *cuvée* is prepared by blending wines of different origins (vineyards classified 100%, 86%, etc.), qualities, and, possibly, vintages. This is indispensable to maintain the quality and character identified with the producer from year to year and is still mainly determined by tasting. There is a single *Appellation d'Origine Contrôlée* for all Champagnes, while the other two appellations in the region, Côteaux Champenois and Rosé de Riceys, are only applicable to still wines. The hierarchy among Champagnes is mainly dependent on the selection of base wines used to blend the *cuvée*.

Once the *cuvée* has been blended, the base wine is cold-stabilized to prevent tartrate precipitation. In some cases, it may be fined just before or after cold stabilization.

Preparation for bottling may also include filtration. Barrel-fermented wines are adequately clarified by simple settling of the lees, but wines fermented in vat always require filtration, especially immediately after cold stabilization.

During bottling, the tirage liqueur (syrup containing 500 g/l saccharose) is added for the second fermentation, calculated to produce the carbon dioxide required for a pressure of 5–6 bars at 10–12°C. Theoretically, 20 g/l saccharose must be fermented to produce 5 bars pressure. Table 14.7 indicates the quantities of sugar required to produce the desired pressure in the bottle after fermentation.

An active yeast starter, consisting of selected strains of *Saccharomyces cerevisiae*, is added at the same time to ensure that the second fermentation ("prise de mousse") will be successfully completed

Table 14.7. Sugar content of the tirage liqueur according to the pressure required (Valade and Laurent, 2001)

Pressure required (bars)		Sugar (saccharose) (g/l)	Sugar (glucose + fructose) (g/l)
At 10°C	At 20°C		
4.0	5.4	16.0	17.0
4.5	6.1	18.0	19.1
5.0	6.8	20.0	21.2
5.5	7.5	22.0	23.3
6.0	8.2	24.0	25.0
6.5	8.8	26.0	27.6

in the bottle. This starter may consist of fermenting must but is better prepared with active dried yeasts. Dried yeasts develop well in must that does not contain alcohol, and may be inoculated directly, particularly in white winemaking (Section 13.7.2). However, in a medium containing alcohol, e.g. when required to restart a stuck fermentation (Section 3.8.3), they must be reactivated prior to use, so that they are in a suitable physiological condition. Laurent and Valade (1994) recommended an effective method for preparing dried yeasts for use in second fermentation in bottle. It is advisable to seed the bottle with an initial population of 1.5×10^6 cells/ml. Below that amount, fermentation is slower and some sugar may remain unfermented, while above that level (e.g. 2.10^6 cells/ml), fermentation is faster but some yeast strains may produce yeasty off-odors.

The tirage liqueur and yeast starter can be added in vat prior to bottling, or to each bottle individually during the bottling process, e.g. via two measuring pumps.

Other substances may also be added at the time of bottling (e.g. 3 g/hl bentonite or 0.2–0.7 g/hl alginate) to facilitate elimination of the yeast sediment when the bottles are disgorged.

(b) Second Alcoholic Fermentation and Aging on the Lees

The bottles are closed with crown stoppers made airtight by plastic seals. They are stored in a horizontal position in box pallets or stacks, interspersed with laths of wood to steady the layers. It is important that the bottles be placed horizontally, firstly to ensure that they remain airtight during fermentation, and, secondly, to provide a maximum interface for exchanges between the wine and its lees. Fermentation takes one month, or sometimes longer, at the constant temperature of 11–12°C in underground cellars in Epernay and Reims. This slow, even, low-temperature fermentation is another quality factor in producing fine Champagne, especially the finesse and persistence of the bead when the bottle is opened. Carbon dioxide pressure increases gradually (Valade, 1999), inhibiting yeast growth and slowing the fermentation rate, especially at low pH and high alcohol levels. Attempts have been made to enhance the fermentation rate by adding nutrients, with inconsistent results. It is more effective to increase the initial yeast inoculation, as well as to adapt the yeast strain and starter preparation conditions to the type of base wine.

The wine is still not proper "Champagne", even when the second fermentation in bottle is completed, i.e. all the sugar has been fermented. The wine spends a long period aging on its yeast sediment, which gradually decreases in volume and becomes more compact. The bottle must remain in a horizontal position to provide a maximum wine-sediment contact surface. The yeasts release substances (Section 14.3.4) into the wine, initially by excretion, then by diffusion from the dead yeast cells. These are mainly amino acids, either synthesized by the yeast or previously absorbed from the wine. Autolysis involving cell wall enzymes has also been observed. All these complex phenomena play a significant role in Champagne quality. The improvement in quality during this stage is correlated to the composition of the base wine, which explains why other sparkling wines benefit less from aging on the lees. Non-vintage Champagne is aged on the lees for a minimum of 15 months, while the minimum for vintage Champagne is 3 years, but they may stay on their lees for up to 8 years, or even longer for some top *cuvées*.

As long as the sparkling wine remains in contact with the yeast sediment under anaerobic conditions, the lees act as a redox buffer and the wine is perfectly preserved. Champagne from bottles several decades old tasted in Champagne cellars were found to be in perfect condition, as they had never been disgorged. Once the bottles have been disgorged, not only does the Champagne stop improving, but there is also a risk of defects developing due to redox phenomena.

The main risk during storage, especially if bottles are exposed to light, is the development of off-odors in the wine. These reduction odors are due to the formation of thiol groups by photodegradation of sulfur amino acids naturally present in Champagne. The reaction, photosensitized by riboflavin (vitamin B2), produces methanethiol and dimethyldisulfide, which are responsible for "sunlight flavor" ("goût

de lumière") (Volume 2, Section 8.6.5) (Maujean and Seguin, 1983). Maujean *et al.* (1978) showed that this drop in redox potential due to light exposure only occurred in disgorged Champagne. The formation of thiol groups and the resulting "sunlight flavor" also depend on the reduction conditions of the wine prior to light exposure. This defect may be prevented by using glass bottles with low transmission values at wavelengths below 450 nm. Adding ascorbic acid, together with SO_2, just before the bottles are finally corked (when the dosage is added) is an effective preventive measure.

(c) Riddling and Removing the Yeast Sediment

The next stage consists of gathering the yeast sediment on the inside of the cap. This is traditionally done by riddling the bottles on special racks, which hold the bottles neck down, at a variable angle. Riddling consists of turning the bottles with a slightly jerky movement, bringing them gradually to a vertical position, completely upside-down, over a period of a month or more.

This operation takes a variable amount of time, generally from three weeks to one month, depending on the type of wine and its colloidal structure, as well as the type of yeast and its capacity to form clumps. Riddling is an awkward stage in the Champagne production chain, due to the space required for the riddling racks, the labor-intensive process, and the fact that the bottles are immobilized for a relatively long period of time. A great deal of work has been done to simplify this operation. The first approach consists of adding various substances to the wine in vat, prior to bottling, intended to facilitate settling of the yeast sediment. While the results have not been negligible, this technique has not made any great improvement in the process.

More significant progress has been made by reproducing the intermittent movements of riddling on the scale of a box pallet (several hundred bottles). Each box pallet is installed on a movable base, which is tilted manually to change the angle of the bottles, gradually bringing them into the vertical, neck-down position. This system may be mechanized and programmed (gyropallet) to riddle the bottles much more efficiently, completing the cycle in one week instead of the one month required for manual riddling. This system is now widely used, in spite of the high initial investment required.

Another approach to simplifying riddling consists of using yeast enclosed in tiny calcium alginate beads for the second fermentation in bottle (Duteurtre *et al.*, 1990; Valade and Rinville, 1991). The sediment settles on the cap almost immediately when the bottle is turned upside down and riddling is no longer necessary. Of course, this assumes that fermentation and aging on the lees continue normally with the enclosed yeasts. The second condition is that yeast cell growth does not burst the beads, producing a powdery deposit that is difficult to eliminate. This problem is now avoided by using a double coat of alginate on the beads. Several million bottles have now been processed with enclosed yeast, and work is continuing to monitor the aging and development of the Champagne. Once this technique has been demonstrated not to affect quality, it will be possible to envisage its use in large-scale production.

(d) Disgorging and Final Corking

Once the sediment has settled on the cap, the wine is disgorged. In the past, this operation was done manually by removing the cap quickly while raising the bottle slightly so that the few milliliters of wine containing the sediment would be expelled without emptying the bottle or losing too much carbon dioxide pressure.

Nowadays, the bottle necks are almost always frozen prior to disgorging, in an automated system that also adds the dosage liqueur, corks the bottles, and fits the wire closure. The bottles are held upside down and the necks plunged into a low-temperature salt solution that freezes about 2 cm of wine above the cap, trapping the sediment. When the bottles are turned upright, the cap is removed and the frozen plug is expelled.

The bottle is then topped up with dosage liquor ("liqueur d'expédition"), a syrup made of reserved wine containing approximately 600 g/l of sugar, used to adjust the final sugar level of the Champagne. "Brut" Champagne generally has 10–15 g/l (1–1.5% dosage), while "Demi-sec" has 40 g/l (4% dosage).

The dosage liqueur can be acidified with citric acid, if necessary. It also contains the quantity of sulfur dioxide required to eliminate any dissolved oxygen, and may be supplemented with ascorbic acid (50 mg/l). This offsets the sudden oxidative effect of disgorging: the redox potential may increase by 150 mV, or even more, depending on the redox buffer capacity of the wine.

According to E. Peynaud (cited by Ribéreau-Gayon *et al.*, 1976), "Dosage is not simply a matter of sweetening the wine, but of improving it. The quality of the dosage liqueur, the way it is aged, the types of wine used, the quality of the sugar, and the preparation formula all play a major role in the quality of the finished product." The dosage liqueur contributes to the overall flavor balance.

14.3.4 Composition of Champagne Wines

(a) Analysis of Champagne Wines

The analysis results in Table 14.8 show the effect of bottle fermentation on the wine's composition. The alcohol content increases by 1.3% vol. during the second fermentation and may drop by a few tenths during preparation for shipping, depending on the composition of the dosage liqueur. If the base wine is not properly cold-stabilized, total acidity may decrease slightly during the second fermentation due to the precipitation of potassium hydrogenotartrate and by the breakdown of small amounts of residual malic acid under the action of the yeasts. Otherwise, there is little variation in the acid content unless a small amount of citric acid is added in the dosage liqueur. This decrease in overall acidity results in an increase in pH.

One of the most significant characteristics of Champagne must and wine is their high nitrogen content, especially in the form of amino acids (Desportes *et al.*, 2000) (Table 14.9), which facilitates the initial and secondary fermentations. The amino acid content of Champagne is twice or thrice as high as that of Bordeaux wine (Ribéreau-Gayon *et al.*, 1976). The same authors gave the following analysis results for Champagne and Bordeaux: 462 and 184 mg/l of total nitrogen, 11.2 and 6.3 mg/l of ammoniacal nitrogen, and 216 and 100 mg/l of amino acid nitrogen, respectively.

Table 14.8. Comparison of wine composition before and after second fermentation (1993 vintage, mean analysis values) (Tribaut-Sohier and Valade, 1994)

	Blend at the time of bottling	After second fermentation and addition of dosage for *brut* quality
Mass density at 20°C (g/dm^3)	990.5	993.9
Alcohol at 20°C (% vol.)	11.0	12.2
Sugars (g/l)	1.3	12.7
pH	3.02	3.05
Total acidity (g/l H_2SO_4)	4.7	4.7
Volatile acidity (g/l H_2SO_4)	0.27	0.3
Free SO_2 (mg/l)	8	8
Total SO_2 (mg/l)	38	58
Tartaric acid (g/l)	3.5	3.2
Malic acid (g/l)	0.2	0.2
Potassium (mg/l)	330	325
Calcium (mg/l)	85	70
Copper (mg/l)	0.17	0.13
Iron (mg/l)	2.1	2.8
Sodium (mg/l)	8	12
Magnesium (mg/l)	60	70
Total nitrogen (mgN/l)	303	410
Ammoniacal nitrogen (mg/l)	13	20
OD 520 nm	0.038	0.028
OD 420 nm	0.087	0.106
Color intensity	0.125	0.134
Shade	2.59	3.89
Conductivity (mS/cm) at 20°C	1.32	1.32
Saturation temperature at 20°C	10°C	12.1°C

The total nitrogen content of Champagne varies from 150–600 mg/l (Maujean *et al.*, 1990) and that of the must is considerably higher. Chardonnay and Pinot Noir/Meunier grape varieties have a high nitrogen content, and Champagne is the wine-growing region where it reaches the highest levels. Table 14.9 compares the mean amino acid content of base wines made from different Champagne grape varieties.

According to the literature, Champagne must contains 25–100 mg/l of proteins (in BSA equivalent), while the level is considerably lower in base wine: 14–32 mg/l (Tusseau and Van Laer, 1993). In harvested grapes, 75% of total nitrogen is in amino acid form, while it accounts for 95% in new base wine (Tusseau *et al.*, 1989).

Several proteins have molecular masses between 20 and 30 Kda, while one with a molecular mass of 62 Kda is probably combined with sugars (glycoproteins). They have isoelectric points between 2.5 and 6.5 (Brissonnet and Maujean, 1993).

Besides proteins and polypeptides from the must, the sparkling properties of Champagne also involve carbohydrate colloids (polysaccharides and glycoproteins) (Marchal *et al.*, 1996; Berthier *et al.*, 1999) released from the yeast cell walls during aging on the lees (Feuillat *et al.*, 1988; Tusseau and Van Laer, 1993). This yeast autolysis is certainly accompanied by more radical transformations. The amino acid content of some sparkling wines has been reported to increase depending on the contact time, and stirring the yeast back into suspension has been recommended to enhance this phenomenon.

Boidron *et al.* (1969) compared the volatile fermentation compounds involved in Champagne aromas with those found in other sparkling wines. Champagnes characteristically have lower concentrations of methanol, higher alcohols, propanol,

Table 14.9. Average amino acid content of base wines made from different Champagne grape varieties (Assayed on sulfenic resin and detected using ninhydrin) (Results in mg/l) (Desportes *et al.*, 2000)

	Chardonnay	Pinot Meunier	Pinot Noir
Aspartic acid	6	28	11
Threonine	14	69	16
Serine	14	43	10
Asparagine	N.I	27	24
Glutamatic acid	38	55	30
Glutamine	67	82	36
Proline	777	165	222
Glycine	27	15	13
Alanine	118	209	95
Citrulline	38	15	11
Valine	32	18	9
Cysteine	24	N.I	NI
Methionine	7	9	9
Isoleucine	4	12	5
Leucine	14	22	21
Tyrosine	41	28	20
β-alanine	11	2	3
Phenyl-alanine	14	18	13
γ-N-butyric acid	67	}78	}74
Ethanolamine	4		
Ornithine	7	11	10
Lysine	8	11	16
Histidine	8	4	6
Arginine	20	321	105
Total	1360	1242	759
Proline/Arginine	38.85	0.51	2.11

ethyl butyrate, and isoamyl acetate, which have a negative effect on aroma. This is probably due to the winemaking conditions (e.g. temperature). Other more positive, aromatic compounds such as ethyl caprate and ethyl lactate (related to malolactic fermentation) are more abundant in Champagne.

In the past, it was relatively common to find large residual yeast populations in Champagne bottles. Yeast counts between 0.16 and 4.8 × 10^3 cells/ml have been found in Champagnes on the market (Ribéreau-Gayon *et al.*, 1976). Current riddling and disgorging techniques, particularly sedimentation additives and fine-tuned riddling programs, have made substantial progress in eliminating residual yeasts.

(b) Effervescence in Champagne Wines

The excess carbon dioxide pressure responsible for effervescence is an essential characteristic of Champagne.

When Champagne is poured into a glass the foam, which is an important quality factor, appears even before the liquid. It is well known that while tasting a poor initial visual impression has a negative impact on the overall assessment, and this is certainly the case with the bead of a sparkling wine (Robillard, 2002). A good quality bead consists of tiny bubbles that remain separate and spherical in shape. Large bubbles produce an unattractive, grayish bead that usually disappears very rapidly.

Effervescence also reveals the wine's aromas, as the bubbles contain odoriferous compounds in addition to carbon dioxide (Maujean, 1996).

It is, therefore, important to consider the criteria for the formation and stability of bead in sparkling wines. The following analysis is based on a 1997 review by A. Maujean (Laboratory of Enology, Reims University) and B. Robillard (Moët et Chandon Research Laboratory), as well as several other publications (Maujean, 1989; Robillard, 1993; Liger-Belair and Jeandet, 2002).

The bubbles in sparkling wine are due to carbon dioxide, formed during the second fermentation and dissolved in the wine. A bubble of CO_2 must push the surrounding molecules apart before it can emerge. A great deal of energy is required to form a liquid/CO_2 interface, but this is minimized by nucleation phenomena.

Bubbles may be formed directly from dissolved gas (induced homogeneous nucleation). When Champagne is shaken up, e.g. during shipment, parent bubbles produce smaller bubbles, some of which are stabilized by contact with proteins and float on the surface. The drop in pressure when the bottle is opened causes them to explode, producing other smaller bubbles, which explode in turn, and so on. This chain reaction is responsible for a violent gush of wine, which may leave the bottle half-empty (Maujean, 1996).

Bubbles are more usually formed by adsorption of the gas on a solid particle (induced heterogeneous nucleation). It has been demonstrated that a minimum radius of 0.25 μm is required for the bubble's internal pressure to be sufficiently low in relation to that of the wine, to enable the bubble to grow and rise through the liquid. Plastics have a higher surface energy than glass, creating a greater affinity for CO_2, so that bubbles coming off a plastic surface will be larger than those released in a glass vessel.

Several factors are involved in effervescence kinetics following degassing (Casey, 1987; Liger-Belair, 2002 and 2003; Liger-Belair *et al.*, 2000).

The first factor is the physical nature of the solid surface (particles in suspension or vessel wall), particularly the number and radius of microcavities on which the bubbles are formed, detaching themselves once they have reached a certain diameter. This produces a bead, or line of bubbles, which always rise from the same spot. Of course, the microcavities must be hydrophobic or they would be filled with wine.

Other factors, such as viscosity and chemical composition, are inherent to the liquid. It is quite probable that some carbon dioxide molecules are immobilized by binding with other substances. Electrostatic interactions may also lead to the adsorption of CO_2 on the surface of macromolecules, as shown by the significant changes in effervescence kinetics when proteins or polysaccharides were added to synthetic wines (Maujean *et al.*, 1988).

The instability of the bead is defined by three parameters:

1. Swelling bubbles: The gas from small bubbles is absorbed into larger bubbles, etc. This results in a coarse irregular foam with an unattractive appearance.

2. Draining: This refers to the liquid that drains out of the foam over time. It leads to a reduction in foam volume and a distortion in the shape of the bubbles. The foam gradually dries out (e.g. as on the head of a beer glass).

3. Coalescence: A break in the film between two smaller bubbles produces a larger one, resulting in a coarse foam that disappears quickly.

Several experimental processes have been proposed for measuring the spontaneous or forced degassing kinetics in sparkling wines (Maujean *et al.*, 1988), as well as for assessing the persistence of foam (Maujean *et al.*, 1990; Robillard *et al.*, 1993).

The Mosalux apparatus (Maujean *et al.*, 1990) is used to determine three characteristics of sparkling wine bubbles:

1. Foamability or maximum foam depth expresses the liquid's capacity to contain gas once it starts effervescing visible in the foam formed when it is poured into a glass.

2. Foam depth describes the constant depth of the foam when the liquid is bubbling in the glass and corresponds to the head of foam.

3. Foam stability measures the time required for the foam to disappear once the liquid stops effervescing. This parameter is only of theoretical interest in laboratory work.

Measurements showed that foamability and foam stability are mutually independent—wines may produce a lot of foam but it is not necessarily very stable. A close correlation has been observed between foamability and protein content. A decrease in protein content of a few mg/l can lead to a 50% drop in foamability (Malvy *et al.*, 1994). However, Maujean *et al.* (1990) did not find any simple correlation between protein content and foam stability.

Protein solubility affects its impact on foaming in sparkling wine. Hydrophobic proteins may also be adsorbed at the gas-liquid interface, on the "bubble skin", stabilizing it by decreasing surface tension. Proteins with lower molecular weights are more rapidly adsorbed at the interface. Proteins that have an effect on effervescence have isoelectric points in the vicinity of wine pH (2.5–3.9). This characteristic does not promote solubility, but makes the protein more hydrophobic. Thus, proteins affect foamability by changing the surface tension when they are adsorbed at the liquid-gas interface of the bubbles. Glycoproteins have an even greater impact on foaming, as the hydrophilic osidic fraction increases the viscosity of the liquid film between the bubbles and reduces the draining of the liquid phase. Although yeast mannoproteins are less hydrophobic than plant glycoproteins, they are present in large quantities in Champagne wines and apparently contribute to their stability (Feuillat et al., 1988).

Foam depth decreases during aging on the lees, but is largely compensated by the improvement in stability.

It is well known that the various stages in the winemaking process have an impact on foam quality. Robillard et al. (1993) examined the impact of filtering base wines. This operation removes solid or colloidal particles that provide a base for bubble formation (nucleation), considerably reducing the intensity of effervescence and thus, the foam stability of the corresponding sparkling wine. The smaller the pores of the filter medium, the more marked the impact on foam stability.

Treatment with plant charcoal or bentonite also causes a considerable decrease in foamability, related to the reduction in protein content. On the contrary, fining with gelatin, combined with silica gel or tannin, improves foaming qualities.

14.3.5 Other Second Fermentation Processes

(c) Transfer Method

The aim of this method is to benefit from the advantages of second fermentation in small bottles and aging on the yeast lees, while avoiding the problems associated with riddling and disgorging. Once the second fermentation and aging are completed, the wine is filtered and transferred to another bottle. This process is not permitted for Champagne, although there is a tolerance for quarter-bottles which are filled after filtration, following second fermentation in full-size bottles. This process is still occasionally used to prepare half-bottles, but its use is due to be prohibited in the near future.

After second fermentation and aging, the bottles are simply taken to the racking area. They are emptied automatically into a metal vat, under carbon dioxide pressure equivalent to that created in the bottles by fermentation, to prevent degassing.

The wine in the vat is refrigerated to $-5°C$ by circulating liquid coolant through a suitable heat-exchanger. This makes the CO_2 more soluble. Dosage liqueur is also added in the vat and the wine is left to rest for a few days. It is then plate-filtered to remove all the yeasts and bottled. As the wine is kept at low temperatures under pressurized carbon dioxide, it retains all the dissolved CO_2.

This system has a number of advantages. It eliminates the labor costs of riddling and disgorging, as well as the time the wine is immobilized on the riddling racks. Dosage liqueur is much more evenly distributed. It is also possible to blend several batches of wine after the second fermentation to obtain the desired quality. Cold-stabilization prevents tartrate precipitation and filtration ensures that the yeasts are completely eliminated, leaving the wine perfectly clear.

If these operations are properly conducted, they give satisfactory results. However, wines made by the Champagne method were always preferred in comparative tastings, probably due to the fact that small amounts of oxygen were dissolved in the wine during transfer operations, however carefully they were controlled. It has also been demonstrated that exchanges occur between the carbon dioxide molecules resulting from the second fermentation and the industrial gas used to protect the wine during the transfer process. Finally, filtration may modify the wine's foaming qualities.

(d) The Charmat (Cuve Close) Method

Second fermentation in bottle is technically demanding and is, therefore, only justified for high-quality products made from fine base wines that are likely to benefit from aging on the yeast lees.

As long aging is not economically viable for cheaper products, a simpler, less expensive process (the Charmat method) has been developed to produce sparkling wine from lower quality grapes.

Figure 14.1 shows a simplified diagram of a system for second fermentation in a sealed vat. The various base wines are blended and transferred to the second-fermentation vat (C) and yeast starter (vat A) as well as syrup (vat B) is added to provide the quantity of sugar required for the second fermentation and the dosage of the finished

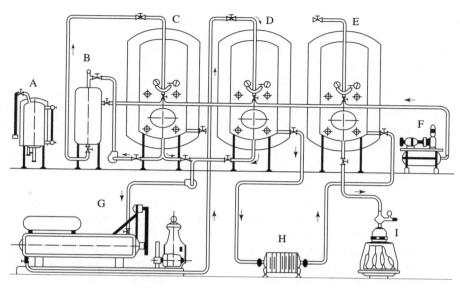

Fig. 14.1. A bulk-method installation for sparkling wine: A, yeast starter preparation tank; B, sugar addition tank equipped with a mixing system; C, second fermentation tank; D, refrigeration tank; E, bottling tank; F, air compressor for isobarometric bottling; G, refrigeration group; H, filter; I, bottler

product. The fermentation vat (C) is equipped with heating and cooling systems to maintain a temperature of 20–25°C. When pressure in the vat reaches 5 bars, fermentation is stopped by reducing the temperature and sulfiting slightly. The wine is then transferred to a refrigerated vat (D) and kept at −5°C for several days for cold stabilization.

The wine is filtered and then transferred to another vat (E), connected to the bottling line. The entire operation is carried out under pressurized carbon dioxide (F), to prevent degassing.

These large-volume processes certainly cannot achieve the same quality as bottle fermentation. This is partly due to the fact that the wine is not aged in contact with the lees, as a sufficient level of interaction can only be achieved in a small container. The quality of the grapes used and the speed of the process also have an impact. In the Charmat method, the yeast is often eliminated after only a few days' fermentation to reduce costs. In view of the other factors involved, it is by no means obvious that aging on the lees would improve quality. Systems have, however, been developed for maintaining the wine in vat on its lees, stirring them into suspension to accelerate

exchanges. The success of this operation depends on the quality of the base wine and keeping fermentation temperatures low to slow down the reactions.

The Charmat method may give better results than bottle-fermentation in hot climates, as it preserves the base wine's freshness and fruit. Finally, the Charmat method is most appropriate for producing sparkling wines from aromatic varieties such as Muscat, as aging on the lees attenuates the Muscat character, without significantly improving quality.

(e) Asti Spumante

This is probably the most famous sparkling Muscat. Unfortunately, when the must is fermented to produce a completely dry wine, it loses all the distinctive grape aromas and has an unpleasant, bitter taste. Long experience has led to the development of a low-temperature fermentation process that is interrupted every time it starts speeding up. The must is clarified, fined, and centrifuged as many times as necessary until the yeast and available nitrogen have run out. Analysis results show clearly that the total nitrogen, particularly available nitrogen, decreases every time the fermenting must is filtered, probably as it was fixed on the yeast. The resulting wine is relatively stable, due to nitrogen deficiency, with 5–7% alcohol by volume and 80–120 g/l sugar. This wine used to be put into bottles for second fermentation, but it was irregular and uncontrollable. Second fermentation now takes place in a sealed vat (Charmat process), using a blend of wines from different vineyards, clarified by fining with gelatin/tannin and filtration. Fermentation starts at 18–20°C and is then slowed down by reducing the temperature. When the pressure reaches 5 bars, the wine is chilled to 0°C and clarified again. The temperature is then reduced to −4°C for 10–15 days to stabilize the wine. Following further filtration or centrifugation, bottling takes place in an environment pressurized with CO_2 to prevent degassing. Some producers use sterile filtration and others pasteurize the wine to prevent it from fermenting again in bottle. The finished Asti Spumante contains 6–9% alcohol by volume and 60–100 g/l sugar. A number of other sparkling wines are produced using similar methods.

14.4 FORTIFIED WINES

14.4.1 Introduction

Fortified wines are characterized by their high concentrations of alcohol and sugar. They are derived from the partial fermentation of fresh grapes or grape juice; The addition of alcohol prematurely stops the fermentation. This fortification can be effected in one step or in several.

These wines were evidently created in the past in response to technical problems encountered in warm regions. Sugar-rich grapes and elevated temperatures resulted in explosive fermentations, easily leading to stuck fermentations. The partially fermented wine was unstable, especially since sulfiting was far from mastered at the time. Lactic acid bacteria subsequently developed, causing lactic disease and the production of volatile acidity (Section 3.8). The addition of alcohol during fermentation was a simple means of stabilizing the wine and produced an alcoholic and sweet product

with an agreeable taste. As late as the 1960s, these wines represented a significant part of Californian and Australian production.

Today, other means can be used to produce standard types of wines in these climatic conditions. Grapes are harvested at sugar concentrations compatible with complete fermentations, even in relatively hot climates. Fermentations are better controlled through sulfiting, aeration and temperature control. They are also complete. Malolactic fermentation can now occur without bacterial spoilage.

Due to greater demand, traditional dry red and white wines have replaced fortified wines at many wineries. Today, only the most famous fortified wines remain. Specific natural factors and well-adapted technology permit these wines to develop their fine aromas and rich flavors. French *vins doux naturels* and port wines are certainly among the most prestigious fortified wines, but other fortified wines from Greece, Italy and other Mediterranean countries also exist.

The Office International de la Vigne et du Vin (OIV) defines fortified wines as 'special wines having a total alcohol content (both potential and actual) above 17.5% vol. and an alcohol content between 15 and 22% vol.' Two types of fortified wines exist:

1. Spirituous wines receive only brandy or rectified food-quality alcohol during fermentation;

2. Syrupy sweet wines can receive concentrated must or *mistelle* in addition to brandy or alcohol.

In both cases, the natural alcohol potential of the grape juice must be at least 12% vol. At least 4% vol. of the alcohol in the final product must come from fermentation.

Storage conditions, up to bottling, vary depending on the type of fortified wine. Due to their high alcohol content, these wines are very resistant to oxidative phenomena. Some actually develop their desired characteristic through a certain degree of oxidation. These wines undergo a true barrel aging. Other finer and more delicate fortified wines are protected from air and are bottle aging.

14.4.2 French Fortified Wines (*Vins Doux Naturels*)

(a) Definition

These famous wines (VDN) are found in a dozen *appellations* across three regions in the south of France (Brugirard *et al.*, 1991). Banyuls, Rivesaltes, Maury and various Muscat *appellations* are among the best known. These wines, fall under the OIV definition of fortified wines but French legislation taxes the two types of fortified wines mentioned in Section 14.4.1 differently. With the spirituous wines (VDN), only the added alcohol is taxed. The other fortified wines, are taxed on the total alcohol, including the alcohol from fermentation of the must.

Production conditions are more constraining in France than as specified by the OIV. Not only is the area covered by each appellation clearly defined, but the grape varieties are also specified. Non-Muscat varieties are Grenache, Macabeu, and Malvoisie, while only Muscat of Alexandria and Muscat à Petits Grains are permitted in fortified Muscat wines.

Crop yield limits are set at 40 hl/ha, with only 30 hl/ha allowed to be used for making VDN. Grape juice must contain at least 252 g of sugar per liter (approximately 14.5% vol. potential alcohol). The proportion of alcohol added at the time of fortification must comprise between 5 and 10% of the must volume. The must is fortified when the fermentation has already transformed a little more than half of the natural sugar. The final product must contain between 15 and 18% vol. alcohol content and at least 21.5% vol. total alcohol. Total alcohol includes the alcohol and potential alcohol, which corresponds with the quantity of alcohol that the residual sugar could produce by fermentation:

$$\text{Residual sugar content (g/l)}/16.83$$
$$= \text{potential alcohol (\% vol.)}$$

The minimum residual sugar content varies from 59 to 125 g/l, depending on the appellation.

The initial must concentration and the percentage of added alcohol are verified by the following relationship P/α, where P = residual sugar weight (g/l), and α = polarimetric deviation which depends on the proportion of glucose and fructose, itself related to the quantity of sugar fermented.

In grape juice, the glucose/fructose (G/F) ratio is equal to 1. Glucose diminishes more rapidly than fructose during fermentation. A French fortified wine must have a P/α of between -2.00 and -3.00. Fraud is suspected below -3.5. A fortified wine artificially made from dry wine, alcohol and saccharose or concentrated must (G/F = 1) would have a P/α of -5.23.

Table 14.10 provides supplemental information concerning the chemical composition of French fortified wines.

(b) Vinification

Several types of French fortified wines (VDN) exist. The white VDN are made from white or gray Grenache or Macabeu grapes. They do not generally undergo a maceration, but are occasionally lightly macerated. They are light, fruity non-oxidized wines made to be drunk young.

Red VDN are macerated. The juice and pomace are generally separated after several days of vatting. Fortification most often occurs on the separated juice but in certain cases the alcohol is added to the pomace and the maceration is continued for 10–15 days. Richly colored fortified wines with high concentrations of dry extract are obtained by this alcoholic maceration. These wines are capable of being aged for a long time.

After separating the must, Muscat wines are made similarly to white wines. However, maceration increases aromatic extraction; making these wines therefore requires a lot of care to respect the finesse of the aromas.

Grape maturity is regularly assessed to determine the harvest date according to the variety. During overripening, the maturity must be carefully followed because sugar concentrations may increase sharply to attain 250–270 g/l (15–16% vol. potential alcohol). The acidity also diminishes considerably. The full aromatic potential of muscat wines is obtained at a sugar concentration of around 225 g/l. *Botrytis cinerea* negatively affects fortified wine quality, especially in the

Table 14.10. Analytical characteristics of French fortified wines (VDN) (Brugirard *et al.*, 1991)

	Minimum values	Average values	Maximum values
Alcohol content (% vol.)	14.8	15 to 17	18.9
Total alcohol (% vol.)	21.5	21.5 to 22.5	23.0
Density at 20°C	1.010	1.015 to 1.030	1.035
Sugar (g/l)	45	70 to 125	150
P/α (15°C)	−1.5	−2.0 to −2.5	−3.5
Total dry extract (g/l)	90	110 to 140	170
Reduced dry extract (g/l)	18	20 to 26	40
Ashes (g/l)	1.4	1.8 to 2.5	3.5
Ashes alkalinity (g/l K_2CO_3)	1.3	1.2 to 2.2	3.1
Total acidity (g/l H_2SO_4)	2.0	0.30 to 3.5	5.0
Volatile acidity (g/l H_2SO_4)	0.15	0.30 to 0.60	legal
pH (20°C)	2.90	3.60 to 3.80	4.20
Tannin (Folin index):			
white VDN	15	25 to 40	55
red VDN	20	30 to 50	70
Aldehydes (mg/l)	25	60 to 120	150
Higher alcohols (mg/l)	50	70 to 90	150
Glycerol (g/l)	3.0	6.0 to 10.0	12.0
Butylene glycol (g/l)	0.30	0.50 to 0.80	1.20
Lactic acid (g/l)	0.19	0.30 to 0.4	0.63
Free SO_2 (mg/l)	0	0 to 15	20
Total SO_2 (mg/l)	traces	100 to 150	legal

case of Muscat. Botrytized grapes should not be macerated.

The first steps of winemaking with maceration consist of moderately crushing and destemming the grapes. The grapes are then transferred to the tank and sulfited at 5–10 g/hl. The fermentation temperature is set at approximately 30°C to favor maceration. Maceration times vary from 2 to 8 days, if the fortification occurs after must separation. In this case, the fermentation speed should be reduced beforehand. Wines are macerated for 8 to 15 days when continuing the maceration after fortification.

When there is no maceration, the grapes are drained and pressed to extract the juice, according to traditional white winemaking methods (Section 13.3). Immediately after extraction, the juice is stabilized by sulfiting at 5–10 g/hl and preferably refrigeration. The must is then clarified by natural settling and racking or centrifugation. Yeast starter may be used and fermentation temperatures are kept relatively low, 20–25°C (and even 18°C for Muscat), to avoid loss of aroma.

(c) Fortification (Mutage)

The addition of alcohol to fermenting must stops yeast activity, increases the dissolution of phenolic compounds during maceration and provokes the precipitation of insoluble substances. A near-neutral wine brandy is used. The addition of non-wine spirits is not permitted. The alcohol addition may be done in several steps to slow and spread out the fermentation phenomena.

The moment of fortification is chosen according to the density, which decreases during fermentation. The density must not drop below a certain established limit, called the fortification point. Choosing the correct fortification point is essential to wine quality. The wine must have a sugar concentration corresponding to the type of product desired and conforming to legislation.

Fortification tables are used to achieve the exact alcohol content required, using either wine spirits at a minimum of 96.0% alcohol by volume or a blend of spirits and must. The addition of wine spirits is effected with either 90% vol.alcohol or with varied blends of alcohol and must. The second form of addition arose from the need to have a tax official present for tax purposes when using alcohol. In the past, wines were stabilized with high SO_2 doses to stop the fermentation, when waiting for the authorization to use alcohol. Nowadays, the wine spirits are denatured by mixing with must that has just started fermenting, in the presence of a government inspector.

It is recommended to stop fermentation (*mutage*) before fortification by refrigerating the must or eliminating the yeasts by centrifugation or filtration.

Sulfiting destined to neutralize the ethanal formed and to block oxidations definitively stabilizes the wine. A free SO_2 concentration between 8 and 10 mg/l should be maintained. Approximately 10 g of SO_2 per hectoliter should be added, considering the high pH (3.5–4.0), sugar concentration and alcohol content of these wines.

(d) Conservation and Aging

Due to their diversity, many storage and aging methods exist for these wines. All are generally aged for a year in tank, undergoing repeated rackings to assure clarification. Different methods specific to each type of wine are used after this period.

Muscat wines are stored in tanks until bottling. Precautions are taken to avoid oxidation and to protect aromas: 15–17°C temperature, sufficient humidity, use of inert gas, etc.

After a selection based on tasting, many red VDN, having undergone maceration, are placed in 6 hl casks exposed to the sun. Oxidation phenomena cause these wines to take on an amber tint and characteristic aromas. The wines are often fined and cold stabilized before being placed in casks. Carrying out these operations at the time of bottling may thin the wine. A simple filtration at this time is preferable. A once-traditional method to obtain the same oxidative transformations consists of leaving slightly emptied glass carboys outside, exposed to natural climatic variations, but it is now rarely used.

The finest and most delicate white and red VDN can be matured in 225 l oak barrels in cellars at moderate temperatures (15–17°C) without any

particular oxidative phenomena, according to traditional fine winemaking methods. The wine is matured for approximately 30 months and bottled after fining with gelatin. Reduction phenomena after the wine is bottled are responsible for the actual aging process.

Rancio wines are made traditionally and locally. The production of these wines is not codified. The method consists of maintaining a 6 hl barrel partially filled. Each year, wine is removed from the barrel to be bottled and replaced with newer wine.

VDN are subject to the same clarification and stabilization problems as other wines. Iron casse, proteic casse, tartrate deposits and colored matter can cloud the wines. Standard preventive measures can help to avoid these problems. Oxidasic casse is another accident linked to grape rot.

The high alcohol content of these wines gives them a certain level of microbial stability, but accidents are still possible due to their high sugar concentration and elevated pH. Some yeasts tolerate 16–17% vol.ethanol and are capable of causing refermentations. In addition, these yeasts resist elevated concentrations of free sulfur dioxide. Particular strains of *Lactobacillus hilgardii* have been identified in certain French VDN. They are apt to develop, provoking lactic disease, responsible for abundant deposits and gustatory flaws.

Standard operations lower the risks of deviation—hygiene, fining, filtration, sensible use of sulfur dioxide, etc.—but pasteurization is the only treatment that completely eliminates germs and stabilizes wine. The correct use of this method does not cause organoleptic modifications, even with Muscat. Sterile filtration can also be used.

14.4.3 Port Wines

(a) Production Conditions

Port wines come from the steeply sloping Douro region in Portugal (Ribéreau-Gayon *et al.*, 1976; Barros, 1991). The schistous soil, the jagged relief, the high temperature variations between seasons, low rainfall and intense sunlight characterize the Douro. These conditions lead to highly aromatic pigmented grapes with high concentration of sugar and phenolic compounds. The *terroirs* are classed in a decreasing scale from A to F according to soil nature, grape variety, vine age, altitude, exposition, etc. There is a great diversity of cultivated varieties in this region (15 red and 6 white). The grapes are picked very ripe but are not vine dried. They are sorted very carefully to eliminate bad grape clusters and spoiled grapes. The must, with a minimum of 11% vol. potential alcohol by volume, but which usually contains 12–14% vol., is sulfited (9–10 g/hl) and may be acidified, if necessary.

A relatively slow partial fermentation is sought. Extraction of skin components occurs during a concurrent maceration. The wine was traditionally fermented in *lagares*, 80 cm-high granite vats containing 2.5–110 hl, an ideal shape for ensuring that all the grapes would be perfectly crushed. The grapes were trodden for several hours each day until the third day of maceration. The fermentation occurred simultaneously. The pomace cap was immersed by mechanical means at regular intervals. When the desired density was attained, the *lagare* was opened and the wine flowed into the casks. Brandy was added to the wine in casks to stop the fermentation and raise the alcohol content to 18–19% vol.

Today, most wines are made in modern wineries and the crushing and maceration operations are mechanized. The open or closed tanks are equipped with automatic pumping-over and mechanical mixing systems. The manual work has all but disappeared. These perfectly controlled technical modifications have improved and regularized the quality of port wine while increasing profitability.

Upon arrival at the winery, the grape crop is destemmed and carefully crushed to facilitate maceration. The must is sulfited but generally not inoculated with yeasts, to avoid explosive fermentations. The temperature is maintained at around 30°C during fermentation.

After reaching 4–5% vol.alcohol, the fermenting must drained from the tank is clarified, possibly by a rotating filter, before being fortified. Correctly choosing the fortification point is essential to the quality of port wine and to obtaining the level of sweetness desired. The quality of the port also depends on the quality of the brandy used for fortification. All brandies used are submitted to analytical and taste tests; they contain 77–78% vol.alcohol. Pneumatic and mechanical horizontal presses are currently replacing traditional vertical presses, since the former are easier to use. Pressing is moderate. During fortification, a fraction of the tannin- and color-rich press wine is added to the free run wine.

(b) Maturation and Characteristics of Port Wines

Figure 14.2 summarizes the maturation and aging process of the different types of port wines. During the winter following the harvest, after the first racking, the wines are classed according to taste. The best batches, during an exceptional year, may be reserved to be declared as vintage port, but most wines are blended.

The blends are aged in 5–6 hl oak barrels (*pipas*) for several years in oxidative conditions that maintain an elevated oxidation–reduction potential. Metal ions, in particular copper and iron, play an essential role in polyphenol oxidation. Even in the bottle, these wines conserve a high oxidation–reduction potential. The iron remains in its ferric state, as if all of the reducing components had been destroyed by oxygen. Their prolonged oxidation and intense esterification give these wines a rich and complex bouquet.

During aging, the tannins become softer as they polymerize or combine with anthocyanins, while coloring matter precipitates and the color changes. The less oxidized "ruby" ports maintain the fruitiness and robustness of young wines. They have a more or less dark red color. The older, more oxidized "tawny" ports are golden red or golden.

White ports undergo a certain level of maceration and are aged in the same oxidative conditions as blends. With certain exceptions, the wines are not oxidized, to maintain their fruity aroma and pale color.

Superior quality products (10-year-old, 30-year-old ports, etc.) also undergo oxidative aging.

At the time of bottling, these oxidized wines are stable in the presence of air. They improve very little during bottle aging.

"Vintage" ports are the best quality wines. After a brief aeration to stabilize the color, they are

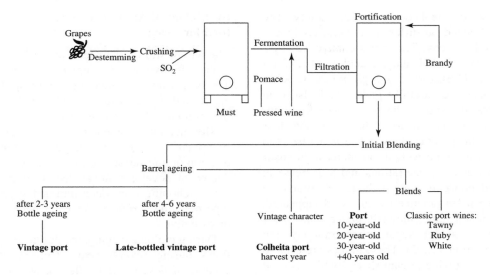

Fig. 14.2. Flow diagram for the production of various styles of port wine (Barros, 1991)

aged in full barrels, like many great red wines. Vintage ports are bottled after 2–3 years' barrel-aging, while "Late Bottled Vintage ports" are bottled after 4–6 years (Figure 14.2). They then continue to improve considerably in bottle. A reduction bouquet develops: it is linked to the low oxidation–reduction potential that maintains the iron ion in the ferrous state. Vintage ports have a considerable aging potential and can be aged for 20 years or more in the absence of air, due to their high polyphenol concentration. These wines are very robust when young and, after years of aging, maintain a high extract concentration, a characteristic fruitiness and relatively high color intensity, with red-mauve tones dominating. Once bottled, these wines are sensitive to oxygen: when the bottle is opened, the wine rapidly loses its qualities.

The year can still be mentioned on the bottle of non-vintage, quality port wines. These vintage character wines are called "colheita ports".

14.5 FLOR WINES

14.5.1 Definition

The Office International de la Vigne et du Vin defines 'flor wines' as:

wines whose principal characteristic is to be submitted to a biological aging period in contact with air, by the development of flor yeasts (film-forming yeasts) after alcoholic fermentation of the must. Brandy, rectified alcohol or agricultural spirits can be later added to the wine. In this case, the alcohol content of the finished product must be equal to or greater than 15% vol.

Sherry in english and german—Jerez in spanish—Xérès in french is the best-known flor wine. In *oloroso*-type sherry, the aging process is essentially physicochemical and biological development is limited. The *oloroso* method will, however, be described in this section. Jura yellow wines are another well-known example of flor wines.

14.5.2 Sherry Wines

(a) Production Conditions

The sherry production zone is situated in the south of Spain, near the city of Xérès de la Frontera. The production of this prestigious wine was described by Casas Lucas (1967), Goswell (1968), Gonzalez Gordon (1990) and Jeffs (1992). This section is based on the work of E. Peynaud (in Ribéreau-Gayon *et al.*, 1976), updated by J.F. Casas Lucas in 1994.

The Palomino cultivar constitutes nearly 95% of the grape production for this wine. The remaining

5% consists of the Pedro Ximenez variety. The vine is cultivated on different *terroirs*, creating a production hierarchy. The musts contain 12–14% vol. potential alcohol and an acidity of 2–3 g/l expressed as H_2SO_4 (or 3–4.5 g/l in tartaric acid).

The grapes are carefully picked and placed in 15 kg cases. In the past, the grapes were traditionally exposed to the sun for a day on a mud floor (*almijar*). This practice, known as *soleo*, results in a 10% loss in grape weight, an increase in sugar and tartaric acid concentrations and a decrease in malic acid concentration. Although favorable to quality, the *soleo* practice has all but disappeared. Pedro Ximenez grapes may still undergo this practice, attaining a high sugar concentration within 15 days (Section 11.2.2). These grapes are used to prepare a sweet wine (cream sherry) which is used in variable proportions to sweeten dry wines.

The sherry winemaking method is based on white winemaking principles without maceration. The juice extraction conditions are consequently of prime importance: moderate crushing, no contact with metal, slow and light pressing, juice selection after pressing. The clarification and refrigeration of these juices tend to be generalized.

Plastering (adding calcium sulfate to must) is a traditional practice in this region. This operation permits the suspended solids to settle more rapidly and the wines obtained are more limpid and their color more brilliant. There is a decrease in pH, a diminution of ash alkalinity (due to the precipitation of acids in the form of salts) and an increase in total acidity and buffering power. Approximately 2 g of $CaSO_4(2 H_2O)$ are added per liter, lowering the pH by 0.2 units. The wine may also be acidified with tartaric acid (1.5 g/l maximum).

Sulfur dioxide is used during winemaking and storage to disinfect the barrels, but concentrations must be limited so as not to hinder the development of flor yeast. Taking hygienic measures avoids undesirable microbial contamination.

In the past, the fermentation occurred in 516 l oak barrels (*botas de extraccion*) filled with 450–467 l of juice. Today, relatively low-capacity stainless-steel containers are increasingly used, to limit excessive temperature increases.

(b) Biological Aging Principles for Flor Yeasts

The wines, still on their lees, are tasted during the months following the completion of fermentation. The best wines, considered the most apt for aging, are racked, fortified to 15–15.5% vol. and stored in a container filled to 5/6ths of its capacity.

The alcoholic content of these wines prevents microbial spoilage, but flor yeasts spontaneously develop on the surface of the wine. After a certain degree of development, another tasting results in a new classification, determining the appropriate type of aging (*crianza*) for each *bota*: biological or oxidative.

During biological aging, the flor develops, sometimes during several years. Certain yeasts are capable of developing on the surface of 15–16% vol.alcohol content wine in contact with air. This film is produced by yeasts specific to the region, coming from either grapes or previously used barrels. These yeasts develop in aerobiosis by oxidizing ethanol. They belong to the *Saccharomyces* genus. Over the years, taxonomy has included these yeasts in the *S. cerevisiae* species or has identified them as *Saccharomyces oviformis*, *Saccharomyces bayanus*, etc. These are therefore not ordinary mycodermal yeasts, responsible for various wine diseases due to poor storage methods.

Martinez *et al.* (1997) identified the following yeast strains in Sherry *flor*:

74% *Saccharomyces cerevisiae beticus*
14% *Saccharomyces cerevisiae montuliensis*
8% *Saccharomyces cerevisiae cheresiensis*
0.3% *Saccharomyces cerevisiae rouxii*
3% strains not typical of Sherry *flor* (*Pichia*, *Hansenela*, and *Candida*).

The different strains develop and form *flor* at different rates, as well as having different metabolic effects, e.g. *montuliensis* produces the highest concentrations of ethanal.

The yeast film is called flor and the biological aging process has been known as *crianza de flor* for a long time at Xérès. The more or less rapid and intense flor formation and its aspect and color (white, cream, golden, burnt) depend on many factors, especially the nature of the yeast and the chemical composition of the wine. The flor rarely forms above 16.5% vol.alcohol and it is impossible above 17% vol. The presence of a little sugar is favorable; the presence of phenolic compounds is unfavorable and darkens the wine color. Sulfur dioxide, nitrogen compounds and other substances are also involved. All of these factors exert an influence that is reflected in future wine aroma and quality.

During this type of aging technique, the wine does not remain permanently in the same container: it is periodically transferred to different *botas*. These transfers are fractional, following the solera system. *Botas* are piled in rows. The barrels (*botas*) in each horizontal row (*escola*) are full of wine from the same crianza (that is, same degree of aging). During aging, the wine is moved around, blended, and redistributed to obtain the most uniform wine possible at the time of bottling. This system also permits new wine to be added regularly, which helps to maintain the flor.

Figure 14.3 summarizes the *solera* system. This example contains three *escolas*: 720 l are taken from the six *botas* of the lowest row, or *solera*, for bottling. The *solera* barrels are filled with 720 l coming from the five *botas* of the preceding *criadera*. Finally, 720 l from the four *botas* of the highest row fill the last five *botas*. These four *botas* are filled with new wine.

The *bota de asienta* intended for sherry aging has a volume of 600 l. The wine volume is 5/6ths, or 500 l.

In practice, the *solera* systems are much larger—they usually contain several hundred

barrels. Transfers are made in groups of 12 to 18 *botas*.

Wine is transferred three to four times per year in the *solera* system. The transfer volume depends on the type and age of the wine desired. The ratio of total system volume to annual volume removed determines average wine age.

(c) Wine Transformations During Biological Aging

The biochemical transformations provoked by the *crianza de flor* have been studied. As oxygen is consumed by the flor, its proportion decreases in the barrel-head space and is replaced by carbon dioxide. The wine transfers, however, aerate the wine. The oxidation–reduction potential of the wine (250–300 mV) indicates a moderately reduced state. The flor acts as an isolating layer, protecting the wine from excessive oxidation. In this manner, the wine ages normally (fino type), acquiring a pale yellow color.

Volatile acidity diminishes to 0.10 g/l in H_2SO_4 (0.12 g/l in acetic acid). Ethanal formation is an essential characteristic of the *crianza de flor*, slowing during aging to produce a total of 220–380 mg/l. Due to its chemical reactivity, ethanal is a precursor of many chemical substances that contribute to the bouquet of sherry wine (diethyl acetal, 50–60 mg/l). Sotolon (Section 10.6.3) is a characteristic element of the aroma of *fino* wines.

The glycerol concentration attains 7–9 g/l immediately after alcoholic fermentation. During the first *crianza* phases, it is significantly depleted. After three years, its concentration can fall to several tenths of gram per liter.

Lactic acid is also formed, reaching 22 mEq/l. This production cannot be explained by alcoholic (7 mEq/l) and malolactic (6 mEq/l) fermentation alone. Malolactic fermentation is nevertheless complete and contributes to wine quality.

Free amino acid concentrations diminish during flor aging, but the evolution of each amino acid varies, depending on the situation. After 7 years *flor* aging, the concentration of proline, the most abundant amino acid in *fino*-type sherries, representing 70% of the initial amount of total amino

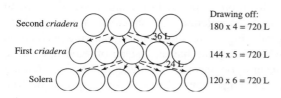

Fig. 14.3. Solera system, showing the partial drawing off and redistribution of wine into the next lowest row of barrels of an older *criadera* stage. This operation was begun by removing 120 l from barrels on the lowest level (or solera) for bottling

acids had decreased to only 31% (Botella *et al.*, 1990).

The *crianza de flor* in sherry-region wineries produces *fino* wines. *Manzanilla* is produced according to the same principle in the Sanlucar de Barrameda region. This style of wine is aged for at least 3 years. At the end of this long aging process, the *crianza de flor* disappears. The aging process can, however, be continued chemically (*oloroso* wines) for 6 years or more. These products have the following names according to their age: *fino*, *amontillado*, *amontillado viejo* and *amontillado muy viejo*.

(d) Oxidative Aging of Oloroso Wines

A flor film develops on nearly all sherry wines. This flor develops several months after the completion of alcoholic fermentation. An initial fortification at 15–15.5% vol.alcohol is practiced on wines most suited to biological aging. After a few months of being aged under the film-like growth, the wines are tasted and classified, confirming those that are destined to be aged biologically and deciding which wines must undergo oxidative aging. The latter are selected according to film growth conditions. If the filmlike growth is not established in suitable conditions, the growth is completely stopped by an additional fortification to 17.5–18% vol.

From this moment on, the wines receiving the additional fortification will age in the absence of a yeast film, without flor yeast activity. Only physicochemical phenomena occur. During the aging process, some of the substances responsible for the fruity character in the wine are oxidized by oxygen. The barrel wood also plays a role in the oxidation process. Its texture acts as a type of tissue of semipermeable membranes. The reactions involved in these phenomena are slow and poorly understood. Basic wood substances are extracted and oxidized.

Oloroso wines can be aged according to the *solera* system or in a more static manner without blending. This static method produces vintage wines (*anadas*). *Oloroso* wines are generally richer in color and more robust than *fino* wines. *Raya*

olorosa wines correspond to a lower class of wine than *oloroso*.

Before bottling, the different types of sherry are clarified by fining with albumin or powdered blood. They are also sometimes stabilized by a bentonite addition.

There is a risk of bacterial spoilage during both aging processes (biological and chemical). For this reason, each barrel is regularly tasted during the process. At the slightest quality doubt, the wine is transferred to sherry vinegar production.

14.5.3 Yellow Wines from Jura

Although they have their own character (Chevennement *et al.*, 2001), there are a certain number of parallels with Sherry. Like sherry, they undergo an aging process with flor development, but no alcohol is added to them.

These wines are made from Savagnin grape. The base wine contains approximately 12% vol.alcohol. The wine is placed in small barrels; the barrels are topped off and sealed. They are aged in a cellar for 6 years, without topping off, producing a head space in the barrels. A filmlike growth progressively forms on the surface of the wine. This flor is composed of aerobic film-forming yeasts which develop by respiration and cause various transformations—in particular, the oxidation of ethanol into ethanal. The yeast most often encountered in yellow wines belongs to the *S. cerevisiae* genus. The inoculation is spontaneous with these wines, causing the film growth to be irregular and sometimes resulting in spoilage. The risks of increased volatile acidity are greater than in sherry wines and increase as the wine alcohol content decreases.

To improve production conditions, film growth can be accelerated by inoculation with a film-producing yeast culture and by leaving head space when filling the barrel, instead of waiting for it to occur spontaneously by evaporation. Maintaining a low temperature (12–13°C) also limits bacterial spoilage.

In practical terms, *vin jaune* is aged for over six years, and is subjected to alternating cold winter (5–10°C) and hot summer (25–30°C) temperatures. These variations cause the development and elimination of a series of *flor* blooms over the years, resulting in the coexistence of live yeast in the surface *flor* and dead yeast cells that are deposited in the lees and autolyzed. The yeasts have an intense metabolic activity at 25°C, but are much less active at 10°C (Charpentier *et al.*, 2002).

The yellow wines from Jura are characterized by their high ethanal concentration (600–700 mg/l), their deep color and their particular organoleptic characteristics.

REFERENCES

André P., Aubert S. and Pelisse C. (1970) *Ann. Technol. Agric.*, 19, 323 and 341.

André P., Aubert S. and Pelisse C. (1971) *Ann. Technol. Agric.*, 20, 205.

André P., Benard P., Bourgeois M. and Flanzy C. (1980) *Ann. Techn. Agric.*, 29, 497.

Barros P. (1991). *La Technologie des vins de liqueur.* Office International de la Vigne et du Vin, Paris.

Bely M., Rinaldi A. and Dubourdieu D. (2003) *Biosci. Bioeng.*, 96 (6), 507.

Berthier L., Marchal R., Debray M., Bruet E., Jeandet P. and Maujean A. (1999) *J. Agric. Food Chem.*, 47, 2193.

Bidan P. (1966) *Bull. OIV*, 42, 34.

Botella M., Pérez-Rodriguez L., Domecq B. and Valpuesta V. (1990) *Am. J. Enol. Vitic.*, 41, 12.

Blouin J. and et Peynaud E. (2001) *Connaissance et Travail du Vin*, Dunod, Paris.

Boidron J.N.B., Avakiants S.P. and Bertrand A. (1969) *Conn. Vigne Vin*, 3, 43.

Brissonnet F. and Maujean A. (1993) *Am. J. Enol. Vitic.* 44 (3), 297.

Brugirard A., Fanet J., Seguin A. and Torres P. (1991) La dégustation et le service des vins doux naturels à Appellation d'Origine Contrôlée, Université des vins du Roussillon, 66300 Tressere (France)

Casas Lucas J.F. (1967) *Fermentation et vinification.* Institut National de la recherche Agronomique Paris.

Casey J.A. (1987) *The Australian Grapeflower and Winemaker*, 55, "Effervescence in sparkling wines".

Castino M. (1988) *Vigne Vini*, 12, 31.

Charpentier C., Dos Santos A.M. and Feuillat M. (2002) *Rev. Fr. Oenol.*, 195, 3.

Chauvet S., Sudraud P. and Jouan T. (1986) *Revue des OEnologues*, 39, 17.

Chevennement R., Cibey R., Grispout P., Levaux J. and Sintot D. (2001) *Rev. Fr. Oenol.*, 127, 26.

Desportes C., Charpentier M., Duteurtre B., Maujean A. and Duchiron F. (2000) *J. Chromatogr. A.*, 9893, 281.

Duteurtre B., Ors P., Charpentier M. and Hennequin D. (1990) *Le Vigneron Champenois*, 7–8, 21.

Dubourdieu D. (1982), Recherches sur les polysaccharides sécrétés par *Botrytis cinerea* dans la baie de raisin. Thèse Doctorat, Université de Bordeaux II.

Feuillat M., Charpentier C., Picca G. and Bernard G. (1988) *Rev. Fr. Oenol., Cahier Scientifiques*, 13, 463.

Flanzy C. (1998) *Œnologie. Fondements Scientifiques et Technologiques*, Lavoisier, Tec et Doc, Paris.

Garcia-Jares C.M., Rozès N. and Médina B. (1993a) *J. Int. Sci. Vigne Vin*, 27 (1), 35.

Garcia-Jares C.M., Médina B. and Sudraud P. (1993b) *Rev. Fr. Oenol.*, 140, 19.

Gerbaux V., Villa A., Monamy C. and Bertrand A. (1997) *Am. J. Enol. Vitic.*, 48, 49–54.

Gonzalez Gordon M.M. (1990) *Sherry, the Noble Wine.* Quiller Press, London.

Goswell R.W. (1968) *Process Biochem.*, 3, 47

Jeffs J. (1992) *Sherry*, 4th edn. Faber and Faber, London.

Lafon-Lafourcade S. and Ribéreau-Gayon P. (1977) *CR Acad. Agric.*, 551.

Lafon-Lafourcade S., Blouin J., Sudraud P. and Peynaud E. (1967) *CR Acad. Agric.*, 1046.

Laurent M. and Valade M. (1993) *Le Vigneron Champenois*, 6, 5.

Laurent M. and Valade M. (1994) *Le Vigneron Champenois*, 1, 7.

Leygnier A., Torres P. and Goyhexex J.M. (2000) *Les vins doux naturels de la Méditerranée*, Aubanel, Editions Minerva, Genève.

Liger-Belair G. (2002) *Ann. Phys. Fr.*, 27 (4), 101.

Liger-Belair G. (2003) *Sci. Am.*, 280 (1), 68.

Liger-Belair G. and Jeandet P. (2002) *Revue Française d'œnologie*, 193, 45.

Liger-Belair G., Marchal R., Robillard B., Dambrouch T., Maujean A., Vignes-Adler M. and Jeandet P. (2000) *Langmuir*, 16, 1881.

Malvy J., Robillard B. and Duteurtre B. (1994) *Sci. Alim.*, 14, 87.

Marchal R., Bouquelet S. and Maujean A. (1996) *J. Agric. Food Chem.*, 44, 1716.

Marchal R., Sinet C. and Maujean A. (1993) *Bull. OIV*, 751–752, 691.

Masneuf S., Muzet M.L., Choné X. and Dubourdieu D. (1999), *Viti*, 249, 20.

Martinez P., Pérez Rodriguez L. and Benitez T. (1997) *Am. J. Enol. Vitic.*, 48, 160.

Maujean A. (1989) *Rev. Fr. Oenol.*, 120, 11.

Maujean A. (1996) *L'Amateur de Bordeaux*, Dec., 32.

Maujean A. and Seguin N. (1983) *Sci. Alim.*, 3, 589 and 603.

Maujean A., Haye M. and Feuillat M. (1978) *Conn. Vigne Vin*, 12, 277.

Maujean A., Gomérieux T. and Garnier J.M. (1988) *Bull. OIV*, 61, 683.

Maujean A., Pinsaut P., Dantan H., Brissonnet F. and Cossiez E. (1990) *Bulletin OIV*, 63, 405.

Moncomble D., Valade M. and Pernot N. (1991) *Le Vigneron Champenois*, 5, 14.

Pernot N. and Valade M. (1994) *Le Vigneron Champenois*, 6, 619.

Pilatte E., Nygaard M., Gaigao Y., Krentz S., Poweer J. and Lagarde G. (2000) *Revue Française d'Oenologie*, 185, 26.

Ribéreau-Gayon J., Peynaud E., Ribéreau-Gayon P. and Sudraud P. (1976) *Sciences et Techniques du Vin*, Vol. III: *Vinifications—Transformations du vin.* Dunod, Paris.

Robillard B. (2002) *Revue Française d'œnologie*, 193, 49.

Robillard B., Delpuech E., Viaux L., Malvy J., Vignes-Adler M. and Duteurtre B. (1993) *Am. J. Enol. Vitic.*, 44, 387.

Sudraud P., Bar M. and Martinière P. (1968) *Conn. Vigne Vin*, 2, 349.

Tribaut-Sohier I. and Valade M. (1994) *Le Vigneron Champenois*, 9, 16.

Tusseau D., Benoit C. and Valade M. (1989) In *Actualités Oenologiques 89.* Dunod, Paris.

Tusseau D. and Van Laer S. (1993) *Sci. Alim.*, 13, 463.

Valade M. and Blanck G. (1989) *Revue Française d'œnologie*, 118, 23.

Valade M. and Laurent M. (1999) *Le Vigneron Champenois*, 6, 67.

Valade M. and Laurent M. (2001) *Le Vigneron Champenois*, 3, 40.

Valade M. and Pernot M. (1994) *Le Vigneron Champenois*, 113, 6.

Valade M. and Rinville C. (1991) *Le Vigneron Champenois*, 3, 22.

Index

Abscisic acid   241, 249
Abscission   243
Acetaldehyde   63, 67
Acetaldehyde dehydrogenase, *see* ALDH
Acetic acid
  accumulation of, by yeast   64–7
  formation of, by yeast   64–7
  see also Bacteria, acetic acid
Acetobacter spp.   183–92, 202
Acetoin   68, 189
Acid rot   292
Acidification   307–8, 345
Acidity   107, 217, 307–10, 370, 374–5
  volatile   66, 91, 108–10, 112, 188, 190, 345, 360,
    371, 451
Acrolein   151
Actinomycetes   123
Additive techniques   313–5
Adenosine diphosphate, *see* ADP
Adenosine triphosphatase, *see* ATPase
Adenosine triphosphate, *see* ATP
ADH   188
ADP   53–4, 62
Aeration   102–4, 341–4, 456
  malolactic fermentation and   375
  momentary/permanent   103
  see also Pumping over
Aerobic respiration   140
Aerobiosis   103, 104, 140, 190
Aging
  acetic acid bacteria during   191
  barrel   433–5, 470
    reduction odors during   439–41
  bottle   470
  flor yeasts   475–9
  fortified wines   472–3
  on-lees   428–9, 432, 434–5, 462–3
  oxidative, *oloroso* wines   478
Ahmeur Bou Ahmeur   263

Alanine   71, 72
Albarino   401
Alcohol dehydrogenase, *see* ADH
Alcohol, potential   80–1, 261
Alcoholic fermentation   57–8
Alcohols, higher   74–6
ALDH   188
Alginate balls   464
Allergies   196
Almijar   476
Amertume   156, 174, 370
Amino acids
  assimilation mechanisms   72–3, 86
  catabolism of   73–4
  classification   71
  composition by grape variety   277
  deamination of   74
  in grape   246, 253–4
  permease, general, *see* GAP
  as stimulants   85–7
  synthesis pathways   70–1
  transamination   73
Ammonium
  assimilation mechanisms   72–3
  in grape   246, 254
  salts, as must stimulant   86–7, 111, 430–1, 455
Anabolism   53
Anaerobic respiration   140
Anaerobiosis   104, 140, 386–8, 391
Anthesis   243
Anthocyanins   246, 256, 265, 268, 323, 328, 345,
    347–8, 351, 388, 446–7
  extraction coefficient   356
Anthranilic acid   257
Antibacterials   211
Antibiotics   224
Antifungals   210–1
Antioxidants   194, 237
Antioxidasics   194
